PLASTICS
Materials and Processing

Second Edition

A. Brent Strong

Brigham Young University

Prentice Hall

Upper Saddle River, New Jersey   *Columbus, Ohio*

Library of Congress Cataloging-in-Publication Data

Strong, A. Brent.
    Plastics : materials and processing / A. Brent Strong.—2nd ed.
      p.   cm.
    Includes bibliographical references and index.
    ISBN 0-13-021626-7
    1. Plastics.   I. Title.
TA455.P5S74 2000   99-38587
620.1'923—dc21   CIP

Editor: Stephen Helba
Assistant Editor: Michelle Churma
Production Editor: Louise N. Sette
Production Supervision: Clarinda Publication Services
Design Coordinator: Karrie Converse-Jones
Cover Designer: Jason Moore
Cover art: Proof Positive/Farrowlyne Assoc., Inc.
Production Manager: Deidra M. Schwartz
Marketing Manager: Chris Bracken

This book was set in Clearface by The Clarinda Company and was printed and bound by R.R. Donnelley & Sons Company. The cover was printed by Phoenix Color Corp.

©2000, 1996 by Prentice-Hall, Inc.
Pearson Education
Upper Saddle River, New Jersey 07458

Printed in the United States of America

10 9 8 7 6 5 4

ISBN: 0-13-021626-7

Prentice-Hall International (UK) Limited, *London*
Prentice-Hall of Australia Pty. Limited, *Sydney*
Prentice-Hall of Canada, Inc., *Toronto*
Prentice-Hall Hispanoamericana, S. A., *Mexico*
Prentice-Hall of India Private Limited, *New Delhi*
Prentice-Hall of Japan, Inc., *Tokyo*
Prentice-Hall (Singapore) Pte. Ltd., *Singapore*
Editora Prentice-Hall do Brasil, Ltda., *Rio de Janeiro*

Contents

Chapter 1 Introduction to Plastics   **1**

Definitions of Plastics and Polymers   1
History of Plastics   5
Raw Material Supply and Pricing   9
Strategic Materials   12
Plastics Industry   12
Uses of Plastics in Modern Society   15
Case Study 1.1—The Development of Nylon   20

Chapter 2 Polymeric Materials (Molecular Viewpoint)   **25**

Introduction   25
Fundamentals of Matter   26
Bonding   31
Basic Concepts in Organic Chemistry   39
Polymers   48
Formation of Polymers   51
Thermoplastics and Thermosets   62
Copolymers   63
Case Study 2.1—Modifications to Improve Teflon® Processing   65

Chapter 3 Micro Structures in Polymers   **75**

Introduction   75
Amorphous and Crystalline   76
Solids, Liquids, and Gases   79
Thermal Transitions of Polymers   81
Effects of Thermal Changes on Polymers   90
Polymer Length   97
Molecular Weight   97
Melt Index   106

Shape (Steric) Effects   108
Case Study 3.1—Mechanical Properties of Polyethylene (PE) as Functions of Density and Melt Index   110

Chapter 4 Mechanical Properties (Macro Viewpoint)   121

Introduction   121
Mechanical Properties in Solids (Elastic Behavior)   123
Mechanical Properties in Liquids (Viscous Flow)   125
Viscoelastic Materials   131
Plastic (High-Strain) Stress-Strain Behavior   135
Creep   141
Toughness and Impact Strength   143
Reinforcements   146
Fillers   147
Toughness Modifiers   148
Case Study 4.1—Testing of Trash Containers to Predict In-use Performance   148

Chapter 5 Chemical and Physical Properties (Macro Viewpoint)   155

Introduction   155
Environmental Resistance and Weathering   155
Chemical Resistivity and Solubility   159
Permeability   169
Electrical Properties   173
Optical Properties   178
Flammability   182
Plastics Identification   185
Case Study 5.1—Using Carbon Black to Protect Polyethylene from UV Degradation   187

Chapter 6 Thermoplastic Materials (Commodity Plastics)   193

Introduction   193
Polyethylene (PE)   194
Polyethylene Copolymers   205
Polypropylene (PP)   208
Polyvinyl Chloride (PVC)   211
Polystyrene (PS)   217
Alloys and Blends   219
Case Study 6.1—Typical PVC Formulation   224

Chapter 7 Thermoplastic Materials (Engineering Plastics)   231

Introduction   231
Polyamides or Nylons (PA)   235
Acetals or Polyoxymethylenes (POM)   239

Thermoplastic Polyesters (PET/PBT)   241
Polycarbonate (PC)   244
Acrylics (PAN, PMMA)   246
Fluoropolymers (PTFE, FEP, PFA)   248
High-Performance Thermoplastics   252
Cellulosics   254
Case Study 7.1—Making Nonstick Electrosurgical Blades   255

Chapter   8   Thermoset Materials   263

Introduction   263
Crosslinking   265
Thermoset Types, General Properties, and Uses   272
Phenolics (PF)   274
Amino Plastics (UF and MF)   276
Polyester Thermosets (TS) or Unsaturated Polyesters (UP)   279
Epoxies (EP)   287
Thermoset Polimides   291
Polyurethanes (PUR)   292
Case Study 8.1—Thermoset Composites for Wrapping Utility Poles   295

Chapter   9   Elastomeric (Rubber) Material   303

Introduction   303
Aliphatic Thermoset Elastomers   307
Thermoplastic Elastomers (EPM and EPDM)   313
Fluoroelastomers   314
Silicones   315
Processing of Elastomers   317
Case Study 9.1—Elastomeric Lining for a Pump   320

Chapter 10   Designing with Plastics   327

Design Methodology   327
Layout/Drawing   330
Constraints   331
Material Choice   336
Prototyping   341
Case Study 10.1—Design of Plastic Stakes for Concrete Tilt-up Walls   344

Chapter 11   Extrusion Process   351

Introduction   351
Equipment   355
Normal Operation and Control of the Process   370
Extrusion Problems and Troubleshooting   378

Material and Product Considerations   386
Postextrusion Forming   403
Coextrusion   403
Case Study 11.1—Extrusion of Irrigation Tubing   406

Chapter 12 Injection Molding Process   419

Introduction   419
Equipment   421
Material and Product Considerations   447
Operations and Control   454
Special Injection Molding Processes   459
Modeling and Computer-aided Mold-flow Analysis   446
Case Study 12.1—Estimating the Cost of an Injection Molded Pocket Knife   468
Case Study 12.2—Mold Costs and Selection   474

Chapter 13 Blow Molding   483

Introduction   483
Molds and Dies   494
Plant Concepts   497
Product Considerations   499
Operation and Control   501
Case Study 13.1—Making Soda Pop Bottles   503

Chapter 14 Thermoforming Process   509

Introduction   509
Forming Processes   510
Equipment   524
Product Considerations   528
Operation and Control   534
Case Study 14.1—Continuous Thermoforming   536

Chapter 15 Rotational Molding Process   543

Introduction   543
Equipment   549
Product Considerations   551
Operation and Control of the Process   558
Case Study 15.1—Trash Cart Manufacturing   561

Chapter 16   Casting Processes   567

Introduction   567
Casting Processes   569
Equipment   578
Product Considerations   580
Operation and Control of the Casting Process   582
Case Study 16.1—Casting a Polyester Thermoset Part in a Silicone Mold   584

Chapter 17   Foaming Processes   589

Introduction   589
Processes to Create Foams in Resins   591
Processes to Shape and Solidify Foams   592
Rebond   603
Product Considerations   603
Control and Operation   608
Case Study 17.1—Foam Insulation   609

Chapter 18   Compression and Transfer Molding Processes   615

Compression Molding   615
Transfer Molding   624
Product Considerations   629
Control and Operation   629
Reaction Injection Molding (RIM)   633
Cold Forming, Sintering, and Ram Extrusion   634
Case Study 18.1—Manufacture of Automobile Body Panels   636

Chapter 19   Polymeric Composite Materials and Processes   643

Introduction   643
Matrix Materials   651
Reinforcements   653
Manufacturing Methods for Composite Parts   658
Plant Concepts   671
Case Study 19.1—Filament Winding of the Beech Starship Airplane Fuselage   674

Chapter 20   Radiation Processes   681

Introduction   681
Equipment and Process   685
Properties, Materials, and Applications   690
Plasma Polymerization and Reactions   693
Case Study 20.1—Making Shrink-Tubing Using Electron Beam Crosslinking   697

Chapter 21 Finishing and Assembly   703

Introduction   703
Runner System Trimming and Flash Removal   703
Machining   705
Nontraditional Machining   708
Shaping (Postmold Forming)   709
Mechanical Joining and Assembly   710
Adhesive Bonding   714
Nonadhesive Bonding   720
Joint Design   725
Coating and Decorating   726
Case Study 21.1—Comparison of Adhesive-Bonded and Metal Attachments   732

Chapter 22 Environmental Aspects of Plastics   739

Introduction   739
Source Reduction   740
Recycling of Plastics   742
Regeneration   749
Degradation   749
Landfills   750
Incineration   752
Total Product Life Cycle   753
Future   757
Case Study 22.1—Recycling Solid Wastes   758

Chapter 23 Operations   765

Introduction   765
Safety and Cleanliness   765
Plastic Resin Handling, Conveying, and Drying   768
Plant Layout   770
Quality Assurance   771
Case Study 23.1—Establishing QC for a PET Bottle Plant   772

Appendix 1 Cost Estimating Form for Injection Molding   777

Appendix 2 Plastics Design/Selection Matrix   778

Answers to Selected Questions   779

Index   801

Preface

This edition retains the general objectives and format of the previous edition with some important additions and reorganizations to clarify some topics. The principal objective of *Plastics: Materials and Processing,* Second Edition, is to introduce plastics to a broad cross section of readers who have a need to gain, improve, or refresh their knowledge of plastics. The book is intended for students of technology, engineering technology, and engineering, and for professionals in the plastics industry (such as technical and nontechnical managers, staff in plastics companies, foremen, and operators). The text emphasizes the fundamentals of plastics materials and processing, yet it is detailed enough to be a valuable resource for future reference. This combination of fundamentals and details makes the book ideal as a textbook for an introductory course in plastics. The instructor can emphasize those topics that have special application for the class and can also assign additional reading to enhance the overall knowledge of the student in the entire field of plastics technology. The book is also an excellent resource for seminars in plastics technology, as well as for company courses and personal study.

The book is not, however, a reference for design data and plastic properties. That role is fulfilled adequately by the several encyclopedias, handbooks of plastics, and computer databases that are published regularly and therefore can present more up-to-date data.

The text parallels an introductory plastics course taught for many years at Brigham Young University. (Hence, the text itself, the objectives, problems, and format have been tried in practice and have been shown to help students succeed.) This is the only plastics course available for most of these manufacturing engineering and technology students, who have reported its value during later work experience in the plastics industry. The text provides a proper foundation for advanced courses in polymer synthesis, polymer properties, and plastics processing.

A background and basic understanding of high school or freshman chemistry, physics, and mathematics is suggested. A few important mathematical formulas are presented and used to show how the various variables are related, to enable important operational calculations to be made, and to illustrate the mathematical theory of key plastic properties. Molecular (chemical) formulas for many of the plastics materials are given, along with an introduction of basic organic chemistry that provides the reader the necessary background to readily understand molecular formulas. As the reader gains experience in plastics, these chemical formulas will serve as valuable references to a deeper understanding of the relationships among plastic structure, properties, and processing.

Plastics is a category of materials that traditionally includes commercial and engineering thermoplastics and thermosets. If a broad view of plastics is taken, elastomers and highly modified natural polymers can also be included. This book takes the broad view, thus allowing comparisons of similar concepts and principles within all these related materials.

Plastics are introduced at three levels of focus: (1) the molecular, (2) the micro (polymer chains and crystals), and (3) the macro (physical properties). Through knowledge of all three levels, readers can understand and predict the properties of the various plastics and their performance in products. Manufacturing methods for plastics and the changes in plastics properties that result from manufacturing are also related to the three levels.

Each chapter in the book has an introductory section that describes the major concepts of the chapter. The chapter then expounds the subject in qualitative and limited (no derivations) quantitative terms. Extensive figures and tables give visual and comparative understanding to the concepts. At the end of each chapter, a case study highlights in detail some important aspect of the chapter in a specific circumstance. Also at the end of the chapter is a summary of the major concepts and objectives. Questions then follow to test the reader's *understanding* (rather than mere recollection) of the principles presented in the chapter. A list of references is provided to assist the student in finding additional material on the subject of the chapter.

The learning of plastics is directly connected with the vocabulary of plastics. Not only are the concepts often expressed in unique terms, but the industry communicates in these terms. Therefore, terms that have unique meanings in plastics technology are italicized when they are introduced in the body of the text and defined briefly when they are used. Furthermore, all of these new terms are included in the index for easy reference. A valuable cost estimating form for injection molding parts is also included in Appendix 1.

Plastics has many highly interrelated topics. Ideally, topics such as molecular interactions, crystallinity, thermal transitions, steric effects, processing methods, and product applications should all be perceived simultaneously in order to gain the best appreciation of each. Simultaneous perception is, however, very difficult when the topics are new. This book, of necessity, presents the material in a linear fashion. However, for best understanding, the book should be reexamined in a rapid, overall reading so that the whole picture of plastics can be appreciated. The structure of the book—with the chapter outlines and summaries, case studies, questions, and appendix—is intended to assist in gaining that overall view.

New Features

In this new edition, some concepts have been reorganized so that the flow is easier for the student. For instance, the tooling chapter of the first edition has been distributed into tooling sections in each of the processing chapters, thus integrating tooling and processing for each process. Similarly, the testing section has been put into the chapters where properties of plastics are discussed. The chapter on design has been moved to immediately follow the chapters on properties, thus giving an immediate example of how the properties can be used in specific examples.

The second edition has several new charts and figures, which not only improve on previous charts but also allow some concepts to be understood in a broader overall view, often better than was done with text only. A glossary of new terms has been added to each chapter.

A form (Design Matrix) that can be copied and used for designing new plastic parts has been added in Appendix 2. A case study in the chapter on design illustrates, in detail, how to use this form.

Brief characteristics of the major plastics are printed inside the front and back covers for easy reference and comparison.

Acknowledgments

I am grateful for the insightful help from the following reviewers: John D. Colluccini, Fitchburg State College; Gary San Miguel, Texas State Technical College; Barry G. David, Millersville University; James T. Johnson, Sinclair Community College; David H. Devier, Ohio Northern University; Mark L. Nowak, California University of Pennsylvania; and Charles L. Hamermesh, Society for the Advancement of Material and Process Engineering. I also thank the staff of the Manufacturing Engineering and Engineering Technology Department at Brigham Young University for their assistance in the preparation of the text. Special thanks to Dannie King Graves, Brenda Baker, Alisa Corfield, Lindsey Dickson Tobler, Camille Call Whiting, Janelle Wakefield, Chad Woolf, Jan Martindale, and Ruth Ann Lowe for help with typing and illustrations. Thanks also to Kent Kohkonen, Roger Turley, Norman Lee, David Sorensen, Ra'ed Al-Zubi, Brian Mansure, and Scott Hansen for their contributions. I also appreciate the encouragement and understanding given by my family and friends during the writing of this book.

Thanks to the reviewers who assisted in the preparation of the second edition: John D. Colluccini, University of Massachusetts, Lowell (adjunct); David H. Devier, Owens Community College; James T. Johnson, Sinclair Community College; and Roy D. Thornock, Weber State Unviersity.

Thanks also to Holly Henjum, Michelle Churma, and, especially, Stephen Helba, an editor who has helped me with this and several other books.

CHAPTER ONE

INTRODUCTION TO PLASTICS

CHAPTER OVERVIEW

This chapter examines the following concepts:

- Definitions of plastics and polymers
- History of plastics
- Raw material supply and pricing
- Strategic materials
- The plastics industry
- Uses of plastics in modern society

DEFINITIONS OF PLASTICS AND POLYMERS

Plastics is not a uniformly defined term. Some prefer to define plastics in a relatively narrow sense, focusing on specific properties (such as formability). Others prefer to define plastics more broadly, viewing collectively properties, processing, and design characteristics of a group of related materials. This book uses a relatively broad definition, with the objective of assisting the reader to appreciate the fundamental similarities between a large group of related materials.

***Plastics* are materials composed principally of large molecules *(polymers)* that are synthetically made or, if naturally occurring, are highly modified.** This definition of plastics can be illustrated in a systematic classification diagram, as shown in Figure 1.1. In addition to their similar nature as synthetic polymers, **all plastic materials have the property that at some stage, they have been or can be readily formed or molded into a useful shape.** (The word *plastic* comes from the Greek *plastikos,* which means to form or mold.)

As Figure 1.1 shows, all materials can be classified as gases, simple liquids, or solids, with the realization that most materials can be converted from one state to another through heating or cooling. If only materials that are solids at normal temperatures are examined, three major types of materials are encountered: metals, polymers, and ceramics. The polymer materials can be further divided into synthetic polymers and natural

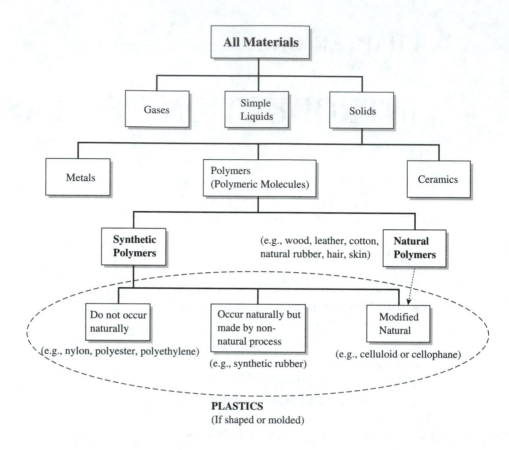

Figure 1.1 Diagram illustrating the definition of plastics.

polymers. Most synthetic polymers are those that do not occur naturally and are represented by materials such as nylon, polyethylene, and polyester. Some synthetic polymers could be manufactured copies of naturally occurring materials (such as synthetic rubber) or even natural polymers that have been so radically modified that they no longer possess the general properties of the original natural polymer, such as celluloid or cellophane, which are derived from cellulose. Therefore, natural rubber is not a plastic but is considered as a reference material in the chapter on elastomers. (Some narrow definitions of plastics exclude all elastomers from the plastics group.) Hence, by our definition, **plastics include all non-naturally occurring polymers, all synthetic elastomers, and all highly modified natural polymers,** as shown in the circled area within Figure 1.1.

Definition of Polymers

A detailed explanation of polymers is given in the chapter on the molecular nature of materials. However, a simple understanding of polymers can be gained by imagining them to be like a chain or, perhaps, a string of pearls where the individual pearls represent small molecules that are chemically bonded together. Therefore, **a *polymer* is a molecule made up of smaller molecules that are joined together by chemical bonds.** The word *polymer* means many parts or units. The parts or units are the small molecules that combine. The result of the combination is, of course, a chainlike molecule (polymer). Usually the polymer chains are long, often consisting of hundreds of units, but polymers consisting of only a few units linked together are also known and can be commercially valuable.

All *molecules,* whether the small type or the large type that result when particular small molecules join together, are made up of atoms (such as carbon, hydrogen, oxygen, or nitrogen). When any small molecule is formed, the atoms join together into a specific arrangement that is characteristic of the particular molecule. The types of atoms and their arrangement determine the properties of the molecule. For instance, the small molecule called methane (natural gas) always has one carbon and four hydrogens, which are arranged in a tetrahedral (pyramid) shape. Another small molecule, ethylene, is a gas derived from petroleum that always has two carbons and four hydrogens arranged so that two hydrogens are connected to each carbon and the carbons are also linked to one another in a planar arrangement. Ethylene is one type of small molecule that can be combined into very long chains to make polymers (polymeric molecules), whereas methane cannot be readily combined to form polymers. Plainly ethylene and methane are different small molecules and have different chemical properties.

Therefore, the *chemical properties* of a molecule determine the types of reactions into which the molecule can enter. In the example of ethylene and methane, the ethylene is more chemically reactive under the conditions needed to form polymers. The reasons for this chemical reactivity and the nature of the chemical reactions that take place are a major study of chemists and are beyond the scope of this book. However, some specific examples are given in the chapter on the molecular (chemical) nature of polymers, especially concerning the reactions that take place when small molecules combine to form polymers. For now, the polymer-forming process can be understood in general terms by examining Figure 1.2, where the combining of small molecules to create a polymer is depicted. Note that each small molecule has two reactive forces, thus allowing each small molecule to be bonded to two others and a long chain to be formed.

The chains formed are called *polymers,* or *polymeric molecules.* Another term that is widely applied to polymeric molecules is *macromolecules* (from the Greek *makros,* meaning long or large). The chains become new molecules with properties that are different from those of the original small molecules, even though the individual units might all be the same. These molecular chains can be short, in which case the molecule is likely to be a liquid at room temperature. These short molecular chains are sometimes called *oligomers.* An example of a short-chain molecule would be cooking oil. Long-chain molecules are usually solids or viscous liquids. When the chains are long, often containing thousands of units, the polymer could be a plastic (provided other definitional conditions are met) and might be called a giant molecule. All plastics are giant molecules, although some giant molecules are naturally occurring and are not, therefore, plastics by the definition used in this book.

Chemical reactions are usually represented as two or three molecules reacting together to form a single, or perhaps a few, new molecules. In reality, a very large number of identical molecules react to form a great number of identical, new molecules. When

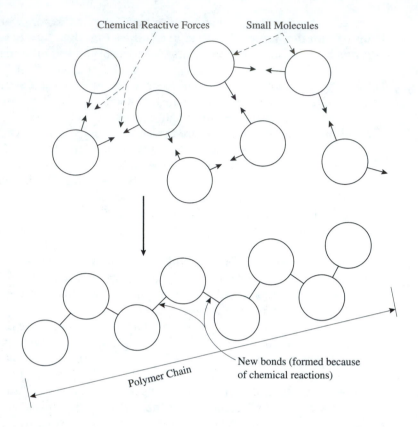

Figure 1.2   Illustration of small molecules combined into a polymer chain.

many molecules of the same type are combined together, the properties that are usually measured are for the large collection of molecules. These properties are called *collective* or *bulk properties* of the material. The bulk properties are determined by both the molecular properties (properties that depend upon the molecular nature of the material—such as chemical reactivity) and the collective properties (properties that depend upon the interaction of the molecules—such as crystal formation). Subsequent chapters examine both the molecular and bulk properties of plastics because understanding many properties of plastics, such as mechanical strength, melting point, and solvent reactivity, depends upon understanding both the molecular and the bulk nature of plastic materials.

Another term that is often associated with polymers and plastics is *resin*. Although no definition for resin is universally accepted, a convenient definition is: **a *resin* is a polymer that has not yet been formed into its final useful shape.** When initially made, polymers are usually viscous liquids or, if solid, are granules (powders) or flakes. In some cases the granules or flakes are formed into some intermediate shape (such as small pellets), but these can also be called resins because they are later formed into a shaped plastic part.

Because of this close connection between the terms resins, plastics, and polymers, these terms are sometimes used interchangeably, although correctly used there are differences. To summarize: polymers are any material made up of molecular chains; plastics are synthetic, long-chain polymers that can be or have been shaped; and resins are solids or liquids that are subsequently shaped into a plastic part.

HISTORY OF PLASTICS

The history of humankind's use of polymers and eventual development of plastics has followed a general pattern of events.

1. Discovery of the polymer. (This usually implies a naturally occurring polymer, but some discoveries of synthetic polymers were made in the laboratory unintentionally.)
2. Use of the polymer. (The early applications were usually based upon the obvious properties of the polymer and required little modification of the material.)
3. Realization of deficiencies of the material and attempts at modification, usually by trial and error.
4. Investigation of the properties of the material and development of a conceptual view or model of the material's basic nature. (This step may take many years.)
5. Systematic modification or synthesis of materials that might mimic the properties of the natural polymer or, in latter cases, development of synthetic materials that do not have natural analogues but have useful properties in their own right.

A time summary of the history of plastics is given in Table 1.1.

Since the beginning of history, people have benefited from naturally occurring polymers. These polymers have provided the raw materials for satisfying basic needs such as clothing (cotton, wool, silk, flax, fur), shelter (lumber, asphalt), and food (starch, protein) and many higher needs such as communication (papyrus, wood pulp), music (strings, glues, reeds, lacquers), decoration (amber), defense and war (arrows, spears, bows), and recreation (rubber). Most of these polymers could be used with only minor modifications, such as weaving the wool or cutting and shaping of wood.

Ancient people found that some natural polymers could be made more useful by making slightly greater changes to the polymer material. For instance, the flax plant could be beaten with rocks or between rollers to crush the cell structure and allow the long fibers to be separated from the rest of the plant. These long flax fibers were then woven into linen cloth. Even when more extensive modifications were made in a natural polymer, such as the soaking of hides in tannic acid (tanning of leather) to prevent hardening when they dried, little change was made to the resultant material except cutting, shaping, sewing, and other changes in physical shape. Many of these polymers are still important today.

Although most of these natural materials would not be considered plastics (they are not synthetic), some natural materials were molded in ways similar to modern plastics. For instance, the sap or resin from resinous trees like pines and firs was found to harden if left to stand in the air and could therefore be placed into a mold of some desired shape and allowed to solidify. The solid part could be removed and it would retain the shape of the mold. Jewelry, amulets, and idols were made by this method. The modern term *resin* comes from this tree-sap analogy.

An ancient natural polymer used in plasticlike processes is lac, which is a resin from certain shrubs that forms the basis of shellac or lacquer. The use of lac as a wood coating material was known and reported in about 1000 B.C. and was described in detail by explorers to India

Table 1.1   History of Plastics

Year	Event
Pre-1800	People discovered and used natural polymers (examples: wool, cotton, leather, wood, silk, flax, lacquers, rubber) with little modification of the polymer.
1839	Charles Goodyear discovered the process for vulcanizing natural rubber.
1868	Celluloid was invented by John Wesley Hyatt. This is considered to be the first plastic, because it was made by substantially modifying a natural polymer and then molding the resulting material into new shapes.
1877	Fredrich Kekulé proposed the chain model for polymers.
1893	Emil Fischer and Hermann Leuchs proposed a chain structure for cellulose and then synthesized the molecule, thus confirming their proposed structure.
1909	Leo Baekeland announced the invention of phenolic resin, the first polymer made from purely synthetic materials.
1924	The polymer chain structure for synthetic polymers was proposed by Herman Staudinger.
1925–1940	Several polymers made by the addition polymerization method were introduced (examples: PVC, PMMA, PS, PE, PVAc, PAN, SAN).
1934	Nylon and the condensation polymerization process were invented by Wallace Carothers.
1940–1950	Several polymers made by the condensation polymerization method were introduced (examples: PET, unsaturated polyester).
1950–1955	Low-pressure catalysts were developed by K. Ziegler and G. Natta.
1955–1970	Many polymers made by a variety of methods were introduced (examples: PC, silicones, PPO, acetal, epoxy/polyurethane).
1955–1970	Composite materials were developed using synthetic resins and strong, stiff reinforcements such as fiberglass, carbon fiber, and aramid fibers.
1970–1990	Production and development of new manufacturing methods for plastic parts were expanded, resulting in lower prices, improved quality, and applications where other materials, such as wood and metal, had previously been used.
1990–2000	Resins with sophisticated structures were developed, giving special properties such as high thermal resistance, low flammability, light sensitivity, electrical conductivity, biodegradability, and biocompatibility.
1990–2000	Several new catalysts that significantly improved the properties of many resins were developed, thus further expanding the capabilities of plastics.

in the sixteenth century. Modern paints employ the same basic principles of drying from a solvent base as did ancient lac.

Natural rubber is another polymer that was described by sixteenth-century explorers. The natives of Central and South America had found that by coagulating the latex sap from certain trees, a flexible, bouncy material was produced. In 1839, Charles Goodyear discovered that natural rubber heated with sulphur retained its elasticity over a wider range of temperatures than the raw rubber and that it had greater resistance to solvents. This process came to be called *vulcanization*. If very large amounts of sulphur were added, the rubber stiffened significantly. This material is known as hard rubber. Later, others extracted several materials from natural rubber and characterized these, eventually breaking down the rubber into its basic chemical constituents. These were then recombined and, in 1897, an elastic, rubberlike material was synthesized. Hence, in the case of natural rubber, the steps in the pattern of polymer discovery—use, characterization, modification, and synthesis—were followed. Natural rubber is not strictly considered a plastic, but highly modified natural rubber and synthetic rubbers would be plastics.

In the nineteenth century, wood pulp, plant fibers, or cotton fibers (all made of cellulose, a natural polymer) were treated with nitric acid to form a highly explosive material called gun cotton or more commonly today, nitrocellulose. It was used as a substitute for gun powder in both the American Civil War and World War I. If the nitrocellulose had a lower nitrogen content, it was less explosive and could be treated with camphor to become pliable and moldable. This material was known as Celluloid. Celluloid was used for early motion picture films, waterproofing coatings, combs and other molded items, and making billiard balls (slightly explosive if impacted very hard). Celluloid, invented in 1868 by John Wesley Hyatt, is considered to be the first commercial plastic. It was soon discovered that treatment of nitrocellulose with other acids and solvents resulted in quite different materials that could also be pressed into films or forced through small holes to form continuous fibers. These became known as cellophane and rayon. By the definition of plastics given at the beginning of this chapter, these *highly modified* natural polymers are viewed as plastics, that is, polymers that are substantially made (or modified) by synthetic (nonnatural) processes.

Near the end of the nineteenth century and the beginning of the twentieth century, key postulates on the molecular structure of polymers were made that gave impetus to the development of new, wholly synthetic polymers. The synthetic fabrications then led to other, improved or expanded structural postulates. For instance, in 1877, Fredrich Kekulé, a pioneer in modern organic chemistry, proposed that natural organic substances consist of very long chains of molecules from which they derive their special properties. In 1893, Emil Fischer proposed a chain structure for cellulose that was followed shortly thereafter by the synthesis of a long, linear molecule based on sugar by Hermann Leuchs, an associate of Fischer. This synthesis confirmed many of the features of the Kekulé and Fischer structures of natural polymers.

Chemists also began to synthesize and explore the properties of polymers that were built up from small molecules rather than derived from natural polymers, although the syntheses were largely done by trial and error. One of the earliest developed (early 1900s) wholly synthetic polymers was phenolic (named Bakelite by Leo Bakeland, the inventor). It was formed by mixing and heating phenol and formaldehyde, two easily obtained, widely used chemicals. The process resulted in a resin that could be shaped (molded) and then, with time and elevated temperature, solidified into a hard material with excellent thermal and electrical insulating capabilities. The material is still used as handles for cooking pans and electrical switches although other plastics now compete for these applications. Shortly thereafter other polymers based upon formaldehyde were synthesized, some of which found use as coatings for paper and wood and are still widely used in kitchen countertops (Formica®) and for the adhesive in particle board lumber and plywood.

Several other polymers were then found by mixing simple gases under extreme conditions (usually high heat and pressure) to form powdery solids. This synthesis method is today called the *addition, chain-growth,* or *free radical polymerization,* which are described in detail in Chapter 2, Polymeric Materials (Molecular Viewpoint). The most common example of this was the reacting of ethylene gas to form polyethylene. Other polymers made during this time and by similar processes were polyvinyl chloride (PVC), polystyrene (PS), and poly-methyl methacrylate (PMMA). The processes were poorly understood with successful results often coming only because of fortuitous accidents. For instance, the discoverers of polyeth-ylene had great difficulty duplicating their original successful synthesis. After much investi-gation they discovered that a trace amount of oxygen was necessary for the reaction to pro-ceed and that in the original experiment a small leak in the apparatus had provided this small oxygen source. In most of these cases the polymeric natures of the products were not well un-derstood, despite the early work of Kekulé and Fischer. (Many scientists believed that the solid products of small molecules were simply small molecules held tightly together by physical, not chemical, forces and were structurally different from the naturally occurring polymers.)

The structural model for modern, wholly synthetic plastics can be traced to the proposed structure of a polymer by Herman Staudinger in 1924. He proposed that a polymer was a lin-ear structure consisting of small units held together by normal chemical bonds. This struc-ture model was disputed by many leading chemists of those days. However, the emergence of X-ray diffraction and of the ultracentrifuge were key analytical tools that were used to inves-tigate the structure of polymers and eventually confirmed the Staudinger structure.

This well-defined structural model led to a decision by the DuPont Company in the early 1930s to make a polymer entirely from small molecules with specific, and preconceived, properties. DuPont hired Wallace Carothers, from Harvard University, to make a material that had properties similar to natural silk. After several years of experimentation and devel-opment, the polymer that we now call nylon was created. Even more important, Carothers developed a new process for making polymers and clarified the supporting molecular struc-ture model. Within a short time, this knowledge of the nylon polymerization process (called *condensation polymerization* or *step-wise polymerization*) led to the development and un-derstanding of additional processes for fabricating other polymers by simply varying the type of small molecules used as the starting materials. These polymerization processes will be discussed in more detail in the chapter on Polymeric Materials (Molecular Viewpoint).

The combination of good structural models and several polymerization methods (addi-tion and condensation), coupled with the needs of World War II and the post-war consumer boom, resulted in rapid developments of many new polymers and hundreds of diverse new applications for plastics. Polytetrafluoroethylene, or PTFE (Teflon®), was discovered (by ac-cident) and then synthesized and became an important material for wire insulation, chem-ical-resistant, and nonstick applications. Other developments in plastics included synthetic rubber; scratch-resistant, weatherable, transparent plastics for aircraft, buildings, and pack-aging; lightweight, low-cost, nonflammable, nonwrinkling fibers for clothes, carpets, and other textile applications; sophisticated coatings and packaging materials for food wrap-pings and containers as well as paints; unique electrical devices and insulators; and struc-tural materials for buildings, aircraft, and automobiles. Plastics successfully replaced tradi-tional materials (such as wood and metal) in many applications, but occasionally with less than acceptable results before the long-term behavior of the plastic was well understood. Hence, plastics developed a reputation for being "cheap" and short-lived. In some applica-tions, however, plastics were indispensable and filled the requirement better than any other material. An example would be wire insulation, which combined flexibility with the good electrical insulative capability of plastics. Improvements in plastics materials and use in more appropriate applications have largely removed the negative image of plastics.

Catalysts developed by Ziegler and Natta allowed low-cost production of high-density polyethylene and polypropylene, which began to be produced in large volumes. Investigations surrounding the use of these catalysts further illuminated the processes of polymerization and the resulting polymeric structure was refined. The concepts of crystallinity in polymers and the dependence of properties on molecular structure were understood at a much deeper and more meaningful level. These concepts led to other catalysts, improved processing methods, and stronger, longer-lasting, and less expensive polymers. Some of these polymers were new materials, but in some cases new plastic materials were made by blending or combining previously known polymers.

The combining of plastics with fiber reinforcement materials, usually ceramics or metals, but occasionally very strong and stiff polymers, provided further advances in properties available for society. These combinations of materials, called *composites*, were easily formable because of the plastic material but were very strong and stiff because of the reinforcements. Fiberglass-reinforced plastic, carbon fiber epoxy, and carbon-carbon composites are typical examples that have been used extensively in automotive, aerospace, and sports applications.

Other areas of polymer development include electrically conductive polymers, light-sensitive polymers, biodegradable plastics, biocompatible plastics for medical applications, and derivation of plastic feedstocks from sources other than petroleum, such as coal and plant materials.

While the developments in plastic materials have been astounding, the developments in the processing, fabrication, and analysis of plastics have been equally important. A multitude of plastic-manufacturing methods are now available, each with one or more particular advantages. These processes allow plastic materials to be shaped inexpensively and accurately while achieving desired functionality or desired material characteristics and optimize performance. New equipment and new methods are still the subject of extensive development efforts.

This history of plastics began with people's use of and curiosity about natural polymers. Improvement of these natural polymers was inevitable. Over time fortuitous events led to the discovery and characterization of synthetic polymers. Eventually the underlying concepts of polymer synthesis, structure and properties were so well developed that polymers could be made to fill specific end-use requirements. Today, the development, characterization, and fabrication of polymers involves chemists, mathematicians, statisticians, design engineers, and manufacturing engineers, among others. In many cases the tasks of these experts overlap, which may facilitate the advancement of the field of plastics as the skill levels required for further advancements continue to increase.

RAW MATERIAL SUPPLY AND PRICING

Although the history of plastics begins with natural polymers, modern plastics are generally made from small molecules and built up into polymeric chains rather than by converting existing natural polymers into plastics. The small molecules used to make most plastics are derived chiefly from crude oil. (The major exception is plastics based upon the molecule ethylene where about half of the ethylene is derived from natural gas, although the rest of the ethylene is from crude oil.) The plastics component of crude oil is approximately 2% of the total volume of crude oil consumed (Figure 1.3). Plastics can also be obtained from coal, trees and other plant products (corn, nut husks, soy, oats, tree sap), as well as from other naturally occurring materials (fish, animals, algae), although crude oil is currently the least expensive and most widely used source.

The prices of raw materials for plastics are dependent on the price of oil, but not to the same extent as fuels, which are the largest use component of crude oil. Other factors such as production costs, capacities, and demand generally have a greater influence on costs. For instance, the costs of polyethylene (PE) and polyvinyl chloride (PVC) are low (approximately one-quarter) compared to the costs of polycarbonate (PC) and polyethylene terephthalate (PET), yet the amount of crude oil contained in all of them is not much different. The differences in price arise from differences in the processes and, eventually, in the amount of the plastics produced (price-volume relationship). The process complexity is dependent upon cost of raw materials, costs of non-petroleum components, costs of equipment, variations and complexity of the processing conditions, labor, environmental protection costs, and energy consumed. Table 1.2 gives the prices of various plastics in 1998, showing the high variability

U.S. Petroleum Consumption, 1988

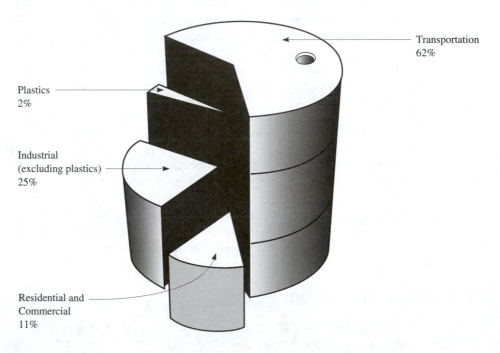

Figure 1.3 Plastics as a component of petroleum consumption in the United States.

of prices with a range of $0.38 per pound for polyethylene to $6.70 per pound for polytetra-fluoroethylene (PTFE) and $33.00 per pound for polyetheretherketone (PEEK).

Table 1.2 Prices of Various Plastics (1999)

Plastic Type	PE	PS	PET	nylon	PC	PTFE	PEEK
Price ($/pound)	0.38	0.40	0.45	1.30	1.70	6.70	33.00

The pricing pattern for a new plastic material has usually been to sell the new plastic for uses that place a high value on the properties of the plastic so that high prices can be justified. As time goes on, the manufacturer of the plastic develops more economical methods of manufacture and the price drops, which allows the development of other uses. Eventually the volume and market stability warrant the erection of a full-size manufacturing facility rather than the semiworks or pilot plant that are initially used. The economies of scale and inevitable competition usually result in a dramatic price reduction for the plastic over the subsequent few years and then a leveling off of the price as the process and consumption (demand) become stable and the major uses are identified and demonstrated. The long-term price then becomes a function of the complexity of the process and the volume (as already discussed). The price-volume relationship is illustrated in Figure 1.4 for several plastic and nonplastic materials. The general linearity of the curve in this figure confirms the strong price-volume dependence of all these materials. The deviations from the line, which are generally small, confirm that the process complexity is a factor in the price, although it is a secondary factor.

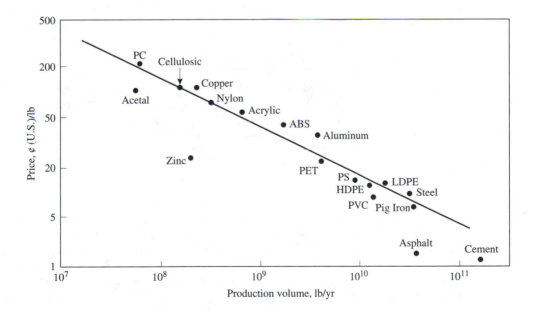

Figure 1.4 Price-volume relationship for plastics and other selected materials (1998 data).

STRATEGIC MATERIALS

One consideration often overlooked in the manufacture of materials is the location and long-term stability of the source of the principal raw material and other critical ingredients. In times of social unrest (such as wars or economic embargoes) or when the supply of a material is controlled (such as diamonds or oil), the availability of the raw materials may diminish. Alternative starting materials or alternative finished goods must then be found, especially for applications that are strategic to national goals. Germany, for instance, was cut off from the world's supply of natural rubber during World War II. Because rubber was critical to the German goals, a substitute had to be found. The result was synthetic rubber (a plastic), which filled most of the applications previously filled by natural rubber. In the United States the silk supply was cut off. This was an impetus for the creation of nylon.

Those times of social instability were also periods that fostered the development of new uses for existing materials not previously envisioned. Nylon was designed as a silk substitute but became widespread as a rope material during WWII, even though hemp, the previously dominant rope material, was not a major target for the original nylon development.

Because plastics are so ingrained in modern life and fulfill so many critical applications for national defense and other national goals (in almost every country), the availability of the raw materials should be examined from a strategic point of view. As already discussed, the principal raw material for most plastics is oil. That can certainly be a problem raw material for most of the world. However, because plastics make up only a small amount of the oil consumed, residual oil from old oil fields could be used as an interim source for plastics in the event normal crude oil supplies were interrupted. Alternate sources of raw materials such as agricultural products, coal, and natural gas could also be expanded. The major impediments to the use of these alternate sources are raw material cost compared to oil and the difficulty of converting some of the manufacturing sites for the chemicals to accept the new raw material feed stocks. Therefore, plastics can usually be made from raw materials that are generally available, although at a somewhat higher cost to produce.

This situation of relatively stable raw material sources also exists for paper, glass, and some metals, especially iron and copper. Other metals, such as chrome (used in stainless steel), are far less available and have fewer alternate materials, and must, therefore, be stockpiled or otherwise secured for strategic purposes.

PLASTICS INDUSTRY

A diagram of the plastics industry is given in Figure 1.5, with each of the major functions represented in a box. The interactions between the various functions are illustrated by the arrows. The heavy arrows are the most common paths. Generally these interactions are sales, and they are made in the directions shown by the arrows, but sometimes the interactions are simply intercompany transfers when a company has integrated more than one function internally. The integration of several steps is quite common in the plastics industry.

The *resin manufacturers* convert chemicals (derived from crude oil, natural gas, coal, and other sources) into the basic polymer materials. Hence, these processes are called *polymerization processes.*

Figure 1.5 Diagram of the plastics industry.

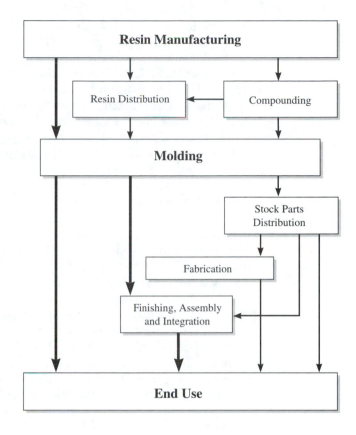

These resins, which require further processing to be useful, are generally made in large, highly integrated manufacturing facilities that resemble oil refineries in their size and scope. The enormous investments required to build such facilities have resulted generally in large petrochemical companies becoming resin manufacturers. Some of the most well-known resin manufacturers are DuPont, ICI, Exxon, Mobil, Hoechst, Shell, BASF, and Allied Signal, among many others. There are approximately 40 major resin types, with each resin type differing from all others according to the fundamental chemical nature of the polymer. Most of the resin types are available from more than one resin manufacturer. Some of the most common of these resin types are: polyethylene, polypropylene, polystyrene, polyvinyl chloride, nylon, polycarbonate, polyurethane, and polyester.

Most of the processes used to make resins are flow processes that require careful control over temperature, pressure, flow rate, catalyst, and other associated parameters in order to obtain the desired properties and to minimize the production of unwanted side products. This text considers only the most basic concepts in polymerization (in the chapter on polymeric materials) so that the reader will be able to appreciate the way polymers are made and understand the changes in polymer properties that arise from the polymerization processes. The emphasis of this book is on the molding of the resins after they have been polymerized.

The resins usually exit from the polymerization reactor in one of three physical forms: liquids, granules, or flakes. The liquids are roughly the consistency of honey. The granules resemble laundry soap powders in texture, size and consistency. The flakes resemble uncooked oatmeal or instant potato flakes in texture and size. If sold directly from the reactor, the liquid resins are generally packaged in 5-gallon, 55-gallon, or tank car containers. The granules and flakes are generally packaged in 50-pound bags, palletized cartons or gaylords (1,000 pounds), in hopper trucks, or in rail cars.

In many cases the resin manufacturers send the granules and flakes through one more process in order to make a more consistent product for later processing and/or to remove contaminants, especially solvents, that might be present. In this additional processing step the granules and flakes are converted into pellets that are shaped like small rods about 0.1 inch (2 mm) in diameter and 0.2 inch (4 mm) long. (They are like spaghetti that has been chopped into very short pieces.) This additional processing step also gives the resin manufacturers an opportunity to adjust the average pellet properties by blending resins of the same polymer material from different batches. Blending is often much easier than trying to fine-tune the resin-making operation itself for each of the minor differences in products that might be offered. Every resin type is available in many varieties, each of which is made by a slightly different set of polymerization process parameters or by blending of the materials from different polymerization batches. These varieties differ slightly from each other in physical properties but are similar in overall properties within each resin type. Minor additives could also be included in the pelletizing step.

Most resins are sold from the resin manufacturers directly to the *molders*. If a molder does not buy in large enough quantities to be supplied directly from the resin manufacturer or, perhaps, desires some special services not provided by the resin manufacturers, the molder might buy from a *resin distributor,* who typically buys in large quantities and then ships in smaller lots. The molder may also need services such as color matching, addition of processing aids, or grinding. In these cases the molder would buy from a *compounder.*

The molders convert the resins (liquids, granules, flakes or pellets) into the desired shapes by using one or more plastic-molding processes. Typical molding processes are extrusion, injection molding, compression molding, and casting. These and other plastic-molding processes are discussed in detail in this text.

Molding companies that have plastic parts as their principal products usually have many plastic processing machines and make parts in large quantities. Typical examples of these companies are plastic-pipe extruders, injection molders of automotive parts, or plastic-toy manufacturers.

Other molders are companies whose principal product is something other than a plastic, but in which some plastic part or parts are used. These companies would typically have only a few plastics processing machines. Examples of this type of company might be a manufacturer of medical devices that uses special plastic fittings, airplane manufacturers who use plastic parts within the airplane, or a milk processor who uses plastic milk jugs.

These essentially nonplastics companies may elect to have an outside company do the molding for them. Such outside companies are called *job shops* or *custom molders,* which typically would have several injection molding machines, often of different sizes, and would make parts for companies who use these parts as components in their products. These job shops often provide other services for the companies such as part design assistance, mold making, and, perhaps, some assembly.

Still other molders make standard-shaped parts that are usually sold to companies that perform additional shaping operations. These shapers of already-molded parts are called *fabricators*. A typical fabricator buys extruded sheet and vacuum forms this sheet into a finished product, such as a case for some instrument. Another typical fabricator buys PVC plastic film and laminates the film to cloth. If the fabricators choose to buy in small quantities or desire special services, they may also buy from *stock parts distributors* that warehouse standard plastic shapes, such as sheets, rods, and blocks.

Molders, fabricators, and stock parts distributors may also transfer their parts to a function that is concerned with *finishing, assembling, or integrating* the plastic part into a larger assembly. These operations are distinguished from fabricating because they are focused on mechanical operations such as cutting, bonding, and painting rather than forming from standard shapes by secondary molding. Finishing, assembling, and integrating are often done in-house by the molder or fabricator but can be done by companies dedicated to this type of specialized operation. Companies that use plastic parts in their products and buy the parts from molders are often involved in this functional step. For instance, furniture companies will cut and shape foam for use in their products. Likewise, aircraft companies could buy molded parts and then rivet or otherwise join them to major components of the airplane.

Molders, fabricators, stock parts distributors, and finishers may all sell directly to end users, but they may sell to some service provider that may perform some finishing step. For instance, a physician might use epoxy resin and fiberglass to mold a cast for a patient's broken arm. A landscaper might bond together components of a sprinkler system. A homeowner might install a plastic roof on a carport or make repairs to a damaged plastic automobile part.

USES OF PLASTICS IN MODERN SOCIETY

Most consumers have relatively little insight into the characteristics of specific plastics unless educated by advertising. For instance, many people are unfamiliar with nylon except as a textile fiber for hosiery, tents, and carpets. Even then, they generally do not understand the properties of nylon that make it a good choice for these applications. They may, however, be more familiar with natural products such as cotton or wool, which might compete with nylon. Similarly, many consumers classify all plastic materials into a general category, unaware of the differences in performance that are possible. Such a comment as "It broke because it is just plastic" might be typical of such a characterization.

The selection of a material for a particular design is based on careful analysis of engineering properties as they relate to the design function coupled with a consideration for the manufacturing capabilities and costs of the material. Since plastics are solid materials when in final use, some characteristics of plastics, as compared to other solid materials, chiefly metals and ceramics, are listed in Table 1.3. Note that some characteristics can be an advantage in one application but might be a disadvantage in another application. For instance, the low melting point of plastics is an important advantage in processing because the molding can be done at lower temperatures than are common with the shaping of metals and ceramics. However, this same low melting point also leads to a narrower useful thermal range because plastics lose many of their beneficial mechanical properties at lower temperatures than do most metals and ceramics. Therefore, the list of characteristics should be reevaluated for each application. Another caution in interpreting Table 1.3 is that there is much diversity within each type of material, which leads to overlaps in properties between plastics, metals, and ceramics. The table gives general properties associated with the type of material and is not necessarily representative of all materials in that type. In some applications, the need for a particular property or properties may be so dominant that other, less desirable properties can or ought to be accepted. For instance, even though plastics may wear out much more rapidly than steel, the high elongation and good recovery of plastics and the ability to form them into various shapes has led to a preference for plastics (rubber) for automobile tires. The ride comfort is worth the loss in long-term wear.

Table 1.3   Characteristics of Plastics When Compared to Metals and Ceramics

Characteristic	Advantage/Disadvantage
Low melting point	Ease of processing/Lower useful thermal range
High elongation	Low brittleness/Higher creep and lower yield strengths
Low density	Lightweight products/Low structural strength
Low thermal conductivity	Good thermal insulation/Dissipates heat poorly
Electrical resistance	Good electrical insulation/Doesn't conduct electricity
Optical clarity (some types)	Usefulness as a clear material/Degradation by sunlight
Easily colored	Use without painting/Difficult to match colors
Solvent sensitivity	Can be applied as a solution/May be affected by solvents
Flammable	Waste can be burned/May cause fumes or fire hazard

Plastics are, by their basic nature, capable of being formed. In general, therefore, the use dictates the form of the plastic part. An infinite variety of forms are achievable. Some applications have traditionally used standard forms, including sheets, films, rods, tubes and pipes, fibers, blocks and slabs.

Because plastics are so prevalent in the world today, only a partial list of applications can be given. Even a complete and comprehensive list of current applications would quickly be outdated since the application of plastics is such a dynamic segment of industry. The following list is intended to show the breadth of these applications across many industries and to illustrate some applications that may be little known.

Packaging

- Wrapping for thousands of different food items. These wraps can be a single plastic or multilayered films using up to seven different plastics and may be flexible or rigid.
- Bottles and other containers for items such as milk, soda pop, shampoo, medicine, liquid laundry products, cleaning agents, insecticides, distilled water, microwavable food, motor oil (often with see-through windows to tell how much is left in the bottle), and blood.
- Blister packs for tools and small hardware items.
- Trash bags and grocery sacks.
- Shrink wrap for palletized containers, small card-display items, and protective shipping coatings.
- Foam packing for impact-sensitive equipment such as computers, electrical instruments, and furniture.

Medical

- Catheters and other tubes that assist in providing entry to the body's organs.
- Hip joint replacement parts.
- Artificial legs, feet, and arms.
- X-ray tables and other items that must be transparent to X rays.
- Disposable surgical clothes and instruments.
- Artificial hearts, lungs, and other organs and synthetic blood vessels, valves, and foundation material for skin grafts.
- Eyeglass frames and lenses.
- Toothbrushes, combs, and other personal care items.
- Dental fillings, bridges, and coatings.
- Diagnosis equipment and tools.

Recreational

- Boat hulls, masts, kayaks, surfboards, canoes, and sails.
- Rackets, golf clubs, poles for vaulting, and oars.
- Bobsleds, go-carts, racing cars, dune buggies, and snowmobiles.
- Athletic shoes.
- Skis, jet skis, ski poles, ski boots, and ski-lift chairs.
- Golf ball covers, golf club shafts, and golf club heads.
- Bicycle parts, helmets, and pads.

Textiles

- Clothing.
- Carpets.
- Nonwoven fabrics.
- Diapers and other disposables.
- Fibers used for clothes either by themselves or blended with natural fibers.
- Netting for sports (basketball, soccer).
- Upholstered fabrics for furniture, draperies, and wall paper material.

Furniture, Appliances, and Housewares

- Telephones and other communication equipment.
- Computer and small appliance cabinets or housings.
- Foam cushioning, molded structural parts.
- Luggage.
- Paint
- Floor coverings (carpets, vinyl flooring)

- Institutional seating.
- Components of washers, dryers, refrigerators, table tops, picture frames, clothes hampers, and lawn chairs.

Transportation

- Automotive components, including bodies (such as the Corvette), body panels, interior trim, seats, and engine parts.
- Airplane components.
- Missiles and rockets.
- Train, monorail, and light-rail cars.
- Seat covers and dashboard covers.
- Truck bed liners.
- Gas tanks.
- Recreational vehicle interior components.

Industrial

- Pipes, valves, and tanks (especially for corrosion protection).
- Gears, housings, activation rods.
- Adhesives and coatings.
- Vibration damping pads.
- Electrical circuit boards.
- Wire insulation and devices (connectors, plugs, sockets).
- Gaskets, sealants.

Entertainment

- Musical instruments (guitars, violins, drums).
- Stereo and TV components.
- Cases for radios, tape players, cassette tapes, and CDs.
- VCR tapes and housings.
- Toys.

Construction and Home Owner

- Moldings.
- Sprinklers and pipes.
- Countertops.

- Sinks, shower stalls, plumbing fixtures.
- Flooring (vinyl and carpeting).
- Cups, plates, and storage containers.
- Paint.
- Protective tarps, tie-down belts, cords, and rope.
- Outdoor signs and lighting covers.
- Trash containers and other pails, buckets, and tanks.

Plastics have ever-increasing application in our modern world (Photo 1.1). Some new applications of plastics continue to be replacements for other, more traditional materials (such as replacement of aluminum sheets by carbon/epoxy composites in airplane structures). Other applications are possible because of the unique properties of plastics (such as lubrication-free bearings). The possibilities for plastics use seem endless.

Photo 1.1 Molded plastic industrial and consumer parts (Courtesy M. A. Hanna Company)

CASE STUDY 1.1

The Development of Nylon

In 1928 the DuPont Company hired a brilliant 32-year-old research chemist, Wallace Carothers, who had graduated with a Ph.D. from the University of Illinois and then taught organic chemistry at Illinois and Harvard Universities. DuPont agreed that Carothers would be given the finest staff and facilities in order to develop a concept that he had been exploring. He believed that he might be able to create a giant molecule (a large polymer) by a new technique that involved a series of condensation reactions. (In condensation reactions molecules combine together to create larger molecules and a by-product molecule, like water, that condenses out of the reaction like dew condensing out of the atmosphere on a humid evening.) Carothers' idea was to choose reactant molecules that had two reactive ends so that each molecule could be chemically connected (bonded) to two others. If reacted in series and if continued long enough, the result would be a very long chain made up of the starting molecules all bonded together (a polymer).

DuPont management realized that such a material made up of bonded molecules would be similar to naturally occurring polymers such as silk. The research objectives for Carothers were, therefore, to explore the concepts of this new reaction (pure research) but to keep in mind that the end result should be a usable product, ideally a synthetic silk (applied research).

The soundness of Carothers' concept was demonstrated in 1930 when his chief associate, Dr. Julian Hill, observed that the compound he had been treating in the molecular still (a device that removed the byproduct and thereby encouraged the condensation reaction to continue) had become tough and semirigid, yet elastic. These were properties long associated with natural polymers. The change of the reactants into these polymeric materials was clear evidence that Carothers' team was on the right track. Carothers named the new materials "superpolymers."

About 2 weeks later, Hill made a second highly important observation. While investigating the properties of the solid material, Hill found that when melted, the thick, molasses-like material could be pulled into a fiber shape that was not brittle but was tough and surprisingly strong. Some of the basic research objectives had been met—a reaction to make a polymer, the equipment to extend the chain, and a strong, pliable fiber—but the material still melted at too low a temperature, softened in hot water, and was sensitive to normal cleaning fluids. Obviously it was not yet a viable substitute for silk. The research team had been working on a reaction system that made a class of organic chemicals known as esters. Although these would later become a large and important group of polymers, further research was needed.

Carothers then turned his attention to reactions of molecules containing the organic chemical group acetylene. He felt that this system could lead to a synthetic rubber. This work paralleled the work of other chemists and, although many important chemicals were developed, the synthetic rubber he sought was not formulated. (One of the discoveries he made was an oil that had the odor of musk, a highly prized ingredient in perfume that was, up to then, obtained only from the rare musk-ox of Asia at almost prohibitive prices.)

At the urging of his associates, Carothers returned to the concept of a synthetic silk. He conceived a more practical synthesis route and focused on a different class of organic chemicals, the amides. On May 23, 1934, another "superpolymer" was synthesized by this new technique. This time, Carothers himself demonstrated how the "superpolymer" could be made into a fiber. He drew the hot viscous substance into a syringe and from the needle squirted a tiny stream of it into the air. The floating stream cooled into a wispy filament, as fine as those of a spider's web. The lustrous filament had excellent heat stability, withstood normal washing and cleaning, and was equal in strength and pliability to silk and other natural fibers such as cotton and wool. "Here," said Carothers, walking into the office of the DuPont chemical director in charge, "is your synthetic textile fiber."

The DuPont Company began an intensive effort to build an experimental or pilot plant to produce sufficient quantities of the material so that commercial tests could be run. The material labeled by Carothers as 66 polymer was selected as the most promising for immediate investigation from the strictly practical standpoint. Eventually the pilot plant produced the materials and small-scale manufacturing was demonstrated.

On October 27, 1938, almost 11 years after the hiring of Carothers, DuPont publicly announced the development of "a group of new synthetic superpolymers" from which, among numerous applications, textile fibers could be spun of a strength-elasticity factor surpassing that of cotton, linen, wool, rayon, or silk.

This new group of synthetic materials based on amide reactions was given the name nylon and quickly captured the imagination of the world. Newspapers hailed the discovery as "one of the most important in the century of chemistry." DuPont exhibited nylon stockings at the 1939 World's Fairs of New York and San Francisco. Nylon hosiery (quickly dubbed "nylons") created a sensation when, on May 15, 1940, the first limited quantities were placed on sale. This began the "nylon riots" and during the next 12 months approximately 64,000,000 pairs of all-nylon hose were bought by American women, with a demand that far exceeded the supply.

Few, if any, major inventions have been so immediately successful as nylon. By the end of 1941, DuPont had yarn plants in operation or being built capable of producing more than two million miles of nylon yarn daily. By 1941, most of the toothbrushes made in the United States, half of the hairbrushes, and scores of industrial and household brushes were being bristled with nylon filaments. Nylon was being used for tennis and badminton racquet strings, catheters, surgical sutures, fishing lines, musical strings, wire insulation, hot-melt glues, self-lubricating bearings, and many wartime applications such as parachutes and ropes.

Carothers' sudden death in 1937 cut short a career that might have added many other contributions. It would be difficult, however, to surpass the contribution he made to plastics technology with the development of nylon.

SUMMARY

Plastics are very large molecules that are synthetic and can be or have been shaped. This definition generally excludes unmodified or slightly modified natural polymers. If the natural polymer has been substantially modified, the material can be considered to be a plastic, assuming it is formable. (Obviously some arbitrariness or opinion is contained in the assignment of some materials, but that represents the situation in practice where some people and organizations consider some borderline materials to be plastics and others do not. On most materials, however, widespread agreement exists.)

The history of plastics illustrates the principles of scientific investigation and development for materials. The material is first discovered and a use identified. Then some deficiencies are identified and modifications are attempted, usually based on trial and error. The material is systematically investigated and a basic model for the structure or nature of the material is developed. The material is then systematically modified or synthesized based upon the model.

The basic chemicals used to make plastics are generally derived from crude oil, although other sources are also important. The prices of plastics are, therefore, somewhat dependent upon the price of crude oil.

The plastics industry has three major functions and several minor functions that are fulfilled by thousands of companies. The most basic function is resin manufacture. The resins are transferred to molders who use plastics processing methods to shape or mold the plastic into desired forms. Often, the plastic part is then transferred to a finisher, assembler, or integrator who makes mechanical changes to the plastic part or combines the plastic part with other parts to make up an assembly or the total product.

Plastics have a wide variety of applications in modern society. The applications touch almost every facet of our lives, and have become virtually indispensable, often without our being aware that a plastic part has been used. The versatility and moldability of plastics suggest that the use of plastics will continue to grow.

GLOSSARY

Atom   The smallest entity of an element (such as carbon, oxygen, nitrogen, hydrogen).

Bulk or collective properties   Material properties that are characteristic of the material when a large amount of the material is present.

Chemical properties   Material properties that depend upon the chemical nature or reactivity of the molecule.

Compounder   A processor of plastic resins who adds minor constituents (such as color) to the plastic and then resells it.

Custom molder   A plastics manufacturer who will do molding for other manufacturers on a contract basis.

Fabricator   A finisher of molded plastic parts, often performing some finishing, assembling, integrating, or decorating of the plastic.

Job shop   Another name for a custom molder.

Molecule   Groups of atoms joined together into specific arrangements that impart certain properties to the group.

Natural polymers   Those polymers that occur in nature and are, therefore, not synthetic.

Oligomer   A relatively short-chain polymer.

Pelletizing process   The manufacturing step that converts raw polymer material into small pellets, often used to purify or add materials to the raw polymer.

Plastic   Materials composed principally of very large molecules *(polymers)* that are synthetically made or, if naturally occurring, are highly modified.

Polymer   A long molecule made up of many smaller molecules that are joined together by chemical bonds.

Polymerization process   The chemical reaction process that converts small molecules into polymers.

Resin   A polymer that has not yet been formed into its final useful shape.

Resin distributor   A company that buys resin in large quantities and resells in smaller quantities.

Resin manufacturer   A company, usually very large, that conducts polymerization processes and therefore makes polymer resins.

Stock parts distributor   A company that warehouses plastic parts already formed into simple shapes such as sheets, rods, and tubes.

Synthetic polymers   Those polymers that do not occur naturally and are, therefore, manufactured; they may also include those natural polymers that are highly modified.

Vulcanization   The process of adding sulphur to natural rubber while heating the batch, thus resulting in improved properties in the rubber; a cross-linking reaction.

QUESTIONS

1. What are the three necessary and sufficient criteria that must be satisfied by all plastics materials?
2. What is a resin?
3. Identify and describe the forms of plastic resins.
4. Identify gun cotton and indicate why it is important in the development of modern plastics.
5. What is the first modern plastic that was synthesized with a specific set of properties in mind, who sponsored the work, and when was it done?
6. Give several reasons for the development and use of plastics.
7. Define the term *polymer* and relate it to the term *plastics*.

REFERENCES

Baird, Ronald J. and David T. Baird, *Industrial Plastics,* South Holland, IL: The Goodheart-Wilcox Company, Inc., 1986.

Brydson, J. A., *Plastics Materials* (2nd ed.), New York: Van Nostrand Reinhold Company, 1970.

Dutton, William S., *DuPont: One Hundred and Forty Years,* New York: Charles Scribner's Sons, 1942.

Mark, Herman F., "A Century of Polymer Science and Technology," in *Applied Polymer Science* (2nd ed.), ed. Roy W. Tess and Gary W. Poehlein, Washington: American Chemical Society, 1985.

Muccio, Edward A., *Plastic Part Technology,* Metals Park, OH: ASM International, 1991.

Richardson, Terry L., *Industrial Plastics: Theory and Application* (2nd ed.), Albany, NY: Delmar Publishers, Inc., 1989.

Seymour, Raymond B. and Charles E. Carraher, *Giant Molecules,* New York: John Wiley and Sons, Inc., 1990.

Ulrich, Henri, *Introduction to Industrial Polymers* (2nd ed.), Munich: Hanser Publishers, 1993.

CHAPTER TWO

POLYMERIC MATERIALS (MOLECULAR VIEWPOINT)

CHAPTER OVERVIEW

This chapter examines the following concepts:
- Fundamentals of matter (periodic table, electron configuration, atomic properties, valence)
- Bonding (ionic, metallic, covalent, secondary)
- Basic concepts in organic chemistry (carbon atom bonding, carbon-carbon molecular orbitals, functional groups, naming organic compounds)
- Polymers (general concepts)
- Formation of polymers (addition polymerization, condensation polymerization, comparison of addition and condensation polymerization, other polymerizations
- Thermoplastics and thermosets
- Copolymers

INTRODUCTION

Although this chapter discusses atoms, molecules, reactions, bonding and other concepts basic to chemistry, the intent of the chapter is not to teach chemistry but, rather, to give the fundamental knowledge necessary to understand the basic nature of polymers. That basic nature of polymers (and therefore plastics) is largely dependent upon the types of atoms used to make up the polymers and the ways in which the various atoms are arranged along the polymer chains. This is called the *molecular viewpoint*. A larger view that considers the effects of clusters of molecules is called the *micro viewpoint* and will be considered in Chapter 3, Micro Structures in Polymers. An even larger view, which looks at the bulk properties without regard to substructures, is explored in the chapter called Mechanical Properties (Macro Viewpoint).

The reader is urged not to be worried or intimidated when chemical formulas are used. These formulas are simply pictorial representations of the arrangements of atoms in molecules where the bonds between the atoms are represented by lines and the atoms are represented by their chemical symbols, which are the symbols given in the periodic table of elements. Simple rules are used to create these symbolic representations of molecules, much like those used to create algebraic formulas using symbols such as $+$, $-$, $=$, and ($\times$). Just as mathematical symbols are the "language" of mathematics, these chemical symbols are the "language" of chemistry. Having a basic understanding of this language will enable the reader to communicate with others in discussing chemistry, polymers, and plastics and will allow important concepts affecting the properties of plastics to be explained in an efficient and unambiguous way.

FUNDAMENTALS OF MATTER

Matter is anything that has weight and occupies space. All gases, liquids, and solids are matter. Heat, light, and electricity, however, are not matter; they are forms of energy. The conversion of matter (mass) to energy releases enormous amounts of energy.

Some ancient Greek philosophers maintained that matter was continuous, like energy, but others believed that matter was composed of small units or building blocks that were called *atoms*. Over many centuries the atomic model of matter has fit the experimental data of normal chemical reactions better than the continuous model. The atomic model, which says that matter is composed of small, discrete particles called atoms, has been adopted by scientists today.

Scientists have found over 100 different types of atoms; about 90 occur naturally and the remainder are synthetic, that is, manufactured. The atoms have been arranged in a table that illustrates how these atoms are formed and assists in understanding the properties of these different atoms. The table is called the *periodic table* and is shown in Figure 2.1.

Periodic Table

All material is composed of various combinations of these basic atoms. Chemistry and materials science are studies of how the atoms are combined and the properties that come from the various combinations. The potential combination patterns of the atoms are infinitely numerous, but knowledge of some basic fundamentals will lay the foundation required for an adequate study of plastic properties.

Materials that consist of only one type of atom are called *elements* (that is, there are 103 elements in Figure 2.1). Elements are some of the most common materials known, such as gold, iron, carbon, and helium. Therefore, the periodic table is not only a listing of the atoms, it is also a listing of the elements. Each element (or atom type) is symbolized on the periodic table by the capitalized initial letter of the element's name, such as H for hydrogen, O for oxygen, and S for sulphur. To avoid confusion, a second, lowercase letter is sometimes used to distinguish elements beginning with the same letter, such as calcium (Ca), and chlorine (Cl). Some symbols were derived from the Latin names, such as Na for sodium (natrium) and Fe for iron (ferrum).

Atoms were originally thought to be the smallest form of matter. Scientists now realize that atoms are actually complex structures made up of many smaller subatomic parti-

Period 1	I A																	VIII A
	1 H 1.01	II A											III A	IV A	V A	VI A	VII A	2 He 4.00
Period 2	3 Li 6.94	4 Be 9.01											5 B 10.81	6 C 12.01	7 N 14.01	8 O 16.00	9 F 19.00	10 Ne 20.18
Period 3	11 Na 23.00	12 Mg 24.31	III B	IV B	V B	VI B	VII B		VIII		I B	II B	13 Al 26.98	14 Si 28.09	15 P 30.97	16 S 32.06	17 Cl 35.45	18 Ar 39.95
Period 4	19 K 39.10	20 Ca 40.08	21 Sc 44.96	22 Ti 47.90	23 V 50.94	24 Cr 52.00	25 Mn 54.94	26 Fe 55.85	27 Co 58.93	28 Ni 58.71	29 Cu 63.55	30 Zn 65.37	31 Ga 69.72	32 Ge 72.59	33 As 74.92	34 Se 78.96	35 Br 79.90	36 Kr 83.80
Period 5	37 Rb 85.47	38 Sr 87.62	39 Y 88.91	40 Zr 91.22	41 Nb 92.91	42 Mo 95.94	43 Tc 98.91	44 Ru 101.07	45 Rh 102.91	46 Pd 106.4	47 Ag 107.87	48 Cd 112.40	49 In 114.82	50 Sn 118.69	51 Sb 121.75	52 Te 127.60	53 I 126.90	54 Xe 131.30
Period 6	55 Cs 132.91	56 Ba 137.34	57 La 138.91	72 Hf 178.49	73 Ta 180.95	74 W 183.85	75 Re 186.2	76 Os 190.2	77 Ir 192.22	78 Pt 195.09	79 Au 196.97	80 Hg 200.59	81 Tl 204.37	82 Pb 207.2	83 Bi 208.98	84 Po 210	85 At 210	86 Rn 222
Period 7	87 Fr 223	88 Ra 226.02	89 Ac 227	104 Rf 261	105 Ha 262													

Lanthanides	58 Ce 140.12	59 Pr 140.91	60 Nd 144.24	61 Pm 147	62 Sm 150.4	63 Eu 151.96	64 Gd 157.25	65 Tb 158.93	66 Dy 162.50	67 Ho 164.93	68 Er 167.50	69 Tm 168.93	70 Yb 173.04	71 Lu 174.97

Actinides	90 Th 232.03	91 Pa 231.04	92 U 238.03	93 Np 237.05	94 Pu 244	95 Am 243	96 Cm 247	97 Bk 247	98 Cf 251	99 Es 254	100 Fm 257	101 Md 258	102 No 255	103 Lr 256

atomic number............ 1
symbol of the element.. H
atomic weight.............. 1.01

Figure 2.1 Periodic table.

cles. The most important of these are protons, neutrons, and electrons. Protons are positively charged identical particles, neutrons are neutral identical particles, and electrons negatively charged identical particles. The variations in combinations of protons, neutrons, and electrons make up the differences between the atom types.

The most fundamental differences between the atoms are the number of protons and the arrangement (configuration) of the electrons. Therefore, the periodic table is a sequential listing of the atoms according to the number of protons, with grouping of the atoms to reflect the similarity of electron configurations. The number of protons is shown on the periodic table as the *atomic number,* and each element is unique in its atomic number.

The atomic weight is also shown for each of the elements in the periodic table. The atomic weight is the average of the number of protons and the number of neutrons. The number of protons is fixed for each element, but the number of neutrons can vary. (When the number of neutrons is different in two atoms, the atoms are called *isotopes* of one another.) The atomic weight and, therefore, the number of neutrons are relatively unimportant in determining atomic properties of interest in this text.

The protons and neutrons are bound closely together in a mass that is called the *nucleus* of the atom. The nucleus of each element is different because the number of protons in the nucleus of each element is different. Each nucleus also has a positive charge, with the magnitude of the charge depending upon the number of protons. (Each proton is assigned an atomic charge of $+1$.)

The electrons are not part of the nucleus but, rather, are moving rapidly around the nucleus in what has been called an *electron cloud*. If the number of electrons surrounding the nucleus equals the number of protons, the arrangement is neutral in atomic charge, because the total of positive and negative charges is the same. (Each electron has a charge of -1.) The neutrons have no charge and so the number of neutrons makes no difference in determining charge balance. The neutral condition is the situation for each of the elements as depicted in the periodic table. Hence, as the number of protons increases across the periodic table, the number of electrons in the element also increases in order to maintain electrical neutrality. (Note that the number of electrons does not change the atomic weight of the element because electrons are so much lighter than the protons and neutrons that their mass is neglected in the atomic weight calculation.)

Electron Configuration

The arrangement of the electrons about the nucleus of an atom greatly influences the chemical and physical properties of the element. This arrangement of electrons is called the *electron configuration* of the atom. The electrons are arranged around the nucleus in *electron orbitals,* each of which represents the location in space where each electron is most likely to be found. The electron orbital is, in essence, a representation of the many pathways that an electron is likely to take when it surrounds the nucleus.

A fundamental rule of nature states that atoms try to be in their lowest-possible energy state, thus requiring that the electrons always try to take the lowest-possible energy orbital. Another rule of nature states that only two electrons (which must be paired) can exist in any single electron orbital. If the atom has more than two electrons (as do all atoms with atomic numbers greater than helium), the subsequent electrons must go into different orbitals. These additional orbitals are progressively higher in energy than the earlier ones. Therefore, as the number of electrons increases in an atom, the energy of the electrons is progressively higher.

Scientists have determined that the energies of the electrons in all the known elements can be grouped into seven major energy levels (or shells). These levels correspond to the seven periods (rows) of the periodic table. The shell with the lowest number is closest to the nucleus and also has the lowest energy.

Scientists have also found that within each of these energy levels the actual electron orbitals can have four different shapes, depending on somewhat more minor differences in electron energy. Beginning with the lowest energy these electron orbitals are called s, p, d, and f. The s orbitals are spherical, and the p orbitals are shaped like dumbbells, with the

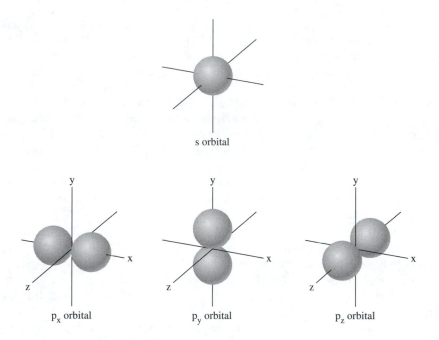

Figure 2.2   Electron cloud representations of s and p electron orbitals.

lobes of the dumbbells lying along the *x, y,* and *z* axes. The s, p_x, p_y, and p_z orbitals are represented in Figure 2.2. The shapes of the d and f orbitals are more complicated and are not represented.

The grouping of the elements into columns reflects the electron orbital that is being filled by the last electron in the atom. The first two columns are created when the s orbitals are being filled. The six columns on the right of the periodic table reflect the filling of p orbitals. The ten columns in the middle of the table are for d-orbital filling, and the two rows at the bottom of the table are formed when the f orbitals are filled.

Atomic Properties

One atomic property that is revealed from the periodic table is chemical stability. All the elements in the last column (VIIIA in the periodic table) are very stable. They are sometimes referred to as the inert gases. This stability is understood by realizing that the s and p orbitals are totally filled in each shell when the atoms in this group are reached in the electron-filling sequence. Independent tests have confirmed that the completion of s and p orbitals gives a low energy and, therefore, very stable state. This stable configuration is sometimes called the *octet rule* because the number of electrons in the s and p orbitals is eight (2 plus 6).

If completion of the s and p orbitals gives a very stable configuration, then having seven electrons (one short of the octet) suggests that the element would have a strong tendency to attract another electron and achieve the stable eight-electron (octet) configuration. This has been found to be true. Fluorine, which is in the next to the last column (VIIA), is very reactive and will withdraw electrons from other atoms. This tendency to attract electrons is called *electronegativity*. Fluorine has the highest electronegativity of any element. Chlorine (below fluorine) also has a high electronegativity as would be expected from the octet rule. Oxygen and nitrogen (in the two columns adjacent to fluorine) also have high electronegativity because of the combination of their need for only a few electrons (two or three) to complete the octet and their small size (which accentuates the strong tendency to attract electrons). In general, F, Cl, O, and N are the atoms with the strongest electronegativity.

All the atoms on the right side of the periodic table lack only a few electrons to complete an octet and therefore generally have a tendency to take on additional electrons. These atoms are called *nonmetals*.

As might be expected, having only a few electrons **more** than the stable octet gives rise to a tendency of the atom to give up electrons. The atoms with these properties are found on the left side of the periodic table. Groups IA and IIA have the strongest tendency to give up electrons. An atom that has a tendency to give up an electron is called a *metal*. The line that separates the metals and nonmetals runs generally along the line (top to bottom) beginning with Si, As, Te, and At.

Those atoms that readily take on or give up electrons are said to have high *chemical reactivities* because the changing of the electron configuration of an atom is the essence of a chemical reaction. Therefore, we can say that when the electron configuration of an atom is changed by interaction with another atom, the two atoms have entered into a chemical reaction. The resulting materials are, therefore, chemically different from the atoms before the electron change occurred.

Valence

The *valence* of an atom is a number that reflects the number of electrons that an atom will usually attract or give up. If the atom attracts one electron, it will have one more negative charge than the neutral element and the valence will be -1. Fluorine and all of the group VIIA elements would be an examples of atoms with this valence (-1). Oxygen and the other group VIA elements would attract two electrons and would have a valence of -2. Atoms from the nonmetal side of the periodic chart (the right-hand side) will normally attract electrons and will have negative valences.

Atoms on the metal side of the periodic table (the left-hand side) will normally give up electrons. For instance, sodium (Na) and the other IA would have a valence of $+1$ because they would give up (lose) an electron and therefore have one fewer electron than the neutral element. The elements in group IIA, such as calcium (Ca), would lose two electrons and therefore have a valence of $+2$. In some cases an element may have more than one valence because several semistable electron configurations can exist for that element. This is especially true when the d and f electrons are being added. For example, iron (Fe) can have both $+2$ and $+3$ valences. These atoms are located in the middle of the periodic table.

BONDING

Bonding occurs when two or more atoms having appropriate chemical reactivities come into close proximity resulting in an attraction between the atoms. In energy terms, the attraction is the result of a lowering of the energy of the system. This phenomenon is illustrated in Figure 2.3. If the atoms are at a large distance from each other, there is little interaction and the system is neither stable nor unstable. As the distance between the atoms decreases, the energy begins to decrease and the system becomes more stable. Eventually the atoms reach an optimal separation which is at the bottom of the energy curve. This separation is the *bond length* and the depth of the energy well is a measure of the *bond energy,* or the energy that would be needed to break the bond and separate the atoms completely. If the atoms approach closer than the bond length, repulsion occurs that leads to an unstable energy condition and the atoms separate slightly to again achieve the optimum energy condition that exists at the bond length.

Hence, attractions between atoms or other phenomena that result in lower energies are likely to cause bonding between atoms. Interactions between the electrons of the various atoms are the origins of most of the attractions between the atoms and therefore control the bonding. Several types of electron interactions are known and each gives rise to a different type of bonding, including ionic bonding, metallic bonding, covalent bonding, and secondary bonding.

Figure 2.3 Bonding energy as a function of distance.

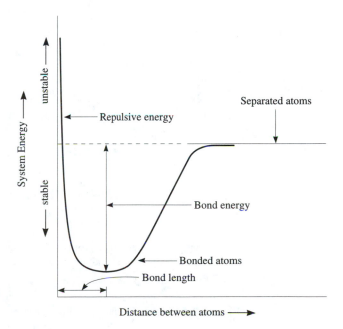

Ionic Bonding

Ionic bonds are formed when an atom that has a strong tendency to give up electrons (a metal) is in close proximity to an atom that has a strong tendency to accept electrons (a nonmetal). This nearness of the atoms allows a **transfer of one or more electrons** (depending on the valence of the atoms) from the outer electron shell of the metal atom to the outer electron shell of the nonmetal. This transfer results in both atoms having stable octet electron configurations. The metal atom will have a positive charge and the nonmetal will have a negative charge. These charged atoms (or any other charged particles), whether positive or negative, are called *ions*.

When a large number of these metal and nonmetal ions are grouped together, they often enter into an arrangement where each of the positive ions is surrounded by negative ions and vice versa. The initial electron transfer and the resulting arrangement when many ions are in close proximity are illustrated in Figure 2.4 where sodium (Na) is the metal and chlorine (Cl) is the nonmetal. The resulting product is symbolized NaCl, common table salt. This type of highly organized and regular structure is called a *crystal*. Crystals are not the only structure that ionic materials may form, but they are common, especially in ionic materials that are solids. This will be discussed further in the section on solids, liquids and gases.

Notice that in a crystal the ions have attractive forces (bonds) to all of the surrounding ions of different charge. This attraction is much like the attraction between north and south poles on magnets and is typical of all ionic materials. In general, positively charged species and negatively charged species of any kind will have this type of attraction. It is called *electrostatic attraction*.

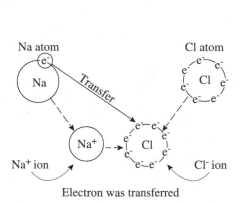

(a) Formation of a metal (Na$^+$) ion and a nonmetal (Cl$^-$) ion as metal atom and nonmetal atom. approach each other

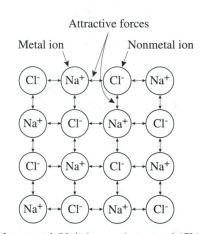

(b) Many metal (Na$^+$) ions and nonmetal (Cl$^-$) ions in a crystal structure showing attractive forces between oppositely charged ions

Figure 2.4   Ionic bonding and the resulting crystal structure.

Most solid materials having ionic bonds are called *ceramics;* typical examples would be table salt and most rocks. Ceramic structures are usually very stable. This stability is shown in the high melting point, high strength, and rigidity of ceramics. The high strength and rigidity are associated with resistance to moving one row of ions relative to the next. Any movement of one row of a ceramic would cause ions of the same charge to come close to each other which would result in a repulsion. This repulsion would be seen as a resistance to movement, thereby requiring more energy to be put into the structure to get it to move.

Metallic Bonding

Metallic bonding occurs when two metal atoms are in close proximity. In this case both atoms have a tendency to give up their electron(s). No transfer of electrons from one atom to the other atom can occur as was the case in ionic bonding. If bonding is to exist, however, some change in electron configuration must occur. In this case, since both metal atoms have a tendency to give up electrons, the energy is lowered in the system when a **releasing of electrons** occurs. The number of outer shell electrons released by each atom is the number that will result in a stable, low-energy electron configuration and is reflected in the valence of the atom. In metallic bonding the electrons given up by the atoms (which become positively charged ions) are not accepted by any other atom but are free to move between all the metal ions. This group of freely moving electrons has been called the *sea of electrons* and is characteristic of metal bonding. Metal bonding is depicted in Figure 2.5. The metal atoms (iron, in the case depicted) approach each other and give up their electrons when in close proximity, thus forming a sea of electrons.

The initial assumption when contemplating a structure of positively charged ions is that the metal ions would repel each other and no bonding would occur. However, the

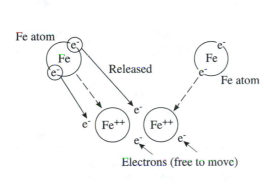

(a) Formation of metal (Fe^{++}) ions and sea of electrons as metal atoms approach each other

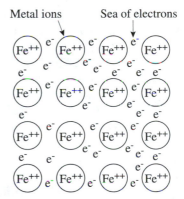

(b) Many metal (Fe^{++}) ions in a crystal structure surrounded with a sea of electrons

Figure 2.5 Metallic bonding of iron and the resulting crystal structure.

electrons are able to cancel the repulsive forces by moving between the positive ions. Hence, metal ions can also form regular crystal structures, although the forces between the metallic ions are generally not as strong as are the forces in ionic crystals. The case depicted in Figure 2.5 is normal metallic iron, which is written simply as Fe. In this case, the +2 valence state of Fe is depicted. In general, materials having metallic bonding are called *metals*. The physical properties of metals are somewhat different from ceramics because of the differences between ionic and metallic bonding. In metals, generally less energy is required to move the rows of the crystal relative to each other. This freer movement is facilitated by the sea of electrons which reduces the repulsion between positive ions as they move near each other. Metals are, therefore, less strong and less brittle (more ductile) than ceramics, all other things being equal. Some metals are known to have higher strengths than would be anticipated from purely bond energy considerations. This additional strength is caused by changes at the micro (multiple crystal) level and in the macro (bulk) level. Further investigation of these changes is made in metallurgy and materials science and is beyond the scope of this text.

It is possible for two different metal atoms (such as Fe and Cr) to also form a structure similar to the one depicted. Again, each atom would give up its electrons and form a positively charged ion. The electrons would form a sea of electrons that would move freely about the ions. The specific arrangement of the atoms would depend on several factors including the size of the ions, their charge, and the relative amounts of each ion. Materials of this type that are composed of two or more metals are called *metallic alloys*.

Covalent Bonding

Covalent bonding, the third type of bonding and the type of bonding most important in plastics, occurs when two nonmetal atoms are in close proximity. In this case both atoms have a tendency to accept electrons. These tendencies to accept electrons are satisfied for both atoms by a **sharing of electrons** between the outer electron shells of the two atoms. The number of electrons shared is dictated by the need for each of the atoms to form a stable electron configuration and, usually, to satisfy the octet rule. (Hydrogen is an exception because its stable configuration is two electrons only.) In most cases of bonding between nonmetal atoms, a nonmetal atom (especially carbon) will share some electrons with one atom and simultaneously share other electrons with one or more different nonmetal atoms. Hence, covalent bonds would be formed from one atom to several other atoms. A sharing of electrons between nonmetal atoms is depicted in Figure 2.6.

Every covalent bond is made up of two electrons. The electrons in the bond move rapidly around both of the atoms making up the bond rather than being statically stationed between the atoms. This encirclement holds the atoms together and constitutes the bonding force between the atoms. In other words, the electrons form a cloud about the two bonded atoms. In this case (see Figure 2.6) C (carbon) shares its four outer shell electrons with four different H (hydrogen) atoms. The resulting structure gives C the required octet of electrons. H needs only two electrons to reach its stable electron configuration, and that is also accomplished through the sharing. A representation showing the distribution of the electrons to accomplish this sharing and a representation showing just the bonds (two electrons per bond) are both given in Figure 2.6.

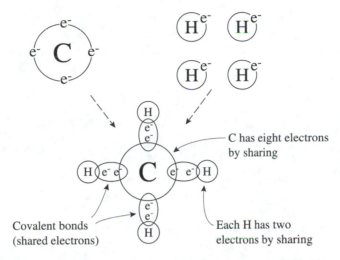

(a) Formation of covalent bonds between nonmetal atoms (one C and for H atoms)

$$H \!-\! \underset{\underset{H}{|}}{\overset{\overset{H}{|}}{C}} \!-\! H$$

Each (–) is a pair of
shared electrons

(b) Representation of the resulting CH_4 molecule

Figure 2.6 Covalent bonding between one carbon (C) and four hydrogens (H) to form methane (CH_4).

The resulting structure of one carbon (C) and four hydrogens (H) is symbolized CH_4 and is called methane. The properties of the new entity, CH_4, are substantially different from the properties of either C or H separately as a result of this covalent bonding. The new entity, which is composed of two or more atoms, is called a *molecule,* the smallest entity that contains more than one atom and possesses all of the properties of the combined unit. The resulting molecule, CH_4, is represented as meaning that one carbon and four hydrogen atoms combined into one unit.

Another example of covalent bonding is seen in Figure 2.7. In this case carbon (C) bonds with two oxygen (O) atoms. In order to obtain its stable octet electron configuration, the C must share two of its electrons with **each** of the O atoms. The O atoms must each share two of their electrons with the C as well. Hence, four electrons are shared between the C and each of the O atoms. Since each two (and only two) electrons constitute a bond, the four electrons shared by each O and C must result in two covalent bonds. This bonding pattern is also represented in Figure 2.7. Nonmetal atoms may share two, four, six, or, in theory, eight electrons to form, respectively, one, two, three, or four covalent

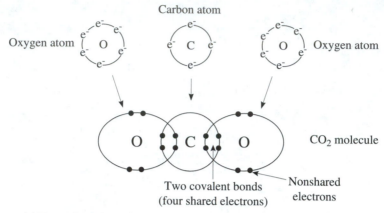

(a) Formation of covalent bonds between non-metal atoms (O and C)

(b) Bond representation of the CO_2 molecule showing polarity of $C{=}O$ bonds

Figure 2.7   Covalent bonding between one carbon (C) atom and two oxygen (O) atoms to form carbon dioxide (CO_2).

bonds. These bonding patterns are logically called single, double, triple, and quadruple bonds.

Because the electrons are shared between atoms, neither of the atoms develops a charge in the same sense as in ionic and metallic bonding, where the electrons are transferred rather than shared. Hence, the atoms when in covalent bonds are not called ions. However, because of differences in electronegativity between the atoms in a covalent bond, one of the atoms may have a stronger attraction for the electrons than will the other. When this difference in electronegativity occurs, the electrons are unequally shared and the electrons spend more time near the atom with the higher electronegativity. This phenomenon produces a bond that has one end slightly negative compared to the other end. Such a bond is called a *polar bond* and is depicted in Figure 2.7, where the oxygens are more electronegative than the carbon and will, therefore, be the negatively charged ends of the polar bonds. A polar bond is also called a *dipole* because there are two poles, like a north and a south pole on a magnet.

To symbolize that the charge is of lower magnitude than that developed by the full transfer of an electron, or in other words that the charge is only partial, the slight positive charge is represented as a δ^+ for the positive end and a δ^- for the negative end, rather than a $(^+)$ or a $(^-)$ as was done in ionic and metallic bond representations.

The electronegativities of C and H in Figure 2.6 are almost the same; hence the covalent bonds between them are nonpolar—that is, the shared electrons spend about as much time near the H as they do near the C.

Secondary Bonding

Several types of atomic attractions exist that are much weaker than the ionic, metallic, and covalent bonds that have already been described. These weaker attractions are called *secondary bonds* and exist chiefly **between** molecules, rather than **within** molecules. One of the most common of these secondary bonds is *hydrogen bonding* which is depicted in Figure 2.8.

As already explained, some covalent bonds are polar; that is, the density of the electron cloud is greater around one atom in the bond than around the other atom (because of the difference in electronegativities of the atoms). This polarity causes one site on the molecule to be somewhat negative and another site on the molecule to be somewhat positive. For instance, the oxygens in Figure 2.7 are slightly negative and the carbon is slightly positive. A similar situation is illustrated in Figure 2.8, where the oxygens are negative and the hydrogens are positive. Note in Figure 2.8 that the negative locations (oxygens) on one molecule can have an attraction for the positive locations (hydrogens) on another molecule. These attractions between molecules are called hydrogen bonds.

Not all attractions between positive and negative locations on different molecules involve hydrogen atoms, but so many cases do involve hydrogen that all attractions of this type are called hydrogen bonds. Perhaps a better name would be *dipole bonds,* but these types of bonds have been called hydrogen bonds for many years and so the name persists. Remember that these bonds can exist only between molecules in which polar covalent bonds are the primary bonds.

Another type of secondary bond is called *van der Waals bonding* and has been attributed to the attraction that all molecules have for each other, somewhat like the gravitational attraction that exists between all bodies. The gravitational attraction is dependent upon the masses of the bodies and the distance between them. Since molecules are so small, the van der Waals attraction will only occur at very short distances. Even then, it is quite weak, several times lower in strength than ionic, metallic or covalent bonding and even lower in strength than hydrogen bonds.

Some scientists have suggested that a third type of secondary bonding occurs. This third type is called *induced dipole bonding* and occurs when one end of a polar bond approaches a nonpolar portion of another molecule.

Figure 2.8 Hydrogen bonding between water molecules.

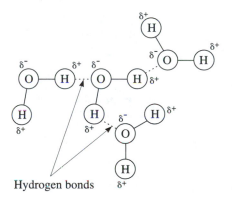

Hydrogen bonds

A momentary and very slight polarity is induced in the nonpolar molecule (CH_4) because of an attraction of the electrons in the vicinity of the H by the negative end of the polar molecule (CO_2). This will occur only at extremely short distances.

The effect of all types of secondary bonding is to give a slightly lower energy to a system of molecules than they would have independently. Hence, the grouped arrangement is slightly more stable and energy must be added to the system to get the molecules to separate. This important consideration, that is, the addition of energy to cause clustered molecules to separate, forms the basis for an understanding of the difference between solids, liquids, and gases.

Nonbonded Interactions Between Polymers

Secondary bonding can have a major effect in polymer systems. These bonds are less strong than covalent bonds but still require energy to be overcome. If one atom in each polymer were to participate in a secondary bond, the force of attraction would be undetectably small. However, with thousands of atoms all participating, as might occur in long polymer chains, these secondary forces can become large enough to influence some of the properties of polymers.

A typical polymer that illustrates this strong secondary bonding is nylon (polyamide); it is shown in Figure 2.9.

The hydrogen bonds in the amide polymer occur in every repeating unit between the partially positive and the partially negative centers. In this case the oxygen (which has a high electronegativity) shares electrons with the carbon but shares them unequally. The electrons are around the oxygen more than the carbon, thus making the oxygen slightly negative and the carbon slightly positive. Likewise, the nitrogen has the electrons for more time than the hydrogen, making the nitrogen slightly negative and the hydrogen slightly positive. Therefore, two regions in each amide unit (the oxygen-carbon region and the nitrogen-hydrogen region) can have hydrogen bonding.

These regions can have an attraction with oppositely charged regions on other polymer molecules. For instance, the negative oxygen atom can form an attraction with a hydrogen atom on another molecule.

The number of secondary bonds formed between polymers can be increased and the overall energy of the system lowered when a polymer folds back and forth, thus bringing many sections of the polymer into close contact and facilitating the formation of sec-

Figure 2.9 Hydrogen bonding in polyamide (nylon).

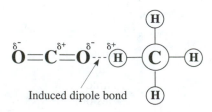

Induced dipole bond

ondary bonds. The stable structure formed has some similarities to the regular crystal structures of metals and ceramics and is therefore called a *crystalline region* in the polymer. In many polymers the physical shape of the polymer prevents the polymer chains from getting close enough to form secondary bonds; thus in these polymers, crystalline regions are not present. More discussion about these crystalline structures and the types of polymers that can form crystalline regions is given in later chapters of this book.

In general, the effect of crystalline regions on the properties of polymers is to increase the melting point of the polymer, because more energy is needed to impart totally free movement to the polymer. These thermal interactions are discussed in the chapter on microstructures.

Secondary bonding will also increase the strength and the creep resistance of plastic materials because the molecules require more energy to slide relative to each other. These properties are associated with the input of energy into the polymer mass. When any energy is input, the tendency for materials to move relative to each other is increased, but this movement is resisted by the intermolecular attractions. These interactions and other mechanical properties will be discussed in detail in the chapter on mechanical properties.

Still other intermolecular forces have been identified. One of the most important is chain entanglement. These entanglements are less dependent on the forces between atoms than on forces involved with entire molecules. They are also discussed in the chapter on mechanical properties.

BASIC CONCEPTS IN ORGANIC CHEMISTRY

One common scientific view that was prevalent before the age of modern chemistry classified all materials as either living or nonliving. (Anything that had ever been living or that was derived from anything that had ever lived was considered part of the living classification.) These two classifications were called, respectively, organic and inorganic. Chemists studying organic materials noted that these materials were generally based on the chemistry of the carbon atom. Because of that connection between organic materials and carbon atom chemistry, the study of carbon chemistry became known as *organic chemistry*. (The study of the chemistry of all the other atoms was called *inorganic chemistry*.) Today, many materials based on carbon are totally synthetic and may never have been alive, but they are still studied within the broad category called organic chemistry.

What is so special about the chemistry of carbon that it should be classed separately from all the other elements in the periodic table? In part, at least, the answer seems to be this: There are so very many compounds of carbon, and their molecules can be so large and complex. Most plastics are based on carbon and are, therefore, part of organic chemistry.

The number of molecules containing carbon is many times greater than the number that do not contain carbon. These organic molecules have been divided into families based upon the arrangement of carbon and other common atoms into various characteristic grouping patterns. They generally have no counterparts among the inorganic compounds. Much of organic chemistry is devoted to investigating the reactions of these families of organic compounds, including the ways the atoms rearrange and combine, and the properties of the resultant, new organic molecules.

Carbon Atom Bonding

Carbon has an atomic number of six and therefore has six protons in its nucleus and, in its elemental state, six electrons surrounding the nucleus. Two of these electrons are in the first shell and do not enter into normal chemical bonding. The remaining four electrons are in the outer shell and are available for bonding. The electron clouds of the four outer electrons for atomic carbon were depicted in Figure 2.2. Normally, in elemental (atomic) carbon, each of the orbitals (s, p_x, p_y, p_z) would have one electron. Carbon needs four more electrons to complete its outer shell and obtain the stable octet configuration. Therefore, carbon will always form four bonds and the valence is four. Because carbon is a nonmetal and most of the other atoms forming compounds with carbon are also nonmetals, the type of bonding between carbon and these atoms would be covalent, as discussed previously in this chapter.

The electron configuration of an atom is changed when the atom becomes part of a molecule. This change occurs because of interactions between the bonding atoms and all other atoms. The case of carbon bonding to four hydrogens illustrates the type of changes in atomic orbitals that occur when the atoms bind together in a CH_4 molecule. This process is illustrated in Figure 2.10.

In the bonding of the carbon with the four hydrogens, three rules must be satisfied in order to achieve the lowest-energy, most stable state. These same rules apply to the bonding of all atoms.

- The number of electrons involved in the bonding must satisfy the octet rule.
- The number of bonds is dictated by the rule that each bond has two electrons.
- Each pair of bonding atoms moves as far away from all other pairs as possible, within the constraint of maintaining the proper bond distance with the central atom. This requirement arises from the repulsion that exists between atoms in which the stable electronic configuration has been satisfied. Spatially this separation can best be done for the CH_4 case if the hydrogens move to the corners of a tetrahedron with the carbon in the center. This idea is illustrated in Figure 2.10.

As indicated in Figure 2.10, the carbon atom orbitals change shape in order to accommodate the preferred spatial arrangement of the hydrogens. The one s and three p atomic orbitals change into four new orbitals which are oriented about the carbon so as to point towards the corners of the tetrahedron. Because these new orbitals were formed by mixing different atomic orbital types (in this case s orbitals and p orbitals), the new orbitals are called *hybrid orbitals*. After the molecule has been formed, all of the orbitals involved in bonding will be called *molecular orbitals*.

The electron configuration about carbon can also be represented with electrons as dots, as shown in Figure 2.11 for CH_4 and another organic molecule, CH_3Cl.

Both molecules demonstrate how covalent bonds are represented by the electron-dot method. In CH_3Cl (Figure 2.11b) note that a chlorine atom has taken the place of one of the hydrogens (that is, one of the four H atoms is replaced by a Cl atom). The chlorine atom has seven electrons in its outer shell prior to bonding. One of these is used in the covalent bond while the others continue to surround the chlorine atom. After the covalent bond is formed between the chlorine and carbon atoms (each atom contributing one electron to the bond), chlorine will have a total of eight electrons in its outer shell, thus

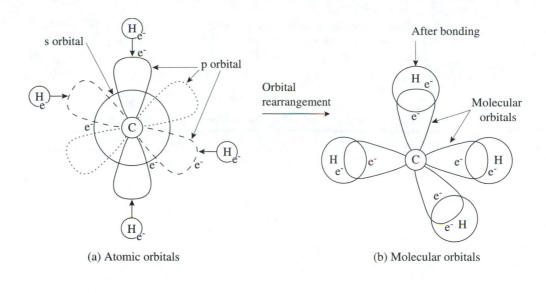

(a) Atomic orbitals   (b) Molecular orbitals

(c) Tetrahedron representation

Figure 2.10   Carbon atom bonding with four hydrogen atoms to form CH_4.

forming a stable configuration with eight electrons in the outer shell. The carbon in CH_3Cl also has eight outer shell electrons so it, too, is stable. Thus, the carbon-chlorine bond is stable when electrons are shared as shown. The electron-dot formulas are somewhat easier to write than the drawings of Figure 2.10, but the spatial orientation is lost. Both representation systems are commonly used.

(a) Methane (CH_4)   (b) Chloromethane (CH_3Cl)

Figure 2.11   Carbon atoms bound by covalent bonds.

Many other atoms can combine with carbon in configurations similar to hydrogen and chlorine to form various molecules. Each atom has a characteristic bonding pattern that is dictated by the number of electrons in its outer shell. (The valence of the atom.) Table 2.1 lists the most common atoms in organic chemistry and the number of bonds each can form.

Table 2.1   Number of Bonds for Typical Atoms (Important in Plastics)

Atom	Number of Bonds (Typically)
C	4
H	1
Cl	1
N	3
O	2
Si	4
F	1
Br	1

Carbon-Carbon Molecular Orbitals

Another key to the large number and complexity of carbon-containing molecules is that carbon can form bonds with itself. Such a molecule (C_2H_6) is illustrated in Figure 2.12a.

In C_2H_6 (called ethane) the covalent bonds with the hydrogens were formed in the method that has previously been described for CH_4 except, in this case, only three hydrogens are bonded to each of the carbons. That leaves each of the carbons with only seven outer shell electrons. One way for the carbons to obtain the stable octet configuration is for the carbons to bond with each other. The electron-dot representation and the simpler representation where each pair of electrons is represented by a single line are shown in Figure 2.12a.

Suppose that a new molecule, C_2H_4, is made by removing one hydrogen and its electron from each of the carbon atoms in C_2H_6. This action will create two unbonded electrons (called *free radical electrons*), as shown in Figure 2.12b. This is an unstable condition because the carbons no longer have eight outer electrons. The unbonded electrons must find other electrons with which to bond. One possibility is that the two free radical electrons, one on each of the carbons, could bond to each other. If that were to occur, two bonds between the carbons would result, as illustrated in Figure 2.12c. The bonding pattern formed between the carbons would then be a *double bond*. The resulting molecular formula would be C_2H_4 (called ethene or ethylene). Molecules containing a carbon-carbon double bond are said to be *unsaturated*. Molecules with only carbon-carbon single bonds are called *saturated*.

Although the two bonds in the double bond appear to be near each other and identical in shape in the simplified drawing of Figure 2.12c, they are really quite different. This difference comes about because of the previously discussed rule that bonded atoms must try to move as far away from each other as possible, while still maintaining the proper bond distance with the atom to which they are bonded. In the case of C_2H_4, only three

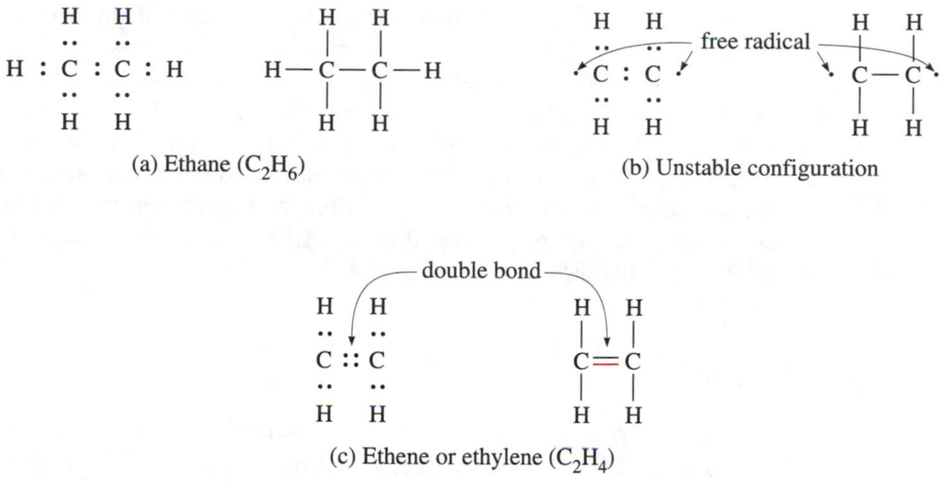

(a) Ethane (C$_2$H$_6$)

(b) Unstable configuration

free radical

(c) Ethene or ethylene (C$_2$H$_4$)

double bond

Figure 2.12 Bonding in carbon molecules.

atoms are bonded to each of the carbons. Hence, instead of the tetrahedral arrangement used when four atoms are bonding to the carbon, a flat trigonal arrangement is used that gives the three atoms bonding to carbon their farthest separation.

The pair of electrons in C$_2$H$_4$ (one on each of the carbons), which form the second bond between the carbons, have not been forced to spatially rearrange by the approach of a bonding atom. They have retained their atomic orbital configuration. In this case, both of these electrons were in atomic p orbitals, which are dumbbell-shaped. These two electrons can then form a new bond and create a new molecular orbital by an overlapping of their orbitals. Such a configuration is illustrated in Figure 2.13.

When two electrons in a molecular orbital are spatially arranged in a cylinder shape along the axis of the carbons, the electrons make up a *sigma (σ) bond*. This is the type of bond that is initially formed between the carbons in C$_2$H$_4$, as shown in Figure 2.13. It is also the type of bond that is formed between the carbons and the hydrogens in C$_2$H$_4$, although these are not shown in the figure, and between all of the atoms in C$_2$H$_4$. The other two electrons in the double bond between the carbons are located above and below the axis of the

Figure 2.13 Types of bonding in double bonds (ethylene).

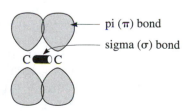

pi (π) bond

sigma (σ) bond

carbon atoms in dumbbell-like lobes which are characteristic of atomic p orbitals. The overlap of these p orbitals makes up a new molecular orbital, which is called a *pi (π) bond*.

The sigma (σ) bond is more stable than the pi (π) bond because in the sigma bond the electrons are held more tightly by the carbon atoms. The relative stability of the σ and π bonds is one of the important factors in determining the chemistry of molecules that contain double bonds. Molecules that contain only carbon-carbon single bonds (such as ethane) and molecules that contain at least one carbon-carbon double bond (such as ethylene) are two examples of functional groups that serve as the basis of grouping organic compounds into families of molecules.

Functional Groups

Over the years, organic chemists have noted certain chemical characteristics associated with various groups of atoms, called *functional groups,* and have given these groups names so that whenever they are encountered in a larger molecule, the characteristic chemistry of the group can be anticipated. The division of organic molecules into families has been made largely on the basis of these functional groups. The science of organic chemistry is largely the science of learning how these functional groups are formed and the characteristic chemistry they exhibit. The most common of these functional groups are represented in Figure 2.14.

These functional groups can be attached to basic groups of carbon atoms, generally by replacing one of the hydrogens attached to the carbon, to form new molecules, much as chlorine was attached in CH_3Cl illustrated previously. These groups impart characteristic chemical and physical properties to the new molecule. The chemistry of these groups is beyond the scope of this text and is covered in textbooks on organic chemistry. Only in a few cases will the chemistry of a particular group be discussed in this text, and these cases will be explained in a simple context.

The variety of molecules that can be made from combinations of these functional groups and various arrangements of molecules can be seen in Figure 2.15. Each molecule is composed of only 3 carbons with hydrogens and oxygens, but with different arrangements. The properties of these molecules vary greatly because of these arrangement differences. For instance, isopropyl alcohol, the first molecule shown in Figure 2.15, is the chemical commonly used as rubbing alcohol. By simply rearranging the atoms (preserving the same number of carbons, hydrogens and oxygens), the material becomes methylethyl ether, which can be used as an anesthetic. By changing the arrangement and the way that the oxygen is bonded to the carbon, two hydrogens are lost and the material becomes acetone, a common solvent used in applications such as fingernail polish remover. Three other rearrangements of the atoms in acetone (keeping the same number of atoms) result in the formation of methyl acetate (a sweet-smelling chemical used in perfumes), propionaldehyde (a sharp-smelling chemical related to cinnamon), and propanoic acid (a chemical related to vinegar). These arrangements of three carbons with hydrogens and oxygens illustrate the many properties that are possible with only minor changes in the arrangement of the atoms in these various molecules. By adding other atoms, in-

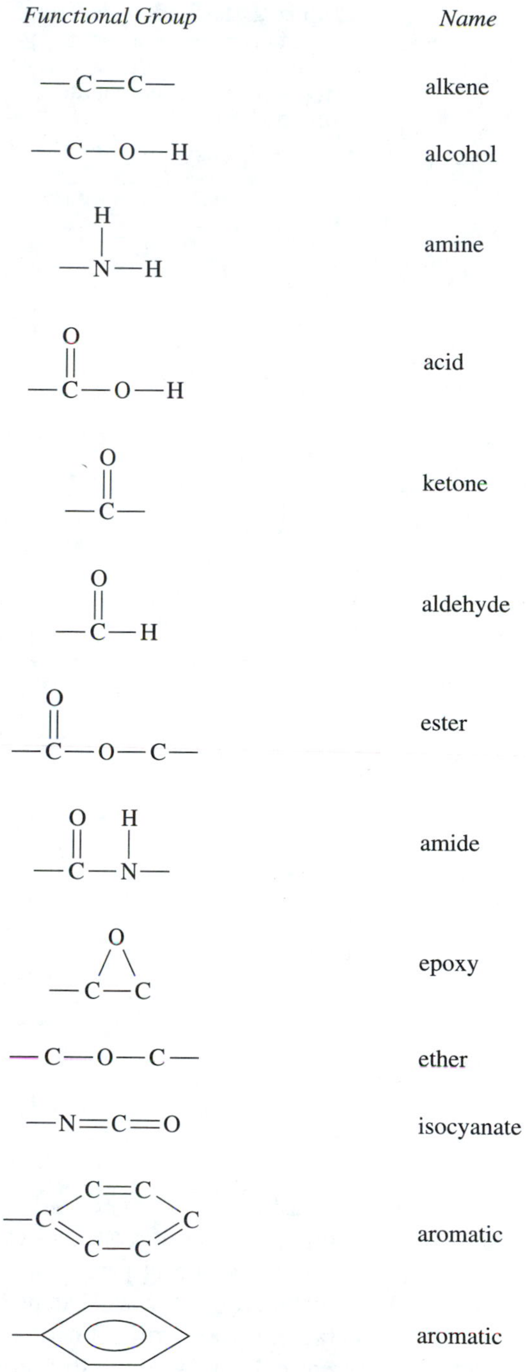

Figure 2.14 Functional groups in organic chemistry.

Functional Group	Name
$-C{=}C-$	alkene
$-C{-}O{-}H$	alcohol
$\overset{\displaystyle H}{\underset{}{-N{-}H}}$	amine
$-\overset{\displaystyle O}{\underset{}{\overset{\|\|}{C}}}{-}O{-}H$	acid
$-\overset{\displaystyle O}{\underset{}{\overset{\|\|}{C}}}-$	ketone
$-\overset{\displaystyle O}{\underset{}{\overset{\|\|}{C}}}{-}H$	aldehyde
$-\overset{\displaystyle O}{\underset{}{\overset{\|\|}{C}}}{-}O{-}C-$	ester
$-\overset{\displaystyle O}{\underset{}{\overset{\|\|}{C}}}{-}\overset{\displaystyle H}{\underset{}{N}}-$	amide
epoxy	epoxy
$-C{-}O{-}C-$	ether
$-N{=}C{=}O$	isocyanate
aromatic	aromatic
aromatic	aromatic

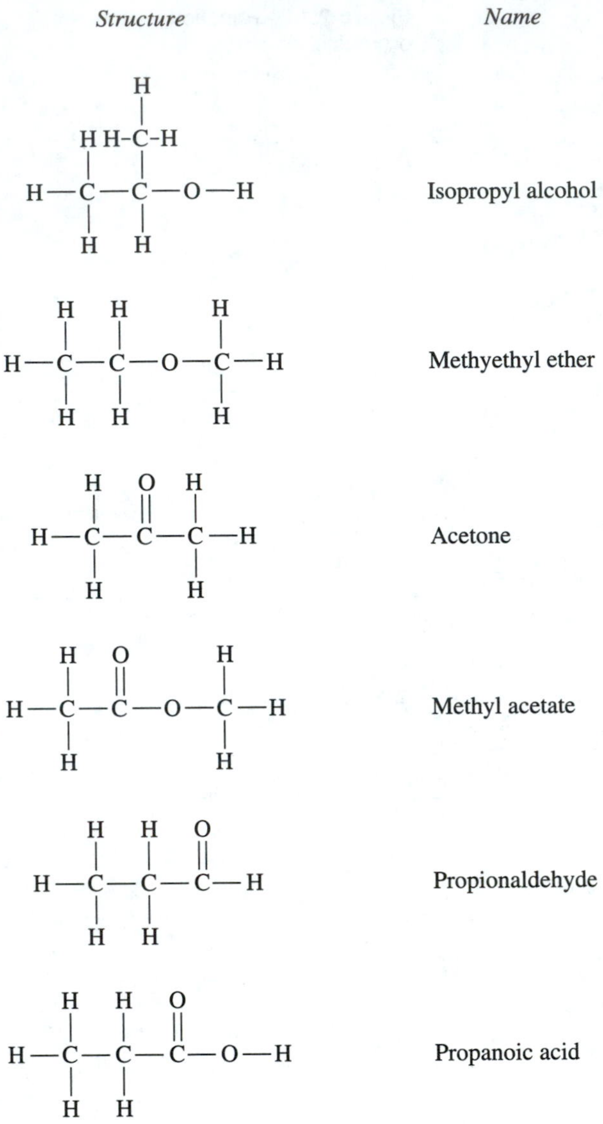

Figure 2.15 Various molecules of carbon, hydrogen and oxygen illustrating the differences in properties with different atomic arrangements.

Structure	Name

Isopropyl alcohol

Methyethyl ether

Acetone

Methyl acetate

Propionaldehyde

Propanoic acid

creasing the number of carbons and making other arrangements of the atoms, a nearly infinite number of molecules can be formed.

One group of particular interest that should be discussed in more detail is the aromatic group. Two equivalent representations are given in Figure 2.14. The first representation shows a ring of carbon atoms with three double bonds. The pi (π) bond electrons not only overlap with the carbons that share the double bond, they also overlap with the pi (π) bond electrons in adjoining carbons, allowing these pi electrons to move in a circular pattern around the ring, which results in a very stable structure. The second, equivalent, representation of the aromatic molecule eliminates the carbons (they are assumed to be at the vertices) and shows the electrons as a circle inside the hexagon of carbons.

Molecules that contain the aromatic group are said to be *aromatic*. Molecules that do not contain the aromatic ring are called *aliphatic*. All the organic molecules in Figure 2.14 except those labeled "aromatic" are aliphatic. Hence, carbon-carbon single and double bonds, carbon bonded to oxygen, nitrogen, or any other element would be aliphatic. Only molecules where six carbons are bonded together in a ring structure and the ring has three double bonds that alternate around the ring would be considered aromatic. If several of these aromatic rings are attached to the molecule, it is said to be highly aromatic. Molecules can, therefore, range from highly aromatic to highly aliphatic. The characteristics of aromatic and aliphatic molecules will be important in the discussions of several key polymers in subsequent chapters.

Naming Organic Compounds

A detailed discussion of the naming of organic compounds is beyond the scope of this book. However, some basic concepts will allow the reader to understand the methods used in naming polymers, which are derived from the system used to name organic chemistry compounds.

The basis for the naming system of organic chemistry is to indicate the family of organic compounds to which a molecule belongs. This family classification is usually dependent upon the functional group that predominates in the chemistry of the molecule. For instance, the molecule might have an alcohol group, in which case the molecular name would be alcohol (such as ethyl alcohol). A newer naming system would indicate that the molecule was an alcohol by ending the root name of the compound in -ol, which is the suffix for alcohol, such as ethanol. Sadly, neither of the naming systems has been universally adopted, especially for molecules which were named many years ago.

The second step in naming an organic molecule is to indicate the number of carbons in the molecule. Prefixes are used for this purpose. The prefixes are given in Table 2.2. The third step is to indicate the number of functional groups in the molecule. Prefixes are also used for this purpose. To avoid confusion, the prefixes for indicating the number of carbons are different from the prefixes used to indicate the number of functional groups. (Note that the prefix "mono" is omitted from the name of the compound unless that omission would cause confusion.)

Table 2.2 Prefixes Used in Naming Organic Chemistry Molecules

Number	Counting Carbons	Counting Functional Groups
1	Meth	Mono
2	Eth	Di
3	Pro	Tri
4	But	Tetra
5	Pent	Penta

Some examples may help in understanding the system. Take, for instance, the two molecules shown in Figure 2.11. The first molecule (CH_4) has only one carbon and so will have "meth" as the prefix. This molecule has no functional groups and so is assigned to the alkane family. The name is, therefore, "methane" where the "ane" ending indicates alkane. The other molecule in this same figure (CH_3Cl) is also part of the alkane family but in this case one functional group (Cl) is present. Hence, the name of this molecule is chloromethane.

The molecules in Figure 2.12 also illustrate the basic naming principles. The first molecule (C_2H_6) has two carbon atoms and will therefore have the prefix "eth." All of the bonds are carbon-carbon single bonds and so the molecule is part of the alkane family. The name is, therefore, ethane. The last molecule in the same figure (C_2H_4) also has two carbons and will have the prefix "eth." However, this molecule has a carbon-carbon double bond and therefore is part of the ethene family. (Notice that the ending of ethane and ethene indicate the differences in these families.) The last molecule will therefore be named ethene. The older name is ethylene, which is a slightly different name but preserves many of the same characteristics of naming the family type, the number of carbons, and the number of functional groups.

The science of polymers and plastics utilizes these names for the naming and categorization of polymer materials. It is for this reason that the functional groups and their names are introduced in this text.

POLYMERS

Just as two carbon atoms are bonded together in ethane (C_2H_6), three, four or many more can also bond in a chainlike arrangement, sometimes thousands of atoms long. These long chains of atoms are called *polymers,* from the word origin meaning *many units or parts.* The polymers are formed by building them up from many very small units or individual groups called *monomers,* meaning *single units.* A three-dimensional perspective representation of monomers built into a polymer chain is illustrated in Figure 2.16.

The number of atoms to which any one atom can bond is determined according to the rules for stable configurations (four bonds in the case of carbon) as discussed previously in this chapter. The proper number of electrons for a stable configuration is shown for each atom in Figure 2.17a. Although the electron-dot notation is important for a basic understanding of the bonding, this method of representing polymers is somewhat tedious to write. Therefore, a shorter method (called standard notation), in which only the atoms and the bonds are shown, is illustrated in Figure 2.17b. It is understood that each dash represents two bonding electrons and that the number of electrons in the outer shell of each atom is appropriate for the type of atom considered, although only bonding electrons are indicated. (Note that each carbon has eight electrons and that between one carbon and one oxygen there are four electrons. When this occurs, a double bond is formed, just as with carbon-carbon double bonds.) An even shorter notation system is illustrated in Figure 2.17c. In this system the hydrogen atoms bonded to carbon atoms are eliminated to permit focusing on the other atoms, which are usually more important in the properties of the polymer. (Remember, however, that the hydrogens are really still attached; they are just not shown.)

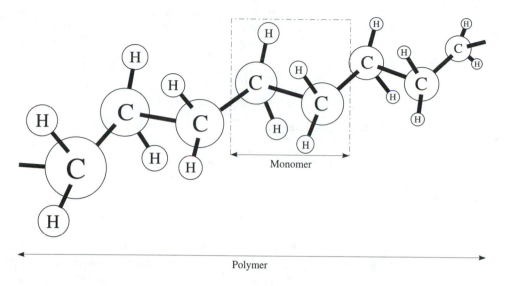

Figure 2.16   Three-dimensional perspective representation of ethylene monomers built into a polymer chain (polyethylene).

The parentheses going through the bonds at the ends of the polymer (Figure 2.17c) indicate that the atoms shown are only a segment of a much longer polymer chain. The n is a whole number that represents the number of such units contained in the entire polymer. An important feature of polymers is that the atoms along the polymer chain can bond both to other atoms along the chain and to atoms off the chain. Those atoms along the chain are sometimes called the backbone atoms and those off the chain are called pendant

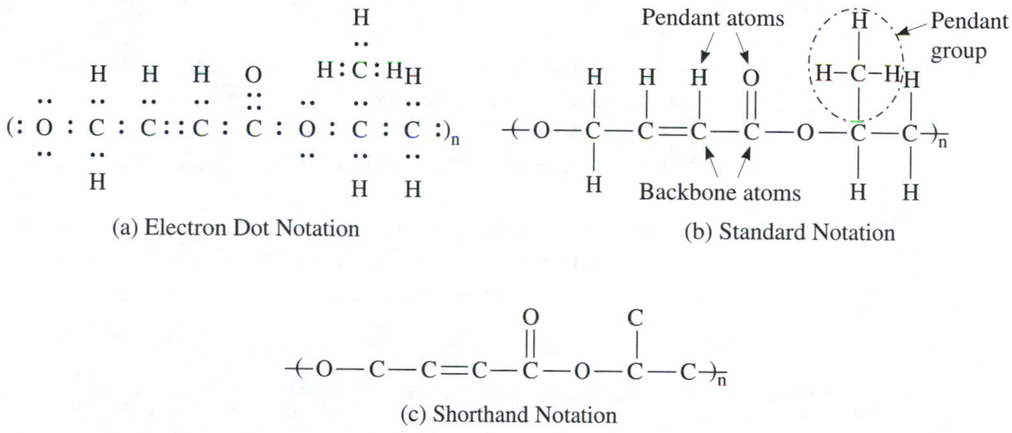

(a) Electron Dot Notation

(b) Standard Notation

(c) Shorthand Notation

Figure 2.17   Polymer chain showing various notation systems.

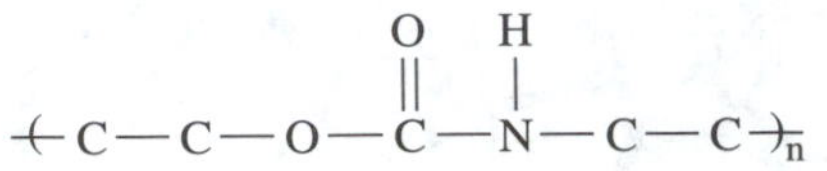

$$\begin{array}{c} \quad\quad\quad O \quad H \\ \quad\quad\quad \| \quad | \\ {+}C{-}C{-}O{-}C{-}N{-}C{-}C\overrightarrow{}_{n} \end{array}$$

(a) Carbon-based polymer

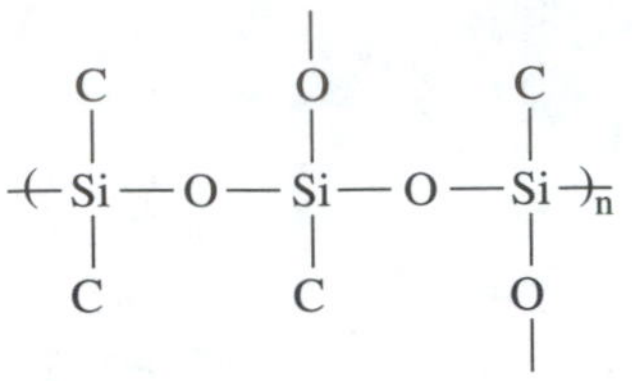

$$\begin{array}{ccc} C & O & C \\ | & | & | \\ {+}Si{-}O{-}Si{-}O{-}Si\overrightarrow{}_{n} \\ | & | & | \\ C & C & O \\ & & | \end{array}$$

(b) Silicon-based polymer

Figure 2.18 Polymers with atoms other than carbon in the chain.

atoms or pendant groups, if several atoms are grouped together. The functional groups shown in Figure 2.14 can be either pendant groups or part of the backbone chain.

Some polymers can also have atoms along the chain itself which are different from carbon, although carbon is by far the most common. These noncarbon chain atoms can significantly affect the properties of the polymer. Some polymers with noncarbon atoms are illustrated in Figure 2.18. The polymer in Figure 2.18a is a carbon-based polymer (that is, carbon is a principal atom along the chain), but nitrogen and oxygen are also part of the chain. The polymer illustrated in Figure 2.18b is not based on carbon but rather on silicon. Although silicon-based polymers have some important applications, the carbon-based polymers are far more important and are the principal focus of this text. (Silicon-based polymers are discussed in the chapter on elastomers.)

Just as was discussed previously in the section on the periodic table, whenever two atoms are bonded which have different electronegativities, the sharing of the electrons may be unequal, that is, the electrons may spend more time in the vicinity of one atom than in the vicinity of the other atom. When unequal sharing of electrons occurs, the bond is a polar bond indicating that it has a positive end (pole) and a negative end (pole). Another way of expressing this concept is that the covalent bond has some *ionic character,* meaning that the electrons are unequally shared and therefore slightly ionic. The limit of unequal sharing is, of course, electron transfer from one atom to another, and that would result in ionic bonding, as discussed previously. A gradation in bonding is therefore possible, from pure covalent to pure ionic. In general, if an appreciable sharing of the electrons exists, as occurs with all bonds between two nonmetal atoms, the bond is covalent. Polymers consist of covalent bonds, although some polarization (ionic character) often exists.

For example, in the molecule represented in Figure 2.19, the pendent oxygen is more electronegative than the carbon which is double bonded to it and so the oxygen would be partially negative and the carbon partially positive. (The term *partially* is used here to in-

Figure 2.19 Polar bonding in polymers.

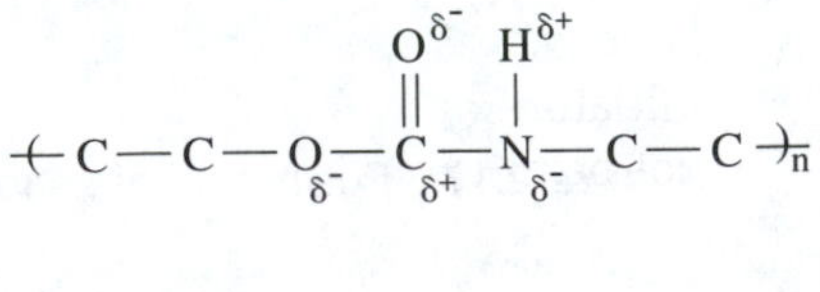

$$\begin{array}{c} O^{\delta^-} \quad H^{\delta^+} \\ \| \quad\quad | \\ {+}C{-}C{-}\underset{\delta^-}{O}{-}\underset{\delta^+}{C}{-}\underset{\delta^-}{N}{-}C{-}C\overrightarrow{}_{n} \end{array}$$

dicate that the charge is less than would occur if an electron were transferred completely.) This same carbon is also bonded to an oxygen which is in the chain. The bond between these two atoms is also polar, again because the oxygen is more electronegative than the carbon. The polar nitrogen and hydrogen bond is also shown. In this bond the nitrogen is partially negative and the hydrogen is partially positive. Carbon and hydrogen have approximately the same electronegativity and so C-H bonds are nonpolar.

FORMATION OF POLYMERS

Polymers are formed by causing small units (monomers) to chemically bond together and build up into long polymeric chains. Two principal methods are used to cause this bonding of monomers to occur: *chain-growth polymerization* or *addition polymerization* and *stepwise polymerization* or *condensation polymerization*. Each of these polymerization methods will be described in detail and then they will be compared and contrasted.

Chain-Growth Polymerization or Addition Polymerization

The addition polymerization mechanism, sometimes called the chain-reaction mechanism, proceeds by several sequential steps, whose essential features are illustrated in Figure 2.20.

The steps for addition polymerization are:

Step 1. Introduce the monomer containing a carbon-carbon double bond into the reaction vessel. (The simplest molecule containing a carbon-carbon double bond is ethylene, which was also depicted in Figure 2.12 and Figure 2.13 and will be used as the example to depict addition polymerization in Figure 2.20.) A high concentration of monomer is usually introduced. (Because the monomer is a gas, the concentration is usually increased by increasing the pressure of the monomer inside the reaction vessel.) Monomers can also be liquids and solids, although these are less common.

Step 2. Inject an *initiator* (usually a small amount is sufficient) into the reaction vessel such that the initiator mixes with the monomer. An initiator is any material that will start the polymerization reaction. In the most common case, the reaction is started by the formation of *free radicals,* that is, molecules that contain an unpaired electron. Peroxide molecules are common initiators for addition polymerization because when heated, they break apart to form free radicals. Peroxide molecules contain two oxygens that are bonded by a single bond and two organic functional groups that are bonded to each of the oxygens. Heated peroxides break apart at the oxygen-oxygen single bond to form the free radicals. These peroxide free radicals are unstable (because of the unshared electrons) and immediately try to join with some nearby electron to form a covalent bond. (Note that initiators are sometimes called catalysts. This is an improper use of the term catalyst because catalysts are not "used up" in the reaction, but initiators are reactants in the polymerization process and are used up as the reaction proceeds. Nevertheless, the use of catalyst to describe these initiators is widespread. In this text, however, a distinction between initiator and catalyst will be made and only when a catalyst meets the strict definition of not being a reactant will that term be used.)

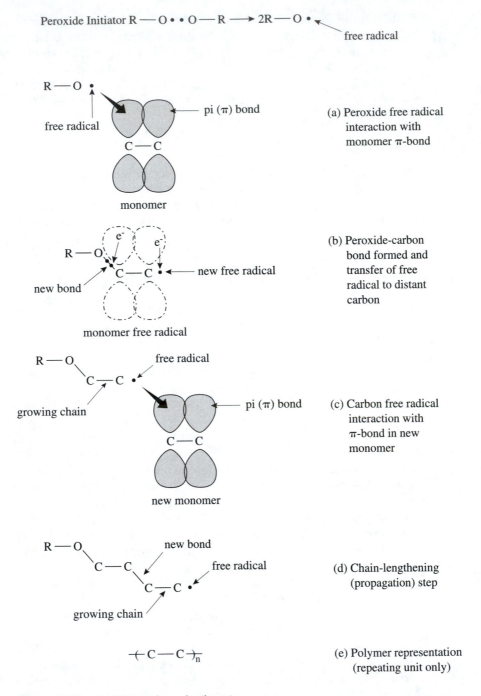

Figure 2.20   Addition polymerization steps.

Step 3. A convenient source of relatively available electrons is the π-bond in the carbon-carbon double bond of the monomer. As previously discussed, π-bond electrons are quite far removed from the center of the carbon-carbon bond. These electrons are, therefore, rather loosely held and readily available for interaction with the free radicals. This interaction of the peroxide free radical and the π-bond is illustrated in Figure 2.20a. This step is sometimes called the initiation step.

Step 4. The peroxide free radical extracts one of the two electrons in the π-bond, thus breaking the π-bond and forming a new bond connecting the peroxide with one of the carbons previously having the π-bond. The other electron in the π-bond is now unpaired. It is, therefore, a free radical. This new free radical moves to the other carbon, which previously had the π-bond. This sequence is shown in Figure 2.20b.

Step 5. This new free radical is also unstable and will try to form a bond. A convenient source of relatively available electrons is the π-bond of another monomer molecule. (This is especially true at high monomer concentrations.) This interaction is depicted in Figure 2.20c.

Step 6. The new free radical extracts an electron from the π-bond of the new monomer molecule and forms a new bond which links the atom on the growing chain with the new monomer atom. The π-bond in the new monomer is broken by this step and a new free radical is formed. This sequence results in a lengthening of the chain by the length of the new monomer (two carbons in this case) and the creation of a reactive site at the distant end of the chain. This step is referred to as the propagation step. (See Figure 2.20d.)

Step 7. This process of creation of a free radical, attack on a π-bond, formation of a bond with a new monomer, and creation of a new free radical can continue as long as a monomer is available to react, provided some other readily available electron source does not interfere with the process by reacting with the newly formed free radical. The chain can, therefore, become very long, often growing to several thousand units. The chain quickly becomes so long that the presence of the peroxide on the end becomes insignificant in the total picture (except for some analytical techniques which examine the end groups). At this point, the polymer is usually represented simply by the *repeating unit*, that is, by the unit that is added to the chain in each step. This unit is usually the monomer, but without the π-bond because it has been lost in the polymerization process. In its place, two bonds have been formed, one on each of the carbons which originally held the π-bond. This representation is given in Figure 2.20e. The n represents the total number of units in the chain. The name of the polymer is usually the name of the monomer preceded by the prefix "poly."

Step 8. Eventually the polymer chain must be ended. Several reactions can result in chain termination. One of the simplest is that the carbon free radical meets another free radical (either peroxide or carbon). In this case the two unpaired (free radical) electrons join together forming a covalent bond between the two free radicals. If two carbon free radicals join, the chains will be combined. Another chain termination mechanism is that the free radical will combine with some other electron-rich molecule. This molecule could be a contaminant in the reaction vessel or could be a molecule intentionally introduced to stop (quench) the polymerization reaction. This step is called the termination step and is the last step in the addition polymerization process.

Still another possibility is that the carbon free radical could extract an electron from a bond other than a π-bond. This is far less likely to occur because the electrons in other bonds are generally much more tightly held. However, it can happen, especially at high temperatures, where all electrons become excited and more active. When this occurs, the most likely place for the free radical to extract another electron would be from a carbon-hydrogen bond in another polymer chain. Should this happen, a bond between the two carbon atoms in the two chains would be formed and a hydrogen free radical would be released. The carbon-carbon bond would result in a connection between the two chains that would be like two different branches of a chain. This process is, therefore, called *branching* and is illustrated in Figure 2.21. The released hydrogen free radical would be available to react with other π-bond electrons or for any other reaction common for free radicals. It could, for instance, terminate some other chain. Other reactions which result in termination are also possible. These are, in general, of less importance than those already discussed.

Many different monomers can be polymerized by addition polymerization. All of them must have a carbon-carbon double bond but can have several side chains attached to the carbons containing the double bond. Several of the most important are listed in Figure 2.22.

Initiation of the addition polymerization reaction can also occur with materials other than peroxides. Almost any material that forms free radicals or that attracts an electron from the π-bond will work. Both positive and negative ions (charged atoms or molecules)

Figure 2.21 Branching reaction in addition polymerization.

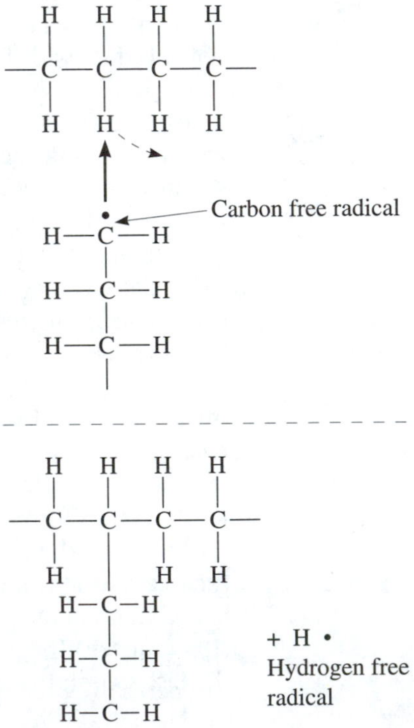

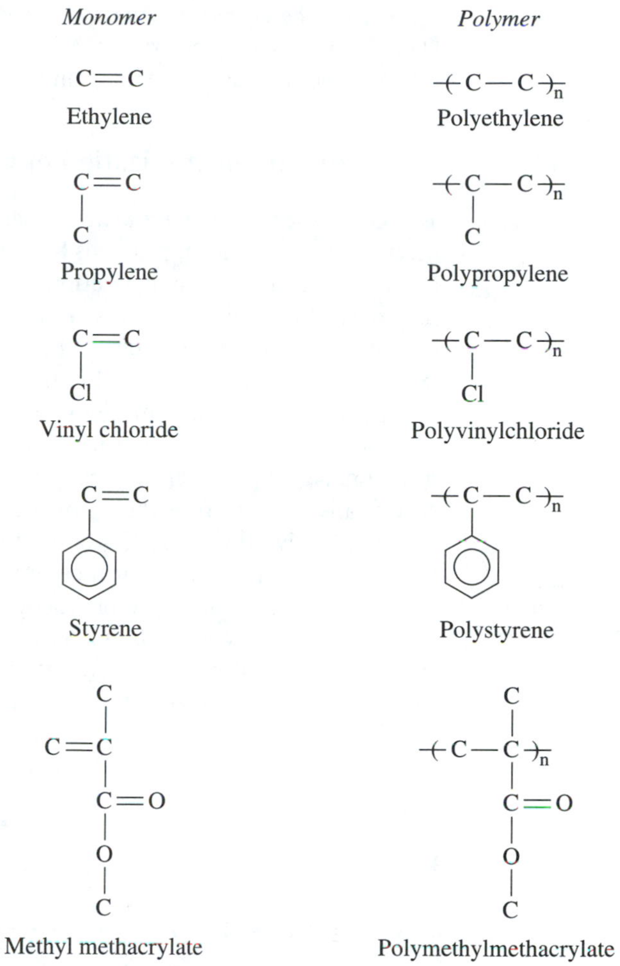

Figure 2.22 Common monomers that polymerize using addition polymerization.

| Monomer | Polymer |

Ethylene — Polyethylene

Propylene — Polypropylene

Vinyl chloride — Polyvinylchloride

Styrene — Polystyrene

Methyl methacrylate — Polymethylmethacrylate

work in some cases. Ultraviolet light, X rays, and other energy sources can also cause free radicals to form, either directly with the π-bond or with an intermediate atom (such as oxygen).

Another important consideration in addition polymerization is the use of a true catalyst. As indicated previously, catalysts are materials that promote the effectiveness of a chemical reaction but are not consumed in the reaction itself. A typical catalyst for addition polymerization is a metal, such as titanium, platinum, or various metal-organic molecule combinations that form a solid surface within the reaction vessel. These metals attract the growing polymer chain and hold it in a particular orientation, thereby facilitating the chain-lengthening reaction with the monomer, which is also attracted by the catalyst. In some cases the holding of the growing polymer in a specific orientation can permit the formation of polymers with specific orientations of any side groups which may be attached to the double-bond carbons. A well-known catalyst system of this type is called the *Ziegler-Natta catalyst,* named after the developers of the system. These catalysts (1) promote the polymerization, (2) can permit specific orientation reactions to occur, and (3) can give the added benefit of initiating the reactions.

Step-Growth Polymerization or Condensation Polymerization

The *condensation polymerization* mechanism, sometimes called the step-growth polymerization mechanism, proceeds by several steps, which are quite different from the steps involved in addition polymerization. Before considering the specific steps in condensation polymerization, some concepts regarding the reactions of functional groups (active reaction sites) should be considered, because these form the basis of condensation polymerization. The most fundamental of these basic concepts is the reaction between functional groups that is illustrated in Figure 2.23.

Each step in a condensation polymerization involves a reaction between dissimilar functional groups which are part of monomer molecules. These functional groups are specifically chosen from the many available in organic chemistry to react in the desired way and to give the properties desired in the resulting material. (Figure 2.14 lists several functional groups used to form commercially important polymers).

The most common type of reaction to form condensation polymers involves the formation of a new covalent bond between the functional groups and the simultaneous formation of a small molecule, such as water, which is a by-product of the reaction. (The joining of two molecules with the formation of a small by-product molecule is called *con-*

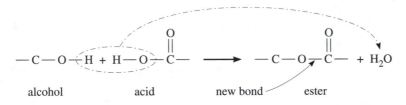

(a) Reaction of an alcohol and an acid to make an ester and water

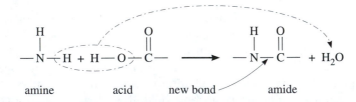

(b) Reaction of an amine and an acid to make an amide and water

Figure 2.23 Reaction between dissimilar functional groups, which is the basis of condensation or step-growth polymerization reactions.

Monomer	Name

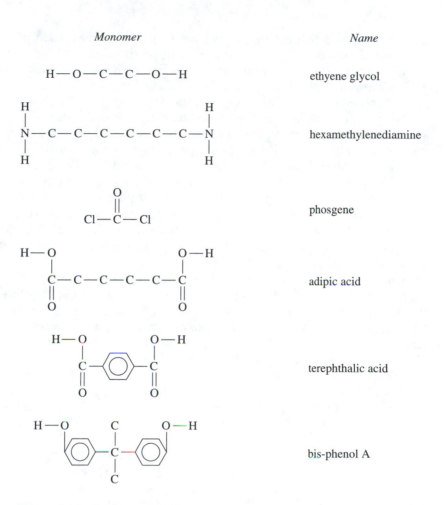

Figure 2.24 Typical molecules having two functional groups, which are, therefore, bifunctional.

densation and is the process from which the polymerization method takes its name.) For instance, an alcohol can react with an acid to form an ester with water as the condensate; or an amine can react with an acid to form an amide, again with water as the by-product. These reactions are represented in Figure 2.23.

For polymerization to occur each monomer molecule must have two reactive functional groups (otherwise the monomers would react once and then stop, with no chain being formed). Monomers that have two groups are *bifunctional*. In most monomers the reactive sites are on either side of a short chain of carbon atoms. Some typical examples of bifunctional monomers are given in Figure 2.24.

The reactive ends are identical on both ends of each of these monomers which are therefore called *symmetric bifunctional monomers*. These are the usual, but not the only kind of monomers that can be used for condensation polymerization. (The use of non-symmetrical monomers will be illustrated in appropriate sections later in this text.)

While the general mechanism for the condensation reaction was represented in Figure 2.23, a complete step-by-step illustration of a condensation reaction to form a specific product is useful and is shown in Figure 2.25.

The particular condensation polymerization reaction illustrated is of great historical and commercial significance. The reaction shown is used to make nylon, which launched the use of condensation polymerization as a method for making polymers. This reaction uses two symmetric bifunctional monomers—hexamethylene diamine (a monomer with two amine groups on either side of six carbons) and adipic acid (a monomer with two acid groups on either side of four carbons). The resulting material is a polyamide, usually known as nylon (6/6), where the numerical designation in the name represents the number of carbons in each monomer; in this case each monomer has six carbons. For simplicity, the hexamethylene diamine monomer will be referred to as "monomer A" and the adipic acid monomer will be called "monomer B."

The steps in the condensation polymerization process are as follows:

Step 1. The two different monomers (A and B), each having two active sites, are introduced into the reaction vessel. The materials are heated and vigorously stirred or agitated to facilitate the reaction. (No initiator or catalyst is normally needed in condensation reactions.)

Step 2. One end of monomer A reacts with one end of monomer B and a new bond is formed linking the two monomers which begins the polymer chain. A molecule of water condensate is formed as a by-product, as shown in Figure 2.25a. The reaction occurs because of the inherent tendency of acids and amines to react. (Reactions of some functional groups occur without the creation of a by-product molecule, although these reactions are far less common.)

Step 3. The beginning polymer chain has active functional groups on each of its ends. (This is because both monomers had two active sites before they reacted.) In the case shown in Figure 2.25, one end of the beginning polymer chain is an amine group and the other end is an acid group. The amine group can react with one of the adipic acid monomers (monomer B) and the acid group on the other end can react with one of the diamine monomers (monomer A). The reaction of the acid end of the beginning polymer with monomer A is illustrated in Figure 2.25b. In this reaction monomer A is added to the beginning polymer which extends the chain and forms a longer polymer. Again a molecule of water is created as a by-product. Note that this reaction step is not dependent upon any previous reaction step, in contrast to the chain reaction mechanism of addition polymerization. For this reason, this reaction method is called stepwise polymerization. The water is removed by a special distillation process as the reaction proceeds.

Step 4. The longer polymer chain still has reactive groups at both ends. An amine can react with the acid monomer (monomer B) or an acid end can react with an amine monomer (monomer A) to produce new bonds and add to the chain to create an extended polymer, as shown in Figure 2.25c. Water is again made as a by-product. This process of reaction with an ever longer polymer can proceed until all the monomers have reacted. Long polymer chains can be formed but are typically not as long as with addition polymerization. Although the actual dimensional length of the chain is very small (because

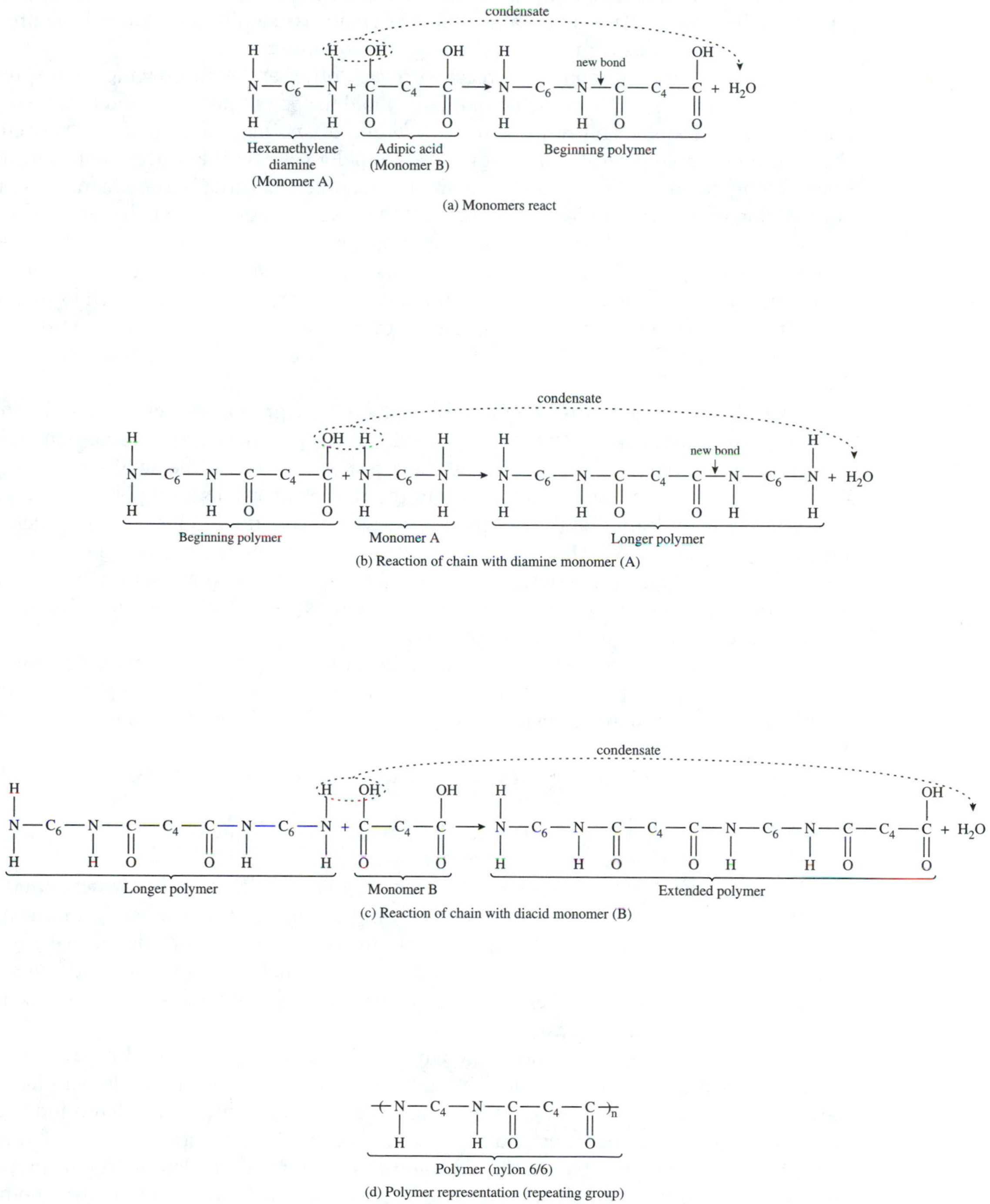
condensate
new bond
H_2O
Hexamethylene diamine (Monomer A)
Adipic acid (Monomer B)
Beginning polymer
(a) Monomers react
condensate
new bond
Beginning polymer
Monomer A
Longer polymer
(b) Reaction of chain with diamine monomer (A)
condensate
Longer polymer
Monomer B
Extended polymer
(c) Reaction of chain with diacid monomer (B)
Polymer (nylon 6/6)
(d) Polymer representation (repeating group)

Figure 2.25 Condensation reaction steps to form a polyamide (nylon 6/6).

atoms are so small), it is significant that the material now has thousands of bonded units and the collection of these units into a polymer chain has significantly altered the properties of the material as compared to the original monomers.

Step 5. Not only can monomers react with each other and with growing chains, two growing chains can react with each other. This would happen if the amine end of one polymer were to encounter and react with the acid end of another polymer. If this happens, the length of the polymer chain can grow very rapidly because the process joins already created long chains instead of adding a single monomer. Eventually the chains will become so long that further movement enabling the reactive sites on the polymers to come into close proximity for bonding is no longer possible. The reaction is then terminated simply by cooling which further reduces the movement and slows the tendency of the functional groups to react. The polymerization reaction can also be terminated by adding a material that has only one active end. This is called *quenching the reaction* and is done after the polymers have reached the desired length. The quenching material is said to form *end-caps*.

When the polymer chains become very long, the end groups are relatively unimportant as far as polymer properties are concerned and the polymers are best represented by a generalized polymer structure showing the repeat unit, that is, the smallest unit that reflects the basic structure of the polymer. In the case of a condensation polymer, this repeat unit is a combination of the two monomers which were used to form the polymer, as indicated in Figure 2.25d. These polymers are named by using the prefix "poly" with the name of the new type of functional group created. For instance, the reaction between an amine and an acid creates an amide and so the polymers are called polyamides. The name nylon is just a common name for polyamides.

In some cases the by-product is not water, but it is almost always a small, stable molecule. An example of a by-product other than water is the reaction between phosgene and bis-phenol A to form polycarbonate. Hydrochloric acid (HCl) is the by-product.

Comparison of Addition and Condensation Polymers

The methods of addition polymerization and condensation polymerization differ in several key areas. These differences are summarized in Table 2.3.

One difference between addition polymerization and condensation polymerization is the chain-growth mechanism. In addition polymerization the polymer grows in a *chain reaction* manner, that is, once started it is self-propagating. An initiator is normally used to start the reaction. In condensation polymerization the polymer grows in a *step-by-step* manner, that is, the polymer chain increases in discrete steps. Initiators are not needed for condensation reactions to occur.

Each of these two major types of polymerization is also characterized by the type of monomer used to form the polymer. In addition polymerization, the monomers have a carbon-carbon double bond which is the single active reaction site. In condensation polymerization each monomer must have two active sites that are functional groups but are not carbon-carbon double bonds. The typical pattern is for one condensation polymerization monomer to have two identical functional groups on each of its ends. This monomer is mixed with a different monomer that also has two identical functional groups on each

Table 2.3 Characteristics of Addition and Condensation Polymerization Methods

	Addition or Chain-Growth Polymerization	Condensation or Step-Growth Polymerization
Polymer growth mechanism	Chain reaction	Step-by-step reactions
Dependence on previous step	Yes—sequential dependent events	No—independent events
Initiator needed	Yes	No
Type of monomer	Contains carbon-carbon double bond	Has reacting bifunctional groups on the ends
Number of active sites (functional groups) per monomer	1	2
Number of different types of monomers needed to form polymer	1	2 (usually)
By-product formed	No	Yes (usually)
Basic (polymer repeat unit) representation	Monomer without the double bond and with bonds on either side	Two monomers joined together
Polymer chain characteristics	A few, long chains	Many not very long chains
Branching	Possible	Unlikely

of its ends, but not the same functional groups as the first monomer. For condensation polymerization to occur, the functional groups of the first monomer must react with the functional groups of the second monomer to form the polymer. Therefore, condensation polymerization generally requires that two types of monomers be present, whereas only one type of monomer need be present for addition polymerization. (Condensation polymerization can also be done when a single monomer has two different end groups which will mutually react. These are, however, rare.)

In view of these differences, some monomers polymerize by the addition mechanism while others polymerize by the condensation mechanism. No polymer can be formed by both mechanisms, although a polymer may be formed by one mechanism and then enter into a later reaction that uses the other mechanism. (This is seen in crosslinking, to be discussed later.)

The formation of a by-product condensate is typical in a condensation polymerization but does not occur in addition polymerization reactions. Some minor products, such as very short-chain polymers, may be formed in addition polymerization, but these are the result of collisions between various free radicals formed during the course of the reaction rather than a specific product from each polymerization step.

The polymer formed from addition polymerization can be represented by the monomer without the double bond and with two bonds extending on either side, thus indicating that a chain is present. The basic polymer formed from condensation polymerization can be represented by two monomers joined together as they would be in the polymer after the condensate has been extracted.

Some differences in the polymers produced by the addition polymerization and condensation polymerization processes can be seen from an examination of their mechanisms. Because of the relatively small amount of initiator used in addition polymerization, only a few chains begin to grow. The chain reaction mechanism proceeds very quickly, especially in the presence of a catalyst, and results in a few very long chains in a short period of time. Eventually the monomer is entirely combined into the chains and the reaction stops.

In condensation polymerization any two different monomers which meet can initiate a chain. Therefore, many chains are growing simultaneously. The growth of these chains depends upon their ability to encounter monomers or other chains and react effectively. Eventually the monomer is essentially all reacted, but because these chains still have active ends, the growing chains can continue to react among themselves, thereby increasing the length of the chain. This chain combination continues until the chains become so long that further movement is difficult, even at the elevated temperatures normally used in the polymerization process or until a quenching agent is added. The likelihood of very long chains is, therefore, more common with addition polymerization than with condensation, and the likelihood of having many polymer chains being formed is greater with condensation than with addition.

Polymerizations Other Than by Addition and Condensation Mechanisms

Some polymerizations use reactions other than addition or condensation. For instance, a di-isocyanate can react with a di-alcohol (diol) to produce a polyurethane. No condensate is formed in this reaction, but some rearrangement of the atoms occurs when the bond between the monomers is formed. Atom rearrangement also occurs in the formation of acetal polymers from formaldehyde.

Another method for making acetal polymers involves the opening of a trioxane ring. Epoxy resins also use the opening of a ring to create two reactive ends, which can bond to other monomers and form a polymer.

These and a few other methods have been used both experimentally and commercially to form polymers, but by far, the greatest number of polymers are formed by either the addition or the condensation polymerization method.

THERMOPLASTICS AND THERMOSETS

All plastics, whether made by addition or condensation polymerization, can be divided into two groups called *thermoplastics* and *thermosets*. Thermoplastics are solids at room temperature that are melted or softened by heating, placed into a mold or other shaping device, and then cooled to give the desired shape. The thermoplastics can be reshaped at any time by reheating the part. Thermosets can be either liquids or solids at room temperature that are placed into a mold and then heated to cure (harden), thus giving the desired shape and solid properties. Thermosets cannot be reshaped by heating.

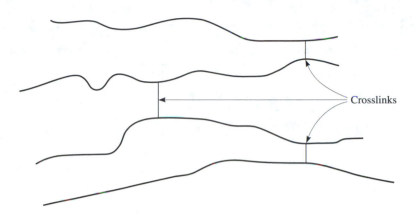

Figure 2.26 Crosslinking of polymer chains.

Two such very different behaviors would be expected to arise from a basic difference between thermoplastics and thermosets. In addition to the normal covalent bonds which join the atoms together in the polymer chain, the thermoset plastics also have covalent bonds which join the chains one to another. These bonds between chains have been given a special name, *crosslinks,* and are illustrated in Figure 2.26.

When the appropriate sites for reactions exist, crosslinks are normally formed by heating the polymer materials, a process called *curing.* The heating gives sufficient energy to excite the molecules and cause them to move close enough together that attractions between the bonding sites can occur, causing the bonds to form. Hence, thermoset materials become firm, cured, or "set" with thermal energy. The crosslinks can either be formed as the polymers themselves are being formed or can be formed between specific sites along existing polymer chains that can join together in normal covalent bonds. Thermoplastic materials, which are characterized by not being crosslinked, will soften and become more "plastic" or pliable or melt with thermal energy. Details of the nature of thermoplastics (commodity and engineering) and thermosets are given in later chapters.

COPOLYMERS

Some unique polymers can be created by mixing together more monomer types than the minimum required to effect normal polymerization, such as mixing ethylene monomer and propylene monomer. Addition polymerization requires only one type of monomer to create a polymer chain, whereas condensation polymerization normally requires two monomers, each with two active ends. When more than these minimum numbers of monomer types are mixed, the polymer will contain some combination of the monomer types present. Polymers with mixed monomers are called *copolymers,* and they are, therefore, characterized as having more than one type of repeat unit. (When only the minimums of monomer types are present, that is, when the normal polymer is made, the polymer is called a *homopolymer.*)

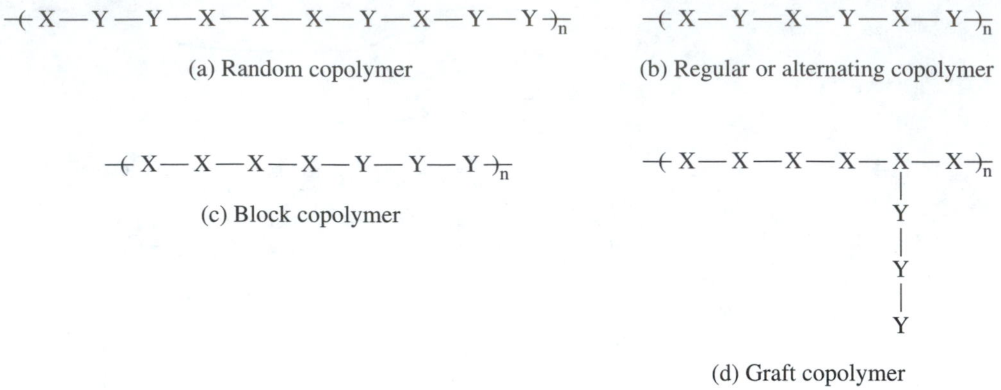

$$\left(\text{X—Y—Y—X—X—X—Y—X—Y—Y}\right)_n$$

(a) Random copolymer

$$\left(\text{X—Y—X—Y—X—Y}\right)_n$$

(b) Regular or alternating copolymer

$$\left(\text{X—X—X—X—Y—Y—Y}\right)_n$$

(c) Block copolymer

$$\left(\text{X—X—X—X—X—X}\right)_n$$
$$\quad\quad\quad\quad\quad\quad\quad\quad | $$
$$\quad\quad\quad\quad\quad\quad\quad\quad Y $$
$$\quad\quad\quad\quad\quad\quad\quad\quad | $$
$$\quad\quad\quad\quad\quad\quad\quad\quad Y $$
$$\quad\quad\quad\quad\quad\quad\quad\quad | $$
$$\quad\quad\quad\quad\quad\quad\quad\quad Y $$

(d) Graft copolymer

Figure 2.27 Copolymer patterns for addition polymerization.

Copolymers in four general patterns can be made by both the addition and the condensation polymerization methods. These are illustrated in Figure 2.27 for the simpler case of addition polymerization. Some commercial copolymers, all to be discussed later in this text, are acrylonitrile-butadiene-styrene (ABS), styrene-acrylonitrile (SAN), and ethylene-vinyl acetate (EVA). (Some copolymers can be referred to as *terpolymers,* which reflects that three monomer types are polymerized together.)

The determination of which type of copolymer would be formed is dependent upon the basic nature of the monomers and their mutual reactivity, the conditions of the reaction (such as concentrations, temperatures, catalysts, etc.) and any specific modifications that might be made to the monomers or polymers allowing the copolymerization step to be activated as desired. All these conditions are used for various copolymerizations.

One monomer, X, which could polymerize alone, is mixed with another monomer, Y, which could also polymerize alone. The mixture of monomers can produce a *random copolymer* structure (Figure 2.27a) where there is no pattern in the order of the monomers along the polymer chain. This is the most common copolymer pattern.

Another pattern could be a *regular or alternating copolymer* pattern, as shown in Figure 2.27b. In this pattern the monomers have a regular, alternating sequence that can be represented as X-Y-X-Y, etc. This polymer should not be confused with a normal condensation polymer although the alternating pattern is the same. In the copolymer case illustrated, two addition monomers were present (one more than the number needed to polymerize) so a copolymer was made. Because copolymerization requires **more** than the minimum number of monomers, a copolymer in a condensation polymerization requires **three** monomers and the resulting polymer would require at least the pattern A-B-C whereas the regular polymer pattern is just A-B.

A third type of copolymer pattern occurs when long sequences of one monomer join long sequences of the other monomer to form the chain. This pattern is called a *block copolymer* and is shown in Figure 2.27c. This pattern is characterized by several monomers of one type in a row along the backbone followed by several monomers of the second type, also along the backbone, such as X-X-X-Y-Y-Y-X-X, etc.

The fourth pattern is a *graft copolymer* (Figure 2.27d). In this type of copolymer a polymer chain is formed by one monomer (X) and then a chain of the other monomer (Y) is attached as a branch to the main backbone.

In general, the properties of random and alternating copolymers are averages of the properties of the two homopolymers. The properties of block and graft copolymers are not averages but have some characteristics like one homopolymer component and some properties like the other homopolymer.

CASE STUDY 2.1

Modifications to Improve Teflon® Processing

Teflon® is a registered trademark of the DuPont Company for its polymers with a carbon backbone and fluorine molecules attached to the carbons at all possible bonding sites. Molecules containing fluorine and carbon are called fluorocarbons. This molecule is represented in Figure 2.28a.

The PTFE material, polytetrafluoroethylene, was discovered rather than intentionally made. A chemist at DuPont (Roy Plunkett) was working with tetrafluoroethylene (TFE), a gas, and found on one occasion the gas cylinder pressure was zero, suggesting that the tank was empty. He lifted the tank and noted that it was unusually heavy for an empty tank and desired to understand why. He cut open the tank and found a white, waxy powder in the bottom. He then realized that the TFE was a monomer that had spontaneously

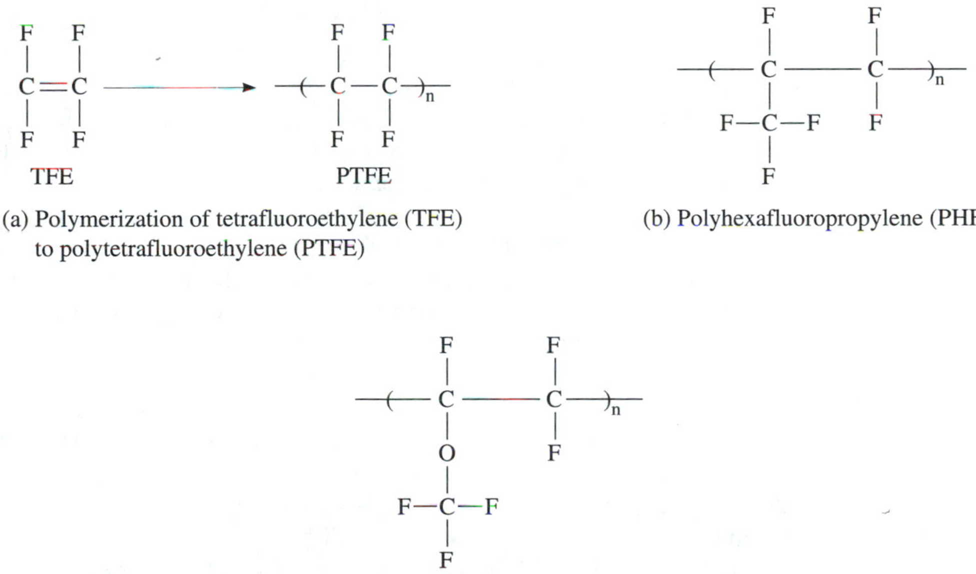

(a) Polymerization of tetrafluoroethylene (TFE) to polytetrafluoroethylene (PTFE)

(b) Polyhexafluoropropylene (PHFP)

(c) Perfluoroalkoxy (PFA)

Figure 2.28 Fluorocarbon polymers.

polymerized to PTFE, the resulting solid material. Because TFE had a carbon-carbon double bond, the addition polymerization method was suspected, and was later confirmed.

The DuPont Company named the new polymer material Teflon® and began to explore its properties. The polymer was found to be more inert and stable than the analog hydrogen-carbon material, polyethylene. (Polyethylene is the same structure as PTFE except that all the hydrogens in polyethylene are replaced by fluorines in PTFE). The differences in properties are because of the nature of the carbon-fluorine bonds.

Fluorine is the most electronegative of all the atoms. Therefore, the carbon-fluorine bond is highly polar, with the negative end toward the fluorine. Hence, the electrons are held very tightly by the fluorine molecule. The bonds are very strong and this pulls the fluorine atoms close to the carbons along the backbone. The fluorines act as shields to the carbons and do not allow the carbons to be attacked by other chemicals, nor will the fluorines be extracted. Hence, the polymer is very inert.

This inertness contributes to other characteristic properties of PTFE. For instance, PTFE will not combine with oxygen and so it will not burn nor can it be corroded. Because the electrons are held so tightly, it will not conduct electricity. Because PTFE resists forming bonds with other materials, very few things stick to it and it has been used extensively in making nonstick surfaces for cooking pans. (The PTFE flows around the roughened edges of the pan surface during manufacture to give a mechanical bond. No chemical bond is formed; hence, the PTFE can be removed by scraping.)

PTFE chains are very long, straight, and stiff, allowing PTFE molecules to pack together tightly. PTFE has the highest density of any plastic. Furthermore, temperatures have little effect. PTFE is unaffected by cold, remaining unchanged at temperatures as low as −268°C, only a few degrees above absolute zero. Likewise, it resists high temperatures. PTFE does not begin to melt until 327°C and then does not become a liquid, but a translucent gel. At temperatures above this point, PTFE decomposes. This resistance to elevated temperature has been both a benefit and a detriment for the use of PTFE. While it allows PTFE to be used in many high-temperature applications, it also means that PTFE is difficult to process by normal thermoplastic processing techniques that require melting of the polymer.

Originally PTFE was made into useful parts by packing the powder into a mold and then heating the mold, under pressure, to just below the melting point. This treatment caused the polymer particles to fuse together into a solid mass, a common process in metals and ceramics called *sintering*. The fabrication of large parts was therefore difficult. Also, the time required to make a part was very long compared with the traditional thermoplastic manufacturing processes.

Therefore, the DuPont Company sought a way to make a polymer that had the beneficial properties of PTFE but would be easier to process by traditional plastic manufacturing methods. The key to success was the discovery of polyhexafluoropropylene (PHFP), a polymer related to PTFE. As seen in Figure 2.28b, the PHFP is based on propylene (a three-carbon group) rather than on ethylene. The resultant polymer has a pendent carbon atom and fluorines at all possible bonding locations. This pendent group means that the chain has far more flexibility and will, therefore, melt at a lower temperature. The pendent group also means that the material loses some of the beneficial properties characteristic of PTFE. PHFP has greater adhesion for other materials, is more electrically conductive, and is slowly attacked by some chemicals.

DuPont found, however, that a copolymer formed from mixing TFE monomer and HFP monomer had properties that were intermediate between PTFE and PHFP. It proved to be a good compromise. The new copolymer material was called fluorinated ethylene propylene (FEP) and is currently in wide use. This material can be molded with normal thermoplastic processing methods.

Another possibility for improvement of PTFE is a modification that gives a slightly higher affinity for bonding with other materials. This can be done by adding oxygen to the side chain, as is shown in Figure 2.28c. The oxygen allows the fluorocarbon material to bond with other materials, such as with the metal in a cooking pan. This material (PFA) is the basis for the new, nonstick coatings that have been produced. Again, the ideal properties of PTFE are diminished, but other properties such as adhesion, perhaps more advantageous, are enhanced.

Other useful polymers have been made by mixing pure polymerized PTFE with other polymers. These are not copolymers because the monomers are not mixed but, rather, the already polymerized polymers are combined. A good example is the mixing of PTFE with polyamideimide (PAI), a very strong and heat-resistant polymer. This mixture was found to be useful in applications where mechanical wear resistance was important, such as non-lubricated bearings. The PAI was resistant to the high temperatures typical of this application (even higher temperature capability than PTFE) and also resisted the abrasion. The PTFE was found to enhance the sliding of the parts against each other and further enhance the abrasion resistance.

Therefore, if the ultimate chemical resistance or nonstick behavior is desired, pure PTFE can be used. If some compromise in properties is acceptable, FEP, PFA or a mixture of PTFE with another polymer such as PAI might be acceptable. The result is a wide range of fluorocarbon materials, each having some unique and beneficial properties which can be tailored to fit each application.

SUMMARY

Matter is composed of atoms which are, themselves, made up of protons, neutrons, and electrons (along with other sub-atomic particles whose description is beyond the scope of this text). The protons and neutrons exist at the center of the atom in small, dense groups called nuclei with the electrons arranged around the nucleus. Neutral atoms will have the same number of protons (+ charges) and electrons (− charges). (Isotopes, which are not discussed in this text, are materials with the same number of protons, but with different numbers of neutrons.) Changes in the number of protons define different types of atoms. Therefore, when one proton is present, a particular type of atom is created (hydrogen). Two protons define a different atom (helium), three define another (lithium), and so on. Each atom has a unique atomic number that corresponds to the number of protons in the atom. At this time over 100 different atom types have been defined. These atom types are called elements.

The elements can be arranged in a simple table wherein they are grouped according to their chemical natures and atomic numbers. This table is called the periodic table, because similar chemical properties repeat in periods.

The chemical properties of the atoms are primarily dependent on the arrangement of the electrons. The electron arrangement, called electron configuration, is a function of the energy of the electrons and the inability of more than two electrons to occupy the same energy level. These energy levels are characterized by their location (shell) and their shape (orbital). The chemical nature of the atoms depends upon whether the atom has a tendency to lose or gain electrons.

If an atom that tends to lose electrons comes close to an atom that tends to gain electrons, one or more electrons are transferred between the atoms and the atoms become charged, one positively and the other negatively. The charged atoms are called ions. The attraction between these positive and negative ions forms a bond that is called an ionic bond. Ceramic materials often have ionic bonds. If an atom that tends to lose electrons comes close to another atom that tends to lose electrons, both atoms will give up their electrons and a loosely-held sea of electrons is formed. This results in a metallic bond. If an atom that tends to gain electrons comes close to another atom that tends to gain electrons, the atoms will share their electrons. This is a covalent bond, and molecular materials, including polymers, have this type of bond.

To the extent that the atoms share the bond unequally, because one atom may have a greater affinity for the electrons than the other, the bond is polar. The affinity for electrons is called electronegativity. Secondary bonds between molecules can be formed that are much weaker than the ionic, metallic, or covalent bonds. Some secondary bonds can occur because of the attractions of polar bonds between molecules. These are called hydrogen bonds. In the case of water, the hydrogens are positive, compared with the oxygen, and will form secondary bonds with the oxygens in other water molecules. Another type of secondary bond is a van der Waals bond, and that results from induced polarity between molecules.

Carbon atoms are able to form many molecules of differing size and complexity. Therefore, the study of carbon-based molecules has been given a special attention and is called organic chemistry. Carbon atoms have six electrons, with two in an inner shell and four in an outer shell. The outer shell electrons are generally the only ones that enter into chemical reactions. These four outer shell electrons will form four covalent bonds and achieve a very stable electron configuration. Since each covalent bond has two electrons, the stable configuration consists of eight electrons around each carbon atom and is called a stable octet. Most other atoms also gain stability when they are surrounded by eight electrons in their outer shell.

Some groupings of atoms that occur frequently in organic chemistry have characteristic properties and are called functional groups. The study of these group characteristics is largely the focus of organic chemistry and is the basis for the naming of organic molecules, including polymers.

Carbon atoms can bond to themselves, often creating long chains of carbons with other atoms attached to the chain. These long atom chains are polymers and are formed by the building up of the chain from smaller units, called monomers. Two principal methods for forming polymers are the addition polymerization method and the condensation polymerization method. Addition polymerization generally requires the presence of a carbon-carbon double bond in the monomer. Condensation polymerization generally requires that two monomers be present, each with two active sites that can react between the monomers.

Polymers which are synthetic and which are formed in order to produce a useful product are called plastics. The raw, formable material is called a resin. Plastics can be divided into two large groups—thermoplastics and thermosets. Thermoplastics are solid materials that are heated to melt and then placed in a mold and cooled to solidify. Thermosets are either solid or liquid materials that are placed into a mold and then shaped and heated. This heating causes the material to crosslink.

Thermoplastics are convenient to use because the polymerization step has been accomplished by the plastic resin manufacturer, usually in large facilities resembling oil refineries. The product is a fluffy powder or flake which is often processed into small polymer strands, about the size of spaghetti, which are chopped into small pellets. These resins are shipped to the plastic fabricators, who then (1) remelt the material, (2) mold it into the various desired shapes, and (3) resolidify it to gain the solid properties desired in the finished part.

The length of the polymer chains produced by the addition polymerization and the condensation polymerization processes has a profound effect on the properties of the polymers. Whereas the monomers are usually gases or liquids, the polymers are solids, often with melting points of several hundred degrees. In general, the longer the chain length, the higher is the melting point of the polymer.

Thermoset materials undergo a crosslinking reaction while in the mold. This reaction takes longer than the cooling of thermoplastics. For some applications, such as the coating of fibers or encapsulation of inserts, thermosets are easier to process than thermoplastics. The crosslinked materials cannot be remelted after they are formed.

Occasionally a thermoplastic material will be polymerized into such a large molecule (with no crosslinks) that it also will reach its decomposition temperature before it melts. Although the behavior is like a thermoset, the plastic is still considered to be a thermoplastic because it lacks the essential feature of thermosets—covalent bonds (crosslinks) between the polymer chains.

When several different types of monomers are reacted together, a polymer with mixed monomers can occur. These polymers are called copolymers and usually have properties that are intermediate between the two types of non-mixed polymers.

GLOSSARY

Addition polymerization   Chain-growth polymerization.
Aliphatic   Molecules that do not contain the six-membered benzene ring.
Aromatic   Molecules containing the six-membered benzene ring.
Atomic model   A model of the fundamental nature of matter, which states that all matter is made up of small, discrete particles called atoms.
Atomic number   The number of protons in an atom.
Atomic orbital   The normal region in which an electron will be when surrounding the nucleus of an atom.

Atomic weight   The average sum of the number of protons and neutrons in the nucleus of an element, reflecting the fact that the same element can have different numbers of neutrons.

Backbone atoms   Those atoms along the main chain of a polymer.

Bifunctional   A molecule having two reactive functional groups.

Block copolymer   A copolymer in which groups of monomer types occur together along the backbone.

Bond   An attraction between atoms or molecules.

Bond energy   The energy needed to break a bond apart and separate the atoms completely.

Bond length   The optimal (normal) separation distance for two bonded atoms.

Branching   When a polymer has a side chain.

Catalyst   A molecule or material that facilitates a chemical reaction but does not, itself, become part of the reaction products. Initiators are sometimes erroneously called catalysts.

Ceramics   Solids that typically have ionic bonds.

Chain-growth polymerization   A method for forming polymers that proceeds by a free-radical mechanism (also called addition polymerization).

Chemical reactivity   The tendency of an atom to change its electron configuration through interaction with another atom.

Condensation polymerization   Stepwise polymerization.

Copolymer   A polymer formed from more than the minimum number of monomers, thus creating a polymer with properties of two or more normally separate polymers.

Covalent bond   Attraction between two atoms as a result of their sharing electrons, with each bond consisting of two and only two electrons.

Crosslinks   Covalent bonds between polymer chains.

Crystal   A highly organized and regular group of atoms or molecules.

Curing   The process of hardening a polymer by the formation of crosslinks.

Dipole   A polar bond.

Dipole bond   Another name for hydrogen bond.

Double bond   When the same two atoms share four electrons.

Electron cloud   The space occupied by electrons surrounding the nucleus of the atom.

Electron configuration   The arrangement or distribution of electrons about the nucleus in an atom or throughout the space surrounding the nuclei in a molecule.

Electron orbital   A region surrounding the nucleus of the atom where a particular electron is most likely to be found.

Electronegativity   The tendency of an atom to attract electrons.

Electrostatic attraction   Attraction between positive and negative centers or particles.

Element   A material that consists of only one type of atom.

End-cap   Reactions or entities which occur at the ends of polymer chains and result in the stoppage of the polymerization reaction.

Fluorocarbon   A polymer based on carbons, with fluorine atoms in all or almost all of the bonding sites.

Free radical   An atom containing an unpaired electron.

Functional group   Certain arrangements of atoms (usually containing carbon) that have characteristic properties.

Graft copolymer   A copolymer in which a chain of one polymer is attached to the chain of a different polymer type.

Homopolymer   A polymer made from only the minimum number of monomer types.

Hydrogen bond   A secondary bond caused by the attraction between a positive location on one molecule and a negative location on another molecule or between widely separated parts of the same molecule.

Initiator   A molecule that reacts (usually by decomposing to free radicals) to start a reaction.

Ionic bond   Attraction between negative and positive ions.

Ionic character   A relative measure of the amount of polarity in a molecule.

Ions   Atoms which have either lost or gained electron(s) and therefore have a net positive or negative charge.

Isotope   Two atoms having the same number of protons but a different number of neutrons and, therefore, different atomic weights.

Macro viewpoint   Properties of polymers which depend upon the bulk nature of the material without regard to the substructures.

Matter   Anything that has weight and occupies space.

Metal   Atoms which tend to give up electrons.

Metallic alloys   Mixtures of two or more different metals.

Metallic bond   A type of attraction in which positive ions are held together by a surrounding sea of electrons.

Micro viewpoint   Properties of polymers which depend upon the nature of clusters or groupings of molecules.

Molecular orbital   The normal region where electrons will be within a molecule.

Molecular viewpoint   Properties of polymers which depend upon the nature of bonding between atoms or molecules.

Molecule   The smallest entity that contains more than one atom and possesses all the properties of the combined unit.

Monomer   A single unit that can be combined with others to form a polymer.

Nonmetals   Atoms which tend to accept electrons.

Nucleus   The dense central portion of the atom containing protons and neutrons.

Octet rule   A stable electron configuration achieved when eight electrons are in the outer shell.

Pendant atoms   Those atoms attached to but not part of the backbone of a polymer.

Periodic table   A systematic listing of the elements arranged according to the number of protons in the nucleus of each atom and the configuration of electrons surrounding the nucleus.

Peroxide   A common initiator molecule.

Pi bond   A dumbbell-shaped molecular orbital.

Polar bond   A covalent bond in which the electrons are unequally shared between the atoms.

Polymer   A molecule made of many parts.

Quench   To add a material that stops a polymerization reaction.

Random copolymer   A copolymer in which the arrangement of different monomer segments along the backbone is random.

Regular copolymer   A copolymer in which the arrangement of different monomer segments occur in a predictable repeating pattern.

Repeating unit   A representation of the polymer in which only the portion added in each step is shown along with the total number of units in the chain.

Ring opening   A chemical reaction that can be used in some polymerization methods.

Saturated molecule   A molecules containing no carbon-carbon double bonds.

Sea of electrons   A group of freely moving electrons that are typical of metallic bonding.

Secondary bond   Attraction between atoms or molecules that is much weaker than ionic, metallic, or covalent bonds.

Sigma bond   A cylindrically shaped molecular bond.

Single bond   When only two electrons are shared between two atoms.

Stepwise polymerization   A method for forming polymers that proceeds by the elimination (condensing out) of a small molecule.

Symmetric bifunctional monomer   A molecule having two identical functional groups that can be used to form a polymer.

Terpolymer   A polymer made from three types of monomers.

Thermoplastic   A polymer, solid at room temperature, that can be melted, put into a mold or other shaping device, and then cooled to solidify in the desired shape.

Thermoset   A polymer that may be either a liquid or a low-melting solid at room temperature that can be placed into a mold and then heated to cure (harden).

Triple bond   When the same two atoms share six electrons.

Unsaturated molecule   Molecules containing a carbon-carbon double bond.

Valence   A number that reflects the number of electrons that an atom will usually give up or attract in chemical reactions.

van der Waals bonding   A type of secondary bonding (very weak bonds) due to the attraction that all molecules have for each other, effective only at very short distances.

QUESTIONS

1. Describe the differences in the carbon and oxygen atoms.
2. Why is an octet of electrons a stable configuration?
3. Identify the type of bond and the product formula expected between potassium (K) and bromine (Br) and explain the basic nature of this bond. Show the resulting electron configurations of K and Br after the bond is formed.
4. Identify the type of bond and the product formula expected between magnesium (Mg) and chlorine (Cl) and explain the basic nature of this bond. Show the resulting electron configurations of Mg and Cl after the bond has formed.
5. Describe the type of bonding between carbon and chlorine.
6. What is an initiator and why is it important in a polymerization reaction?
7. Why is it harder to make very long polymer chains using the condensation polymerization method than by using the addition polymerization method?

8. Describe the bonding in a carbon-carbon double bond. Include in this description an explanation of the mechanism by which the various bonds are formed. What does the existence of a double bond tell about the other atoms bonded to the carbon atoms?

9. Define monomer and polymer. Write typical polymeric repeating unit structures for both addition and condensation polymerization and explain the various symbols contained therein.

10. Which polymerization method, addition or condensation, is expected to result in branched molecules? Why?

11. Describe crosslinking and the resultant properties that it will create.

12. What is a copolymer and how are copolymers formed?

13. What is the molecular difference between thermoset and thermoplastic materials?

REFERENCES

Brydson, J. A., *Handbook for Plastics Processors,* Oxford: Heinemann Newnes, 1990.

Brydson, J. A., *Plastics Materials* (2nd ed.), New York: Van Nostrand Reinhold Company, 1970.

Daniels, C. A., *Polymers: Structure and Properties,* Lancaster, PA: Technomic Publishing Co., Inc., 1989.

Gruenwald, G., *Plastics: How Structure Determines Properties,* Munich: Hanser Publishers, 1993.

Mark, Herman F., *Giant Molecules,* ed. René DuBois, Henry Margenau, and C. P. Snow, Life Science Library, New York: Time Incorporated, 1966.

Quagliano, James V., *Chemistry,* Upper Saddle River, NJ: Prentice Hall, Inc., 1958.

Richardson, Terry L., *Industrial Plastics: Theory and Application* (2nd ed.), Albany, NY: Delmar Publishers, Inc., 1989.

Seymour, Raymond B. and Charles E. Carraher, *Giant Molecules,* New York: John Wiley and Sons, Inc., 1990.

CHAPTER THREE

MICRO STRUCTURES IN POLYMERS

This chapter examines the following concepts:

- Amorphous and crystalline
- Solids, liquids, and gases
- Thermal transitions
- Effects of thermal changes on polymers
- Polymer length, molecular weight (average molecular weight and molecular weight distribution (MWD))
- Physical and mechanical property implications of molecular weight and MWD
- Melt index
- Steric (shape) effects

INTRODUCTION

This chapter focuses on the shapes of polymer chains and the interactions between polymer chains rather than on individual atoms or small groups of atoms along the polymer chains. This level of focus is called the **micro** view, as opposed to the **molecular** view of the previous chapter. The **macro** (an even larger) view, which looks at bulk properties of polymers, is discussed in the following two chapters.

The length of the polymer chain, the ways polymer molecules or chains interact with other chains, and the shape of the chain, especially as the shape is dictated by pendent groups, are key parameters in determining many polymer characteristics, including solid/liquid behavior and crystallinity.

The chain interactions can be of two types. The first type is a coiling of several polymer chains about each other with entanglement, much as a nest of snakes would intertwine. The second type of interaction is a systematic closely packed folding of sections of a chain into a regular pattern. Each of these interactions is discussed in detail in this chapter.

AMORPHOUS AND CRYSTALLINE

Amorphous and crystalline describe the randomness or regularity of the entanglement of polymer chains. *Amorphous,* which means "without shape," describes those polymers or regions within a polymeric structure in which the entanglement between polymer chains and within a chain is random. In many polymers, however, the entanglement can be very regular with large regions of the chains packing together into regular, repeating structural patterns. When this occurs, these regularly packed regions are called crystals and the polymer is said to be *crystalline.* A polymer with some crystalline regions is illustrated in Figure 3.1. Some polymers which are crystalline include high-density polyethylene (HDPE), polypropylene (PP), acetal, and nylon.

The areas of crystallinity are composed of folded chains which are held together by crystal bonds (secondary branch). These bonding forces between the chains are localized to the tightly packed, crystalline areas. This bonding occurs because the crystal structures, when they occur, represent structures with lower energy than a random, noncrystalline arrangement of the molecule. The lower energy occurs because the molecules form bonds that release energy. These crystalline sections are scattered throughout the polymer with some nonstructured (amorphous) areas between them.

The amount of crystallinity in the polymer depends upon several factors. Perhaps the most important is the size and regularity of the pendent groups (the groups attached to the main polymer backbone). If these pendent groups are relatively small and are regularly spaced along the polymer chain, they will not interfere with each other and the polymer chains can pack closer together. Forces of attraction and other similar interactions between polymers, such as hydrogen bonding, also increase crystallinity. Still another influence on crystallinity is the manner in which the polymer is cooled from the melt. Slow cooling allows the crystalline regions to organize while there is still movement among the chains and, therefore, favors crystallinity, provided the molecular structure is appropriate.

The amount of crystallinity, that is, the total number of atoms involved in a crystalline structure as opposed to the number in amorphous regions, can vary widely. In some polymers, no crystallinity is formed. In others, if all of the conditions are favorable, the crystallinity can approach 100%, but is more likely to be in the 60 to 70% range. Polymers

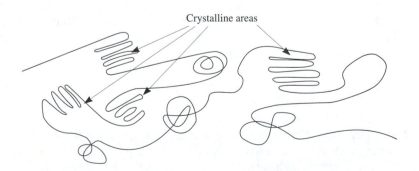

Crystalline areas

Figure 3.1 Crystalline and amorphous regions in a polymer structure.

whose structures favor crystallinity, such as HDPE, PP, acetal and nylon, are referred to as crystalline polymers, even though the actual crystallinity in the particular material may be only moderate.

The *degree of crystallinity* is the amount of the structure of the polymer that is crystalline, as opposed to amorphous. Three methods have principally been used to determine the degree of crystallinity of a polymer. These three are: specific volume (specific gravity or density), X-ray diffraction, and infrared spectroscopy.

As the material becomes more crystalline, it also becomes more dense. Especially in polyethylene and other plastics with a wide range of possible crystallinities, the density is the most common method of expressing crystallinity. For instance, polyethylene with a density of 0.97 g/cc would be high density (HDPE) while a density of 0.92 g/cc would be low density polyethylene (LDPE). The following methods are commonly used to measure specific gravity and density.

- **Specific Gravity (ASTM D 792).** This test determines the weight of a sample in air and then immersed in water. The ratio of the weights is the specific gravity, or the density of the material compared to the density of water. Most plastics have specific gravities in the 0.9 to 3.0 range. Small samples (about 1 inch^3) are used in the test and are suspended in the water by a thin wire.

- **Density by Density-Gradient Technique (ASTM D 1505).** The density of materials is determined by comparing the point at which a small sample (usually a pellet) will be suspended in a fluid of varying density with the suspension points of small glass floats of known densities. This method uses a density-gradient column prepared by adding two miscible liquids (usually water and ethanol) in varying concentrations so that more of the dense liquid is near the bottom of the column and more of the less dense liquid is near the top. The glass floats of known density are then carefully added to the column. The floats sink according to their densities and, by noting the depths, a plot of depth versus density can be established for the column. Then the plastic samples are added to the column and, by noting the depth at which they are suspended, their density can be read from the calibration chart. After some time (usually weeks), the liquids in the column mutually diffuse and the column must be refilled and recalibrated. A density-gradient column is illustrated in Figure 3.2.

- **Bulk Density (ASTM D 1895).** The bulk density is the apparent density of the material, that is, the density of the material without compaction or modification. This property is important when the plastic material is a powder or a flake. The test is conducted by carefully allowing the plastic to flow into a beaker or other container of known volume. The excess material is scraped off the top, and the container with the material is weighed. The weight of the container is subtracted from the total weight, and then density is calculated as the weight per volume. A related property is the *bulk factor,* which is the bulk density divided by the density of the plastic part after molding. Another related property is *pourability,* which is a measure of the time required for a standard quantity of material to flow through a funnel of specified dimension. Results are given as g/cc or pounds/foot^3.

- **Sieve-Analysis (Particle-Size) Test (ASTM D 1921).** The size of the particles and the distribution of the particle sizes in a particular batch can be important in

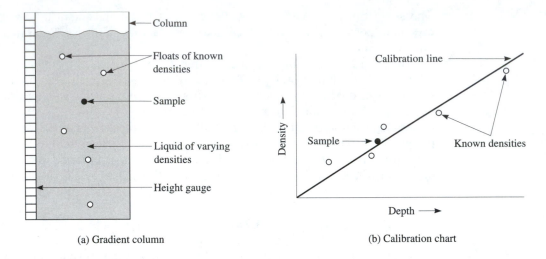

(a) Gradient column

(b) Calibration chart

Figure 3.2 Density-gradient column method for determining density, showing the column and a plot of the column calibration.

some processes and in compounding. For instance, rotational molding fusion can be highly dependent on the sizes of the particles. Also, when large and small particles are mixed together, melting tends to be uneven and can result in nonuniform mold filling and surface defects such as orange peel. The test to determine the size distribution of the particles is simply to pour the powdered material through a series of sieves with various opening sizes. The distribution is then determined by weighing the different sieves before and after the test. A shaker usually is employed to facilitate passage of the material through the sieves.

X-ray diffraction is useful in determining the degree of crystallinity because X-rays will develop a characteristic pattern when diffracted through a crystal structure. (This same technique is used to investigate crystal properties in metals and ceramics.) Infrared spectroscopy can also be used to investigate crystallinity because the vibrations and rotations of the atoms that are detected by infrared spectroscopy are affected by the crystal structure and, therefore, appear at slightly higher energy levels than do freely rotating and vibrating atoms. Hence, a shift in spectrographic pattern is detected when crystalline regions are present. Differential scanning calorimetry is another method that has been used but not as frequently as the others.

Because of the bonding forces within a crystal, crystallinity affects many physical properties in ways that are similar to other intermolecular attractions already discussed. Tensile strength, for instance, is increased by crystallinity because of the high resistance to movement in the crystalline regions and the need to overcome the intermolecular (crystalline) forces. This resistance can be very high in some cases, resulting in a marked increase in these properties over amorphous polymer analogs.

The effect of crystallinity on impact toughness is somewhat more involved. The crystalline sections of a polymer are not as effective in absorbing and dissipating impact energy as are the amorphous regions because the atoms in the crystalline regions are not as free to rotate, vibrate and translate. This restriction on atomic movement causes highly crystalline materials to be stiffer and more brittle. Therefore, even though the strength increases, the impact toughness often decreases for highly crystalline materials.

As will be discussed further in the chapter on chemical and physical properties, these properties are also affected by crystallinity. For instance, solubility of the polymer is generally reduced in crystalline materials because of the compactness of the crystalline structure compared to the amorphous region. This compactness retards the access of solvent molecules to the bulk of the structure.

Crystallinity is a basic property of plastics that should be considered in the selection of a polymer for any application. Many polymers can be obtained in a range of crystallinities, thus allowing the designer a wide choice of material properties. Another major effect of crystallinity is in thermal changes.

SOLIDS, LIQUIDS AND GASES

Matter can be described in terms of its physical state, that is, its physical condition at any one moment. The possible states are known as solid, liquid, and gas. When in its solid state, matter has a fixed volume and a fixed shape. An example would be an ice cube. The ice cube will occupy the same volume of space, no matter how much larger than the ice cube the container may be. Furthermore, the ice will maintain its shape (a cube) regardless of the shape of the container, provided the container is larger than the cube and no other changes are made, such as temperature.

When in a liquid state, matter has a fixed volume but not a fixed shape. Water is an obvious example of a liquid. When placed in a container, such as a glass, the water will flow to take the shape of the container. If placed in a container of a different shape, such as going from a glass to a cup, the water will change its shape accordingly. The water will, however, maintain its volume, even though it might be placed in a much larger container.

When in a gaseous state, matter has neither a fixed volume nor a fixed shape. Steam or water vapor is the corresponding gas to ice and water. When placed in a container, the gas will take the shape of the container and will occupy the entire space available within the container. If the size or shape of the container is changed, the gas will change accordingly.

Most types of matter can be converted from one state to another by changing the temperature or, less commonly, changing the pressure. For instance, ice will change into a liquid at the melting point, which will, in turn, change into steam at the boiling point.

The single most important property that controls the state of a material is the freedom of movement of its atoms or molecules. If the particles can move independently, the material is likely to be a gas. If the particles are highly restricted in their movements, the material is likely to be a solid. A liquid is between the two in ease of particle movement.

For instance, in ice, the molecules are locked in a three-dimensional crystal structure with relatively strong bonds holding each water molecule in place relative to its neighbors. As heat is added to the ice crystal, the atoms gain more energy and begin to vibrate and move slightly. More heat will result in more vibrations, rotations and other localized movement until, eventually, the molecules have more energy than the energy associated with the crystal structure bonds which hold the molecules in place. (These are secondary bonds acting in three dimensions.) When this energy point is reached, the ice melts. However, even after melting there are still some secondary bonds acting between water molecules. These secondary bonds are generally not three dimensional and are, besides, weaker than in the solid case because the molecules are farther apart and secondary bonds are very distance dependent. With added energy, the molecules continue to gain greater energy which imparts further increases in vibrations, rotations, and other movements until eventually the energy of the remaining secondary bonds are exceeded and the molecules begin to act completely independently. This is the boiling point. The molecules with energies above this boiling point energy are no longer associated with the rest of the molecules and are free to move about within the container. They will, therefore, fill the container in a random fashion, characteristic of a gas. Additional energy input will cause all of the liquid to boil and become a gas.

Additional energy could still be added to the gaseous system. This will continue to cause the molecules to vibrate, rotate and move until eventually sufficient energy is present in the system that the bond energy is reached. When that occurs, the atoms in the molecule will separate and the material will no longer be water but will revert to hydrogen and oxygen atoms. The point where the covalent bonds are broken is called the decomposition point. (Further energy addition could conceivably break the atoms apart but these energies are associated with nuclear reactions, such as occur in atom bombs, and are beyond the scope of this text.)

As the number of atoms in a molecule increases, the number of sites for secondary bonding also increases. Hence, the amount of energy required to break these bonds and convert the material from a solid to a liquid also increases. Polymers, which have very long chains of atoms, will generally have higher melting points than smaller, nonpolymeric covalent materials. Other characteristics within the molecules that could increase secondary bonding (such as having polar sites that can form strong hydrogen bonds) will also increase the melting point of covalent materials.

Ionic solids (ceramics) and metallic solids (metals) can also be melted if sufficient energy is added to the system to cause the bonds between the particles in the solid structure to separate. In these cases, the particles are atoms and the bonds that hold the materials together are the ionic or metallic bonds which were discussed in the chapter on molecular structures. These bonds are much stronger than the secondary bonds that hold covalent solid structures together. Hence, the energies required to melt ceramics and metals are generally much higher than the energies required to melt covalent solids. The melting points of ceramics and metals are also higher than covalent solids. (The energies to break the covalent bonds in covalent molecules are roughly the same as the energies to break apart the atoms in ceramics and metals. But, in the case of covalent molecules, these energies are associated with the decomposition point rather than the melting point.)

All the changes in state (solid to liquid to gas) are reversible with the subtraction of energy from the system. Gases can be condensed to liquids and liquids frozen to solids by cooling. Most materials can be changed from one state to another and back repeatedly, with some important exceptions which will be discussed later.

The changes between the solid and liquid states with the addition of energy are important in the understanding of plastics. As discussed in the Introduction to Plastics chapter, the definition of a plastic material implies that it is used as a solid but at some time has been shaped or molded into a useful shape. The shaping or molding is usually done as a liquid. Hence, almost all plastics have made a transition from the liquid to the solid state. Many of the processes discussed in the later chapters in this text will discuss different methods of making these transitions.

THERMAL TRANSITIONS OF P0LYMERS

Responses to Heat Inputs

The responses of polymer materials to heat inputs are similar in some ways to those of small molecules, as has been described in the section on solids, liquids, and gases. The input of thermal energy (heat) is both common and very important for polymeric materials. When polymers are heated, the basic nature of the polymer is to move internally to absorb the energy input. This internal motion can be molecular twisting, vibrating, stretching, translation, and other movement. The extent of these motions is detected by a rise in the temperature of the polymer. Some of these motion modes occur at lower energy inputs than others (for instance, vibrating occurs with little energy input), but as the amount of energy input increases, all the motion modes become active and the temperature increases greatly as a measure of this increased molecular motion.

At low temperatures, polymers are solids. The motions of atoms within a solid polymer are initially limited to small movements (usually vibrations) of a few atoms. As the amount of heat input increases, these motions become both larger in amplitude and involve more atoms. Because of the limited amount of motion allowed in a solid material, the easiest way to absorb the energy is to involve more of the atoms along the polymer chain but to limit the motions to those that take up the least space and energy—vibrations, rotations, and twisting. Hence, with moderate heat input, the majority of the atoms are in a mode of increased vibration, rotation, and twisting. The properties of the polymer during this stage of heating are little changed except for an increase in the temperature of the polymer and a minor increase in the space occupied by the polymer (volume), because the motions take slightly more space and the atoms are forced apart. This situation is represented in Figure 3.3, which graphs the thermal transitions of an idealized plastic.

The minor increase in volume is measured by the *coefficient of thermal expansion.* The size of the coefficient of thermal expansion is dependent on the amount of energy required to force the atoms apart. This quantity is discussed in more detail later in this chapter.

Creep

As the polymer is heated further, translational movements become more important. Translational movements occur when atoms move from one place to another in space, beyond the movements of vibration, rotation, and twisting, which are centered about a stationary reference position. These increased movements allow the polymer chains to slowly disentangle and to move apart, thus increasing the space between the atoms. This increase in space results in a decrease in the strength of any secondary bonds that may be present and a decrease in the entanglement between polymer chains.

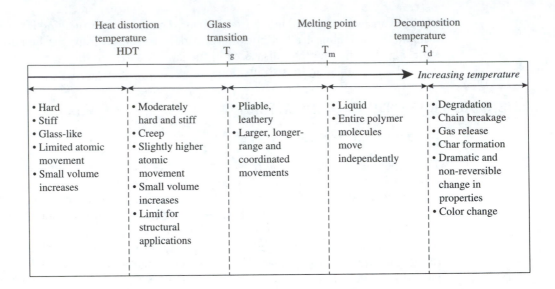

Figure 3.3 Idealized plastic.

With these changes, the polymer's movement under applied external loads is increased. In other words, the solid polymer slowly moves under applied loads. This phenomena is called *creep*. Creep becomes greater as the temperature of the polymer increases and as the amount of secondary bonding and entanglement decreases. Therefore, polymers that have high secondary bonding, such as crystalline polymers, have inherently less creep than do amorphous polymers, assuming other factors are equal. Creep is also reduced by high entanglement, by the presence of large pendant groups that inhibit movement, by crosslinking, and by the inherent stiffness of the backbone of the polymer. These effects are discussed later in this chapter in the section on steric (shape) effects and in the chapter on mechanical properties.

Heat Distortion Temperature (HDT)

At some temperature the plastic will become so pliable and so easily distorted under load that it may not perform the function that is intended, especially if that function is structural. The temperature at which this happens varies widely among different plastics and among different applications. For instance, although one plastic may be "dishwasher safe," that is, it will not distort at dishwasher temperatures (about 120°F, or 50°C), another plastic will curl and deform. Therefore, the first plastic material is thermally stable at dishwasher temperatures, but the second is not. In another instance, two plastics may be very suitable for cassette tape cases in normal use but one will distort when left inside a car in the summer. The designer of plastic parts should, therefore, be aware of the maximum structural use temperature. Aromatic content, crosslink density, crystallinity, and secondary bonding can all raise the temperature at which distortion occurs.

A convenient test, the *deflection temperature under load* (ASTM D648), can be used to measure this upper use temperature. In this test, illustrated in Figure 3.4, a sample of the plastic material (5 × ½ × ¼ inch) is placed in a heated bath of mineral oil or some other liquid that is thermally stable at the temperatures to be used and will transfer heat readily to the sample. A specific weight (designated by the test procedure and dependent upon the type of plastic material being tested) is then placed on the sample. The entire apparatus is then heated, usually with stirring to insure good heat evenness in the liquid. The temperature at which the sample deflects (bends) a specified distance under these conditions is the *heat distortion temperature (HDT)*. The HDT is often considered the maximum structural use temperature, especially for any application in which the part will be loaded mechanically.

For some applications, the maximum use temperature may be lower than the HDT because of excessive creep of the plastic material. Plastics with high creep may, given enough time, eventually distort, even under only moderately elevated temperatures, or even at room temperatures. The maximum structural use temperature is, in these cases, determined from the **lowest temperature at which change in the plastic material occurs which would be detrimental to the particular application.**

The Vicat softening temperature (ASTM D 1525) is also used to measure the maximum structural use temperature of a plastic. The Vicat test measures the temperature at which a flat-ended needle penetrates a sample to a specific depth.

Another test used to determine a maximum use temperature is the UL temperature index (UL 746B). Underwriters Laboratories (UL), an independent organization concerned with consumer safety, has developed a temperature index to assist UL engineers in judging the acceptability of individual plastics in specific applications involving long-term

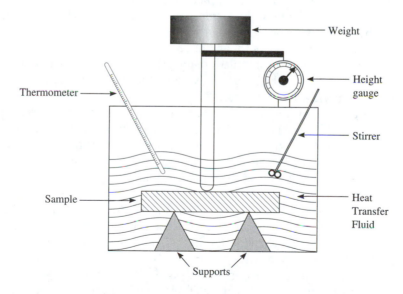

Figure 3.4   Deflection under load test to determine heat distortion temperature.

exposure to elevated temperatures. The UL index lists a temperature for each plastic that is considered the maximum long-term use temperature. The index temperature is determined by exposing samples to circulating air at various temperatures for 10,000 hours. That temperature that causes a sample to lose 50% of its mechanical property (usually strength or toughness) is the index rating temperature.

Glass Transition Temperature

When one atom in a polymer chain moves in translation, it has a tendency to pull surrounding atoms with it. Hence, these translational movements require more energy than do vibrations, rotations, or twisting. With further heating beyond the HDT, several adjacent atoms (perhaps five or six) will eventually have enough combined energy to move (translate) as a unit, perhaps with a semirestricted motion somewhat like that of a jump rope or a sinusoidal wave, because some segments of atoms in the polymer are very free to move, whereas other adjacent segments remain more restricted. This coordinated, long-range translation results in a significant increase in the flexibility of the material in the region where the long-range motion occurs.

When some of the atoms begin to exhibit this phenomenon, added thermal input will cause other polymers to begin similar motions, thus absorbing the input thermal energy and causing the temperature to remain constant. Temperatures such as this, where added heat does not increase the temperature but causes some change within the material, are called *thermal transitions*. (The melting of ice is a thermal transition.) In this case, the thermal transition that occurs when the polymer molecule begins to make coordinated long-range movements is called the *glass transition temperature (T_g)*. The relationship between the HDT and the T_g is shown in Figure 3.3.

Note that the HDT is not a formal thermal transition in the sense of T_g or, as we shall see, a melting point. The temperature does not stop when the HDT is reached, but some internal change in state occurs within the polymer. HDT is highly dependent on the weight applied and the shape of the sample, as well as the temperature, and should therefore be considered only as a convenient reference temperature. The HDT is much easier to determine experimentally than is T_g, thus leading to the widespread use of HDT in characterizing polymer use temperatures.

The most important method for determining T_g is the differential scanning calorimetry (DSC) test (ASTM D3417 and D 3418). This test involves measuring the heat absorbed by a sample when that sample goes through thermal transitions. The DSC test allows these transitions to be identified for a plastic material by noting the absorption of heat from a plot of heat versus time as the sample is gradually heated. Results identify the temperatures of the transitions.

Where previously (below T_g) the polymer was rigid and hard, with the long-range movements of several adjacent atoms that occur above the glass transition temperature, the polymer becomes pliable and leatherlike. (T_g is called the glass transition temperature because glass behaves similarly.)

The changes in physical properties at T_g allow this transition to be determined using a test called thermomechanical analysis (TMA). The TMA method places a probe into the sample and measures the changes in the size or mechanical behavior of the sample as the temperature is progressively raised. TMA can determine the coefficient of volume expansion of a sample and can also determine the T_g of a sample because the sample becomes more pliable above the glass transition temperature. TMA data can also be used to determine the maximum use temperature of the material. Results are a plot of mechanical changes versus temperature.

If the polymer is (1) tightly entangled, (2) extremely stiff, or (3) highly restricted in some other way, T_g will be high, indicating that considerable energy must be imparted in order to induce the characteristic long-range movements. In fact, the backbones of some polymers are so stiff that the polymers appear to be hard and stiff even above T_g. It is, therefore, sometimes difficult to tell from just feeling the polymer whether it is above or below its glass transition temperature.

Melting

As more and more heat above T_g is put into the polymer, the thermoplastic polymer continues to soften and become more pliable, because larger and larger segments of the polymer become excited and gain coordinated movements. The polymers continue to disentangle from each other. Eventually, the polymer has so much internal energy that entire polymer molecules are moving freely relative to all the other polymers in their vicinity, and melting occurs. (If the polymer is joined to those neighboring polymers by covalent bonds, it is a thermoset, and melting does not occur, as explained later.) Hence, melting is simply the process of polymer chains gaining sufficient energy to move independently. Initially only a few polymers have sufficient energy to move independently, but with increased thermal input, all the polymers will gain this freedom. The temperature at which this occurs is called the *melting point* or *melting temperature, T_m*, and it is a thermal transition as defined previously for T_g. Figure 3.3 shows the relative position of the melting temperature and the other thermal transitions which have already been discussed.

The thermal transitions are also depicted in Figure 3.5, where the differences between thermoplastics (both amorphous and crystalline) are shown and compared with a thermoset. Note that in a typical **amorphous thermoplastic** the polymer is hard and stiff below the T_g and passes through the HDT and then to T_g. Above T_g the polymer would be leathery up to the melting temperature, where it would, of course, melt to a liquid.

In the amorphous thermoplastic polymer, the melting point, T_m, is shown as a range over which melting would occur. This melting range is caused by the wide variations in entanglement, chain length, and secondary bonding between polymer chains that often occur in amorphous plastics.

As can be seen in Figure 3.5 for the case of **crystalline thermoplastics,** there is no T_g. In highly crystalline polymers where the crystal bonding energies are strong, resulting in tightly held molecules, almost no translation motions occur until the input energy is sufficient to overcome the crystal bonding energies. The crystalline structure prevents the long-range, coordinated atomic motions that are characteristic of the change from rigid structures to leathery structures. When the energy reaches the threshold energy equal to the crystal bonding energies, the crystal lattice breaks apart and the molecules become free to translate. Therefore, crystalline thermoplastics remain quite rigid up to the

Thermoplastics (amorphous)

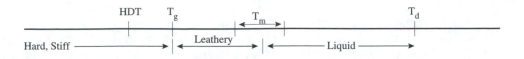

Thermoplastics (crystalline)

Thermosets

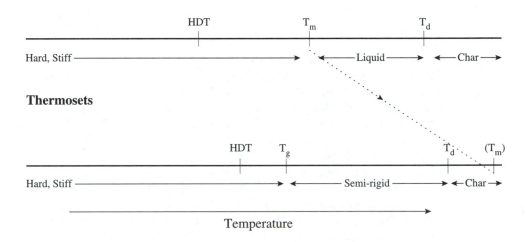

Temperature

Figure 3.5   General thermal behavior for thermoplastic and thermoset plastics.

crystalline melting point, which is quite sharp (narrow temperature range). The temperatures at which the crystalline areas break apart are usually high enough that other effects, such as secondary bonding and differences in chain length, have little effect on the melting point sharpness.

This behavior is obviously important in certain applications where the plastic material must give structural support. Hence, crystalline materials are more likely to be used in structural applications than are amorphous polymers. The melting point for a crystalline thermoplastic is typically somewhat higher than for an amorphous thermoplastic (all other things being equal), because of the higher energy required to break the crystalline bonds.

Most real polymers are mixtures of crystalline and amorphous regions. The two types of regions act independently in thermal transitions. Therefore, the crystalline regions have sharp melting points and the amorphous regions have glass transitions and broad melting temperatures.

For **thermoset materials,** the presence of crosslinks restricts the movements of the atoms in the polymers, thus increasing T_g (and HDT) over the values that would occur in otherwise equivalent amorphous or even crystalline thermoplastics. The restrictions from the crosslinks stiffen the thermoset material above T_g so that it is less leathery and more rigid than thermoplastics. This pattern is depicted in Figure 3.5, where this region is labeled *semirigid*. The behavior of Formica™ countertop material (a thermoset) is a good example of how the properties of a thermoset will change in the region of T_g. It is hard in normal use but can be softened somewhat by heating so that it can be bent and shaped to fit the contour of the counter. As with all thermosets, further heating of the countertop material will not cause it to melt, but rather to degrade or, at extremely high temperatures, to burn.

Furthermore, crosslinks largely prevent the formation of crystal regions. Most thermosets will, however, exhibit a glass transition temperature. The T_g is related to the number of crosslinks formed, because higher crosslinking will give further restrictions to the molecules and require higher temperatures to effect the long-range movements that are characteristic of the glass transition. This direct relationship between T_g and crosslink density has led to the use of T_g as an indicator of the extent of cure (crosslinking). Because T_g is one measure of the maximum use temperature of the plastic for structural applications, the use range can be increased by increasing the number of crosslinks, which in turn raises T_g.

In crosslinked thermosets the melting point is dramatically increased relative to thermoplastics, as shown in Figure 3.4 with the dotted line. As discussed later in this chapter when molecular size effects are examined, the larger the molecule, the greater the energy that needs to be input to melt it, thus raising the melting point. For thermosets, the increase in molecular weight is so great that the melting point is raised above the decomposition temperature, thus creating a situation where there is no real melting point because the thermoset material will start to decompose before it will melt.

Characteristics of the polymer that raise T_g or T_m have a tendency to raise its maximum structural use temperature as well. Therefore, in general, thermoplastics have lower thermal stability temperatures than thermosets because the crosslinks in thermosets raise T_g and T_m. Other polymer characteristics that raise the amount of energy required to impart internal movements and, therefore, raise the maximum use temperature are higher degrees of aromatic character, hydrogen bonding, and the stiffness of the polymer backbone.

Decomposition

In its melted state, the polymer contains a high amount of energy; this is manifested in translation, vibration, rotation, and twisting motions. If additional energy is input, the amplitude and speed of the motions increase. This increase in amplitude is especially important in the case of vibrations, because eventually the vibration amplitude will become large enough that the bonds will break. At this point, a sufficient amount of the input energy, which randomly moves through the molecule, has localized in the bond and equals the bond energy. This breaking of covalent bonds causes a loss of properties and a change in the basic nature of the polymer. This is called *decomposition* or *degradation*, and the temperature at which it occurs is the *decomposition temperature (T_d)*. For thermoplastics, decomposition would generally occur in the liquid state; for thermosets, it generally occurs in the solid state, as shown in Figure 3.3 and Figure 3.5.

When thermoplastics degrade, they often release a gas and may form crosslinks, thus becoming thermoset materials at high temperatures. When thermosets are heated, either those formed from overheated thermoplastics or from intentional crosslinks, a *char* is formed. These chars are similar to charcoal, which is, in fact, the char of wood. When a char is formed, by-product gases are often released and the polymer may begin to change color, often yellowing or blackening.

At very high temperatures when oxygen is present, the polymer or the gases that may be given off by decomposition may ignite. This is called *combustion,* a rapid decomposition.

The decomposition temperature can be determined using DSC or TMA but is probably easier to obtain using a test called thermogravimetric analysis (TGA). TGA is a procedure in which the sample is progressively heated and changes in the weight of the material are recorded. The weight changes are usually associated with the volatilization or decomposition of components of the sample. Often some portions of the sample, such as mineral fillers, are not volatilized or decomposed and the concentrations of these materials can be determined. (To get the weight of these materials without a char being present, the char is usually subjected to heat in an oxygen atmosphere, thus causing the char to burn away.) The results of the TGA test are a plot of weight changes versus temperature.

In small molecules (smaller than polymers), additional heating of a melted material will cause the material to gain enough translation energy that all the attractions present in the liquid will be overcome and the material will evaporate into a gas, as was discussed earlier in this chapter and illustrated by the example of water turning to steam. But, because polymer chains are so large, the temperature at which the liquid attractions will break is higher than the decomposition temperature. Hence, polymers degrade before they evaporate.

Processing Temperature

One other important temperature for plastics is the processing temperature. This is the temperature at which a plastic material can conveniently be molded. The processing temperature is determined experimentally and is somewhat different for different processing equipment and conditions. Nevertheless, an approximate value can be useful as a starting place for new processing conditions. Some examples of the important temperatures associated with common plastics are given in Table 3.1.

Non-Thermal Energy Inputs

In addition to thermal sources of energy, energy input could come from a mechanical source (such as impact) or from any other energy source (sound, light, X rays, etc.). With all these energy inputs, the polymer moves internally to absorb the energy. If the internal

Table 3.1 Thermal Properties of Selected Plastics

Polymer	T_g	T_m	Processing Temp
Polyethylene—low density (LDPE)	−130 to −13°F (−90 to −25°C)	208 to 240°F (98 to 115°C)	300 to 450°F (149 to 232°C)
Polyethylene—high density (HDPE)	−160°F (−110°C)	266 to 280°F (130 to 137°C)	350 to 500°F (177 to 260°C)
Polypropylene (PP)	−103 to −94°F (−25 to −20°C)	320 to 356°F (160 to 180°C)	374 to 550°F (190 to 288°C)
Acrylonitrile butadiene styrene (ABS)	212°F (100°C)	230 to 257°F (110 to 125°C)	350 to 500°F (177 to 260°C)
Nylon (6,6)	120°F (49°C)	470 to 500°F (243 to 260°C)	500 to 620°F (260 to 327°C)
Polyethylene terephthalate (PET)	150 to 175°F (66 to 80°C)	413 to 509°F (212 to 265°C)	440 to 660°F (227 to 349°C)
Polycarbonate (PC)	300°F (149°C)	284 to 300°F (140 to 149°C)	520 to 572°F (271 to 300°C)
Polyphenylene oxide (PPO)	375 to 428°F (190 to 220°C)	500 to 900°F (260 to 482°C)	400 to 670°F (204 to 354°C)

motions can dissipate the energy sufficiently, then no breakage of bonds will occur. However, if the energy input is large or very rapid, the polymer may not be able to dissipate the energy sufficiently and in some area the localized energy can exceed the bond strength. When this occurs, the polymer breaks. This localization of energy can occur with any type of energy input, although mechanical impacts are a very common source of such energy concentrations because these impacts give very large, rapid energy inputs at a single location.

The energy inputs can add together. Sometimes these additions will create localized energy concentrations that are sufficient to exceed the bond energies, even when the individual inputs would not be sufficient to cause breakage. This may happen, for instance, when internal stresses are introduced into a plastic material as part of the molding operation, and then later thermal stresses, such as rapid cooling, may cause the material to crack.

The internal motions that accompany all energy inputs can become quite extensive and may result in changes in mechanical or other properties because they may facilitate movement of one polymer chain past another. For instance, the origin of tensile strength is the resistance to motion of one polymer chain past another. If the internal motions are high, less resistance of movement between polymers would be experienced and the tensile strength would decrease. On the other hand, some mechanical properties may be improved by some energy input. For instance, elongation would be greater when the molecules have more internal freedom of motion. Toughness, which is a combination of both strength and elongation, could be either raised or lowered, depending on the specific conditions.

EFFECTS OF THERMAL CHANGES ON POLYMERS

Flexibility

The properties of plastics are quite different below and above these thermal transitions. An understanding of these changes is important in predicting how and under which conditions the polymers can be used. Generally a highly crystalline material will soften slightly but retain its shape and generally stiff or brittle characteristics up to temperatures near the melting point. This behavior indicates that most of the energy is in vibrations rather than translations, as would be expected from a compact, crystalline structure. Therefore, the maximum structural use temperature (to maintain stiffness and strength) for crystalline materials is reasonably close to T_m.

Amorphous materials are relatively rigid, stiff and brittle at temperatures below T_g, the *glassy region,* with mechanical properties somewhat like crystalline polymers. Above T_g the material becomes significantly softer and takes on many physical characteristics that are much like leather (tough, pliable, flexible) and therefore it is called the *leathery region.* These relationships are illustrated in Figure 3.6.

The glass transition temperature is a convenient tool for predicting the useful temperature range for using a particular polymer. In some cases a polymer must be flexible when in use. (Some examples would be rubber, a plastic strap, and a plastic flexible hinge.) This flexible behavior is characteristic of the leathery region. Therefore, materials of this type will be largely amorphous and will have a T_g at a very low temperature so that the temperature of use will be above T_g. When the temperature of use is above T_g, the material is pliable. However, as the temperature is lowered, the temperature may drop below T_g and the material will change to its glassy state and become brittle. Examples of plastic materials becoming brittle at low temperatures are common and include such items as embrittled cold rubber balls, toys which break when used outside in the winter, and indoor

Figure 3.6 Glass transition and melting point relationships for thermoplastics.

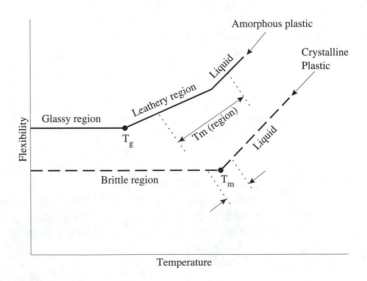

trash carts which break from being dropped when used outside in the winter. If you lived in a cold location that had an extremely low temperature, you might need to worry about the temperature dropping below the T_g of the tires on your car. If you tried to drive on the tires when they were below their T_g, you might break the rubber and ruin the tires.

In other uses, a polymer must be rigid and stiff, such as for an electrical wall outlet or a compact disk box. In these cases the polymer must be used below the glass transition temperature. Because most amorphous polymers have a gradual onset of pliability near T_g, a convenient rule of thumb is that the maximum use temperature for a polymer to stay rigid (that is, safely in the glassy region) is approximately 75% of the glass transition temperature. For instance, if the T_g for a resin is 300°F (149°C), the safe-use temperature would be 222°F (107°C).

In summary, if the desired properties of the plastic material are **pliability** and **resiliency** rather than structural support, then the plastic should be used **above** the T_g. For instance, rubber materials would almost always be used above their T_g. If the desired properties are **rigidity** and **strength,** then the material should be used **below** the T_g.

Thermal Degradation

When degradation occurs, most mechanical and physical properties of the plastic material are seriously altered. Mechanical strength, stiffness, and elongation often drop precipitously.

The accumulation of thermal energy, that is, the exposure to high heat, can occur in many ways. For instance, the polymer could be a component of an oven, or a frying pan handle, or could be near an exhaust duct for high-temperature gases. But the most common situation in which a plastic material will be exposed to high temperatures is during processing. Thermosets are heated to cure them so that crosslinks will be developed. Thermoplastics are heated to melt them so that they can be molded. Neither type of plastic material should be subjected to excessive heat during these processing steps or thermal degradation can occur. For thermosets this usually means not extending the thermal cure cycle for longer times or at higher temperatures than are necessary. For thermoplastics this usually means not heating to a higher temperature than is required for melting and proper viscosity control for molding.

With thermoplastics, however, another problem arises because thermoplastics can be processed several times. The most common instance occurs when scrap material is reprocessed, usually after chopping or grinding this material into small pieces so that it can be more easily and uniformly fed into the processing equipment. The scrap material being reprocessed is called *regrind*. Experience has shown that when regrind material is processed, the evidence of thermal degradation is much more apparent than is seen with nonregrind *(virgin)* material.

A method for limiting the effect of regrind problems is to mix the regrind with a large amount (usually over 60%) of virgin material. In this way the amount of regrind in any scrap will be continuously diluted and the effect of thermal degradation on part properties will be minimal.

Each thermal process causes some thermal degradation to occur, even at temperatures well below the decomposition temperature. This degradation is the result of random local buildups of energy, which can be sufficient to break one or more of the covalent bonds in the energy concentration area. In most cases this degradation is limited to the breaking of a few bonds and the effects are not detected in the overall behavior of the polymer.

If the plastic material is only processed (melted) once, this degradation is usually minimal, but with each subsequent reprocessing the effects become more evident. Therefore, plastics have an accumulated thermal degradation that is a function of the number of times the material has been heated or melted or, in other words, a function of the accumulated time at elevated temperature. This phenomenon is called a *thermal history* of the plastic material. These effects are most evident when high temperatures are used for processing. Some plastics, such as polyvinyl chloride (PVC), are especially sensitive to thermal degradation and minimization of their thermal history is important to insure that good performance is maintained.

Another problem with thermoplastic melts is their tendency to form small solid masses that cannot be melted. These masses, which are called *gels,* result from crosslinks formed when side-chain bonds break (allowing the by-product gas to form) and then recombine to form a bond with an adjacent polymer chain. (This is another form of thermal degradation because the desired properties are altered by heat.) These crosslinked masses form most often in continuous processes such as extrusion where some thermoplastic material may get caught in a crevice or on a nonsmooth part of the system and remain there for a long period of time at a high temperature. These gels may eventually break free, causing a defect in the part or may remain in place and, when cooled, cause great difficulty in disassembling the various parts of the system.

In some thermoplastics, the degradation and melting temperatures are close to each other and substantial degradation could occur before melting. Thermoplastics of this type are, generally, not processable by traditional thermoplastic processing methods like extrusion or injection molding. An example of this type of material is polytetrafluoroethylene (PTFE), which is processed by methods more akin to metal processing, such as powder fusion. (Recent modifications in PTFE have changed the relative positions of the melting and degrading temperatures so that when modified, usually by copolymerization, this polymer can be normally processed.)

In thermosets the degradation product is a char. This char material usually resists further heating and is, therefore, an excellent thermal insulator. Some plastics (such as phenolic) take advantage of this property in applications requiring a high thermal insulation after exposure to high heat, such as rocket exhaust nozzles and high-temperature insulation.

Aging

When a relatively low amount of heat is applied to the polymer over a long period of time, the cumulative effects can be similar to high-temperature degradation. That is, the polymer can reach its thermal stability limit and could eventually begin to degrade. This effect is called *thermal aging* or simply *aging.*

Because of the long-term nature of aging, the effects of this process are much more difficult to detect than the more rapid types of thermal-induced change. The consequences are, however, potentially catastrophic. With aging, the part will gradually lose mechanical strength and elongation resulting in an embrittlement that can easily lead to failure.

Aging cannot really be prevented, but the useful life of plastic parts which may be especially subject to long-term, slightly elevated temperatures can be extended by the use of thermal stabilizers.

Oxidation is a similar long-term process in which oxygen reacts with the polymer. This reaction causes changes in the polymer's properties similar to aging. Oxidation can also be slowed through the use of additives.

Thermal-protective Additives and Processing Methods

Several materials have been developed to assist in the reduction of thermal degradation in thermoplastics. These materials, which preferentially absorb heat, are added to the plastic material and therefore reduce the amount of thermal energy absorbed by the polymer chains. These materials are commonly called *thermal stabilizers*. Some of the most effective are powdered inorganic minerals such as limestone, talc, and alumina. These materials absorb heat because of their large heat capacities and therefore "protect" the plastic material. In some cases, however, the presence of these inorganic materials is detrimental to other desired properties of the plastic material. For instance, they add weight and decrease tensile strength. Organic heat stabilizers are also available, and although more expensive, are preferred where strength, weight, or other problems with inorganic materials may exist. Materials such as PVC that have relatively low bond energies are especially sensitive to thermal degradation. The most commonly used heat stabilizers in PVC are mixed metal salt blends (such as Ba/Cd/Zn stearates), organotin compounds (such as mercaptides), and lead compounds (such as stearates).

Another technique that is used to protect heat-sensitive plastics is the use of *processing aids*. These are materials that lubricate the process, often by melting at low temperatures to facilitate the melting of the remainder of the polymer. Waxes are some of the most common processing aids. When they are added to a polymer mix, the processing temperature can be reduced 60 to 90°F (20 to 30°C). Some materials, such as the stearate salts, can serve as both thermal stabilizers and processing aids.

Thermal degradation can also be reduced if energy is input to the polymer by methods other than heat. Therefore most plastics processing equipment use both mechanical and thermal energy to melt the plastic material. The mechanical energy is often put into the system by screw mixing, which has the added benefit of conveying the polymer through the heating zone. (These processing machines will be described in detail later in this text in chapters which discuss the various processing methods.) Some equipment, such as twin screw extruders, have a very high mechanical input and are, therefore, especially useful in processing heat-sensitive materials such as PVC.

Thermal Conductivity

Thermal conductivity is a measure of how quickly or easily heat moves through or along a material. As already discussed, when plastics are heated, they have a tendency to move internally and therefore absorb input energy. If the energy is absorbed, it is not transferred along. Hence, plastics have low thermal conductivities in comparison to metals. Typical thermal conductivities for plastics and metals are given in Table 3.2. These values are determined using the thermal conductivity test (ASTM C 177). Thermal conductivity is defined as the rate at which heat is transferred by conduction through a unit cross-sectional area of a material when a temperature gradient exists perpendicular to the area. The coefficient of thermal conductivity is sometimes called the *K factor*. The primary technique for measuring thermal conductivity of plastic materials is the guarded hot-plate method. This method is complex and the equipment is expensive. Great care must be taken to ensure that operator technique does not enter into the test results.

Table 3.2 Typical Thermal Conductivities, Heat Capacities, and Coefficients of Thermal Expansion for Plastics and Metals

Material	Thermal Conductivity	Heat Capacity	Thermal Expansion (CTE)
Plastic	0.03 to 0.06 Btu/h-ft °F (0.05 to 1.0 W/m·K)	0.4 to 0.9 Btu/lb °F (0.4 to 0.9 cal/g °C)	9 to 12 in./in. °F $\times$ 10^{-5} (5 to 7 cm/cm °C $\times$ 10^{-5})
Metal	30 to 60 Btu/h-ft °F (50 to 500 W/m·K)	0.1 to 0.3 Btu/lb °F (0.1 to 0.3 cal/g °C)	2 to 8 in./in. °F $\times$ 10^{-5} (1 to 4 cm/cm °C $\times$ 10^{-5})

Plastics are, therefore, ideal materials for insulating applications, provided the temperature does not get higher than the thermal stability temperature. Some plastics, especially thermosets, have a combination of high thermal stability temperatures, high degradation temperatures, and low thermal conductivities. Such materials are excellent for thermal insulators at moderately elevated temperatures (up to about 600°F or 300°C) and can be found in products such as handles for frying pans and toasters.

A property related to thermal conductivity is *heat capacity*. This is the measure of the temperature rise in a given weight of material for a given amount of heat input. As might be expected, the tendency of polymers to absorb energy internally results in rather high heat capacities, as can be seen in Table 3.2.

Thermal Expansion

Most materials expand when heated and contract when cooled. Because plastics absorb input energy through internal motions, the polymer chains usually move apart to allow for this motion. Hence, plastics are likely to expand more than metals or ceramics when heated. This change in dimension of a material with input heat is called *thermal expansion* and is usually measured as the ratio of the change in a linear dimension (such as overall length) to the original dimension per unit change in temperature. The resulting quantity is called the *coefficient of thermal expansion (CTE)*. The tests for coefficient of thermal expansion (ASTM D 696 for linear and D 864 for volume) are conducted by placing a sample inside a tube that is submersed in a temperature-controlled bath. A quartz rod is placed against the sample and held in place by a low-force spring. The movement of the quartz rod is sensed by a dial gauge or by an electronic measuring device. The sample

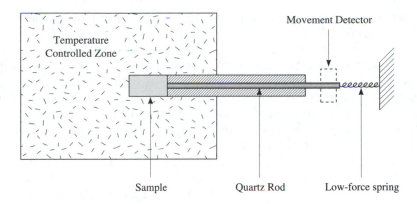

Figure 3.7 Test apparatus (dilatometer) used for measuring coefficient of thermal expansion (CTE).

is heated and the movement of the quartz rod against the sample is noted over the temperature range of interest. The apparatus used is called a *dilatometer*. A diagram of a test apparatus for making CTE measurements is given in Figure 3.7. Test specimens can be from 2 to 5 inches long and round so as to fit easily within the testing device. Results are given as length/length °F.

Typical CTE values for plastics and metals are given in Table 3.2. Ceramics have CTEs in the range 0.06 to 2 in./in. °F, which are much lower than for either plastics or metals. Therefore, the addition of ceramics (such as fillers) into a plastic resin can lower the overall CTE of the mixture. The addition of fiber reinforcements, especially ceramic fibers, will generally decrease the CTE for plastics because the polymer chains are not allowed to move freely, being restricted in their movement by the fibers.

This expansion of plastics when hot must be taken into consideration when designing a part. For instance, if a tight fit is desired between a metal screw in a plastic part, the use temperatures should be kept in a narrow range, otherwise at high temperatures the plastic material will expand much more than the metal and the screw will become loose.

The CTE for plastics must also be considered when designing a mold for plastics. When a plastic part is formed, it is usually at an elevated temperature and will shrink considerably when cooled. Consequently, the mold cavity should be precisely oversized so that the finished part can shrink to the correct dimensions.

Thermal Stresses

If a plastic part is constrained so that it cannot expand or contract with changes in temperature, internal energy called *thermal stresses* develop. The magnitude of these stresses will depend upon the temperature change, the method of shape confinement, and the CTE of the plastic material. These stresses (which represent retained energy) can result in lower overall mechanical and thermal properties because the total energy threshold for a particular property is reached at a lower applied energy level. If other stresses are introduced, such as from drilling holes, internal imperfections, or mechanical stresses, the part may fail prematurely.

In some cases, these thermal stresses can be relieved by heating the material in a manner that allows free chain movement. This stress-relieving process is called *annealing*. Generally, annealing is done at a moderately elevated temperature for a long time rather than at a high temperature for a short time. When annealed, some materials may actually contract because the molecules are allowed to move to a more energy-favorable configuration which may be closer together. Metals are annealed in much the same way as described here for plastic materials.

Plastic materials can build up stresses from energy inputs other than thermal. For instance, a part could be stretched and then quickly cooled so that the molecules cannot move to return to their original positions. Alternately, the molecules may be forced into a highly oriented state, as is often done in the extrusion process, and be cooled before the natural randomization of the molecules can occur. Both of these instances result in internal stresses which can generally be relieved by annealing.

Thermal Effects on the Rate of Chemical Reactions

Some processes that have been discussed involve chemical reactions (such as thermal degradation and aging) whereas others are changes in physical state (such as thermal transitions, thermal expansion, and thermal stresses). The difference between the chemical reactions and the physical changes is that in chemical reactions the bonds between the atoms in a polymer are broken and reformed in new configurations, whereas in physical changes the molecules are merely separated from neighboring molecules.

The dependence of physical changes on thermal input is additive and linear. That is, the transition is dependent upon the amount of heat input directly. This can be described mathematically as a linear function between heat input and temperature and the transitions are directly related to the temperature.

With chemical reactivity, the relationship between temperature and the rate of the chemical reaction is more complicated. The rate was shown to be an exponential function of temperature. This relationship was formalized in 1886 by Svante Arrhenius and is called the *Arrhenius equation,* shown as Equation (3.1):

$$Rate = Ae^{-(E/RT)} \tag{3.1}$$

where A is called the collision factor and is a measure of the effectiveness of collisions between reacting species, e is the natural logarithm base, E is an activation factor which indicates the amount of energy required to make the reaction occur, R is the gas constant, and T is the temperature in absolute units (Kelvin). The rate of the reaction increases as A increases, decreases as E increases, and increases as T increases. The rate of the reaction approximately doubles with every 10 K (or 10°C) rise in temperature.

The Arrhenius equation is very important for predicting the rate of any chemical reaction. For instance, oxidation and degradation by ultraviolet light can be described by the Arrhenius equation. Several other phenomena which do not involve chemical reactions, such as diffusion, viscous flow, and electrolytic conduction, can also be described by the Arrhenius equation.

POLYMER LENGTH

The length of the polymer chain is a key factor in determining the nature of the interactions between polymers. To understand this, examine the general formula of a polymer which can be represented by the following expression:

$$+ A +_n$$

where A is the polymer repeat unit and n is the number of repeat units in the chain. This is the same representation, in a general form, that was introduced in the chapter on molecular structures for various specific thermoplastic and thermoset polymers. The repeat unit (A) was previously defined as the basic representation of the polymer unit. The repeat unit is usually just the molecular formula of the monomer or monomers which make up the polymer, with minor adjustments in the bonding to reflect the new bonds which were created to form the polymer and, perhaps, the separation of a condensate. The value of n can be very large, often in the hundreds or thousands of units for some polymer types. The length of the chain can, therefore, be found if the parameter n is known. Rather than attempting to measure a value for n directly, a much simpler experimental measurement of the size of the polymer is to determine the *molecular weight,* which can be directly related to the number of repeat units in the chain (n) and therefore to the chain length. Therefore, the molecular weight and the polymer chain length are closely related and are often discussed interchangeably.

MOLECULAR WEIGHT

The molecular weight of a polymer is defined as the sum of the atomic weights of each of the atoms in the molecule. The atomic weights can be found by referring to the periodic table. For instance, the atomic weight of carbon is 12 g/mole, so each carbon in the molecule would be counted as 12. The atomic weight of hydrogen is 1 g/mole, of oxygen is 16 g/mole, and so on. When the exact molecular formula is known, the molecular weight is easy to calculate. For example, water (H_2O) would have a molecular weight of 18 g/mole $(16 + 1 + 1)$, methane (CH_4) would be 16 g/mole $(12 + 1 + 1 + 1 + 1)$, and benzene (C_6H_6) would have a molecular weight of 78 g/mole $[(6 \times 12) + (6 \times 1)]$. The molecular weight of a polymer could be likewise calculated if the exact formula is known. For instance, if the polymer is polyethylene with n equal to 1000, the molecular (repeat unit) formula would be as follows:

$$+ C_2H_4 +_{1000}$$

The molecular weight would be 28,000 g/mole $\{[(2 \times 12) + 4] \times 1000\}$ since each C_2H_4 would have a weight of 28 g/mole and there would be 1000 of these units.

Average Molecular Weight

In all real polymer systems the nature of the polymerization process results in chains with many different lengths. In other words, the polymer molecules (chains) are usually different molecular weights. This variation in the molecular weight of molecules is a characteristic of polymers that is not found in small molecules. For instance, all methane molecules (CH_4) have the same molecular weight—16. But all polyethylene molecules do not have the same molecular weight. In one batch of polyethylene polymers the molecular weights could vary over several thousands. It is not possible to examine each of the polymer chains to determine its precise weight. Therefore, an exact molecular weight for each cannot be determined and the value of n is not precisely known. The polymer chains must be treated as a group and a reasonable representation of the polymer must be some average group representation. Statistics are a convenient tool to allow determination of representative group values of the polymer chains.

A useful statistical concept that can be used to characterize polymers is called the *distribution of values*. A value can be any quantity that is to be examined. For instance, a value could be the height of students in a class. The distribution would be determined by counting the number of students in the class of each height. The distribution can be visualized by plotting the number of students in each group on one axis and the various heights on the other axis, resulting in what is commonly referred to as a *histogram*.

For polymers, an important value to be examined is the molecular weight. Therefore, the *molecular weight distribution* is simply a count of the number of molecules of each molecular weight. The molecular weight distribution can be envisioned by plotting the number of molecules of each molecular weight against the molecular weights as shown in Figure 3.8.

Rather than plot every single molecular weight along the x-axis, the molecular weights are often grouped so that similar molecular weights can be counted together.

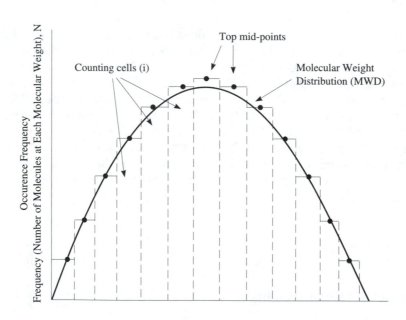

Figure 3.8 Plot of number of molecules at each molecular weight versus molecular weight thereby defining a molecular weight distribution.

(This could be like grouping all the students having heights between 65 and 70 inches into one group and all those with heights between 70 and 75 inches into another group, and so on). These groups are called *counting cells*. The counting cells would, therefore, be plotted along the *x*-axis with the width of the cell corresponding to the range of values to be put into that cell. (In the student height example, one cell would have a width from 65 to 70 and the next cell would be from 70 to 75.) The number of molecules in each counting cell would be plotted on the *y*-axis and would determine the height of the cell. The counting cells are, therefore, rectangles with a width the same as the spread in values and the height the number of values within that spread. The counting cells are given a subscript (*i*) so that each can be identified in a general representation, as shown in Figure 3.8. When this is done, the histogram has been completed.

A curve can be drawn representing the overall shape of the plot by connecting the tops of each of the cells at their midpoints. This curve between top midpoints can then be modified slightly to give a continuous and smooth curve, as is represented by the line labeled Molecular Weight Distribution (MWD) in Figure 3.8. (The implications of the shape and other variations in the molecular weight distribution curve are discussed in detail later in this chapter.)

An *average molecular weight* can then be determined by summing the weights of all the chains and then dividing by the total number of chains. The average molecular weight is an important method of characterizing the polymer.

The actual calculation of the average molecular weight can be done in three slightly different ways. One way calculates a quantity known as the *number average molecular weight* (M_n) and can be calculated by Equation (3.2):

$$M_n = \frac{\sum N_i M_i}{\sum N_i} \tag{3.2}$$

where M_i is the molecular weight of species *i*, that is of the weight of molecules in counting cell *i*, and N_i is the number of molecules of a particular molecular species *i* (comprising molecules in the same counting cell). This method of calculating an average molecular weight gives equal value to all chains, large or small. A sample problem illustrates the value of a number average molecular weight distribution.

Calculation of Number Average Molecular Weight

What is the number average molecular weight of a polymer sample in which the polymer molecules can be divided into 5 groups with the following molecular weights (in g/mole):

Group 1	50,000
Group 2	100,000
Group 3	200,000
Group 4	500,000
Group 5	700,000

The ratios of the number of molecules in each group are 1:4:5:3:1.

Solution:

$$MW_n = [1(.5 \times 10^5) + 4(1 \times 10^5) + 5(2 \times 10^5) + 3(5 \times 10^5)$$
$$+ 1(7 \times 10^5)]/(1 + 4 + 5 + 3 + 1)$$
$$= 2.6 \times 10^5 \text{ g/mole} = 260{,}000 \text{ g/mole}$$

The distribution of samples given in the sample problem is plotted in Figure 3.9. This distribution is obviously nonsymmetrical. Such nonsymmetrical distributions are called *skewed distributions*. In this case, the *tail* of the skew, that is, the small region at the end of the curve, is toward higher molecular weights.

The number average molecular weight (M_n) can be determined experimentally by analyzing the number of end groups (which permits the determination of the number of chains) or through some other property of the polymer that depends on the number of molecules (called *colligative properties*) such as boiling point elevation, freezing point depression or osmotic pressure (pressure through a membrane). Therefore, if the number average molecular weight (M_n) of the polymer can be determined experimentally, the average number of repeating units (n) can be calculated by dividing by the number average molecular weight by the molecular weight of the repeating unit (M_o), as shown in Equation (3.3).

$$n = \frac{M_n}{M_o} \tag{3.3}$$

The molecular weight of the repeating unit is simply the weight of the unit shown in the repeating unit representation of the polymer. For instance, M_o for polyethylene would be 28 g/mole.

The number of repeating units (n) is sometimes called the *degree of polymerization* (*DP*) and relates to the amount of monomer that has been converted into polymer. If the average molecular weight (M_n) is large, n will also be large, since M_o is constant for each type of polymer. Hence, in this case, many groups have joined together and the polymerization has proceeded to a high degree. Due to the length of most polymer chains, polymer molecular weights of several hundreds or even thousands are not unusual.

Another way of determining average molecular weight favors large molecules more than small. This average is particularly useful in understanding polymer properties that relate to the weight of the polymer, such as permeation through a membrane or light scattering. In this method the molecular weight can be represented by the *weight average molecular weight* (M_w) which is given by Equation (3.4).

$$M_w = \frac{\sum_i N_i M_i^2}{\sum_i N_i} \tag{3.4}$$

The parameters are the same as those used in Equation (3.2). A calculation of the weight average molecular weight using the same data as used in the calculation for number average molecular weight would result in a weight average molecular weight of 4.2×10^5 g/mole. The higher value for the weight average molecular weight, when compared with the number average, is a reflection of the fact that the distribution used in the sam-

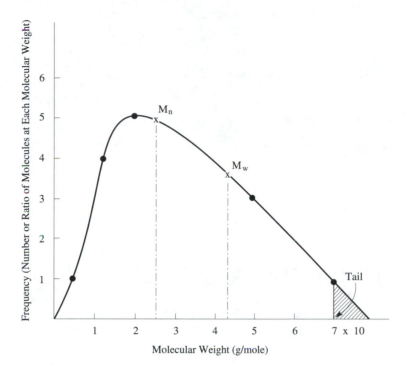

Figure 3.9   Molecular weight distribution illustrating the sample problem.

ple calculation is skewed, having a long tail toward the higher molecular weights. The weight average molecular weight is also shown in Figure 3.9.

A third way to calculate the average molecular weight emphasizes large molecules even more than the weight average calculation. This third method is called the *z-average molecular weight* and is useful for some calculations involving mechanical properties. The experimental method relating to this method would be ultracentrifuge, a separation method that uses the weight differences of the polymers to separate them when they are spun in a centrifuge. This average would be even more toward the high molecular weight tail than the other averages.

Molecular Weight Distribution

As already indicated, the nature of the polymerization process results in chains of varying size (molecular weight) and a distribution of the size and weight of the polymers. In some polymerization processes growing chains would successfully react with many monomers to form very long chains, and in other processes the chains would be considerably shorter. This variance in chain size or molecular weight is called the *molecular weight distribution (MWD)* which is another important method of characterizing polymers.

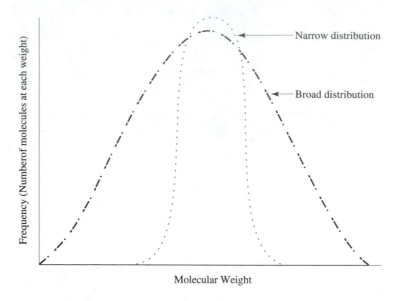

Figure 3.10   Broad and narrow molecular weight distributions.

In some cases the molecular weight distribution would be very narrow indicating that most of the polymers were nearly the same length. Other molecular weight distributions would be quite broad indicating that the polymer batch has some short and some long polymer chains. Examples of narrow and broad molecular weight distributions are given in Figure 3.10.

The molecular weight distribution as well as the average molecular weight are controlled by the conditions of the polymerization. A detailed discussion of these conditions is beyond the scope of this text, but several of the books in the bibliography treat this subject in detail.

Molecular weight distributions can be symmetrical, as in Figure 3.8, or skewed, as in Figure 3.9. When the distribution is symmetrical, the average molecular weight can be easily approximated graphically. For skewed distributions this approximation is much more difficult and the formulas or experimental methods described earlier in the text must be used.

The molecular weight and molecular weight distribution of a plastic material can be determined by the difference in absorption of small and large molecules on special particles that line a tube through which the plastic is caused to flow on a substrate (such as paper) on which the samples are placed. This test is called gel permeation chromatography (GPC) and follows procedures given in ASTM D 3593. Materials are separated by the differences in the absorptivity that are associated with molecular weight. After time, these differences can be seen and quantitatively measured.

Physical and Mechanical Property Implications of Molecular Weight and MWD

Both the average molecular weight and the shape of the molecular weight distribution can significantly affect some key physical and mechanical properties of the polymer such as tensile strength, impact toughness, creep resistance, and melting temperature. In general, a higher average molecular weight increases all of these properties. The reason is primarily explained by *entanglement*. Entanglement is simply the mutual wrapping of polymer chains around each other. Higher molecular weights imply longer polymer chains and longer polymer chains imply more entanglement. When the polymers are highly entangled, they resist sliding over each other. Entanglement, therefore, affects the plastic much like the secondary or intermolecular bonding (hydrogen bonding) discussed in the chapter on polymer structure—molecular viewpoint. Both entanglement and secondary bonding create forces (*intermolecular attractions*) that tend to keep the molecular chains associated together. These forces are less strong than primary bonds (crosslinks) between the molecules. The forces from entanglement and secondary bonding can, therefore, be broken with much lower energy input than would be required to break crosslinks. All of the mechanical properties will be discussed in greater detail in the chapter on mechanical properties, where the effects of both microstructure and macrostructure on these properties can be seen. However, some general relationships between molecular weight, molecular weight distribution, and various mechanical properties are discussed here as a background for other discussions.

One of the most basic mechanical properties is *tensile strength,* which is the resistance to two forces pulling on a sample of the polymer in opposite directions. Entanglement and other intermolecular attractions have a strong effect on tensile strength. As the molecules are pulled in opposite directions (tension), the energy levels of the molecules increase. This increase in energy level creates more molecular movement (vibration, rotation and translation), which has the effect of gradually disentangling the molecules. Ever more tensile force will create more disentanglement until the molecules are free to slide relative to each other. The force required to cause this sliding of molecules is called the tensile force. Hence, the more entangled the molecules, the more tensile force that is required to cause them to slide.

The effect of molecular weight distribution (MWD) on tensile strength is more subtle than the effect of molecular weight. A broad MWD, especially when skewed to the small end, implies that some of the molecular chains are much shorter than the average. These shorter chains become disentangled much more readily than would a chain of average length. The net effect of the broad MWD is, therefore, a lowering of the resultant tensile strength.

Impact toughness (or impact strength) is another important mechanical property of polymers that is significantly affected by molecular weight. *Impact toughness* measures the ability of a material to withstand an impact blow, that is, to absorb energy. The case of impact toughness is, however, somewhat different from tensile strength because impact toughness depends on the ability of the material to absorb energy through vibrating and minor movements rather than just resist input forces. If the material cannot move, the impact energy will be concentrated in one area and rupture will occur. Materials that do not allow this movement or vibration are said to be brittle. (Crosslinking and, to a lesser extent, secondary bonding, restrict movement and therefore embrittle the plastic.) Long chains, however, mean that the energy can be transmitted along the chain and shared over more atoms, eventually being dissipated through vibrations, minor translations, and heat. With entanglement, some of the energy can also be transferred to other molecules in the mass, thus further reducing the concentration of the energy. Therefore, impact toughness is generally increased by increasing molecular weight up to the point where embrittlement becomes important.

The effect of MWD on impact toughness is much the same as the effect on tensile strength. A broad MWD results in more short molecular chains that have a diminished ability to entangle and, therefore, a diminished ability to transmit the energy between chains. This means that the impact toughness is reduced by a broad MWD.

Other mechanical properties, such as elongation, creep, and stiffness, show the same trends as tensile strength and impact toughness. In general, higher molecular weights will increase these mechanical properties and broader MWD will decrease the properties.

The most important thermal property for polymers (the melting point) has already been introduced. The melting temperature (or melting point) is a measure of the amount of thermal energy that has to be put into the polymer mass to induce the molecules to slide freely relative to each other. When the entanglement is great, the amount of input energy required to get free movement is high and the melting temperature goes up. A high melting temperature can be useful in allowing a plastic to be used for high-temperature applications, but can also be a detriment in processing because the processing temperature will be higher. Hence, low molecular weights generally reduce melting points and improve ease of processing.

The net effect of increasing molecular weight is, therefore, an increase in many mechanical properties and an increase in melting point. In most cases the increase in mechanical properties is desirable and the increase in melting point is acceptable, perhaps even desirable.

It would be possible, at least in theory, to continue to increase mechanical properties and melting point by increasing molecular weight until the chains were so long that the energy required for melting would exceed the energy of the bonds. At this point decomposition would occur. This situation is, in fact, the case for thermoset polymers. The crosslinks increase molecular weight by linking the molecules together. This increases mechanical properties but also increases the melting point. Generally, there are enough crosslinks formed in these thermoset materials that the materials cannot be realistically melted.

In actual practice, increases in mechanical properties are *not linearly* related to increases in molecular weight, as can be seen in Figure 3.11.

Initially the mechanical properties increase rapidly with increases in molecular weight. Eventually, however, the rate of increase in mechanical properties slows with increases in molecular weight and a practical maximum value for mechanical properties appears to be reached. Therefore, increases in molecular weight beyond a certain limit will result in only small increases in mechanical properties. Increases in molecular weight have a much stronger effect on melting point. Therefore, for thermoplastic polymers that

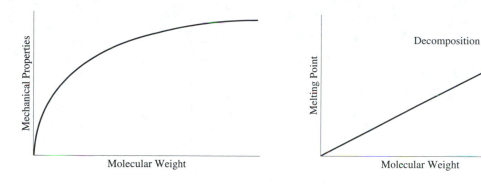

Figure 3.11 General mechanical and melting properties as a function of molecular weight.

are to be processed by melting, a practical limit of molecular weight is dictated by a diminishing of the increases in mechanical properties versus the increase in melting temperature and a desire to melt the polymer for ease of processing.

Plastic processing can also be affected by the molecular weight distribution. In some types of processes, a narrow distribution is preferred while in others a wide distribution gives better processing. A narrow distribution means that the material will melt over a narrow range of temperatures. (This is because the molecules are almost the same size and will, therefore, require about the same energy to cause them to move freely.) Uniform melting can be useful in processes, such as injection molding, that depend upon a rapid freezing of the molten polymer. The freezing, which is just the reverse of melting, will be done over a narrow temperature range that can be easily controlled. On the other hand, some processes, such as extrusion, work better with a wide molecular weight distribution. Processes of this type require a high *melt strength,* which is a measure of the ability of the molten material to be shaped. Honey and taffy, for instance, have high melt strengths because they can be pulled and shaped while molten, as opposed to water, which has little melt strength. Like honey or taffy, plastics with a broad molecular weight distribution have high melt strengths. The small molecules melt first and lubricate the entire mass, thus giving some ease of sliding to the very large molecules, even while they are still slightly entangled. This lowers the effective melt temperature. The large polymers give strength to the melt because of their residual entanglement.

An interesting use of modifying the molecular weight distribution to give selected processing capabilities is seen in ultrahigh molecular weight polyethylene (UHMWPE). This material, as the name implies, has extremely long polymer chains and, consequently, excellent strength, impact toughness, and a high melting point as compared to other polymers of the same family. The nature of the polymerization process usually results in a narrow MWD. Except for the high melting point, this material works well for injection molding. However, the material is not good in extrusion and related processes requiring a high melt strength. The processing temperature is, of course, very high, but even more important for these processing methods, the material has little melt strength because by the time all of the molecules are melted, they are too free moving. Any attempt to lower

Figure 3.12 Bimodal molecular weight distribution.

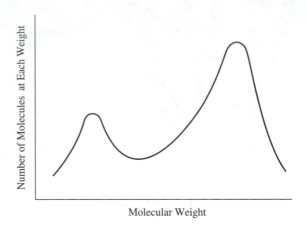

Number of Molecules at Each Weight

Molecular Weight

the temperature slightly to get more melt strength results in freezing. The temperature range for operating is just too narrow.

This problem has been effectively solved by adding a lower molecular weight polyethylene. This second material acts as the low-melting lubricant which is similar to the behavior of broad-MWD material. In this case, however, two rather distinct distributions exist, as shown in Figure 3.12. This type of distribution is called *bimodal*. The case cited is an example of a high-bred material which has been customized for specific properties by combining two polymers.

MELT INDEX

Rather than attempt to determine the average molecular weight and the MWD, either by calculation or by using some of the experimental methods already mentioned, many important flow characteristics of the polymer, which are strongly dependent on these properties, can be found using a simple test called the *melt index*. This method is so simple that it is the preferred method for determining these molecular weight and molecular distribution characteristics in industry. In fact, the melt index is one of the most common parameters specified when describing a polymer. A typical device for conducting this test is shown in Figure 3.13.

In the melt index test, a large block of metal in which heating coils have been imbedded surrounds a tube into which several grams of the polymer material to be tested are placed. This apparatus allows a constant, preset temperature to be maintained on the polymer. The desired temperature for the test is selected and the apparatus is brought to this temperature. (The temperature depends upon the type of polymer to be tested.) A weighted piston (with the weight specified by the test method for each type of polymer) is then placed in the tube so that it applies a constant force on the polymer. To avoid confusion, the temperature and weight conditions are reported with the results of the test. Tables of suggested values for these parameters are given in the testing literature. (For instance, this test is outlined in ASTM D 1238 as *Standard Test Method for Flow Rates of*

Figure 3.13 Melt index test.

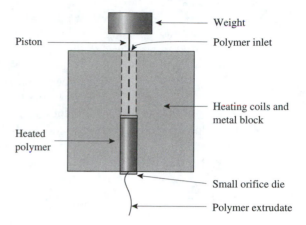

Thermoplastics by Extrusion Plastometer.) Typical examples of temperature/weight conditions are: polyethylene 190°C/1.0 kg, nylon 235°C/1.0 kg, and polystyrene 200°C/5.0 kg.

Under the constant temperature and weight that is appropriate for the type of polymer to be tested, the polymer eventually begins to flow slowly (extrude) out the small orifice at the bottom of the tube. When this flow is constant, the polymer extrudate is removed (that is, wiped away or cut off at the orifice) and discarded. A stop watch is started and after a set amount of time (usually from about 15 seconds to 5 minutes) as specified by the test method according to the type of polymer, the extrudate that has come out during the set time is carefully removed and weighed. The weight of the material extruded during the specified time is the melt index expressed in grams per 10 minutes.

The melt index is not an intrinsic or fundamental property of a polymer. It is, rather, a convenient method for expressing important flow characteristics of the polymer which clarifies the way in which the polymer can be processed. The melt index has been found to relate closely with average molecular weight, and, to a lesser extent, with MWD.

Note that if the melt index is a large number, that means that much material flowed through the orifice in the allotted time. Material with this behavior, that is, high flow, would have short chains and, therefore, low molecular weight. **Hence, a high melt index indicates a low molecular weight. The opposite is also true; a low melt index means a high molecular weight.** The viscosities of the polymers also provide a general comparison. Polymers with a high melt index flow quickly through the test apparatus and are low in viscosity. Polymers with a low melt index flow slowly and have a high viscosity. Typical values for melt index and molecular weight for polyethylene samples are given in Table 3.3.

Table 3.3 Comparison of Melt Index and Average Molecular Weight

Weight Average Molecular Weight (g/mole)	Melt Index (g/10 min)
100,000	10
150,000	0.3
250,000	0.05

The values given for melt index and molecular weight in Table 3.3 are approximate and could vary as much as 20% depending upon the shape of the MWD. For instance, if the MWD were skewed to the high end, the weight average molecular weight would be moved toward higher values. Therefore, Table 3.3 should only be considered as indicative of the types of results that can be obtained.

The use of melt index values for processing can be illustrated for polyethylene. In general terms, melt index values of above 15 for polyethylene would be considered to be quite low in molecular weight and would be used when ease in processing would be highly desirable. Fractional melt indices (below 1.0) would be more difficult to process and would be used when mechanical properties performance is very important.

As with the relationship between molecular weight and mechanical properties (see Figure 3.11), the relationship between molecular weight and melt index is also nonlinear so that in the region of high molecular weight, large increases in molecular weight result in only small changes in melt index. Note that with molecular weight and melt index, however, the relationship is an inverse. (High molecular weight results in a low melt index.)

The underlying purpose for determining melt index is to estimate the ease of melting of the polymer. As already noted, the melting characteristics of a thermoplastic polymer almost always indicate the ease of processing the polymer, particularly the amount of heat and/or mechanical and shear energy which must be added to the polymer to melt it. The melt index test approximates the melting conditions for the polymer, although with less shear and mechanical energy than would normally be used for commercial plastics processing. Even though it is only approximate in duplicating the melting conditions, the melt index has proven to be a convenient approximation to the melting characteristics of the plastic. Hence, high melt index numbers indicate ease of melting, lower energy input required and, often, easy processing.

SHAPE (STERIC) EFFECTS

The effects of the shape or size of the atoms or groups of atoms are called *steric effects*. These steric effects are very important in determining some of the microproperties of polymers. Because the shapes and sizes of polymer molecules may be difficult to envision, chemists have developed several different methods to represent the shapes and sizes of molecules when written. Some of these representation methods are illustrated in Figure 3.14 for the simple molecule 2-chloropropane (C_3H_7Cl).

As suggested in Figure 3.14, the shape and size of the pendent groups has a major effect on some of the important properties of plastics. When the pendent groups are large, crystallinity is almost always reduced or eliminated simply because the molecules cannot get close enough together to form crystalline bonds or because no stress-free or low-hindrance packing pattern exists. Because of this lack of crystallinity, the initial assumption would normally be that the properties most enhanced by crystallization, such as tensile strength and thermal properties, would be lower. This expected decrease in properties does not necessarily occur because the bulky pendent groups of one molecular chain interfere with the bulky pendent groups on another chain. The effect of this hindered movement

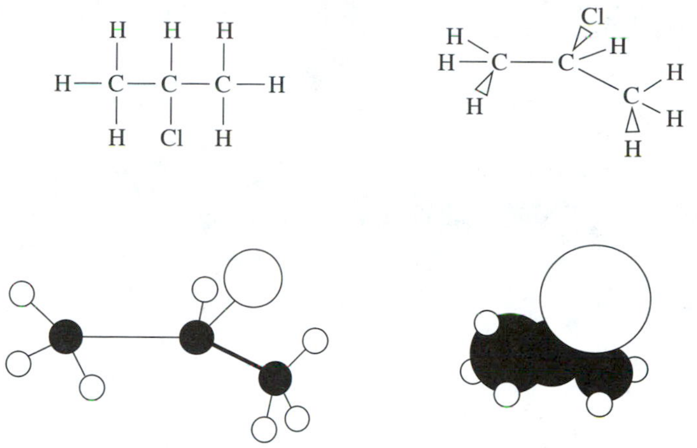

Figure 3.14 Various methods for representing the 2-chloropropane (C₃H₇Cl) molecule and to illustrate steric effects.

would be to increase mechanical and thermal properties just as would the intermolecular bonding and other interactions such as entanglement, which have previously been discussed. Therefore, the two effects arising from steric hindrance—low crystallinity and hindered movement—counteract each other and the net effect on properties depends on whichever of the factors happens to dominate for the particular polymer.

Bulky groups may hinder rotations or flexibility, which can affect properties such as T_g. If the movements are hindered, T_g would be expected to increase.

Many large molecular groups can be bulky and cause the effects described. In particular, aromatic pendent groups are very bulky and have the effect of increasing mechanical and thermal properties. An example of this behavior is seen in polystyrene, a very hard and brittle polymer which has a pendent aromatic group.

On the other hand, some pendent groups are very flexible and, although they prevent crystallization, they do not inhibit translational, rotational, or flexing movement. (The lack of hindrance of movement of one molecule relative to another can be explained by the ability of these flexible side chains to easily move to avoid interfering with another molecule). An example of this rather large pendent group arrangement that does not increase stiffness is the branching that occurs in low-density polyethylene, a very flexible material that has long pendent groups. The net effect in these cases is a reduction of crystallinity without the hindrance-of-movement effect encountered in the case of bulky pendent groups. Therefore, in these cases, the mechanical and thermal properties all decrease. An example of this pendent type would be an aliphatic branch. (Aromatic and aliphatic are discussed in the chapter on polymeric materials (molecular viewpoint) and are illustrated in Figure 3.15.)

Therefore, no general rule can be stated as to whether a bulky pendent group will increase or decrease the properties except that the general trend for pendent aromatic groups is to increase and for pendent aliphatic groups is to decrease these properties.

The steric nature of the backbone also affects the physical properties. If the backbone itself is stiff, the strength, impact toughness and temperature properties will generally

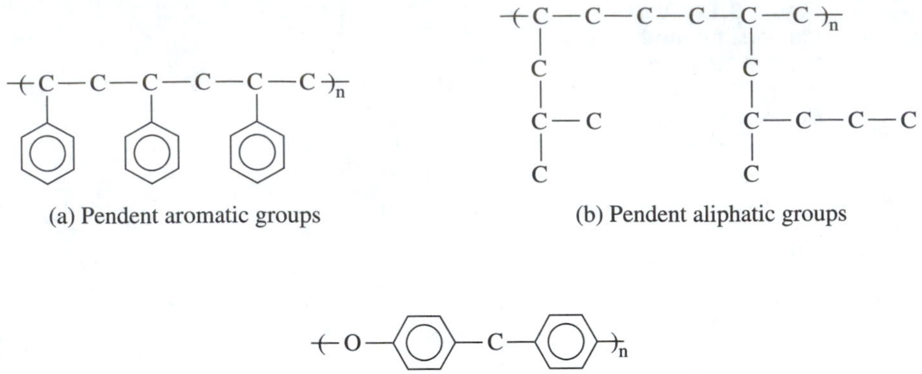

(a) Pendent aromatic groups

(b) Pendent aliphatic groups

(c) Chain containing aromatic groups

Figure 3.15 Aromatic and aliphatic pendent groups and backbone constituents.

increase. The atoms or groups contained within the backbone itself can have a major effect on backbone stiffness. The inclusion of aromatic groups in the backbone will cause the backbone to be stiffer because of steric interferences with other backbone atoms and also with pendent atoms. Aliphatics along the backbone result in more flexibility in the backbone.

Some of the highest strength and stiffest plastics are those with extensive aromatic groups along the backbone. These materials have high melting points, high glass transition temperatures, and often excellent solvent resistance. Two groups of polymers of this type are the *aramids* and the *liquid crystal polymers*. One aramid is Kevlar®, which is made into a fiber that is used for bulletproof vests and other composite reinforcements, and another aramid is Nomex®, which is used for a super strong plastic and coatings. The polymer backbones are so stiff that the crystal structure is partially retained, even in the liquid phase, and are called liquid crystals. This results in a directional solid material that can be very strong. These polymers are often used for high-temperature and high-strength applications, such as aerospace. Examples of liquid crystal polymers include Xydar® and Vectra®.

======= CASE STUDY 3.1 =======

Mechanical Properties of Polyethylene (PE) as Functions of Density and Melt Index

The relative effects of polymer density and molecular weight on physical properties can be complicated and sometimes confusing. This potential confusion arises because density and molecular weight have similar effects on some mechanical properties (such as tensile strength) but have opposite effects on other properties (such as brittleness). To further complicate these property dependencies, in some cases the effects of density will dominate and in others, the effects of molecular weight will be stronger.

During the manufacture of a polymer, the processing variables can usually be selected to vary density and molecular weight independent of each other, at least within some general ranges of operability. Therefore, many polymers can be made with all of the possible combinations of high density/high molecular weight, high density/low molecular weight, low density/high molecular weight, and low density/low molecular weight. These variabilities were found in the case of some polyethylene (PE) resins which were furnished for this case study by Paxon Polymer Company. The pertinent data for these polymers is given Table 3.4. As is usual with polymers, the molecular weight is measured as the melt index. Because the melt index and molecular weight have an inverse relationship (as molecular weight gets larger, melt index gets smaller), the reciprocal melt index is also given in the table. Use of the reciprocal melt index will allow some comparisons to be made more simply than could be done with the melt index itself.

Table 3.4 Polyethylene Resin Density and Melt Index Data
(Courtesy of Paxon Polymer Company)

Resin	Density (g/cm^3)	Melt Index (MI) (g/10 min)	Reciprocal Melt Index (1/MI)	Tensile Strength (MPa)	Tensile Toughness (J/cm^2)
AA60-003	0.960	0.3	3.33	30	24
AA45-004	0.945	0.35	2.86	22	34
BA50-100	0.949	<.01	10	25	36
AA55-600	0.953	55	0.02	27	9

As was previously discussed in this chapter, the tensile strength of a polymer should increase with increases in density and with molecular weight (or reciprocal melt index). These quantities are plotted in Figure 3.16 for the four resins in this case study. Note that the relationship between the tensile strength and density is, as predicted, a straight line. Hence, as density increases, so does tensile strength. The relationship between tensile strength and reciprocal melt index (1/melt index) is not a straight line. In this case, other factors seem to be dominating the relationship. These relationships can be summarized as follows for polyethylene:

- Density is a strong determiner of tensile strength, perhaps the dominant factor. Tensile strength has a linear relationship with density.
- Tensile strength does not always have a linear relationship with molecular weight (reciprocal melt index). Other factors such as molecular shape, secondary bonding, and crystallinity, seem to dominate.

As previously discussed, toughness should increase with increasing molecular weight and may decrease with increasing density because of the enhanced brittleness associated with crystalline structures common to high-density resins. These relationships are plotted in Figure 3.17.

The relationship between toughness and density is not linear. There may be some reduction in toughness at higher densities, as would be expected from brittleness considerations but the data is scattered (see Figure 3.17a). The relationship between toughness

Figure 3.16 Plots of tensile strength versus (a) density and (b) reciprocal melt index for four polyethylene resins.

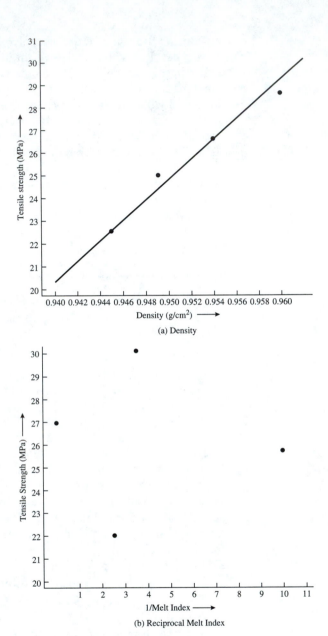

(a) Density

(b) Reciprocal Melt Index

Figure 3.17 Plots of toughness
versus (a) density and
(b) reciprocal melt index for
four polyethylene resins.

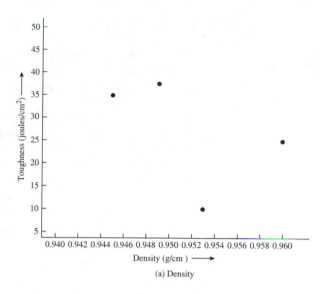

(a) Density

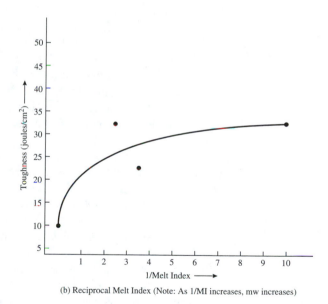

(b) Reciprocal Melt Index (Note: As 1/MI increases, mw increases)

and molecular weight (reciprocal melt index) is much stronger. The toughness increases with increasing molecular weight, but the relationship is not linear. At high molecular weights, increases in molecular weight seem to have a diminishing effect on toughness. This relationship was described previously in this chapter and explains why molecular weight cannot continue to be increased in order to obtain ever higher polymer physical and mechanical properties.

===================== SUMMARY =====================

The intent of this chapter was to examine and explain polymer properties that depend upon the interrelationship of several molecular chains. This perspective is called the micro view (as compared to the molecular view of the previous chapter and the macro view of ensuing chapters).

When the polymers are solids, the polymer chains become entangled, and portions of the polymer molecules can pack closely together. In some polymers this packing becomes very regular, often because a polymer will fold back and forth on itself, but other times because it aligns tightly with neighboring molecules. When this close packing occurs, secondary bonds form between the polymers. These areas of close packing are called crystalline regions because of the analogy with metallic and ceramic crystals, which are also regular and held together by bonds between the elements of the crystal. In other molecules, however, the molecules do not pack closely and the overall regions of the molecules are random. These regions are called amorphous. Therefore, regions within molecules can be either crystalline or amorphous, and the molecules themselves are often referred to as being crystalline or amorphous, depending on which type of region dominates. Crystallinity in a polymer increases the density of the polymer.

Polymer strength and stiffness are increased by crystallinity. Therefore, a direct relationship between density and strength is commonly found.

When polymers are heated, the heat is absorbed by the polymer through increases in molecular movement. Initially this movement is small and results in only minor increases in volume with very little additional change in polymer properties. With increased heating the motions become very large and the molecules begin to slide and disentangle. When this occurs, the polymer is less able to support loads. A common test in which polymers are subjected to loads at elevated temperatures is called the HDT and it measures the point at which the ability of the polymer to support loads is diminished beyond a set amount.

Further heating of the polymer will cause increased movement until several atoms within the molecule move in long-range, coordinated movements. The temperature at which this happens is called the glass transition temperature, T_g. This is a thermal transition in the polymer because further heating will not raise the temperature until all of the molecules in the batch are able to move in similar long-range motions. Below T_g the polymer is rigid and hard. Above T_g the polymer is pliable and leathery.

The next transition of the polymer is at the melting point. Here the polymer molecules can move freely from each other. If the polymer is amorphous, the melting occurs over several degrees because of the various lengths and other interactions of the polymers. When the polymer is crystalline, the melting point is quite sharp and corresponds to the breaking of the secondary bonds that hold the crystalline structure together.

Beyond the melting point, further heating will eventually cause the molecule to degrade because the thermal energy is sufficient to break the primary bonds of the molecules.

Imagine a polymer of enormous chain length. As this polymer is heated, the energy causes both vibrational and translational motions, as with the normal-sized polymer. The enormous polymer, however, is so large that very high energies are required to reach the level of activation needed to melt it. Before the entire polymer can be totally activated, the vibrational energies in the bonds might exceed the bond strengths and the polymer could begin to degrade. The melting point of this polymer has been increased (due to the long chain length) such that the melting temperature has become higher than the decomposition temperature. This situation, in which the enormous chain length causes the melting temperature to be higher than the decomposition temperature, can often occur if fully formed polymer chains are bonded together, as occurs in the curing of thermosets. When a bond occurs between two chains, the total length of the resultant chain would be approximately double the length of the separate chains. Further bonds between this double-length chain and other chains would give further rapid increases in total chain length until, conceivably, the entire structure could be bonded together with these interchain bonds. These interchain bonds are called crosslinks and the process of forming these bonds is called crosslinking or curing. Polymers that have these crosslinks are called thermosets; their structures have become rigid (set) with temperature because these bonds between the polymer chains are formed with heating. Because of the very high cumulative length of these crosslinked polymers, these thermoset materials decompose before they melt.

Plastics can also degrade over long periods of time. This degradation is from the accumulated effects of heat, often from repeated processing or from exposure to high temperatures in use. Hence, elevated temperatures for long periods of time are to be avoided.

Thermal conductivity is a measure of the ability of a material to conduct thermal energy. Most plastics are poor thermal conductors compared to metals. This lack of thermal conductance allows plastics to be used extensively as insulative handles for cooking pans and for other devices that may get hot.

When materials get hot, they generally expand. This is called thermal expansion and is generally higher for plastics than for metals and ceramics. Thermal expansion is an important consideration in designing a mold for plastics to ensure that the finished part is the proper size, even after the subsequent normal contraction as the part cools. If the part is constrained so that it cannot change its size with thermal changes, some stresses arise within the part. These are thermal stresses, and they can lead to premature failure in some circumstances if not relieved. They can easily be relieved by annealing.

The molecular weight of polymers is an important measure of the size of polymer chains. Because polymer chains can be of several different lengths, statistics are used to describe the molecular weight. Hence, molecular weight is described as an average of the polymers contained within the sample. Several types of averages can be found. The most common of these types are the number average and the weight average molecular weights. The number average molecular weight gives equal weight in the average calculation to every polymer chain, regardless of its size. The weight average molecular weight gives greater emphasis to the large molecular weight chains. If the molecular weight averages can be found experimentally, the degree of polymerization (DP) can be calculated. The degree of polymerization is the number of monomer units which have been linked together in the average polymer chain.

Molecular weight distribution (MWD) is another statistical measure of polymers. This property describes how similar in size the polymers are. If the MWD is small (narrow), most of the polymers are of the same length, which is usually desirable when injection molding the polymer. A wide MWD indicates that the polymers have a wide variety of sizes and is often desirable in extrusion processes.

Melt strength is a property that is related to MWD. Melt strength is a measure of the ability of the polymer to be shaped while in a molten state. Melt strength arises when some of the molecules have become free to move extensively (as they would in a liquid) while other molecules are still strongly interacting with surrounding molecules. Hence, some molecules are melted and others are not. The melted molecules give fluidity to the polymer sample and the non-melted molecules give strength. Melt strength is enhanced by wide MWD because small molecules can become melted while long molecules are still not melted.

The melt index is a method of measuring the effects of molecular weight. In this test the tendency of a polymer to melt is measured. High molecular weight materials are more difficult to melt and therefore have a low melt index. Hence, molecular weight and melt index are inversely related.

Mechanical and physical properties of polymers are often increased (non-linearly) with increasing molecular weight. This relationship is especially true for toughness.

The density of a polymer is a measure of how tightly the polymer chains can pack together. High density means that the packing is close. Often, close packing is the result of the formation

The shape of a polymer can strongly effect the polymer's crystallinity. If pendent groups are large, these groups will tend to interfere with the groups on other molecules and prevent close packing. Hence, crystallinity is less likely. The presence of large groups will also effect mechanical and physical properties of the polymer. These effects are complicated and not easily predicted except that the presence of aromatic groups will generally increase strength and toughness.

GLOSSARY

Aging   Long-term, low-temperature degradation.

Amorphous   Without shape; refers to areas within polymers or to types of polymers dominated by areas in which no regular structural pattern (such as crystallinity) occurs.

Annealing   Moderate heating to allow relief of internal stresses.

Arrhenius equation   An exponential function that relates, among other quantities, the rate of chemical reactions to the temperature.

Bimodal   A molecular weight distribution in which two peaks occur, often because low and high molecular weight materials have been mixed.

Char   The material remaining after a polymer decomposes.

Coefficient of thermal expansion (CTE)   A quantitative measure of the increase in volume or length of a material as it is heated.

Colligative properties   Properties that depend on the number of particles in the material.

Combustion   The burning of a polymer or decomposition at very high temperatures in the presence of oxygen.

Creep   The tendency of a polymer to distort under external loads, especially as the temperature increases.

Crystallinity   Areas within a polymer or types of polymers in which the polymer molecules fold into a tight, regular structure.

Decomposition   The breaking of primary bonds in a molecule.

Decomposition temperature (T_d)   The temperature at which decomposition begins to occur.

Degradation   The decomposition of a material.

Degree of crystallinity   The amount of the structure that is crystalline as opposed to the amount that is amorphous.

Degree of polymerization (DP)   The number of repeat units in a polymer molecule.

Differential scanning calorimeter (DSC)   A test that determines thermal transitions (such as T_g, T_m, and T_d) by measuring the heat gained by a sample as it is heated at a specific rate.

Distribution of values   A statistical method for relating quantities that are not all exactly the same.

Entanglement   Wrapping of polymer chains about each other.

Gel permeation chromatography (GPC)   A method for determining molecular weight and its distribution based on the differential absorptivity of solvated polymers on strips of paper.

Gels   Small solid or semisolid masses that resist melting, usually caused by minor crosslinking of thermoplastic resins.

Heat capacity   The measure of the temperature rise in a given weight of material for a given amount of heat input.

Histogram   A graphical representation of the number of occurrences of a particular value (or parameter) at each possible occurrence of that value.

Impact toughness   A measure of the ability of a material to withstand an impact blow.

K factor   A measure of thermal conductivity.

Melt index (MI)   A test measuring the weight of a polymer that can be extruded through a small orifice in a particular time.

Melt strength   The ability of a molten material to be shaped.

Melting point (T_m)   The temperature at which a material changes from a solid to a liquid (or vice versa).

Molecular weight   The sum of the atomic weights of all atoms in a molecule and a measure of the size of the molecule.

Molecular weight distribution (MWD)   A count of the number of molecules of each molecular weight; it measures the variability in the molecular weights of the molecules in a sample.

Number average molecular weight (M_n)   The average molecular weight as normalized by the total number of chains.

Processing aids   Materials that lower the thermal requirements to process a polymer, usually done by lubricating the material.

Regrind   Scrap material that is made during the molding process that is ground into flakes and reused in the process.

Steric effects   The influence of molecular shapes on the properties of a material.

Tensile strength   A measure of the force required to pull a sample apart.

Thermal conductivity   The measure of how quickly or easily heat moves through or along a material.

Thermal expansion   The increase in dimension of a material upon heating.

Thermal history   The accumulation of the effects of elevated temperature processing.

Thermal stabilizers   Materials that assist polymers in withstanding the effects of heat.

Thermal stresses   Internal energy built up due to thermal effects.

Thermal transition   A temperature at which some basic change occurs in the nature of the material, such as melting, as evidenced by a holding of any temperature rise with increased energy input until the change is complete.

Thermogravimetric analysis (TGA)   A test that measures the weight loss of a sample as it is heated at a specific rate; often used to determine T_d.

Thermomechanical analysis (TMA)   A test which measures changes in the mechanical properties of a material (as sensed by penetration of a flat needle) as the material is heated at a specific rate.

Virgin material   Resin that has not been previously processed (that is, nonregrind).

Weight average molecular weight (M_w)   The average molecular weight, as normalized by the total weight of each molecule.

z-average molecular weight   A measure of molecular weight that heavily weights the influence of large molecules.

QUESTIONS

1. Contrast the interatomic or intermolecular forces present in solids, liquids, and gases. Explain the consequences of these forces and how they are normally overcome.
2. Can all materials exist as solids, liquids, and gases at various temperatures? Explain.
3. Name three methods of reducing the amount of thermal degradation that might occur in a heat-sensitive plastic. How does each reduce thermal degradation?
4. What is a heat history, as applied to polymers, and why is it important in polymer technology?
5. What are thermal stresses and how are they caused in plastic materials? How can they be relieved?
6. What is the difference between a number average and a weight average molecular weight?
7. Why must average molecular weights be used for polymers rather than exact molecular weights based upon the molecular formula as is done for small molecules?
8. Discuss the relationship between melt index and molecular weight.
9. Contrast and discuss the difference between amorphous and crystalline regions in a polymer.
10. Explain why the properties of a polymer below the glass transition temperature are different from the same properties in the same polymer above the glass transition temperature.
11. Discuss how you would expect the glass transition to be affected if a large, pendent group were added to the monomer unit.

12. Explain the effects of crosslinking on the glass transition temperature.
13. Discuss the implications on MWD when the number average molecular weight and the weight average molecular weight are widely different.
14. Explain bimodal distributions of polymers and their usefulness.
15. Discuss how copolymerization would affect crystallinity.
16. Given the following data, calculate the number average molecular weight and the weight average molecular weight, and discuss whether the MWD distribution is wide or narrow compared with a normal commercial ratio of 20.

$$N_i \ (\times \ 10^{24}) \ 5 \ 7 \ 8 \ 9 \ 10 \ 8 \ 6 \ 3 \ 2 \ 1$$
$$M_i \ (\times \ 10^4) \ 1 \ 2 \ 3 \ 4 \ 5 \ 6 \ 7 \ 8 \ 9 \ 10$$

17. Why does processing of a crystalline material like nylon require more critical thermal control as opposed to an amorphous material like ABS?

REFERENCES

Billmeyer, Fred W. Jr., *Textbook of Polymer Science* (3rd ed.), New York: John Wiley & Sons, 1984.

Clegg, D. W., and A. A. Collyer, *The Structure and Properties of Polymeric Materials,* London: The Institute of Materials, 1993.

Cowie, J. M. G., *Polymers: Chemistry & Physics of Modern Materials* (2nd ed.), New York: Chapman & Hall, 1991.

Daniels, C. A., *Polymers: Structure and Properties,* Lancaster, PA: Technomic Publishing Company, Inc., 1989.

"Resins and Compounds," *Modern Plastics Encyclopedia,* Hightstown, NJ: McGraw-Hill (1993), 178–229.

Rosen, Stephen L., *Fundamental Principles of Polymeric Materials,* New York: John Wiley & Sons, 1982.

CHAPTER FOUR

MECHANICAL PROPERTIES (MACRO VIEWPOINT)

CHAPTER OVERVIEW

This chapter examines the following concepts:

- Mechanical properties in solids (types of forces, elastic behavior and definitions)
- Mechanical properties in liquids—viscous flow (viscous behavior and definitions, Newtonian and non-Newtonian flows, measurements of viscosity)
- Viscoelastic materials (viscoelastic behavior and definitions, time dependence, long-range and short-range interactions)
- Plastic stress-strain behavior (plastic behavior and definitions, interpretation of plastic behavior, mechanical model of plastic behavior, non-tensile forces)
- Creep
- Toughness and impact strength
- Reinforcements
- Fillers
- Toughness modifiers

INTRODUCTION

Previous chapters have considered the properties of polymers from the **molecular** view and from the **micro** view. This chapter and the chapter on Chemical and Physical Properties will consider polymers from the **macro** view, that is, from considerations of the bulk material. This view is the view of actual use. It is, therefore, an important view, but is best understood in conjunction with the molecular and micro views. This chapter will examine the mechanical properties of polymers. Mechanical properties describe how polymers behave when subjected to various forces such as tension, compression, shear, and impact.

The mechanical properties of traditional, nonplastic materials (such as metals and ceramics) have been used for centuries to understand materials, scientifically investigated for decades, and described quite well using classical mechanics theory. This classical theory often models the mechanical behaviors of these materials as systems of simple mechanical devices, such as springs and dashpots, which can be described using conventional mathematics (usually calculus). Various mechanical conditions, such as impositions of different types of forces, have been studied and adequately solved over the years. The classical theory has been found to work well for traditional materials, which are generally divisible into two categories—*elastic solids* (such as iron, concrete and copper) and *viscous fluids* (such as water, oil, and molten steel). Elastic solids are materials that completely recover their shape and restore the energy imparted to them after an imposed force has been removed. Viscous fluids are materials that flow when exposed to an imposed shear force but when the force is removed, immediately stop their flow and do not return any of the imposed energy.

Some other materials, such as polymers, do not follow these classical equations and definitions. These nonclassical materials have properties that seem to be combinations of viscous fluids and elastic solids and are therefore called *viscoelastic*. For instance, they may flow when exposed to an imposed force (like a viscous fluid) but return energy and recover some shape when the force is removed (like an elastic solid). The degree to which a viscoelastic material behaves like a classical elastic solid or a classical viscous liquid depends upon the basic nature of the material (crystallinity, interchain interactions, chain stiffness, and other molecular and micro view characteristics). All of these characteristics are influenced by many of the conditions under which the test is conducted, but temperature and time are especially important.

Example of Time and Temperature Dependence of a Polymer

The rate (time) and temperature dependence of the mechanical properties of "silly putty" (a familiar viscoelastic material) can be seen from a few simple experiments. If a piece of "silly putty" were rolled into a ball and then dropped to the floor, it would bounce. This behavior would be like an elastic solid, wherein the original shape was recovered and the energy that was put into the ball was returned in the form of a bounce. However, if the same material were left on a table overnight, it would flow into a puddle shape. Hence, at high rates, that is short times for the experiment (like the dropping impact), the material behaves as an elastic solid, while at low rates (like sitting overnight) the material behaves like a viscous fluid.

The temperature dependence of "silly putty" can be seen by performing the same experiments at different temperatures. If the ball of "silly putty" were heated to a temperature considerably higher than room temperature and then dropped, the ball would bounce very little but would stick to the ground and quickly flow into a puddle. On the other hand, if the ball were cooled to a temperature significantly below room temperature and then left on the tabletop, the ball would retain its shape with little flow, even over several days (provided the temperature remained low). Therefore, short reaction times and low temperatures favor elastic properties, and long reaction times and high temperatures favor viscous properties. The "silly putty" is, therefore, a viscoelastic material.

Most plastics are viscoelastic. In order to understand viscoelastic behavior, a background in classical elastic and viscous behavior is helpful. Therefore, these two subjects are introduced first, and then viscoelastic behavior is examined.

MECHANICAL PROPERTIES IN SOLIDS (ELASTIC BEHAVIOR)

Types of Forces

Whenever a force is applied to a solid material, that material will deform in response to the applied force. The most common types of force are: a pulling force (called *tensile force*), a pressing or pushing force on the end of a columnar sample (called *compressive force*), a pressing or pushing force on the middle of a supported, long sample (called a *bending* or *flexural force*), a rotational or torsional force (called a *torsion force*), or a combination pushing force with a sliding force (called a *shear force*). These forces are illustrated in Figure 4.1. Tensile forces are often the easiest to visualize and will, therefore, be

Figure 4.1   Types of forces.

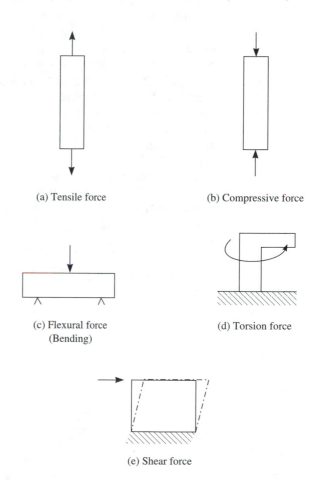

(a) Tensile force

(b) Compressive force

(c) Flexural force
(Bending)

(d) Torsion force

(e) Shear force

used to develop the concepts needed to understand mechanical properties. The other forces will be examined later in this chapter. Most of the concepts derived from a consideration of tensile forces are fundamental and will be directly applicable to the other forces, although in a slightly different form.

Elastic Behavior and Definitions

If only small deformations are considered and the solid material will return to its original shape when the force is relieved, then the deformation is called *elastic*. In elastic deformations all of the mechanical energy that was put into the material by the applied force to cause the deformation was held within the material and was then used to cause the material to return to its original shape and position. A common example would be a spring that is deformed slightly, thus imparting potential energy to the spring, which is then available within the spring to cause it to return to its original shape. Another way of saying this is that energy was returned or recovered. Energy is always recovered in elastic deformations.

So that materials of different sizes can be directly compared, the force is usually divided by the area of the sample to give units of pascals (newtons per square meter) or pounds per square inch. (The length dimension is not important in this calculation because it is assumed that the applied force is evenly distributed over the entire length of the sample between the pulling forces.) The force divided by the area is called the *stress* (σ) and the units are force per area (typically pascals or pounds per square inch). The displacement (movement) of the material is called the *strain* (ϵ). Normally the strain is given as the change in length (Δl) divided by the original length (l_O) and the units are dimensionless. The elongation is a measure of the strain when the force is tension. Elongation is usually expressed as a percentage increase in length compared to the original length of the test specimen. Plots of the stress versus the strain such as the type shown in Figure 4.2 are called *stress-strain curves* or *stress-strain diagrams*.

Figure 4.2 Stress-strain relationships for elastic solids.

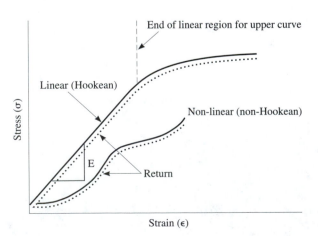

The relationship between stress and strain can be given by an equation such as (4.1),

$$\frac{F}{A} = \sigma = E\epsilon \tag{4.1}$$

where F is the force, A is the cross-sectional area, σ is the stress (force divided by area), ϵ is the strain, and E is the proportionality factor which is called the *modulus,* sometimes referred to as *Young's modulus* for the tensile stress case. The modulus is the slope of the stress-strain curve. If the modulus is large (corresponding to a steep angle of the curve), the material resists deformation strongly. Such materials are said to be *stiff.*

If the stress-strain curve is linear, regardless of its slope, the stress and strain are directly proportional and E is a constant over the elastic region (the straight portion of the curve). Such materials are said to follow Hooke's law and to be *Hookean.* Hooke's law applies to mechanical springs and can be written in the form

$$F = kx \tag{4.2}$$

where F is the force, x is the displacement, and k is the proportionality factor which is called the spring stiffness and is constant for small displacements. This equation is similar to Equation (4.1) with only minor modifications for area and original length to fit normal forms of the two equations. Hence, the two equations are, in essence, the same.

If the stress-strain curve is nonlinear, the material is said to be *non-Hookean.* The non-Hookean modulus is not constant and is defined only at specific points on the curve using calculus as the derivative of the stress to the strain. In both the linear and nonlinear cases the material returns to its original shape and position when the force is relieved, so the material is elastic. (The return to original shape is indicated in Figure 4.2.) Most metals and ceramics are Hookean solids.

MECHANICAL PROPERTIES IN LIQUIDS (VISCOUS FLOW)

Viscous Behavior and Definitions

The stress-strain relationships discussed for solids do not apply to liquids. This can be easily seen by imagining a sample of water being pulled in tension. The water is incapable of being pulled in this way. Hence, a tensile stress on water will not produce a displacement (strain). Therefore, different mechanical behavior relationships must be developed from those used for elastic solids. The most common mechanical formulation involves the relationship between stress and the velocity gradient (that is, the change in velocity with distance) as illustrated in Figure 4.3.

In Figure 4.3 a fluid is contained between two plates, one stationary and the other movable. The moveable plate has an area A and is subjected to a force F. This force causes the layer of liquid next to the moving plate to be displaced at a certain velocity. The movement of that layer of fluid influences the next lower layer to move, but its velocity is less than the first layer and is therefore shown as a shorter arrow in Figure 4.3. This reduction in velocity from layer to layer is caused by the tendency of the second layer of fluid to resist movement. This resistance to movement is much like inertia and is called the *viscosity*

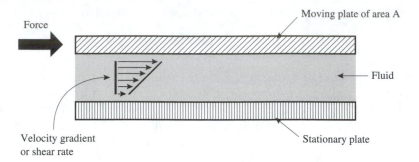

Figure 4.3   Stress and velocity gradient relationships.

of the fluid. Higher-viscosity fluids have higher resistance to flow. (Honey has a higher viscosity than water.) The same scenario holds true for succeeding lower layers of fluid, as indicated in Figure 4.3. Therefore, the stress (*F/A*) causes a *velocity gradient* (change in velocity with distance) in the fluid. The velocity gradient can also be thought of as a change in the shear with respect to time, a *shear rate* ($\dot\gamma$). (This alternate notation is possible because the force is a shear force and the velocity is both a function of distance and time.)

Note that a critical assumption of this analysis is that each layer has a resistance to the movement induced by the previous layer. This resistance to flow, viscosity, is defined by Equation (4.3)

$$\frac{F}{A} = \eta\,\frac{dv}{dx} = \eta\dot\gamma \tag{4.3}$$

where *F/A* is the stress, *dv/dx* is the velocity gradient (expressed as a derivative), $\dot\gamma$ is the shear rate and η is the viscosity. Only fluids which have this resistance to flow (viscosity) will be discussed, but almost all normal liquids possess viscosity. The unit of viscosity is the *poise* (i.e., 1 poise = 1 g cm^{-1} s^{-1}). The study of the flow of viscous fluids is called *rheometry* and devices which assist in determining the viscosity are called *rheometers*.

Note the similarity of Equations (4.3) and (4.1). The viscosity is a proportionality constant for viscous liquids much as the modulus is a proportionality constant for elastic solids. An important difference between the mechanical behavior of viscous fluids and elastic solids is in the recovery of original shape, position and energy in each of the systems. In elastic solids, the solid material returns to its original shape and position after the force is relieved. This means that whatever mechanical energy was put into the system to deform it is still in the material and can be used to cause the material to return to its original form, thus returning the energy. In viscous fluids, when the force is relieved, the system does not return to its original shape and position. Instead, the fluid simply stops moving. The energy that was put into the system by the force on the system (the sliding of the plate in Figure 4.3) is not recovered but has been dissipated in the fluid as heat. This heat was caused by the internal forces needed to overcome the viscosity of the liquid and cause it to flow.

Newtonian and Non-Newtonian Flows in Fluids

When the stress-velocity gradient or shear rate curve is linear, the liquid system is said to be *Newtonian* and the viscosity is constant. Water is an example of a Newtonian liquid. When the stress and velocity gradient or shear rate are not linear, the fluid is *non-Newtonian*. Two types of non-Newtonian behavior are possible and are illustrated in Figure 4.4.

Materials which become thinner (less viscous) at high shear rates are called *shear-thinning* or *pseudoplastic,* whereas materials which thicken at high shear rates are called *shear-thickening* or *dilatant*. Most polymer liquids are shear-thinning. An example of a shear-thinning material is the ink in a ballpoint pen. The ink flows readily when sheared (such as during writing), but not otherwise (such as when the pen is in your pocket.) Dilatant behavior is not observed in polymers.

Because shear thinning is so important in polymer processing, some additional discussion on this point is needed. The key to understanding Newtonian and non-Newtonian flows is to keep in mind the basic nature of liquids. Initially, we can consider simple molecules, such as water. At the micro level, liquids are composed of molecules that have enough energy to move about easily (as compared to solids, which move only under strong forces), but not so much energy that the molecules are flying in all direction (as do gases). Using the analogy that simple molecules are like marbles, we can understand how they behave when they are subjected to a shear force. The solid molecules don't move and are like marbles locked in a gelatin matrix; the liquid molecules slide over each other like several marbles being rolled inside a bag; and the gas molecules bounce against each other like marbles being shot at each other during play.

The resistance to movement of the small liquid molecules is usually minimal (that is, the viscosity is low) and is linear in its dependence on the rate of shearing, that is, the rate at which the bag of marbles is squeezed.

The shear-thinning (pseudoplastic) nature of polymeric liquid materials requires a change in the analogy. Rather than being like rigid marbles, each polymer molecules is like a miniature octopus. Therefore, as the molecules move past each other, they can lightly entangle, like the arms of two passing octopuses might entangle. If the molecules

Figure 4.4 Newtonian and non-Newtonian liquids.

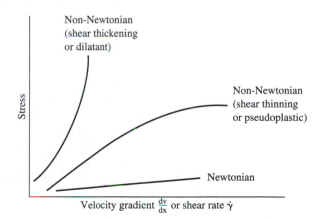

pass slowly, this entanglement can be quite strong, thus resulting in high viscosity. If the molecules pass quickly (high shear rates), the entanglement is minimal and the viscosity is lower. During high shear, the arms of octopuses, which are moving quickly, would be streaming out behind them and would not interact with other octopuses in the neighborhood.

Types of shear forces that are encountered in polymer processing include: squirting the polymer through a small hole at the injection port of an injection molding machine, pushing the polymer along a thin channel in a mold, and squeezing the polymer between a turning screw and the walls of a barrel during extrusion. All these shearing forces cause the viscosity of the polymer to decrease.

Note, however, that temperatures can also have a major effect on the viscosity at the same time. Changes in temperature indicate that thermal energy (heat) is being added or subtracted from the polymer. If heat is added at the same time that shear occurs, as is done in an extruder, the polymer viscosity will drop greatly. If heat is withdrawn (cooling) at the same time that shear is occurring, the viscosity may increase in spite of the shearing forces because energy is leaving the polymer through heat loss at a faster rate than it is entering the polymer by shearing. This situation is encountered in an injection mold where shear occurs at the injection nozzle and through the runners and gates, but so does cooling, resulting in a thickening (increase in viscosity) of the polymer. Care must be exercised that the polymer is not cooled so quickly that it will become too thick to fill the mold cavity.

A property closely related to shear-thinning is *thixotropic* behavior that is also a thinning of the material. The differences between the two properties are that shear-thinning is dependent on shear rate, whereas thixotropic thinning is independent of shear rate but dependent on time, at a constant shear rate. For example, many paints and gels are thixotropic. Imagine the process of painting. If the painter increases the rate at which the paint is applied (increasing shear rate), the paint will not necessarily get thinner. Trying to spread the paint by brushing faster is not likely to work. However, if an area is brushed for a long time without increasing the rate of brushing, the paint is likely to thin and spread more evenly. Then, after the brushing has stopped, the paint will again thicken. This thinning with shearing over time and subsequent thickening when the shear has stopped is typical of thixotropic materials.

Measurements of Viscosity

The flow of viscous materials can be measured in several ways, both in solutions and as melts. These methods vary in the range of viscosities over which they are applicable. Because of the wide range of viscosities possessed by polymers (viscosities typically range from 10^{-2} poise to 10^{12} poise), measuring the viscosity in one range and then extrapolating that data to a widely different use range should be avoided.

For low viscosities, the method most commonly used to determine viscosity is the capillary pipette (ASTM D2857). In this method the flow of the fluid (usually a solution) through a pipette is timed. The time for the pure solvent to flow through the pipette is also measured and then the two values are compared to give a *relative viscosity* (η_r), which is the ratio of the viscosity of the polymer solution to the pure solvent.

The relative viscosity can be used to obtain a quantity known as the *intrinsic viscosity* [η]. The procedure for obtaining the intrinsic viscosity is to divide the relative viscosity by the concentration (C) at each point that the relative viscosity is obtained. This quantity (η_r/C) is then plotted versus the concentration. A linear plot is obtained which can be extrapolated to zero concentration. The value of the quantity η_r/C at zero concentration (where the line intercepts the y axis) is the intrinsic viscosity, [η].

Because its value is determined by extrapolating the polymer solution to infinite dilution, the intrinsic viscosity represents the viscosity that a polymer would have if the polymer chains were isolated from each other. In other words, it is the viscosity of the polymer when it is dependent only on the basic nature of the polymer molecules, the temperature, the pressure, the solvent effects, and (most important in this context) the polymer molecular weight. This fundamental picture of viscosity gives rise to the idea of an "inherent" or "intrinsic" viscosity, which is, one hopes, independent of effects outside the solution itself. The intrinsic viscosity is therefore useful as an experimental method for determining the average molecular weight (M). The relationship between intrinsic viscosity and molecular weight can be given by Equation (4.4),

$$[\eta] = KM^a \tag{4.4}$$

where K and a are constants that are determined from plots of the log of [η] versus the log of M. The type of molecular average determined in this calculation, M_w or M_n, is determined by the specific procedures used in the test, but conversions between these two types of molecular weight averages can be made using the methods discussed in the chapter on polymeric microstructures.

A widely used method for determining viscosities in the middle of the viscosity scale is the rotating spindle or cylinder method. The Brookfield viscometer is of this type and has been used extensively for many years. This testing device uses a rotating spindle or cylinder that is immersed in the test liquid. The torque is measured and converted into units of viscosity on a convenient dial indicator. By measuring the torque required to spin the spindle or cylinder, the stress can be found. The shear rate is determined from the rotational speed and the gap distance between the spindle or cylinder and the walls of the vessel containing the liquid. The ratio of the stress to the shear rate is the viscosity. The size of the spindles and cylinders is changed to expand the range of viscosities over which the viscometer can be used. In some cases a "T" spindle (a spindle with a cross-member) that rotates and descends into the liquid is used to give more accurate measurements for thixotropic materials. This test apparatus is illustrated in Figure 4.5a.

The cone-and-plate method determines the viscosity of a plastic by shearing it between a rotating cone and a plate as indicated in Figure 4.5b. The angle of the cone is selected to impart a near-constant shear rate on the polymer across the entire surface of the plate. (Because the cone is traveling further at the edge, the shear rate is higher there and so the gap distance between cone and plate is greater to compensate.) The angle is typically about 5°. The determination of the viscosity is similar to that for the rotating cylinder, with appropriate adjustments for geometries.

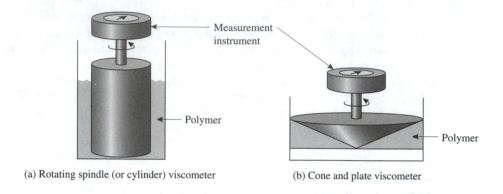

(a) Rotating spindle (or cylinder) viscometer   (b) Cone and plate viscometer

Figure 4.5   Spindle-type and cone-and-plate viscometers.

Another method to determine viscosities uses Zahn and Ford cups, which are small cups with known-sized holes in the bottom. These are filled with the liquid to be tested and then the drain time is measured. These cup tests are not as accurate as other methods and are principally used for quality control measurements within a plant where fluids of similar viscosities are compared. For additional accuracy, the liquid can be pressed through a capillary with a known weight. This method, called capillary extrusion, is similar to the melt index test discussed in the chapter "Microstructures in Polymers."

For viscosities at the high end of the viscosity range (for instance, with melted thermoplastics or elastomers), heated viscosity measurements are often used. These measurements require a machine that is capable of slowly heating the polymer at a very precise rate and simultaneously shearing the polymer. Two types of machines are very common. One is called a rheometer (from the name for the study of viscosity), which uses either a capillary with a pressure system or two plates to shear the polymer (ASTM D 1238, D 1823, and D 3835). The other instrument is a Mooney viscometer (ASTM D 1646, D 1417, and D 3346), which uses a toothed plate inside a die to shear the polymer. They can both be used on a wide variety of materials, but the Mooney is most often used for elastomeric materials. The rheometer is somewhat more flexible in its applications and is used for thermoplastics and, occasionally, for elastomers.

In classical mechanics, the behavior of viscous liquids is related to a *dashpot,* much as the behavior of elastic solids was modeled by a spring. A dashpot is a piston moving in a Newtonian (viscous) fluid. The dashpot behavior depends upon the force applied and the change in the movement (displacement) of the piston, which is in a slow-reacting material. This behavior is described by Equation (4.5),

$$F = b\dot{x} \tag{4.5}$$

where F is the force, $\dot{x}$ is the change in displacement (distance) with time and b is the proportionality factor. This equation for viscous liquids is analogous to Equation (4.2) for elastic solids, in that both describe the behavior of the materials in terms of simple mechanics.

VISCOELASTIC MATERIALS

Viscoelastic Behavior and Definitions

Most polymer materials have some characteristics that are similar to viscous liquids and some that are similar to elastic solids. These materials are therefore known as *viscoelastic*. Viscoelastic materials can be either liquid or solid, although the distinction between liquids and solids in these materials is not a clear one. A representation of this continuum of properties between pure viscous liquid, viscoelastic liquid, viscoelastic solid, and elastic solid is given in Figure 4.6. In this figure the simple mechanical devices which have been previously used to describe viscous liquids (dashpot) and elastic solids (spring) are shown with their appropriate equations. Because the viscoelastic materials are combinations of the viscous liquids and elastic solids, the two simple methods for combining the mechanical elements (in series and in parallel) are also shown. The series arrangement of the spring and dashpot elements behaves much like a viscoelastic liquid *(Maxwell model)*,

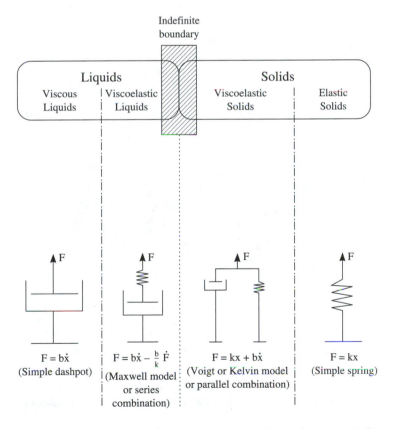

Figure 4.6 Continuum of viscoelastic properties and representations using simple mechanical devices.

whereas the parallel arrangement is more like a viscoelastic solid *(Voigt* or *Kelvin model).* Both of these simple mechanics models can be useful for simple analysis and conceptual purposes but should be considered as approximations to actual systems.

The use of the various models is seen more readily in the different mathematical formulations which arise from the way the models are expressed. These formulations are beyond the scope of this text, but are useful for the advanced student in understanding, predicting, and comparing the behavior of the various materials.

Plots of the responses of a typical solid-like material with elastic behavior, a liquid-like material with viscous flow, and a viscoelastic material with plastic behavior to a given steady force are given in Figure 4.7.

The plot of the applied force (stress) is shown in Figure 4.7a. The plot of response of the solid-like material (Figure 4.7b) indicates a direct and linear response to the applied force. The solid-like material moves (strains) instantaneously to the application of the constant force and continues at this position so long as the force is applied. At the moment the force is relieved, the solid-like material immediately returns to its original posi-

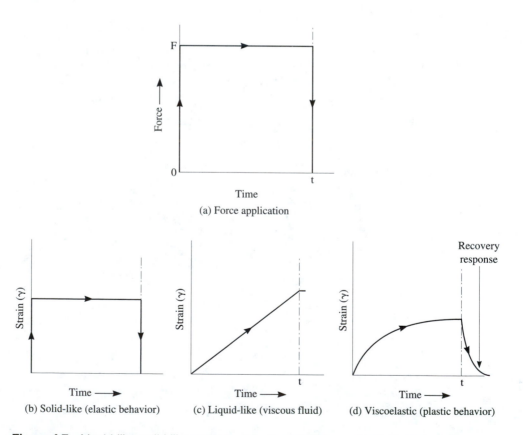

(a) Force application

(b) Solid-like (elastic behavior)   (c) Liquid-like (viscous fluid)   (d) Viscoelastic (plastic behavior)

Figure 4.7 Liquid-like, solid-like, and viscoelastic materials mechanical response to an imposed steady force.

tion. This behavior is like a spring that is instantaneously stretched and held in an elongated position. At the moment the force is relieved, the spring returns to its original, non-stretched shape.

The plot of the response of the liquid-like material possessing viscous flow is shown in Figure 4.7c. In this material the imposition of the steady force begins to deform the material, but the movement gradually increases as time proceeds. This movement will continue so long as the steady force is applied. This behavior is like honey being subjected to a block sliding against it. The honey will begin to flow when the force is first applied, and will continue to move as long as the force is applied. When the force is relieved, the honey will stop moving but will not return to the original position.

The response of a viscoelastic material to an applied steady force is illustrated in Figure 4.7d. The viscoelastic material begins to move immediately upon application of the force, but not as much as the solid-like material, although more than the liquid-like material. The response is, therefore, intermediate between the two other materials. The viscoelastic material will continue to move so long as the force is applied. When the force is stopped, the viscoelastic material will attempt to recover to the original position, but will be slowed in this recovery. That slower recovery response is illustrated in the plot as a sloping, nonlinear curve which continues past the time the force is relieved. Hence, the viscoelastic material is like both a solid and a viscous liquid.

Time Dependence of Solid and Liquid Viscoelastic Materials

The time dependence of viscoelastic materials is an important consideration that significantly affects their behavior. The shear rate (time) dependence of viscoelastic liquids has already been discussed. Most polymer viscoelastic liquids exhibit shear-thinning, and many are thixotropic.

The time dependence of viscoelastic solids is illustrated in Figure 4.8 where the slope (viscosity) of the stress-strain curve is seen to rise with increasing shear rate. At the micro level this time (rate) dependence can be understood in polymeric solids by noting that forces on polymeric systems are resisted by the substantial entanglement of the polymer

Figure 4.8 Time dependence of viscoelastic (polymeric) materials showing the changes in stress-strain with shear rate.

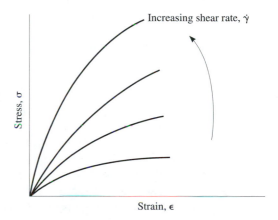

chains and by secondary bonding. (Remember that these intermolecular forces are quite small in liquids and are mostly momentary entanglements that give rise to non-Newtonian behavior, as previously described and illustrated in Figure 4.4.) In solids, however, as energy is added through mechanical forces, the chains will gradually disentangle and the secondary bonds will eventually dissolve, thus allowing the molecules to translate (move) relative to each other to relieve the stress. However, this untangling and sliding takes time to accomplish. Therefore, at higher shear rates, the polymer molecules do not have as much time to move and the disentanglement is forced, resulting in a greater resistance to movement. The result is an apparently stiffer (higher-viscosity) material.

Glass Transition Temperature (Viscoelastic View)

The glass transition has been previously defined as the point (temperature) at which the energy in the system is sufficient to initiate long-range molecular motions, thus imparting a pliable and leathery nature to the solid material. These motions were considered from a thermal viewpoint, but they can also be understood from a viscoelastic viewpoint.

As has been previously explained, solid viscoelastic materials can be thought of as having two behaviors—viscous and elastic. The elastic behavior is typified by a relatively minor deformation of the solid when subjected to a force with a full return of the material to its original shape when the deformation force is stopped. The solid material remains hard and rigid. The energy put into the system is recovered completely when the force is removed. The mechanical analogy is a spring.

Polymer molecules that behave like elastic solids can be thought of as being many small springs. When they are stretched and then released, they return to their original positions. At the micro level, the small springs are the molecules themselves, which are stretched and bent but not displaced relative to their neighbors. No disentanglement occurs. This behavior can exist only if the movements are small. Hence, these are called *short-range movements*.

The viscous nature of a polymer solid can be associated with *long-range movements*. The viscous material will move more freely than an elastic solid when a force is imposed, and all the energy input into the material may not be returned because of permanent deformations (such as disentanglement) or the creation of internal heating.

The long-range movements require more energy to activate than do the short-range movements. Hence, above a certain level of internal energy, the behavior of the material will more likely be dominated by long-range movements, and below this characteristic energy level, the material will exhibit only short-range movements. The point where this occurs is, of course, the glass transition temperature (T_g).

The time for the long-range movements to occur is also longer than the time involved in short-range movements. Therefore, by subjecting the system to stresses over very short times, the long-range effects can be avoided. Hence, the long-term effects can be circumvented by either lower temperatures or shorter times. This is called the *time-temperature superposition*. The reverse of the experiments just explained is also true, that is, higher temperatures can be related to longer times. The earlier example of "silly putty" clearly reflects the time-temperature superposition phenomenon.

PLASTIC (HIGH-STRAIN) STRESS-STRAIN BEHAVIOR

Plastic Behavior and Definitions

Up to this point, the discussion of mechanical properties in solids has focused principally on situations of small strain (small movement) where behavior is elastic. The deviation from ideal elasticity in viscoelastic materials was addressed, but not the behavior of these materials at high strain (large deformations). In testing and in actual use high strains are often encountered. Therefore, the behavior of viscoelastic materials at large deformations is significant. A plot of stress versus strain over the entire range of strains is presented in Figure 4.9.

The elastic region in the stress-strain curve of Figure 4.9 is the initial part of the curve, identified as the linear region where the stress and strain values are small. In this region the modulus is given by the Hooke's law relationship, as shown in Equation (4.2). The point where the stress-strain curve begins to deviate from linearity is called the *proportional limit*. Another key feature of the stress-strain curve (Figure 4.9) is the onset of permanent deformation, which is marked as the *yield point*.

In most materials, the onset of nonlinearity (proportional limit) and the onset of permanent deformation (yield point) are very close to each other. (They are shown quite far apart in Figure 4.9 just to illustrate clearly that they are actually different points.) The onset of nonlinear behavior in polymer materials need not necessarily result in permanent deformations. For instance, elastomeric materials can have highly recoverable properties

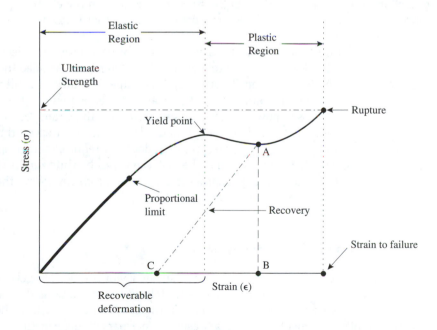

Figure 4.9 Stress-strain behavior over the entire strain range for a typical polymeric material.

(that is, they return to their original shape) but still be nonlinear in their stress-strain behavior.

In structural materials the yield point is important because it marks the upper limit of applied stress from which full recovery of shape occurs when the stress is removed. Imagine a coat hanger that deflects when a normal load (a coat) is placed on it and then, when the coat is removed, the hanger recovers (elastically) to its original shape. However, if an excessively heavy coat is put on it, the yield strength may be exceeded, in which case the hanger would be permanently deformed (although some elastic recovery would be seen, but not enough to regain the original shape). In nonstructural materials the yield point is less critical except when shape retention is important. If the stress is removed before the stress-strain curve reaches the yield point, full recovery will occur by either retracing the same path (full energy recovery) or by returning to the origin by a lower path (energy loss or *hysteresis*). The energy loss path is characteristic of viscoelastic materials, as previously discussed.

Beyond the yield point, the material is permanently deformed. That is, even when the stress is removed, the material will not recover to its original shape. The region beyond the yield point is called the *plastic region*. The nonrecovery of shape is indicated in Figure 4.9 using point *A* as the point where the stress is removed. Point *A* is in the plastic region, that is, beyond the yield point. The loading follows the curve past the yield point and finally reaches point *A*. The strain at point *A* is noted as point *B*. When the stress is removed, the material will recover part of its original shape and move from a strain of *B* to a strain of *C*. (The recovery actually follows the dotted line from *A* to *C*. The dotted line is parallel to the slope of the stress-strain curve in the elastic region.) The distance from the origin to point *C* is the amount of strain that is not recovered or *permanent deformation*.

At some point on the normal stress-strain curve, a maximum stress level is reached. This is the highest stress that can be tolerated by the system and is called the strength or the *ultimate strength*. (For tensile stresses, this value is called the ultimate tensile strength, or, by analogy, the ultimate flexural, compressive, or torsion strength, depending on the type of stress imposed.) If the stress is continued, the strain gets higher but the stress decreases because of the ability of the molecules to slide over each other. This gives movement (strain) in the polymeric structure but does not require increasing amounts of stress. Eventually the maximum strain of the material is reached and *rupture* (breakage or failure) occurs. The strain value at rupture is called the *strain to failure,* that is, the accumulated strain or movement to the point of failure.

Interpretation of Plastic Behavior

The molecular interpretation of elastic and plastic behavior is that in the elastic region the strain results from recoverable movement—stretching out of the twisted molecules in the amorphous regions and minor deformations in the crystalline regions. At the yield point nonrecoverable movements begin that result in permanent deformation. Some of the most common are: disentanglement of the molecules, slip of one molecule past another in a way that the slip could not be recovered, slip along a crystal plane, and formation of a crack. Further input of mechanical energy will lead to continued movement and increased

Figure 4.10   True and engineering stress-strain.

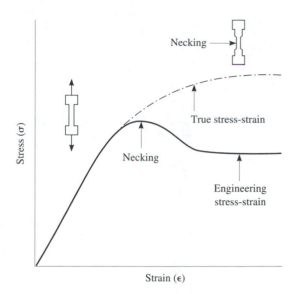

energy in the molecule until the amount of energy in the molecule reaches the level of bond energies. At that point, further increases cause bond rupture and the sample breaks.

The behavior described at the molecular level is illustrated in Figure 4.10 as the *true stress-strain* line. True stress-strain, as opposed to *engineering stress-strain* (the other line in Figure 4.10), is defined so that the strain is the instantaneous value of the stress-strain curve ($d\sigma/d\epsilon$) rather than the change in strain divided by the original strain ($\Delta\epsilon/\epsilon_0$), which is the definition of the engineering stress-strain. The true stress-strain is not used in common practice, but the situation depicted in Figure 4.10 illustrates its usefulness. The true stress-strain curve shows the increasing resistance to the stress as the material deforms plastically. The engineering stress-strain, however, shows a reduction in the stress after the yield point. This reduction is a result of the thinning of the test sample. Because the sample thins (called *necking*), the force required to elongate the sample is greatly reduced and appears as a reduction in the stress-strain curve. In summary, the true stress-strain curve reflects what is actually happening at the molecular level—a continuous increase in the stress—while the engineering stress-strain curve reflects the necking of the sample and better indicates what a plastic part might do in actual use.

Therefore, when the engineering strain-to-failure is reported, the value is often given as a ratio of the amount of deformation compared to the original length. In plastic materials, a related quantity is even more commonly reported. This quantity is the *elongation,* which is simply the engineering strain-to-failure reported as a percentage of the original length of the sample.

Mechanical Model of Plastic Behavior

A mechanical model can be built from classical mechanical elements for the plastic deformation of viscoelastic materials, much like those models already discussed for the elastic

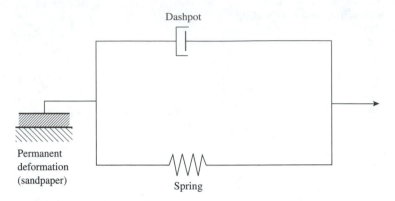

Figure 4.11 Mechanical model for plastic deformation of viscoelastic materials.

and viscous cases. This model, which is shown in Figure 4.11, uses the parallel-element model of the spring and the dashpot and adds an element which represents a permanent deformation. The new element can be thought of as two pieces of sandpaper, one stationary and the other pressed against the first. A minimum force is required to cause these two pieces of sandpaper to slide relative to each other so some small movement can occur in the dashpot and spring without any movement in the sandpaper. If the minimum force that causes the sandpaper to slide is not exceeded, the movement of the system is reversible (returns to the original position when the force is relieved) and corresponds to elastic behavior. When the minimum force is exceeded, sliding of the sandpaper will occur. This sliding is irreversible because the sandpaper will not slide back when the force is removed. This behavior corresponds to plastic behavior.

Tensile Forces

The tensile properties are determined using procedures such as those given in ASTM D 638. Normally the sample is shaped so that the part will break between the grips regions of the tensile test machine. The most common shape for plastic samples is called a "dog-bone" shape and is illustrated in Figure 4.12. The increased width of the sample in the gripping zones ensures that the sample will break within the failure zone.

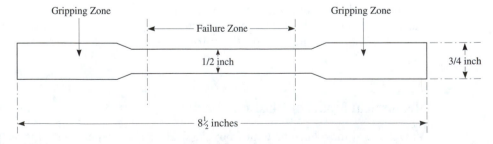

Figure 4.12 Typical tensile specimen shape (dog bone) for plastic samples.

The major plastic tensile properties determined by this test are tensile strength, tensile yield, tensile modulus, and elongation. Because of the viscoelastic nature of plastic materials, care should be taken to ensure that sample results that are to be compared are taken at the same rate of pull and at the same temperature, as these parameters can strongly affect the tensile results. The results of the tensile test are usually expressed as tensile strength (in psi or Pa), tensile modules (in psi or Pa), and elongation (in inch/inch, mm/mm, or %).

Nontensile Forces

The mechanical behavior, elastic and plastic, has been described up to this point in terms of responses to a tensile force applied to the material. However, in Figure 4.1 several forces besides tensile forces were illustrated. All these forces are also important in plastic materials.

Compression Properties.   The compressive force gives information about the strength and stiffness of a columnar sample that is supported vertically, then pressed on its ends as shown in Figure 4.13a. The equations and symbols for stress, strain, and modulus (σ, ϵ, E) developed for the tensile force also apply to the compressive force, but are named compressive strength, compressive modulus, etc. The compressive strength and modulus can be quite different from the tensile strength and modulus because of the difference in the ability of the polymer material to support a columnar load versus a pulling load. In general, compressive strengths and moduli are lower than tensile values for polymers.

Compression properties are measured following procedures such as those given in ASTM D 695. Compressive strength is not as significant in most plastics as are some of the other mechanical properties. Because of the low modulus of most plastics, the compression samples must be quite thick or be supported. The dependance of the test results on sample geometry has led to a recommendation of the testing association that compression values be used for structural studies only if the test samples are the same shape as the parts will be in actual use.

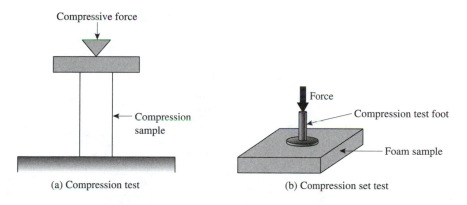

(a) Compression test   (b) Compression set test

Figure 4.13   Compression test methods.

For foam samples, the compression test is very important but is usually run under quite different conditions than would be used with a rigid plastic sample. The compression tester for foam samples uses a compression foot with a large diameter to limit the penetration, and registers quite low loads. The compression test for foams is therefore somewhat unique, as are many other tests associated with foam materials such as open-cell content, compression set, tear-resistance, and the various cushioning tests described in the chapter on foam materials. Test results are compression strength (psi or Pa), compression modules (psi or Pa) and strain-to-failure (inch/inch, mm/mm, or %). This test is illustrated in Figure 4.13b. When the yield point is exceeded in compression, a permanent deformation or compression set occurs. This compression set is imporatnt in characterizing the behavior of flexible foams.

Flexural Properties.   Flexural strength and modulus also use the same symbols as those associated with tension and compression forces. In fact, the bending forces which are used to determine flexural strength and modulus induce tension and compressive forces in the sample as illustrated in Figure 4.12b. It is not surprising, therefore, that flexural strength and modulus are often thought of as combinations of tension and compression.

The flexural properties are determined using procedures like those given in ASTM D 790. Many plastic parts are used in applications where flexural properties are important. For instance, plastic seating must have a minimum flexural strength and modulus or the seat will bend (sag) excessively or break. The flexural sample is a simple rectangular-shaped beam that is placed over two rests or supports and then loaded in the middle of the beam between the supports (see Figure 4.14). The beam is typically $\frac{1}{8} \times \frac{1}{2} \times 4$ inches, although this size can vary for different materials. Results are given as flex strength (psi or Pa), flexural modulus (psi or Pa), and flexural elongation (inch/inch, mm/mm, or %). The flexural modulus is the most important flexural property for plastics.

Torsion and Shear Properties.   The equations describing the torsion and the shear forces illustrated in Figure 4.1 do not use the same symbols as the tensile, compressive and flexural forces, although the equations are of the same form. The stress-strain equation for torsion and for shear, analogous to Equation (4.1), is given in Equation (4.6) as follows:

$$\tau = G\gamma \tag{4.6}$$

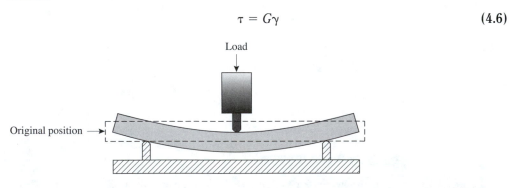

Figure 4.14   Flexural tensile and modulus test schematic.

where τ is the shear stress, G is the shear modulus, and γ is the shear strain. The shear strain, and its time derivative, the shear rate, have already been introduced in this chapter in discussing the shears induced in viscous flow. While the values for the shear strength, shear modulus, and other shear-related values may be quite different from the tensile-related values, the basic concepts regarding the microstructure origin of these parameters are generally analogous.

Blushing.   Plastic deformations from any type of stress will act first on the amorphous portions of the plastic. During the stretching and unfolding of the molecules in these areas, adjacent molecules can be brought into close alignment. This alignment can result in the formation of some crystallinity in that region, especially in simple molecules that align easily. This increased crystallinity can be noted by a change in the index of refraction of the polymer and the newly crystalline area takes on a cloudy appearance. This phenomenon is called *blushing*. The most common occurrence of blushing is when a plastic is severely bent. The blushing appears at the outer edge of the bend where the tensile forces are the strongest and the molecules are stretched the most. The increased crystallinity at this site causes the local modulus of the material to increase. This phenomenon is, therefore, somewhat analogous to work-hardening in metals. If the material is subjected to repeated bending at the same place, the material will eventually break because of the embrittlement from the higher crystallinity.

CREEP

The deformations discussed up to this point have occurred over a relatively short span of time and with relatively large forces being applied. Most viscoelastic materials will also show significant deformations with small loads but over long periods of time. This property is called *creep* and is defined as the gradual deformation of a material under a load that is less than the yield strength of the material. (When loads greater than the yield strength are applied to any material, it will permanently deform.) Creep is illustrated in Figure 4.15 for a simple beam. A load is applied over a long period of time and this load causes the material to deform. When the load is removed, a small amount of the deformation will be immediately recovered (the elastic or short-range portion of the deformation). However, the long-range portion is not immediately recovered. This deformation is much slower to recover and in some cases will not be completely recovered *(permanent set)*.

The amount of creep is strongly dependent upon the amount of the load, the time the load is applied, and the temperature of the material. Increases in any of these parameters will cause the creep deformation to be greater. These relationships can be expressed by the modified stress-strain equation, (4.7),

$$\sigma_0 = E(t,T)\ \epsilon(t,T) \tag{4.7}$$

where σ_0 is the applied stress (usually a constant over the course of the experiment), E is the creep modulus and ϵ is the amount of deformation (strain due to creep). Both the modulus and the deformation are functions of time and temperature. Furthermore,

Figure 4.15 Creep in viscoelastic materials.

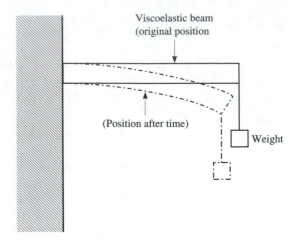

increases in the amount of stress (σ) from experiment to experiment will also increase the amount of deformation.

If a viscoelastic material is loaded for a long period of time, at a high temperature, with a substantial load, the sample will eventually break. This is called *creep rupture,* and it is useful in developing allowable stress limits in plastics applications. If the stress-to-rupture test is performed at several loads, the times for rupture will be different (assuming constant temperature). A plot of the stress level versus the time to failure (usually plotted using a log-log scale) gives the allowable stresses for design considerations with the additional allowance for safety factors. Such a plot is given in Figure 4.16.

Creep results from the stretching and gradual uncoiling of the molecules as they are subjected to the constant stress. Initially only the stretching of the polymer occurs, but with longer times, the molecules gradually uncoil and eventually begin to slip past each other. All of these phenomena result in deformation (strain). Higher applied stress levels

Figure 4.16 Creep-rupture data on log-log plot.

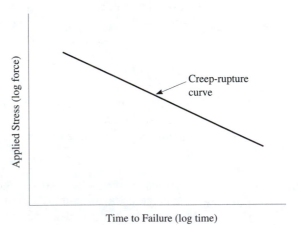

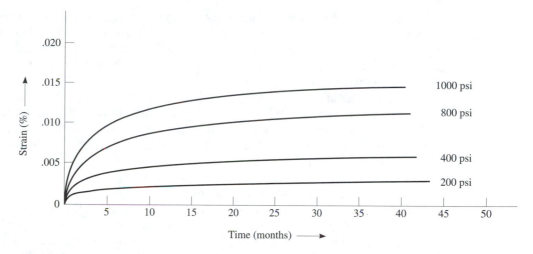

Figure 4.17 Creep data showing increasing creep (strain) with higher applied stresses.

result in more strain because there is more force to give the energy to cause the molecular movement. Likewise, higher temperatures supply more energy that can be used to facilitate molecular movement.

Properties of the plastic that retard molecular motion will reduce the amount of creep. For instance, crystallinity and crosslinking, because of the intermolecular forces, will reduce creep. High molecular weight will also reduce creep because the additional entanglement with higher chain length requires greater disentanglement before movement can occur. Creep is also reduced by large and stiff polymer repeating units because of the greater energy needed to cause these materials to disentangle.

No single ASTM standard applies to creep measurements. The creep tests usually are done by subjecting the sample to the type of force that it will be subjected to in actual use, and then measuring the change in strain or elongation as a function of time. For instance, tensile creep is measured by placing the sample under a specified tensile load and then measuring tensile elongation as a function of time. Because creep is so sensitive to temperature, creep measurements often are conducted at specified temperatures other than room temperature. The results are usually presented as a plot of strain (%) versus time for various loads. The creep results are often presented in a plot like that shown in Figure 4.17.

TOUGHNESS AND IMPACT STRENGTH

The ability of a material to absorb energy without breaking (rupture) is called *toughness*. Toughness is related to the area under the stress-strain curve (such as the area under the engineering stress-strain curve in Figure 4.9) because energy absorption is the summation of all of the force resistance effects within the system. Using calculus, this summation is accomplished by integrating the equation describing the stress-strain curve. Generally, the toughness of a material is much more important when the force is applied suddenly in an impact rather than over a relatively long period of time as it is done in the stress-strain experiment. Therefore, a property called *impact toughness* (sometimes called *impact strength*) is defined as the maximum force that a material can withstand upon sudden impact without rupture.

Impact toughness is strongly dependent upon the ability of the material to move or deform to accommodate the impact. This movement is related to elongation or strain. Therefore, materials which exhibit high elongation are often tough, especially if they also have good strength. However, the modulus is usually low in these materials (low stiffness). Hence, materials with high elongation and low modulus are called tough, and materials with low elongation and high modulus are called brittle.

High molecular weight favors high toughness. This results from the combination of higher strength and better sharing of the impact force along the polymer chain by causing more atoms to rotate, vibrate, and stretch to absorb the energy of the impact. Crystallinity gives higher strength, but lower toughness unless the nature of the backbone changes. For instance, even though nylon is crystalline, the ability of this material to absorb energy by passing the energy from one molecule to another through intermolecular forces makes it very tough. Crosslinking of a brittle polymer will usually decrease toughness because of the increased limitation of motion within the polymer mass caused by the intermolecular bonds, but crosslinking of a flexible polymer will increase impact toughness because of the greater ability of the polymer to transfer the energy.

The Izod and Charpy tests (ASTM D 265) utilize a pendulum-impact testing device (illustrated in Photo 4.1 and Figure 4.18). The sample is clamped in the sample holder either vertically (for Izod) or horizontally (Charpy). The samples may have a notch cut in them to initiate the rupture.

In the Charpy and Izod tests, the pendulum will generally break the sample and continue its path beyond the impact point. Some of the energy in the pendulum will be absorbed in the breaking of the sample; therefore, it will not swing as high after the impact as it started before the impact. The reduction in height (potential energy) can be related directly to the energy absorbed by the sample. A pointer, which is pulled by the pendulum and then stops at the maximum height, records the maximum height of the swing.

Hence, the energy absorbed from the impact can be measured from the height of the pointer. This energy absorbed is a measure of the toughness of the sample. The samples are typically $1/8 \times 1/2 \times 2$ inches but can be of other thicknesses up to $1/2$ inch. The thickness strongly influences the test results, so precise thickness measurement is required. The results are usually expressed as impact toughness (Izod or Charpy) with units of foot-pounds/inch or J/m.

Another test to determine impact toughness is the *falling dart test* (ASTM D 3204 for rigid sheets, D 1709 for PE film, and D 244 for thermoplastic pipe and fittings). In each of these tests the sample is placed under a column that guides a metal dart, tup, or other impacting device onto the sample. The impactor has an accelerometer attached that measures the energy absorbed on impact and thus determines the toughness of the sample. The weight of the impactor, the drop height, and the size of the impactor point can all be varied to accommodate different plastics and different sample thicknesses.

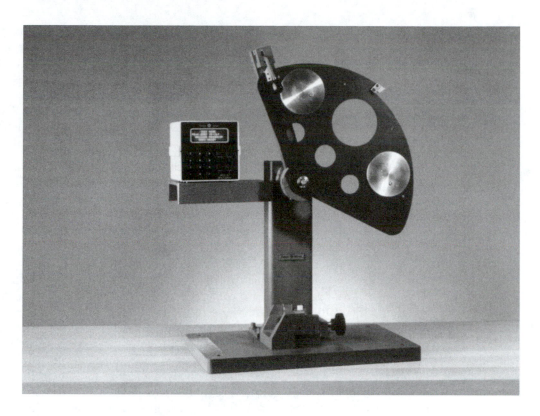

Photo 4.1   Izod/Charpy impact test machines. (Courtesy: Tinius Olsen Company)

Figure 4.18   Charpy/Izod impact toughness tests.

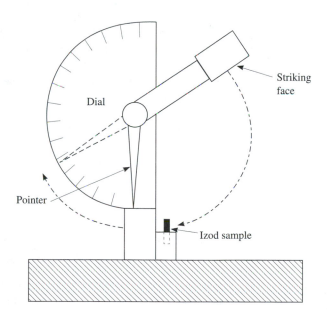

Some tests lack the accelerometer and measure the toughness by noting the drop height and weight of the impactor that cause sample rupture to occur. The test results are expressed as foot-pounds/inch or J/m. In some cases the results are given as the height at which 50% of the samples break. This is called the stair-step procedure.

In some dart-impact testers, the weights of the dart and the heights can both be varied so that subtleties of impact breakage might be explored in more detail. Accelerometers can be attached to the sample to aid in understanding the energy absorption.

Abrasion and Hardness.   Abrasion resistance (ASTM D 1044 and D 1242) is defined as the ability to withstand mechanical action (such as rubbing, scraping, or erosion) that tends progressively to remove material from a surface. Plastics are often placed in environments where surface abrasion occurs. This abrasion can be a performance problem and so the ability of a plastic material to withstand abrasion must be measured. Abrasion, however, is a very complicated phenomenon. Many theories support claims that abrasion is closely related to the frictional force, load, and true area of contact. An increase in one or more of these parameters will usually increase abrasion. Other factors are the test method, type of abrasion, and heat generated. This complex behavior, depending on several factors, makes abrasion difficult to measure. Abrasion resistance is most often measured as the loss in weight of a material when subjected to a standard testing condition. The Taber test is a common abrasion test in which a wheel of known abrasiveness is placed on a sample, weighted, and then rotated around the surface of the sample for a set period of time or set number of cycles (usually 5000). Other methods to measure abrasion include measurement of the loss of optical clarity for transparent samples and loss of volume for samples of fixed dimensions. The samples are generally 4-square-inch plates. The sample is weighed to the nearest mg before and after the test. Results are expressed as weight loss in mg/1000 cycles.

Hardness is defined as the resistance of a material to surface deformation, indentation, or scratching. These several hardness tests (ASTM D 785 for Rockwell, D 2240 for Durometer or Shore, and D2583 for Barcol) are designed for materials of widely differing hardnesses. All of the tests use an indentation method to determine the hardness of a sample, although the type of indentor and the pressures involved vary among the tests. The Rockwell test is for rigid plastics like acetal, polystyrene, and polycarbonate. The Barcol test also measures hardness in very hard plastics and was designed specifically for reinforced and nonreinforced rigid plastics. The Durometer or Shore hardness test is mainly used for measuring the hardness of soft materials. The indentor of the Durometer test is somewhat larger than the indentors of the other tests and so penetration into the softer samples is restricted by the indentor. Tests can be run on actual parts so no special sample is required. Results are given as a number read from the scale on the tester. Reports indicate the tester used and the scale read, such as "20 Shore A hardness."

REINFORCEMENTS

The strength and modulus of many plastic materials can be substantially increased by the addition of strong, high-modulus materials, called *reinforcements*. These reinforcements are often in the shape of fibers (such as fiberglass) because fibers can be made very stiff, primarily in their long direction. In some cases the orientation of the fibers can be controlled so that a substantial number of the fibers are in one or more directions, thus allowing the strength and modulus to be greater in one direction than in others. The elongation of these materials is much less in that same, high-modulus and -strength direction, and toughness is usually also reduced. These materials, which combine plastics with reinforcements, are called *composites* and are discussed in greater detail in the chapter on composite materials and processes.

When the relative volume of the reinforcement fibers is high compared to the volume of the binder or resin (plastic) in the composite material, the mechanical properties of the fibers tend to dominate the mechanical properties of the combined material. This is especially true when the fibers are long because the applied forces tend to be transmitted along the fibers in these cases. When the fibers are short, the volume percentage of fibers is low, or the mechanical properties of the fibers are not significantly higher than the plastic, the properties of the composite will be much more dependent upon the mechanical properties of both the reinforcements and the plastic.

When the reinforcements are long, the processes for manufacturing the composite parts are often quite different from traditional plastic manufacturing processes. Most often, thermoset materials which are not yet crosslinked (and are therefore liquids) are combined with the fibers and then placed into a mold where the materials are heated to form the crosslinks and solidify the part. Plastics with short fibers can, on the other hand, often be incorporated into the plastic material that is in traditional forms (such as pellets) and processed in conventional thermoplastic molding equipment.

FILLERS

Many inert and active materials can be added to a plastic mixture including fillers, plasticizers, colorants, lubricants, and stabilizers. All can be called *additives*. Each of these will be discussed in this and later chapters. When inert materials that do not have any particular reinforcement capability are added to plastics, the materials are called *fillers*. Fillers usually have no particular length-to-width ratio and are therefore not fibers but, rather, particles. Typical examples would be clay, ground limestone, and other powdered or granulated minerals.

The effect on mechanical properties of adding reasonably low concentrations of fillers to the plastic is generally not large, although some minor increased stiffness or reduced strength and reduced elongation is common. The principal purpose for adding fillers is most often to reduce the cost of the overall material. Fillers usually cost much less than the plastic and therefore any combination of the plastic and filler will be less costly in terms of total weight or total volume than the pure plastic material itself. Fillers can also be added to impart special properties to the plastic part, such as flame retardance, color, and opacity (no light transmission).

Fillers are usually mixed into the liquid or molten plastic material. This can be done by the plastic manufacturer, a special compounder whose expertise is the addition of materials to plastics for resale, or at the site where the plastic part is actually made. Extrusion is the most common method for incorporating fillers into molten plastics.

TOUGHNESS MODIFIERS

Just as materials can be added to plastics to increase mechanical strength and modulus (reinforcements), materials can also be added to increase toughness. These *toughness modifiers* or *toughness enhancers* are materials that have high toughness themselves, typically with very high elongations. The most common are rubbers and when added as toughness modifiers, they are usually in small particulate form and would be mixed into the melted plastic material just as a filler might be. Other rubbers can be added as monomers or as short polymers in the polymerization reaction of the base polymer.

When thoroughly mixed with the plastic material, these toughness modifiers give additional movement when the plastic is impacted, thus improving the ability of the material to absorb the impact energy and, consequently, improving the toughness. The effect of toughness modifiers is, therefore, roughly opposite that of reinforcements.

An example of copolymerizing toughness modifiers is ABS. This plastic is based on polystyrene but has two rubber modifiers added—polybutadiene and polyacrylonitrile. The resulting performance of ABS is much tougher than non-modified polystyrene.

Other materials can be added to plastics to cause a change in their mechanical properties. The action of these materials is, however, generally chemical or physical and not specifically mechanical. These are discussed principally in the chapter on chemical and physical properties.

=============== CASE STUDY 4.1 ===============

Testing of Trash Containers to Predict In-use Performance

Making large trash containers (carts) with wheels that allow the container to be rolled from a storage location to a pickup point is a major business in the United States and Europe. These carts are generally made of polyethylene by several different manufacturing processes, including blow molding, injection molding, and rotational molding. (Each of these methods is discussed later in this text.) These trash carts are usually purchased by the municipality or by a contract trash hauler and must last for several years to be economically justified. The normal expected life is usually rated as greater than 10 years.

The purchasers and manufacturers of these carts need to know whether a particular cart design, resin type, or manufacturing process will perform properly during the entire expected life. Therefore, a set of tests has been developed to investigate whether the carts can meet the minimum use standard. These tests are designed to examine the properties of the carts that are most likely to be affected by prolonged use. In many cases, these tests specifically address the important issues that have been discussed in this chapter such as viscoelastic behavior, yield strength and permanent set, toughness as a function of time and temperature, and blushing which can lead to brittle failure.

A list of the standard tests for trash carts and a brief explanation of the test is given in Table 4.1.

The **squeeze/lift test** examines the ability of the plastic material to withstand repeated flexural bending. Some carts could develop blushing from this test and that could result in embrittlement and premature failure. Other components of the cart such as wheels, axles, and lids are also checked by this test.

Table 4.1 Trash Cart Performance Tests

Test	Procedure	Comments
Squeeze/lift	Load the cart with a standard weight of sand bags. Use a standard squeezing/lifting device to lift and dump the cart 520 times.	520 cycles represent a lifetime of normal use. (1 lift per week for 10 years.) The cart should remain serviceable during and after the test.
Drop (sand)	Fill the cart with standard sand and drop from 5 feet.	Simulates dropping a cart off a porch.
Drop (water)	Fill the cart with water and drop from 5 feet.	Simulates performance beyond normal use but provides information on resins and designs.
Dart impact	Drop a 20-pound dart on cart samples that have been frozen at $-40°$.	Investigates the toughness of the material at low temperatures as might occur when struck by a snow plow.
Crush	Place the cart against a solid wall and then push the cart to within 1 foot of the wall and hold for 1 minute and release. Then measure the shape recovery.	Examines the tendency of the material to take a permanent set.
Abrasion	Fill the cart with a standard amount of water and drag along the ground until a leak appears.	Examines the toughness of the resin.
Tensile and flex strength, modulus and elongation	Cut samples from a cart that has been squeeze/lifted for 520 times and then determine the tensile and flexural properties. Compare to a nonexercised cart.	Determines the changes in properties that occur after a normal life of use.
Other tests	Various tests examine other properties of the carts such as shape, ease of movement, etc.	These are generally related to performance other than mechanical behavior.

The **drop (sand) test** is a check on the impact toughness of the cart during normal use. The loading is moderate and the height is also moderate. If the resin is too brittle, cracking can occur in this test. Designs which induce stress in the part are also subject to premature failure because the stresses from the drop will often localize in these high-stress areas. Wheels and axles are also tested.

The **drop (water) test** investigates the toughness of the cart under extreme conditions. Failures in this test are most often corrected by lowering the melt index or by significant changes in cart design. In some cases the density of the resin can be so high that the cart will fail at any reasonable melt index.

The **dart impact test** is also a toughness test but in this case the samples are frozen. Even plastics that have a glass transition temperature below $-40°$ will rupture when subjected to sudden impacts at low temperatures. This test measures the ability of the material to withstand such impacts. High-density materials are both stronger and more brittle than low-density materials. Hence, this test effectively investigates which property dominates at low temperature.

The **crush test** simply determines whether the sample will recover from a significant deformation. Amorphous materials recover better than crystalline because the amorphous regions are more flexible and will not have their yield deformation exceeded as readily.

The **abrasion test** looks at toughness at a fairly high and continuous rate. As the polymer heats, the abrasion resistance will often decrease.

Tensile and flexural properties for exercised and nonexercised carts reveal the amount of change that has occurred during the squeeze/lift test. When the change is large, it indicates that the material is near its performance limit. In some cases the modulus of the exercised material will be higher than the modulus of the nonexercised material. This situation indicates that some embrittlement has occurred. A loss of tensile elongation indicates that the polymer chain structure has begun to break down.

SUMMARY

Although polymers do not behave exactly like classical elastic solids or exactly like classical viscous fluids, the principles used in classical mechanics can be modified slightly for use with polymers. Polymers have characteristics both of elastic solids and of viscous fluids and are therefore called viscoelastic materials. The equations that describe viscoelastic materials give insight into the mechanical behavior of polymers and allow their behavior in mechanical tests to be predicted and understood.

The viscoelastic nature of polymers can be seen when a force is applied to a polymer sample. Upon imposition of the force, the polymer will deform much like an elastic solid. But when the force is removed, the polymer may recover its shape immediately, as would an elastic solid, or may recover its shape slowly or, perhaps, not recover the shape completely, as would a viscous fluid. This latter behavior indicates that some of the energy input to the sample by the force is not recovered, but is dissipated in the polymer sample by internal heating.

Mechanical properties for all five of the common forces (tensile, compression, flexure, torsion, and shear) are usually determined using standard tests (most often with procedures developed by national testing agencies such as ASTM). Tests have also been developed by companies, government agencies, and various researchers to meet specific performance requirements.

Another example of viscoelastic behavior is seen when a stress-strain test is performed at different stress rates. As the stress rate increases, the polymer molecules have less time to move and accommodate the applied stress. Therefore, the polymer becomes stiffer and stronger as the stress rate increases.

Viscosity is an important property of liquid, molten, or dissolved polymers and it is measured by a variety of methods, some for low-viscosity situations and others for high viscosities. The viscosity of polymers will often change in non-Newtonian ways, that is, the stress will change nonlinearly with changes in shear rate.

Creep is another property of polymers that is explained by considering the nature of polymer materials. When a small load is applied to a polymer for a very long time, especially at a high temperature, the polymer molecules are forced to gradually disentangle or gradually move relative to one another in a nonrecoverable way. This results in a gradual nonrecoverable elongation of the polymer called creep. If the force is applied long enough, rupture can occur.

Toughness is a measure of the ability of the material to absorb energy when a force is applied. When that force is applied quickly, the impact toughness is measured. The viscoelastic nature of polymers suggests that faster impacts are not tolerated as well as slower impacts. Hence, polymers which are able to move and adjust to absorb the input energy are tougher than those stiffer polymers which cannot adjust. The speed of adjustment depends upon temperature as well as the nature of the polymer chains themselves.

Polymers can be modified to increase strength by the addition of reinforcements. These reinforcements are usually very strong and stiff fibers. When a load is applied to a polymer in which reinforcements are present, much of the load is transferred by the polymer material onto the reinforcement and the net result is an increase in the strength and stiffness of the entire mixture.

Materials can also be added to polymers to increase toughness. In this case the materials added are those which will absorb energy. Rubbers are often the materials chosen as impact enhancers.

Fillers are materials added to polymers which are neither reinforcements nor impact enhancers. These materials are generally added to reduce the cost of the total material. In many cases the changes in mechanical properties due to the addition of the fillers is not important in the application.

GLOSSARY

Abrasion   Loss of material from rubbing or scraping its surface.

Additives   Fillers, plasticizers, lubricants, stabilizers, colorants, and other materials included in the plastic material mix.

Bending force   A flexure force.

Blushing   A localized whitening of a plastic material indicative of alignment of molecules from local stresses.

Charpy toughness   The toughness of a material as determined by using the Charpy test (falling pendulum on a horizontal sample).

Composites   Plastics into which reinforcements have been added.

Compressive force   A pushing force on the end of a columnar sample.

Creep   Long-term plastic deformation.

Creep rupture   Breaking of the material due to creep.

Dashpot   A mechanical device that is used in classical mechanics as the model for a viscous fluid.

Dilatant (shear-thickening)   Non-Newtonian materials that deviate from the linear viscosity equation by increasing viscosity with an increase in shear rate.

Elastic solids   Materials that completely recover their shape and restore the energy imparted to them after an imposed force has been removed.

Elongation   The strain for a tension measurement, usually expressed as the strain at failure as a percentage of the original length of the sample.

Engineering stress-strain   The strain defined as the distance elongated divided by the original length.

Fillers   Particulate additives to plastics.

Flexural force   A pressing force on the middle of a supported long sample.

Hardness   Resistance of a material to surface penetration.

Hookean   A material in which the displacement is linear with an applied force, thus obeying Hooke's law.

Impact toughness (impact strength)   The ability of a material to absorb the energy of an impact.

Intrinsic viscosity [η]   The viscosity of an infinitely dilute solution of a polymer; can be used to calculate molecular weight.

Izod toughness   The toughness as determined using the Izod test (falling pendulum on a vertically held sample).

Long-range movements   Long displacements of the molecules in a solid polymer that would be typical of viscous behavior and occur above the T_g.

Maxwell model   A representation of viscoelastic materials using a dashpot and a spring in series.

Modulus   The ratio of stress to strain in the elastic region.

Necking   A thinning of the sample during tensile testing.

Newtonian   Fluids that have a linear relationship between stress and velocity gradient or shear rate.

Permanent deformation or set   The amount of strain not recovered when a stress is removed past the yield point of the material.

Plastic region   The region of the stress-strain curve beyond the yield point.

Poise   The unit of viscosity, expressed as $g \cdot cm^{-1} \cdot s^{-1}$.

Proportional limit   The point in the stress-strain curve where nonlinear behavior begins.

Reinforcements   Additives, usually fibers, that increase the strength and/or stiffness of materials.

Relative viscosity   The ratio of the viscosity of the polymer solution to the pure solvent.

Rheometry   The study of the viscosity of a material, especially as related to time.

Shear   A force that causes one portion of a material to move past another portion of the material.

Shear rate   The change in shear with time.

Shear-thinning (pseudoplastic)   Non-Newtonian materials that deviate from the linear viscosity law by decreasing viscosity with an increase in shear rate.

Short-range movements   Small displacements of the molecules in a solid polymer that are characteristic of the elastic region of the stress-strain curve.

Stiff   A material characteristic related to modulus.

Strain   The displacement (stretching) of a sample due to an imposed force.

Stress-strain curves (or diagrams)   Plots of the stress (y axis) and strain (x axis).

Tensile force   A pulling force.

Thixotropic   Materials that decrease in viscosity with time even though the shear rate is constant.

Torsional force   A force that presses on a sample, causing one portion of the sample to rotate about a fixed portion of the sample.

Toughness   The ability of a material to absorb energy without rupture.

Toughness modifiers   Material (often elastomers) added to a plastic to increase its toughness.

True stress-strain   The behavior of a material during a stress-strain experiment, where the strain is defined as the instantaneous value of the stress-strain curve.

Ultimate strength   The highest stress that a material can withstand without breaking.

Velocity gradient   The change in velocity with distance.

Viscoelastic   Materials that possess properties of both viscous fluids and elastic solids.

Viscosity (η)   The resistance of a fluid to movement.

Viscous fluids   Materials that flow when exposed to an imposed shear force and do not return any of the imposed energy when the force is removed.

Voigt or Kelvin model   A representation of viscoelastic materials using a dashpot and a spring in parallel.

Yield point   The point in the stress-strain curve where permanent deformation begins.

Young's modulus   A tensile modulus.

Weighting factor   A number assigned to a property based upon the importance of that property in the performance of a part; used in the design process for choosing a material.

QUESTIONS

1. Explain what is meant by a viscoelastic material and relate its response to applied stresses by comparing the material with an elastic solid and a viscous fluid.
2. Explain plastic deformation in terms of molecular chains.
3. Using a molecular view, explain why compression strength is generally less than tensile strength in polymers.
4. How is modulus likely to be affected when toughness modifiers are added to a plastic? Why?
5. Discuss how creep is likely to be affected by aromatic pendant groups.
6. Describe the Izod toughness test.
7. What is ASTM and why is such an organization important?

8. Why is the flexural modulus so important for many plastic applications?

9. A stress-strain experiment was done on a plastic and the following data was noted on the plot of the experiment: ultimate tensile strength = 9000 psi, yield strength = 5000 psi, proportional limit = 4000 psi, strain-to-failure = .025, strain at yield = .020, strain at proportional limit = .015. The original length of the sample was 4 inches. What was the modulus?

10. What is the purpose of notching Izod or Charpy samples?

11. What is the likely affect on modulus of adding a filler? Why?

REFERENCES

Driver, Walter E., *Plastics Chemistry and Technology*, New York: Van Nostrand Reinhold Company, 1979.

Rodriguez, Ferdinand, *Principles of Polymer Systems*, Washington: Hemisphere Publishing Corporation, 1982.

Rosen, Stephen L, *Fundamental Principles of Polymeric Materials*, New York: John Wiley & Sons, 1982.

Young, R. J., and P. A. Lovell, *Introduction to Polymers*. London: Chapman & Hall, 1991.

CHAPTER FIVE

CHEMICAL AND PHYSICAL PROPERTIES (MACRO VIEWPOINT)

CHAPTER OVERVIEW

This chapter examines the following concepts:

- Environmental resistance and weathering
- Chemical resistivity and solubility (physical property effects on solvent-solute interactions, thermodynamics of solvent interactions, plasticizers, solvent welding, environmental stress cracking and crazing)
- Permeability (definitions and general principles, diffusion constant, Fick's laws of diffusion, barrier properties of plastics)
- Electrical properties (resistivity, dielectric strength, arc resistance, dielectric constant, and dissipation factor)
- Optical properties (light transmission, colorants, and surface reflectance)
- Flammability

INTRODUCTION

In contrast to the previous chapter, the properties discussed in this chapter will not relate directly to the reaction of plastics to imposed mechanical forces, but will relate to the response of plastic materials to various environments. The environments may be unusual, such as high heat or high electrical voltage, or quite common such as sunlight or water. The view will principally be the macro view, although in all of these properties the molecular nature, the microstructure, and the macrostructure of the plastic material will have some effect on the material's behavior. In each case the response of the plastic material to each environment and the reasons behind that response will be discussed.

ENVIRONMENTAL RESISTANCE AND WEATHERING

Many applications require that plastics retain critical properties, such as strength, toughness, or appearance, during and after exposure to natural environmental conditions.

Environmental agents that may damage plastic materials include solar radiation, microorganisms, ozone and oxygen, thermal energy, pollution and industrial chemicals, and water.

Weathering is the degradation of the material from the long-term effects of the environment. Ultraviolet (UV) degradation from sunlight is, perhaps, the most significant type of environmental degradation for plastic materials, although oxidation and the phenomenon called stress cracking are also important. Attack by water and other solvents, which is also important, will be discussed in the section on solubility and resistance to chemicals.

Ultraviolet (UV) Light Degradation

The absorption of UV light, chiefly from sunlight, degrades polymers in two ways. First, the UV light adds thermal energy to the polymer, just as any heating would do, causing thermal degradation. Second, the UV light excites the electrons in the covalent bonds of the polymer. This excitation raises the electrons to a higher energy level. These higher energy electrons are less restricted to the particular covalent bond in which they are located. Hence, the bond is weaker and can be broken more readily.

One major difference between the energy from sunlight and the energy from thermal heating is that sunlight is made up of a wider energy spectrum of different intensities. These different intensities are characterized by different frequencies of vibration of the light energy. The energy of light is directly proportional to the frequency according to Equation (5.1):

$$E = hv \qquad (5.1)$$

where E is the energy, h is a constant (called Planck's constant) and v is the frequency of the light vibrations. The UV frequencies are near the higher end of the sunlight spectrum and therefore have more intense energy than the other parts of sunlight (visible and infrared light). UV light is especially damaging to plastics because the energy intensity closely matches the energy levels in the bonds between most of the atoms. Hence, exposure to UV light can result in enough energy to cause many bonds to rupture. This rupture can result in many of the same phenomena associated with thermal degradation—discoloration, loss of mechanical and thermal properties, especially embrittlement, and occasionally, evolution of gaseous by-products.

Some materials absorb UV light more readily than polymers and can, therefore, be added to polymers to improve the resistance of the polymer to UV light degradation. The most common of these materials is *carbon black,* which is made by the incomplete combustion of oils or other organic fuels and is similar to soot. This powdery black material has the advantage of relatively low cost and high UV absorption efficiency. The principal negative effect is the intense black color that is imparted to any material containing even small amounts (such as 2%) carbon black. Therefore, other UV-absorbing materials (such as Irganox, a commercial energy-absorbing additive) have also been developed for applications that require colorless or nonblack colored plastics. Many UV-absorbency materials are eventually consumed because they absorb the UV light by combining with it. Most of the chemical materials, such as Irganox, are of this type.

The tendency of plastics to degrade with exposure to UV light varies widely. For instance, polyethylene film that has no UV absorbent added, if left in the sunlight, will degrade to a powder in 2 or 3 months. On the other hand, acrylics, especially PMMA, will show almost no effect from UV degradation even after years of exposure. Tests to measure the effects of UV exposure are usually related to the application of the plastic material. For instance, acrylics are often used for outdoor signs and therefore must remain clear and unyellowed. A typical test is to measure the loss of light transmission after UV exposure.

The method of exposure is also a matter of concern. Ideally the material would be exposed to normal sunlight and then be tested. However, this type of exposure often takes long periods of time (years) for significant effects to be seen. Therefore, methods to accelerate the exposure have been developed (ASTM G 53 for exposure to UV lamps, D 1499 for exposure to carbon arc lamps, D 2565 for exposure to xenon arc lamps, and D 1435 for outdoor weathering). In these methods, special lamps are used which emit UV light in a spectrum closely matched to that of normal sunlight. The plastic material samples are, therefore, exposed to this artificial sunlight in chambers which can also simulate temperature and humidity conditions appropriate for the particular application. Some caution should be exercised in extrapolating the results from accelerated tests to normal outdoor exposure. The variations of outdoor tests are so broad that good statistical correlations have rarely been accomplished. Therefore, accelerated UV testing should be viewed as an indication of the actual weathering performance, but not as a simulation of that performance. A typical commercial UV exposure chamber is shown in Photo 5.1.

Photo 5.1   A commercial UV exposure chamber (Courtesy Q-Panel Lab Products)

The tests are generally run by measuring some critical performance property (such as elongation or optical clarity) on samples before exposure. Then, either the same samples or equivalent samples are subjected to weathering and then are removed and retested. The loss in property value is noted as the test result. It is not unusual to make several measurements after different exposure periods so that a plot of property loss as a function of exposure time can be made and, hopefully, extrapolated to values of the property considered critical for performance. Many attempts have been made to correlate exposure in an accelerated test and actual outdoor exposure. These correlation attempts have been largely unsuccessful, chiefly because of the high variability of actual outdoor exposure. Nevertheless, approximate correlation values are often given.

Resistance to Microorganisms

Some plastic materials, especially those that have their origins in natural products or that have natural products mixed with the plastic, are potentially susceptible to degradation by microorganisms. Other plastics, although not degraded by the microorganisms, serve as a site for growth of the organisms (such as the mold that grows on bathroom grout). In either case, the presence of the microorganism is objectionable.

Tests have been developed for monitoring the resistance of plastics to microorganisms (ASTM G21 for fungi and G 22 for bacteria). In these tests, plastic samples are placed in petri dishes containing an appropriate nutrient salts agar. The samples are sprayed with either a fungi dispersion or a bacterial dispersion and then allowed to incubate, usually for 21 days. The samples are then removed from the petri dishes and washed free of growth material. The key performance properties are then measured and results are given as losses of performance relative to the original values.

Oxidation

Degradation occurs when the electrons in a polymeric bond are so strongly attracted to another atom or molecule (called a "foreign" atom or molecule) outside the bond that the polymer bond breaks. The foreign molecules are often part of the environment surrounding the plastic material and their effects are, therefore, part of weathering. The most common foreign molecule that has a significant effect on plastics is oxygen (due to its high reactivity and relative abundance). The process of reaction with oxygen is called *oxidation*.

The results of oxidation are similar to other instances of polymer degradation—loss of mechanical and physical properties, embrittlement, discoloration, etc. The degradation from oxidation generally occurs over long periods as the foreign molecule (oxygen) diffuses into the polymer and attacks the bonds. However, raising the temperature will often significantly increase the rate of oxidation, as will increases in the concentration of the foreign molecule surrounding the plastic material. This is another example of the effect of temperature as described by the Arrhenius equation.

Some plastics are more susceptible to oxidation than others. Reasons for greater tendency for oxidation are (1) the lower energy of the bonds in the plastic material, (2) the greater attraction of the electrons for the foreign molecule (because of the inherent nature of the other polymer bonds), and (3) the more open nature of the polymer structure (that is, its greater ability to allow the foreign molecule to enter its structure).

As with thermal and UV degradation, materials have been developed which preferentially absorb oxygen or other foreign molecules and can be added to polymers to reduce oxidation of the plastic material. These materials are especially important during processing because of the higher temperatures encountered at that time which accelerate the oxidation as described in the Arrhenius equation. In cases where the plastic material is especially susceptible to oxidation or where prevention of oxidation is especially important, processing is sometimes done in a less reactive atmosphere. For instance, the high temperatures and high pressures often used for the curing of thermosets for aerospace applications would normally result in high oxidation. Therefore, these materials are often cured in a nitrogen atmosphere.

Testing for oxidation of plastics is often closely associated with thermal stability because the oxidation is performed at high temperatures in order to accelerate the times involved.

CHEMICAL RESISTIVITY AND SOLUBILITY

Chemicals affect plastics in a variety of ways. The lowest interaction level is no perceptible interaction with the plastic material at all, such as water with polyethylene. In other cases the chemical may swell or soften the plastic material such as water with nylon or ketones with PVC. In a few cases the chemical may dissolve the plastic material such as water and polyvinyl alcohol. Finally, the chemical may react with the plastic material to permanently alter the nature of the plastic material and form new bonds, as with strong acids on cellulosics. These are only examples of the many interactions possible between plastics and various chemicals, and all shades of these various plastics/chemical interactions are possible across the spectrum of plastics and chemicals which might be in their environment. A chart indicating some of these interactions is given in Figure 5.1.

The nature of the interaction between a plastic material and a chemical is usually discussed in terms of solvents and solutes, where the chemical is the *solvent* and the plastic

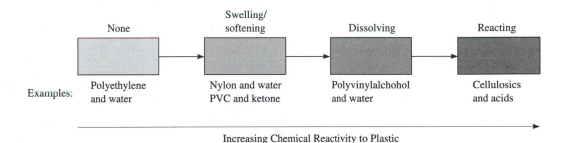

Figure 5.1   Types of solvent effects on various plastics.

material is the *solute*. (A solute is the material that is being dissolved by the solvent.) These terms are applied even when no actual dissolving takes place, such as with swelling.

The interactions are dependent upon both the chemical and physical natures of the solvent and of the plastic material, with the chemical natures the most important.

Chemical Nature and Solvent-solute Interactions

A strong interaction between a plastic material (solute) and a chemical (solvent) is favored when the chemical natures of the materials are similar. For instance, if the solvent is highly polar, such as water, and the plastic material has many polar groups, such as nylon, the plastic material is likely to be strongly affected by the chemical. This observation of chemical similarity in dissolving phenomena has been expressed in traditional chemistry as "like dissolves like" which implies that polar solvents strongly affect polar plastics and nonpolar solvents have a strong interaction with nonpolar plastics. Even though the nature of the interaction between plastic and solvent is often less than total dissolution, the rule is convenient for remembering the basic nature of the interactions.

The lowest level of interaction is, of course, no interaction at all. This occurs when the chemical natures of the plastic and the solvent are vastly different. As the chemical natures of the plastic and solvent become more alike, the first level of interaction is usually a swelling of the plastic material. The similar chemical natures favor the formation of secondary bonds, such as hydrogen bonds or van der Waals forces between various sites on the polymer and the solvent. Often many solvent molecules will crowd around these sites within the plastic material structure that are chemically favorable for bonding with the solvent. The energy of solvation gained from this process is sufficient to open up the physical structure of the plastic material and cause it to swell. Interactions of this type are represented in Figure 5.2, which shows several water molecules surrounding the polar locations in a nylon polymer. The forces that expand the structure of the polymeric material because of these many solvent molecules is obvious. (Refer to the chapter on polymeric materials (molecular structure) for a discussion of polar molecules and an illustration of polar atoms in nylon.)

Figure 5.2 Solvent interactions with a polymer.

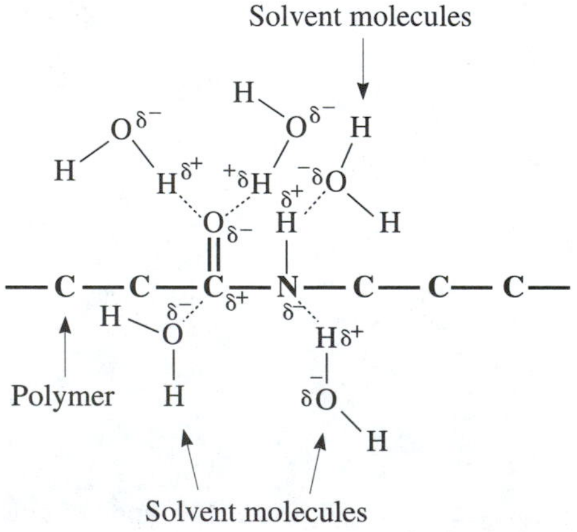

This initial level of interaction between solvent and plastic, which is labeled in Figure 5.1 as swelling/softening, is also called *plasticizing* and is very important for some plastic materials. PVC is a good example of such a plastic. Although virgin PVC is stiff and hard, when certain chemicals are added a distinct softening of the PVC occurs, rendering the plastic pliable and soft. This important property of plastics is discussed in greater detail later in this chapter, where the nature of the chemical additives and the nature of their interactions is examined.

As the nature between solvent and solute becomes more similar, the interactions between a polymer and a solvent can lead to even more solvent molecules crowding around the favorable sites and can eventually result in a breaking of the secondary bonds between the polymer chains and forcing disentanglement. When this occurs, the polymer chains can move freely with respect to each other, and the polymer is said to be in solution. This requires, of course, that a sufficient amount of the solvent be present to surround the polymer chains so that they "float."

An example of complete polymer dissolution is polyvinyl alcohol in water. This property has been used for many years by using polyvinyl alcohol as a protective coating on fibers, thus allowing them to be woven, knitted, and otherwise made into fabrics without excessive breakage on the textile processing equipment. Then, after the fabric has been formed, the polyvinyl alcohol coating is removed by simple hot-water washing of the cloth. This allows the fabric then to be colored or printed. The same principle is used in swimming pool chlorination packets, which are made by bagging the chlorination chemical inside a polyvinyl alcohol bag. The pool is easily chlorinated by tossing the bagged chlorine into the pool, where the polyvinyl alcohol plastic bag dissolves and releases the chlorination chemical.

The nature of these polymer solutions is quite different in many ways from solutions of small molecules (such as the much higher viscosity of the polymer solutions and their non-Newtonian flow). In most actual plastic applications, complete solvation of the plastic material is rare, because the amount of solvent is usually not sufficient to complete the solvation and because the process may take considerable time. Therefore, when the solvent and solute are similar in polarity, swelling of the plastic material is the most common solvation effect.

In a few cases, the chemical will react with the plastic material and form permanent, covalent bonds between the chemical and the plastic material rather than the secondary bonds characteristic of ordinary solvation. In some cases the reactivity of a polymer to a chemical is part of the process for forming some plastics. Some examples are the reaction of cellulose with nitric acid to form cellulose nitrate or the reaction of cellulose with acetic acid in the formation of cellophane.

If the plastic material is in use and if it comes into contact with a chemical that reacts with it, the chemical reaction between the plastic material and its chemical environment can be catastrophic because the nature of the plastic material can be altered greatly. The cases of chemical reaction are, however, rare in actual use except for some incidental occurrences such as staining of carpets or splattering of chemicals on various plastic material surfaces.

In general, plastic materials are dominated by nonpolar carbon-hydrogen bonds. Therefore, plastics are usually resistant to polar solvents such as water. The resistance of plastics to water is a decided benefit in resisting environmental effects because water is so prevalent in the environment. Even when the polymer has polar groups, such as nylon, the interaction with water is limited to minor swelling and softening. These interactions are stronger if temperatures are increased because the solvent can penetrate faster and farther into the plastic material. The interactions are more evident in applications where a property that might be degraded by the presence of the solvent is important. An example of this is the excessive abrasion of nylon guide strips in a soft drink bottling plant. The nylon strips are used to guide the cans or bottles along a conveyor. However, the conveyor system, including the nylon guide strips, is often under hot water which swells and softens the nylon. The rubbing of the bottles or cans against the swelled and, therefore, softened nylon strips causes excessive wear.

The general resistance to water and the nonconductivity of plastics (which will be discussed in more detail when the electrical properties of plastics are examined) means that galvanic corrosion is not a problem in plastics as it is in metals. Galvanic corrosion is responsible for rusting, pitting, and many other degradation phenomena common in metals. To this extent, plastics perform much like certain alloyed metals such as stainless steel.

Physical Property Effects on Solvent-solute Interactions

Although generally not as important as the chemical similarities or dissimilarities of the polymer and solvent, the physical natures of the polymer and solvent can also influence how the polymer might be affected by chemicals. The most important physical characteristic of the solvent is size. Generally smaller molecules are more aggressive as solvents than are larger molecules. This is simply a reflection of the ability of the small solvents to penetrate the polymeric structure. For instance, consider the differences in solvating capabilities in hexane, wax, and polyethylene, which have the same basic chemical nature but differ in the size of the molecules. Hexane will attack many plastic materials, causing them to swell. Melted wax molecules are approximately 10 times larger than hexane molecules and have little solvating effect. Melted polyethylene, which has molecules that are hundreds or thousands of times larger than hexane, has no solvating effect. The same effects are seen when these materials are the solute. Hexane is easily dissolved by other nonpolar solvents, waxes are dissolved slowly, and polyethylene is almost insensitive to solvation.

The swelling phenomenon is especially sensitive to the physical nature of the plastic material. Swelling occurs because the structures of plastic materials are relatively open (that is, there is more void space) in comparison to metals and ceramics, as evidenced by the low density of plastic materials. (Plastic material specific gravities [densities] are typically 0.9 to 2.0 g/cc whereas steel is 7.8 g/cc.) This open structure in plastic materials allows the solvent molecules to penetrate and become attached to the appropriate sites along the polymer chains. Furthermore, the low density also implies that there are relatively few secondary bonds to be broken or stretched in order to accommodate the swelling.

Even though plastic materials have low densities compared to many other materials, some plastics have significantly higher densities than others because of higher crystallinity. If the plastic is highly crystalline, the structure is more compact and the solvent cannot penetrate as readily. Moreover, the secondary bonds in crystalline plastics are more numerous and stronger than in amorphous plastics, thus making penetration by the solvent more difficult. Hence, highly crystalline plastics are more resistant to solvents than are largely amorphous plastics.

The swelling from solvent absorption results in a decrease in the tensile strength and modulus of the plastic material because the molecules are further apart, the secondary bonds are weaker or in some cases broken, and the polymer chains can, therefore, slide more easily over each other. Other properties, such as elongation and toughness, are often increased by the addition of a small amount of solvent but will eventually decrease with further increases in absorbed solvent.

Because of the importance of assessing the chemical resistance of plastics, especially in applications such as process pipes, liners for processing equipment, and environments where particular chemicals are common, several tests have been developed to measure the chemical resistance of plastic materials. Three tests are especially common:

- Immersion test (ASTM D 543). The simplest and most widely used test for determining the resistance of a plastic to a chemical is the immersion test. The test is conducted by carefully weighing and measuring the dimensions of a plastic sample and then immersing the sample in a solvent under specified conditions. After seven days the sample is removed, reweighed, and remeasured. Changes in weight and dimensions are noted. The sample is also examined for other effects of chemical attack such as dissolved plastic, plastization (usually weakened mechanical properties), cracking, warping, swelling, embrittling, etching, discoloring, crazing and other changes in appearance. Generally, the results are given in a comparative fashion with several different plastic materials being reported.

- Stain resistance (ASTM D 2290 and D 1712). Plastics are used extensively in household applications, such as countertops and carpets, where incidental contact with staining agents can occur. The ability of the plastic material to resist staining is measured in D 2290. In the test, staining material is applied to a horizontal test surface and then placed in an oven for 16 hours at 50°C. Excess staining solution is removed, and then the material is examined for residual staining. The other test (D 1712) is specifically aimed at plastics that have pigments or other additives containing salts of lead, copper, and antimony, which are especially susceptible to staining from sulfide-containing liquids.

- Water absorption (ASTM D 570). The tendency of some plastics to absorb moisture is very important in the processing of those plastics and also in some electrical, optical, and mechanical properties. The tendency of a plastic material to absorb water is chiefly dependent on the chemical nature of the material. Polymers composed only of carbons and hydrogens (such as PE, PP, and PS) are highly resistant to water absorption. Polymers that have highly polar groups, especially those containing pendent oxygens, are susceptible to water absorption. The test for water absorption is conducted by conditioning a sample in a dry atmosphere for at least 24 hours at a known, elevated temperature. Next, the sample is cooled in a desiccator and then immediately weighed. The sample is then immersed in water for a specific period of time (usually 24 hours) at a specific temperature (usually either room temperature or at the boiling point). Following this immersion, the sample is reweighed. The moisture absorption is the difference between the two weights divided by the dry weight, expressed as a percentage weight gain.

Thermodynamics of Solvent Interactions

The interactions of solvents on plastic materials can be expressed conceptually by examining changes in the *free energy* of the plastic material. Chemical and physical changes will occur naturally when the change in free energy is negative. A negative ΔG is associated with energy leaving the system. When energy leaves the system, a lower energy state is achieved. Low energy states are stable, so stability is obtained whenever energy leaves. This stability is the conceptual basis of why the reactions occur if ΔG is negative. The change in free energy is given by Equation (5.2),

$$\Delta G = \Delta H - T\Delta S \tag{5.2}$$

where ΔG is the change in free energy, ΔH is the change in *enthalpy* or bonding energy, T is the absolute temperature (that is, the temperature on the Kelvin scale) and ΔS is the change in *entropy* or randomness. In this representation, the solvent will have a strong interaction with the polymer when ΔG becomes negative.

The sign of ΔG is obviously dependent upon the signs and relative magnitudes of the enthalpy term, ΔH, and the entropy term, $T\Delta S$. Each of these terms will be examined separately so that the general trends in each can be seen.

The change in entropy, ΔS, is always positive when going from a solid to a solution because the material increases its randomness. Therefore, since T is also always positive, the entropy term, with its negative sign, will always contribute to a negative free energy. Entropy always favors solvation. If ΔH is small, then entropy will dominate and the solvation will occur.

The enthalpy change, ΔH, is associated with changes in the primary and secondary bonding energies. If bonds are formed, ΔH is negative, that is, energy leaves the system and the total system drops to a lower energy state. If bonds are broken, ΔH is positive, that is, energy input is required to break the bonds. If bonds are both formed and broken, the algebraic sum of the energies of the formed and broken bonds will dictate the sign of ΔH. Strong bonds or a large number of bonds will dominate. Hence, if weak bonds are broken and strong bonds or many weak bonds are formed, ΔH is negative and the net result is energetically favorable.

In open polymer structures many solvent molecules can approach the polymer and, therefore, many solvent bonds can be formed. When the solvent-polymer bonds are strong, ΔH is negative. Strong solvent-polymer bonds are favored with polar solvents when the polymer chain has polar sites because hydrogen bonds are formed.

Crystalline polymers have high intermolecular energies which must be overcome to form the solvent bonds and hence, even if secondary bonds (which are weak) are formed between the solvent and the polymer, the net result is a small ΔH term. Also, the tight nature of the crystalline structure reduces the number of solvent molecules that can approach the polymer and therefore reduces the number of solvation bonds that can form. As a result, crystals resist solvent interactions from a thermodynamic point of view.

In the case of nonpolar solvents, even with nonpolar polymers, the ΔH term is not large because the secondary bonds between the solvent and the polymer are not strong. The bonds are generally only due to van der Waals forces which are much less strong than even hydrogen bonds. Hence, nonpolar solvents and nonpolar molecules generally result in a small ΔH or might even be positive if heat is required to break the intermolecular attractions. Therefore, for nonpolar solvents to dissolve plastic materials, the $T\Delta S$ must be large enough to dominate over the ΔH term. One condition in which the $T\Delta S$ is large is when the polymer is short, that is, polymers with low molecular weights. (This is because small molecules can be highly ordered and therefore have a larger ΔS when they go to the disordered state of a solution.) As a result, interactions between nonpolar solvents and plastic materials are favored by low molecular weights.

The $T\Delta S$ term can also dominate over the ΔH term as the temperature is increased. This effect can be visualized at the micro view level by considering that plastic materials naturally expand with increased temperature and are, therefore, more open. Hence, the solvent can penetrate more easily. This is a molecular model confirmation of what is seen from thermodynamic considerations.

An interesting point regarding thermodynamics is that the melting of polymers also follows the free energy equation. Hence, melting will occur when the ΔG term is negative. The ΔH term is always positive (indicating an input of energy) because in melting bonds are being broken but new bonds are not formed. The entropy ΔS is always positive because disorder is increased in going from a solid to a melt, so the entire entropy term carries a negative sign. Because the ΔH and $T\Delta S$ terms have opposite signs, the free energy will be negative only when the entropy term $T\Delta S$ is larger than the enthalpy term ΔH. This situation is most likely to occur as T increases. Therefore, the melting point, T_m, can be considered as the point where the $T\Delta S$ terms becomes larger than the ΔH term.

Some insight into the relative values of ΔH and ΔS can be gained by noting when T_m is high or low. If T_m is low, the ΔS term must be high relative to the ΔH term, as it would be in low molecular weight polymers. If T_m is high, the ΔS term is small compared to ΔH, as occurs with high molecular weight polymers. The situation of a small ΔS relative to ΔH also occurs with polymers in which the bonds are strong, such as with crystals. Both high molecular weight polymers and highly crystalline polymers have a high T_m.

Plasticizers

Plasticizers are chemicals that have strong solvent effects on certain plastic materials but are only added in moderate concentrations. Therefore, rather than dissolve the plastic material, the plasticizer will just cause the polymer structure to swell. This swelling permits increased chain movement, especially locally, which makes the plastic material softer and more flexible. This greater chain movement means that the material changes from the glassy state (hard and brittle) to the rubbery state (flexible and soft), a process called *plasticization*. Another way of thinking about this change is that the T_g of the plastic material is lowered. The greater flexibility also means that the plastic material becomes easier to process and usually melts at a lower temperature.

The amount of plasticizer that is added to the plastic material and the method by which the plasticizer is added can be quite critical to the determination of the properties. If the plasticizer concentration is too low or the plasticizer is poorly distributed, the plastic material will not be flexible enough. If too much plasticizer is added, the plastic material will have general chain movement (as opposed to local chain movement) and the strength of the material will be lost. Many plastic materials can be plasticized by contact with a solvent, but this is usually inadvertent and would generally be viewed as undesirable, such as the softening of nylon in a water environment.

To be commercially viable, a plastic/plasticizer system should give easily controlled property changes in the plastic material over a reasonably wide range of plasticizer concentrations. The plastic material should soften but should still retain most of its strength. Furthermore, the plasticizer should meet most of the characteristics of the "ideal" plasticizer that are given in Table 5.1.

In only a few plastic materials can inexpensive plasticizers be added to give precisely controlled properties. Most plastic materials are either insensitive to plasticizers or have such complicated dependencies that precise control over resulting properties is not possible. Hence, only a few plastic materials are plasticized commercially. Although only few plastics are ideally suited for plasticization, those that are tend to be widely used.

The most important commercial application for plasticization is in polyvinyl chloride (PVC). This polymer is hard and brittle in its nonplasticized state and is used for applications such as sprinkler pipe. When plasticized, PVC is often referred to as "vinyl" and is soft and pliable. It is used for applications such as car seats, car dashboard covers, covers for three-ring binders, and some flexible bottles.

The case of PVC illustrates both the advantages of using a plasticizer and many of the difficulties in obtaining all of characteristics of the ideal plasticizer in one system. A plasticized PVC can be processed at a much lower temperature than rigid PVC. Hence, the problem with thermal breakdown that is so prevalent with unplasticized PVC is largely avoided in vinyl. Processing can even be done at room temperature as when pressing *(cal-*

Table 5.1   Characteristics of an "Ideal" Plasticizer

Low cost

Not volatile, that is, it should remain in the plastic material

Effective over a wide range of concentrations

Does not increase flammability

Nontoxic

Not detrimental to the environment

Not water extractable

Stable through processing and use

Does not migrate to the surface of the plastic material

Stable to ultraviolet (UV) light

endaring) the vinyl onto a fabric to make vinyl cloth for automobile seats. Some of the difficulties with vinyl plasticizers are the volatilization of the plasticizer, which is seen in the oily residue that coats the inside of car windows, and the embrittlement with time that is evident on dashboards that crack when they get old.

The problem of plasticizer migration is especially difficult to solve. All materials will migrate to areas of lower concentration. For a plasticized plastic material the surface of the material is usually the area of lowest concentration because the molecules on the surface evaporate or are wiped away. Small molecules generally migrate faster than larger molecules, but lower-weight plasticizers are generally more effective in softening the plastic material. If a heavier, less volatile plasticizer is used, it will migrate slowly to the surface and evaporate slowly, thus staying as an oily residue. Obviously some compromise must be reached between a light-weight plasticizer that migrates quickly but evaporates cleanly and a heavier plasticizer that migrates more slowly but leaves an oily surface. Reasonable compromises have been made to meet the needs of a large number of specific applications.

Plasticizers for PVC are generally phthalate esters and other fairly sophisticated, high-boiling-point organic solvents. Although PVC is inherently difficult to burn, when additional flame retardance is needed, phosphate esters are used. In some applications mixtures of plasticizers are used, especially when one of the plasticizer molecules is not very effective alone but works well when another plasticizer is present to give initial swelling.

Great care is usually required in blending the plastic material with the plasticizer. This blending is done in mixers which give uniform dispersion and thorough coating of the powdered plastic material and the liquid plasticizer. The resultant material may be soft and doughlike or it may be a nearly dry powder. In either case, the molder usually buys the preplasticized resin rather than doing the plasticizing in-house.

Solvent Welding

Some advantages not readily available in metals and ceramics are available in plastic materials because of their sensitivities to certain solvents. An example of this is *solvent welding* where only the surface is exposed to the solvent. In this process the solvent is usually applied to the surface of the plastic material to soften it just prior to the welding process. When two such surfaces are brought into contact and held together, they join and become bonded as the solvent evaporates. This process is commonly seen in the joining of PVC pipe and fittings. Solvent welding and other joining processes will be discussed further in the chapter on finishing and assembly.

Environmental Stress Cracking

Some plastic materials develop cracks when simultaneously placed into certain environments and subjected to mechanical stresses. For instance, if a plastic material is in contact with a soapy solution and is also kinked, cracks can occur. Often, the plastic material would not crack (at least not in the same amount of time) if exposed to either the hostile environment or to the mechanical stresses separately. This phenomenon is called *environmental stress cracking*. Examples are polyethylene tubing which cracks when bent in the presence of surface-active solvents such as detergents, nylon in the presence of salt solutions, or polycarbonate in the presence of chlorinated solvents.

This cracking is different from the polymer degradation phenomena which have been previously discussed because stress cracking does not break the primary polymer bonds, either along the main chain or on side chains. Instead, it breaks the secondary linkages between polymers (hydrogen bonds or van der Waals forces). These secondary linkages are broken when the mechanical stresses cause minute cracks in the polymer that are enlarged by the swelling action of the solvent. These cracks can propagate rapidly throughout the entire plastic material structure like a tear or a split if the combination of mechanical stresses and environmental swelling persists. The result can be a catastrophic failure of the plastic part.

The resistance of plastics to stress cracking varies widely depending on (1) the effects of the environmental chemical, (2) the tendency of the plastic to resist the stresses applied, and (3) the ability of the plastic to stop the propagation of the cracks. This overall resistance is called *environmental stress crack resistance (ESCR)*.

The resistance of plastics can be improved by the obvious methods of reducing the applied stresses or eliminating the environmental chemical agent. Avoiding the presence of internal stresses is also effective in improving ESCR. For instance, reducing the speed of extrusion, reducing the amount of necking down that occurs due to pulling the extruded part slightly faster than it is pushed out of the extruder (drawdown), and increasing the temperature of injection molding can reduce internal stresses and improve ESCR. Other methods to improve ESCR include annealing, which reduces any residual stresses, and crosslinking, which creates strong covalent bonds in the place of some of the secondary bonds. Crosslinking of polyethylene by treatment with electrons has been extensively used to increase the ESCR of tubing and wire insulation and is discussed in the chapter on radiation processes.

Modifying the molecular structure has also been effective in some cases. For instance, linear low-density polyethylene is much more resistant to stress cracking than is conventional, branched polyethylene. This molecular modification seems to change the nature of the secondary bonding. It may also affect the way in which the cracks are initiated and/or propagated.

The standard test for ESCR is called the *bent-strip test.* In this test small strips of plastic material are bent (to cause a severe stress) and then placed in a solution of surface-active agent (such as a soap, although a commercial material called *Igepal* is the standard). The strips are inspected periodically to see if cracks or other evidence of degradation occur. The solution may be heated to accelerate the cracking. The results are expressed as the time to failure (in hours) or, if no failure occurred, as a "pass" at the number of hours tested (usually 1000 hours).

Crazing

In some cases an environmental chemical will embrittle the plastic material or stress the plastic material and cause small surface cracks, even when no other stress is applied. This stressing can, of course, be caused by the swelling of the plastic material, as the solvent molecules penetrate the polymer matrix.

Cracks may also appear when the plastic part is stressed (usually in tensile) with no apparent environmental solvent present, although the appearance and the growth of the cracks are almost always accelerated by the presence of a solvent. These phenomena are called *crazing* or *solvent crazing* and differ from environmental stress cracking in both the direction of the cracks and the extent of the cracking. The crack direction in environmental stress cracking is in the direction of molecular orientation in the part (usually the machine direction). The crack direction in crazing is perpendicular to the molecular orientation direction. In crazing the cracks are usually much more numerous in a small area but are much shorter than environmental stress cracks.

The many small cracks often result in a whitening of the part because of a change in the way the light is reflected. This whitening is sometimes called *blushing*. Blushing can also occur when the molecules in a particular area become highly oriented because of the formation of local internal stresses that change the local crystallinity and possibly cause internal voids (such as when a plastic part is severely bent). Repeated stressing of a plastic part in the region where blushing has occurred will likely result in embrittlement and fracture.

This phenomenon of embrittlement and fracture is similar to metal embrittlement resulting from the effects of cold working. In both plastics and metals, the part develops stresses in the worked (bent) area which will lead to cracks and, eventually, fracture. Just as with metals, the effects of the bending can be reversed by annealing, provided cracks have not yet formed.

PERMEABILITY

Definitions and General Principles

Permeability is a measure of how easily gases or liquids can pass through a material. A low permeability means that the gases or liquids pass through with difficulty, usually requiring a long period of time and/or high pressures. Hence, the material is said to be a barrier to the passage of gases or liquids and permeability is often called a *barrier property*. Many plastics applications, especially in packaging, require that the plastic material be a barrier to the passage of gases and liquids. These plastic materials would, therefore, have low permeabilities.

The permeability of a plastic material is governed by many of the same properties that determine the susceptibility of the plastic material to solvents. For instance, just as a polymer with many polar groups is sensitive to a polar solvent, that same polymer would be permeable to a polar gas or liquid. Hence, "like is permeable to like." Conversely, a nonpolar polymer would be a barrier to polar gases and liquids. As with solvents, the openness of the plastic material structure is also a factor. If the structure is largely amorphous with few areas of dense packing of the atoms, gas and liquid molecules can move more easily into and through the plastic structure and the material will have a high permeability. Therefore, in two plastic materials with similar polarities, the higher-density, more crystalline material would be the better barrier resin.

In comparisons of the openness or permeability of plastic materials an assumption is logically made that the permeating gas or liquid is the same for the materials compared.

The nature of the permeating gas or liquid will obviously have an effect, as has already been mentioned, with respect to the polarity of the permeating material relative to the polarity of the polymer. The size of the gas or liquid molecule is also extremely important. Small molecules can work their way through the polymeric structure much more easily than can large molecules. This size effect is so strong that the permeation rate of a small molecule (helium) can be 10^{16} greater than that of a large gaseous molecule (pentane). Size effect can, therefore, outweigh all other permeation effects.

Diffusion Coefficient

The permeability characteristics of a material (plastic, metal, or ceramic) can be expressed by a quantity called the diffusion coefficient or *diffusivity constant, D*. Diffusion or diffusivity is the susceptibility of a material to allow permeation through it in relation to a particular gas or liquid. For instance, if the plastic material is solvent-sensitive to a particular gas or liquid, the *D* for that system will be large. If the plastic material has an open structure, *D* will be larger than for a dense, crystalline plastic, assuming the same diffusing gas. Therefore, a plastic material with a high diffusion coefficient will have high permeability.

The environment under which the diffusivity is measured can also affect *D*. Any condition that will cause the plastic structure to swell will significantly increase *D*. One critical environmental effect is temperature. Raising the temperature will impart flexibility to the plastic system, which will make it more open and thus allow the gas or liquid to permeate more easily. Therefore, the temperature at which the diffusivity was determined must be specified. Tables of diffusion constants for many plastic materials with various common gases or liquids are available in several plastic materials handbooks. The diffusivity is given for a standard temperature (usually room temperature, 73°F or 23°C). Conversion to another temperature is done using Equation (5.3), which is one form of the Arrhenius equation.

$$D = D_o e^{-(A/RT)} \qquad (5.3)$$

D is the diffusivity at the environmental temperature, D_o is the diffusivity under standard temperature conditions, *A* is the activation constant and measures the energy required for the gas or liquid to pass through the molecules of the plastic material, *R* is the gas constant, and *T* is the absolute temperature (usually measured in kelvins). A useful rule of thumb is that a temperature increase of approximately 5 K can double the diffusivity.

Other environmental effects, such as the presence of a second gas or liquid, can also affect the permeability. For instance, a plasticizer will significantly increase diffusivity because the polymeric structure has been swelled and softened, thus allowing the permeating agent to pass more easily. These combinations of factors are, however, so numerous that diffusivities for all these conditions would be impractical to list. Therefore, for conditions such as these, specific diffusivities would be measured in each case.

Fick's Laws of Diffusion

It has been shown that the permeability of a particular gas or liquid through a barrier material depends not only on the diffusion coefficient of the barrier material, but also upon the difference in concentrations of the diffusing material from point to point within the material, and inversely upon the thickness of the material. If a steady-state condition exists, such as when the concentrations on either side of the barrier material are not allowed to change with time, this relationship can be expressed as given in Equation (5.4), which is called Fick's first law of diffusion:

$$J = -D\frac{dC}{dx} \qquad (5.4)$$

where J is the flow of gas or liquid through the material and is expressed as a flux (flow per unit area), D is the diffusion coefficient of the material, dC is the change in concentration of the gas or liquid from one side of the plastic part to the point of measurement, and dx is the distance from one edge of the material to the point of measurement. The negative sign is introduced because dx is usually chosen to be opposite in sign to the direction of flow. These relationships are illustrated in Figure 5.3.

The permeability of materials can also vary with time. This case has been shown to follow a law called Fick's second law of diffusion. This law can be written in mathematical form as given in Equation (5.5),

$$\frac{\delta C}{\delta t} = D\frac{\delta^2 C}{\delta x^2} \qquad (5.5)$$

where C is the concentration of the gas or liquid, t is the time to permeate, D is the diffusion coefficient, and x is the distance from one edge to the point of measurement. The partial differentials simply mean that the rate of change of concentration is dependent upon both the thickness, x, and the time, t. If the constant D and any two of the variables in Fick's law are known (C, t or x), the other variable can be found by using mathematical solution tables for the "error function" which fits the form of Fick's second law.

Figure 5.3   Diffusion variables.

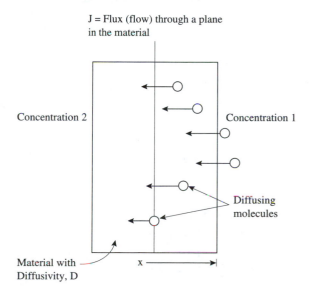

These permeability equations allow the barrier natures of various plastic materials to be expressed quantitatively and compared. They also permit the calculation of the amount of gas or liquid that will permeate through a particular thickness of a known barrier material in a specified time, assuming that the concentration of the gas or liquid is also known.

Barrier Properties of Plastic Materials

In practice, most applications require plastic parts with low diffusivity, which are, therefore, barrier plastics. For instance, packaging would normally try to exclude the permeation of air or water. The permeation of oxygen and of water through various plastic materials is illustrated in Table 5.2 where the dependence of permeation on the properties of the plastic material and the gas or liquid that is permeating through can be seen. The relative humidity (RH) is noted in the conditions because it affects the concentration gradient of water.

The differences in permeabilities of various plastics as seen in Table 5.2 can be readily understood from a basic understanding of the natures of the polymers. For instance, ethylene vinyl alcohol has polar groups along the chain and has, therefore, a relatively high permeation rate for water (which is polar) and a low permeation rate for oxygen (which is nonpolar). Other polymers with polar groups along the chain are nitrile barrier resin and nylon which also have high permeation rates for water. Polyethylene, on the other hand, has no polar groups along the chain and has, therefore, a low permeation rate for water but a much higher permeation rate for nonpolar oxygen than do the polar polymers. Note

Table 5.2 Barrier Properties of Selected Commercially Available Plastics (from *Designing with Plastics and Composites* by D. V. Rosato, D. P. DiMatta, and D. V. Rosato, Van Nostrand Reinhold, 1991, page 280)

Polymer	Permeability of Oxygen at 25°C, 65% RH (cc, mil/100 in^2/24 h)	Permeability of Water at 40°C, 90% RH (cc, mil/100 in^2/24 h)
Ethylene vinyl alcohol	0.05–0.18	1.4–5.4
Nitrile-barrier resin	0.80	5.0
High-barrier PVDC	0.15	0.1
Good-barrier PVDC	0.90	0.2
Moderate-barrier PVDC	5.0	0.2
Oriented PET	2.60	1.2
Oriented nylon	2.10	10.2
Low-density polyethylene	420	1.0–1.5
High-density polyethylene	150	0.3–0.4
Polypropylene	150	0.69
Rigid PVC	5–20	0.9–5.1
Polystyrene	350	7–10

also that high-density polyethylene has a lower permeation rate than does low-density polyethylene, illustrating the more compact structure of the high-density material.

A difficulty arises when a barrier to both polar and nonpolar gases or liquids is desired because most plastics materials are either one or the other. The packaging industry has solved this problem by coating one material with another, or by combining two or more materials together into a multi-layered film wherein each of the layers is a barrier to a type of gas or liquid to be excluded. Layers can also be added for improving other properties such as strength or ability to heat-seal. Seven-layer barrier films of this type are currently available and are used chiefly for wrapping meat.

Some common coated plastic materials are cellophane coated with polyethylene or PET coated with metal, which is widely used for helium-filled balloons. Metals generally have much lower permeabilities than plastics because of the highly dense, crystalline nature of metals in comparison to plastics. Therefore, the metal-coated PET material will resist the permeation of helium, even though helium has the highest permeation rate of any material since it is the smallest of all atoms. (These metal-coated balloons are sometimes known by the PET trade name, as Mylar^{TM} balloons.)

Another method that has been employed to modify the permeability of plastic materials is to chemically bind polar groups onto the chain of an otherwise nonpolar polymer. By careful regulation of the conditions under which the chemical reaction is conducted, the polarity of the polymer can be chosen and its permeability controlled. An example of this procedure is the creation of DuPont's Nafion® which is a modified Teflon® to which polar groups have been added. Nafion is used for membranes in chemical cells that require the passage of water and some ions to work, but still require the overall chemical resistance of Teflon®. Unmodified Teflon® would not permit the passage of the molecules.

Tightening the structure of a plastic material is another method of improving barrier properties. As indicated previously, this can be done by increasing the density of a material. Thermosetting plastics can be crosslinked to tighten their structure and thus increase their barrier properties. This crosslinking not only tightens the structure (higher density and less open), it also makes the molecules less mobile and, therefore, more resistant to the passage of a penetrating molecule.

ELECTRICAL PROPERTIES

The basic electrical properties of all materials are related to the ease of formation of a conductive path through or along the surface of the material. In some materials, such as metals, the sea of electrons which bind the metal ions together can move freely throughout the metal structure and thus provide a conductive path. Metals are, therefore, highly conductive. In other materials, charged atoms (ions) can move freely within a solution, such as salt dissolved in water. These, too, are *conductors*.

The electrons and ions in plastics and ceramics are much more restricted. Therefore, these materials are not usually conductive to electricity but are, instead, *insulators*. Some caution should be exercised, however, in using the terms conductor and insulator to describe any particular material because the terms are really comparative in nature and not absolute. For instance, most plastics would normally be considered as insulators but that is only as compared to common conductors such as metals. A specific plastic material could be considered a conductor if compared to some other plastic material with much lower conductivity. PVC is, for instance, an insulator when compared to copper but is a conductor, or at least is more conductive, when compared to PTFE.

Several electrical properties have been devised to measure the tendency of a material to form a conductive path under different conditions. The most common of these properties are: resistivity, dielectric strength, arc resistance, dielectric constant, and dissipation factor.

Resistivity

The *resistivity* of a material is the resistance that a material presents to the flow of electrical charge. This is sometimes called *ohmic resistance* because it can be expressed or quantified by the simple Ohm's law, that is, the voltage (V) is equal to the current (I) times the resistance (R) as shown in Equation (5.6).

$$V = IR \qquad (5.6)$$

The resistivity can be measured through the thickness of the sample and is then called the *volume resistivity,* or it can be measured along the surface of the sample and is then called the *surface resistivity.* Plastics have much higher volume resistivities than do metals, as illustrated in Table 5.3.

The resistivities measured by standard (ASTM) tests may be considerably different from resistivities in actual use. The standard test values are measured using controlled specimen size and shape, surface cleanliness, and moisture content. In actual use the parts are almost always of a different shape, will often have surface contaminants such as mold release, solvents, or oils, and may have absorbed considerable moisture, especially if the polymer is moisture sensitive such as nylon or cellulosics. The higher conductivity (lower resistivity) of the plasticized PVC compared to unplasticized PVC illustrates that the pres-

Table 5.3 Volume Resistivity for Various Materials

Material	Volume Resistivity Range ($\Omega \cdot$ cm)
Fluorocarbons (PTFE)	10^{18}
Polystyrene	10^{17}
Polyethylene	10^{15}
PVC	10^{15}
PVC (plasticized)	10^{11}
Phenolics	10^{11}
Antistatic polymers	10^7–10^{13}
Static-dissipator polymers	10^3–10^6
Conductive polymers	10^0–10^3
Metals	10^{-6}

ence of nonplastic plasticizers, fillers, pigments, or other additives can also make a significant difference in the electrical conductivity of the base plastic material. The values given in Table 5.3 are, therefore, only guidelines for relative resistivities.

The test for volume and surface resistivity is ASTM D 257. In this test a low voltage is applied across the thickness or along the surface of a sample under standard conditions.

The reciprocal of the resistivity is the *conductivity*. Therefore, metals are seen to have many times higher conductivities than plastics, even those few plastic materials which are considered to be conductive. Conductive polymers are used in situations in which the nonelectrical properties of a plastic are desired, but some conductivity is required, such as in certain electromagnetic interference (EMI) shielding applications. Conductive and semiconductive polymers can also be used to dissipate static charges. These static-dissipative polymers are used for applications such as carpets, where the buildup of static charge is undesirable.

The most common method to make a plastic material conductive is to simply add a conductive filler. Many materials will work, including most metal powders, many plasticizers, and many inorganic fillers. The properties of the plastic material when these fillers are added would be consistent with the properties expected when other traditional fillers are added.

Conductive polymers are made by using chemical groups that allow the electrons to move relatively freely along the polymer chain. These electrons are said to be *delocalized*. One method of achieving delocalized electrons is to have a series of carbons with double bonds between every other pair of carbon atoms, as illustrated in Figure 5.4. This arrangement of alternating double bonds is called *conjugated bonds*. The effect of this conjugation is that the electrons are free to move along the polymer chain, thus imparting electrical conductance along the polymer, as illustrated in Figure 5.4b.

Conjugated double bonds are also found in aromatic molecules such as benzene, as described in the polymeric materials (molecular viewpoint) chapter. The conjugation in benzene is limited to only the ring atoms and therefore the electrons are delocalized only over the ring itself. Some additional minor delocalization can be achieved by linking linear conjugated double bonds onto the benzene molecule.

An arrangement that has significant delocalization is graphite. Graphite is a structure in which the carbons are bonded in flat sheets that are then stacked upon each other. The flat sheets have delocalization across the sheet, but not from one layer to the next.

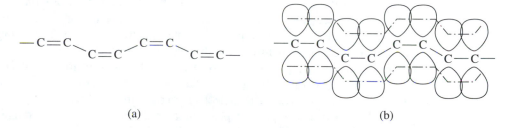

(a)   (b)

Figure 5.4   Example of a conductive polymer showing bond view (a) and atomic orbital view (b).

The polymer illustrated in Figure 5.4 can be made by the polymerization of acetylene gas and is therefore called polyacetylene. It is conductive along the chain of a polymer molecule but is not able to transfer the electrons from one molecule to another. Hence the electrons cannot move in all directions throughout the material and it is not as truly conductive as a metal. Polyacetylene can be made truly conductive by incorporating molecules that have free electrons that can move in directions other than along the polymer chain. Metals have this property, as do some organic molecules with atoms such as nitrogen and sulphur, especially when they are associated with a conjugated ring. Polymers made from acetylene and these electron-donating molecules possess true conductivity.

The primary function of plastic materials in electrical applications is that of an *insulator*. This insulator, or *dielectric,* separates two field-carrying conductors. Typical electrical applications of plastic materials include plastic-coated wires, terminals, connectors, industrial and household plugs, switches, and printed circuit boards. The major requirements of an insulator are: (1) have a high enough dielectric strength to withstand an electrical field between the conductors; (2) possess good arc resistance to prevent damage in case of arcing; (3) have high insulation resistance to prevent leakage of current across the conductors; (4) maintain integrity under a wide variety of environmental conditions; and (5) be mechanically strong enough to resist vibrations, shocks, and other mechanical forces. The key electrical properties embodied in these requirements are dielectric strength, dielectric constant, dissipation factor, volume and surface resistivity, and arc resistance.

Dielectric Strength

As indicated previously, the resistivity of a material is determined under a relatively low voltage. If the voltage is steadily increased, a point will be reached when the electrical force on the electrons within the material is so great that there is an electrical breakdown and a conductive path is formed. The voltage at which this occurs is called the *breakdown voltage* and when divided by the thickness of the sample is called the *dielectric strength*. The dielectric strength is an important measure of the stability of a plastic insulator in various electrical environments, especially high-voltage applications where plastic insulators are used to separate high-voltage wires. (Ceramic insulators are also used for this application.)

The most common test for measuring the dielectric strength of a material is ASTM D 149. In this test a voltage is applied across the thickness of a sample, which has been carefully conditioned to a standard temperature and moisture content. The voltage is then increased until the sample is forced to conduct, usually by the internal heating caused by the high electrical field imposed. Because the thickness is an important factor in determining the voltage required to cause conduction, the results of the test are normalized to a unit thickness by reporting the dielectric strength in volts/mil.

Typical dielectric strengths of plastic materials would be in the range of 20 to 50 kV/mm which are approximately 2 to 10 times higher than ceramic materials and, of course, hundreds of times higher than metals, which are not insulators, but conductors.

Arc Resistance

The *arc resistance* is the property that measures the ease of formation of a conductive path along the surface of a material (rather than through the thickness of the material as is done with dielectric strength). In ASTM D 495 a constant, high voltage is applied to the surface of a clean, dry sample and the time to form a conductive path is measured. This property is important in switches and various electronic housings where the rapid opening and closing of electrical contacts can cause sparks that result in surface erosion or breakdown. Another important application requiring good arc resistance is for stand-off insulators that keep the high-voltage wires insulated from the towers that suspend them.

Aromatic materials (those containing the benzene ring) have a tendency to char and thus form graphite-like structures which are far more conductive than the original material. These types of materials would have arc resistances of 10 to 150 seconds. On the other hand, nonaromatic (aliphatic) materials have arc resistances of 150 to 200 seconds. PTFE has the highest arc resistance at over 200 seconds. Some fillers, such as aluminum trihydrate, have been shown to increase the arc resistance of plastic materials.

Dielectric Constant

Dielectric constant or *permitivity* is a measure of how well the insulative material will act as a dielectric in a capacitor. This constant is defined as the capacitance of the material in question compared (by ratio) with the capacitance of a vacuum. A high dielectric constant indicates that the material is highly insulative, thus permitting the use of small (thin) dielectric material in capacitors. Dielectric constants are dependent upon temperature and the frequency of the alternating electric field that is applied to the sample.

The test for dielectric constant or permitivity is ASTM D 150. In the test, a carefully conditioned sample is placed between plates which are then charged, as they would be in a capacitor. The leakage current is measured and compared to the situation that would exist were a vacuum to exist between the plates.

Nonpolar polymers would generally have a dielectric constant in the 2–3 range while polar polymers can range up to 7. (Pure water, by comparison, has a dielectric constant of 80). Some applications require a high dielectric constant (such as a microwave dish) while others require a low dielectric constant (such as a capacitor). In either case, the ability of the plastic material to dissipate the charge placed upon it is also important. This property is called the *dissipation factor*.

Dissipation Factor

The dissipation factor of a material measures the tendency of the material to dissipate internally generated thermal energy (heat) resulting from an applied alternating electric field. This heating is caused by the movement of the electrons and atoms within the material in response to the changing polarity of an alternating electrical field. Polymers with highly polar groups tend to have high dissipation factors because they move more in an imposed electric field (assuming all other factors, such as size, are constant). Amorphous structures would be more easily moved than crystalline structures and would be expected to have higher dissipation factors. Another factor affecting the dissipation is the temperature of the material. The dissipation factor for plastic materials increases dramatically above their glass transition temperature because the ability of the materials to move is greatly increased.

The dissipation factor is related to the dielectric constant because of the tendency of materials to discharge a capacitor when they heat. Hence, the charge is dissipated easily if the dissipation factor is high. The dissipation factor test is ASTM D 150.

Summary of Electrical Properties

All the electrical properties of plastic materials are used to determine the behavior of parts under various electrical environments and applications. Plastic materials are, in general, highly insulative (nonconductive) when compared to metals. Plastics are therefore convenient materials for capacitors and for other dielectric applications. Hence, plastics are used extensively as wire insulation, circuit boards, capacitor dielectrics, and EMI shielding when other plastic properties, such as formability or low radar detectability, are desired. Ceramics are also insulators and can be used in many of these applications. Ceramics are, however, brittle materials and so applications which require flexibility, such as wire insulation, would not be appropriate for ceramics.

Plastics are not often used as conductors. Metals are, of course, the preferred material for conductors. Some plastics have conductivities high enough that they are used as conductors in special applications where metals are not desired, although even these plastics are not as conductive as metals. Plastics are not used as semi-conductors because the energy required is too great. Semi-conductors must be "almost" conductors, that is, the electrons in semi-conductors must be able to be excited with little energy into conductive pathways.

Ohmic conductance in plastics can be achieved by delocalization of the electrons. This is usually done by conjugating double bonds and by the addition of electron-donating molecules in the chain. Capacitive conductance is achieved by adding polar groups to the polymer and by increasing the ability of the polymer chain segments to move, such as in an amorphous, open structure or by increasing the temperature of the material.

OPTICAL PROPERTIES

The optical properties of a plastic part generally involve the way in which the plastic materials interact with light. These interactions can be grouped into three different types of interactions: (1) how the light passes through or is diffused by the plastic material, (2) the way light is reflected off the surface of the material, and (3) the color of the plastic material. These optical properties are important in most applications and are critically important in some.

In many plastics, optical properties in combination with other properties, such as toughness, flexibility, and moldability, are the crucial combination that gives real value.

Some applications in which optical properties are especially important are outdoor signs, optical fibers, automotive taillights, safety glasses, window glazing, merchandise display cases, instrument panels, contact lenses, prisms, low-cost camera lenses, magnifiers, and numerous boxes, packages, and coatings where clarity is desired.

Light Transmission

Plastics differ greatly in their ability to transmit light. Some plastics allow light to pass through them with little change. Images are distinct when viewed through these materials. These materials are called *transparent*. Good examples of transparent materials (even though they are often colored) are plastic lenses for eyeglasses, bulletproof glass, and automobile back-up lenses. Window glass and clear water would, of course be other non-plastic transparent materials. The plastic materials most often associated with high transparency are acrylics (Plexiglass[TM], Acrylite[TM], and Lucite[TM]), polycarbonate (Lexan[TM] and Sparlux[TM]), and polystyrene (Dylene[TM], Opticite[TM], and Santoclear[TM]).

Other plastic materials do not allow any light to pass through them. These *opaque* polymeric materials can be used for car tires, football helmets, and computer housings. They are often polymers to which fillers or pigments (inorganic colorants) have been added. A few polymers can be naturally opaque.

Some plastic materials have light transmission properties that are intermediate between transparency and opacity. These materials are semitransparent materials, sometimes called *translucent*. They allow light to pass through them but the materials appear to be cloudy or shadowy. Images are detected but are not distinct. Some common translucent materials are nylon gears, plastic milk bottles, and covers for fluorescent lights that have been surface roughened to diffuse the light.

Plastic materials can also be semiopaque and allow a small amount of light to pass but do not permit the detection of images through the plastic material.

The boundaries between the various categories of light transmission are not well defined. Hence, some convenient rules of thumb in discussing the various degrees of light transmission are useful. If light will pass through a plastic part such that a newspaper can be easily read through the plastic, then the material can generally be considered to be transparent. If a newspaper cannot be easily read through the part but general shapes can be perceived, such as holding your hand on the plastic part and detecting the outline of the hand, the material is translucent. Semiopaque material allows only vague shadows to be perceived through the part. Nothing is perceived through an opaque plastic material.

Even in highly transparent materials, some light is absorbed by the plastic material. This absorption causes heating of the material and, in the case of the high-energy ultraviolet light, can cause degradation. Eventually the bonds will break down and the plastic part will become less transparent, often with an accompanying yellowing of its color. (The yellow color is caused by a preferential absorption of blue light by many degraded plastics.) The transmission can be measured by comparing the amount of light that passes through the plastic part with the amount of light passing through clean air or a vacuum. In many plastics these light transmission changes are gradual and so a plastic material can be transparent with little perceived change for long periods of time.

Polymethylmethacrylate (PMMA) has been found to lose less than 20% of its light transmission over 20 years if properly formulated. This is the best plastic material in retention of optical clarity.

The standard test for measuring light transmittance and haze is ASTM D1003. Luminous transmittance is defined as the ratio of transmitted light to the incident light. The value is generally reported as a percentage of the transmitted light. Haze is the cloudy appearance of an otherwise transparent specimen caused by light scattered from within the specimen or from its surface. Both luminous transmittance and haze are measured by using a hazemeter.

Transmission depends upon the relatively unobstructed passage of light through the plastic part. If the light is scattered by inhomogeneities in the part, the amount of light transmitted is reduced. Relatively small reductions or minor scattering cause the material to become translucent. A typical cause of this phenomenon is the scattering of light by the crystal structure in a crystalline polymer as in polyethylene. Low-density polyethylene that has few crystals is transparent. However, as the number of crystals increases, the transparency decreases until, in high-density polyethylene, the material is translucent (as with plastic milk bottles). This scattering by the polymer crystal structure occurs because the size of the crystal is approximately the same as the wavelength of the visible light. If the crystals were significantly smaller or larger, little scattering would occur. In general, polymer crystals are about the right size to cause scattering, and so a useful rule of thumb is that **transparent polymers are noncrystalline and translucent polymers are crystalline.** However, some crystalline polymers such as PET, are transparent because the crystal size is not in the wavelength range of visible light.

The effect of additives can significantly alter this rule. Additives will almost always **decrease** the light transmission capability of the plastic material. This is especially true for many fillers that not only scatter light but absorb light.

Just as fillers tend to make plastics opaque, so also do additives which create separate phases. For instance, ABS is a mixture of three types of plastic materials (polyacrylonitrile, polybutadiene, and polystyrene). The mixtures of these materials will often have small, separate phases where one or two of the materials are dispersed through the other. Light will reflect off the phase boundaries where the phases touch and be scattered or absorbed, thus resulting in an opaque material.

Colorants

The color of the plastic part is affected by the way the light is absorbed or diffracted by either the polymer itself or by additives in the plastic material. Additives which cause specific light to be absorbed are called colorants. These colorants can be either semi-transparent or opaque; they can have either a small effect on the transmission of light or a major effect. The materials which are most likely to have a small effect are organic liquids which are added to the plastic material and are called *dyes*. The materials which have a major effect, by scattering the light or absorbing the light, or some combination of both, are usually inorganic materials and are called *pigments*. So, a plastic part that is clear was likely colored by a dye, whereas a plastic part that is dark and opaque was likely colored by a pigment. The most common pigment used to color plastics is carbon black which also acts as a ultraviolet light absorber and therefore provides additional weathering protection.

Materials with highly delocalized electrons tend to absorb light because the frequency of light is often the same as or close to the frequency needed to excite these electrons. This absorption property is readily apparent when the molecular structure of dyes is examined. These are almost all aromatic molecules with added carbon-carbon double bonds to increase the delocalization of the electrons. Hence, they absorb certain light frequencies and let other pass through. The absorption of part of the visual light spectrum results in a colored material. Conductive and semi-conductive polymers have highly delocalized electrons and are almost always highly colored. Graphite, for instance, which is conductive, is black, indicating that light of almost all visible frequencies is absorbed.

Color matching is an important property in some plastic applications. This matching can be done analytically by using the light reflectance or transmittance spectrum of the plastic material in comparison with the desired color match partner. The amount of pigment or dye is carefully adjusted to give the same reflectance or transmittance between the two materials to achieve the proper match. Care should be taken to monitor this color match after the dye or pigment has been thoroughly mixed into the plastic material as additive dispersion has a major effect on the color.

These color matches are obtained visually by using specific light sources to view the color and then comparing the color with standard color samples. The test method for this procedure is ASTM D 1729. Automated color matching machines indicate when a color match has been achieved (with a selected variance percentage) and, when the color match is not correct, suggest combinations of standard pigments that can be added to the mix to obtain the match.

Surface Reflectance

The reflection of light off the surface of a plastic part determines the amount of gloss on the surface. The reflectance is dependent upon a property of materials called the *index of refraction* which is a measure of the change in direction (angle) of an incident ray of light as it passes through a surface boundary. If the index of refraction of the plastic part is near the index of air, light will pass through the boundary without significant change in direction. If, on the other hand, the index of refraction between the air and the plastic material is large, the ray of light will significantly change direction causing some of the light to be reflected back toward its source. This reflection emphasizes the presence of the surface. For instance, glass is a material with nearly the same index of refraction as air, therefore glass window surfaces are not readily apparent. However, if the glass is coated with a shiny metal, as is done in a mirror, the surface becomes very visible. (A mirror will reflect the light because the silver surface on the back of the mirror has a very high index of refraction compared to air or glass.)

The refractive index is measured by ASTM D 542 and is actually defined as the ratio of the sine of the angle of the incident light to the sine of the angle of refracted light. Refractive index values are important to designers of lenses for optical instruments such as microscopes and binoculars. The optical index of most transparent plastics is near that of glass. The test is conducted using a refractometer or a microscope to compare the actual thickness of a specimen with the apparent thickness when viewed through the material. The result is a number (index) without units.

Surface reflectance is also affected by geometric features of the boundary that may cause the light to change directions. For instance, roughing the surface or coating the surface with irregular materials will cause the light to change direction and therefore cause surface reflectance.

A convenient test for assessing the amount of light reflected off the surface is the specular gloss test, ASTM D 523. Specular gloss is defined as the relative luminous reflectance factor of a specimen in the specular (observed) direction. In other words, the specular gloss is the amount of shininess exhibited by a surface. Specular gloss usually is measured by a glossmeter, which shines a light on a sample, measures the reflected light at a particular angle, and compares this reflected light with the amount that would be reflected from a shiny black surface.

In some cases, surface reflectance is desired. For instance, the cover over a light fixture is often roughened or lightly pigmented so that the light is scattered, thus diffusing the light over a wider area and decreasing direct shadows. The glare of the light, which is a measure of the amount of light that is moving directly to the viewer, is reduced because of the diffraction caused by the change in surface reflectance. For good transparency, a matching of the indices of the plastic part and of air is generally desired. Some of the plastics which give this good match are PMMA, polycarbonate, and polystyrene.

FLAMMABILITY

Most untreated plastic materials will burn. This flammability is a result of the chemical nature of the polymers (carbons and hydrogens) which are readily oxidized (burned) to carbon dioxide and water vapor. Other atoms and insufficient heat or oxygen can also result in other by-products of the burning of polymers, but the basic tendency to burn remains with only a few exceptions, one being highly aromatic polymers. Many of these materials will burn only slowly and only if an independent flame source is directly in contact with the plastic part. If the flame source is removed, the polymer will stop burning within a specified distance or time in a standard test. This property is called *self-extinguishing*. Some examples of polymers that are highly aromatic and can be naturally self-extinguishing are phenolics, aramids, and polyimides. When burned, these materials form solid, ash-like materials that are largely inert and hence resistant to further burning. These types of burned, inert materials are called *char*, which are also formed when thermosets decompose.

Another important exception occurs when the polymer contains significant amounts of fluorine, chlorine, bromine or iodine atoms. (These atoms are all in one chemical group that is called the *halogens*.) Typical examples of polymers containing halogens are PVC (contains chlorine) and PTFE (contains fluorine). When a polymer containing halogen atoms begins to burn, the bonds are broken and the halogens are released, along with hydrogens, carbons, and other atoms which may be part of the polymer. The halogens have a strong affinity for the hydrogen atoms and will readily combine to form dense gases (HF, HCl, HBr, and HI). These gases are so dense that they exclude oxygen from the area of combustion of the polymer and essentially smother the flame. Hence, highly halogenated polymers are self-extinguishing.

The self-extinguishing property can be imparted to polymers that do not contain halogens by incorporating halogen-containing additives into the plastic material. For instance, halogen salts (such as magnesium bromide and phosphorous pentachloride) are solid materials that can be added to many plastic mixes to make them self-extinguishing.

A serious problem with these halogenated polymers and additives is that the hydrogen-halogen gas that is formed can be fatal. Hence, great care should be taken when using halogenated materials in applications where people might be trapped inside a closed area with the gas from the burning polymer (such as inside an airplane). Antimony compounds have been shown to enhance the flame-retardant effectiveness of the halogen compounds and are often also added.

Another material that smothers flames and can be used as an additive for plastic mixes is phosphorous. Just as with halogens, these materials are added to the plastic mix and can render the resultant plastic material self-extinguishing.

Water can also smother a flame and is often used as a method of improving the non-burning nature of a polymer. Water molecules are bound to some inorganic materials such as alumina (Al_2O_3) which are then called *hydrated*. Hydrated alumina or other hydrated materials can be added to polymers to impart self-extinguishing characteristics. As with the other cases of flammability additives, higher concentrations of the additives will improve flame resistance. An advantage of the hydrated materials in flame retardance applications is that the gas liberated by the additive (water vapor) is not harmful, whereas the halogenated and phosphorous materials often are.

The nature of the gaseous by-product from burning plastic parts is dependent on both the polymer and the additives. As already discussed, many additives and polymers can create toxic smoke. To help suppress this smoke, chemicals such as zinc borate, molybdenum trioxide, or ferrocene can be added to the plastic.

Plastics are used in many applications where flammability requirements are dictated by law or standard, and in many other applications where low flammability is desired simply because of performance specifications. Flammability is quite complex and often several different flammability properties are measured in order to characterize the flammability of a particular plastic material.

- **Flammability Test (ASTM D 568 for flexible plastics and D 635 for self-supporting plastics).** In the test for flexible plastics the plastic sample is hung vertically; in the test for self-supporting samples the sample is clamped in a horizontal orientation. After the sample is clamped in the appropriate orientation, it is marked with a gauge mark (usually 15 inches, 38 cm from the bottom of the sample). A Bunsen burner is ignited, and the flame is adjusted to the proper height. Then, the tip of the flame is applied to the end of the sample. The time it takes for the sample to burn to the gauge mark is reported as the average time of burning (ATB). Materials that do not burn to the gauge mark within the allotted time are called *self-extinguishing*. This test is not well accepted because of the wide variation in possible results and the difficulty in determining the reasons for the variations.

- **Oxygen Index Test (ASTM D 2863).** The oxygen index is defined as the minimum concentration of oxygen in a mixture of oxygen and nitrogen that will support flaming combustion of a material. This test is especially useful because it allows

Figure 5.5 Test apparatus for the limiting oxygen index (LOI) test.

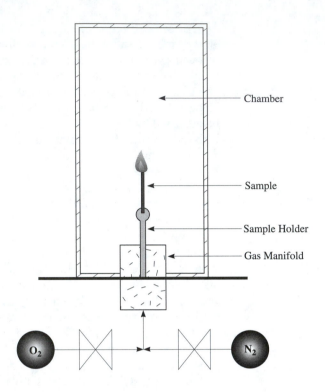

many types of materials to be rated on a numerical basis. The test is conducted by placing the sample in a holder so that the sample is vertical (like a candle). The sample and holder are placed inside a glass column mounted on a manifold, which distributes the gas evenly around the sample. The specimen is ignited, and the concentration of oxygen in the gas surrounding the specimen is varied until the combustion is just maintained. The level (%) of oxygen is called the limiting oxygen index (LOI). The test apparatus for the LOI method is shown in Figure 5.5.

- **Radiant-Panel Test (ASTM E 162).** This test involves a radiant panel (maintained at a temperature of 1238°F) as a heat source to ignite the plastic specimen. The specimen (a 6- × 18-inch sheet) is mounted at a set distance from the radiant panel, with the top of the specimen tilted at 30° toward the panel. The panel is ignited and the specimen begins to burn. The rate of burning and the amount of heat evolved in the burning (sampled in the flue stack above the specimen) are combined to give a flame-spread index for the specimen. The index is defined from the test procedure.

- **Smoke-Density Test (ASTM D 2843).** This test measures the loss of light transmission through a collected volume of smoke produced under controlled, standardized conditions. The specimen is burned inside a smoke-density chamber in which a light is caused to pass between two photoelectric cell plates. The light transmission is plotted against time. The total smoke produced is the area under the transmission/time plot.

■ **UL 94 Flammability Tests.** The Underwriters Laboratories have developed several burn tests that are all classed as UL 94 tests. These tests measure the burning characteristics of various materials in horizontal and vertical burning modes. The tests establish various maximum burn distances and rates that must not be exceeded for a material to be in compliance with the particular UL standard.

PLASTIC IDENTIFICATION

Plastic products are manufactured using a variety of processing methods and raw materials. With only a few exceptions, visual identification is difficult, if not impossible. Two different methods for identification are in common practice. One method uses sophisticated spectroscopes that identify the plastic based on some inherent pattern of energy absorption. The other method uses some systematic step process, which dictates a path to new tests depending on the results of previous tests, for identifying plastics based on the peculiar characteristics of each plastic.

Spectroscopy Identification

The most common spectrophotometric method is *infrared spectroscopy.* In this method the sample is placed in a machine that shines infrared light either through or reflected off the sample. The light transmitted through or off the sample is compared with the nondisturbed light to view various absorptions. These absorptions correspond to internal molecular motions of the polymer. The motions are characteristic of a polymer type and so, by comparing the spectrum obtained for a sample with a library of spectra obtained from many different materials, the sample can be identified. To enhance some signals, the spectra are processed using a Fourier transform method. These spectra are called *Fourier transform infrared (FTIR).* Other spectrophotometers work in a similar manner but use light of different energies, such as ultraviolet light.

Another identification spectrometer uses *nuclear magnetic resonance (NMR),* which measures absorptions of a material based on atomic movements within a magnetic field. Again, experience with interpretation of the spectrum allows identification. Yet another spectroscopic method is *mass spectroscopy,* which involves breaking a sample into molecular fragments that are then separated by momentum as they pass through a magnetic field. The charge and mass numbers of the fragments can tell an experienced operator what the material might be.

Systematic Stepwise Identification

This method uses a series of tests to sequentially eliminate groups of polymers from consideration until, finally, a test reveals the identity of the polymer based on some unique set of characteristics that only a particular polymer may possess. Several such stepwise schemes exist. One, for instance, makes an initial separation of the material into thermoplastics and thermosets by placing the material on a hot plate to determine if the sample melts. If it melts, it is a thermoplastic. Assuming that the material is a thermoplastic, for purposes of illustration, the sample could then be tested for density by dropping a small amount of the plastic in water. If the material sinks, it has a specific gravity greater than 1.0. Polyethylene and polypropylene would thus be eliminated because they both have densities less than 1.0. The sample could then be burned, and its burning characteristics (no burn, self-extinguishing, continues to burn) used to further group the plastic.

Again for illustrative purposes, assume that the plastic is self-extinguishing. The common plastics with all of the characteristics thus far identified include nylon, PC, PPO, polysulfone, and PVC. These can be further separated on the basis of whether they drip when burned. PPO and PVC do not. Assuming that our sample does not drip, then PPO and PVC can be distinguished from each other on the basis of the color of the flame, the odor from burning, the smoke type, and the ease of ignition. PVC burns with a yellow flame having green edges, has the odor of hydrochloric acid, and has white smoke. PPO burns yellow orange, has the smell of phenol, and is difficult to ignite. Obviously some experience with the characteristics is useful, as is a known sample of material to use as a comparison. Where results are still in doubt, some other characteristic of the material, such as chemical reactivity or melting point, could be used to make the separation. The burning characteristics of some common plastics are given in Table 5.4.

Table 5.4   Burning Characteristics of Some Common Plastics

Material	Burning Characteristics	Smoke or Vapor Characteristics
ABS	B. Yellow, sooty.	Acrid with cinnamon odor.
Acetals	B. Blue to colorless.	Slight smoke. Formaldehyde odor.
Acrylics	B. Bright with crackling.	Fruity odor.
Cellulosics	B and RB. Yellow.	Burnt paper, vinegar, or rancid butter smell (depending on type).
Epoxies	B. Small yellow.	Black with some soot.
Fluorocarbons	O.	—
Nylons	SE and B. Yellow. Drips.	Odor of burnt hair or horn.
PC	SE. Drips or chars.	Slight phenol odor.
PE, PP	B. Blue with yellow tip.	Parafin wax odor.
Phenolics	O.	Chars.
Polyesters	SE and B. Yellow with blue edges.	Sweet but irritating.
Polystyrenes	RB. Yellow and sooty.	Choking.
Polyurethanes	B. Yellow.	Black with stinging.
PPO	SE. Yellow-orange. No drips.	Phenol smell. Hard to ignite.
PVC	O. Yellow with green. No drips.	Acrid. White.
Silicones	O.	Glows in flame.
Urea formaldehydes	O and SE. Difficult to ignite.	Formaldehyde odor.

Code for burning characteristics:

   O = Difficult to ignite (essentially nonburning)

   SE = Material stops burning when ignition source is withdrawn

   B = Continues to burn with moderate speed after ignition source is withdrawn

   RB = Rapid or vigorous burning

===== CASE STUDY 5.1 =====

Using Carbon Black to Protect Polyethylene from UV Degradation

While there are many chemical UV stabilizers available today, by far the most common UV stabilizer is carbon black. The benefit of carbon black in plastics to inhibit UV degradation has been known for many years. However, the amount of carbon black to ensure adequate UV protection needs to be better defined. Therefore a study was conducted to define this amount and to clarify the relationship between polyethylene degradation from UV and carbon black content.

The most convenient method of adding carbon black to plastic materials is to create a *master batch* of plastic with a high carbon black content. The master batch is made in special locations which are adapted to handle the powdery, messy raw carbon black. This raw carbon black is mixed with plastic material to give approximately 30% carbon black. The plastic/carbon black material is then extruded into pellets. This is the master batch which is then sold to plastic processors who want to add carbon black to their virgin plastic material. The master batch is simply metered into the virgin material at a rate that will give the desired carbon black concentration in the final plastic product.

The test samples were made by extruding thin-walled polyethylene tubing and varying the carbon black content from 0% to 3.0% by adjusting the metering screw that feeds the master batch into the virgin polyethylene. At each setting of the metering screw, sufficient tubing was made to ensure that the carbon black concentration in the final product had stabilized and that the product was consistent.

The manufacturer of the tubing occasionally treats the tubing with electron beam irradiation to induce crosslinking for the purpose of improving the environmental stress-crack resistance (ESCR). This capability provided the opportunity of investigating the relationship between carbon black content and crosslinking in polyethylene tubing by examining the effects of carbon black content in crosslinked and noncrosslinked samples.

The extrusion process permitted samples of two different thicknesses to be made. Therefore, the relationship between thickness and carbon black content could also be determined.

In order to evaluate the UV degradation in the polyethylene tubing, the samples needed to be exposed to UV light. In actual use, the UV exposure would be from natural sunlight. However, this was impractical for this study because of the long exposure times required. Therefore, exposure to artificial sunlight (artificial weathering) was used to accelerate the testing. Artificial weathering methods have been used in plastics research for decades. These methods not only accelerate the weathering process, they also give improved predictability over natural weathering. With the popularity of artificial weathering, many studies have tried to draw a direct correlation between hours of exposure in a weatherometer and years of outdoor exposure. These studies have resulted in a wide range of correlations which depend upon the artificial weathering apparatus, location of the natural weathering, seasonal variations, and yearly weather trends. The best that can be hoped for is a rough estimate of the actual weathering performance after a machine and location have been specified.

Elongation is a good measure of the extent of degradation in polyethylene. Studies have shown that polyethylene may retain much of its strength while elongation has all but disappeared, leaving the material brittle. The elongation is especially sensitive in the machine direction, that is, the direction of flow of the material as it exists from the extruder. The samples were judged to fail when the sample had lost 60% of its original elongation.

Plots of the failure points (60% criterion) for various exposure times and carbon black contents were then prepared for two thicknesses (0.012 and 0.016 inches) and for noncrosslinked and crosslinked samples. (Crosslinked is abbreviated as "x-linked" in the plots.) These plots are given in Figure 5.6. The results indicate that carbon black concentrations above 2.4% were relatively safe, that is, the degradation curve was quite flat, indicating that the performance of the material was not very sensitive to the carbon black concentration in this region. However, concentrations below 1.0% were in the fatal zone which is characterized by a rapidly decreasing curve indicating that a small change in concentration could result in little UV protection.

Figure 5.6 Exposure versus carbon black content for polyethylene tubing samples at two thicknesses and with and without crosslinking (x-linking). Failure was at loss of 60% of original elongation.

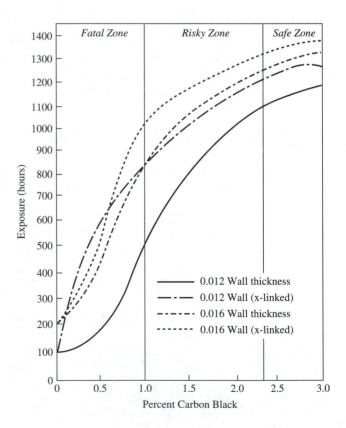

SUMMARY

This chapter has examined the effects of various environmental conditions on plastic materials, some of which are detrimental to the properties of most plastics. Other environmental conditions, while not favorable to plastics, are more unfavorable to other materials, thus making plastics preferred for these environments.

Plastics are affected by sunlight and oxygen. These are generally referred to as weathering of the plastic (along with heat aging). Sunlight (specifically the ultraviolet component of sunlight) will often degrade plastics by rupturing the covalent bonds. Ultraviolet (UV) light has about the same energy as a covalent bond, thus enhancing the potential for interaction between the UV light and the electrons in the bond. When bonds are ruptured, the mechanical and physical properties of the plastic material are adversely affected. Chemicals which preferentially absorb ultraviolet light can be added to the plastics to retard UV degradation.

The high chemical reactivity of oxygen promotes reactions with various parts of the polymer chain with accompanying changes in polymer properties. As with other degradation mechanisms such as thermal and ultraviolet light, chemical additives are available to inhibit the action of oxygen and prolong the useful life of the plastic material.

The effects of solvents on plastics can vary from no effect at all to softening of the plastic and on to complete dissolving of the plastic in the solvent. In general, the effects of solvents are strongest when the chemical nature of the solvent (especially its polarity) is similar to the chemical nature of the plastic. Solvent effects are usually reduced by properties of the plastic material that tend to make the structure more dense or more rigid, such as crystallinity and crosslinks, and by low temperatures.

Plasticizers are an important class of solvents which are used to soften certain plastic materials such as PVC. When no plasticizer is present, PVC is a rigid, brittle plastic that has many uses, for instance, sprinkler pipe. When plasticizer has been added, PVC is called vinyl. It is soft and pliable and commonly used for seat covers and dash covers in automobiles, among many other applications.

A property closely related to solvent sensitivity is permeability or the tendency of a gas or liquid to pass through a material. When the permeability is low, that is when passage through a material is difficult, the material is called a barrier. Barrier properties are especially important in some applications of plastics such as packaging. The laws governing permeability were given and explained, along with some considerations for decreasing the permeability of plastics such as increasing density, choosing a plastic that is different chemically from the gas to be passed, and utilizing multiple layers of plastic to take advantage of low permeabilities in combination.

Most plastics are more electrically resistive than are metals; hence they are usually referred to as insulators. This insulating ability is useful for wire and cable insulation as well as many special electrical devices, such as capacitors and resistors.

The optical properties of plastics range from transparent, to translucent, to opaque. The tendency of the material to absorb and refract light determines which level of optical clarity is achieved. In general, plastics with fillers are opaque and plastics without fillers are either translucent or transparent, depending upon the amount of crystallinity.

Most plastics will burn to some extent. Therefore, if the plastic is to be used in an application requiring nonflammability, some special considerations will often be required. Additives containing halogens, phosphorous, antimony or hydrated fillers are effective in reducing flammability by producing a gas that smothers the flame. Polymers in which halogens are bonded to the polymer chain, such as PVC and PTFE, are often inherently nonflammable. Materials are called self-extinguishing if they burn when a flame is applied but are extinguished when the flame is removed.

GLOSSARY

Arc resistance   The ability of a material to resist conduction across its surface from the formation of an electric arc.

Barrier property   Characteristic of materials having a low permeability.

Blushing   Whitening of a plastic due to crazing.

Breakdown voltage   The voltage at which an insulating material begins to conduct electricity.

Calendering   A process in which a plastic material is placed between rollers and flattened.

Carbon black   A sootlike residue from the incomplete combustion of hydrocarbons that is used as a black colorant and a protectant against UV light degradation.

Char   The carbonaceous material formed when some materials incompletely burn.

Conductivity   The reciprocal of the resistivity and a measure of the ability of a material to conduct electricity.

Conductors   Materials that conduct electricity.

Conjugated bonds   Alternating single and double carbon-carbon bonds that result in electron delocalization.

Crazing   The formation of small surface cracks as a result of stressing, often in the presence of a solvent.

Delocalized   An electron configuration in which the electrons are free to move among the atoms in the material.

Dielectric   An insulating material.

Dielectric constant   The ability of a material to serve as a capacitive insulating material, expressed as a ratio to the capability of a vacuum.

Dielectric strength   The voltage at which an insulating material is forced to conduct electricity.

Diffusivity constant   A measure of the susceptibility of a material to permeation.

Dissipation factor   The tendency of a material to allow current to leak across it when the material is the insulative material in a capacitor.

Dyes   Organic compounds used to impart colors to plastics or other materials.

Enthalpy   Total bonding energy content of a material.

Entropy   A measure of the randomness of a material.

Environmental stress cracking   The formation of cracks in a plastic from the simultaneous application of a solvent and a mechanical stress.

ESCR   The resistance of a material to environmental stress cracking.

Free energy A thermodynamic property, usually designated as ΔG, that corresponds roughly to the change in energy content of a material from the combined changes in bond energies, temperature, and entropy.

Halogens Elements of group VII-A (fluorine, chlorine, bromine, iodine), which are often added to plastics to reduce flammability.

Hydrated Compounds that contain loosely bonded water molecules.

Infrared spectroscopy (IR or FTIR) A spectrographic method for determining the chemical composition of a material by passing infrared light through or off the material.

Insulators Materials that do not conduct electricity.

Mass spectroscopy A spectrographic method in which the molecules are fragmented and then passed through a magnetic field that separates the fragments by mass, thus permitting identification of the original material.

Master batch A blend of polymer and some additive, such as carbon black, in which the additive is in very high concentration, thus permitting the master batch to be used as a concentrate to incorporate the additive into regular polymer material.

Nuclear magnetic resonance (NMR) A spectrographic method in which molecular motion in a magnetic field is used to identify the molecules.

Opaque Materials that do not allow light to pass through them.

Oxidation Combining of gaseous oxygen with a polymer resulting in a degradation of the polymer.

Permeability A measure of how easily gases or liquids can pass through a material.

Plasticization The softening and swelling of a plastic by a solvent or chemical additive.

Pigments Inorganic compounds used to impart colors to plastics and other materials.

Resistivity The resistance that a material presents to the flow of electricity, sometimes called ohmic resistance.

Self-extinguishing The ability of a material to stop burning within a specified time or distance.

Solute Material that is dissolved in a solution (such as the sugar in a sugar-water solution).

Solvent A chemical that acts to dissolve another material (such as the water in a sugar-water solution)

Solvent crazing Crazing when solvent is present.

Solvent welding The joining of plastic materials together by using the surface softening imparted by a solvent.

Specular gloss The amount of light reflected off a surface; the shininess of the surface.

Surface resistivity The resistance of a material measured along its surface.

Translucent Materials that allow a portion of the light to pass through them but appear cloudy or shadowy.

Transparent The ability of a material to allow the passage of light without significant change to the light.

Ultraviolet (UV) light degradation Degradation from UV light, chiefly sunlight.

Volume resistivity The resistance measure through the thickness of a material.

Weathering Degradation of material from long-term effects of the environment.

QUESTIONS

1. Describe how ultraviolet light degrades plastics and why UV light does not generally degrade metals.
2. Describe the processes of solvent/solute interaction that lead to swelling and, in some polymers, dissolving of a plastic by a solvent.
3. Explain the relationship between plastic crystallinity and sensitivity to solvent attack.
4. Explain why plasticizers increase flexibility and elongation in plastics. What is the relationship between plasticizers and the glass transition temperature?
5. Explain in thermodynamic terms and molecular structure terms why small molecules are more readily solvated than large molecules.
6. Discuss the physical factors in a plastic that affect permeability.
7. Describe the key features of a polymer that would make it electrically conductive.
8. Identify three tests for thermal properties of plastic materials.
9. What is the limiting oxygen index (LOI)?
10. Explain why laboratory identification of a plastic so often involves burning the plastic.

REFERENCES

Charrier, Jean-Michel, *Polymeric Materials and Processing,* Munich: Carl Hanser Verlag, 1991.

Gruenwald, G., *Plastics: How Structure Determines Properties,* Munich: Carl Hanser Verlag, 1993.

Moore, G. R., and D. E. Kline, *Properties and Processing of Polymers for Engineers,* Englewood Cliffs, NJ: Prentice Hall, Inc., 1984.

Rosato, Donald V., David P. DiMattia, and Dominick V. Rosato, *Designing with Plastics and Composites, A Handbook,* New York: Van Nostrand Reinhold, 1991.

CHAPTER SIX

THERMOPLASTIC MATERIALS (COMMODITY PLASTICS)

CHAPTER OVERVIEW

This chapter examines the following concepts:

- Polyethylene (LDPE, HDPE, LLDPE, UHMWPE, relationship between density and molecular weight, crosslinked polyethylene)
- Polyethylene copolymers (EVA, EAA, EPM)
- Polypropylene (stereoisomerism, properties, applications)
- Polyvinyl chloride (heat sensitivity, plasticization, properties, other vinyl polymers)
- Polystyrene (properties, expandable, constrained geometry)
- Alloys and blends (general properties, HIPS, SAN, ABS)

INTRODUCTION

This chapter discusses thermoplastic materials that are used in high-volume, widely recognized applications and are known as *commodity thermoplastics*. Some resin manufacturers have objected to the commodity designation because that term can imply that the materials are interchangeable from supplier to supplier without differences in properties. Some differences can be seen, but within a product classification, they are not great. The resin suppliers are far more likely to sell service and reliability than resin properties that can generally be duplicated by other manufacturers as needed.

All of the commodity thermoplastics that will be considered in this chapter are made by the addition polymerization method. As discussed in the chapter on polymeric materials (molecular viewpoint), this method requires that the monomer have a carbon-carbon double bond and all the monomers meet that requirement. The differences between the monomers used to make these commodity thermoplastics are in the functional groups attached to the carbons. Although functional group substitution can be made at four locations on a carbon-carbon double bond, only one site is used for substitution in all of the major types of commodity thermoplastics which will be considered. Therefore, all these commodity thermoplastic monomers and polymers can be represented by the general formulas given

Figure 6.1 General representation of commodity thermoplastics.

in Figure 6.1, where X represents a functional group of the type discussed in previous chapters. Note that in one case hydrogen is considered to be a functional group in this representation. In all cases, three hydrogens are also attached to the carbon-carbon double bond.

The differences between the commodity thermoplastics arise, therefore, from the differences caused by the substitution of one functional group on the carbon-carbon double bond. One of the most important effects is *steric,* that is, the consequences of differences in the size of the functional groups. Steric effects were discussed in the chapter entitled "Micro Structures in Polymers." When the functional groups are small (such as hydrogen), then little *steric hindrance* (interference because of size) is encountered. Without steric hindrance the polymers are relatively free to rotate, bend, and pack together. The steric hindrance increases as larger functional groups are substituted onto the carbon-carbon double bond, with the results of restricted polymer motion, less ability to pack densely, and changes in mechanical, physical, and chemical properties. Each of these effects will be discussed when the specific thermoplastics are presented.

Other effects of the substitution of a functional group can be important. For instance, electronegativity and the chemical properties of the functional group can make a significant change in properties. Each of these will be considered when the particular plastic is discussed.

The first major polymer types (polyethylene and polypropylene) are often given a special name, *polyolefins,* because of their similarity in chemical properties. This term means "oil-like" and refers to the oily or waxy feel that these materials have. Polyolefins consist of only carbons and hydrogens without other atoms in the polymer. Furthermore, they are all aliphatic (nonaromatic) groups. Oils and parafin waxes are also aliphatic with only carbons and hydrogens but are much shorter molecules, having a maximum of about 20 carbons. Chains of this length are not long enough to entangle and cannot, therefore, exhibit the properties characteristic of polymers. A few other polyolefins exist, such as polybutadiene rubber, but these will not be considered here as their unique mechanical properties suggest that they be grouped separately. Others may have such low sales volumes that they are not generally considered to be commodity resins.

POLYETHYLENE (PE)

The polymer unit for polyethylene (PE) is given in Figure 6.2. The functional group (X in Figure 6.1) is simply hydrogen. PE is the simplest of all polymers with just two carbons and four hydrogens in the basic polymer repeating unit.

Figure 6.2 Polymeric representation of polyethylene (PE).

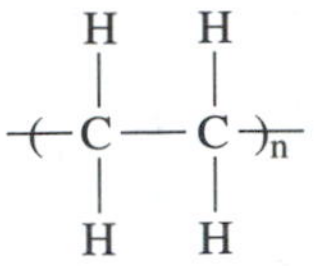

$$\begin{array}{ccc} & H & H \\ & | & | \\ \left. \hspace{-0.3em}\right(\hspace{-0.3em}C & - & C \hspace{-0.3em}\left.\right)_{\overline{n}} \\ & | & | \\ & H & H \end{array}$$

Many properties of PE can be predicted from its basic polymer representation. For instance, PE consists of only carbons and hydrogens, usually with high molecular weights, and so it is relatively insensitive to most solvents. This is an advantage when PE is used for applications such as chemical reaction vessels or pipes where inertness of the container is critical. However, the solvent insensitivity is a problem when inks, paints or other solvent-based materials are used to mark or decorate PE. The inks and paints will generally not adhere to PE. This problem can be overcome by treating the surface of PE where the ink or paint is to be placed with a flame or electric spark, thus changing the chemical nature of the PE surface. The problem with this solution is that it requires an extra step in the manufacturing process and, furthermore, the adherence of the ink or paint is still not very good.

Joining or bonding PE is another process that is made difficult because of the inherent solvent resistance of PE. Many adhesives for plastics depend upon the ability of their solvent base to soften or partially dissolve the surface of the plastic material to form a good bond. This will not work well with PE. This problem is overcome by using bonding techniques that melt the surfaces of the PE parts to be joined and then pressing them together. Processes of this type are more difficult than solvent welding and require special joining equipment. They will be discussed in further detail in the chapter on finishing and assembly.

High electrical resistance is another property that results from the basic chemical nature of PE. The carbon and hydrogen have approximately the same electronegativity, resulting in little polarity. As a result, electrical charge is not easily transferred and PE is, therefore, an excellent insulator, used extensively for insulating wires and cables and in many electrical devices.

For many of the same reasons, PE is also a good thermal insulator. However, the melting point of PE is quite low and so its use is limited in applications where high temperature is present.

Perhaps the most important applications for PE are based upon its low cost and ease of manufacture. PE is polymerized from ethylene gas that is easily and inexpensively obtained from either natural gas (methane) or from crude oil. Furthermore, the processes that are used to make the PE are easily scaled to make the polymer in very large quantities. To further reduce its cost, the temperatures required for processing PE into final shapes are also the lowest of any of the common, high-use thermoplastic materials. This means that comparatively little energy is required in the molding operations. The molding operations are further simplified because PE is stable during processing and poor quality parts can be reground and reprocessed with very little difficulty. PE applications that require this low cost and ease of processing include trash bags, packaging and other films, containers (such as milk bottles), many children's toys, and various housewares.

Some properties of PE depend more on the way the PE molecules interact with each other and these interactions are most apparent in the micro and macro views of the polymer that were discussed in previous chapters. The interactions of the PE molecules are strongly dependent on the shape (steric effects) of the molecules. The differences in shapes could not be reasonably predicted from the simplified view of addition polymerization presented earlier, which showed the basic chain extension mechanism. That view would not have suggested major differences between the PE molecules, except perhaps in molecular weight. The differences in shapes can result in changes in PE properties that are often very important in choosing the type of PE for a particular application. However, it should be remembered that the basic properties of PE arise from its basic nature and are largely unaffected by the changes in shape. In other words, the property differences between PE types arising from shape are relatively minor when compared to the differences between any of the PE types and other nonpolyethylene polymers.

The major differences in the shape of PE molecules arise from changes in the conditions that exist in the polymerization reactor during the polymerization reaction. Reactor conditions such as temperature, pressure, and catalyst type can have a major effect on the shape by either creating or suppressing the formation of molecular branching. *Branching* is the formation of side chains off the basic polymer backbone.

These side chains can form when a hydrogen-carbon bond is broken during the polymerization reaction. (In the basic view of addition polymerization presented in a previous chapter, only carbon-carbon double bonds were assumed to be broken.) However, when the polymerization is carried out at high temperatures, there is often sufficient energy in the molecules that some carbon-hydrogen bonds break, thus creating a free radical on the carbon. (The hydrogen leaves with one electron from the bond and leaves the other electron localized on the carbon, thus forming a carbon free radical.) This carbon free radical can then serve as a site for chain growth to begin. When this occurs, the chain can grow at two locations simultaneously. The net result is a branch off the main carbon backbone. The mechanism for the breaking of carbon-hydrogen bonds and the formation of side chains (branches) is illustrated in Figure 6.3.

Changes in the amount of branching (that is, the number of side chains and the length of the side chains) result in major differences in the interactions between PE molecules. Some of the properties affected by branching are given in Table 6.1.

Branching causes strong steric interference between molecules and thus forces an open noncrystalline structure. Many effects of branching on properties can be tied directly to the openness of the molecular structure. The melt temperature of highly branched PE material is lower than close-packed, crystalline materials because fewer intermolecular attractions exist in the open structure and, therefore, the energy to allow the molecules to move independently is lower. Lower creep resistance, lower tensile strength, lower stiffness, and lower hardness (scratch resistance) also result from the lower intermolecular forces in the open polymer structure of the branched material. Impact toughness is higher because the open structure can move more readily and absorb the energy of the impact. Transparency is higher because of the absence of crystal structures that often cause light to diffract. The oxidative resistance, UV stability, and solvent resistance are all lower because the oxygen, UV light or solvents can more easily penetrate the structure. This ease of penetration also increases the permeability. The decrease in shrinkage occurs because the open structure is less likely to contract to a highly packed structure when it is cooled.

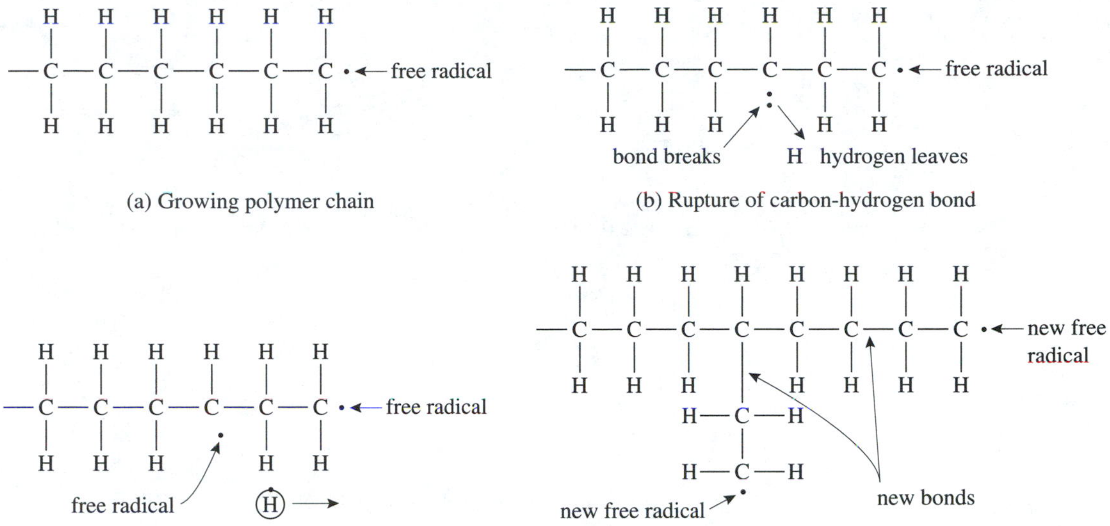

(a) Growing polymer chain

(b) Rupture of carbon-hydrogen bond

Figure 6.3 Branching mechanism for polyethylene.

Three general types of commercially made PE differ chiefly in the way the molecules interact, caused by the amount and type of branching and are illustrated in Figure 6.4.

The three PE materials are distinguished on the basis of density rather than branching because density is a property that is easily measured and is directly dependent on the amount and type of branching. The differences in polymer shapes represented in

Table 6.1 The Effect of Branching on Several Polymer Properties

Property	How Increased Branching Affects the Property
Density/crystallinity	Decreases
Melting point	Decreases
Creep resistance	Decreases
Tensile strength	Decreases
Stiffness	Decreases
Hardness	Decreases
Impact toughness	Increases
Transparency	Increases
Oxidative resistance	Decreases
UV stability	Decreases
Solvent resistance	Decreases
Permeability	Increases
Shrinkage	Decreases

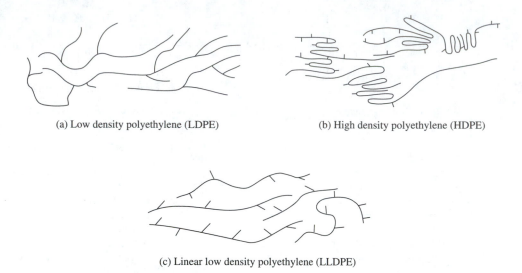

(a) Low density polyethylene (LDPE)   (b) High density polyethylene (HDPE)

(c) Linear low density polyethylene (LLDPE)

Figure 6.4   Different types of polyethylene showing the effects of branching.

Figure 6.4 are idealized. There is some overlap in the nature of the materials and so the densities of the materials are normally given in ranges. These normal density ranges are given in Table 6.2. Each of the three major types of PEs will be discussed separately.

Low-Density Polyethylene (LDPE)

The type of PE formed when high-temperature and high-pressure polymerization conditions are used is called *low-density polyethylene (LDPE)*. The density is low because these polymerization conditions give rise to the formation of many branches, which are often quite long and prevent the molecules from packing close together to form crystal structures. Hence, LDPE has low crystallinity (typically below 40%) and the structure is predominantly amorphous.

The low density and highly amorphous nature of the structure affects the physical properties of LDPE, as reflected in Table 6.1. These properties lead to uses for LDPE that emphasize its flexibility, impact toughness, and stress crack resistance. This material is used extensively in films and flexible tubing. Furthermore, LDPE is the lowest melting

Table 6.2   Densities of Polyethylene Types

Polyethylene	Density (g/cm^3)
Low-density polyethylene (LDPE)	0.910–0.925
High-density polyethylene (HDPE)	0.935–0.960
Linear-low density polyethylene (LLDPE)	0.918–0.940

and easiest to process of the PE types so it is used extensively in high-volume applications, such as packaging films, toys, and squeeze bottles for food and other household applications, especially when strength and other mechanical properties are not critical.

High-Density Polyethylene (HDPE)

If polymerization conditions are used that result in limited branching (that is, low temperature and pressure), the result is a PE that is more linear, with only a few, short branches. This type of PE is called *high-density polyethylene (HDPE)*. As the name implies, the polymer chains in HDPE can easily pack tightly and crystalline structures are formed, thus increasing the density. The properties of HDPE relative to LDPE can be assessed from Table 6.1 with the realization that branching is much lower in HDPE than in LDPE. In general, HDPE is stiffer, stronger, and more abrasion resistant than LDPE.

HDPE is made in a process that requires much lower temperatures and lower pressures than the process used to make LDPE. In order to get long polymer chains under the HDPE conditions, a catalyst is required. The first catalyst for this process was developed by Karl Ziegler in 1952 and was then applied to polymerizations of other monomers by Giulio Natta. The catalyst is called a *Ziegler-Natta catalyst*, which is a general name applied to all similar catalysts even though some more recent types may be covered by different patents from different inventors.

HDPE is used in preference to LDPE when greater stiffness or strength is required. For instance, milk, water, detergent and bleach bottles are HDPE because they are usually made with very thin walls to save material and cost, yet still must retain their shape. HDPE gives sufficient stiffness to accomplish this, whereas LDPE would tend to sag. The improved stiffness and strength are even more important as the size of the container increases. Therefore, barrels, trash carts, and chemical storage tanks are usually made of HDPE in part because of their superb chemical resistance. The use of HDPE for automotive fuel tanks relies upon its strength, chemical resistance, and low permeability.

The HDPE molecules are essentially linear with little entanglement in the melt, at least compared to LDPE. Therefore, when processed in the melt, HDPE molecules tend to be aligned in the direction of flow, especially when the flow path is highly restricted. This orientation also leads to rapid crystallization and high shrinkage upon cooling. Hence, the cooling rate of HDPE is faster than LDPE which can be an advantage in very high-volume processes such as the manufacture of margarine tubs. This orientation in the melt also adds to the strength of the melt, which is useful in blow molding very large parts. This advantage is explained in more detail in the chapter on blowmolding.

While both LDPE and HDPE are used in extruded pipe, the HDPE pipe is generally used in higher value, more critical applications such as pipe for high-pressure delivery of natural gas. In any pipe application using PE the joining of pipe to fittings must be accomplished by some method other than with solvent adhesives. For the natural gas application, special joining techniques and equipment that melt and press the pipe and fittings are used. These special techniques ensure that no leaks occur. For low-pressure, noncritical water pipe and tubing applications where LDPE is used, mechanical joints can be used, although small leaks are common. (The pipe used in sprinkler systems, where pressures are relatively high but leaks cannot be tolerated, is made of PVC. The pipe used for drain, waste, and vent in houses where stiffness and high impact strength are needed are made of ABS. Both types are discussed later in this chapter.)

The optical properties of HDPE reflect the increased crystallinity. HDPE is less optically clear than LDPE, all other factors being equal. Hence, HDPE cannot be used when optical clarity is an important consideration. HDPE is used for packaging but is most often used for applications such as grocery bags, where visual clarity is unimportant and strength is at a premium.

HDPE has the disadvantage of increased brittleness compared to LDPE. In applications where the high strength of HDPE and high impact toughness are required, a very high molecular weight grade of HDPE has been produced. This material is called *ultra high molecular weight polyethylene (UHMWPE)* and is really a subgroup of HDPE, since it is made by a similar process. The use of UHMWPE is much smaller than the use of the other types of PE. UHMWPE will typically have molecular weights in the range of 3 million to 6 million versus typical HDPE molecular weights of 50,000 to 300,000. In addition to the increased impact toughness of UHMWPE, the abrasion resistance is also significantly improved over other types of PE. Hence, UHMWPE is used for liners in coal cars, guides in mechanical equipment where rubbing is expected, such as gears, and prosthetic devices.

The density of UHMWPE is generally slightly higher than conventional HDPE. Therefore, the material is even higher in solvent resistance and lower in permeability than HDPE and this has led to some unique applications in the chemical industry.

The major problem with UHMWPE is the difficulty of melting the material. The molecular weight is so high that decomposition will often occur before melting. Hence, the material cannot be processed in traditional plastic molding equipment. It is generally *sintered*. (A process where a powdered material is packed into a mold, heated to just below the melt temperature, and held for an extended period under pressure. The powder particles fuse together and take the shape of the mold.) Sintering is not practical for producing complicated parts and so the shapes obtainable in UHMWPE are limited.

Linear Low-Density Polyethylene (LLDPE)

A third type of PE, *linear low-density polyethylene,* LLDPE, is made by a low-pressure catalyst process similar to the HDPE process, but with longer and more branches. LLDPE would typically have 16 to 35 branches per 1000 backbone carbons, whereas HDPE would typically have 1 to 2 branches per 1000 backbone carbons. This branching in LLDPE is sufficient to prevent close-packing of the molecules. Therefore, LLDPE has a low density like LDPE but a linear structure much like HDPE.

The side chains are actually made by adding another monomer (called a *comonomer*) to the ethylene monomer during the polymerization process, along with an appropriate catalyst. The comonomer must contain a carbon-carbon double bond and then a few (two, four, or six) additional carbons. (Organic molecules of this type are called α-olefins, where the α indicates that the double bond is between the first and second carbons.) The additional carbons become the side chains and are two, four, or six carbons long, depending on the comonomer used (butane, hexane, or octane). Longer side chains generally give improved physical properties because of increased chain entanglement and stronger secondary bonding. The number of side chains is determined by the concentration of the comonomer relative to the amount of ethylene and is typically 8% to 10% in most commercial grades.

Although technically a random copolymer (because ethylene and the comonomer are polymerized together), conventional usage has referred to LLDPE as a homopolymer because the chemical and physical properties are so similar to the other PE homopolymers (LDPE and HDPE). Common usage of the term copolymer when associated with PE refers to copolymers with significantly different properties from homopolymer PE. These are discussed later in this chapter.

The LLDPE process has proven to be less expensive than the process used to make conventional LDPE. Since the introduction of the LLDPE process in the later 1970s, all subsequent plant constructions for making LDPE have used this technology rather than the high-pressure/high-temperature process used to make LDPE. The advantages of the LLDPE process are shown in Table 6.3.

The molecular interactions between LLDPE molecules are different from either LDPE or HDPE and yet are related to both. The effects of lack of crystallinity and density are obviously similar in LLDPE and LDPE, while the linear shape of the LLDPE molecule is similar to the shape of HDPE molecules. These similarities and differences result in LLDPE properties which are generally between the properties of LDPE and HDPE. For instance, the strength of LLDPE is about 15% higher than LDPE. The stiffness of LLDPE can be as much as 25% greater than LDPE, and impact toughness is about 10% higher in LLDPE over LDPE. These property differences can often result in about a 25% reduction in the

Table 6.3 Comparison of High-Pressure and Low-Pressure Processes for Making Low-Density Polyethylene

High-Pressure Process (LDPE)	Low-Pressure Process (LLDPE)
Operating pressures as high as 50,000 psi (350 GPa)	Pressures of less than 1 psi (6 KPa)
Temperatures of 600°F (300°C)	Temperatures of 200°F (100°C) or less
Long construction lead time	Reduced construction lead time by 8 to 12 months
Mammoth space requirements	Occupies 1/10 the space of LDPE process
Huge capital outlay	Capital outlay reduced by as much as 50%
High energy demands	Reduced energy demands by 75%
Limited to low-density polyethylene	Can produce both high- and low-density PE
Costly and complex maintenance	Easy to maintain
Production rates vary with PE grade	Same production rate for all resin grades
Meets environmental requirements with difficulty	Environmental pollution minimal
Rapidly inflating operating costs	Operating costs reduced
Limited catalyst system choice	Wide catalyst flexibility
Catalyst removal required	Catalyst removal not needed
Acceptable resin properties	Superior resin properties

weight of plastic used when changing an application from LDPE to LLDPE. The liabilities of LLDPE include the following: melt processing temperature is about 20°F higher, shrinkage is about 8% greater, it is less clear (optically), it is less flexible, and increases in its melt index will often produce lower ESCR and higher densities, whereas in LDPE the melt index and density are more independent of each other.

An LLDPE material that is occasionally identified as a separate product is *ultralow density polyethylene (ULDPE)*. This material has a density range of 0.880 to 0.915 g/cm^3 and is made by using only the longer co-monomers such as 1-octene and adjusting the polymerization conditions so that crystallinity is very low. These materials are very flexible yet have good tear strength. Their heat sealability is excellent. Applications would include food packaging, shrink-wrap, heavy-duty film, and heat-seal layers. ULDPE can also be blended with other polymers, such as polypropylene and HDPE, to improve tear and impact toughness.

Relationship Between Density and Molecular Weight in Polyethylene

When purchasing or designing with a PE material, both the density and the molecular weight (or the melt index) are generally specified. The effects of density and molecular weight are interrelated for some polymer properties and are nearly independent of each other for other properties. Therefore, the relationship between molecular weight and density is complex. These relationships were previously discussed in the chapter on micro structures in polymers, but a review here is appropriate because molecular weight and density are so important in understanding the properties of PE.

A graph illustrating the relationships between crystallinity, molecular weight and density for PE is presented in Figure 6.5.

This graph shows the entire range of molecular weights for PE-like materials, from liquids (oils), through greases, waxes, and finally polymers. Each type of material has a range of variation of crystallinity with molecular weight. Above molecular weights of 10,000 the materials would be considered polymers. In that region the soft materials are identified with relatively low crystallinity and conventional types (LDPE) are identified. The linear materials (HDPE) are also noted. Note that the combination of high crystallinity and low molecular weight for these conventional and linear polymers results in brittle materials. If the molecular weight is increased, the materials become stiff and tougher. Increasing density will increase hardness and abrasion resistance in the plastics. The increase in crystallinity with increases in density is, of course, expected.

The overall trend in the polymer region of Figure 6.5 is that increases in molecular weight result in increases in crystallinity. Hence, in a general sense, crystallinity and density increase with molecular weight. Therefore, these properties are not totally independent, although the existence of regions for the various materials allows some variation in this relationship.

Another important and related characteristic of PE that affects properties is molecular weight distribution (MWD). The major effect of MWD is in processing. When the MWD is narrow, melting occurs over only a few degrees (called a *sharp melting point*) and the resulting viscosity is generally lower than for a material with an equivalent molecular weight and a broad MWD. This sharp melting point and low viscosity is ideal for injection

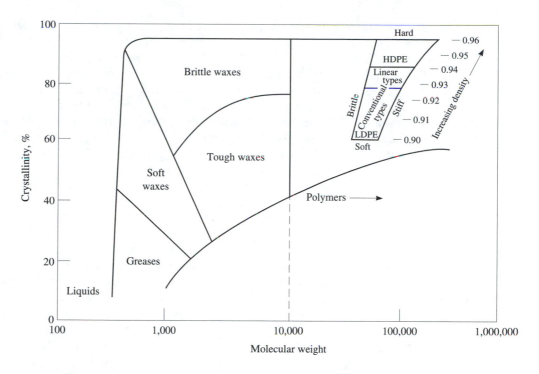

Figure 6.5 Relationships between crystallinity, molecular weight, and density for polyethylene.

molding. In contrast, broad MWD materials melt over a wide range of temperatures. The low molecular weight materials melt first and act as lubricants for the higher molecular weight molecules, facilitating their flow. Wide MWD materials will have higher viscosity in the melt because of the presence of high molecular weight molecules that are only partially melted. These high molecular weight molecules are still entangled and they give strength to the melted polymer mass. Hence, broad MWD polymers have high melt strength, a property of value in processes where the melt must retain its shape, such as extrusion, blow molding, and thermoforming.

As the MWD broadens, the toughness of the material decreases, the ESCR increases, part shrinkage decreases, and the tendency to warp decreases. All of these changes in properties result from the ability of the low molecular weight polymers to move into the spaces (amorphous regions) of the longer materials and reduce the effects of having only long polymers.

Crosslinked Polyethylene

Some applications have taken advantage of the capability of PE to be crosslinked. This crosslinking can be done either by electron irradiation or by chemical means. When done by electron irradiation the bonds are formed by passing the molded material through a chamber where high-energy electrons are created and accelerated through the material. When the high-energy electrons hit the PE molecules, some of the carbon-hydrogen and carbon-carbon bonds are broken. The breaks create free radicals in much the same way as high temperatures created free radicals in the branching of LDPE. In electron irradiation, the free radicals are more likely to react with another free radical in the vicinity because the polymer chains are already fully formed and somewhat restricted in their movement. (In contrast to the branching phenomena in LDPE where the free radicals are created during the polymerization process and the molecules are still highly mobile). This tendency to react locally will often result in a rearrangement of the bonds between atoms, which can cause crosslinks to form between nearby polymer molecules. As a result, the PE is converted from a thermoplastic to a thermoset material. Electron irradiation is more effective in amorphous regions than in crystalline regions because in crystalline materials the structures are so rigid that electron penetration is more difficult and when a free radical is formed, less rearrangement is likely to occur. Hence, highly amorphous LDPE is the material that is most commonly crosslinked by irradiation.

In a chemical crosslinking process, a special type of peroxide is used to initiate the addition polymerization reaction. This peroxide has the capability of forming multiple free radicals on the ends of a large organic molecule. (Triallylcyanurate is one of the most common of these chemical crosslinkers.) These special peroxides begin several polymerization chains that are all joined together by the multiended peroxide molecule and are, therefore, crosslinked. This method is effective in both LDPE and HDPE.

The crosslinking of PE by either the electron beam bombardment or by the chemical method results in a polymeric material that cannot be melted prior to its decomposition. It becomes, therefore, a thermoset. However, the number of crosslinks formed is usually not sufficient to change the fundamental nature of the material, except in some selected properties. In most cases, crosslinked and noncrosslinked PE are difficult to distinguish without subjecting the material to testing for those specifically modified properties, which will be discussed later. This testing often requires the use of sophisticated equipment. However, a simple method of distinguishing crosslinked from noncrosslinked PE is to place a sample of the material in question on a hot plate. Both materials will soften and become tacky as they heat. As the temperature continues to increase, the noncrosslinked material will eventually melt into a liquid. The crosslinked material will continue to soften but will never completely liquify. In the crosslinked sample, some of the polymer chains may not have been linked to other chains and these chains will liquify. However, the bulk of the material in the crosslinked sample will not liquify. Eventually the bulk of the noncrosslinked material will char.

Some properties are, of course, strongly affected by crosslinking. Both the electron and the chemical crosslinking can give significant increases in the impact toughness and the environmental stress crack resistance of PE. Applications in which impact toughness is important include trash containers and storage barrels, especially those where cold weather impact is expected. Wire and cable coating is an application in which the environmental stress crack resistance is important. To be effective, the crosslinking should usually link over 80% of the molecules. (This level is determined by solvent extraction.)

Another application for crosslinked PE is shrink tubing. This application relies upon the *memory* of the material when the crosslinks are formed. The memory of the material reflects the fact that when the crosslinks are formed they are in a stable energy state. If the material is subsequently distorted, the internal energy of the system is raised and the system becomes less stable. The material has a strong internal energy that will favor a return to the original, lower-energy shape. This contributes to the elasticity in plastics. Therefore, if the material is first formed into a tube and then crosslinked, the energy of the crosslinked material will be set at a level that favors returning to the original form. If the material is subsequently heated, expanded, and then cooled in the expanded form, the internal energy of the system will be higher. (This is like an elastic band that is stretched and then not allowed to spring back.) If the material is later heated sufficiently to allow the molecules to move, they will return to their shape when first crosslinked. In other words, the tubing that is crosslinked in one size and then expanded will shrink to its original, crosslinked size when heated. Additional information about radiation processing of polymers will be discussed in the chapter on radiation processes.

POLYETHYLENE COPOLYMERS

The addition polymerization method allows copolymers to be made by mixing two or more types of monomers that each contain a carbon-carbon double bond. The polymerization conditions, such as temperature, pressure, monomer concentrations, catalyst and initiator amounts all affect the nature of the copolymer, chiefly by determining the number and sequence of arrangement of the monomers along the chain. The types of copolymers illustrating the various monomer sequence possibilities were discussed in the chapter on polymeric materials (molecular viewpoint).

The natures of the monomers, that is their chemical, steric, mechanical and other properties, as well as the interactions which may occur between the monomers or, after they have copolymerized between the segments of the polymer, also have a profound effect on the nature of the copolymer.

The many variables that can affect the properties of copolymers can result in a wide range of copolymer materials. These effects can be classified into two groups of copolymers. In the first group the natures of the copolymer are substantially different from the natures of any of the monomers used to form the copolymer. The second group of copolymers are those in which the copolymer properties are similar to those of one of the monomers when it is a homopolymer. LLDPE is one of these latter types of copolymer. As was previously indicated, these materials are usually referred to as modified homopolymers. The second monomer is called a comonomer. The comonomer is added for some specific purpose which gives only minor changes, and the modified materials are used in applications that would be typical for the homopolymer. In the case of LLDPE the comonomer was chosen principally to prevent crystallization. In other cases, comonomers have been added to improve the thermal stability of polymers, to add greater flexibility, or to give improved impact toughness, and so on. If these changes are small, the copolymer is considered as a homopolymer. When the changes are large, the new material is called a copolymer.

Copolymers of ethylene (major changes in properties) are numerous. Many of the properties of these can be predicted based upon the principles discussed in the molecular and micro views of polymers that were discussed previously.

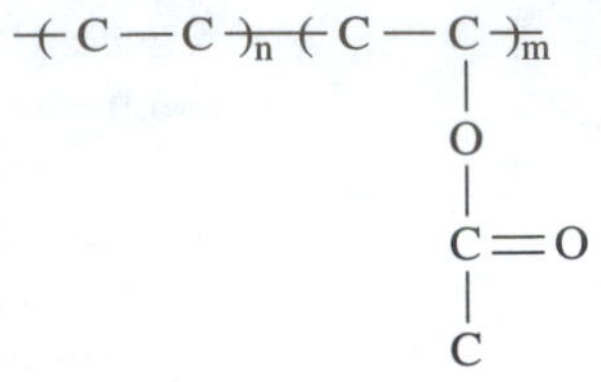

Figure 6.6 Generalized representation of ethylene vinylacetate (EVA) copolymer showing separate repeating units.

$$+\!\!\!\begin{array}{c}\text{C}\!-\!\text{C}\!\end{array}\!\!\!\Big)_{\!n}\!\!+\!\!\!\begin{array}{c}\text{C}\!-\!\text{C}\!\end{array}\!\!\!\Big)_{\!m}$$

Ethylene Vinylacetate Copolymer (EVA)

The repeating unit formula for EVA is given in Figure 6.6. In this representation the polymer repeating unit for each of the monomers is shown separately, with each usually having a different subscript.

The vinylacetate monomer contains a carbon-carbon double bond and a pendent group with several atoms, some of which have significantly different electronegativities so parts of the pendent group are highly polar. The vinylacetate monomer reduces crystallinity in the copolymer (because of steric interactions with the bulky pendent group) and increases the chemical reactivity of the copolymer because of the regions of high polarity. The net result is a very flexible copolymer that bonds well to many other materials. This combination of properties makes an excellent adhesive and that is a major application for EVA. (Elmer's Glue™, a Borden Chemicals trademark, is a well-known example of EVA with other additives used for this application.) In films, the higher polarity of the copolymer increases the tendency of the material to attract itself and other polar materials. This attractive nature is called *cling*. Some stretch films and other packaging materials make use of this property.

The polarity also changes the permeability of the material. EVA films are much more permeable to water and other polar materials but are better barriers to oils and other nonpolar substances when compared to PE homopolymer. The optical clarity of EVA is better than PE film because of the reduced crystallinity.

Commercial concentrations of vinylacetate groups in the copolymer are 5 to 50%, with 5 to 20% being the largest product group. At these concentrations, the costs are still low, bonding properties are excellent, and flexibility is good but the material is not too soft.

Other applications for EVA include flexible packaging, shrink-wrap, automotive bumper pads, flexible toys, tubing, and flexible thread, cords, and cabling.

If the manufacturing equipment is designed correctly, EVA can be made in the same polymerization reactors as LDPE. Hence, the resin manufacturer can shift production from one product to the other to meet fluctuating market demands. The increasing number of uses for EVA, coupled with the higher price and higher profitability, have resulted in a preponderance of EVA production in many of the facilities having this dual product capability.

Ethylene Acrylic Acid Copolymer (EAA)

Acrylic acid is a monomer used in copolymerizations with ethylene. Acrylic acid contains a group that is called an organic acid that will easily give up a hydrogen ion (H^+) to a chemical that will accept hydrogen ions. (This is the typical behavior of most acids.) The groups that accept the hydrogen ions are called bases. The acid group is pendent off the backbone.

The backbone is formed from a carbon-carbon double bond by addition polymerization. The two monomer types (ethylene and acrylic acid) react to give copolymers that are similar in structure to the ethylene vinylacetate copolymers except that the acrylic acid group is even more polar than the vinylacetate group.

An appropriate base, such as a metal hydroxide or oxide, can be added to the copolymer material and cause an acid-base reaction to occur. In the common acid-base reaction with EEA and a metal oxide or hydroxide, the metal forms an ion that is attracted to the copolymer by ionic bonding. The metal ions can have positive charges of +2 and so they will be able to form ionic bonds with more than one copolymer molecule. When this multiple bonding occurs, the two polymers are linked together by ionic bonds in a crosslinking fashion. This is illustrated in Figure 6.7, where the metal ion is Zn^{+2}.

These ionic crosslinks are not nearly as strong as the covalent crosslinks that are formed in thermoset materials. Ionic crosslinks can be broken as the material increases its movement when heated. Therefore, ionic crosslinked materials do not become thermosets. They can be processed in normal thermoplastic processing equipment. These copolymers that have ionic crosslinks have been given the name of *ionomers*. (These materials were originally developed by DuPont and are trade named Surlyn^{TM}, a DuPont trademark.) Typical concentrations of acrylic acid groups in EAA are 3 to 20%.

As would be expected from the presence of a polar pendent group, the bondability of EAA is much higher than for PE. Crystallinity is lower in EAA than in PE and so film clarity is higher. EAA films are more resistant to oils and greases than PE but are more permeable to water vapor. They are, therefore, another choice for food packaging films and shrink-wrap.

The presence of the ionic bonds increases significantly the impact toughness, tensile strength, puncture strength, and abrasion resistance of the ionomer compared to PE. This leads to applications such as golf ball covers, bowling pin coatings, and automotive bumper pads.

Figure 6.7 Generalized representation of ethylene acrylic acid copolymer (EAA) showing the formation of ionic crosslinks.

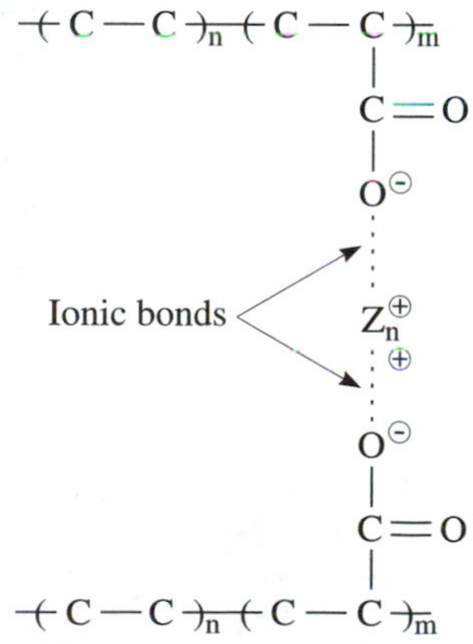

The gradual reduction of ionic bonding with temperature increases the melt strength of ionomers in comparison to PE homopolymers. Therefore, applications in which high melt strength is an advantage, such as blow molding large parts or some coating applications, will often specify ionomers. The increased bondability of ionomers also increases the paintability and printability of these polymers, thus further increasing their use in coating applications.

Ethylene Propylene Copolymers (EPM)

When ethylene and propylene are copolymerized such that the copolymer is random, the irregular structure prevents crystallization from taking place. The copolymer has a T_g that is intermediate between those of HDPE (about $-110°C$) and polypropylene (about $-20°C$). The copolymer is, therefore, in the rubbery region (above T_g) when at room temperature. The flexibility is further increased because of the low crystallinity. The flexibility and chemical inertness make these materials attractive as alternatives to other elastomeric materials for applications such as o-rings, gaskets, hoses, mats, weather strips, wire coatings, and fabric coatings.

Ethylene and propylene can also be copolymerized with small amounts (3 to 9%) of a monomer containing two carbon-carbon double bonds. Substances containing two carbon-carbon double bonds are called *dienes*. ("Di" means two and "ene" is the name for a carbon-carbon double bond, as in ethylene.) When three monomers are polymerized, the resulting material is sometimes called a *terpolymer*. ("Ter" means three.) The terpolymer formed from ethylene, propylene and a diene is called EPDM.

EPM and EPDM are elastomers and will be discussed in greater detail in the elastomeric (rubber) materials chapter. These materials and blends of these materials with PE and polypropylene are collectively called *thermoplastic olefin elastomers (TPO)*.

POLYPROPYLENE (PP)

The repeating unit for polypropylene (PP) is given in Figure 6.8.

The presence of the pendent CH_3 group permits the formation of three different types of PP. These three types of molecules differ in the way the atoms are spatially arranged about the backbone carbons and are called *stereoisomers*. These three arrangements (stereoisomers) are illustrated in Figure 6.9.

The differences in these spatial arrangements can be seen by focusing on the backbone carbon to which the pendent group is attached. Note that this backbone carbon has three

Figure 6.8 Polymer repeating unit for polypropylene.

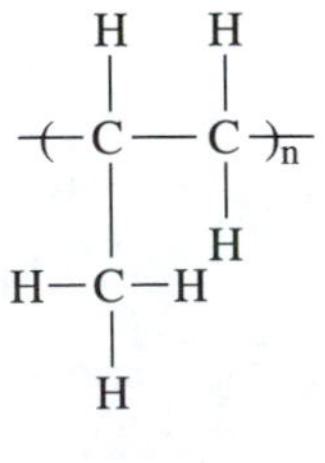

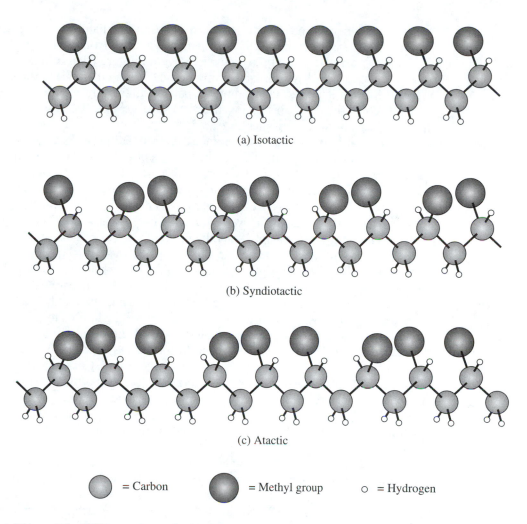

(a) Isotactic

(b) Syndiotactic

(c) Atactic

⬤ = Carbon ⬤ = Methyl group ○ = Hydrogen

Figure 6.9 Different types of polypropylene that depend upon the arrangement of groups attached to the carbon (stereoisomerism).

other carbons attached to it—the carbon in the pendent group and the two adjacent backbone carbons. Carbons to which three other carbons are attached are called *tertiary carbons*. The other backbone carbon in the PP repeating unit is not a tertiary carbon because it has only two carbons attached to it. Carbons to which only two other carbons are attached are called *secondary carbons*. (The terms tertiary and secondary are introduced to assist in identifying which of the two types of backbone carbon is being referred to in the discussion.) The tertiary carbon is key in understanding stereoisomerism.

In the *isotactic* configuration which is represented in Figure 6.9a, the pendent group is always attached to the tertiary carbon on the same side. This results in a very regular structure. (This arrangement can be compared to a line of people who are all facing the same direction, each holding a balloon in his/her right hand.) Isotactic actually comes from words meaning "same" (iso) and "hand or touch" (tactic).

Another arrangement of the atoms is shown in Figure 6.9b. In this arrangement the pendent group regularly alternates from one side to the other side of the tertiary carbon. (This is like the line of people holding the balloon alternately right hand and left hand all the way down the line.) This arrangement is called *syndiotactic,* which implies that a set pattern exists in the arrangement.

In the third arrangement of the atoms, the pendent group is attached to the tertiary carbon in a random fashion. (This is like the line of people where some have the balloon in their right hand and some have it in their left hand, but no particular order is established.) This arrangement is called *atactic,* which implies that there is no pattern.

The differences in properties between isotactic, syndiotactic, and atactic PP are quite pronounced and arise from the way in which the polymer molecules can pack together. Only the isotactic arrangement allows the molecules to pack tightly into crystalline structures. In the syndiotactic and atactic arrangements, the methyl pendent group (CH_3) is too large to allow for tight packing and crystalline regions are not formed. Hence, isotactic PP is much more rigid and strong in comparison to the rubbery nature of syndiotactic and atactic PP. The only PP of commercial importance is the highly crystallized isotactic arrangement and the remainder of this discussion will be limited to this type.

In order to obtain the regular arrangement of atoms required to make isotactic PP, a catalyst is used to force this arrangement during the polymerization of the polymer. Such catalysts are called *stereoregular.* The Ziegler-Natta catalyst used to produce HDPE is of this type. Other types of stereoregular catalysts (called *high-selectivity catalysts*) have now been developed that are increasingly used to produce PP, in part because the ability to control the shape and length of the polymer is even better with the new catalysts. Therefore, commercial grades of PP are made using Ziegler-Natta or some other stereoregular catalyst.

It is not surprising that PP and PE, especially HDPE, have similar properties and compete for many of the same applications. However, PP and PE differ in some important respects and these differences have led to preferences for one or the other in various applications. PP is stiffer than PE, so in applications requiring flexibility (such as wire coating), one of the PE materials would be used. On the other hand, if greater stiffness is needed, PP is the preferred resin. This is especially true if the application also requires abrasion resistance or hardness, such as for gears, toys, automotive battery cases, and seats for stacking chairs.

The resistance to environmental factors is similar for PP and PE. PP is somewhat more susceptible to UV and oxidative degradation than is PE but is more resistant to stress cracking than PE. Hence, crosslinking of PP for improved ESCR is not practiced commercially, partially because the electron beam radiation degrades the PP.

PP has a higher glass transition point and a higher melting point than PE (except for UHMWPE). This means that processing temperatures are generally higher, but it also means that service temperatures are higher. Sterilizable medical devices, dishwasher-safe food containers, and appliance parts are often made of PP for this reason.

A very important property difference that has led to many applications for PP is its superior resistance to cracking from mechanical stresses. PE materials will readily blush and craze when subjected to bending, but PP will not. Applications requiring this property include carpets, ropes, strapping tape, and molded items incorporating integral hinges. (*Integral hinges* or *living hinges* are formed when the hinge of a container is formed from the same material as the container itself and is molded as one piece. The hinge is usually a region of the part that is thinner or narrower so that bending will preferentially occur in that region.)

The superior stiffness of PP over PE and the low price of PP compared to the engineering plastics have led to its use in some structural applications. If additional stiffness or strength is needed, reinforcements can be added to PP. For instance, the addition of 30% short fiberglass reinforcements can double the tensile strength and impact resistance of PP. Impact modifiers can be added to PP to further improve impact strength, especially for low-temperature applications where PP is less impact resistant than HDPE. EPDM, the copolymer of PP, PE, and a diene monomer, has improved impact properties and much greater elongation than either PP or PE.

Fillers (such as calcium carbonate or talc) are often added to PP up to about 30% concentration by weight. The filled plastic has improved stiffness, lower mold shrinkage, and lower cost. Many molded automotive parts have been converted from thermoset materials to filled PP because PP can be molded into very complex shapes using fast molding cycles and still retain the dimensional stability that was previously provided by the thermoset materials.

POLYVINYL CHLORIDE (PVC)

The vinyl chloride monomer consists of a carbon-carbon double bond and a pendent chlorine atom and three hydrogen atoms (see Figure 6.10). This monomer polymerizes by the addition (free radical) polymerization method. This polymerization is done commercially using suspension, emulsion, bulk, or solution techniques and produces polymers that are nonstereospecific (atactic), although isotactic, syndiotactic, and atactic are all known. Vinyl chloride monomer (a gas) is a suspected *carcinogen* (may cause cancer) and has been shown to be toxic in large doses. Therefore, stringent standards have been established for exposure limits to vinyl chloride monomer, including exposure to the monomer during polymerization and a maximum permissible level of monomer residue in the polymer. These standards have been in place since the early 1970s in the United States. Therefore, polymerized PVC is free of monomeric vinyl chloride and is not considered a potential carcinogen. In fact, several food grades of PVC packaging material have been approved by the FDA. The polymerized product is usually a clear or white flake or powder, which is called polyvinyl chloride.

The presence of the chlorine pendent atom causes significant property changes in PVC compared to the polyolefins (PE and PP). The chlorine atom prevents close packing of the

Figure 6.10 Polymer repeating unit of polyvinyl chloride (PVC).

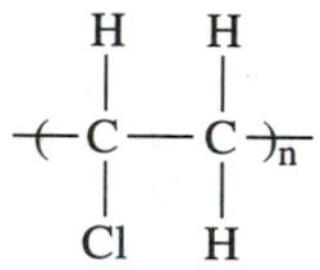

polymer and also gives a solvent sensitivity that is not seen in PE or PP. The solvent sensitivity is so important that two types of PVC have arisen based upon whether the PVC is modified with a solvent (plasticizer) or not. The unmodified PVC is called *rigid PVC* and the solvent-modified PVC is called *plasticized PVC,* or *vinyl.*

An important property of PVC that is common to both rigid PVC and vinyl is flame retardance. When PVC burns, HCl gas is produced. This gas is more dense than air and therefore smothers the flame by excluding oxygen from the flame vicinity. As a result, PVC will burn with difficulty if an externally fueled flame (such as a bunsen burner) is maintained in contact with the material but will extinguish if the fueled flame source is withdrawn. Materials with this characteristic are called *self-extinguishing.* Many applications for PVC, both rigid and vinyl, depend upon this self-extinguishing property. For instance, vinyl is used for wire and cable coating (such as Romex™), chiefly because it is self-extinguishing rather than because of its insulating capability, which is not as good as PE or PP.

Another inherent property of both rigid PVC and vinyl is thermal decomposition. PVC has a tendency to decompose by giving off HCl gas and forming crosslinks among the polymer chains when the HCl leaves. The color of the PVC generally changes to a yellow or, at advanced stages of decomposition, a brown. This decomposition can occur at temperatures near the melting point. Once started, the decomposition process tends to cause other nearby H and Cl atoms to combine, thus accelerating the decomposition. When sufficient decomposition sites have been formed within the molecule, the decomposition pro-

Photo 6.1 Various types of rigid PVC and plasticized PVC (vinyl) (Courtesy of Elf Autochem North America, Inc.)

ceeds rapidly. This process is called *autocatalytic decomposition*. It is important, therefore, to prevent decomposition by minimizing the amount of the polymer's thermal exposure. Polymers which are sensitive to accumulated heating are said to have a *heat history*. This tendency for thermal degradation is very important in the processing of PVC and complicates how PVC can be processed by heating.

Molten PVC has a high viscosity that further complicates the problem of thermal decomposition because PVC cannot be heated more to decrease the viscosity and allow easier processing. Processing would be extremely difficult except for several thermal stabilizers and lubricants that preferentially absorb thermal energy, reduce the viscosity of the melt, and thermally protect the PVC, as discussed in the chapter on chemical and physical properties.

PVC is also sensitive to UV and oxidative degradation, again with the evolution of HCl, indicated by yellowing. PVC can be protected by the addition of UV and oxidation stabilizers.

Rigid PVC

When compared with PE and PP, unmodified PVC is more rigid, stronger, and more solvent sensitive. The chlorine atom is approximately the same size as the CH_3 group, as shown among the resin comparisons in Figure 6.11. The size of the chlorine atom is sufficient to interfere with close packing and crystallization resulting in a largely amorphous PVC polymer. Commercial PVC typically is less than 10% crystalline.

Even though PVC is largely amorphous, the size of chlorine atoms causes significant intermolecular interference and the polarity of the Cl atom results in intermolecular attractions, thus increasing the tensile strength and modulus compared to PE and PP. The intermolecular interactions and general stiffness also increase the glass transition and melting point of PVC. The glass transition of PVC varies somewhat with the polymerization conditions used to make the polymer, but is generally about 140° to 180°F (60° to 80°C), a temperature that is significantly higher than room temperature (73°F, 23°C). Since most applications for PVC are at room temperature PVC is commonly used **below** its glass transition temperature, whereas PE and PP are used **above** their glass transition temperatures. It is not surprising, therefore, that PVC is much more rigid and brittle than the polyolefins in its unmodified (rigid) applications.

Rigid PVC is used in many applications where cost is a major factor. As a result, fillers are often used to reduce the cost of the product. These fillers also add stiffness and may give some thermal stability benefit. The fillers usually reduce toughness. Typical fillers are

Figure 6.11 Size representations of common pendent groups in commodity polymers.

talc, calcium carbonate, and clay. Impact modifiers can also be added to PVC so that the toughness will be improved.

When properly formulated, rigid PVC can be processed by most conventional thermoplastic processing methods. In each of these processes care should be taken to reduce the heat history of the resin. This is done by (1) processing at the lowest possible temperature, (2) using additives that protect the resin, (3) utilizing equipment that gets good mixing without excess heating (such as twin-screw extruders), and (4) ensuring that the concentration of regrind in the mix is low so that the regrind material (which has a high heat history) is surrounded by virgin resin.

Care should be taken to carefully inspect the metal dies and tooling used in the process because of the likelihood of corrosion from HCl that is unavoidably given off by PVC, at least in small amounts, in any of the resin-heating processes. This corrosion phenomenon can be very severe in some configurations and formulations. In these cases, replacement of molds, tools, fixtures, and screws should be considered and added into the overall cost of any PVC part. Corrosion-resistant coatings and metals are available to minimize the effects of HCl, but these also add cost to the tooling.

Extruded PVC products include house siding, pipe for sprinkler systems and electrical conduit, rain gutters, and window frames. Rigid PVC bottles can be made by blow molding, although plasticized PVC is more commonly used for bottles because it is less rigid. PVC is a commonly thermoformed material for low-cost, semirigid applications such as tote bins. Innumerable injection molded parts are produced of rigid PVC, especially when low cost, rigidity, and strength are needed.

The ability of PVC to be solvent welded or joined is an obvious advantage over PE and PP that generally must be welded with heat or some method that causes molecules to fuse together. Solvent welding employs traditional adhesives and is much easier to use than the fusion or heat seal methods. A common example of the ease of joining PVC is seen in the solvent adhesive joining of sprinkler pipe. The materials to be joined, usually PVC pipe and a PVC fitting, are coated on the joining surfaces with the solvent adhesive. The parts are then placed together and allowed to dry. When the joint is dry (usually in about 24 hours, although 80% of the strength is achieved in 1 hour), the joint is often as strong as the surrounding plastic.

The sales of rigid PVC and vinyl are approximately equal.

Plasticized PVC (Vinyl)

When plasticizer has been added to PVC, the plastic is substantially more flexible than the rigid PVC that has just been described. This plasticized PVC is commonly called *vinyl*. (The term vinyl is occasionally used to describe any molecule containing a carbon-carbon double bond and a noncarbon and nonhydrogen pendent group. This definition, though chemically correct, leads to confusion in plastics where the term vinyl has been associated so closely with PVC. Therefore, throughout this book, and in most common usage, the term vinyl, unless clearly indicated otherwise, will denote the plasticized form of PVC.)

The action of plasticizers was previously discussed in the chapter on chemical properties, where plasticizers were identified as chemicals that were added to plastics to soften them and add flexibility and elongation. The action of plasticizers is like a solvent for the plastic where only enough solvent is added to cause some swelling, disentangling of the molecules, and some breaking of secondary intermolecular bonds, but where sufficient intermolecular interactions still exist that the material is not liquid. The plasticized material is generally a semirigid solid.

Adding plasticizers to PVC is very beneficial for many applications. The rigidity and brittleness of rigid or unmodified PVC can be substantially reduced by adding plasticizers, although this increase in flexibility is accompanied by a decrease in tensile strength. The impact toughness of the plasticized material increases initially and then decreases. The physical properties of a plasticized PVC suggest that the glass transition temperature of the material has been reduced such that at room temperature the material has passed from the rigid zone to the leathery or rubbery zone. This reduction in T_g has been confirmed by experimentation.

The plasticizer is usually infused into PVC flakes, granules or particles before any thermal processing, rather than attempting to add the material to the melt. This pre-processing plasticization has the advantage of avoiding the difficult melting of PVC and its accompanying heat history. Since the plasticizers are solvents of the PVC, this infusion into the solid can usually be done with only a slight rise in temperature that swells the PVC and facilitates solvent entry into the plastic structure.

As discussed in the chapter on chemical properties, an ideal plasticizer would impart good flexibility to the vinyl, would be inexpensive and easy to add, would not add appreciably to the flammability of the plastic, would not be toxic, would not be extractable by sunlight, would not change the color of the vinyl, and would have low permeation rates so that it stays within the vinyl. Few plasticizers meet all of these requirements. Therefore, products have been developed that can be sprayed onto vinyl products and other plastics which can be plasticized to add plasticizer back into the plastic. (Armorall^{TM}, an Armorall Corporation trademark, is a brand of one of these products.) These products seem to work well in practice. The difficulty with these products is that they add the plasticizer to a formed part without the opportunity to heat slightly to open the structure. To get the required penetration, the plasticizer molecule must be small. Consequently, the migration rate out of the vinyl is also high and so repeated applications must be made to ensure continued plasticity.

The most common method of processing vinyl is to melt the plasticized pellets in traditional plastic-processing equipment. This melting can be done at lower temperatures than are used with rigid PVC because of the lower T_g and T_m that the plasticizer creates. This lower-temperature processing reduces the heat history and extends the useful life of PVC over high-heat-history material.

Extruded vinyl is made into many parts including tubing (brand name Tygon^{TM}), which has wide application when flexibility, low cost, and optical clarity are important. Other extruded vinyl products include sheets for floor mats and garden hose. In some of these products the improved flame retardance of vinyl over other plastics is an important characteristic. Extruded vinyl sheet can be readily thermoformed into products such as storage boxes, often with integral hinges.

Blow molded vinyl bottles are widely used where low cost, flexibility and clarity are important. Bottles for cooking oil, bleach, and shampoo are some familiar products made of vinyl.

Vinyl can be made into blown film. It is widely used for shrink-wrap, food packaging, bags for blood plasma and other medical fluids, garment bags, and wall coverings. These film applications often rely on the ease of sealing vinyl with either solvent adhesives or with low-temperature fusion or ultrasonic sealing.

Some processing methods for vinyl do not require melting the vinyl plastic. One of the most common of these is accomplished by adding sufficient solvent to the vinyl material that the vinyl dissolves or becomes suspended in the solvent. The resulting liquid material is called a *plastisol* or *vinyl dispersion*. The plastisol can be applied to other materials and then dried and fused (at moderately elevated temperatures) to form a soft vinyl covering. Many metal parts, such as racks for dishes and parts storage, are sprayed with or dipped in plastisol to provide corrosion protection and cushioning. Handles for screw drivers, pliers, and other tools are often made by dip coating in plastisol. Vinyl gloves are made by coating a mold (in the form of a hand) with plastisol and then drying.

Vinyl can be foamed to produce products such as carpet padding, weather stripping, and backing for various fabrics, and other sheet materials. Vinyl floor covering includes vinyl foam, a reinforcement layer, a printed film and backing material, as well as a vinyl wear surface.

The ability to process vinyl without heating extensively has given rise to an important processing method called *calendering*. This process involves the pressing of sheet materials together, such as vinyl onto cloth, by passing the materials through rollers. Vinyls are usually calendered by softening the polymer with high concentrations of plasticizer or solvent prior to pressing. This process can be facilitated by slightly heating the materials. The calendering technique is used extensively for coating cloth to make vinyl seats and dashboard covers in automobiles and for general-purpose vinyl fabrics. Plasticizer that evaporates out of the vinyl fabrics (bleed-out) can sometimes be sensed as an oily residue on some vinyl fabrics and as an oily coating on the inside of automobile windows. This evaporation is highest when the temperature is high and so locations directly in the sunlight, such as the dashboard, are the most seriously affected. When the plasticizer migrates out of the vinyl, the vinyl material embrittles and often cracks.

Vinyl Copolymers and Related Polymers

An important copolymer of PVC is made by combining vinyl chloride monomer with vinyl acetate monomer. The resultant copolymer is generally about 85% vinyl chloride. A major use of this copolymer is in place of PVC homopolymer to improve the flexibility of finished vinyl flooring material. Other applications include many of the same applications where PVC homopolymer could be used but where rigid PVC is too stiff or where the plasticizer problems of vinyl (such as bleed-out) are unacceptable.

A plastic material that is chemically related to PVC is polyvinylidene chloride. The monomer for this plastic has a carbon-carbon double bond but has two chlorines on one of the carbon atoms. The polymer formed from the addition polymerization of this material has excellent barrier properties, especially against oxygen gas, and is used extensively as a food wrap. Polyvinylidene films have good *cling* properties, that is, a tendency to stick to itself, which is an advantage in food wraps. A common trade name for this material is Saran™, a trademark of Dow Chemical Corporation.

POLYSTYRENE (PS)

The styrene monomer has a carbon-carbon double bond to which a benzene ring and three hydrogens are attached. This monomer is polymerized by the addition polymerization mechanism. The repeating unit for PS is represented in Figure 6.12.

The size of the benzene pendent group is represented in Figure 6.11, where the benzene ring is much larger than any of the other pendent groups (hydrogen, CH_3, or Cl) associated with the other commodity resins. The benzene ring reduces the ability of the polymer chain to bend and interferes substantially with other parts of the molecule. These characteristics prevent crystallization. Therefore, PS is essentially 100% amorphous.

The amorphous nature of PS allows light to pass through the structure without significant refraction and so PS is transparent and clear. Pellets of PS have a certain glitter or sparkle and so this unfilled, clear grade of PS has been referred to as *crystal polystyrene*. The term crystal PS refers to its appearance and not its crystallinity, which is near zero.

The chemical properties of PS are dictated largely by the presence of the benzene ring. Any molecule containing the benzene ring is called aromatic and has certain characteristic chemical properties. Among those properties are a sensitivity to aromatic and chlorinated solvents. Therefore, PS can be dissolved in these solvents and will swell in the presence of small amounts of both aromatic and chlorinated solvents. Solvent adhesives employing these solvents are effective in joining PS. However, PS is resistant to water and has been used extensively for applications, such as food packaging, where water resistance and clarity are important.

PS will burn readily with a yellow flame and dark, sooty smoke. This type of burning is characteristic of polymers with aromatic pendent groups. (When aromatic groups are in the backbone of the molecules, flammability is reduced in comparison to pendent aromatic groups.) Therefore, flame retardant additives must be added to PS to obtain slow burning or self extinguishing properties. These flame retardant additives usually make PS nontransparent and therefore eliminate one of the major advantages of PS.

The long-term clarity of PS is not good because of a tendency of the material to yellow with exposure to UV light and to oxygen. The material is subject to environmental stress cracking, which further limits its long-term use. PS is therefore most appropriate for applications of short duration (such as packaging) and not for long-term use (such as for outdoor signs). The misuse of PS in the 1950s and 1960s was largely responsible for the bad image of "cheap plastics." Light covers that yellowed, appliance covers that burned easily, and outside panels that embrittled are all examples of poor use of PS. Perhaps the most troublesome use of PS was in toys that often broke easily because of PS's brittleness.

Brittleness (low toughness) in PS is a result of the interference of the aromatic pendent groups with neighboring molecules that prevents the molecules from sliding past

Figure 6.12 Polystyrene (PS) polymer repeating unit.

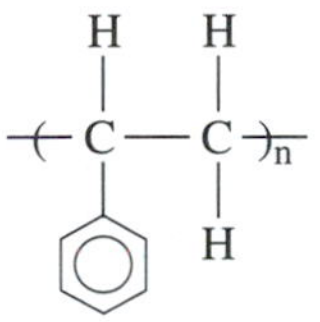

each other. Today, the most important applications of crystal PS utilize the low cost and clarity of the material and are not negatively affected by the brittleness. Most of these are applications that do not require toughness. Examples include cosmetic bottles and packages, drinking cups (such as the clear cups used on airplanes), and thermoformed packaging, such as blister packs, where the rigidity and clarity are premium characteristics.

The strength of PS is strongly determined by the molecular weight. Normal uses of PS do not usually depend upon high strength and so the molecular weight is often kept at moderate levels to facilitate easy processing.

The wide range between the softening point of PS (212°F, 100°C) and its decomposition temperature (500°F, 250°C) allows the viscosity of PS to be reduced by increasing the temperature. PS should not be held at temperatures above 300°F (150°C) for long periods of time because some degradation can occur. However, simple processing is usually of a short duration so the extent of this thermal degradation is rarely a problem. The noncrystalline nature of PS results in low mold shrinkage and minimal part warpage. All these properties simplify the processing of PS so it is readily processed by all of the common thermoplastic processes.

The development of catalysts that hold the monomer and the polymer chain in specific geometrical configurations during polymerization has resulted in the production of a PS with high crystallinity. The catalysts are part of a chemical family called *metallocenes*. These catalysts have been used to make geometrical isomers of several plastics. These are called *constrained geometry polymers* and are like the stereoisomers discussed in relation to PP except that the geometrical restraints using metallocene catalysts are even stronger than with the Ziegler-Natta catalysts. The result of using these new metallocene catalysts can be significant. Constrained geometry PS has a melting point of 520°F (270°C) which is 340°F (170°C) higher than amorphous PS. The mechanical properties are also claimed to be significantly higher for the constrained geometry PS.

Expanded Polystyrene (Polystyrene Foam)

One of the most important applications for PS is as a foamed material. PS foam is widely used for disposable drinking cups, fast-food containers, picnic plates, wall insulation, packing material for delicate instruments, and numerous other applications. (A common brand of PS foam is Styrofoam™, a Dow Chemical trademark.) The widespread use of foamed PS has drawn criticism from environmentalists who are concerned that the material is not being properly recycled, even though recycling of PS is very easy. The plastics industry is addressing these environmental concerns and many of the steps being taken are outlined in the chapter entitled "Environmental Aspects of Plastics." Even with the environmental concerns, the use of PS foam continues to rise. This increase in use is undoubtedly because of the superior performance of this material in comparison to other competitive materials, and the environmental difficulties, which are also present with almost any other material that might be used in place of the PS foam.

The advantages of PS foam over competitive materials include the following:

- Low heat flow, making for good insulation.
- Good energy absorption for packaging delicate instruments and other impact applications.

- High buoyancy.
- High stiffness to weight so that parts can be self-supporting and lightweight.
- Low cost per volume.

The process for making foams will be discussed in detail in the foaming processes chapter, but some general considerations here will assist in understanding how the foam is processed to make PS parts. The most common method for making PS foam involves the use of prefoamed PS beads. The resin manufacturer makes these beads by adding an inert gas, such as pentane, during the polymerization. Under the proper conditions, which usually involve a water suspension environment, polymerization occurs with the formation of small, internally foamed PS beads with the inert gas trapped inside. These beads are then shipped to the part molders. The part molders convey the beads from the shipping container to the mold by air pressure or vacuum. During this conveying step, the beads are often heated and will expand up to 20 times their original volume, but this is only a partial expansion. The beads are conveyed into molds that are then closed and heated, usually by passing steam through the mold. This heating causes the beads to further expand, often doubling their size over the partially expanded size, and to fuse together. The molds are then cooled and the parts removed. A close examination of many parts made from PS will permit identification of the individual beads that have been fused to form a continuous part.

Foamed PS slabs are made by a different technique that is much more typical of other foamed plastics. These slabs are made by introducing a foaming agent into a PS solution or melt that is extruded to form a flat sheet. As the sheet leaves the end of the extruder, the pressure on the melt reduces, thus allowing the foaming agent to expand and foaming of the PS occurs. The dimensions of the slab are set by causing the foamed slab to pass between dimensioning belts or plates with a specified gap between them. The foamed PS slab is then cooled and cut to dimension.

ALLOYS AND BLENDS

The property problems of PS, chiefly brittleness and poor weatherability, have led to the development of several polymeric materials that use PS for the principal ingredient but have substantially improved impact properties or weathering. These improved materials can be made in a variety of ways, each of which gives a different set of properties. The methods can be classified as follows: copolymerization, alloying, and blending. Firm definitions have not been agreed upon for each of these terms, but a working definition of each is:

- *Copolymerization* is the combining of two or more types of monomers such that the resulting polymer contains some of each of the types of monomer. (Copolymerization was discussed in the chapter on polymeric materials [molecular viewpoint] but is reviewed here because it is so important in understanding alloying and blending, and because the methods of alloying and blending often depend upon the use of a copolymer.)
- *Alloying* is the combining of polymers, after they have formed, or of a polymer and a monomer, into a single-phase, *homogeneous* polymer material. Some chemical attraction between the combined polymers is usually required to form an alloy. Metals also form alloys where two different metals combine together.

■ *Blending or mixing* is the combining of polymers, after they have formed, in such a way that the resultant polymer material is two or more phases.

Alloys and blends have become important for many polymers besides PS. Nylon, polycarbonate, polypropylene, acrylics, and many others have some grades that are made by alloying or blending. Estimates of polymer production indicate that blends and alloys represented over 20% of the total resin market. Over 1000 patents per year (an average of about three per day) were issued for polymer alloys and blends during the 1980s. That rate has decreased somewhat because of the problems in creating new alloys and blends with sufficient stability throughout the entire plastic-part-manufacturing process. Still, even with the reduction in intensity of effort into alloys and blends, the potential advantages that can be realized from these materials are enormous. Future developments in this area are inevitable and should prove to be of major importance in the plastics industry.

One of the promises of alloying or blending is that new polymers can be developed without the need for the vast investments of time and equipment that are required to make new polymers by polymerization of monomers. Moreover, alloying or blending can be done by custom compounders as well as by resin manufacturers, suggesting that the competition will tend to keep prices low. Another advantage is that the effort to develop a new alloy or blend can be justified on much smaller potential sales than would be required for the development of a totally new polymerization. The diversity of alloys and blends, that is, the choices of polymers than can be combined and the many ratios of combination, given far more property possibilities than are available from just pure resins. Alloys and blends can therefore be used to give tailored properties for specific applications.

The current and near future work in alloys and blends seems to be proceeding in three directions. First, efforts are under way to improve the properties of the commodity resins, especially PP and PVC through alloying and blending. Second, properties of engineering plastics are being improved for specific applications. Third, alloying and blending are being used to make new high-performance resins.

Some general observations on the methods of making copolymers, alloys and blends and the resulting properties are as follows:

■ Copolymerization generally results in the best properties that can be obtained from two different polymer-forming materials. It is also the most expensive of the processes for mixing two polymer-forming materials.

■ Because molten polymers are more reactive than solid polymers, alloys are almost always formed by using an extruder to melt and mix two or more polymers. The extrusion mixing of polymers has the danger that overheating the polymers can form crosslinks, degradation, or significant increases in molecular weight (which can cause gels to form).

■ If the polymers are *miscible,* that is, the two polymer melts will mix to form one liquid phase, forming alloys when they cool. If the molten polymers are *immiscible,* that is, two phases of liquids are formed when they are mixed, a blend is likely to form when they are cooled. (Two common miscible liquids are water combined with ethyl alcohol which form one phase. Two immiscible liquids are water and oil which form two phases.)

- The ability of polymers to form alloys is increased if there is an attraction between the polymer molecules. This attraction can be enhanced in some polymers by reactions that add reactive sites to the polymer chain. Copolymerization is one reaction method that is used to add reactive sites to PS and improve its ability to react and form alloys.
- An alloy will usually have properties, at least in some respects, that are superior to the properties of either of the materials separately. This enhancement of properties is called *synergy.*
- *Compatibility agents* are materials that can be added to two immiscible polymers to couple them and allow alloys to form.
- Blends can be formed by mixing dry polymers and then extruding.
- A blend will usually have properties that are averages of the properties of the polymers that were combined to form the blend.
- Physical properties of alloys can be as much as 10 times greater than for blends.
- Recycling of alloys and blends is more difficult than for homopolymers because of the potential need to separate the polymer materials in the recycling process.

A simple example of how alloying improves properties can be seen by mixing a crystalline resin with an amorphous resin. The resulting alloy can have the improved dimensional stability and solvent resistance of the crystalline material and the improved warp resistance and melt strength of the amorphous material.

High-Impact Polystyrene (HIPS)

Various elastomers (rubbers) can be added to PS to improve its impact toughness. The two most common rubbers for this purpose are butadiene rubber (BR) and a copolymer of styrene and butadiene rubber (SBR). The SBR copolymer will typically have no more than 25% to 30% styrene. Since these are rubber materials, their specific properties will be examined in the chapter discussing elastomeric (rubber) materials, but here we can note that they have very high elongation, high flexibility, and are quite soft.

The monomer for butadiene and the polymerized form are given in Figure 6.13a and b. Note that after the polymerization has occurred, a double bond remains in the polymer

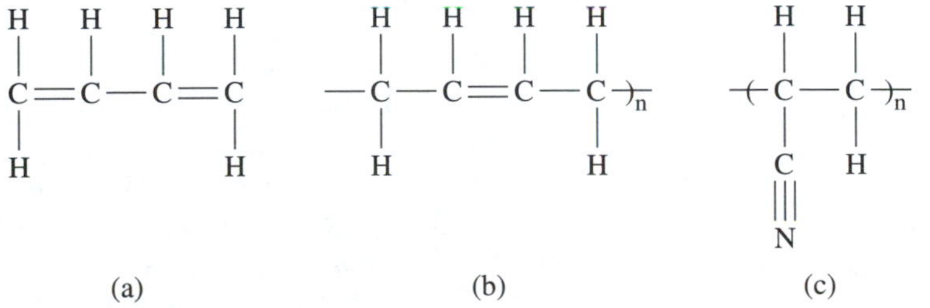

(a)   (b)   (c)

Figure 6.13 Butadiene monomer, polybutadiene repeat unit, and polyacrylonitrile repeat unit.

repeat unit. This is an active site that will allow crosslinking to occur, as described in more detail in the chapter on elastomers.

The original method of making HIPS was to form the styrene into PS and then combine the PS and rubber polymers in a two-roll mill, an internal mixer, or a mixing extruder. The result of these mechanical blends was only a minor improvement in the impact properties over unmodified PS. Today, the preferred method of making HIPS is to dissolve the rubber polymer in styrene monomer and then polymerize the styrene by conventional means. The resulting material contains not only the rubber polymer and PS but a copolymer in which short PS side chains are grafted onto the rubber polymer. These graft copolymers (compatibility agents) improve the compatibility of the PS and rubber polymers and the entire mixture becomes an alloy. The total butadiene rubber content in HIPS made by this method is usually less than 30%. Resin manufacturers obtain the optimum properties by controlling the size of the particles that result from the polymerization process to be in the range of one to ten microns.

Since only one phase is present in this alloy, HIPS made by this method is clear, although it will not pass light as completely as unmodified PS.

The impact toughness of HIPS made by this copolymer method can be as much as seven times greater than nonmodified PS, but tensile strength, hardness, and softening temperature are all lower.

Styrene Acrylonitrile (SAN)

Styrene and acrylonitrile monomers can be copolymerized to form a random, amorphous copolymer that has improved weatherability, stress crack resistance, and barrier properties. The polymer repeat unit for acrylonitrile is given in Figure 6.13c. The copolymer is called styrene acrylonitrile or SAN. The styrene content in these copolymers is about 65 to 70%. The copolymer is clear, although not as clear as crystal PS. SAN yellows more quickly than PS but still has good gloss.

The polar nature of acrylonitrile increases the resistance of the copolymer to nonpolar solvents such as oils and greases. This polarity also increases interactions between the copolymer molecules resulting in higher tensile strengths (about 10% more than PS) and higher heat distortion temperatures (about 10°C more than PS). The higher heat-distortion is accompanied with higher processing temperatures, thus making processing somewhat more difficult.

Applications for SAN include dishwasher-safe containers, oil-resistant containers, inspection ports and gauge covers, packaging for foodstuffs, pharmaceuticals, and cosmetics where oil resistance is needed, and bottles employing the barrier properties.

The acrylonitrile polymer (PAN), without any styrene, is used extensively for fibers (such as Orlon™, a DuPont trademark,), which are easily colored and have a soft texture. Sweaters and some carpets are major applications. Polyacrylonitrile can also be used as a molding material, competing with the engineering thermoplastics. High-quality (high strength and stiffness) PAN fibers are used as the starting material in the formation of carbon/graphite fibers, which are used extensively in composite materials. In this process the PAN fibers are pulled through a series of ovens, where the fibers are treated under high temperature and controlled atmospheric conditions, which convert the fibers into carbon/graphite fibers.

Acrylonitrile Butadiene Styrene (ABS)

The above discussions of HIPS and SAN show that modifications to PS can increase selected properties of PS by creating copolymers, alloys, and blends with modifying polymers. The impact toughness of SAN is slightly higher than PS but not as high as HIPS; however, the weatherability, barrier properties, and heat distortion temperature of SAN are higher than HIPS. In an effort to obtain the best properties of both HIPS and SAN, a combination material called acrylonitrile butadiene styrene (ABS) has been developed.

The primary focus of ABS development has been to improve impact toughness over SAN and HIPS while still providing good solvent resistance, gloss, weatherability, and processability.

The complex relationship of the three materials is illustrated in Figure 6.14, where polyacrylonitrile is at one apex of the triangle, polybutadiene is at another apex, and PS is at the third apex. ABS is in the middle of the triangle. The property changes created by increasing the concentration of one material relative to the others are seen by moving along the sides of the triangle. For instance, barrier properties are increased in PS by moving toward polyacrylonitrile. This represents the formation of SAN. Similarly, by moving from PS toward polybutadiene, the HIPS region is entered and the toughness is increased.

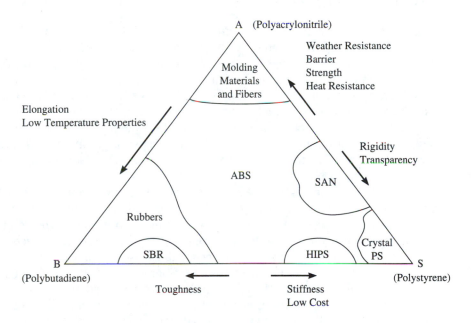

Figure 6.14 Representation of the changing properties among polyacrylonitrile, polybutadiene, polystyrene, and ABS.

Initial ABS combinations were made by blending SAN with butadiene-acrylonitrile rubbers. A second method of making ABS combines styrene and acrylonitrile monomers to either BR or SBR rubbers to form polymers and grafted copolymers and *terpolymers* (three polymers united into one molecular structure). The property variation from these two methods, including variations within each of the methods, is very large, resulting in many different grades of ABS. In general, the desirable properties—toughness, good solvent resistance, gloss, and processibility—have been obtained. The cost of ABS is 50 to 70% more than general-purpose PS, but the advantages of ABS have proven to be worthwhile for many applications.

ABS can be processed by any of the standard thermoplastic processing methods. Injection molded ABS is used for telephones, helmets, steering wheels, and small appliance cases. Extruded ABS is used for pipe (especially for drain, waste, and vent pipe), and many extruded specialty shapes. The solvent adhesive capability of ABS allows for convenient joining of ABS products such as pipe. Thermoformed ABS products include snowmobile covers, luggage, golf carts, and refrigerator liners, among many others. ABS can be electroplated and is used for metallized knobs and handles for radios, television sets, and automobiles. Foamable ABS is made into wall panels, picture frames, handles, ice buckets, and furniture by injection molding.

CASE STUDY 6.1

Typical PVC Formulation

Because of the difficulty of manufacturing/molding PVC, hundreds of different formulations using many different additives have been developed. The number of additives and their purposes are so complex that a typical formulation for making rigid PVC house siding is presented in Table 6.4 to illustrate the purposes of each of the components in the mix. This formulation shows the number of additives and the approximate concentrations which are often added to rigid PVC formulations.

Each of the materials in the formulation has a purpose. Some of the materials are proprietary and are sold based upon specific property improvements that these materials can produce. Other materials are generic and are sold on the basis of supply, service, and cost.

Formulations such as the one given here can be obtained from these suppliers as well as from the PVC resin manufacturers. These formulations are, of course, just guidelines and many PVC molders and extruders develop their own versions. The point of showing this formulation is not to suggest that one recipe for PVC siding is correct, but rather to indicate the complexity of the formulations that are used commercially.

Other resins, such as PE, have far fewer ingredients. A typical formulation for PE might have only three ingredients: resin, UV protectant, and an antioxidant. The complexity of PVC simply illustrates one point—PVC is difficult to manufacture into parts because it is so difficult to melt without degradation.

Table 6.4   Formulation for Rigid PVC Siding Showing the Many Additives Needed
for Processing and Property Protection

Ingredients	Concentration (parts per hundred)	Function	Comment
PVC resin	100.00	Base resin	Relative viscosity of 2.25 (a measure of molecular weight)
Processing aid	0.75 to 1.50	Promotes fusion	Improves mixing
Acrylic impact modifier	4.50 to 7.00	Impact strength	Weatherable, shock absorber
Methyl tin mercaptide	0.75 to 1.50	Heat stabilizer	Retards thermal degradation
Calcium stearate	0.80 to 1.20	Lubricant, co-stabilizer	Lubricates to release from hot metal
Wax	0.60 to 1.20	Lubricant	Reduces processing power needed
Partially oxidized PE	0.00 to 0.45	Lubricant	Adds gloss and release from equipment
Titanium dioxide	9.00 to 11.00	Pigment, UV stabilizer	Adds white color and UV protection
Calcium carbonate	3.00 to 5.00	Filler	Cost reduction and dimensional stability

SUMMARY

Commodity resins are those that are sold in very large volumes and can be obtained, with only minor variations in properties, from several competing resin manufacturers. The resins most commonly referred to as commodity resins are: polyethylene (PE), polypropylene (PP), polyvinyl chloride (PVC), polystyrene (PS), and various modifications of these basic resins. All of the commodity resins are polymerized by the addition mechanism from monomers containing a carbon-carbon double bond, three hydrogens, and one pendent atom. The nature of the pendent atom significantly affects the properties of the polymer. The pendent atoms are hydrogen for PE, CH_3 (which is methyl) for PP, chlorine for PVC, and benzene for PS. PE and PP are called polyolefins because of their oil-like properties (waxy feel, chemical nature, and relative softness).

Three types of PE are made commercially—low-density polyethylene (LDPE), high-density polyethylene (HDPE), and linear low-density polyethylene (LLDPE). These types differ in the amount of branching (side-chain length and number), which is a major factor in determining the intermolecular interactions of the polymer. If the polymer has branches that are long and frequent, the polymers are kept apart and cannot fold into dense crystal structures (LDPE). If the polymer has a few branches that are short, the polymers can interact, fold, and form dense crystals (HDPE). The third type of polyethylene (LLDPE) has some moderately short branches with a structure similar to HDPE but the chains are just long and frequent enough that the molecules cannot pack together into crystals. The density is, therefore, low. HDPE and LLDPE are made with catalysts that facilitate and control the branching length and frequency.

LDPE is the most flexible of the PE types. It is the lowest melting and generally the easiest to process. Products made from LDPE include trash bags, films, wire insulation, and other products where high-volume production and flexibility are important. HDPE is used in applications where strength, stiffness, solvent resistance, and nonpermeability are important factors, such as milk bottles, toys, trash carts, and tear-resistant bags. LLDPE is used in many of the same applications as LDPE but is often preferred because of somewhat higher strength and toughness. A new plant to produce LLDPE costs much less than a plant to make LDPE and so new plants for PE have been built to utilize the LLDPE technology.

Important parameters to specify in PE are type (HDPE, LDPE, or LLDPE), density, molecular weight (specified as a melt index), and molecular weight distribution (MWD). The interrelationship between these parameters is complex but generally, higher molecular weight is needed for increased toughness, higher density for strength (although higher molecular weight will also increase strength), and lower molecular weight for ease of processing. MWD affects processing more than the physical and mechanical properties. A wide MWD is preferred for processes like extrusion and blow molding where a high melt strength and broad melting range are important. Narrow MWD is preferred for injection molding where the material should have a low melt viscosity and rapid resolidification.

PE can be crosslinked by electron beam irradiation and by chemical methods. The crosslinking is done after the part has been formed, as postcrosslink melting is not possible. Crosslinking is used to improve toughness and environmental stress crack resistance (ESCR). PE crosslinked by electron irradiation and by chemical methods becomes a thermoset.

Four major copolymers of PE were discussed. Ethylene vinylacetate (EVA) is used as an adhesive and where a clearer product than PE is needed. Ethylene acrylic acid (EAA) is also useful as an adhesive and as a clear alternative to PE. However, because EAA can form ionic crosslinks, it is much tougher than PE. Polymers that can form ionic crosslinks are called ionomers. Ionic crosslinks can be broken with heat so they do not make the polymer a thermoset. EPM and EPDM are elastomeric materials made by copolymerizing ethylene with propylene and a diene. These copolymers are collectively called thermoplastic olefin elastomers (TPO).

PP has a CH_3 (methyl) pendent group that permits three stereoisomers to be manufactured. Stereoisomers differ in the arrangement of atoms along the chain. The only stereoisomer of commercial importance in PP is isotactic in which all of the pendent groups are on the same side of the molecular chain. This arrangement facilitates dense packing, so isotactic is highly crystalline. The interaction of the methyl group with other chains causes PP to be stiffer and stronger than PE. PP is less affected by severe bending of the molecule and can be used to make integral hinges, whereas PE would blush and craze if bent in a similar manner.

If the pendent group is a chlorine atom, the polymer formed is polyvinyl chloride (PVC). Intermolecular interference of the chlorine group prevents crystallization and adds stiffness and strength to PVC. A major problem with PVC is its tendency to decompose and give off HCl when heated. The amount of degradation increases as the accumulated amount of heat increases, therefore processing is very difficult. To prevent this decomposition, many heat stabilizers and other additives are formulated with PVC resin.

A method of processing PVC without high heat is to add special solvents to the resin. These solvents, called plasticizers, lower the T_g of the plastic and increase the PVC's flexibility, making it easier to process. When highly plasticized, PVC is called vinyl and is used to coat fabrics and to make flexible bottles and film and many injection-molded products in which low cost and flexibility are required.

PVC is inherently self-extinguishing because burning releases HCl, which smothers the flame. PVC is much more solvent sensitive than the polyolefins and can be readily welded with solvent adhesives.

Vinyl chloride can be copolymerized with vinyl acetate to give increased flexibility over rigid PVC and reduced plasticizer problems over vinyl. This copolymer is used extensively in vinyl floor coverings.

A polymer that is chemically related to PVC is polyvinylidene chloride, which has excellent barrier properties against oxygen and is used extensively as a food wrap.

PS has a benzene ring for the pendent group. The large size of this group prohibits any crystallization and also makes the polymer very stiff (brittle). The polymer is very clear and is sometimes called crystal styrene even though it is completely amorphous.

PS is very easy to process and is used for many items that require clarity and stiffness such as drinking cups and food packaging. Foamed PS is a major product that has unusually good stiffness to weight, thermal insulative properties, and low cost.

Efforts to improve the toughness of PS have led to the development of rubber-toughened PS, which is called high-impact polystyrene (HIPS). This material is made by alloying or blending rubber polymers with polystyrene. Alloys and blends are important methods of obtaining plastics with new properties rather than by developing a new polymer through new monomer polymerization. Another styrene-based plastic with improved properties is styrene acrylonitrile (SAN). This plastic has improved weathering and barrier properties over polystyrene. Combinations of acrylonitrile, butadiene, and styrene have led to the plastic material known as ABS, which can be either an alloy or a blend.

The cost of manufacture for the various styrene-based alloys and blends is generally as follows: PS < HIPS < SAN < ABS. Hence, improved physical properties will generally cost more, at least among this group of polymers.

GLOSSARY

α-olefin A chemical containing only carbons and hydrogens in which a carbon-carbon double bond is present between the first two carbons in the molecule.

Alloy (in plastics) Two or more polymers that are combined in such a way that they form a single phase.

Atactic A configuration in which the pendent groups are randomly attached to one or the other side of the tertiary carbon.

Autocatalytic decomposition A polymer decomposition that is accelerated by the presence of the products of the decomposition reaction.

Blend (in plastics)   Two or more polymers that are combined in such a way that they form more than one phase.

Branching   Formation of side chains off the basic polymer backbone.

Calendering   Forming a material by squeezing and flattening with a set of rollers.

Carcinogen   A cancer-causing material.

Cling   The tendency of a polymer film to attract itself.

Commodity thermoplastics   Thermoplastics that are high volume, widely recognized, and have generally generic properties between all manufacturers.

Comonomer   An α-olefin mixed into the ethylene in the making of LLDPE that forms small side chains.

Compatibility agent   A material that enhances the tendency of materials to form a stable mixture or a single phase when mixed.

Constrained geometry polymers   Polymers made in such a way that specific geometric arrangements of the atoms are formed.

Copolymerization   The combining of two or more types of monomers in such a way that they polymerize jointly.

Crystal polystyrene   A clear grade of polystyrene; it has no internal crystal structure.

Diene   A chemical containing two carbon-carbon double bonds.

Heat history   The property of some polymers in which the effects of heating of the polymer are cumulative.

High-density polyethylene (HDPE)   A low-branched form of polyethylene usually made with a catalyst.

High-selectivity catalyst   A catalyst that controls the shape of the molecules made using that catalyst.

Immiscibility   The property of materials (including plastics) that results in two or more phases when the materials are mixed.

Integral hinge   A hinge made from the same material as the container itself.

Ionomer   A polymer in which ionic crosslinks form.

Isotactic   A configuration (such as in one type of polypropylene) in which the pendent groups are always attached to the same side of the tertiary carbon, thus resulting in a very regular structure.

Linear low-density polyethylene (LLDPE)   A type of polyethylene characterized by little branching but low density, usually made under low-temperature and low-pressure conditions but with the use of a comonomer.

Living hinge   A hinge made from material that is flexed in order to create the hinge mechanism.

Low-density polyethylene (LDPE)   A highly branched form of polyethylene usually made under high-pressure and high-temperature conditions.

Memory (in a plastic)   The ability of a plastic to reflect the conditions under which it was originally molded, even though it may have been subsequently shaped or formed.

Metallocene catalyst   A catalyst system that fixes the geometry of the reacting material and therefore creates constrained geometry polymer products.

Miscibility   The property of material (including polymers) that results in a single phase when the materials are mixed.

Mixture (in plastics)   Another name for a blend of polymers.

Plastisol   A vinyl dispersion in which only enough solvent is added to liquify the mixture.

Polyolefin   Plastics containing only carbons and hydrogens and absent of aromatics; polyethylene and polypropylene are part of this group.

Secondary carbon   A carbon atom to which two other carbons are attached.

Self-extinguishing   Polymer materials that burn only with difficulty and stop burning if the ignition source is removed.

Sharp melting point   A narrow melting point range.

Sintering   A forming or molding process in which the raw materials (often a polymer or metal powder) are packed and then heated to below the melting point while under pressure, thus causing the particles to fuse together.

Stereoisomers   Molecules with the same chemical formula that differ in the spatial arrangement of the atoms.

Stereoregular catalyst   A catalyst that induces a repeating spatial arrangement of atoms in the molecules made using that catalyst.

Steric   Having to do with shape or size.

Steric hindrance   Interference because of size or shape.

Syndiotactic   A configuration in which the pendent groups regularly alternate from one side of the tertiary carbon to the other side.

Synergy   Enhancement of properties in a mixture, alloy, or copolymer in which the combined material has benefits beyond those of either material alone.

Terpolymer   A polymer formed from three monomers.

Tertiary carbon   A carbon atom to which three other carbons are attached.

Thermoplastic olefin elastomers (TPO)   Polymers made from the copolymerization of ethylene and propylene, sometimes with the addition of small amounts of diene.

Ultrahigh molecular weight polyethylene (UHMWPE)   A form of HDPE in which the molecular weight is especially high.

Ultralow-density polyethylene   A form of polyethylene in which the density is below 0.85 g/cc.

Vinyl   PVC which has been modified by the addition of a plasticizer.

Vinyl dispersion   A mixture of vinyl resin with a solvent.

Ziegler-Natta catalyst   A chemical system invented by Ziegler and Natta which is used to polymerize HDPE, PP, and other stereoregular polymers.

QUESTIONS

1. Explain the cause of the differences in structure in LDPE, HDPE, and LLDPE.
2. What is meant by a commodity resin?
3. Explain why isotactic polypropylene can crystallize while the other polypropylene stereoisomers cannot crystallize.
4. How does changing the molecular weight distribution (MWD) alter the processability of the plastic?
5. Why should the concentration of regrind PVC be kept low in any rigid PVC extrusion operation?
6. Which material, EVA or PS, would be expected to be a better barrier material against water? Why?

7. Why is crystal polystyrene clear?
8. What is ionic crosslinking and what properties will it produce?
9. Would you expect ABS to be clear or opaque? Why?
10. Explain why low molecular weight polyethylene is often added to UHMWPE.

REFERENCES

Birley, A. W., R. J. Heath, and M. J. Scott, *Plastics Materials: Properties and Applications* (2nd ed.), New York: Chapman and Hall, 1988.

Brydson, J. A., *Plastics Materials* (2nd ed.), New York: Van Nostrand Reinhold Company, 1970.

Chanda, Manas, and Salil K. Roy, *Plastics Technology Handbook,* New York: Marcel Dekker, Inc., 1987.

Charrier, Jean-Michel, *Polymeric Materials and Processing,* Munich: Hanser Publishers, 1991.

Domininghaus, Hans, *Plastics for Engineers,* Munich: Hanser Publishers, 1988.

Milby, Robert V., *Plastics Technology,* New York: McGraw-Hill, 1973.

Ulrich, Henri, *Introduction to Industrial Polymers* (2nd ed.), Munich: Hanser Publishers, 1993.

CHAPTER SEVEN

THERMOPLASTIC MATERIALS (ENGINEERING PLASTICS)

CHAPTER OVERVIEW

This chapter examines the following concepts:

- Polyamides or nylons (PA)
- Acetals or polyoxymethylenes (POM)
- Thermoplastic polyesters (PET and PBT)
- Polycarbonates (PC)
- Acrylics (PMMA)
- Fluoropolymers (PTFE, FEP, PFA)
- High performance thermoplastics—polyphenylenes (PPE, PPO and PPS), polysulfones (PSU and PES), thermoplastic polyimides (PI, PAI and PEI), and polyaryletherketones (PEEK, PEK and others)
- Cellulosics

INTRODUCTION

Engineering thermoplastics were originally identified by their ability to replace metallic parts in applications such as automobiles, appliances, and housewares. While this criterion is still true, these resins have been used in many applications beyond those originally associated with metals. Therefore, a more useful definition focuses on the properties of the resins. Most of the resins identified as engineering thermoplastics possess the following key property characteristics:

- High strength and stiffness, comparable to most metals when adjusted for the difference in weights. Some of the engineering thermoplastics may require reinforcement to match the stronger metals in strength or stiffness.
- Retention of mechanical properties over a wide range of temperatures, especially high temperatures. Most engineering thermoplastics have continuous-use temperatures, suggesting a substantial retention of properties in excess of 175°F (80°C).

- Toughness that is sufficient to withstand incidental impacts that accompany applications in which these plastics may be used, although few are as tough as the ductile metals.

- Dimensional stability throughout the temperature range of normal use. This is especially important when plastic parts are substituted for metals. Dimensional stability can be indicated by measuring the creep and the coefficient of thermal expansion (CTE) or the coefficient of linear expansion (CLE). Creep is the tendency of materials to change dimensions under load while at normal-use temperatures and is a particular problem when the material is used in a structural application. CTE and CLE are measures of the expansion of a material when heated. Low values of creep and expansion are preferred so that the fit of the part into an assembly can remain within tight tolerances.

- The ability of the material to withstand environmental factors such as water, solvents and other chemicals, UV light, and oxygen. Engineering thermoplastics and metals are resistant to most of these factors, but not all. Environmental resistance is, therefore, an important consideration for each intended use. Resistance to all environmental factors is seldom required, although the broader the spectrum of resistance the better.

- As easy to shape and finish as metals because they are often substituted directly for a metal. Plastics are normally superior to metals in this property. The opportunity to combine several metal parts into a single plastic molded part is one of the most important considerations in the substitution of the plastic for a metal.

Other properties can be important for specific applications. Some of the most common are abrasion resistance, extended fatigue life, lubricity, electrical properties (generally high dielectric strength, high dissipation factor, and low dielectric constant), flammability, and overall cost (which may include manufacturing time and complexity, equipment investment, possibilities for parts consolidation, and raw material cost).

Some material applications require a level of performance in one of the special consideration properties that can only be met by an engineering thermoplastic. In these cases the plastic is not replacing a metal but is creating a new application capability. Examples of this would be the electrical resistance and lubricity of fluoropolymers and the clarity and shatter-resistance of polycarbonate bulletproof windows. Photos 7.1 and 7.2 show some typical engineering thermoplastic types.

The commodity resins discussed in the previous chapter can meet some of the property expectations of engineering thermoplastics but will typically have a major deficiency that prohibits their use in an engineering application. The uses of commodities are, therefore, chiefly dependent upon their cost and processibility rather than on their metal-like properties.

The engineering plastics usually are sold as molding resins. They can be molded into near netshape parts that require little finishing work after molding, and into stock shapes (such as rods, tubes, and blocks) that can be machined and finished into the final part shape. Most engineering resins can also be made into fibers, films, and coatings which is beyond the capability of most metals.

The engineering resins considered in this chapter can be divided into families, which are discussed separately. Within each discussion the general family characteristics are dis-

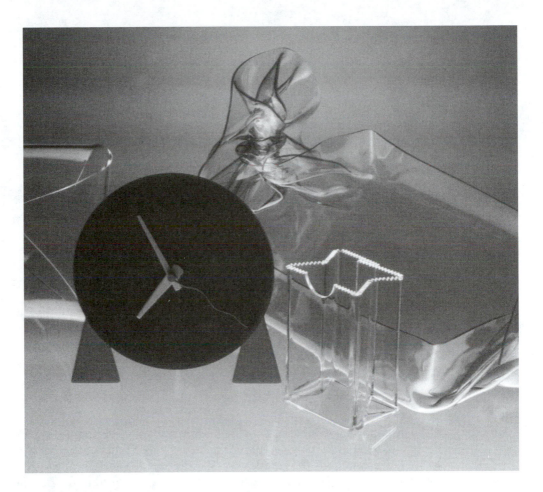

Photo 7.1   Acrylic (PMMA) used for art pieces (Courtesy Atohaus North America, Inc.)

cussed, usually with comparisons to other engineering thermoplastic families and to met-als. Then, the most common specific resin types within each of the families are discussed and the differences between types pointed out. The differences within the members of a family are generally less pronounced than the differences from one family to another or to other types of materials.

The following families of thermoplastic (TP) engineering plastics are considered: poly-amides or nylons, acetals or polyoxymethylenes, TP polyesters, polycarbonate, acrylics, fluoropolymers, high-performance thermoplastics (including polyphenylenes, polysul-fones, TP polyimides, polyaryletherketones), and cellulosics with several specific polymers within each family discussed. The summary of this chapter presents a table comparing the most important properties of representative members of each of these families. The reader may want to consult that table while reading this chapter to assist in distinguishing the differences between the engineering thermoplastics families.

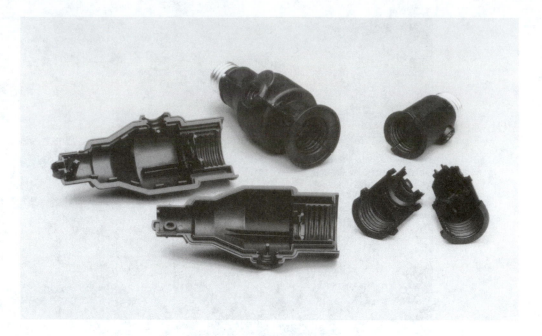

Photo 7.2   Injection molded PET parts with a metal insert (Courtesy: Kestler & Co.)

The family names are based on the most important or distinguishing functional group or type of bond in the polymer. For instance, the amide functional group characterizes the polyamides. Similarly, the acetal bond characterizes the acetals. The distinguishing bond or functional group is often the bond or group that is formed in the polymerization reaction. A prior knowledge of these functional groups or types of bonds is useful, but is not required for a clear understanding of the properties of each of the families. The polymer properties are based upon the principles that have previously been used to understand and predict polymeric properties, namely functional group size, electronegativities within the functional group, chemistry of the group, and intermolecular interactions.

The names of individual polymers, that is, specific polymers within the various families, are based on two systems which depend upon the polymerization mechanism. If the polymer is made by addition polymerization, the name of the monomer is the basis of the polymer name. For instance, polytetrafluoroethylene is made by addition polymerization from the monomer tetrafluoroethylene. (The commodity polymers polyethylene, polypropylene, polystyrene, and polyvinyl chloride are all made by addition polymerization so they are also named from their respective monomers.) On the other hand, if the polymer is made by condensation polymerization, the polymer is named from the distinguishing functional group or bond, just as the engineering thermoplastic families are named. Examples are polycarbonate, which is distinguished by the carbonate bond, and polyetheretherketone, which is named by describing all of the key bonds in the polymer.

Engineering thermoplastics are made by both the addition and condensation polymerization reactions. The key to their performance is not the method of polymerization but the structure and chemistry of the polymer and its interactions.

POLYAMIDES OR NYLONS (PA)

The polyamides were the first engineering thermoplastics produced and are the largest family in both production volume and number of applications. Originally the name *nylon* was associated with and coined by the DuPont company to represent the family. It has since become a synonym for all the polyamides, without regard to the producing company. The major producers of nylon resin have their own brand names for their polymers. Some brand names for nylon include: Zytel™ (DuPont), Ultramid™ (BASF), Torayca™ (Toray), Durethan™ (Bayer), Capron™ (Allied Signal), and Akulon™ (Akzo). The differences in properties for the engineering thermoplastics are more pronounced from company to company than are the differences between commodity thermoplastics, especially because of the wide range of specific resin types within most of the engineering thermoplastic families.

Polyamide General Family Characteristics

The polymer repeating unit that represents the family of polyamides is given in Figure 7.1. The subscripts *a* and *b* in the repeating unit indicate that in the nylon family the number of CH_2 groups can vary from one particular nylon type to another. The variations of the number of CH_2 groups is the principal difference between the different types of polymers within the family. These differences are explored in detail later in this section, after the general family characteristics are discussed.

Nylons are formed by condensation polymerization from reactants (monomers) that combine to make amide groups with water as a by-product. The reactants that form amide groups are the acid and amine functional groups that were identified in the chapter on polymeric materials (molecular viewpoint).

The amide group, which is present in each of the polymer repeating units, is the major determiner of nylon family properties. Both the N—H bond and the C—O bonds are polar

Figure 7.1 Polymer repeating unit for polyamides, where the subscripts *a* and *b* can vary depending upon the type of monomers used and *n* represents the number of repeating units in the polymer chain.

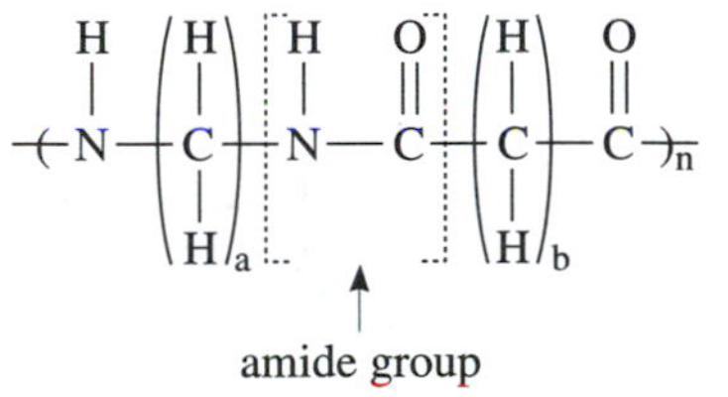

amide group

with the N and the O being the negative ends. This polarity induces the formation of secondary bonds between adjacent nylon molecules. These secondary bonds (hydrogen bonds) restrict the movement of the nylon molecules relative to each other, thus increasing the tensile strength. The secondary bonding also facilitates the close packing of nylon molecules resulting in high crystallinity. This crystallinity leads to high strength, high stiffness, good toughness, low gas and vapor permeability, translucency rather than transparency, a sharp melting point, good abrasion resistance, good fatigue life, and high-temperature processing. Typical nylon parts that utilize these properties are gears, rollers, shafts, impellers, bearings, films for vacuum systems, cooking bags, and zip fasteners (zippers).

The polarity of the amide group makes nylon sensitive to polar solvents such as water. Water absorption of nylon (approximately 2.5% by weight) is higher than most of the other engineering thermoplastics and can have a significant effect on the properties of nylon. For instance, the tensile strength and tensile modulus can decrease by 20% with water absorption. The tendency for nylon to absorb water increases significantly with temperature so applications where the nylon would be in hot water should be avoided. On the other hand, in many applications in which nylon is exposed to only cold water for brief periods the properties are not seriously affected by the water absorptivity. For instance, nylon is widely used for ropes, which can often get wet, and the properties are not seriously affected. Absorptivity of water is a problem in processing because nylon resins must be dried before an injection molding process. This drying requirement is not, however, unusual for the engineering thermoplastics.

The polarity of the amide group increases the ability of nylon to be joined with traditional solvent-based adhesives. Many effective adhesives for nylon are commercially available. Nylon can also be solvent welded by simply subjecting the surfaces of the two parts to a solvent for nylon. The solvent will soften the surfaces, and while still soft, the parts can be joined and clamped and then dried.

Just as the polarity of nylon increases sensitivity to polar solvents, it decreases sensitivity to nonpolar solvents. Therefore, nylon is used extensively in applications in which the plastic part is subjected to oil and grease.

Other materials are usually the preferred choice in electrical applications because the resistivity of nylon depends significantly on the moisture content.

The sharp melting point of nylon with the accompanying low viscosity of the melt are advantages in injection molding but disadvantages in extrusion and blow molding where melt strength is required. When nylon is extruded, wide molecular weight distributions should be used and the temperature profile of the extruder should be lowered at the exit so that the viscosity of the melt can be maximized.

Properties of Specific Nylon Types

The differences between the various types of nylons largely depend on the number of carbons in the molecular sections between the amide groups. These segments are simply the segments that exist in the monomers which are used to form the nylon polymer. Hence, if the nylon is made from a six-carbon diamine, as shown in Figure 7.2a, and a 12-carbon diacid, as shown in Figure 7.2b, the resulting polyamide would be named nylon (6/12),

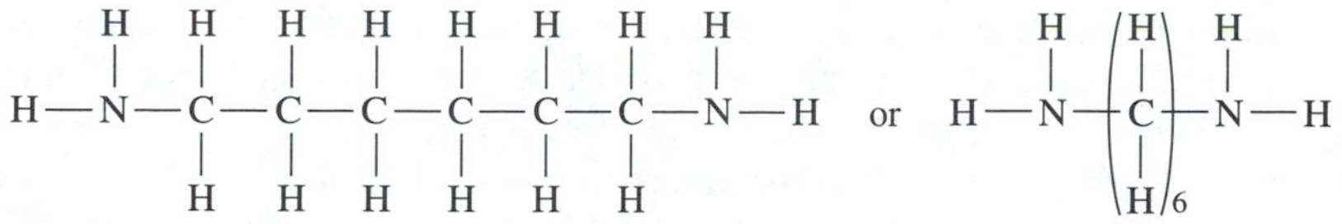

(a) 6-Carbon diamine

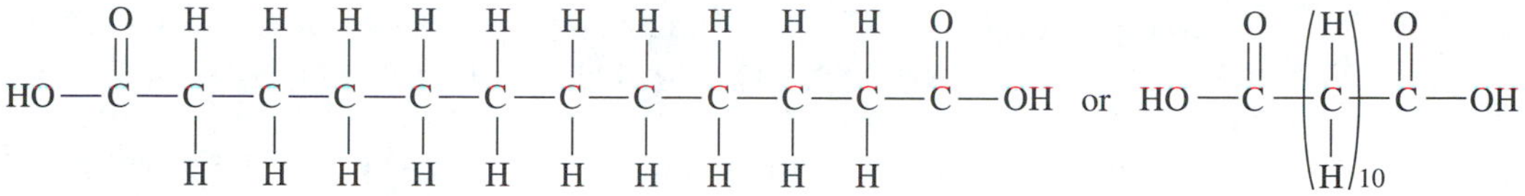

(b) 12-Carbon diacid (Note that the carbons in the acid functional groups count in the overall carbon number.)

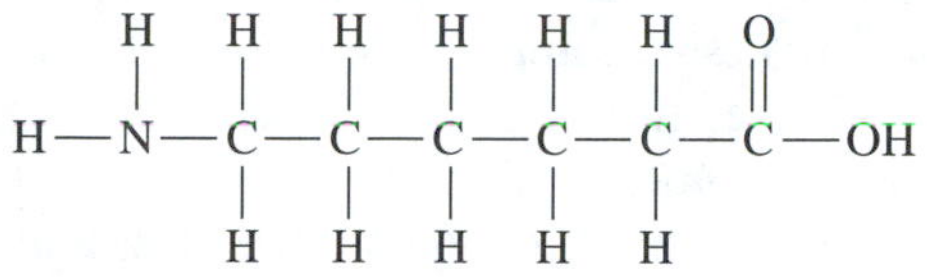

(c) 6-Carbon molecule having both amine and acid groups

Figure 7.2 Monomers which can be used to make polyamides (nylons). Monomer (a) can react with monomer (b), whereas monomer (c) can polymerize by itself.

which is pronounced "nylon-six-twelve." (The number of carbons in the diamine is given first followed by the number of acid carbons in the carbon number designations.) Polyamides can also be formed from just one monomer if that monomer has an amine group on one end and an acid group on the other. Such a monomer is shown in Figure 7.2c. The resulting polyamide would have six carbons between each of the amide groups but, because only one monomer is used, the polymer name would be nylon (6). Nylons of many different types are being sold commercially including: nylon (6), nylon (11), nylon (12), nylon (6/6), and nylon (6/12). The most widely used of the different types is nylon (6/6), which was also the first made.

The choice of six carbons in both the diamine (hexamethylene diamine) and the diacid (adipic acid) to make nylon (6/6) was probably dictated by the low cost and availability of the monomers. This choice has proven to be a good one for achieving a compromise of properties. The six-carbon chains provide enough flexibility to give nylon (6/6) good toughness and yet allow the chains to still be stiff, strong, and crystalline. Longer chains are much more flexible. Shorter chains are stronger and stiffer but more brittle. Applications for nylon (6/6) include bristles, sheets, rods, tubes, coating, molded industrial parts, and many others.

Nylon (6/6) has proven to be nearly ideal for making nylon fibers, which are used for applications such as hosiery, carpets, reinforcement for hoses, belts, and tires; and for other heavy-duty industrial applications. Industrial and carpet fibers must be stiff, strong, tough, and resistant to cracking under severe bending. Other fibers may be softer, more easily colored, or less expensive, but for carpets and industrial purposes, nylon (6/6) fibers are usually superior. One drawback to nylon (6/6) fibers is their strong tendency to build up static charge. Two methods to reduce the static are the inclusion of metallic fibers in the carpet to conduct the electricity to ground and the treatment of the surface with an antistatic agent to prevent or decrease the buildup of the charge.

Water absorption is affected strongly by the length of the CH_2 segments in the nylon molecules. When the number of CH_2 units in the monomers is small, the number of amide groups per unit of length is high and so the overall polarity in the plastic is relatively higher. This results in more water absorption. Therefore, shortening the CH_2 content, such as in nylon (2/2), will add strength and stiffness but will also increase the amount of moisture sensitivity. Conversely, if a nylon is desired that will make a softer, more flexible fabric (perhaps for apparel), the number of CH_2 segments between the amide sections should be increased, such as nylon (6/12).

The second-highest-volume polyamide is nylon (6), which, as the numerical designation implies, is formed from a single monomer. (See Figure 7.2c.) A convenient source of this six-carbon monomer is a ring compound that opens during the polymerization process to form the desired monomer. The ring that is used to form nylon (6) is called caprolactam. This polymerization method is low cost and creates a polymer with properties that are competitive with nylon (6/6).

Several copolymers, alloys, and blends have been developed with nylon as the principal component that have a wide range of physical properties, especially toughness and resistance to environmental conditions such as sustained high temperature and hot water. For applications requiring additional strength or toughness, nylons reinforced with short glass or carbon fibers or with minerals have been developed.

Aramids

A quite different polyamide material is illustrated in Figure 7.3. This material contains the amide groups with benzene rings between them. This highly aromatic polymer has been given a generic name of aramid, which seems to be a contraction of aromatic polyamide. It is not generally called a nylon.

The properties of the aramids are unique among polyamides. The high aromatic content, especially when part of the backbone, stiffens and strengthens the material beyond any of the other nylons. When made into a fiber, the material is called Kevlar®, a DuPont trademark, and is widely used in making bulletproof vests and as a reinforcement in composites. The aramids are nonburning, solvent resistant, and very high melting. If the polymer is changed slightly so that the amide groups attach to a different carbon on the ben-

Figure 7.3 Aramid polymer repeating unit (Kevlar®).

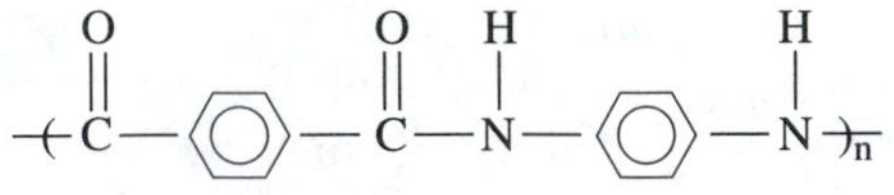

zene ring, a molding plastic resin is created called Nomex®, a DuPont trademark. It possesses most of the same properties as Kevlar® and is more easily processed in standard molding equipment. It can also be used as a coating to strengthen and add flame retardance to many other materials such as paper and cloth.

ACETALS OR POLYOXYMETHYLENES (POM)

The acetals or polyoxymethylenes (POM) are a family of polymers that compete with nylons for many of the same applications. Acetals are not used as commonly as nylons, due in part to an instability in acetals that causes difficulty during processing. However, for some applications, especially in water, acetals are preferred over nylons.

As can be surmised from the simple polymeric repeat unit that has no large pendent groups (Figure 7.4), these polymers are highly crystalline, in fact, they have the highest crystallinity of any thermoplastic polymer family.

Acetal General Family Characteristics

The very high degree of crystallinity gives excellent strength, stiffness, surface hardness, barrier properties, solvent resistance, and a sharp melting point. In these properties acetals are generally equivalent to nylon (6/6). The higher crystallinity of acetal reduces the toughness when compared to nylon, but only slightly, so acetals can readily be used in applications requiring some toughness. Acetals are somewhat more *notch sensitive* than nylons, that is, they fail at a lower impact strength if a notch has been cut in the sample. In spite of this, acetals have good fatigue resistance as is evidenced by their use as switches in vehicle turn signal indicators.

Although the carbon-oxygen bond in acetal is polar, it is less polar than the carbon-oxygen double bond of nylon and is far less available for secondary bonding because it is between atoms along the backbone rather than pendent to the chain. Hence, acetal is normally not sensitive to polar solvents, as shown by a very low moisture absorption (about 0.22%), which is less than one-tenth the amount of water absorbed by nylon. The dimensional and property changes that characterize nylon in a moist environment are not present with acetal. The properties of acetal are, therefore, more predictable over a wide range of environmental conditions. The only common environmental solvents that have strong action on acetals are acids and strong oxidizing agents. Prolonged contact with these should be avoided. Acetals are used widely in water environments such as toilet tanks, shower heads, hose connections, and valve bodies. Very hot water or steam applications should be avoided because the maximum use temperature is about 212°F (100°C), approximately the same as for nylon.

Figure 7.4 Polymer repeating unit for acetal or polyoxymethylene (POM).

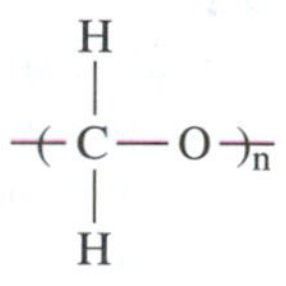

$$\left(\begin{array}{c} H \\ | \\ C - O \\ | \\ H \end{array} \right)_n$$

The low molecular polarity reduces the tendency of acetal to bond with other materials, and most solvent adhesives are not especially effective. The low bonding tendency gives acetal a low coefficient of friction and a glossy molded surface. These properties are useful in applications such as gears, chain links, conveyor product guides, cams, nonlubricated bearings, automotive window cranks, mounting brackets, and zippers. The good abrasion resistance of acetal further enhances its performance in these applications.

Acetal is sensitive to UV light, which causes a chalking of the surface and degradation of the polymer.

Properties of Specific Acetal Types

Two major types of acetal resins are commonly available, a homopolymer and a copolymer. The homopolymer (most common brand name is DuPont's Delrin®) is made by the polymerization of formaldehyde. This polymer is very susceptible to thermal degradation by depolymerizing (unzipping the molecule) with evolution of formaldehyde, a potentially toxic and flammable gas. To reduce the tendency for degradation, a chemical is added to the polymer which reacts with the ends of the polymer molecules. This process is called *end-capping* and is used to stabilize acetals and a few other polymers.

The copolymer (most common brand names are Celanese's Celcon® and Hoescht's Hostaform®) is made by the copolymerization of formaldehyde with ethylene oxide or by the ring opening polymerization of dioxane with ethylene oxide. The resulting structure differs from the homopolymer in that a C—C bond occurs along the backbone along with the C—O—C bonds. The C—C bond interrupts the degradation (unzipping) process and therefore stabilizes the polymer. For additional stability, the copolymer may also be end-capped.

The obvious advantage of the copolymer is improved thermal stability. The copolymer also has more flexibility, twice the elongation, and 20% less water absorption. These are all characteristics to be expected from the addition of the more flexible C—C bond. The homopolymer has the advantages of 15% higher tensile strength, 20% higher modulus, 20% higher impact strength, 20% higher surface hardness, and 20% higher fatigue limit.

Processing of the homopolymer is more difficult than the copolymer because of the degradation problem, although even the copolymer is relatively unstable and caution should still be exercised. Safe practice in extrusion, injection molding, and other melt processes dictates that the barrel of the molding machine be cleared of the acetal before shutdown to ensure that the acetal resin is not in the barrel during the subsequent start-up, which may overheat the polymer. Further, even temporary shutdowns, as for a mold change or even a difficulty with the mechanical function of the molding machine, can result in some evolution of formaldehyde. If the shutdown is for more than a minute or two the barrel should be purged. *Caution should be taken to only use acetal polymers in well-ventilated rooms.*

The high crystallinity of acetal introduces some processing modifications as well. To obtain the best properties, the processing cycle in injection molding should allow full crystallinity to occur by keeping the mold hotter than for other resins (such as nylon). (Note: a cold mold is normally used to hasten solidification and thereby reduce the cycle time.) If the mold temperature is too low, the molecules could be trapped in a partially amorphous state in which the amount of crystallinity could vary widely from part to part. Another disadvantage of the highly crystalline acetal resins in processing occurs because the melt viscosity doesn't change significantly with temperature. This means that the normal process of reducing the viscosity to improve flow by raising the temperature is not practical. The high crystallinity makes film extrusion so difficult that acetal films are not generally available.

Both homopolymer and copolymer are easily machined and fabricated. Threaded acetal parts are often machined out of an extruded rod and used in place of machined brass parts. Parts can be assembled by press-fits, snap-fits, cold-heading, riveting, as well as spin and ultrasonic welding.

Acetal resins can be combined with short fiberglass reinforcements to enhance strength and toughness. Mixtures of acetal with fluorocarbons will enhance the inherently good surface lubricity of acetals.

THERMOPLASTIC POLYESTERS (PET/PBT)

Some care should be exercised in discussing polyester resins because two classes exist: thermoplastics, discussed here, and thermosets, discussed in the chapter on thermoset materials. The differences between the thermoplastic and thermoset natures of these two polyester classes is much greater than the similarities which might arise from the presence of the polyester functional group in both classes. In other words, being a thermoplastic or thermoset is more important in property distinction than is being a polyester.

Within the group of engineering thermoplastics, the polyesters have some unique properties that lead to use in applications that are not readily filled by other engineering thermoplastics. In other applications, the polyesters compete with the other engineering thermoplastics and the choice of resin for some of these applications depends on subtle differences. The polymer repeat unit is represented in Figure 7.5 for the most common of the thermoplastic polyesters. When the parameter m is 1, the polymer represented is polyethylene terephthalate (PET) and when m is 2 the polymer is polybutylene terephthalate (PBT). The general properties of both of these thermoplastic polyesters are discussed and then the differences between PET and PBT are explored.

Thermoplastic Polyester General Family Characteristics

The TP polyesters are the fastest growing of the engineering thermoplastics, due in large part to the acceptance of PET as the resin of choice in the soft drink container market and for high-performance films such as photographic, magnetic tape, electrical insulation, and

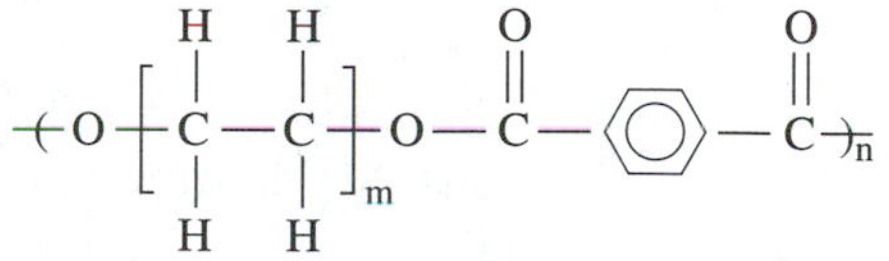

Figure 7.5   Polymer repeating unit for the most common thermoplastic polyesters where m can be 1 (for PET) or 2 (for PBT).

decorative film laminates. Other applications have also grown significantly resulting in wide usage. With tensile strengths and use temperatures comparable to nylon and acetal, the TP polyesters compete for applications such as pump housings, windshield wiper arms, sunroof frames, and light-duty gears.

Crystallization of the TP polyesters is slow and generally reaches no more than 50%. While the excellent mechanical properties of nylon and acetal are due to high crystallinity and/or secondary bonding, the excellent mechanical properties of TP polyesters are attributed to *orientation effects*. When the molecules within a plastic are aligned or oriented in one direction, the plastic material is stronger in that orientation direction. This occurs simply because the molecules are much stronger when they are being pulled along the backbone direction than when being pulled out of a random arrangement. These orientation effects are especially strong with large and complex polymer repeating units such as those of the TP polyesters. After the molecules have been oriented, they can relax into a random state, thus losing the advantages obtained from the orientation. To prevent this relaxation from occurring, the oriented molecules are crystallized to prevent the molecules from moving out of orientation. A typical processing method for film that takes advantage of this property is:

- The TP polyester is melted, shaped into a film, and then cooled quickly so that it is amorphous.
- The film is heated to a soft stage above T_g and stretched in the machine direction while retaining the width so that only orientation occurs and not necking. (This is usually done by heating a small section of the film while it is between rollers running at different speeds. The contact with the rollers maintains the width.)
- The film can then be stretched in the cross direction by clamping and pulling it while warm.
- The film is then crystallized by heating to a temperature that is higher than the stretching temperature while maintaining the machine and cross-direction tension.
- The film is then cooled below the glass transition temperature to set the shape, crystallinity, and molecular orientation.

The good mechanical properties are, therefore, attributed to the orientation received from stretching, which is locked in place by the crystallinity.

Bottles and other TP polyester parts are likewise oriented and then crystallized to lock the orientation in place and obtain superior mechanical properties. The size of the crystals is usually not large enough to interfere with visible light and cause scattering. Therefore, the films and bottles are usually transparent.

For additional strength and toughness, the use of short fiberglass reinforcement is also common with TP polyesters. The inclusion of fiberglass will, of course, make the part nontransparent.

Additives are often blended with the polymer resin to control surface roughness which can have a major effect on frictional properties. Additives are also used to control optical properties such as color, opacity, and surface reflectivity. For example, metalized films with either high reflectivity or dull surfaces can be selected by choice of additive. When used for drafting films, for instance, the surface of the film needs to have a moderate roughness so that it can be written on with a pencil. This factor can also be controlled with additives.

The TP polyesters are composed of both aromatic and aliphatic portions that are joined by a polyester functional group through a condensation polymerization. This polyester group has a polar carbon-oxygen double bond, but as a percentage of the entire molecule, the polar component is rather small. Therefore, TP polyesters have low moisture absorptivity. This low moisture sensitivity gives generally consistent electrical properties that have led to numerous electrical applications such as ignition E-coils, lamp sockets, electric hand tool housings, and numerous electrical switches and connectors that were previously made from phenolic resins. Phenolic resins are thermoset materials that inherently have a longer processing time than do the thermoplastics and so are vulnerable to being displaced by a thermoplastic with acceptable physical properties. Such is the case with replacement of phenolic in many applications by TP polyesters.

TP polyesters have excellent dimensional stability over a wide range of moisture and thermal conditions, but even small amounts of moisture can react with the melt and cause some degradation in mechanical properties. Therefore, TP polyester resins should be dried before molding.

The melts of TP polyesters are of medium viscosity compared with other engineering thermoplastics and do not present serious difficulties in processing, except that crystallization is often slow. Several grades of TP polyester are now offered that speed the crystallization process and improve the cycle time. If full crystallization is not allowed to occur, the resin may crystallize after molding, resulting, perhaps, in shrinkage and cracking of the part.

Another group of thermoplastic polyesters is called the *polyarylesters* because they are made from monomers that form polyester linkages and also contain the aryl or aromatic (benzene) group. These resins have higher mechanical and thermal properties than do the polymers represented in Figure 7.5, but also much higher costs. Processing of the polyarylesters is also much more difficult. Therefore, the polyarylesters are part of the group of thermoplastic resins called *high-performance thermoplastics*. This group is discussed later in this chapter.

Properties of Specific TP Polyester Types

The most important of the TP polyesters is polyethylene terephthalate (PET). This resin has been used extensively in three major product types—fibers, films, and molding resins. The fibers have been used for textile applications, formerly used alone but now usually blended with wool or cotton. The polyester gives wrinkle resistance, permanent pleat capability, and staining resistance to the fabric. Fibers made from polyester are also used as reinforcements in tires, conveyor belts, and hoses, especially where hot water is likely to be present, making nylon inappropriate. Brand names for polyester fibers include DuPont's Dacron® and Kodak's Kodel®.

PET films are used extensively because of their ruggedness and clarity. The films are used as magnetic tapes, substrates for photographic films, release films, drawing foils, and, because of the high operating temperature capability of PET, as sterilizable packaging for medical and other applications. A copolymer of PET with a modified glycol is available as a film that is completely amorphous and has improved processing. This copolymer is called PETG. Typical PET film brand names include: DuPont's Mylar® and Kodak's Kodar®.

The use of PET for soft drink bottles requires that the resin be tough (so that it can withstand a drop), inexpensive, and have a low permeability to carbon dioxide. PET is a reasonable compromise and has performed well in this application. The use of PET for soft drink bottles has increased its use in other bottle applications, some where permeability is not a major issue. The clarity and durability of PET, along with the excellent odor resistance have led to many of these applications.

Polybutylene terephthalate (PBT) is related to PET but has the advantage of easier processibility because it crystallizes more rapidly than PET. The mechanical properties of PBT are somewhat lower than PET. Major applications for PBT include: slide bearings, conveyor chain links, roller bearings, gears, cams, protective headgear, couplings, screws, and various electrical components.

POLYCARBONATE (PC)

This class of engineering thermoplastic resins differs from all the others discussed so far in that only one type of polycarbonate has reached significant commercial success. Therefore, the properties of the class and the properties of the individual polymer are the same. Brand names commonly associated with polycarbonates include GE's Lexan® and Bayer's Merlon®.

Polycarbonate is formed by a condensation polymerization resulting in a carbon that is bonded to three oxygens, a characteristic of carbonates. The polymer repeating unit for polycarbonate is given in Figure 7.6.

The large, complex, aromatic structure of polycarbonate determines the physical and mechanical properties of the molecule. Polycarbonate is noncrystalline and therefore clear, yet is nearly as strong as the highly crystalline nylon and acetal plastics and is somewhat tougher. This mechanical performance is due to the large aromatic content of the polymer leading to backbone stiffness coupled with moderately large pendent groups and the hydrogen bonds that can form between the polar carbonates on adjacent molecules. All these factors increase the resistance to intermolecular movement that is needed for high strength.

Figure 7.6   Polymer repeating unit for polycarbonate (PC).

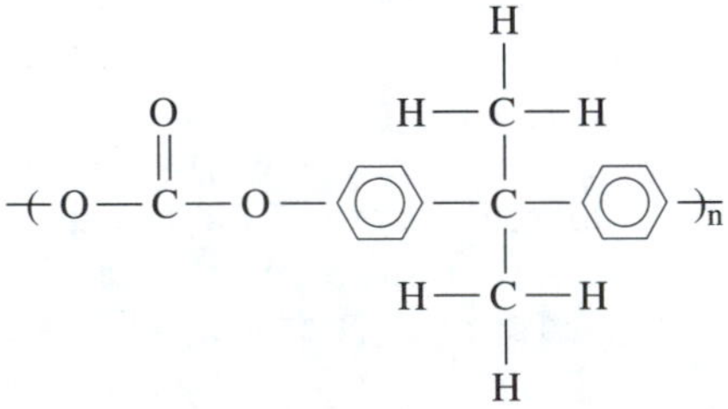

The same factors that give strength also increase the energy needed to cause melting. In the case of polycarbonate, the melting point and the use temperature are much higher than for the competitive nylon, acetal, and TP polyester materials. (Polycarbonate can be used continuously to 275°F, 135°C.)

Creep resistance over an exceptionally wide temperature range and other factors contributing to dimensional stability are important properties for polycarbonate. These properties are also contributed by the high aromatic content, pendent groups, and hydrogen bonding.

Impact toughness depends upon the ability of the material to absorb and distribute localized energy. To do that effectively, the material must have strength so that it won't break and, simultaneously, sufficient resilience or elongation that the force can be transferred throughout the structure. Polycarbonate has the strength, as already described, and also has the resilience because of its interconnected (with hydrogen bonds) yet amorphous structure.

All these superior mechanical and physical properties have created a strong application capability for polycarbonate. Uses include: small tool housings, hot-dish handles, hair dryers, camera bodies, pump impellers, safety helmets, business machine housings, and recreational vehicle bodies. Polycarbonate competes with nylon, acetal, and TP polyesters for these markets.

A unique combination of properties possessed by polycarbonate are its high optical clarity and toughness. These properties have led to applications such as small appliance housings (especially where temperature is a problem), bulletproof windows, break-resistant lenses, compact discs, food processing containers and housings, returnable milk bottles, headlights, taillights, runway markers, microwave containers, medical tubing connectors, and filter housings. One caution is that polycarbonate is sensitive to UV light. It degrades with some yellowing and loss of mechanical properties. This sensitivity is less than with polystyrene but more than with acrylics, which are discussed later in this chapter. Another drawback of polycarbonate in optical applications is the ease of scratching the surface, but coated grades have been developed to solve this problem.

Polycarbonate is sensitive to aromatic, chlorinated solvents and some polar solvents, and has poor resistance to alkali solutions. These solvents can cause crazing and weakness. The advantage of this solvent sensitivity is that polycarbonate can be joined easily with many common solvent-based adhesives.

The polar nature of polycarbonate leads to a modest amount of moisture absorption. This absorptivity is highest and most damaging in the melt, therefore polycarbonate should be dried before processing.

The high melting temperature causes some difficulties in processing polycarbonate but it is not especially sensitive to thermal degradation, so thinning by raising the temperature is usually an acceptable method of improving processing. Polycarbonates can be processed by extrusion, injection molding, blow molding, thermoforming, and structural foaming.

Polycarbonate can be filled with glass fibers to increase strength, especially for metal replacement applications. Polycarbonate is the basis of numerous blends and alloys which use the toughness and strength of polycarbonate and add some other beneficial property from the second polymer.

ACRYLICS (PAN, PMMA)

The acrylics group is dominated by two resins—one used principally for blending with other resins and as a fiber (PAN) and the other used principally for molding (PMMA). The blending and fiber resin, polyacrylonitrile (PAN), which is made from the acrylonitrile monomer, was discussed in the chapter on commodity thermoplastics as part of the discussion of modified styrene. It is a blending and alloying resin for styrene and butadiene. (The acrylonitrile monomer is the "A" in ABS.) When made into a fiber, PAN is called acrylic and is characterized by softness and ease of coloring. The fiber is used to make sweaters and carpets. (DuPont brand name: Orlon®.) PAN is also used as a starting material in making carbon fibers for plastic-reinforced composite materials. A few other acrylic resins are made, but they are less important and are used chiefly as copolymers with commodity resins.

The molding resin, PMMA, is the subject of the remainder of this discussion on acrylics because it is this polymer that competes with the other engineering thermoplastics. Common brand names for PMMA include Atohaas' Plexiglas®, ICI's Lucite®, and Cyro Industries' Acrylite®. The resin is polymerized by the addition polymerization method to form the polymer that is represented in Figure 7.7. The plastic is atactic and therefore amorphous.

The most important property for PMMA is its optical clarity. This plastic has as high as 92% light transmittance, the highest of any plastic material. It also has the lowest sensitivity to UV light of any plastic, very low oxidation sensitivity, and overall weather resistance which together result in a high retention of clarity and light transmittance even after very long periods of time. PMMA has very high gloss. These excellent optical properties have led to applications such as windshields (especially for planes and helicopters), skylights, fluorescent light diffusion panels, outdoor signs, automobile taillights, headlight covers, compact discs, display cases, light fixtures, and general glass substitutes. In these applications PMMA competes with polycarbonate. PMMA has better initial and long-term optical properties but is more brittle and not as tough and strong as polycarbonate. Impact grades of PMMA copolymer have up to twenty times the impact

Figure 7.7 Polymer repeating unit for polymethylmethacrylate (PMMA), which is an acrylic.

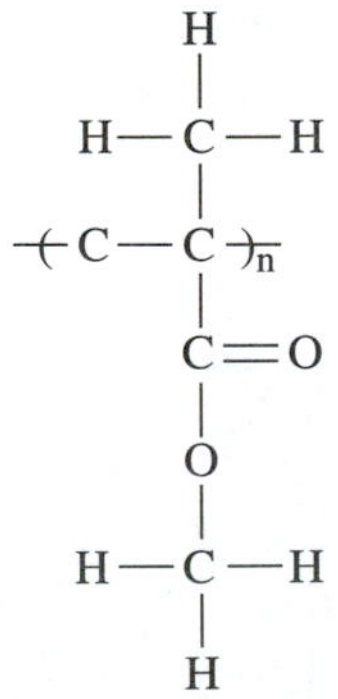

strength, although the optical properties are not as high as for homopolymer PMMA. As with polycarbonate, PMMA is scratched easily, although coated grades are available which significantly reduce the scratch problem. PMMA has a significant advantage in price against polycarbonate.

PMMA is sensitive to chlorinated solvents that can result in some crazing and stress cracking, but also allows the use of solvent cements for joining. The polarity of the pendent group leads to some moisture absorption that does not seriously affect physical properties but can lead to defects in molding, so PMMA should be dried before it is molded. In repeated hot water wash cycles, PMMA will craze, so applications in which this treatment is likely should be avoided. PMMA is easily cleaned with a mild soap, household ammonia, or dilute inorganic acids.

PMMA can be dispersed or dissolved in water or solvents to form floor sealants and polishes. This property, along with the excellent ability to color and weather, has led to its use in paint.

The relatively low processing temperature, low shrinkage, and good dimensional stability make PMMA easy to process in injection molding and extrusion. A major product for PMMA is extruded sheet which can be thermoformed into many of the products mentioned earlier, especially outdoor signs. PMMA is one of the easiest of plastics to machine and join, thus adding to the range of applications for this plastic.

If a sheet of the highest optical clarity is desired, another processing method is used to make acrylic sheet. This process is casting, which is discussed in detail in the casting chapter. The advantage of this process over extrusion is that no thermal stresses are present and optical distortions are minimal. The material that is cast is a solution of the PMMA polymer dissolved in the monomer (MMA) that has been initiated with a peroxide or that is initiated with UV light. This solution of PMMA in its monomer is called *acrylic syrup* because of the obvious similarity in viscosities to maple syrup. Acrylic syrup could also be made by interrupting the polymerization process before the chains get very long.

Other products in addition to sheets are cast from acrylic syrup. For instance, when highly filled with calcium carbonate or alumina trihydrate, acrylic syrup can be poured into molds to form sinks, countertops, and other bathroom and kitchen fixtures that resemble marble. A brand name of these products is DuPont Corian®. Cast acrylic countertops compete against cast thermoset polyesters, which are discussed in the thermoset resins chapter and, of course, against marble. An advantage of the acrylics in this market is their weatherability, low burning rate, low smoke emission, and ease of removal of stains and burn marks. Acrylic is cast without fillers as an encapsulating material, especially when clarity and other optical properties are required. Acrylics are among the easiest plastics to form by mechanical machining and other simple techniques. They are, therefore, widely used by hobbyists and commercially to make parts for displays, and artistic works (see Photo 7.1).

Acrylic blends, alloys and copolymers achieve selected properties that are significantly higher than homopolymer acrylic in toughness, use temperature, and ease of molding. These mixtures are inferior to the homopolymer in optical qualities, but for some applications the decrease is not significant.

FLUOROPOLYMERS (PTFE, FEP, PFA)

The fluoropolymers are different from the other engineering thermoplastics because their usefulness is not primarily based on their mechanical properties but rather on the unique physical and chemical properties that result from the presence of fluorine in the polymer. In the most important of the fluoropolymers, polytetrafluoroethylene (PTFE), which is represented in Figure 7.8, only carbon and fluorine atoms are present in the molecule. In general, the higher the concentration of fluorine atoms to the total, the stronger are the unique properties associated with the presence of the fluorine. The discussion of fluoropolymer properties focuses on the polymer characteristics which arise from the fluorine atom, and then will focus on the relatively minor changes that exist between the various fluorocarbons.

During the life of the original patents on fluoropolymers, the DuPont brand name of Teflon® was synonymous with the fluoropolymer class. DuPont has applied the Teflon® brand name to a variety of fluoropolymers beyond the simple PTFE. Today, several other companies make fluoropolymers and some of the most common brand names are: Allied's Ultralon®, Whitford's Xylan®, and Hoechst's Hostaflon®. All these polymers are made by addition polymerization.

Fluoropolymers General Family Characteristics

Fluorine is the most electronegative of the elements and strongly attracts electrons to it in any bond that it forms. The electrons around fluorine are held tightly thus forming very stable bonds with low chemical reactivity. Therefore, an inherent property of fluoropolymers is that they do not bond readily with other materials, a property that is popularly called *nonstick*. This chemical inertness and lack of bonding was obvious in early nonstick coating applications on pans. The fluoropolymer gave good nonstick performance to prevent cooked materials from sticking to the pan but the fluoropolymer coating was removed with only minor mechanical abrasion. No chemical bonding occurred between the coating and the pan. The coating was held in place by mechanical linkages (that is, the fluoropolymer flowed into cracks, crevices, and craters on the surface of the pan). These mechanical linkages were not strong and so even minor abrasion caused the coating to flake off. More recent grades of fluoropolymer coatings have improved the bonding capability and fluoropolymers are widely used to give nonstick surfaces to pans, medical devices, industrial rollers of many types (such as rollers for printing, photocopying, and food applications), process guides and dies, and a myriad of other products for consumer and industry that need easy-release properties.

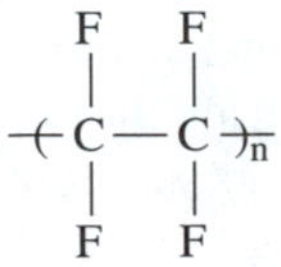

Figure 7.8 Polymer repeating unit for polytetrafluoroethylene (PTFE).

$$\begin{array}{ccc} & F & F \\ & | & | \\ -\!\!\left(\!C\!-\!C\!\right)\!\!-_{\!n} \\ & | & | \\ & F & F \end{array}$$

The chemical inertness gives superior solvent resistance. PTFE is not attacked by any known solvent under normal operating conditions and by only a few solvents under extreme conditions. This has led to applications such as linings for reaction tanks, valves, pipes and chemical storage containers, gaskets, packing, and thread sealants.

Bonding of fluoropolymers is a problem because of the chemical inertness. No solvent bonding system will effectively bond these materials and bonding by other methods such as melt bonding, ultrasonic welding, rf welding, etc., is only partially successful.

The tightly held electrons in fluorocarbons result in very high electrical resistances and the lowest electrical permitivity of any plastic. Hence, fluoropolymers are used extensively as wire insulation, especially for high-value applications where the high cost of fluoropolymers can be accepted.

Lubricity is closely related to nonstick and implies that the material will easily slide against other materials. The lubricity of fluoropolymers is very high. They are used for guide plates, rollers, and bearings, especially when external lubricants are not added. Powdered fluoropolymers are added to other materials to improve their lubricity.

The presence of fluorine atoms makes fluoropolymers inherently nonflammable. This property enhances the value of fluoropolymers in electrical insulation, bearing assemblies, and many electrical and mechanical devices in sensitive aerospace applications.

The presence of fluorine atoms presents a problem if the fluoropolymer decomposes. The resulting products can be toxic. Fortunately, the service temperatures of fluoropolymers are quite high, ranging to 500°F (260°C) for continuous use.

Properties of Specific Fluoropolymer Types

As already indicated, PTFE is the most widely used of the fluoropolymers, with perhaps as high as 90% of total fluoropolymer sales. Its repeating unit, shown in Figure 7.8, is very simple, having only carbons along the backbone and fluorine pendent atoms. PTFE was first discovered, by accident, at a DuPont laboratory when no gas would come out of a supposedly full cylinder of tetrafluoroethylene, sometimes called perfluoroethylene. The cylinder was cut open and found to contain a waxy solid in the bottom. The waxy solid was identified as PTFE, the properties were characterized, and DuPont began scale-up and marketing. The product was an instant success because of the unique nonstick, electrical insulation, chemical resistance, and lubricity properties.

PTFE molecules are long and straight with little branching, thus leading to structures that are over 90% crystallized. This high level of crystallinity gives PTFE the highest density of any plastic material, 2.0 to 2.3 g/cm^3. The denseness of the structure combines with the inherent chemical inertness to increase PTFE's resistance to chemical attack, increase the nonstick property, and decrease the coefficient of friction.

The high crystallinity also increases the mechanical properties of PTFE. Were it not for the bonds within the crystal structures, PTFE would have low strength because the straight, unreactive molecules with small pendent groups would have little intermolecular interactions. As a result, the tensile strength of PTFE (2000 to 5000 psi, 14 to 36 MPa) is considerably lower than the strengths of any of the other engineering thermoplastics and is closer to the range of commodity thermoplastics like polypropylene (3000 to 5500 psi, 21 to 39 MPa) or HDPE (3000 to 5400 psi, 21 to 38 MPa). Hardness is usually related closely to tensile strength and that is true in PTFE where hardness is about the same as HDPE and is less than nylon.

The use of PTFE in abrasion pads is due more to the nonstick properties than to the inherent abrasion resistance. However, PTFE can be filled with fiber reinforcements and is then much more abrasion resistant and still retains good nonstick properties. PTFE is quite tough, about the same as nylon, due to the relatively high elongation, and this also adds to its value for abrasion applications.

In spite of the high crystallinity, PTFE creeps readily. The crystalline bonds do not seem to be sufficient to hold the molecules in place in the near absence of any hydrogen bonding or pendent group interference.

A major problem with PTFE is the difficulty in processing because the melting point of PTFE is close to or above the decomposition temperature. Hence, PTFE has no melt phase that can be used to mold the plastic. Consequently, forming techniques more closely associated with metals and ceramics are used to mold PTFE. The most common of these is *sintering*. In this process a powder of the material is packed into a mold and held under pressure while the material is heated to near its melting point. At these temperatures and pressures, the powder particles *fuse* together, that is they join along their edges, to produce a solid mass. Some voids between the particles will inevitably remain and so the tensile strength of the part is reduced from that of a nonvoid material; however, for most applications not requiring strength, this molding method is satisfactory. Parts made by sintering can be machined and otherwise finished but the shapes cannot be as complex as those possible with molding from a melt. Another method of forming PTFE is *ram extrusion*. In this process the powdered material is heated and pressed through a forming die to make a continuous, uniform cross section part. Ram extruded parts are simple in shape.

Coating of PTFE on surfaces is a major application area. Small polymer particles are usually dispersed in a solvent and then sprayed onto the surface of the item to be coated. The item is then heated to drive off the solvent and fuse the PTFE.

The inability to process PTFE in conventional plastics molding processes like injection molding, extrusion, and blow molding led to the development of a second type of fluoropolymer. What was desired was a lower melting point, less crystallinity, and more flexibility while still retaining the nonstick, chemical inertness and electrical resistance of PTFE. The latter requirement suggested that a *perfluoro* monomer be used, that is, one in which only carbon and fluorine are in the molecule. To give the greater flexibility and lower melting point, the three-carbon monomer was chosen. The resulting polymer is called polyhexafluoropropylene (PHFP) and is shown in Figure 7.9a.

PHFP proved to have a melt temperature that made the polymer processible in conventional thermoplastic molding equipment, but the properties associated with the fluorine character were less like PTFE than anticipated. For many applications, the PHFP would not work. Perhaps a copolymer of perfluoroethylene and perfluoropropylene would give the desired fluorocarbon properties and still be melt processible. This supposition proved to be correct and the resulting copolymer is called FEP for fluorinated ethylene propylene. The relative amounts of perfluoroethylene and perfluoropropylene can be adjusted for specific applications.

(a) Polyhexafluoropropylene   (b) Perfluoroalkoxy

Figure 7.9 Repeating units for (a) polyhexafluoropropylene (PHFP) and (b) perfluoroalkoxy (PFA).

FEP is the second largest fluoropolymer in terms of sales and is used in applications such as tubing and pipes for chemical processes, gaskets, tubing for shrink-wrap electrical connections, shrink tubing for coating rollers and other long, thin parts (rather than coating by spraying on the PTFE), chemical resistant containers, and wire insulation.

Many properties of FEP make it less advantageous than PTFE. Tensile strength is 10 to 15% lower, maximum use temperature is 20% lower, and solvent sensitivity is higher. However, some properties of FEP are improved over PTFE such as toughness (10% higher) and dielectric strength (10 to 20% higher).

Another derivative of PTFE is shown in Figure 7.9b and has a pendent group in which an oxygen links a perfluorinated carbon to the backbone. This polymer is called perfluoroalkoxy (PFA) and has the capability of being melt processable, but it retains more of the nonstick properties of PTFE than does FEP. Hence, PFA is widely used as a coating material. It has also been found to deform slightly when bonding to substrates and is, therefore, more rugged as a coating for pans and industrial products that is PTFE. PFA also retains most of the electrical, chemical inertness, and high-use-temperature characteristics of PTFE. To avoid internal stresses in the product, PFA should be processed at high temperatures (around 700°F, 370°C) and at slow rates.

Several other fluoropolymer products have been developed, largely for special applications and unique combinations of properties. These polymers include the following:

- Nafion® (DuPont), a PTFE derivative that has a sulfonated pendent group on the end. This sulfonated group allows the material to absorb and pass water. The product is used extensively in chemical cell membranes.

- Films made from monomers in which fluorine atoms do not replace all of the hydrogen atoms in ethylene. These films have excellent weatherability and are used for exterior coatings, packaging of corrosive materials, and flame retardance (Tedlar® from DuPont and Kynar® from Pennwalt). Related products are made from copolymers of ethylene and perfluoroethylene (Tefzel® from DuPont).

HIGH-PERFORMANCE THERMOPLASTICS

This set of polymers is distinguished from the engineering thermoplastics because of exceptional performance in one or more properties, usually at a premium price. These polymers are usually chosen because of their high performance in the specific applications in which a plastic may be the only viable choice.

Every polymer in the class of high-performance thermoplastics has a polymer backbone structure that is chiefly composed of benzene rings or substituted benzene rings that are linked by various functional groups or atoms. This highly aromatic nature gives very high strength, high temperature performance, usually high impact toughness, low flammability (because the polymers tend to char), and reasonably good solvent resistance.

Most of the high-performance thermoplastics have been used as the base resin for composite materials in which long fibers (usually carbon or fiberglass) are mixed with the melted resin. These advanced thermoplastic composites compete with thermoset composites for applications in aerospace, high-performance sporting goods, and medical applications.

The high-performance polymers can be subdivided into families that are related on the basis of the type of bonding between the aromatic rings. Each group will be identified and then the variations within the group for specific polymers explained.

Polyphenylenes (PPE, PPO, and PPS)

The phenylene bonding pattern is two benzene rings or benzenes with substituent groups that are joined by an divalent (2-bonding) atom, such as oxygen or sulphur. These materials are thermally stable, solvent resistant, and flame retardant, with somewhat high molding temperatures, but low shrinkage and warpage. The moldability has been further improved in polyphenylene oxide (PPO) by blending with high-impact polystyrene. The impact strength of these polymers leads engineering thermoplastics in applications such as pump impellers and housings, radomes, small appliance housings, video display terminal housings, electrical sockets, valve components, flow meters, wheel covers, and window frames.

PPO has a lower water absorption than any of the common engineering thermoplastics except the fluoropolymers. In spite of the good solvent resistance, these materials can be solvent welded, especially at elevated temperatures with the use of chlorinated solvents. The polyphenylenes can also be metal coated, which has led to some automotive applications where thermal stability, dimensional stability, and fatigue resistance are required.

Brand names of these materials include Noryl® (GE) and Ryton® (Phillips).

Polyaryletherketones (PEEK, PEK, and Others)

This group of polymers has benzene rings joined by either oxygen in ether linkages or by carbon and oxygen in ketone linkages. The properties are, therefore, similar to those of the polyphenylenes that are linked by oxygens or sulphurs. These linkages are illustrated in Figure 7.10.

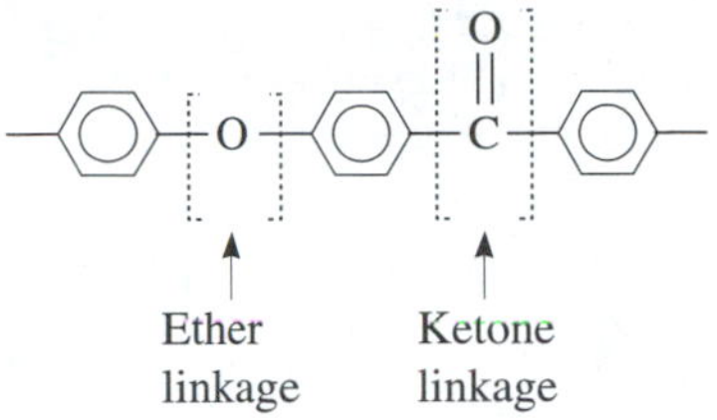

Figure 7.10 Ether and ketone linkages between aromatic groups that are characteristic of the polyaryletherketones.

Ether linkage   Ketone linkage

The name of the polymer is simply a naming of the linkages along the backbone. For instance, if only one ether linkage and one ketone linkage are present, as is the case in the polymer shown in Figure 7.10, the polymer would be called polyetherketone (PEK). If two ether linkages and one ketone linkage are present, the polymer would be called polyetheretherketone (PEEK). PEEK is the most common and widely sold member of the polyaryletherketone group.

The ether linkage favors processibility while the ketone group gives stiffness to the backbone that results in strength and high modulus. Therefore, by adding the ketone group to what would otherwise have linkages like the polyphenylenes, the molecule becomes somewhat stiffer and stronger than the polyphenylenes and has higher use and melting temperatures. PEEK has excellent impact strength and chemical resistance and is used extensively as a resin in carbon-fiber-reinforced composites for the aerospace industry.

Polysulfones (PSU and PES)

The structure of the polysulfones is aromatic groups, usually having more than one benzene ring, joined by an SO_2 group. In some cases, the multiple aromatic rings are directly joined (PSU) and in others they are joined by an oxygen (PES). The SO_2 group adds stiffness and strength to the molecule, much like the ketone group does in PEEK. The polysulfones, therefore have high stiffness, high temperature stability, and high strength.

The polysulfones are more dimensionally stable and tougher than the polyphenylenes, and about the same as the polyaryletherketones. Creep is very low in these polymers over a wide range of temperatures allowing them to be used for hot water pipes, battery cases, circuit breakers, circuit boards, automobile applications near the engine, and interior dishwasher components. The polysulfones compete in many applications against high-performance thermosets, but because the polysulfones can be injection molded, the total part cost can be less because of the reduced time to fabricate by injection molding versus a cure method required with thermosets.

The costs of these resins are higher than the polyphenylenes but somewhat lower than the total cost of a metal when the ease of fabrication and the possibility of combining many metal fabrication steps into one molding step are considered.

The polysulfones are noncrystalline and can be clear in some formulations, thus increasing their applicability in appliance housings and high-temperature view ports. Some caution should be used, however, since these materials will solvent craze, and physical and optical properties could be reduced.

Figure 7.11   Imide group that is part of the polyimides (PI) and the polyamideimides (PAI).

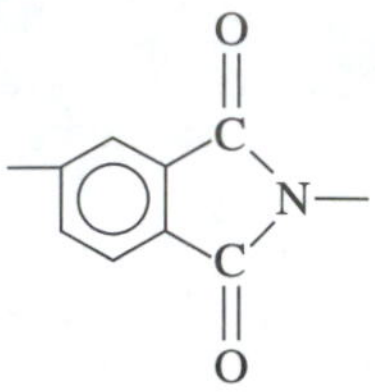

Thermoplastic Polyimides (PI and PAI)

The polyimide group adds considerable stiffness to the backbone because it is composed of a benzene ring to which a ring of carbons, nitrogen, and oxygens is attached as shown in Figure 7.11. This imide group restricts the movement of the backbone and increases the amount of energy required to cause melting. Consequently, polyimides have very high melting points, high use temperatures, and high dimensional stability. They are also very difficult to process with high melting points that are very near the decomposition temperatures. In some polyimides the imide groups are attached to the rest of the chain with amide bonds. These are the polyamideimides (PAI).

Brand names for PAI include Torlon® (Amoco) and for PI, Vespel® (DuPont), Kapton® (DuPont), and Kinel® (Rhone-Poulanc). Torlon is a molding resin that is processible by conventional thermoplastic methods. Vespel is a molding resin that can only be processed by sintering and other solid-phase consolidation methods. Both of these materials are used for high-performance gears, sliding parts such as bearings, and coatings because the coefficient of friction is quite low for the polyimides. In some cases, although not as good, these materials compete with the fluoropolymers for nonstick applications, but the adhesion to other materials and the toughness and scratch resistance are much better. Fluoropolymers or molybdenum disulfide can be added to PAI and PI to make them self-lubricating.

Kapton and Kinel are usually sold as films that are used in electrical and high-temperature applications.

Processing of all these materials is difficult. They have very high viscosities and long times must be allowed for them to flow. Some evidence of skin irritation has been seen with members of this group so precautions against long-term exposure to the liquids should be taken.

CELLULOSICS

The cellulosics are included in the group of engineering thermoplastics as much for historical reasons as for their performance. The strengths, operating temperatures, toughness, and flammability are more like the upper-end products in the commodity plastics than like the engineering thermoplastics. However, in the later 19th and early 20th centuries, cellulosics were used as metal replacements and have, therefore, been traditionally classed with the engineering thermoplastics.

These materials are a wide variety of resins that are all derived from cellulose (wood, cotton, or similar products). The raw cellulose is chopped and then chemically treated to make it soluble in water. The water solution is then further treated with other chemicals to give the desired chemical structure to the polymer and the material is extruded or pressed from the water solution and dried to form the polymer. Many well-known plastics are part of the cellulosics group including cellophane packaging film, rayon fiber, and celluloid, which are some of the oldest of the cellulosics.

In addition to the various treatments possible to change the chemical groups attached to the polymer, many plasticizers are available to give softening and flexibility. Cellulosics are used extensively as modifiers for solutions of other polymers. They are sometimes called "universal thickeners" because of this capability. Their effect is often to *level* paint, that is, to assist the paint in covering the surface uniformly.

In general, the cellulosics are flammable, have poor solvent resistance, and are inexpensive compared with other engineering thermoplastics.

Properties of Specific Cellulosic Types

Cellulose acetate (CA) is the most commonly used of the cellulosics. It is noted for its attractive appearance (gloss and clarity), toughness, and high-impact strength. It is used extensively for tool handles, brush handles, eyeglass frames, pen barrels, caps, and packaging film where moisture passage is desired, as for fruits and vegetables. The major drawbacks of cellulose acetate are poor weatherability, poor solvent resistance, and flammability.

Cellulose acetate butyrates (CAB) are tough, transparent, and water resistant. Some typical uses include data keyboards, cash register keys, transparent dial covers, light covers, and blister packaging. A major use is as an additive to prevent cratering in injection molding, assist pigment dispersion, and reduce solvent crazing in other plastics.

Cellulose nitrates were the first of the cellulosics to be developed commercially. They were originally used for billiard balls and continue today in toilet articles and lacquers and for sheet applications, although their high flammability has dramatically reduced the consumption of the plastic.

=== CASE STUDY 7.1 ===

Making Nonstick Electrosurgical Blades

Electrosurgical knives, sometimes called cautery blades or tips, were designed to meet a specific need in the operating room. These blades are attached to a properly controlled, handheld electrical source and can then be used to perform surgery by cutting and *cauterizing* (closing bleeding vessels) simultaneously. With such a device, the surgeon can make precise cuts without the constant problem of excessive bleeding in the area. Blades of various shapes are available to facilitate operations in different parts of the body and for different purposes. Usually the blades are installed onto the end of an electrical pencil-like holder that generates the radio frequency current necessary to heat the blade. The holder can also provide the surgeon with fingertip control over power level and suction for removal of debris and blood.

Originally the blades were made of uncoated stainless steel. While these blades were capable of cauterizing and cutting, one problem that resulted was tissue sticking to the blade, thereby reducing cutting efficiency and accuracy and ultimately requiring replacement of the blade. One approach to overcoming this problem was to coat the blade with nonstick material to which cauterized tissue and debris would be less likely to adhere. Of course the nonstick coating must be capable of passing current from the blade to the tissue so that cauterization could be effected.

In the first attempts to create a nonstick coating on the electrosurgical blades, small cracks in the coatings were created to insure electrical conductivity along the blade. These cracks provided electrical pathways between the stainless steel and the tissue. Some improvement in the efficiency was shown over bare stainless blades, but in use of the coated knife, the conductive locations or openings typically became coated with charred tissue, soon rendering the knife nonconductive and thus unusable.

A second method of coating the blades was tried by using a successful industrial coating technique. In this method the stainless steel blade was first coated with a known durable binder material that was required to have high strength over a wide range of temperatures and to withstand repeated wiping of the blade by the surgeon to remove the char. To further enhance the nonstick nature of the coating, some inherently nonstick material was added to the binder dispersion. The materials chosen for the binder were high-performance thermoplastics, in one case polyimide, and in the other case polyamide imide. Both of these materials were stable at high temperatures, gave good bonding to stainless steel, and could be mixed with PTFE (up to about 30%) to give additional nonstick capability. This coating method gave some improvement over the previous, uncoated or incompletely coated blades, but still resulted in excessive buildup of char which was difficult to remove. The char was built up because of the conductive nature of the binder material and the low concentration and incomplete coverage of the PTFE at the surface. Hence, a new and somewhat different approach was needed.

The new approach was to coat the stainless steel surface with pure PTFE. This coating would surely provide very high nonstick character to the blade. The question remained, however, if the blade would still operate properly because of the known high electrical resistance of the PTFE material. A few of the pure PTFE-coated blades were fabricated. These proved to be far superior to any other blades in ease of char removal and, somewhat surprisingly, gave cauterizing efficiencies at least as good as the other blades that had been used in spite of the very high electrical resistivity of PTFE. The developers of this latest blade discovered that PTFE provided a secondary method for electricity to be conducted through the coating. This method is called capacitive coupling and occurs because of the leakage of electrical current through the material, much like current will eventually leak though a capacitor in an electrical circuit. In all of the other devices tried, the passage of electrical current was by traditional ohmic methods, like passage through an electrical resistor.

The coating of the blade with the PTFE was, in itself, a difficult problem. The coating must be durable so that it is not easily wiped off when the char is removed, must conduct the current capacitively, and must be cost effective in manufacture. The problem of adherence to the blade was solved by using two techniques. The stainless steel blades were first etched using a sand blast. In this process, the blades were supported vertically in a metal holder and then introduced into a spray box. The fine grit sand was entrained in a water flow and sprayed onto the blades, giving a pitted surface that would provide good mechanical bonding. After sand blasting, the surfaces were blown clean.

The second step was to clean the surface with a primer solution of PTFE suspended in a mixture of chromic and phosphoric acids. The acids ensured that no contaminants remained on the blade surface and allowed the PTFE to flow into all of the interstices of the mechanically abraded surface. This primer coating was then heated to evaporate the solvents and cause some adhesion of the PTFE to the metal surface.

A top coat containing PTFE and a solvent was then sprayed on the blade. This top coat had a much higher concentration of PTFE than the primer and, when baked at 400°F (200°C), the PTFE particles fused together to form a continuous PTFE surface over the blade.

Actual surgical testing of the blades demonstrated excellent cutting efficiency and accuracy with little char buildup. The buildup that did occur could be easily removed by gentle wiping of the blade. This wiping did not remove the PTFE coating because of the superior mechanical anchoring to the pitted blade. The PTFE-coated blades usually lasted through an entire case, thus saving precious surgical time and reducing frustration for the surgeon.

Further refinements in the manufacturing method have resulted in the use of a high-concentration PTFE primer coat that can be fused, thereby eliminating the need for the special top coat. Other fluoropolymers are also being evaluated.

The general techniques discussed in coating the surgical blade are similar to those used in coating cooking pans. Electrical conductivity is not important in pans, but thermal conductivity and abrasion resistance are. To provide improved thermal conductivity, a conductive filler can be added to the fluoropolymer dispersion. Furthermore, because the abrasion resistance of a pan surface must be much higher than that for a surgical blade, the use of PI or PAI binders can give this improved durability, although with some loss in nonstick behavior. The use of PFA in place of pure PTFE has also been explored as a compromise in achieving good adhesion without sacrificing too much nonstick characteristic. Undoubtedly the field of nonstick coatings will continue to evolve as new techniques for coating and new resins are developed.

SUMMARY

The engineering thermoplastics are similar in many respects because they are generally strong, have a high use temperature, have a good impact strength, have good dimensional stability over a wide range of temperatures, have generally good resistance to environmental conditions, and are much more moldable than the metal, ceramic, or wood parts that they often replace. Rather than reiterate the values for each of the major types of engineering thermoplastics, these are presented in Table 7.1.

Table 7.1 Summary of Key Properties for Various Engineering Thermoplastics

Resin	Tensile Strength psi/(MPa)	Max. Use Temp. °F/(°C)	Notched Izod Impact ft-lb/in. (J/mm)	Dimensional Stability (Creep resistance, coeff. of linear exp.) $\times 10^{-6}$°F ($\times 10^{-6}$K)	Environmental Resistance	Moldability, Finishing, and Other Key Properties
Polyamides (PA) nylon (6/6)	9000–12000/ (62–84)	175–250 (80–120)	2:1 (15–20)	Creep: fair CLE: 44 (80)	Flam:slow to SE H_2O:2.5% Acids:poor UV:fair	Properties vary with moisture
Acetals or polyoxymethylenes (POM)	9000–10000/ (62–70)	194–270 (90–110)	1–2.3 (8)	Creep: exc CLE: 50–61 (90–110)	Flam:slow H_2O:0.22–0.25% Acid:poor Others:exc UV:poor	Can easily decompose
TP polyesters (PET/PBT)	5800–10500/ (40–72)	212 (100)	0.8–1.0 (4)	Creep: good CLE: 39 (70)	Flam:burns to SE H_2O:0.08:–0.3% Others:exc	Moderately difficult to mold
Polycarbonate (PC)	8000–9700/ (56–57)	275 (135)	12–18 (20–30)	Creep: exc CLE: 33–39 (60–70)	Flam: SE H_2O: 0.16% Solv: poor UV:good	Clear
Acrylic (PMMA)	7250–11000/ (50–77)	150–194 (65–90)	0.3–0.5 (2)	Creep:good CLE: 39 (70)	Flam:burns H_2O:0.1–0.4% Others:good UV:exc	Best weather. v. clear
Fluoropolymer (PTFE)	3680–5200 (25–36)	480 (250)	3.0 (13–15)	Creep:poor CLE: 56 (100)	Flam:no H_2O:none Others:none UV:exc	Difficult to process, best elect. resist., v. low friction
Polyphenylenes (PPO/PPS)	8000–10800/ (55–75)	175–500 (80–260)	0.3 (–)	Creep:good CLE:26–31 (47–55)	H_2O:none Flam:SE Solv:exc	Difficult to process
Polyaryletherketones (PEK/PEEK)	13300–15000/ (92–103)	266–275 (130–135)	8 (–)	Creep:v.good CLE: 36 (65)	Flam:no Acid/base: good Solv:exc	Difficult to process, high price
Polysulfone (PSU/PES)	7250–14300/ (50–100)	300 (150)	1, 3 (–)	Creep:v.good CLE: 30 (54)	Flam:low Acid/base: good Solv: fair	High price Can be clear
TP Polyimides (PI/PAI)	14800 (102)	460–600 (240–320)	0.5–1.0 (–)	Creep:v.good CLE: 28–35 (50–63)	Flam:no H_2O:none Solv:exc	Difficult to process, high price
Cellulosics (CA/CAB)	3000–11000 (21–76)	175 (80)	1.5 (6–20)	Creep:fair CLE: 67 (120)	Flam:slow H_2O:6% Acid:v.poor Others:var. UV:exc	Clear

GLOSSARY

Acrylic syrup A solution of acrylic polymer and monomer, often used for casting.

Acrylics A class of polymers characterized by the presence of an acrylic bond.

Aramid An aromatic polyamide.

Cauterizing Closing bleeding vessels.

Cellulosics A group of plastics derived from cellulose, usually by acid treatment.

End-capping Reacting of the ends of a polymer to give the polymer some specific property (such as thermal stability).

Fluorocarbon A material containing fluorine and carbon.

Fluoropolymer A polymer containing fluorine.

Fuse Joining of solid particles or parts by partial melting or surface heating below the melting point.

High-performance thermoplastics TP materials, usually highly aromatic, that have extraordinary physical and mechanical properties.

Level To assist paint in covering a surface.

Lubricity The tendency of a material to slide when pressed along the surface of another material.

Nonstick A material property that reflects the material's poor bonding to other materials.

Notch sensitivity Prone to failure in impact tests if a notch or crack is present in the material.

Nylon Another name for polyamides.

Orientation effects Properties that depend on or are related to alignment or other orientations of the molecules.

Perfluoro A molecule with a high fluorine content.

Polyacrylonitrile (PAN) An acrylic resin used principally to make fibers and as a blending resin in ABS.

Polyamide (PA) A polymer in which the amide bond is formed during polymerization; nylon.

Polyarylesters High-performance thermoplastic polyesters that have a very high aromatic content.

Polyaryletherketone A group of polymers in which benzene rings are linked by ether and ketone groups; PEEK (polyetheretherketone) is the most important member of the group.

Polycarbonate (PC) A class of engineering thermoplastics that is characterized by the carbonate linkage.

Polyester A class of polymers characterized by the formation of ester bonds; both thermoplastic and thermoset polyesters are common.

Polyethylene terephthalate (PET) The most common of the thermoplastic polyesters.

Polyimides (PI) A group of polymers in which the imide group is a major constituent.

Polymethylmethacrylate (PMMA) An acrylic resin used for molding; the plastic with the best UV light stability.

Polyoxymethylene (POM)   A polymer formed from formaldehyde or from formaldehyde and polyethylene; acetal.

Polyphenylenes   A group of polymers in which benzene groups are linked by sulphur or oxygen atoms.

Polysulfones (PSU)   A group of polymers in which benzene rings are linked by sulfate groups.

Polytetrafluoroethylene (PTFE)   The most common of the fluoropolymers.

Ram extrusion   A process in which powdered material is heated and pressed through a forming die.

Sintering   A forming technique in which a powdered material is packed into a mold and then heated for an extended period of time to a temperature somewhat lower than the melt temperature, thus causing the particles to fuse.

Unzipping   Degradation by depolymerization back to the monomer.

QUESTIONS

1. Which type of backbone substituents stiffen the polymer chains?

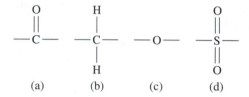

(a)   (b)   (c)   (d)

2. Indicate three advantages of clear polycarbonate sheet over clear acrylic sheet.
3. Compare the water absorptivity of nylon (6/6) to nylon (6/12) and explain the differences that are noted.
4. Discuss the physical, chemical, and mechanical properties of PMMA compared to polystyrene.
5. Discuss why acetals are generally not made into fibers.
6. Identify FEP and discuss its main advantage over PTFE.
7. If nylon is sensitive to moisture absorption, why can it still function well as a tow rope for water skiers?
8. Why is the temperature raised in order to achieve crystallization in PET? Is that temperature higher or lower when the PET has been oriented?
9. Why is a polyimide film ideal in a very high-temperature electrical application, even one in which the film may be slowly degraded?
10. Which of the engineering thermoplastics could reasonably be used as a tire cord material?

REFERENCES

Birley, A. W., R. J. Heath, and M. J. Scott, *Plastics Materials: Properties and Applications,* (2nd ed.), New York: Chapman and Hall, 1988.

Daniels, C. A., *Polymers: Structure and Properties,* Lancaster, PA: Technomic Publishing Company, Inc., 1989.

Domininghaus, Hans, *Plastics for Engineers,* Munich: Hanser Publishers, 1991.

"Engineering Plastics," *Engineered Materials Handbook,* Vol. 2, Metals Park: OH, ASM International, 1988.

"Engineering Resins Primer," *Plastics Engineering,* Dec. 1974.

Richardson, Terry L., *Industrial Plastics: Theory and Application* (2nd ed.), Albany, NY: Delmar Publishers, Inc., 1989.

Rosato, Donald V., David P. DiMattia, and Dominick V. Rosato, *Designing with Plastics and Composites: A Handbook,* New York: Van Nostrand Reinhold, 1991.

Seymour, Raymond B., and Charles E. Carraher, *Giant Molecules,* New York: John Wiley and Sons, Inc., 1990.

Strong, A. Brent, *High Performance and Engineering Thermoplastic Composites,* Lancaster, PA: Technomics Publishing Company, Inc., 1993.

Tess, Roy W., and Gary W. Poehlein, *Applied Polymer Science* (2nd ed.), ACS Symposium Series, 285, Washington: American Chemical Society, 1985.

CHAPTER EIGHT

THERMOSET MATERIALS

This chapter examines the following concepts:

- Crosslinking (reactive groups, crosslinking as the polymers are formed, crosslinking between two existing polymer chains, viscosity and thermal control)
- Thermoset types, general properties, and uses
- Phenolics (PF)
- Amino plastics (UF and MF)
- Polyester thermosets or unsaturated polyesters (including vinyl esters, alkyds, allylics, BMC and SMC)
- Epoxies (EP)
- Imides
- Polyurethanes (PUR)

INTRODUCTION

Thermoset plastic parts are made from polymeric resins that are capable of forming chemical crosslinks. These resins, called *thermosets,* are shaped (usually by placing them into molds) and then are chemically crosslinked. The process of crosslinking is called *curing.* These crosslinking reactions bind the polymer molecules together in a three-dimensional network resulting in a part that cannot be melted. The crosslinking reactions are not reversible. If the thermoset parts are heated, they may soften somewhat, but they will retain their general shape as the temperature is raised, not melting, until decomposition begins. Therefore, as opposed to thermoplastics in which molding and shaping only involves freezing the hot (melted) material, in thermosets the molding and shaping involve a chemical reaction that causes the material to harden and take a permanent shape.

To clarify the nature of the curing of a thermoset material, the following common example, baking a cake, is offered. The ingredients are mixed together to form a batter. One of these ingredients is a polymer that is capable of forming crosslinks (wheat flour). Other ingredients could be colorants or additives that give particular taste, texture, or processing properties. (This batter is like the plastic thermosetting resin that is capable of forming crosslinks, and the other materials are like the various fillers and processing aids that are commonly added to thermoset resin mixes.) The batter/resin is placed into the cake pan (the mold). The cake pan is often coated with butter or grease, which serves to assist in removing the cake when it is finished. (Thermoset resins will often employ mold releases for the same purpose.) The filled pan is then placed in the oven and heated until the cake is formed and all the batter is reacted. The completion of the reaction is usually determined by time, although some specific tests for unreacted batter do exist, such as probing with a toothpick to see if liquid batter adheres. Further heating of the cake will not cause it to melt, but rather to burn. If burned, the material becomes a carbonaceous mass that certainly does not have the properties of the cake. When cooked for the correct time, the cake pan is removed from the oven and the cake is removed after cooling. (In like manner, the thermoset plastics are put into molds, heated to crosslink, and then cooled to facilitate removal of the part from the mold.)

A slightly different scenario, but one that also illustrates the curing of thermoset plastics, can be seen with cookies. In this case the material to be baked is much thicker than the cake batter. It is called a dough. (Likewise, some thermoset resins are quite thick and they are called molding compounds or, sometimes, molding doughs.) The cookies can be placed onto a sheet and cooked without a mold. (This happens in thermoset plastics as well when shape retention is not important.) A modification of the cookie baking process can be imagined if shape is important, for instance, in a commercial cookie factory. If all the cookies are to have the same shape, the cookie dough could be packed into molds of the desired shape and with some added pressure on the molds. The cookies would retain the shape of the mold when they were baked. The filling of the cookie molds might even be done with a device that squeezes a set amount of cookie dough into the mold. (A thermoset plastic process mimics this filling and baking process by having the plastic metered into molds and cured under pressure.) The cookies might even have coconut shreds. (Although coconut shreds are put in cookies for taste, reinforcements of fiberglass are often put into thermosets to hold the part together and give it additional strength.)

One further example of thermoset-like behavior, although not a food analogy, is the mixing of a two-component adhesive. After mixing the two components, the material can then be applied as an adhesive. It reacts over time to solidify (crosslink) and join the parts. No external heat is needed for the reaction to occur.

The features described for the cake, cookies, and adhesive are all common features of the processes associated with thermosetting materials. Thermosets can be batter-like liquids, dough-like molding compounds, or liquids that are mixed. Thermosets can be cured in molds or without molds, can be crosslinked under pressure or without, and can be cured with or without external heating. All of these conditions are dependent upon the requirements of the part and the nature of the thermosetting material used.

CROSSLINKING

Reactive Groups

The characteristic of polymer resins that allows them to be crosslinked is that they contain more reactive groups than are required for just polymerization. In the case of addition polymerization, only one reactive group is needed in each monomer to form a polymer— a carbon-carbon double bond. Therefore, if two or more carbon-carbon double bonds are present in the monomer, crosslinking can occur. One carbon-carbon bond in each monomer would be used in making the polymer but the additional carbon-carbon double bonds would be unused and could then react with carbon-carbon double bonds in other molecules and form crosslinks. Butadiene rubber is crosslinked using this methodology.

In the case of condensation polymerization reactions, each monomer must have two reactive groups to form the basic polymer—one group at each end of the molecule. There-fore, if the reacting molecules have three or more reactive groups, crosslinking can occur by condensation reactions at the extra active sites.

In some cases condensation polymerization can be used to form the polymer and then additional polymerization can be used to form the crosslinks. In this case, each of the re-acting molecules must have two reactive groups for the condensation polymerization and some of the monomers must have a carbon-carbon double bond or some other, similar, group that will bond by addition reactions and form crosslinks. Unsaturated polyesters are crosslinked by this methodology.

Crosslinking reactions that are not strictly addition reactions or condensation reac-tions are also possible. For instance, epoxy resins crosslink by a ring-opening step that cre-ates active sites for bonding to a crosslinking molecule. Also, urethanes can form crosslinks by reactions between reactive monomers even though no condensates are formed.

Depending upon the nature of the reacting groups, the polymerization and crosslink-ing reactions can occur simultaneously or can occur at different times and under differ-ent reaction conditions. For instance, a polymer could be formed by a condensation reac-tion and then, at some later time it could be crosslinked by addition reactions. In this way the monomers, which are often gases or highly volatile liquids (which are difficult to store and ship) can be converted into a polymer (much easier to store and ship) and then, when desired, can be molded into a thermoset part by crosslinking.

The complexity of the thermosetting process implies that some measure of precise control over the crosslinking reaction is desired. Usually some specific, external event is required before crosslinking begins. In some cases, that event is simply the mixing of the reactants. Another event could be the mixing of the thermosetting resin with a material that starts the crosslinking reaction. Such materials are called *initiators* or *hardening agents*. In many cases, initiators are chosen that require heating of the mixture to activate the initiator and "kick off" the crosslinking process. A broad term that is often applied to initiators, hardeners, or any other material that promotes crosslinking is *curing agent*. A term that is widely, though incorrectly, used for these curing agents is *catalyst*. A catalyst is defined as a material that promotes or enhances a reaction but is not consumed in the reaction. A catalyst is recoverable after the reaction is over. The curing agents all react and are consumed in the crosslinking reaction. Therefore, in this text the term catalyst will be used only for those cases that meet the strict definition. In commercial practice, however, catalyst is widely used as a term for curing agents.

Crosslinking as the Polymers Are Formed

Many examples of monomers having more reactive sites than the minimum required for just polymerization are well known. One of these can be illustrated in the polymer known as crosslinked melamine formaldehyde (commonly called Melmac® or Formica®, which is often used as picnic dishes and plastic countertops). The monomers for this reaction and the resultant crosslinked polymer are depicted in Figure 8.1.

In the case of melamine formaldehyde, the polymerization occurs by a somewhat standard condensation mechanism, that is, between an active site on one monomer and an active site on the other monomer, with water formed as a by-product. To satisfy the requirement for crosslinking, one of the monomers (melamine) has **three** active sites, one more than the two needed to form a condensation polymer. Therefore, as the melamine and formaldehyde monomers are heated together, they can react to form a new bond at all three active melamine sites. (Water is formed as a by-product of all the reactions.) Because formaldehyde has two active sites, a subsequent reaction with melamine monomer can also occur on the other end of the formaldehyde molecule. Chains can, therefore, grow simultaneously in many directions. These can combine with monomers or with active sites on other growing polymer chains or parts of the same chain. The eventual result is an interlocking, multidirectional network of polymers. In essence, all of the atoms are joined *together* into a continuous and interlocking network rather than just in a linear chain. Hence, the polymer has an extremely high molecular weight. This massive network caused by the interlocking of the chains is characteristic of thermoset materials.

Crosslinking between Two Existing Polymer Chains

Another way to provide for crosslinking is to form polymers in which reactive sites are present in the polymer chain after the chain has been formed. These reactive sites can then be used as locations to form crosslinks between the polymer chains. For this sequence to occur, several requirements must be met. First, the monomers that react to form the chains cannot disturb or react with the reactive sites that will later be used to form the crosslinks. Second, a mechanism must exist for the active sites along the chain to react and form covalent bonds without destroying the already formed polymer chains. Some of the most common examples of this type of polymer are the crosslinking of polyester resins that are often used as the binding material for fiberglass in recreational boat hulls and whirlpool baths. This crosslinking method is illustrated in Figure 8.2.

In crosslinkable polyesters the polymers are formed in a normal condensation polymerization reaction. The H on one monomer reacts with the OH on the other to form the water by-product and a new bond is formed linking the monomers. This can happen repeatedly because each of the monomers has two active sites for polymerization. In this case, one of the polymers also has another active site, but that site is undisturbed by the

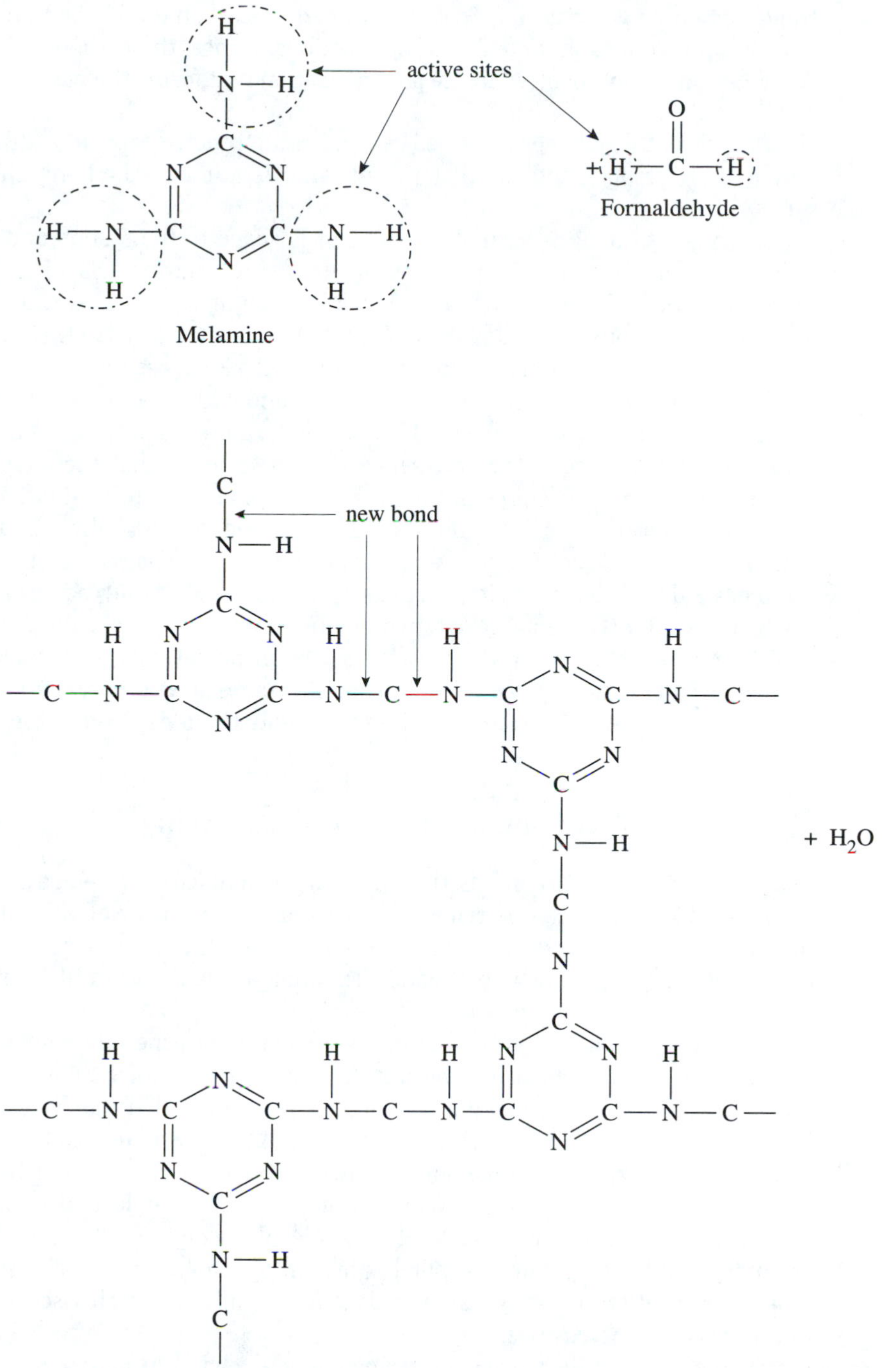

Figure 8.1 Polymerization and crosslinking of melamine formaldehyde.

condensation polymerization. That site is a carbon-carbon double bond that is only active in addition reactions, not condensation reactions. Hence, the carbon-carbon double bond is present but is not involved in the polymerization that forms the basic polymer. This step is illustrated in Figure 8.2a.

Therefore, the polymer is formed by condensation polymerization, although usually to only a short chain length so that the material is still liquid. (Long chains tend to be solids.)

When crosslinking is desired, an initiator is added to the system and mixed. The system is then heated to break apart the initiator, which starts the addition polymerization reaction at the carbon-carbon double bond. (See Figure 8.2b.) The initiator will combine with the π-electrons in the double bond to form a new bond. The other electron in the π-bond will move to the other carbon and a free radical will result.

The resulting free radical can then form a bond with a carbon-carbon double bond in another chain. This new bond would be, of course, a crosslink that ties the two chains together. In actual practice the polymers are often so bulky that the double bonds in two polymers rarely get close enough to form bonds. Therefore, to facilitate the formation of a bond between the polymer chains, another component is usually added that contains a double bond and can act as a *bridge molecule* between the chains. Because the bridge molecule has a double bond, it simply reacts by the same addition reaction mechanism (Figure 8.2c and d). Styrone is the most commonly used bridge molecule.

The resulting structure is characterized by principal chains of polyesters that were formed by condensation polymerization and then subsequently crosslinked using addition reactions, often using bridge molecules to facilitate the formation of the crosslink bonds.

Viscosity and Thermal Control during Crosslinking

Thermoplastic resins are solids that are melted, molded and cooled to form the solid, shaped part. The part can be remelted if desired. The behavior of viscosity with temperature for thermoplastics is seen in Figure 8.3. The solidification by cooling is seen as a retracing of the thermoplastic path, and the double-headed arrows illustrate that this path is reversible.

Even when molten thermoplastics are heated, they generally have much higher viscosities than thermosets, as can be seen from Figure 8.3, which shows the thermoplastic line above the thermoset line during the early part of the thermal cycle, before crosslinking occurs. This is because with thermosets, the molecular weight can be increased through crosslinking, thereby improving many of the physical and mechanical properties, such as toughness and strength. With thermoplastics, the molecular weight cannot be increased during processing; therefore, the molecular weight of the material must already be high in order to obtain acceptable physical and mechanical properties. This higher molecular weight in thermoplastics leads to higher inherent melt viscosities, as compared to the thermoset viscosities.

The behavior of thermosetting resins is also depicted in Figure 8.3. In this case, the thermosetting resin can initially be either a liquid or an easily melted solid. (The transition from a liquid to an easily melted solid is simply the condition of the material at

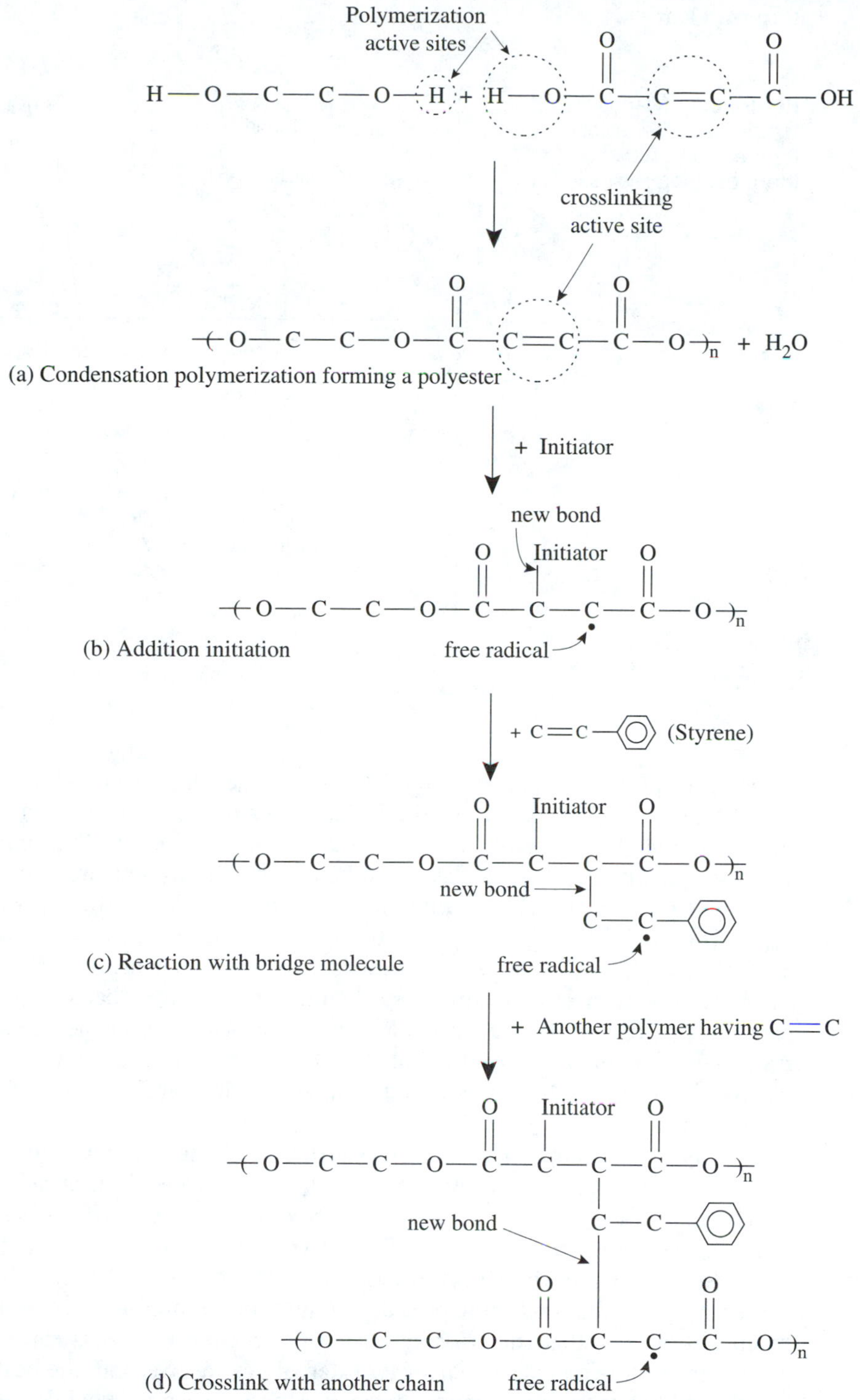

Polymerization active sites

crosslinking active site

(a) Condensation polymerization forming a polyester

+ Initiator

new bond

Initiator

free radical

(b) Addition initiation

+ C=C—⬡ (Styrene)

Initiator

new bond →

free radical

(c) Reaction with bridge molecule

+ Another polymer having C=C

Initiator

new bond

free radical

(d) Crosslink with another chain

Figure 8.2 Crosslinking of already formed chains. The polyester crosslinking system.

269

Figure 8.3 Viscosity-temperature behavior for thermoplastic and thermosetting resins.

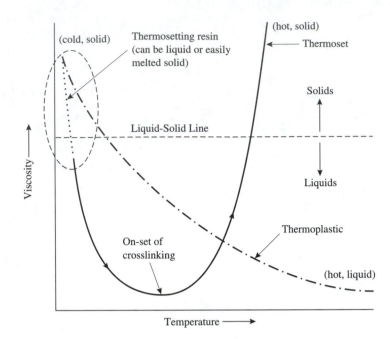

room temperature and is therefore illustrated on the graph as a region that overlaps the liquid-solid line. The position of the line depends on the ambient temperature.) When heated, the viscosity initially drops because of normal viscosity thinning with temperature, just as with the thermoplastics, although the thermosets are usually lower in viscosity at this stage. Further heating of the thermosets causes the crosslink reactions to begin. Eventually the viscosity begins to increase because the formation of the crosslinks binds the molecules together and restricts their movement. The thermoset material curve eventually rises and crosses the liquid-solid line thus becoming a hard finished part. This path cannot be retraced, that is, the process cannot be reversed (as is illustrated by the arrows). Hence, the initial, downward part of the thermoset curve is dominated by temperature thinning while the latter, upward part of the curve is dominated by crosslinking.

Another important temperature consideration of the thermosetting polymers is the heat that is usually given off when the crosslinks are formed. This is called the *heat of reaction* and, because the heat is given off, is called an *exotherm*. (If the heat were taken in, it would be called an *endotherm*.) Several consequences follow from the existence of exotherms in most of the curing reactions. First, external heat may be required only in certain stages of the reaction to provide the molecular movement necessary to line up the reactive groups so that the crosslinks can form. Some thermosets can be crosslinked at room temperature and have no need for external heating. Second, the heat generated may be so high that some temperature control may be required to avoid decomposition or the rapid expansion and contraction that might occur and result in cracks in the part. Temperature control may be made by cooling, keeping the rate of reaction within certain limits, or by limiting the mass (thickness) of the parts so that heat will not build up in one section.

Even though the exotherms of thermosetting reactions may require some limiting of the thickness of the parts, thermosets are much less sensitive to thickness variations across the part than are thermoplastics. In thermoplastics, differences in thicknesses from one section of the part to another will result in differences in heat retention during the molding (cooling) of the part. These heat retention differences result in thermal stresses and differential shrinkage of the part. (Thick sections shrink more than thin sections and this often results in dimples or other sink marks in thick sections.)

One of the advantages of thermosetting resins is that they can be liquids when the process begins. This allows for easy mixing of colorants, fillers, reinforcements, and processing aids, usually with just simple stirring or mechanical mixing. Thermoplastics must be melted to allow these materials to mix and that is usually done in an extruder. The melted thermoplastic usually has a higher viscosity than does the thermoset liquid, which makes the mixing in of additives even more difficult in thermoplastics.

In some cases, the viscosity of the initial thermoset material is too low for good mixing. The resin flows away from the fillers, reinforcements or other additives. To correct this problem, a process has been developed in which some thermosetting resins are partially polymerized which increases the viscosity. When the viscosity is appropriate for mixing and, perhaps, long-term storage, the reaction is stopped or quenched. The materials are mixed and the resinous mixture is called a *molding compound* or *molding resin*. When the final molding and curing of the part is then desired, this molding resin is placed in the mold and crosslinked. Three stages of thermosetting resins are therefore identified. The first, called *stage A,* is when the material is thin and principally monomeric. In the second stage, called *stage B,* the material has been partially polymerized. The third, *stage C,* is when the material has been cured to its final state. The process of partially polymerizing the material is sometimes called *B-staging*. The molecules in the partially polymerized resin also have a special name. They are called *oligomers* or *prepolymers* which means that they are partially formed polymers and implies that full polymerization will occur at a later time.

Another method of controlling the viscosity of the thermosetting resin is to dissolve the resin in a solvent. This is done when the resin is too thick at the initial stages of the process. The solvent is sometimes called a *diluent* because it dilutes and lowers the viscosity of another liquid. The problem with this method is that the solvent must be eliminated during or after the reaction so that the finished part will be a solid and not plasticized or otherwise affected by the solvent. In some cases, the solvent chosen will become a reactant in the curing reactions. It could be a comonomer that extends the length of the polymer chains or the solvent could serve as bridge molecules to help form the crosslinks. Solvents that participate in the reaction and are thereby consumed are called *reactive diluents*.

The time required for the molding and curing step for thermosets is usually much longer than the time for thermoplastics to be molded and solidified. Therefore, the longer molding times of thermoset materials are leading to thermosets being replaced by thermoplastics, especially where there are also improved physical, mechanical, and chemical properties. Still, many applications are uniquely served by thermosets. Properties such as the ability of the thermoset resin to easily wet reinforcement fibers, to have widely varying thicknesses from one section of the part to another, to withstand high temperatures without melting, to be molded without heat, to be mixed without an extruder, and to be molded without high pressure are some of those advantages which some thermosets possess and dictate their continued use.

The long cycle times of thermoset resins can be shortened significantly by preheating of the material before it is placed in the mold. This preheating is often done using a dielectric heater (radio waves). Preheating can also allow many of the trapped gases to escape that can cause poor part appearance. In some applications, this preheating has become a necessity for economic viability of the thermosetting process.

It is possible to stop the polymerization or the crosslinking reactions before completion and then resume it again. This allows monomers to be reacted only until they are converted into short polymer chains, usually with little or no crosslinking. They can be stored and shipped easily in this form and then, when desired, the polymerization and crosslinking reactions can be resumed and completed. The stopping of the polymerization and crosslinking reactions is called *quenching* and the short-chain polymers formed are called *oligomers*. Polymers that have been stopped partway in the curing process are said to be in a *B-stage*. (The A-stage is unreacted monomers and the C-stage is fully cured polymers.)

THERMOSET TYPES, GENERAL PROPERTIES, AND USES

Some properties are characteristic of most thermosets. As the number of crosslinks increases, the stiffness of the material also increases. Thus, many thermosets are typically stiffer and more brittle than the thermoplastic materials. The impact toughness of thermosets can be increased by adding fillers or reinforcements, which can also increase strength. These fillers and reinforcements will also affect the viscosity of the molding resin and adjustments in the manufacturing method may have to be made to allow for the higher viscosities of these filled compounds.

The crosslinks improve the creep resistance of thermosets over most of the engineering thermoplastics. Long-term thermal stability and dimensional stability with temperature are also improved with crosslinking, although some thermosets soften with heat sufficiently that they cannot be used when high temperatures are encountered.

Because thermosets can be formed from generally low-viscosity liquids, some important capabilities and properties arise. For instance, thermosets are much more capable of wetting reinforcement fibers than are the higher viscosity thermoplastics. Also, the low viscosity of thermosets allows them to be molded into very complex shapes. Furthermore, because thermosets can be crosslinked at room temperature, the resin may not need to be heated, even during the molding phase. These advantages have led thermosets to be used for applications such as composite materials (where a fibrous reinforcement is included with the resin), complex electrical switches and assemblies, very large parts that could not be easily heated for molding (such as boats), and many parts where thermal stability is critical.

In each of the groups of thermosets that will be discussed, the first focus will be on the nature of the resin and the crosslinking reaction that allows it to form a thermoset part. Then, the mechanical, physical, and chemical properties characteristic of that thermoset group will be discussed. In some groups of thermoset materials, several different types of specific resins exist. These will then be compared.

The thermosetting resins that will be discussed in this chapter have been grouped into several families. These groupings are chiefly based upon the nature of the crosslinking reaction that is characteristic of the family.

- Phenolics are made by the reaction of phenol (an aromatic molecule) and formaldehyde (a common organic liquid). The reaction of phenol with formaldehyde simultaneously forms polymer linkages and crosslinks. The resulting material is very hard and stiff. The properties of phenolics suggest that this material is highly crosslinked and that the crosslinks are strongly three-dimensional. That is certainly true. This hardness has allowed phenolic to be used as a substitute for other very hard materials, such as ivory (billiard balls).

- Amino plastics are also formed by simultaneous polymer and crosslinking reactions, but use amines in place of phenol as the reactant with formaldehyde. In many cases, the multiple amine sites can be present on one molecule, thus giving a very tightly bonded crosslink structure. These materials, like phenolics, are hard and brittle plastics. The amino plastics have very high surface energies along with the hardness. These surface energies contribute to the widespread use of these materials as adhesives and as surface materials for countertops.

- Unsaturated polyesters and related materials are low molecular weight polymers that contain carbon-carbon double bonds that can be used to form crosslinks by the same free-radical mechanism used to create addition polymers at a time chosen by the molder. The reactive polymers are usually liquids at room temperature that become cured after the addition of an initiator and, occasionally, heat. The resulting materials are less rigid than the phenolic and amino plastics, but are still quite stiff and brittle. The unsaturated polyesters are, by far, the most common of the thermoset materials. Their most predominant application is in composites, where fiberglass is the reinforcement. The vinyl esters are closely associated with unsaturated polyesters and cure by the same general mechanism.

- Epoxies are polymers with three-membered rings on the ends of the polymer chains. These rings are active bonding sites for a wide variety of chemicals (called hardeners) that can react with the polymer and form a bridge to another polymer, thus creating a crosslink. Epoxies are stiff and strong. Epoxies are used extensively as adhesives and as the resin in advanced composite applications. (Those composites in which carbon fiber is used as the reinforcement and that require higher performance from the resin than can be obtained with polyesters.)

- Imides crosslink by a condensation polymerization method between molecules that contain the imide group. This group is somewhat like an aromatic group but is even stiffer and stronger. The imides are stiff, strong materials with very high thermal stabilities.

- Polyurethanes are created by the reaction between polyols and isocyanates. This reaction takes place simply by mixing the two reactants and forms a urethane linkage. No condensation product is made. Urethanes can be both thermoplastics and thermosets, although the thermosets are more important commercially. They are generally flexible materials, although a wide range of flexibilities and other properties are available. The polyurethanes can be easily adjusted for stiffness and strength versus flexibility and toughness by changing the aromatic content of the monomers. This freedom of choice in properties, coupled with the excellent abrasion resistance and durability of the polyurethanes in general, has led to a rapid rise in the uses of polyurethanes. Perhaps the single largest use is as the principal material in athletic shoes.

■ Elastomeric materials are also crosslinked, but they are examined in a separate chapter because of their unique properties.

PHENOLICS (PF)

Phenolics were the first thermoset materials synthesized (under the name of Bakelite™ by Leo Bakeland in 1907). They are still among the most highly used thermosets, undoubtedly because they are some of the lowest-cost engineering materials on a cost-per-volume basis.

Phenolics are formed from the condensation polymerization reaction between phenol, an aromatic molecule, and formaldehyde, a small organic compound often used as a solvent or as a preservative. As discussed previously, crosslinking requires that at least one of the reactants have more active reaction sites than the minimum needed for just polymerization. Phenol has three active sites, as indicated in Figure 8.4, which is one more than condensation polymerization requires and therefore permits crosslinking to occur.

The condensation reaction for phenolics can be carried out under two different conditions which produce very different intermediates (B-stage materials). The intermediates,

Figure 8.4 Phenol showing (a) the reactive sites and (b) the crosslinked phenolic thermoset.

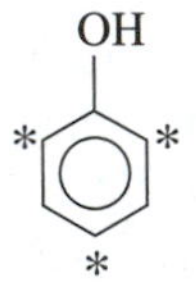

(a) Phenol with active sites marked (*)

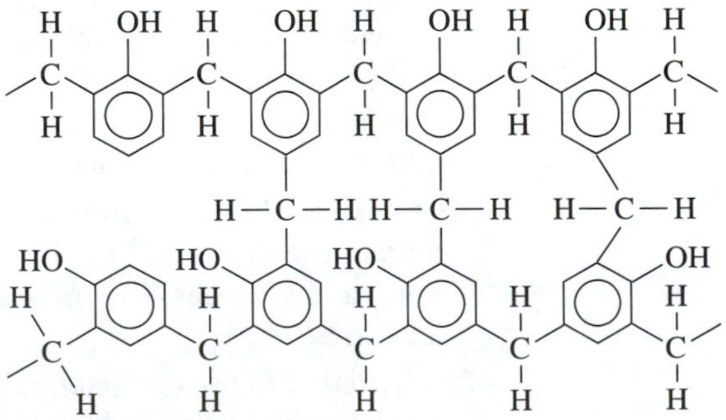

(b) Phenolic crosslinked polymer

called *resoles* in one case and *novolacs* in the other, are the materials that are usually sold to molders. In the resole process the condensation polymerization is performed in an alkali solution with excess formaldehyde and is carefully controlled so that a linear, noncrosslinked polymer liquid, called a resole, is produced. The resole can be stored or shipped for subsequent molding operations. When molding, the crosslinking is accomplished by simply heating the viscous liquid. Since the molder can form the crosslinked part by simply heating without the need for addition of any other materials, resoles are also called *one-stage resins.*

The uncured resole is a viscous liquid that is water soluble and is widely used as a coating and impregnation material to give strength and body to paper when subsequently cured. Resoles can also be stirred vigorously to form a foamed material about the consistency of whipping cream which can then be molded and cured into a rigid foam. These foams will readily absorb and hold water and are used widely by florists as embedding forms for making floral displays.

The novolacs are formed by reacting phenol and formaldehyde in an acid solution and with insufficient formaldehyde to complete the reaction, the opposite conditions from those used to form resoles. The resulting novolac material is a noncrosslinked polymer in the form of a powder. Novolacs will not crosslink with just the addition of heat, but require a curing agent. The most common of the curing agents for novolacs is hexamethylene tetramine or, simply, *hexa.* Hexa is heat activated and so molding is usually done with pressure to compress the powder and with heat. Because a second material (hexa) must be added to novolacs, they are called *two-stage resins.*

Most molded phenolic parts are made from novolacs. If molded without fillers or reinforcements, the parts are brittle and have high shrinkage in the mold as would be expected from the highly aromatic and multiple crosslinked nature of the cured resin. (See Figure 8.4). Novolac powders are, therefore, usually blended with fillers and reinforcements to increase strength and toughness. These blends are called phenolic molding powders. The cost of the part is also reduced by the addition of the fillers. The most common filler is wood flour, which is a purified sawdust. Other common fillers and reinforcements are cotton fibers, fiberglass, and chopped thermoplastic fibers such as nylon.

The high number of OH groups in the polymer gives excellent adhesive qualities. Phenolic adhesives are used for plywood, printed circuit boards, foundry shells and cores, sandpaper, brake linings, and grinding wheels (where the grit is mixed with phenolic powder and molded to the desired shape).

A molding problem is caused by the highly adhesive nature of phenolics. They tend to stick to the mold. Mold releases can be sprayed onto the mold surface or the hot mold can be coated with bee's wax. Alternately, materials such as waxes are also blended into the molding powder. Materials that are added to the resin for this purpose are called *internal mold releases.* A typical formulation for a phenolic molding material is given in Table 8.1.

An important property of phenolics that leads to many applications is nonflammability. When phenolics are subjected to a flame, they char rather than melt or burn. Many applications sensitive to flammability and smoke, such as the interior of airplanes, require that phenolics be used for coatings and many molded parts. Another application for which these properties is important is rocket nozzles. Most materials cannot withstand the tremendous heats of the exhaust gases from rockets. Phenolics are used in this application

Table 8.1 Typical Formulation for a Phenolic Molding Compound

Material in Compound	Parts by Weight
Phenolic resin	60
Filler (usually wood flour)	50
Fiber reinforcement (instead of or in addition to filler)	80
Hardening agent (such as hexa)	15
Plasticizer (if used)	5
Dye or pigment	5
Internal release	1

because they form a rigid and stable char that is only slowly eroded away by the action of the exiting gases. Furthermore, the char has a very low thermal conductivity so that surrounding materials are protected by the decomposed phenolic (char).

Low thermal conductivity is a property characteristic of phenolics that promotes their use as pan handles, bases for toasters, knobs for appliances, and motor housings. Most people recognize the dark handles that are characteristic of phenolics. A dark color is inherent with phenolics and limits the use of phenolics in some applications. A dark pigment is often added to phenolics to standardize the color and to decrease the sensitivity of the material to UV radiation.

Phenolics have very high electrical resistance. They are used extensively for electrical switches, circuit breakers, connectors, and commutators, cabinets for radios, and automotive electrical parts. These latter applications are decreasing because of the more rapid processing and inherent impact toughness of engineering thermoplastics.

Solvent sensitivity of phenolics depends strongly on the nature of the filler. It is generally good for organic solvents but is poor for acids and bases.

Phenolic powders are usually compression molded in matched metal molds, although they can also be transfer molded and injection molded. (When injection molded, care must be taken to not allow the thermoset material to set up in the barrel of the extruder, as might occur during an interruption in the molding process.) Because the crosslinking occurs by condensation polymerization reactions, water is a by-product of the process. If not allowed to escape from the part while crosslinking is occurring, this water can create part defects and surface blemishes. Therefore, when molding phenolics, the mold is often cracked open during the molding cycle to allow the water vapor (usually steam) to escape. This process is called *breathing the mold* or *bumping the mold*. Use of a dielectric preheater could help reduce the amount of water condensate.

After the parts are molded, machining is rather difficult because of the abrasive nature of the filled phenolic resin. Fortunately, most phenolic parts are molded to near net shape so extensive machining is not required.

AMINO PLASTICS (UF AND MF)

This class of thermoset resins is closely related to the phenolics and has properties and applications that are similar. The polymers are formed by the condensation reactions of multifunctional monomers containing amine groups (NH_2) and formaldehyde. The two most important specific types within this group are called the urea formaldehydes (UF) and the melamine formaldehydes (MF). The active sites in one member of the amino plastics group, melamine, is illustrated in Figure 8.5.

Many of the mechanical, physical and chemical properties of amino plastics are similar to the phenolics and result in competition between amino plastics and phenolics for the same applications. Just as with phenolics, thermoplastics are displacing amino plastics, especially in some molded part applications. The nonfilled, nonreinforced amino plastic resins are hard and brittle. Therefore, just as with the phenolics, most molding compounds are filled, often with wood flour, α-*cellulose* (a doubly refined sawdust), talc, clay, fiberglass, and other chopped fibers. Just as with phenolics, amino plastics are molded by compression molding, transfer molding, or injection molding (with caution to not allow premature set-up.) Because phenolics are less expensive than the amino plastics, the majority of applications for amino plastics depend, to some extent, on some property that is superior to phenolics, at least for that particular application.

Amino plastics are lighter in color than phenolics. They can therefore be used in applications where the color must be light. Amino plastic molded parts can even be clear, especially when filled with α-cellulose. Some of these applications include appliance knobs, dials, handles, and electrical parts that are visible. The arc resistance of amino plastics is slightly superior to phenolics, so when this property is critical, amino plastics are preferred. Some related applications are as a substitute for small glass windows, although the amino plastics are susceptible to scratches and will craze with age and with moisture absorption. They will also embrittle over 170°F (77°C). Despite these limitations, another related application for amino plastics is as an enamel coating for kitchen appliances such as refrigerators and dishwashers.

Amino plastics have excellent adhesion, even better than phenolics. The adhesion is from the hydrogens attached to the nitrogen atoms (the active reaction sites), many of which remain as active sites even after the polymerization and crosslinking. Amino plastics are often used when a plastic is to be molded around a metal insert because of this superior adhesion characteristic. A major application featuring this property is as the adhesive in particle board and plywood and for laminating paper to other materials. Amino plastics are also used extensively as bonding materials for furniture. The relatively poor water resistance of the amino plastics dictates that most of these applications be for interior use, unless painted or protected from moisture by some other method.

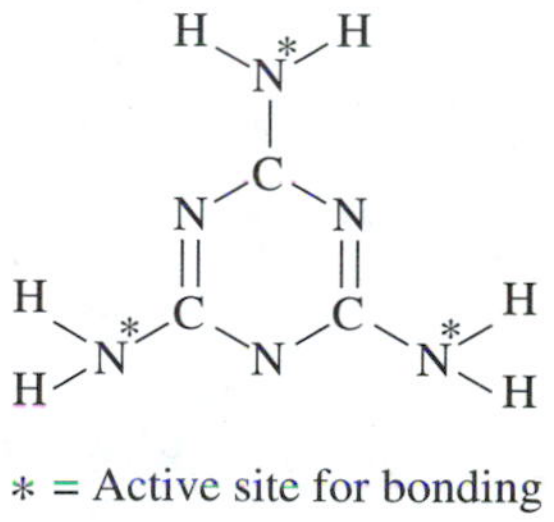

Figure 8.5 Active reaction sites in melamine, a member of the amino plastics polymer group.

* = Active site for bonding

The good bonding capability of amino plastics, especially to wood and cellulose products, has led to an interesting application. Amino plastics are coated onto cotton and rayon fabrics to give crease retention, wrinkle resistance, water repellency, body (stiffness), shrinkage control, and, when filled appropriately, fire retardance.

Urea Formaldehyde (UF) Resins

These resins are intermediate in price between the phenolics and the melamine formaldehydes, which are discussed next. Molded parts from these materials have good grease resistance, hardness, mar resistance, and are easily colored. Urea formaldehydes make up the majority of the molded amino plastic products. Internal mold releases are almost always used. An application that continues to be important is molded bottle caps. For these, the low cost and colorability are important factors.

Urea formaldehydes are used for most of the wood adhesive applications discussed previously because of their lower cost relative to the melamine formaldehydes.

A recently developed application of urea formaldehydes uses cured and uniformly-sized pellets which are sprayed at high pressure against a surface to remove paint and other surface materials without damaging the surface underneath (somewhat like sandblasting). This process is less costly and safer than the use of stripping chemicals and has gained wide approval because of its favorable environmental characteristics. Automobile paint is removed using this technology.

Melamine Formaldehyde (MF) Resins

These resins, sometimes called *melamines,* have the hardest surface of any commercial plastic material. This hard surface along with the excellent grease and water resistance, low flammability, and clarity of the plastic has led to the use of melamine formaldehyde as a countertop material. A common brand name for this material is Formica® (Formica Corporation). These countertop materials are thin sheets composed of a heavy paper or thin cardboard backing that is bonded to the back side of a patterned paper. The bonding material is usually phenolic, which may be foamed, but is not thick. The top side of the paper is then coated with melamine formaldehyde. The entire sandwich, which is roughly 0.008–0.016 inches (2 to 4 mm) thick, is sold as a roll, perhaps 6 feet (2 meters) wide. When used, the material is unrolled, softened slightly with a heat gun so that it will lay flat and, perhaps, shaped to conform to the wooden base that forms the counter. The sheet is then adhesively bonded to the wooden base.

Melamine formaldehyde is compatible with a wide variety of fillers and has been used for many molded parts, although now engineering thermoplastics are replacing it. Many market products, such as molded dinnerware, picnic dishes, and cups are made of melamine resins. The hard, stain-resistant surface and relative low cost make this an excellent application.

Melamines are more resistant to heat and to attack by acids and bases than are the urea formaldehydes and therefore make excellent surface-coating materials. Although now largely replaced by other polymers, melamine formaldehydes were also used extensively as automotive coatings. Although the appearance and physical properties were excellent, the relative long cure times were a problem. Some applications in which surface hardness is an important property still use melamine formaldehyde coatings.

POLYESTER THERMOSETS (TS) OR UNSATURATED POLYESTERS (UP)

Thermoplastic polyesters (PET and PBT) were discussed in the chapter on thermoplastic materials (engineering plastics) and are polymers formed by condensation polymerization with ester bonds linking the repeating units. The thermoset polyesters are also polymerized using condensation reactions to form ester bonds but some significant differences allow these polyesters to crosslink and become thermoset. With thermoset polyesters the polymerization is terminated much sooner so that the thermosets are short-chain, low molecular weight polymers. They are, therefore, viscous liquids or low-melting solids allowing the thermosets to be shaped easily before crosslinking. (The crosslinking reactions which occur later will cause an increase in the molecular weight and the corresponding increase in physical and mechanical properties needed for most applications.) The thermoset polyesters also have active sites in each polymer repeating unit that allows crosslinking to occur. The active site for polyesters is a carbon-carbon double bond.

Crosslinking Mechanism for Thermoset Polyesters

The thermoset polyesters are able to form crosslinks because each repeating unit contains an active carbon-carbon double bond that can react by the addition polymerization mechanism, as described in the chapter on polymeric materials (molecular viewpoint), and form a link to a carbon-carbon bond in another molecule. Because carbon-carbon bonds are called *unsaturated bonds,* the thermoset polyesters containing these bonds are called *unsaturated polyesters,* which is another way of naming the thermoset polyesters. The carbon-carbon double bonds that are used in these crosslink reactions are not created by the condensation polymerization reaction but, rather, are present in one of the monomers and simply go through the condensation polymerization reaction without being affected. The carbon-carbon bonds are therefore available for the crosslinking reactions after the polymer is formed. The separate crosslinking reaction involving the carbon-carbon double bonds is carried out when the polyester resins are molded. A polyester molecule containing a carbon-carbon double bond is shown in Figure 8.6a.

Carbon-carbon double bonds react by the addition polymerization mechanism. This reaction requires that a free radical be formed that attacks the carbon-carbon double bond to initiate the reaction sequence. Typically, a heat-activated or time-activated peroxide or some other free-radical source is added to the polyester to initiate the crosslinking process. The typical concentration of initiator is one to two percent. (Concentrations of initiator that are significantly higher or lower will result in incomplete crosslinking and parts with inferior properties.)

Using the addition polymerization mechanism, the initiator reacts with a carbon-carbon double bond forming a new bond with one of the carbons in the carbon-carbon double bond and creates another free radical on the other carbon. This new free radical

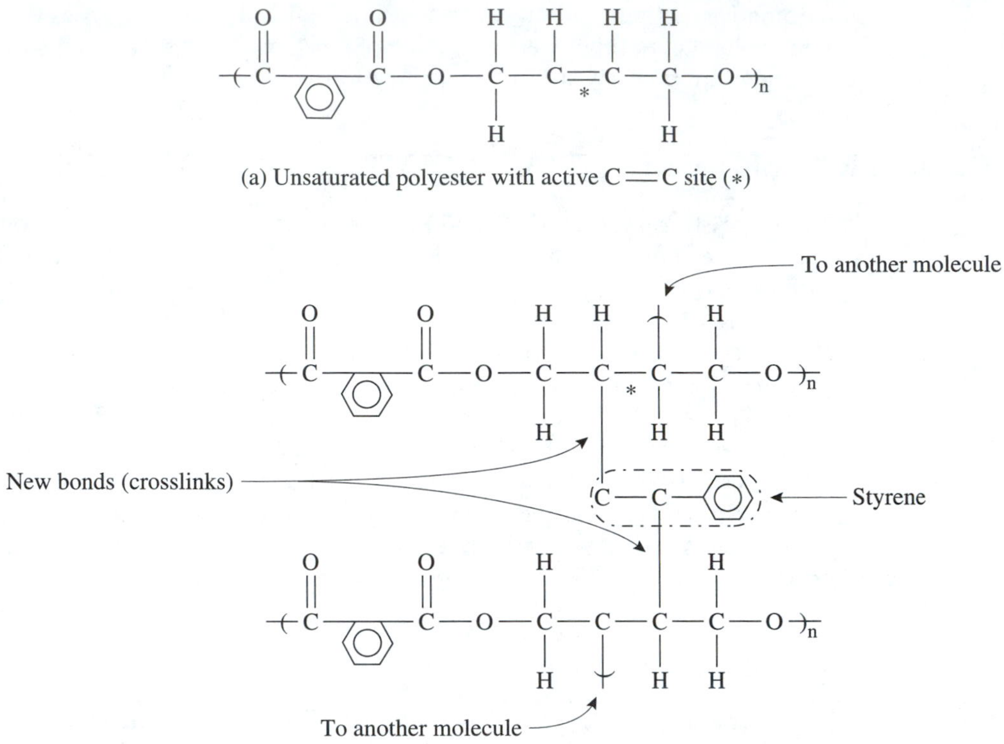

(a) Unsaturated polyester with active C=C site (∗)

(b) Crosslinking of unsaturated polyester

Figure 8.6 Unsaturated polyester showing (a) reactive carbon-carbon double bond and (b) crosslinking reaction.

can then react with another carbon-carbon double bond to form another new bond and another free radical. This process can be repeated in a chain-reaction fashion. If the carbon-carbon double bonds that are linked are on separate molecules, a crosslinked, thermoset structure is created. No condensation by-product is formed during the crosslinking process, thus making the molding of thermoset polyesters much less difficult than phenolics or amino plastics.

In practice, direct bonds between polyester molecules are difficult to form because of the high viscosity of the polymers and the steric (shape) interferences that are often present. Therefore, solvents are often added to the polyesters. These solvents allow the polymer molecules to move freely, thus facilitating the crosslinking. Generally, the solvents chosen contain an active carbon-carbon double bond so that they can also participate in the crosslinking reaction. This further facilitates the crosslinking because the solvent molecules can serve as "bridges" between the polymer chains and make the bonding less hindered. Furthermore, solvents that participate in the crosslinking can be completely used up thus eliminating the need for solvent removal from the crosslinked plastic. A reaction illustrating the use of a solvent molecule is shown in Figure 8.6b. Almost all commercial unsaturated polyesters contain solvents which participate in the crosslinking (reactive diluents). The most common solvents are styrene, vinyl toluene, diallyl phthalate, and methyl methacrylate.

A high concentration of solvent vapors from these solvents usually accompanies the molding of unsaturated polyesters. Most processes for molding or shaping thermoset polyesters allow the vapors to freely enter the workplace, thus causing a potentially flammable environment and, perhaps, unhealthy breathing conditions. Modern manufacturing methods seek to control these vapors, but the potential for difficulty is still present and needs to be carefully monitored.

Additives to Thermoset Polyesters

Free radicals can be formed spontaneously by sunlight, heat, oxygen or contaminants even without the addition of an initiator. So most polyester resins, even those to which no initiator has been added, have a limited shelf life since eventually enough free radicals are formed to crosslink the material. To prevent this premature crosslinking, the materials are often stored at low temperatures. Chemicals that retard the formation of free radicals or that preferentially absorb free radicals can also be added to prolong the shelf life. These materials are called *inhibitors*.

For some applications, the speed of the crosslinking reaction is too slow, either because the mixture has been inhibited or because the reaction is inherently too slow. Under these circumstances, chemicals can be added that speed the reaction. These chemicals, which are usually based on a metal such as cobalt or manganese, are called *accelerators*. Care must be taken that the accelerators and initiators are never mixed together in high concentrations. Each must be added separately to the polyester mixture. Any polyester mixture to which an accelerator and initiator have been added will have a very short shelf life and must be used immediately.

When the molder is ready to form the part, the polyester material is shaped (usually but not always in a mold) and then the crosslinking reaction is allowed to proceed. If a heat-activated initiator is used, the material is heated. If a time-activated initiator is used, the material is simply allowed to sit until the crosslinking reaction is complete.

The unmodified, cured thermoset polyesters are generally hard, brittle materials that do not compete well with either the commodity thermoplastics (because of price) or with the engineering thermoplastics (because of performance). Therefore, thermoset polyesters are almost always modified with fillers and reinforcements to reduce costs and improve properties. If the resin is modified with fillers or reinforcements at the time of molding, the resin is called a *laminating resin*. If the resin is modified before molding, generally by the resin manufacturer, and is later molded as a modified material, the resin is called a *molding compound*. Both of these types of thermoset polyester materials will be examined.

Laminating Resins

Unsaturated polyester laminating resins or resins combined with a diluent are usually sold to the molder as *neat resins,* which strictly means a resin containing nothing but the main polymer (and possibly the diluent), although minor additives such as inhibitors and antioxidants can be included. Initiators are not added to the neat resins or solutions until just prior to use. This procedure prolongs the shelf life of the resin and allows the user to choose the type of activation system most appropriate for the particular application. Shelf life for thermoset materials, even without initiators added, is very short (only a few weeks or months) compared to thermoplastics. Polyesters, for example, should generally be processed in less than 60 days.

Unsaturated polyester resins are used extensively with fiber reinforcements when the purpose of the resin is to bind the fibers together and give shape to the part. Reinforcing fibers are saturated with the resin at the time of molding or shaping. The resin is then cured either by heat or time. The fibers significantly increase the strength of the polyester resin and allow these materials to compete effectively against engineering thermoplastics and metals, especially when low cost and light weight are important. Materials that combine resins with reinforcements are called *composites* and constitute an important class of materials.

Another term for these combinations of resin and reinforcements is *fiberglass-reinforced plastics (FRP)* which some use as a synonym for composites, but is more frequently used as a special type of composite where the resin is polyester and the reinforcement is fiberglass. FRP is the most common type of composite material. Many of the processes employed for making these fiber-reinforced plastics (which will be examined in detail in the chapter entitled Composite Materials and Processes) are suitable for making very large parts. Therefore, common applications for FRP include boat hulls, spas (whirlpool baths), shower stalls, corrugated panels (such as carport roofs), electrical insulators, pipes, storage tanks, and wall panels. See Photo 8.1.

In many FRP products a coating of unsaturated polyester is applied to the mold surface and allowed to partially cure before the wetted reinforcement is put in place. This coating provides a protective layer (called a *gel coat*) that can be colored so that painting is not necessary. (The term "gel" comes from the fact that the coating is allowed to partially cure or gel before the reinforcement layers are added.)

The ability of unsaturated polyesters to form protective coatings has led to their widespread use as coatings for wood, metals, and various other materials. They can be applied as gel coats, which are then cured or as solvent-based paints. Baked enamels are often based upon unsaturated polyesters.

Several types of unsaturated polyester resins can be used for these applications. These resins all have polyester linkages along the backbone and contain a carbon-carbon double bond in each repeating unit but differ in the other parts of the molecules. For instance, some polyesters can be almost entirely aliphatic while others have high aromatic content. The flexibility, toughness, strength, weather resistance, flammability, and other physical properties can be affected significantly by the molecular variations. Some common types of unsaturated laminating resins are called orthophthalic, isophthalic, BPA fumarate, and chlorendic which reflect the structure of a key molecular segment of the polyester molecules. These variations in polyester resins are important in determining part properties and are, therefore, important in specifying the particular polyester resin to be used. For instance, the isophthalic (pronounced eye-so-thal'-ick and sometimes called simply "iso") resin is known to have good environmental resistance and is used extensively in gel coats

Photo 8.1 Fiberglass reinforced plastics (FRP) walls of a house. (Courtesy: Composite Fabricators Association (CFA).

and in applications where resistance to solvents or weather may be important. Orthophthalic (pronounced or-tho-thal'-ick and sometimes called simply "ortho") resin has generally lower properties than does iso, but the ortho is less expensive. Therefore, a common procedure is to use the iso resin as the gel coat and then use the ortho resin for the remainder of the material. This procedure is possible because the resin in laminated parts is often applied in layers, and thus two different materials can be used in adjacent layers.

The chlorendic (pronounced clor-end'-ick) resins give one of the methods for making unsaturated polyesters flame retardant. These resins contain chlorine atoms as part of the resin, which give some natural flame suppression. Bromine-containing resins are also available. Flame retardance can also be achieved by using fillers that contain halogen molecules or that contain water, as was described in the chapter on chemical and physical properties.

Vinyl Esters (VE)

An important class of resins that are closely related to the unsaturated polyesters are the vinyl esters (pronounced vine'-ell ess'-ters). These resins are generally slightly more expensive than the other unsaturated polyesters but have superior toughness and corrosion resistance. Vinyl esters use carbon-carbon double bonds as active reactant sites for crosslinking, but these active sites are on the ends of the vinyl ester polymer chains rather than in every repeating unit. This gives lower crosslinking density but more latitude in choosing the polymer between the active sites.

The polymer between the active sites in a vinyl ester resin is often derived from an epoxy by reacting the epoxy with an acrylic group. This reaction results in the addition of the carbon-carbon double bonds to the ends of the molecule. Therefore, performance of vinyl esters is similar to epoxies except for the lower performance of the styrene crosslinks, as opposed to the epoxy crosslinks, which are discussed later in this chapter.

Vinyl esters cure in much the same way as unsaturated polyesters—that is, they employ peroxide initiators that allow either room temperature cures (usually with an accelerator added) or heated cures. As with unsaturated polyesters, the vinyl esters have a limited shelf life because of the random creation of free radicals, which then cure the resin. Hence, inhibitors are also usually added to vinyl esters.

The main uses for vinyl esters are in corrosion-resistant applications. They are used as high-performance gel coats, often with unsaturated polyesters making up the rest of the part. Vinyl esters are also used in pipes and reaction vessels, where the high corrosion resistance is a strong benefit.

Molding Compounds

Unsaturated polyester molding compounds are made by mixing the resin or resin solution with fillers or reinforcements prior to use. To insure that the initiator is uniformly distributed throughout the mix, the initiator is usually added to the resin or resin solution prior to adding the filler and reinforcement. Therefore, the shelf life of the molding compound is much less than laminating resins. To extend their shelf life, molding compounds are often stored at low temperatures.

The properties of the molded parts are dependent upon the resin type, the solvent, the reinforcement, the filler, and other minor constituents. For instance, the long-term stability of polyester thermosets in hot water depends strongly on the selection of all these constituents in the molding compound. If any constituent is changed, the hot water performance is affected strongly. Even with the best of each type of constituent, polyester thermosets are likely to absorb significant moisture in hot water and lose both mechanical and chemical property performance.

Several types of polyester thermoset molding compounds are commonly used. These types include alkyds, allylics (principally DAP and DAIP), BMC/DMC, and SMC. These differ in the types of unsaturated polyesters that are used, the types of active diluents, and the method of adding reinforcement, if any. Each is discussed briefly.

Alkyds (pronounced al'-kids), are polyester molding compounds that are based on the same types of resins used in laminating. They are prepared by blending the resin with the initiator and then with cellulose pulp, mineral filler, lubricants, pigments, and, perhaps, short fibers. The mixing is usually done in a heated roller mill which gives some curing to the polymer. When the proper degree of polymerization is reached, the semisolid material is removed from the rollers, cooled, crushed, and ground into a molding powder. This powder is usually compression molded, often requiring only low pressures. Alkyd parts have high electrical resistance, low moisture sensitivity, and good dimensional stability.

Alkyds are used principally for paints and molded parts. The paints have high durability and have been used for many years, but their use has been limited by the relatively slow drying rate. Furthermore, the paints and molded parts have some tendency to absorb water, thus restricting their application to low-water environments. The molded alkyd resins have been used extensively in electrical applications, especially in automobiles, where their low conductivity and high dielectric strength have value.

The allylics (pronounced al-lil'-icks) are a group of unsaturated polyesters that are based upon resins formed from a monomer containing the allyl group, a particular organic chemistry functional group that contains a carbon-carbon double bond. The most common of these monomers are diallylphthalate (DAP) and diallylisophthalate (DAIP). When polymerized into a linear, uncrosslinked polymer and then combined with a filler and, perhaps a reinforcement, these materials are called allylic, DAP or DAIP molding compounds. The reinforcements are usually short fibers or mineral fibers so that its paste-like consistency can be maintained.

Allylic pastes are used as body putty for repairing automobiles and for many other applications in which a high degree of shaping is required. Allylics are also used extensively for electrical parts (chiefly connectors, switches, bobbins, and insulators). These parts are made by compression or transfer molding (which will be discussed in a later chapter). Allylics are also used for *potting*, which is a process of encasing an article or assembly in a resinous mass. This is done by placing the article to be potted into a container that serves as a mold, pouring the liquid resin into the mold so that the desired portion of the part is covered with resin, and then curing the resin. After the casting is completed the mold remains as part of the assembly. Potting is used extensively in the electrical industry to encapsulate wire ends and is discussed more fully in the chapter on casting.

Bulk molding compound (BMC) or, alternately, dough molding compound (DMC) is made by combining the unsaturated polyester resin with an initiator, a filler, and reinforcement fibers. The term *premix* can also be applied to these mixtures. The mixing of these components is done at room temperature to avoid premature curing and with as little agitation as possible so that quite long fibers (up to about 2 inches or 5 cm) can be used. The resultant materials are stored at low temperatures to prolong their shelf life. Typical concentrations of BMC components are: resin—30%, filler—60%, reinforcement—10%.

BMC materials are sticky and have a doughlike consistency which allows them to be metered, shot by shot, into an open mold. Therefore, the principal molding method is compression or matched die molding. This permits largely automated and rapid molding of parts having moderately high complexity. The automobile industry has adopted this method for manufacture of numerous parts including body panels, grills, trim, air conditioning ducts, electrical components, and various semistructural members.

One problem inherent in the use of BMC is the limited amount of material movement that is possible within the mold. Generally, the charge of BMC is placed in the center of a mold and then it moves to fill the mold as the mold presses against the material. Even with careful adjustment of the viscosities of resin and the proper amounts of resin, filler and reinforcement, movement over long distances in the mold will tend to cause separation of the components. Therefore, parts made by BMC molding are limited to about 16 inches (40 cm) in their longest dimension.

Larger parts can be made with sheet molding compound (SMC). The composition of SMC is about the same as BMC but the method of mixing the components is quite different. The process for making SMC is depicted in Figure 8.7.

Rather than mix the components in a bulk process, the preinitiated resin and filler material is doctored onto a moving sheet of polyethylene film. The reinforcement fibers are chopped to the desired length (typically 1 to 3 inches, 3 to 7 cm) and sprinkled onto the resin/filler layer. A second polyethylene sheet that has also been doctored with a layer of resin and filler is then placed on top of the chopped fibers so that a sandwich is formed with the reinforcements in the middle, surrounded by the resin/filler mixture, and enclosed by the polyethylene film. This sandwich is passed between rollers that mix the fibers into the resin and filler, then rolled to an appropriate size for easy handling, and removed for storage at a cool temperature.

When the SMC is to be used, the roll is taken to the molding station (usually a large compression molding machine) and is unrolled and cut to the desired lengths. These lengths are typically about the same size as the part to be molded. The polyethylene sheets are removed from the SMC sandwich as the material is placed into the mold. Sufficient layers of SMC sheet are placed in the mold to obtain the thickness desired. If additional thickness is desired in some locations, smaller strips of SMC can be laid into the mold at those locations. When all of the material has been properly placed into the mold, the mold is closed and the part is cured. Because the SMC is placed throughout the mold, little

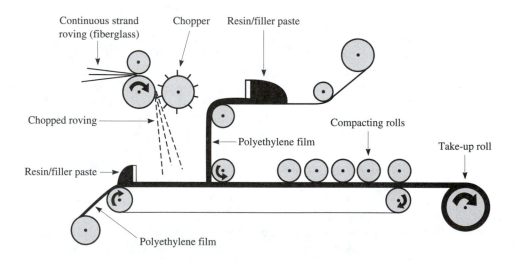

Figure 8.7   Process for making sheet molding compound (SMC).

movement of material occurs within the mold. Therefore, very large parts can be made by this process. A well-known example of a part made from SMC is the body of the Corvette™ (General Motors).

The use of BMC and SMC for exterior body panels in automobiles has required the development of a molding system that will result in a very smooth, defect-free surface (often called a *class A surface*). The inherent shrinkage of polyester materials as they crosslink normally results in small sink marks and dimples that are unacceptable in the automotive panel market. A special additive system has been developed to solve this problem and is called *low-profile or low-shrink system*. The system is based upon the addition of some thermoplastic resins and the addition of fillers containing divalent metals (such as Ca^{+2} and Mg^{+2}). The addition of these materials to polyester resins results in a thickening of the resin/fiber/filler mixture (a process called *ripening*), and depends on the formation of ionic bonds between the ionic groups on the polymer and the positive metal ions. The ripening or thickening of BMC and SMC occurs to some extent even without the presence of the low-profile additives because of the normal curing that takes place with an initiated resin system. Therefore, BMC and SMC systems can be viewed as B-staged resin systems. With these systems, the use of thermoset polyesters in automobiles has increased steadily for many years.

EPOXIES (EP)

These resins are characterized by the presence of the three-membered ring epoxy group. The groups are not typically part of the polymer repeating unit but are, rather, attached to the ends of a polymer as is illustrated in Figure 8.8a. This is a normal structure, although the epoxy groups could be in locations other than on the ends. If elsewhere, the epoxy groups must be reasonably accessible for reactions to occur with them. For crosslinking to occur, at least two epoxy groups must be on each polymer molecule (though not in every repeating unit).

The crosslinking of an epoxy resin is initiated by the opening of the epoxy ring by a reactive group on the end (usually) of another molecule. This type of reaction is illustrated in Figure 8.8b, where an amine is used as the reactive group on the second molecule. Note that two new bonds are formed when the ring opens. One bond is with a carbon atom that was in the epoxy ring and the second bond is between the oxygen of the epoxy ring and the hydrogen that was on the amine. The bond between the amine and the carbon is the key bond in crosslinking. The other bond, which creates an OH group, is important in some of the properties of the epoxy resin, such as bondability. The amine molecule usually has another amine group on the other end of the molecule that can react with a second epoxy molecule. The two epoxy molecules would therefore be joined together by the amine molecule. This is, of course, crosslinking.

Several types of reactive groups will react with an epoxy ring and start the crosslinking sequence. The most common reactive groups for this purpose are amines, anhydrides, amides, and mercaptans. Each type can affect the curing characteristics and final part properties. All of these molecules that have reactive groups and are used to cure epoxies

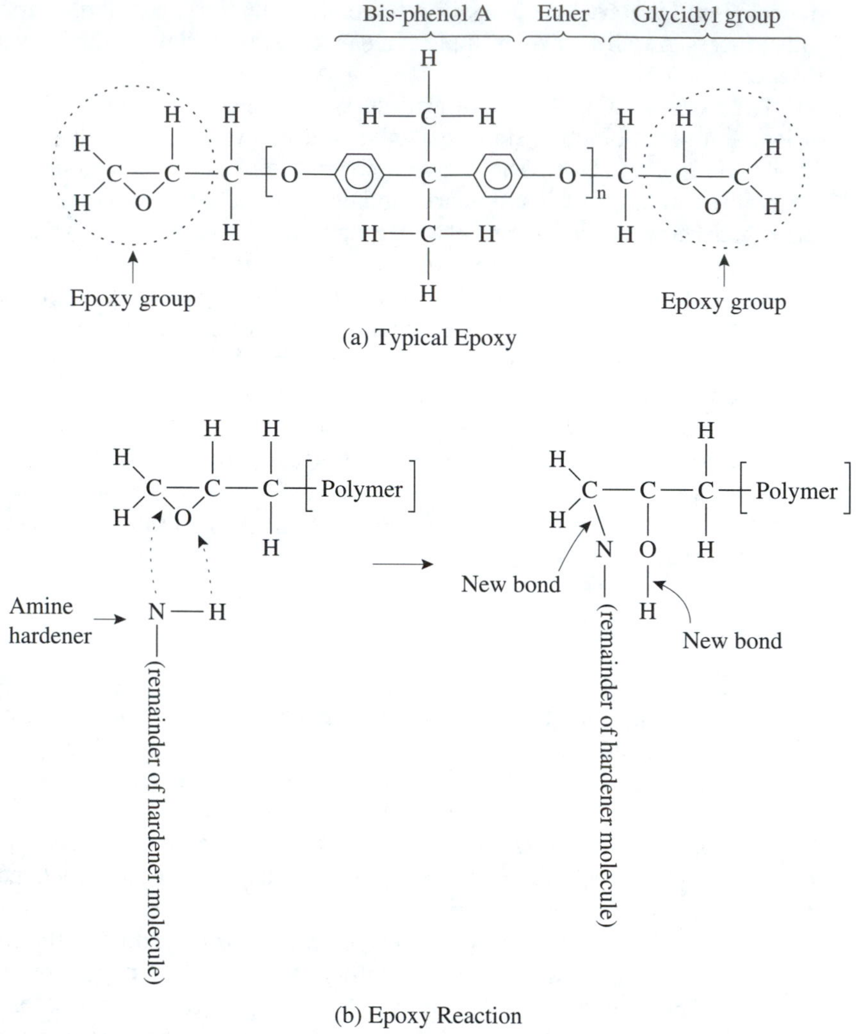

(a) Typical Epoxy

(b) Epoxy Reaction

Figure 8.8 A typical epoxy resin.

are called *hardeners*. The molecules containing the reactive groups can be small molecules or polymers, can have only two reactive groups or many, and can be slow or fast to react. In every case, however, the reaction is started merely by mixing the epoxy with the hardener. (The epoxy and hardener are often referred to as "parts A and B.")

The relative concentrations of the epoxy and the hardener are important in determining the properties of the final part. If the total number of epoxy groups is much higher than the number of reactive groups on the hardener molecules, the hardener molecules will tend to add to the epoxy at one location but little crosslinking will occur. A thermoplastic material with poor mechanical properties will result. If the concentration of hardener reactive groups matches the concentration of epoxy groups, a well-crosslinked structure with maximum properties will result. If the concentration of hardener active sites is greater than the concentration of epoxy sites, a thermoplastic material with poor properties will result. Hence, the number of epoxy sites and the number of hardener active sites must be properly balanced and the addition ratios carefully calculated, at least to within $\pm 10\%$.

Some epoxies are cured at room temperature and others are heated to facilitate the crosslinking reactions. A general rule is that the maximum use temperature is two-thirds of the Fahrenheit or Celsius cure temperature, so high temperature epoxies (with cure temperatures of about 350°F, 175°C) are common. No condensation by-product is formed during an epoxy crosslinking reaction.

Cured epoxies are generally hard and brittle but the nature of the polymer located between the epoxy end-groups can have some effect on many of the properties of the epoxy resin. These polymers are typically short-chain aromatic polymers resulting in precured epoxies that are viscous liquids or low-melting solids. These resins can, therefore, easily be used to coat reinforcements or be mixed with fillers. As with unsaturated polyesters, the physical and mechanical properties required in the end product are achieved by the increase in molecular weight that occurs during crosslinking. If higher molecular weight polymers are desired, the precured epoxy resin will likely be a solid. In these cases, solvents can be added to provide the capability to soak the reinforcement or mix easily with a filler. These solvents are removed from the finished part unless they are an active participant in the crosslinking reaction.

Toughness of the epoxy part is strongly dependent on the length of the polymer chain between the epoxy end groups. Long polymers are generally tougher than a short polymer of the same chemical type. However, having a long polymer also means that there will be fewer crosslinks per unit length. (The number of crosslinks per unit length is called the *crosslink density*.) Lower crosslink density means that the material is less strong, less stiff, more sensitive to solvent attack and has a lower maximum use temperature.

The toughness of the epoxy part is also dependent upon the amount of aromatic character in the polymer and in the hardener molecules. Less aromaticity generally means more flexibility in the molecule and therefore improved toughness but decreased strength, stiffness, and maximum use temperature. Another method to increase the toughness of epoxies is to add rubber polymers. These can be copolymerized with the epoxy or can be alloyed. The problem with adding rubber polymers is that chemical resistance and strength are often negatively affected.

An innovative epoxy resin is made by using a polymer between the epoxy groups that contains carbon-carbon double bonds. In this case, two crosslinking reactions can be carried out simultaneously, one by the traditional epoxy method and one by the addition mechanism. A solvent such as styrene that will react in the addition reactions is typically used so that it is consumed in the curing reactions and solvent removal from the cured part is not necessary. This dual-cure system allows for both a high molecular weight polymer and a high crosslink density.

While epoxies are generally more expensive than unsaturated polyesters, some important advantages of epoxies contribute to the steady growth of this resin class. One advantage is the low shrinkage that occurs during molding. The crosslinking reactions of polyesters tend to draw the molecules together tightly, whereas the crosslinking reactions of epoxies do not. This property improves molding capabilities, especially for parts in which dimensional tolerances are critical. This property is also important when the mold is made of a flexible material and shrinkage could change the shape of the mold itself.

Epoxies are stronger, stiffer, more durable, more solvent resistant, and have a higher maximum operating temperature than polyester thermosets. These characteristics have led to the use of epoxies for high-quality coatings, paints, adhesives, and liners for reaction vessels and tanks. Epoxies are also used extensively as the resin in electronic printed circuit boards and as potting material, applications that utilize many of the superior properties of epoxies.

These properties are also useful in the composites industry where epoxies are combined with reinforcements such as fiberglass and carbon fibers. The use of epoxies in composites is also aided by the superior adhesion properties of epoxies because the bond between the resin and the fiber reinforcement is important to the performance of the part. The aerospace industry uses many epoxy composite parts because of the favorable strength and stiffness to weight that can be obtained. The growth of composite parts, chiefly epoxy-based, in this industry is very rapid.

A form of epoxy combined with reinforcement that is used extensively in the aerospace industry is a *prepreg*. A prepreg is a sheet made of fibers or cloth on which liquid epoxy resin mixed with hardener has been coated. (Sometimes the epoxy is a solution.) The epoxy is allowed to crosslink slightly until it becomes viscous enough to stay on the fibers without running off. This process is called B-staging, much like the partial polymerization that is done with phenolics. The resultant material is tacky. The prepreg is placed on paper to keep the layers from sticking together and is rolled up and stored at low temperatures until use. When the material is to be made into a part, the roll is allowed to return to room temperature so that it is tacky and pliable and is then laid into the mold or otherwise shaped. Several layers are often placed on top of each other until the desired thickness is reached. The material is then cured, often under pressure so that the layers are squeezed together. Autoclaves are conveniently used to cure these prepreg parts because heat and pressure can be applied simultaneously. This molding method is discussed further in the chapter on composites.

Epoxies are easily cast or prepreg molded to form parts. This has resulted in a unique application for epoxies as rigid molds for short production runs. Although only minimal pressure can be applied to these molds, their dimensional stability at relatively high temperatures and their hardness and rigidity have allowed this application to grow for both high-performance aerospace parts and for prototype molds.

The naming of epoxy resins can be quite complicated. One system uses the first letter of the various descriptive terms to identify the polymer. For instance, the polymer shown in Figure 8.8 is known as DGEBPA (or DGEBA), which stands for *diglycidyl ether of bisphenol A*. This name is instructive as a model for naming epoxies. The prefix *di* means two of something, in this case, two glycidyl units. The glycidyl unit is the epoxy unit (the three-membered ring plus one more carbon atom, as seen in Figure 8.8). There are two of these units, one on each end of the molecule. The *ether* term comes from the linkage of the glycidyl units into the large aromatic group in the repeating unit (within the brackets). The linkages are oxygens, which are bonded to the glycidyl units and to the aromatic with single bonds. These are ether linkages. Finally, the name of the large group in the repeat unit is given. In this case it is bis-phenol A. Hence the name is the di-glycidyl ether of bis-phenol A (DGEBPA).

THERMOSET POLYIMIDES

Thermoset polyimides find use almost exclusively in high-temperature applications where epoxies cannot be used. Such applications include aircraft engine exhaust ducts, skin panels for super-high-speed aircraft, and various small parts that might be subjected to temperatures over 300°F (150°C), especially electrical components. Some polyimides can withstand temperatures to 600°F (315°C) and some even to 660°F (350°C) with intermittent service. Just as with thermoplastic polyimides, the thermoset polyimides are stiff, strong, but somewhat brittle materials.

Several types of thermoset polymers, all containing the imide group, are used commercially. These polyimides can be divided into two groups depending upon the crosslink mechanism. One group crosslinks by the condensation mechanism. The most important member of the group is PMR-15. The other group crosslinks using the addition mechanism and employs carbon-carbon double bonds for this purpose. The most important member of this group is bis-maleimide (BMI). In some addition-type polyimides, the carbon-carbon bonds are on the ends of the molecules, much like most of the epoxies.

The polyimides which are crosslinked by condensation are difficult to use. These materials are often polymerized in several steps that are done as part of polymer formation and crosslinking. These chemical steps are difficult to run, requiring carefully controlled conditions over long periods of time. The cures must occur under high vacuum to remove the by-products from the part. The very high viscosity of these resins dictates that most are polymerized in solution, which means that both the by-product and the solvent must be removed. The temperature capabilities of some thermoset polyimides are further raised by reacting the polymer with active groups that attach to the ends of the polymer *(end-capping)*, and this is done just prior to or during crosslinking.

The addition crosslinking resins (BMI) have the obvious advantage that no by-product is formed during the crosslinking reaction so crosslinking is less complicated and no by-product removal is required. Addition crosslinking polyimides can also be polymerized and prepared as prepregs, sheets formed by a resin coated onto fibers and then B-shaped, or other convenient forms before crosslinking, thus simplifying the reactions. Addition polyimides are, therefore, emerging as the preferred thermoset polyimides, except for specific applications where the chemical or thermal properties of one of the other members of the group are needed. Mechanical properties of addition polyimides are similar to epoxies and are, therefore, used in many applications where epoxies would be used but for the temperature requirement.

Some thermoset polyimides are nonflammable, a property that suggests their use in aircraft interiors and other flame- and smoke-sensitive applications. This property is becoming ever more important as the amount of plastic materials in these sensitive locations continues to increase.

POLYURETHANES (PUR)

The polyurethanes are a family of polymers that can have widely differing properties. They can be molded into pliable or rigid parts, formed into soft and resilient or hard and rigid foams, and applied as durable coatings. Polyurethanes can be either thermoplastics or thermosets, although thermosets are more important commercially. Clearly, polyurethanes are some of the most versatile polymers. This versatility and the multitude of properties that can be obtained with polyurethanes result from the many different molecular variations that can be joined together using the urethane bond. The basic nature of the reaction that forms the urethane bond is considered first, and then the types of molecular structures that are commonly joined in polyurethanes are studied.

The basic chemistry for formation of a polyurethane is similar to a condensation (step-growth) reaction in that two monomers, each having at least two reactive groups, unite in a head-to-tail manner to create the polymer. However, in contrast to a condensation reaction, no condensate is formed when the urethane bond is created. The urethane bond is formed from the reaction of a polyol with an isocyanate, as illustrated in Figure 8.9.

A *polyol* (which means multiple alcohols or multiple OH groups) can have from two to many OH groups. Just as with a condensation polymerization reaction, the polyol monomer must have at least two reactive groups (in this case OHs) in order to polymerize. If three or more reactive groups are present, crosslinks can form. The monomer depicted in Figure 8.9 is shown with two OH groups and with an R between these groups. In organic chemistry a letter such as R is often used to indicate that a general, non-specified group of atoms is present. The nature of the R group is understood to be such that the basic reaction illustrated would not be affected by the nature of R. This notation system provides a convenient method of representing a wide variety of different molecules that all enter into the same basic type of reaction.

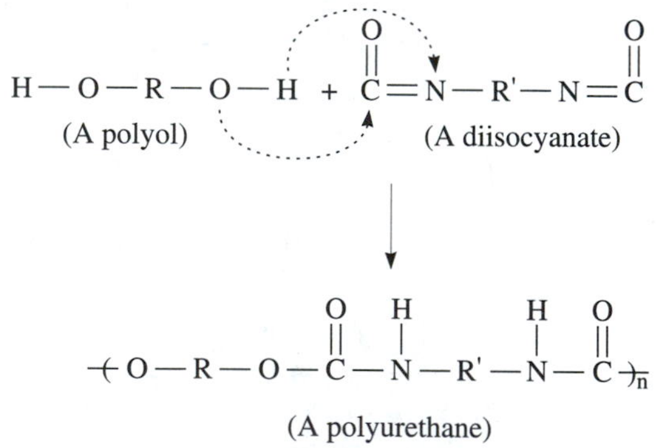

Figure 8.9 Basic reaction to form polyurethanes indicating the flexibility possible through different choices of the substituent groups R and R′.

Note: R is usually a multifunctional polyether or polyester but can also be a small organic group.

R′ is usually a large aromatic group

The other monomer in the reaction to form a urethane bond is an *isocyanate* which is the NCO combination of atoms. This monomer must also have at least two reactive groups in order to form a polymer. The type of chemical compounds that have two isocyanate groups are called *diisocyanates*. Several different atomic arrangements can form the molecular structure between the two isocyanate groups and so the letter R′ has been used to indicate this versatility. The prime (′) is added to the R to reinforce the concept that the unspecified group in the diisocyanate can be different from the unspecified group in the polyol.

When a polyol reacts with an isocyanate, a molecular rearrangement occurs that creates a more stable molecular structure. The hydrogen on the polyol forms a bond with the nitrogen in the isocyanate and the oxygen in the polyol forms a bond with the carbon in the isocyanate. Some previous bonds in the polyol and in the isocyanate break to allow these new bonds to form.

Reactions by molecular arrangement have been previously encountered with epoxies and can give some useful comparisons to polyurethanes. The epoxy ring was opened by the reaction of a hardener and the bonding between the epoxy and the hardener involved a molecular rearrangement. The role of the polyol in polyurethane chemistry is like the role of the epoxy molecule in epoxy chemistry. The isocyanate role in polyurethanes is like the hardener in epoxy chemistry. Epoxy molecules generally have epoxy groups on the ends of the various branches of the molecule, much as polyols would have OH groups on the ends of the branches. Just as the epoxy molecules could have many different chemical groups or arrangements of atoms between the epoxy rings, the polyols can also have many different groups between the OH groups (as represented by R in Figure 8.9). It is this variation in the types and arrangements of atoms between the reactive groups (called R-groups for simplicity) that gives much of the variation between types of epoxies, and that also gives much of the variation between polyurethanes.

Although many types of R-groups can be used in polyols, two general arrangements are common in commercial polyurethanes. Both are relatively short-chain polymers, one based on ether linkages and the other based on ester linkages between the polymer units. The ether-based polyurethanes are generally more flexible than are the esters and the principal use of ether-based urethanes is in foams. Ester-based polyurethanes have higher mechanical properties than the ethers and are more often used in molded polyurethane parts and coatings.

The other component in the urethane polymerization reaction is the isocyanate. Usually the R′-groups (the groups of atoms between the isocyanates) are large aromatic groups. Typical examples of commercial diisocyanates are toluene diisocyanate (TDI) and methylenediphenyl isocyanate (MDI). MDI is darker in color and has lower oxidative resistance and UV stability than TDI. Even greater UV stability and also increased toughness can be obtained by using an aliphatic diisocyanate. Excellent weather resistance and improved UV resistance can also be obtained by using a triisocyanate.

Caution must be exercised in the handling of isocyanates. These chemicals cause respiratory distress and can be toxic. They will react rapidly with water, often with much heat. Therefore, extreme caution should be used in the storage and use of isocyanates materials.

Both the polyol and the isocyanate monomers are generally liquids when they are combined to form the polyurethane, which simplifies the mixing and metering of these materials. In most cases, the reaction between the polyol and the isocyanate is almost instantaneous, even without external heating. A manufacturing method that is especially well suited to this mixing of reactive liquids is called *reaction injection molding (RIM)*. In RIM the two liquid components are pumped through a mixing chamber or tube and then into a closed mold. The polyurethane is formed inside the mold, usually with only moderate pressures required. Hence, the molds can be made of inexpensive materials and can be quite large, assuming that some care is exercised to ensure that the entire mold is filled. When the polyurethane is formed (cured), the molds are simply opened and the part extracted. The RIM process can be automated which has made this process inviting for making automotive parts. The totally enclosed nature of the RIM process has further increased its desirability as a method of producing both foamed and nonfoamed parts that can be rigid and yet have the resiliency of an elastomer. Automotive bumpers and other impact pads are typical examples of parts made by this process. The RIM process and related processes such as RTM are described in more detail in the chapter on composites.

Polyurethanes can also be molded by traditional thermoplastic processes (such as injection molding and extrusion) or by traditional thermoset processes (compression molding, transfer molding) depending upon the product and the nature of the specific polymer.

The versatility of polyurethanes can be seen by briefly considering the different types of polyurethane resins and the products that are often made from each resin type.

- Flexible foams—Foams with densities from 1 to 12 pounds/cubic foot (0.016 to 0.19 g/cc) are commonly made using polyether-based polyols and TDI or MDI. These components are relatively inexpensive, easy to process, and result in foams with excellent resiliency and good water resistance. An advantage to this system is that if water is added to the reactants, it will combine with some of the TDI to create CO_2 gas, which serves as a foaming agent for the system. Other foaming agents can also be added for specific purposes. If more dense foams are desired, a polymeric diisocyanate can be used in place of TDI. The principal products for flexible polyurethane foams are bedding, furniture, vehicle seating, carpet padding, and packaging.

- Rigid foams—Rigid polyurethane foams are highly crosslinked, usually by employing multifunctional polyols and multifunctional, polymeric isocyanates, which are abbreviated as PMDI. The common densities are from 5 to 15 pounds/cubic foot (0.08 to 0.24 g/cc). Major markets for rigid polyurethane foams are in construction and other insulation applications, furniture, packaging, and transportation.

- Thermoset plastics and elastomers—These were originally investigated as substitutes for natural rubber and were formulated to be processed in most of the same methods that are used for natural rubbers. (Elastomers are discussed in more detail in a later chapter). The polyurethane rubbers were found to have superior tear strength and abrasion resistance over natural rubber. Typical applications include automotive bumpers and athletic shoe components. Some polyurethanes compete well with nylon and acetal, having a lower water absorption than nylon (6/6) and about the same water absorptivity as acetal. The polyurethanes have good low-temperature impact strength and good tear resistance, but relatively poor high-temperature stability.

- Thermoplastic elastomers—When the polyol and isocyanate monomers each have only two functional groups, a linear thermoplastic polymer is formed. Because of hydrogen bonding, the isocyanate portions of these polymers tend to aggregate into clusters (called *domains*) which gives additional strength to the product and some advantages in flexibility and durability. Thermoplastic polyurethane elastomers can be injection molded, extruded, or processed in most other typical thermoplastic molding processes. Common commercial products include flexible sheets and films, wire insulation, hoses, tracks for sports vehicles, roller skate wheels, seals, bushings, bearings, small gears, and automotive exterior parts.

- Fibers—A specific type of thermoplastic polyurethane elastomer is the type of fiber that has now been given the generic name of spandex. These fibers were developed by the DuPont Company who have given their brand of spandex fibers the name Lycra®. These fibers have domains, just as the other thermoplastic polyurethane elastomers. In the fiber form, these domains take on the characteristics of "hard" and "soft" segments where the clustered isocyanates are hard and the clustered polyols groups are soft. The isocyanates are hard because they are often highly aromatic, and therefore stiff, and they are held tightly together by hydrogen bonding. The polyols are soft because they can be more aliphatic and flexible and have less tendency to be drawn tightly together. The result of these hard and soft segments is a surprisingly wide range of flexibility with initial elongation coming from the soft segments and then a later, higher-energy elongation coming from the hard segments. Spandex fibers have been used extensively in swimwear, foundation garments, and outer garments where flexibility and stretch are desired.

- Coatings—The choice of isocyanate has a major effect on the UV sensitivity of the polyurethane coating. Highly aromatic isocyanates, such as TDI and MDI, are sensitive to UV light and weather poorly. Aliphatic, often cyclic aliphatics, are used more often when UV stability is desired. The major advantages to polyurethane coatings over other plastic coatings include: toughness, abrasion resistance, flexibility, fast curing, and chemical resistance. Coatings for metal, wood, rubber goods, and aircraft are commercially important. The hardness of the surface is dependent upon the choice of both polyol and isocyanate (higher aromatic content is harder) as well as the amount of crosslinking that is allowed to occur. Higher crosslinking makes the surface harder. Optical clarity is enhanced by reducing the aromatic content so some compromises may be necessary in formulating for specific applications.

===== CASE STUDY 8.1 =====

Thermoset Composites for Wrapping Utility Poles

The average life expectancy of a wooden utility pole is 25–50 years and, with 100 million wooden utility poles in the United States, replacement is a major problem for utility companies because of the cost and the possible interruption of service which might occur. Even after the replacement has occurred, the disposal of the old pole is a problem because of the chemicals, such as creosote and pentachlorophenol, that have been used to treat the wooden poles to reduce rotting, flammability and infestation. In many states, products containing these treatment chemicals require special disposal. Costs for replacement of poles can exceed $10,000 per pole.

An alternate solution to pole replacement that is now receiving wide acceptance is a system called FiberTect®, which has been developed by PoleCare Industries, a division of Timber Products Incorporated. After the pole is cleaned and treated, a fiberglass fabric is wrapped around the pole, spiral fashion, starting at a point 1–2 yards (1–2 m) below the ground and continuing to a point 2–6 yards (2–6 m) above the ground. Resin is then coated onto the fabric. Alternate layers of fabric and resin are added until the total wrap thickness is 0.3–0.5 inches (0.8–1.4 cm). A final pigmented layer is applied. The resin cures in a few hours to complete the restoration.

The choice of resin is a critical decision for the success of the product. The resin must be strong and stiff, compatible with both the fiberglass and the old wooden pole, and flame retardant. Flame retardance is required because of the need for poles to withstand brush fires, house fires, and vandalism. The logical initial choices for resin would be unsaturated polyesters. These have good compatibility with the fiberglass, are strong, and are low in cost. However, additives must be added to polyesters to meet the flammability requirements. Epoxies are another candidate resin but these also require additives to meet the flammability specification. A resin that is inherently flame retardant is phenolic, which was then considered. Phenolics are stiff, strong, and have excellent compatibility with wood. A fiberglass that was compatible with phenolics was identified. The ability to cure phenolics without external heating was also a benefit. Hence, the resin of choice was phenolic.

A special fiberglass weave was created to meet the particular demands of the pole-wrap product.

The final result, a wrapped and refurbished utility pole, is actually superior to the original pole. The wrapped pole has greater strength, better flame retardance, better infestation resistance, and superior aesthetics. The market is huge, perhaps as high as one million poles per year.

SUMMARY

The difference between thermosets and thermoplastics is that thermosets form crosslink bonds between the polymer chains. These crosslinks allow the uncured resin to have a low molecular weight (and, therefore, to often be a liquid) and then to rise in molecular weight as the polymer chains are linked together during cure. As liquids, the thermosets will flow easily into molds without the need for external heating and can coat fiberglass or other reinforcement materials.

Several different crosslinking systems are used with the common thermoset products. These are summarized in Table 8.2.

Phenolics and amino plastics crosslink by condensation reactions. The polymerization and crosslinking reactions can be combined into one step or, by carefully controlling the reaction conditions, the polymerization can be done in one step and then the crosslinking done in a second step. When separated into steps, the second step (polymerized but not crosslinked) is called *stage B*.

Table 8.2 Types of Thermoset Materials and Their Crosslinking Mechanisms

Thermoset Resin Class	Crosslink Mechanism
Phenolics	phenol plus formaldehyde
Amino plastics	amine plus formaldehyde
Unsaturated polyesters	—R—C=C—R'— carbon-carbon double bond plus R—O—O—R' initiator (usually peroxide)
Epoxies	epoxy ring plus hardener (amine shown)
Polyimides	imide condensation or imide and carbon-carbon double bond
Polyurethanes	H—O—R—O—H polyols plus diisocyanates

The crosslinks in unsaturated polyesters are formed by a free-radical addition reaction that usually employs a peroxide initiator and a reactive solvent, such as styrene, to link together polymers that have previously been fully formed. In these unsaturated polyesters the crosslinking step is clearly separate from the polymerization process.

Epoxies combine with hardener molecules to form bridges between the epoxy polymer chains. These hardeners have two ends, each of which can react easily with an epoxy ring by a rearrangement of some atoms, and thereby link two chains.

Polyimides can crosslink in either one or two steps and by either a condensation reaction or by an addition reaction. However, the high molecular weight of polyimides and their stiffness complicates the processing of these materials.

Polyurethanes are formed from two monomers, polyols and isocyanates, which will form crosslinks if either of the monomers has more than two reactive sites. The reaction that forms the urethane bond involves a molecular rearrangement and no condensate is formed.

The properties of the various thermosets are dependent upon the structures of the polymers, the nature of the crosslinks, and the crosslink density. Generally the same rules apply for thermosets as for thermoplastics in predicting these properties. For instance, highly aromatic thermosets usually have high strength, high stiffness, low flammability and poor UV resistance, just as would be the case with thermoplastics. While thermoplastics can be highly crystalline, thermosets are generally not. The crosslinks often interfere with crystal formation. So crystalline effects are rare. On the other hand, thermoplastics do not have any crosslinks. The effects of crosslinking in thermosets are to raise the molecular weight and with it the strength, stiffness, thermal stability, hardness, abrasion resistance, and other properties.

The formation of the crosslink bonds takes time. This time usually results in longer molding cycles for thermosets than for thermoplastics. The economic incentive to decrease molding time has led to a change from thermosets to thermoplastics in many applications. However, the increased use of reinforced materials and the superior ability of thermosets to wet the reinforcement fibers have given rise to a demand for some thermoset resins. Other growth markets for thermosets include coatings, molded parts in which the reaction time is low (such as polyurethanes), and applications where the liquid nature of the resin is an advantage.

GLOSSARY

Accelerator   A material that enhances the speed of a reaction, especially of crosslinking.

Alpha (α) cellulose   A purified form of sawdust.

Amino plastics (UF and MF)   A group of thermoset resins made by the reaction of urea and formaldehyde or melamine and formaldehyde.

A stage   The condition of a polymer that has not yet begun to crosslink.

Bis-maleimide (BMI)   A thermoset polyimide that uses the free-radical reaction at carbon-carbon double bonds as the crosslinking mechanism.

Breathing the mold   A process during compression molding involving the brief opening of the mold to allow condensate (a by-product of the crosslinking reaction) to escape.

Bridge molecule   A molecule that facilitates the formation of a crosslink by connecting two polymer chains.

B stage   The condition of a polymer that has been partially crosslinked and the reaction then stopped, usually to be continued to completion at a later time.

B-staging   The process of beginning the crosslinking of a polymer and then stopping or quenching the process so that the material can be stored until a more appropriate time to fully cure it, perhaps after it has been placed in a mold.

Bulk molding compound (BMC)   A molding compound consisting of an initiated resin (usually unsaturated polyester), filler, and fiberglass that are mixed together and sold as a doughlike mass to be subsequently molded.

Bumping the mold   Another term for breathing the mold.

Catalyst   A material that facilitates a chemical reaction but is not consumed in the reaction; the term is also used (improperly) as a synonym for initiator, especially a peroxide initiator.

Char   A carbonaceous mass that is the result of thermal decomposition of a polymer, something like charcoal.

Class A surface   A very smooth, defect-free surface, usually associated with the type of surface required in the automotive industry prior to painting a part.

Composites   Materials consisting of two or more major components that can be mechanically separated, such as a resin and fibrous reinforcements.

Crosslink density   The number of crosslinks per unit length of polymer molecule.

C stage   The condition of a polymer that has been fully cured.

Curing   The process of crosslinking in a molecule.

Curing agent   An initiator or a hardening agent.

Diisocyanate   A group of chemicals containing two isocyanate groups; used in making polyurethanes.

Diluent   A material that solvates a polymer; a material that lowers the viscosity of a polymer.

Domain   An area within a polymeric structure wherein the molecules have a tendency to aggregate into clusters but are not necessarily crystalline.

Dough-molding compound (DMC)   Another name for bulk molding compound.

End-capping   Reacting a polymer with groups that attach to the ends of the polymer; this will often improve properties of the polymer, such as thermal stability.

Endotherm   The heat that must be supplied to chemicals to make them react, the heat consumed in a reaction.

Ester linkages   Polyurethanes in which the primary linkages in the polyol are esters.

Ether linkages   Polyurethanes in which the primary linkages in the polyol are ethers.

Exotherm   The heat given off by a chemical reaction.

Fiberglass-reinforced plastic (FRP)   A composite material containing fiberglass and resin (usually unsaturated polyester or vinyl ester).

Gel   To partially or completely cure or set up.

Gel coat   A layer of resin, usually unsaturated polyester or vinyl ester, that is sprayed into a mold such that it serves as the outer surface of a molded part when the part is removed from the mold.

Hardening agent or hardener   A material that induces crosslinking or curing.

Heat of reaction   The heat given off or consumed in a chemical reaction.

Hexa   A curing agent used with phenolic novolacs.

Inhibitor   A material that absorbs free radicals or otherwise retards the formation of crosslinks.

Initiator   A chemical that begins a curing or crosslinking reaction, usually a peroxide.

Internal mold release   A material, often waxy in nature, that is added to a molding compound or powder to facilitate the release of the molded part from the mold.

Iso   A short name for isophthalate polyesters.

Isocyanates   Groups of chemicals containing the isocyanate group ($O=C=N-$); used in forming polyurethanes.

Laminating resin   An unsaturated polyester to which fillers or reinforcements are added at the time of molding.

Low-profile or low-shrink system   A resin-filler-reinforcement system containing additives that give a controlled expansion so that the natural shrinkage of the plastic upon molding can be counteracted.

MDI (methylenediphenyl isocyanate)   A chemical used to make polyurethanes, especially foams.

Melamines (MF)   A group of thermoset resins made from melamine and formaldehyde.

Molding compounds   Thermoset materials that contain resin, filler, and, usually, reinforcement and therefore have a doughlike consistency.

Molding dough   A molding compound.

Neat resin   A resin to which no fillers or reinforcements have been added.

Novolac   A B-staged phenolic made by the reaction of an excess of phenol in an acid environment that is a thermoplastic in the B-stage and requires the addition of a curing agent (typically hexamethylenetetramine) to form full thermosetting phenolics; a two-stage resin.

Oligomer   A short-chain polymer.

One-stage resin   A resole phenolic that will cure from the B-stage with heating.

Ortho   A short name for orthophthalate polyesters.

Phenolic (PF)   A thermoset resin made by the reaction of phenol and formaldehyde.

PMR-15   A thermoset polyimide that polymerizes by the condensation method, used in very high-temperature applications.

Polyester thermosets (UP)   A group of thermoset resins characterized by the presence of a polyester bond and a carbon-carbon double bond; crosslinked by the free-radical mechanism.

Polyol   A group of chemicals containing at least two alcohol groups, these are often short-chain polymers on which several OH groups (alcohols) are pendant, often used to make polyurethanes.

Polyurethanes (PUR)   A group of polymers formed by the reaction of a polyol and a di-isocyanate; they can be thermoplastic but are more often thermosets.

Potting   The process of encasing an article or assembly in a resinous mass.

Premix   Another name for a bulk molding compound.

Prepolymer   An oligomer.

Prepreg A sheet formed by a resin coated onto fibers and then B-staged.

Reaction injection molding (RIM) A manufacturing method for polyurethanes that involves mixing the monomers together and injecting them into a mold.

Reactive diluent A chemical that both dissolves and reacts with a polymer, usually in the crosslinking process as a bridge molecule.

Resole A B-staged phenolic resin created in an excess of formaldehyde in an alkaline environment that is a thermoset and will fully cure upon heating, without the addition of a curing agent; a one-stage resin.

Ripening The process of thickening that occurs in BMC and SMC that is characteristic of the partial curing of the resin and of the action of low-profile additives.

Sheet molding compound (SMC) A mixture of initiated resin, filler, and fiberglass that is initially made in a flat sheet, thus allowing molding of large parts because the material does not have to move greatly in the mold.

TDI (toluene diisocyanate) A chemical used to make polyurethanes, especially foams.

Thermosets Polymers in which crosslinking has occurred.

Two-stage resin A novolac phenolic that will cure from the B-stage with the addition of a curing agent.

Unsaturated Containing a carbon-carbon double bond.

Unsaturated polyester thermosets (UP) A name for polyester thermosets that emphasizes the presence of a carbon-carbon double bond in the backbone.

Urea formaldehyde (UF) A group of thermoset resins made from urea and formaldehyde.

Vinyl esters (VE) A group of polymers that have properties somewhat like the epoxies but can be cured like unsaturated polyesters.

QUESTIONS

1. What is meant by a stage B resin?
2. Compare the processing differences of resole and novolac resins. What is meant by one-stage and two-stage resins?
3. What are the structural features in the melamine monomer that leads to the very high hardness of cured melamine resins?
4. Discuss the role of styrene in the crosslinking of unsaturated polyesters.
5. Discuss two methods that might be employed to make an epoxy resin tougher.
6. What do the initials DGEBPA stand for?
7. What is the origin of the variety in the properties of polyurethanes?
8. Discuss the safety precautions that should be observed when handling isocyanates.
9. If a polyol has four active sites per molecule and an isocyanate also has four active sites per molecule, why isn't a 1:1 mixture by weight a likely ideal concentration mixture? What should be the correct ratio?
10. Discuss the choice of materials for use as handles for a frying pan.

REFERENCES

Birley, A. W., R. J. Heath, and M. J. Scott, *Plastics Materials: Properties and Applications,* (2nd ed.), New York: Chapman and Hall, 1988.

Chanda, Manas, and Salil K. Roy, *Plastics Technology Handbook* (2nd ed), Albany, NY: Delmar Publishers, Inc., 1989.

"Engineering Plastics," *Engineered Materials Handbook,* Vol.2, Metals Park, OH: ASM International, 1988.

Milby, Robert V., *Plastics Technology,* New York: McGraw-Hill Book Company, 1973.

Richardson, Terry L., *Industrial Plastics: Theory and Application* (2nd ed.), Albany, NY: Delmar Publishers, Inc., 1989.

Seymour, Raymond B., and Charles E. Carraher, *Giant Molecules,* New York: John Wiley and Sons, Inc., 1990.

Ulrich, Henri, *Introduction to Industrial Polymers* (2nd ed), Munich: Hanser Publishers, 1993.

CHAPTER NINE

ELASTOMERIC (RUBBER) MATERIAL

This chapter examines the following concepts:

- Aliphatic thermoset elastomers (natural rubber, isoprene, polybutadiene, and related polymers)
- Thermoplastic elastomers
- Fluoroelastomers
- Silicones
- Processing of elastomers (compounding, preforming, molding, and dipping)

INTRODUCTION

Plastic materials have been separated into two groups, thermoplastics and thermosets, based upon their behavior at elevated temperatures or, alternately, upon whether the plastic is crosslinked or not, which is the key structural feature that determines behavior at high temperatures. The previous few chapters have explored the most common thermoplastics and thermosets and related specific properties of these plastics to other structural features, such as crystallinity, aromatic character, ability to hydrogen bond, and so on. This chapter considers a group of polymers that all have a common characteristic, very large elastic elongations. For convenience of classification, polymers that have more than 200% elastic elongation are grouped together as elastomers. In other words, *elastomers* are materials that can be repeatedly stretched to over twice their normal length and then immediately return to their original length when released. Elastomers can be either thermoplastics or thermosets, but if thermosets, they are so lightly crosslinked that hardening does not occur. Yet the crosslinking is sufficient that they cannot be melted after shaping.

Thus, elastomers are considered by some to be a separate class of plastics.

All materials have some elastic elongation, but for many materials, especially ceramics and metals, the elastic elongation is very small, typically less than 2%. Some engineering plastics have elastic elongations in the same range as metals; whereas others, such as polyethylene, can have elastic elongations up to 50%. Only a few materials, the elastomers, have elastic elongations higher than 200%. *Elastic elongations* are defined as the elongation of any material when that material is at its yield point, sometimes called the *elastic limit*. Any elongation beyond the yield point would be inelastic (that is, would cause a permanent deformation) and full recovery to the original length would not be realized. Therefore, elongation beyond the yield point would be nonelastomeric. These relationships are indicated in Figure 9.1.

The high elastic elongation of elastomers is usually associated with low stiffness (low modulus) and low strength. Conventional plastics and metals will have higher stiffness and higher strengths than elastomers. Some idealized stress-strain curves for metals, conventional plastics, and elastomers are illustrated in Figure 9.2, which shows the lower modulus and lower strength of elastomers.

As discussed in the chapter on chemical and physical properties (macro viewpoint), high strength and high stiffness in materials are associated with crystalline structures, other strong interactions between the molecules or atoms, and general stiffness of the molecules. These features all prevent easy movement of the molecules relative to each other and decrease elastic elongation. The elastomers, therefore, have the opposite structural features. Elastomeric polymers are highly random, are generally totally amorphous, have few strong interactions between molecules, and are flexible polymer chains (usually aliphatic rather than aromatic). When a tensile force is applied to an elastomer, the mol-

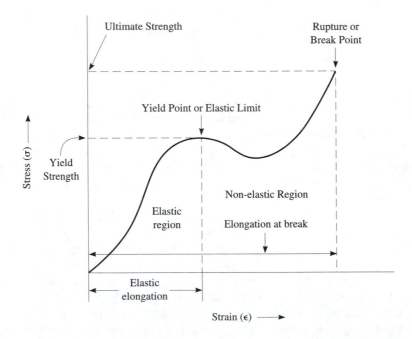

Figure 9.1 Stress-strain diagram showing the yield point or elastic limit, elastic elongation, and elongation at break.

Figure 9.2 Idealized stress-strain curves for metals, conventional plastics and elastomers.

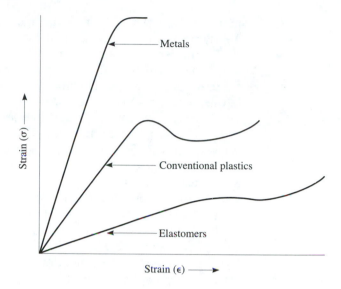

Strain (σ) →

Metals

Conventional plastics

Elastomers

Strain (ε) →

ecules can easily move relative to each other, probably just a simple uncoiling of the tangled molecules. This movement can continue with little additional force until the molecules are totally stretched or some other internal resistance is met. This behavior is represented diagrammatically in Figure 9.3.

When the tensile stress is relieved, the molecules will assume their original, random shape and the entire structure will return to its original shape. This elastic recovery will occur provided the molecules have not been displaced in absolute position relative to one another, that is, they have not slid but have only uncoiled. If sliding happens, the elastic

Figure 9.3 Diagram showing the random, natural state of elastomers when under no stress and the more ordered state when stressed.

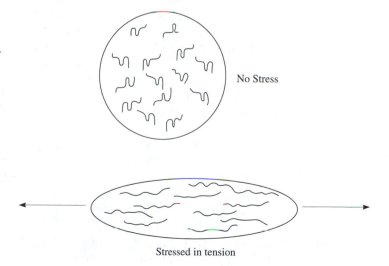

No Stress

Stressed in tension

limit (yield point) has been exceeded and some nonelastic movement has been introduced. This nonelastic movement cannot be recovered.

The tendency of the material to return to its original, random state can be attributed to entropy. (Remember that entropy is a measure of the disorder of a system.) The non-stressed state is one of high entropy because of its high randomness. When the external tensile force is applied to the system, order is imposed and entropy is forced to decrease. This is a less favorable (unstable) state from thermodynamic considerations. Therefore, when the external force is removed, the entropy will naturally try to increase which is the driving energy for a return to the original, random state.

When the molecules are forced into their ordered positions by stretching the polymer and reducing entropy, the total energy of the system is lower, heat is given off and a slight warming will be detected. Upon relaxation of the stretching force, heat goes into the material to create the randomness and a cooling is sensed because the heat is being taken in.

Elastomers will typically also have compressive recovery, provided the compressive elastic limit is not exceeded. This behavior is sometimes called *resilience*. When the material is compressed, the molecules are forced into a more ordered state rather than the preferred, random state. Hence, entropy is decreased by the compression and, when the compressing force is relieved, the entropy will tend to increase again. The molecules push against the surface in order to return to their random positions again.

A material may be elastomeric at room temperature and rigid at lower temperatures. The loss in elastomeric behavior is associated with a decrease in the ability of the molecules to uncoil or to have long-range, extended movements. In most elastomers a transition temperature exists such that above the transition temperature the material behaves as an elastomer and below the temperature the material is rigid. That transition temperature was discussed in the micro structures in polymers chapter when thermal transitions were examined. The transition temperature is the glass transition temperature, T_g. Above the glass transition temperature the material is said to be rubbery (elastomeric) and below the glass transition temperature the material is called rigid or glassy.

Elastomers are sometimes called rubber or rubbery material, although those terms are more often applied to natural rubber than to elastomers in general. Natural rubber was the first elastomer material investigated in depth and is a member of the largest group of elastomeric materials, the aliphatic thermoset elastomers. The term elastomer is, therefore, a general term whereas rubber is somewhat more specific.

The idea of thermoset elastomers, that is, of having crosslinks in an elastomeric material may seem to be counterintuitive because crosslinks will bind molecules together, exactly the opposite concept of elastomers. To some extent, this is true. However, if the crosslinks are few in number, they can give some important benefits to elastomers and not be so restrictive that the elastomeric movement is prevented. Most crosslinked elastomers have about a hundred atoms between crosslinks. This level of crosslinking is called *lightly crosslinked* or *soft crosslinking*. The atoms between the crosslinks can still move with great freedom, uncoil and coil as required by external forces, and exhibit the long-range movements typical of materials above their glass transition temperature. The advantages of the crosslinks are that long-term creep is reduced or limited, especially at elevated temperatures, complete melting is prohibited, and the maximum extent of elongation is limited. Without the crosslinks, an elastomer may be elongated beyond the elastic limit with-

Photo 9.1 Common molded elastomeric parts. (Courtesy: Wabash MPI.)

out any warning, whereas with the crosslinks, a maximum elongation is set safely within the elastic region. A limit can be easily set because of the resistance given by the crosslinks after the desired elongation is reached. If less elongation is desired, the number of crosslinks can be increased. (This is called increasing the *crosslink density*.) If the crosslink density is increased sufficiently, the material will cease to be an elastomer because the elongation will drop below 200%. Some common molded elastomeric parts are shown in Photo 9.1.

ALIPHATIC THERMOSET ELASTOMERS

The family of aliphatic thermoset elastomers is the largest group of elastomeric materials. This family has an important historical development that illustrates several important concepts in elastomers. This historical perspective explains why these materials have developed chiefly as crosslinked, thermoset materials and illustrates the logical development steps that led to the emergence of several important elastomers which are part of this family.

Natural Rubber (NR)

Historical evidence indicates that natural rubber was widely known among native South Americans prior to the arrival of Western Europeans in the 16th century. This natural rubber was obtained by drying the sap of certain plants, chiefly the rubber tree of the Amazon forest *(Hevea braziliensis)*. The sap is an emulsion or suspension of a nonsoluble component in a water solution. Such an emulsion is called a *latex* and the rubber sap is sometimes called latex rubber. (Another well-known latex is milk in which the butter fat component is suspended in water solution.)

Crude natural rubber quickly found many applications when it was introduced into Western Europe. It was coated onto fabric and made into waterproof coats, used as a waterproof coating for shoes, made into balls and coatings for balls for various games, and found to be useful in rubbing out pencil marks from paper (from whence it was given the name "rubber"). In all of these applications some serious limitations were noted, especially the tendency of the crude material to soften and creep at elevated temperatures.

Several experimenters were actively attempting to solve this high-temperature problem of crude natural rubber when Charles Goodyear accidentally discovered that the problems of high-temperature stability could be solved by cooking crude natural rubber with sulphur. The use of heat and sulphur led to the name of this process, *vulcanization,* after the Roman god of fire, Vulcan. Later experimental investigation discovered that vulcanization had created crosslinks between the rubber molecules. Today, vulcanization is a term applied to the crosslinking or curing of any elastomeric material, even those processes that do not utilize sulphur or heat.

Natural rubber is highly elastomeric, even after moderate crosslinking. The elongation is about 1000% for most vulcanized natural rubbers. Compared to other elastomeric materials, natural rubber is also characterized by high tensile strength, high tear strength, high resilience, resistance to cold flow, resistance to wear, and resistance to fatigue. These highly desirable properties have led to the continuous high use of natural rubber, sometimes blended with other elastomers that have now been developed. Most of the supply of natural rubber comes from plantations of *Hevea braziliensis* in Indonesia and Malaysia. Rubber trees still grow and are harvested in the Amazon jungle, but the yields from these natural varieties of trees are much lower than from the carefully bred varieties in Southeast Asia.

If additional sulphur is added and the vulcanization process is continued for a longer period than normal, additional crosslinks will form. As these form, the polymeric material gets harder and the amount of elongation decreases. Eventually a material is made that is nonelastomeric (less than 200% elongation). This highly crosslinked material is called *hard rubber.* Hard rubber is widely used in combs and insulation, and has been used for bowling balls.

Molecular investigations conducted in the early decades of the 20th century discovered that crude natural rubber was chiefly composed of cis-polyisoprene, a polymer chain with a carbon-carbon double bond in the repeating unit. The *cis* prefix means that the two pendent groups (in this case an H and a CH_3) that are attached to the two carbons in the carbon-carbon double bond are on the same side of the carbon-carbon bond. This cis arrangement is illustrated in Figure 9.4a. The alternate configuration where the two

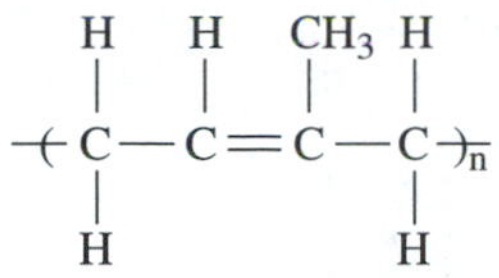

(a) Hevea rubber (cis-polyisoprene) 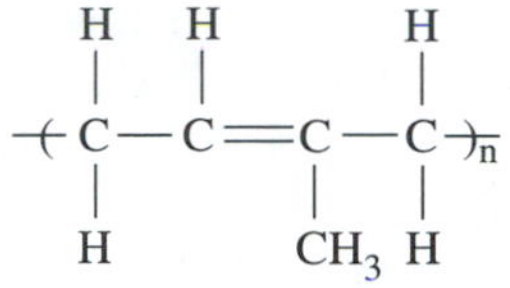   (b) Gutta percha or balatta (trans-polyisoprene)

Figure 9.4 Polymer repeating groups for (a) hevea rubber and (b) gutta percha or balatta illustrating the cis and trans configurations of isoprene.

groups are located on opposite sides of the carbon-carbon double bond is called *trans* and is illustrated in Figure 9.4b.

The properties of the cis and trans polymers are quite different. The cis arrangement is the form of natural rubber (natural isoprene). The cis is highly elastomeric and is sensitive to heat softening. The trans arrangement permits the molecular chains to fit together more easily which increases the interactions between the molecules. These interactions make the trans material harder. In the case of isoprene, the trans material is called *gutta percha* or, occasionally, *balatta*. Gutta percha is much harder than the cis isoprene, natural or latex rubber material that is called *hevea rubber*. Gutta percha is used for golf balls and shoe soles.

Several chemical studies have shown that when either the cis or trans forms of isoprene are vulcanized, the sulphur reacts with the carbon-carbon double bonds in two molecules which, when heated, move close enough that two or more sulphur atoms form a short chain that bridges between the polymer molecules. In this vulcanization process, the carbon-carbon bond disappears when the sulphur reacts. This crosslinking reaction is similar in many ways to the crosslinking reaction of unsaturated polyesters that was discussed in the chapter on thermoset materials. When natural rubber is vulcanized, about 100 atoms separate each of the sulphur crosslinks (lightly crosslinked).

Natural rubber has excellent fatigue resistance, excellent resilience, and low energy loss when stretched and returned to the original shape.

Synthetic Polyisoprene or Isoprene Rubber (IR)

With the disruption of supplies of natural rubber during World Wars I and II and the increasing need for elastomeric materials, several groups began work on synthetic rubber. Synthetic polyisoprene was made in the early 1900s and was used for tires for lightweight vehicles such as bicycles and early automobiles, but was found to have somewhat different properties from latex-derived natural rubber. The synthetic polyisoprene was found to be a mixture of both the cis and trans molecular forms, thus giving a mixture of their properties. After the Ziegler-Natta catalyst system was developed in the 1950s, it was found that 90% pure cis-polyisoprene could be produced using this catalyst system. In fact, by varying the polymerization conditions and the catalyst, mixtures anywhere from 90% pure trans-polyisoprene to 90% pure cis-polyisoprene could be made. The small amount of trans in the 90% cis-polyisoprene made the synthetic somewhat more difficult to process, but synthetic isoprene is widely used. Natural rubber is, however, used more extensively, chiefly because of its low cost.

Butadiene Rubber (BR) and Styrene Butadiene Rubber (SBR)

In the feverish activity to develop a synthetic rubber that occurred from 1930 to 1945, one of the polymers that was developed was polybutadiene or butadiene rubber. The structure of this polymer is similar to polyisoprene but with an important difference. The repeating units of both have a backbone of four carbon atoms including a carbon-carbon double bond so general performance in crosslinking is similar. The difference is that polyisoprene has a hydrogen and a methyl carbon attached to the two carbon-carbon double bond carbons, but polybutadiene has just two hydrogens. The structure of polybutadiene is shown in Figure 9.5. This difference in structure, that is, the absence of the methyl carbon in polybutadiene, has a significant effect on the physical properties of the polymer. The methyl group interferes with movement in the polyisoprene polymer, restricting bending and twisting motions as well as sliding of one molecule relative to another. This interference leads to increased stiffness, higher strength, and higher temperature stability in polyisoprene than in polybutadiene. Polybutadiene would, therefore, have poorer tensile strength, tear resistance, and tack than would polyisoprene. The resilience in both elastomers is about the same. Polybutadiene has poor resistance to solvents.

The advantages of polybutadiene are low cost, improvement of low-temperature flexibility, compatibility with many other polymeric materials, and good adhesion to metals. These advantages have led directly to some applications for unmodified polybutadiene and to the development of ways to modify polybutadiene to improve its physical properties. The major application for unmodified polybutadiene is as a toughener for other materials. Several plastics are sold in a toughened grade where that grade is made by adding polybutadiene to the other polymer. Some of the most common of these plastics that are toughened with polybutadiene are polystyrene, epoxy, nylon, acetal, polycarbonate, and other elastomeric materials that may be too stiff or that need improved low-temperature toughness.

In some plastics butadiene monomer is added to the monomer of the other plastic so that a copolymer is created. This is the case with polystyrene. The copolymer is called styrene butadiene rubber (SBR). This material is used extensively as an improved, tough-

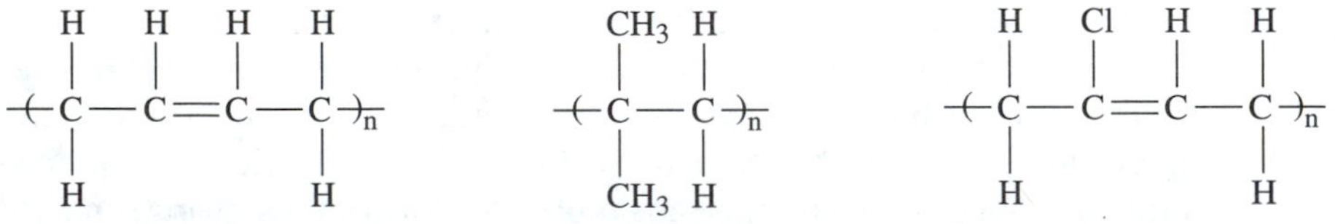

(a) Butadiene rubber (BR)   (b) Butyl rubber   (c) Neoprene

Figure 9.5 Polymer repeating units for (a) polybutadiene (butadiene rubber), (b) polybutylene (butyl rubber), (c) neoprene.

ened styrene (impact styrene) and as an improved butadiene rubber with properties that depend on which molecule is in greater abundance. The bulky styrene molecules add stiffness and intermolecular interference to the butadiene while the butadiene adds flexibility and toughness to the styrene. The styrene acts as a stiffening agent for the butadiene, much like the methyl carbon does in polyisoprene. Therefore, the principle of molecular interference to achieve higher strength and stiffness is clearly at the basis for the improved properties of SBR over unmodified polybutadiene. During World War II, SBR was used extensively as a substitute for natural rubber. It was given the name of Buna-S or GSR (government styrene rubber). Its major applications were for tires, footwear, wire insulation, adhesives, gaskets, and seals. The major drawbacks of SBR were poor oil resistance and sensitivity to oxidation and UV radiation.

SBR is also used as a basis for making ABS where acrylonitrile is also added, either as a monomer or as a polymer, to achieve a tough, low-cost, and weatherable product. The combination of styrene, butadiene, and acrylonitrile to make ABS can be done in many different ways and so a wide variety of ABS products with a wide range of physical properties is available (materials formed from three monomers are called terpolymers). This polymer system was discussed more fully in the chapter on commodity thermoplastics. SBR can also be added as a toughener to plastics, much as unmodified polybutadiene would be added.

Another method of adding intermolecular interference is through the use of fillers. If these fillers are small enough in particle size, the filler particles can fit between the molecules and give interference much like a methyl or styrene group. The filler that has been used most extensively for this purpose is carbon black. The physical strength, tear resistance, and stiffness are all vastly improved by the presence of carbon black in a tire. Another advantage is improvement in resistance to UV radiation.

Another polymer that was developed during the 1930s and 1940s that is closely related to polybutadiene and polyisoprene is polybutylene or *butyl rubber,* which is shown in Figure 9.5b. (Note that butyl rubber and butadiene rubber are not the same material.) Butyl rubber has only two carbons along the backbone in the repeating unit but has two methyl groups attached to one of the carbons. It therefore has more interference than does polyisoprene or butadiene and is stiffer and stronger.

Butyl rubber has high damping capability, low gas permeability, and good UV radiation and oxidation resistance, but has poor compatibility with other rubbers and is therefore not used extensively in rubber blends. Butyl rubber has relatively poor resilience. The principal applications for butyl rubber are as vibration damping pads, inner tubes in tires, and high-temperature hoses (such as steam hoses). Also, butyl rubber does not contain a carbon-carbon double bond. The low permeability of gases through butyl rubber has led to its use as a layer in commercial automobile tires. However, the poor compatibility of butyl rubber with other rubbers that are commonly used in tires requires that a compatibility material be coated to the butyl rubber, thus forming a bond between the butyl rubber and the other rubbers that are layered next to it.

The good weathering characteristics of butyl rubber have led to its use as a sealant or weatherstripping material. In these applications the butyl rubber is mixed with a filler (often ground rocks, such as limestone or calcium carbonate) and with a plasticizer to give the material additional stickiness and a low modulus so that it will be easily shaped to fill in gaps and take the shape of the materials it is sealing between. The butyl rubber mixture is then extruded into a strip shape and wound onto a roll, often with paper between the layers of butyl to keep the layers from sticking together.

If additional toughness, rigidity, or higher-temperature performance is desired in butyl rubber, a small amount of butadiene can be copolymerized with the butyl polymer. The copolymer will have a small number of carbon-carbon double bonds, which can be crosslinked to give the enhanced properties that might be desired.

Oil-Resistant Elastomers

A copolymer of butadiene and acrylonitrile was developed as another substitute for natural rubber and to correct one of the key problems present in polybutadiene, namely, sensitivity to oil. The resultant polymer was originally called Buna-N but is now known as nitrile butadiene rubber (NBR). Nitrile rubber is made by the copolymerization of butadiene and acrylonitrile. This combination is, of course, part of the combination of ingredients that is used to make ABS, as discussed in the chapter on commodity thermoplastics. We see, therefore, that each of the constituents in ABS—acrylonitrile, butadiene, and styrene—has commercial value individually, with each of the other constituents, and as a combination of all three together.

NBR is much more expensive than either butadiene rubber or SBR and is therefore limited to use in applications where oil resistance is required. NBR also has improved resistance to degradation from oxidation and finds some application where that property is important. NBR has good abrasion resistance but poor low-temperature elasticity. Typical applications for NBR include oil and fuel lines, gaskets, seals, conveyor belts, and coatings for printer rolls.

Another oil-resistant polymer is chloroprene rubber (CR) or neoprene, that was developed by DuPont in the 1930s. This polymer has a chlorine bonded to one of the carbons in the carbon-carbon double bond in a structure that is otherwise like polybutadiene (see Figure 9.5c). The effect of the chlorine is to improve oil resistance and, because the chlorine is quite large and will cause intermolecular interference, strength and stiffness are also improved over unmodified polybutadiene and over polyisoprene and natural rubber. Thermal stability is also better in neoprene than in polybutadiene or polyisoprene although neoprene is much more difficult to vulcanize, which must occur to obtain this thermal stability. The vulcanization of neoprene is accomplished by using metal salts with ionic charges of two (such as MgO or ZnO) as additives that will act as bridge molecules between two polymer chains. The reaction is thought to be between the metal ion and the chlorides to leave oxygen bridges between the adjacent molecules.

The presence of chlorine makes neoprene nonflammable, hence applications requiring this property and elasticity are common. An example of such an application is mattresses for the Navy. The mattresses must be nonflammable because of the highly dangerous consequences of a fire on ships, yet the resiliency of the elastomer results in good comfort.

Other applications for neoprene are fuel hoses, boots, shoe soles, and coatings for fabrics where oil resistance and, perhaps, nonflammability are important.

THERMOPLASTIC ELASTOMERS (EPM AND EPDM)

Any thermoplastic material that can be repeatedly stretched to over twice its original length and will return quickly to the original length when the stress is relieved can be considered a thermoplastic elastomer. These materials are not crosslinked (vulcanized) and therefore have some distinct processing advantages over the more traditional thermoset elastomers (such as the natural and synthetic rubbers). Several types of thermoplastic materials qualify by their performance as elastomers. As with the thermoset elastomers, most thermoplastic elastomers are aliphatic and noncrystalline, as would be expected for flexible, highly elongating materials. The advantages of thermoplastic elastomers are broadening their use in many of the applications where aliphatic thermoset elastomers were formerly the material of choice. However, the thermoplastics are more temperature sensitive, especially over the wide range of temperatures expected of modern elastomers and their properties have not yet been as well defined as traditional rubbers. Therefore, thermoplastic elastomers have not yet made major inroads in complicated structures such as tires. Furthermore, the durability, toughness, and adhesion to the reinforcing fibers are better for the crosslinked elastomers than for the noncrosslinked, particularly at the higher service temperatures.

Thermoplastic Olefin Elastomers (TPO)

Copolymers of polyethylene and polypropylene are elastomeric and constitute the basis for this important group of thermoplastic elastomers. The elastomeric behavior results from their amorphous nature, whereas homopolymer polyethylene and polypropylene fit closely together and can, therefore, crystallize. These copolymers are sometimes called ethylene propylene monomer rubber (EPR or EPM). Ethylene-propylene elastomers have improved oxidation resistance and improved acid and alkali resistance over natural rubber but have poor compatibility with other rubbers, poor creep resistance, relatively poor resilience, and poor resistance to hydrocarbon solvents.

Some important applications for thermoplastic olefin elastomers are body and chassis parts, bumpers, hoses, weatherstrips, seals, mats, wire insulation, appliance parts, gaskets, and coated fabrics. In most of these applications, the more rapid processing of thermoplastic elastomers over thermoset elastomers has been a major factor in the choice of thermoplastics.

A modification to the ethylene and propylene copolymer is the addition of polybutadiene or some other monomer that will allow a carbon-carbon double bond to persist after polymerization. These *terpolymers* (copolymers made from three monomers) have the properties of thermoplastic elastomers but also allow crosslinks to be formed if desired. The terpolymers are ethylene propylene diene monomer rubber (EPDM). The carbon-carbon double bond is usually on a side chain that still allows crosslinking with sulphur as with normal thermoset elastomers, but improves the resistance of the material to degradation from oxidation. This improved oxidation resistance occurs because the oxygen or ozone attacks the double bonds that have not been crosslinked. When these double bonds are in the backbone, as they are in normal thermoset elastomers, attack by oxygen or ozone will result in chain breakage *(scission)*, whereas if the double bond is in a side chain, the scission is far less important.

Other Thermoplastic Elastomers

Several other families of thermoplastic elastomers are commercially available. One of the most important is polyurethane elastomers. These were discussed in detail in the chapter on thermoset materials and will not be discussed here. Another family of thermoplastic elastomers is the copolyesters. Several commercial grades are available from DuPont (Hytrel®), Hoechst-Celanese (Riteflex®), and Eastman (Ecdel®). The copolyesters are higher priced than the EPM, EPDM, or polyurethanes but have excellent tear strength, good resistance to flex fatigue at both high and low temperatures, and have excellent heat resistance. Applications include hoses, tubing, belting, sheets, flexible couplings, ball bladders, sporting shoe soles, and wire coverings. A particularly interesting application for copolyester sheets is as linings under solid-waste disposal sites. The toughness and the low permeability of these materials is especially beneficial in this application because of the need to prevent the contamination of groundwater sources under the waste dump by water that may filter through the dump and become contaminated by leaching out toxic or noxious chemicals from the solids in the dump.

Other thermoplastic elastomers can be created by judicial copolymer formation, alloying, or other methods which would result in amorphous, aliphatic materials which have good recovery properties. Such materials have been based upon polyamide alloys, styrene copolymers, and chlorosulfonated polyethylene. Major applications for such materials include liners for potable water systems, liners under landfills and other waste water ponds, diaphragms where long-term flexibility is required, and other more traditional elastomeric uses such as hoses, shoe soles, and gaskets.

FLUOROELASTOMERS

Fluoroelastomers have many of the same desirable properties found in fluoropolymer plastics such as Teflon® PTFE or FEP resins. Both the fluoroplastics and fluoroelastomers have a high content of fluorine substitution on the carbons, which is the feature that gives these materials their unique properties. The fluoroelastomers are generally copolymers of vinylidene fluoride with another highly fluorinated monomer such as tetrafluoroethylene, the monomer used to make PTFE. The vinylidene fluoride monomer is depicted in Figure 9.6. The first commercial fluoroelastomer was developed by DuPont and was given the brand name Viton®. A family of Viton materials have been made by varying the monomers used to copolymerize with vinylidene fluoride.

The copolymers of vinylidene fluoride and other fluorocarbon monomers seem to possess many of the same structural characteristics that give rise to elastomeric behavior in hydrocarbons. Just as the copolymer of ethylene and propylene seems to give sufficient irregularity in the polymer sequence that crystallization is prevented, that same feature exists in the fluorinated copolymers with vinylidene fluoride. Mechanical properties are

(a) Vinylidene fluoride monomer   (b) Tetrafluoroethylene monomer

Figure 9.6 Vinylidene fluoride and tetrafluoroethylene monomers which are the monomers used to make a fluoroelastomer.

comparable to those of many general-purpose elastomers, although somewhat lower than natural rubber.

Added to these general elastomeric characteristics are the special properties that arise from the fluorocarbon nature of the fluoroelastomers. The fluoroelastomers offer excellent service in corrosive fluids and chemicals at temperatures as high as 400°F (200°C). In air, the fluoroelastomers remain usefully elastic for indefinite periods of continuous exposure up to 450°F (230°C) and of intermittent exposure at higher temperatures. Resistance to oils, fuels, lubricants, most mineral acids, and many aliphatic and aromatic hydrocarbons is outstanding. Resistance to ozone, weather and flame is exceptional. Radiation resistance is good and high-vacuum performance excellent.

The high cost of fluoroelastomers limits their application to only those that require the special fluoropolymer properties, but these applications are still numerous. They include: gaskets, O-rings, oil seals, diaphragms, pump and valve linings, flue duct expansion joint seals, hose, and tubing when these materials are used in extremely difficult environments. Fluoroelastomers are coated onto fabrics for chemical- and petroleum-handling equipment and are used for critical automotive and aerospace components, high-vacuum equipment and equipment involving cryogenic temperatures and radiation. Recent applications include those where the environment is less difficult but a long, maintenance-free service life is critical.

SILICONES

Silicones are the largest group of plastic materials that are not based principally on carbon atoms. The silicones are based on the silicon atom which is directly below carbon in the periodic table and therefore is in the same chemical group. Both silicon and carbon form four bonds in their normal bonding pattern and have many similar chemical reactions. There are, of course, also some differences between silicon and carbon. Silicon is less electronegative than carbon which increases the activity of some silicon bonds and allows some reactions that are generally not available with carbon. Also, silicon-silicon double bonds are rare and cannot be used for polymerization or crosslinking as carbon-carbon double bonds would be used. Silicon is also used extensively as a semi-conductor material for making electronic components.

Silicon-based polymers are made by the polymerization of monomers which are called *silanes*. These silane monomers do not polymerize by the traditional addition and

Figure 9.7   Polymer repeating unit of a typical silicone polymer showing the siloxane bond and where the pendent groups are methyls.

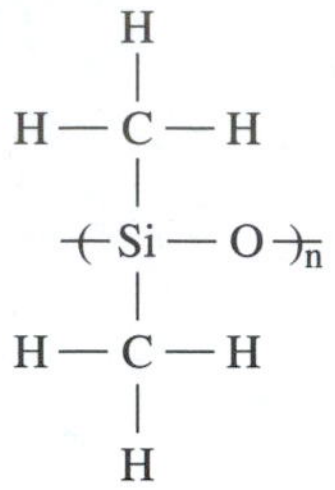

$$
\begin{array}{c}
\text{H} \\
| \\
\text{H} - \text{C} - \text{H} \\
| \\
\left(\!\text{Si} - \text{O}\!\right)_{\!\overline{n}} \\
| \\
\text{H} - \text{C} - \text{H} \\
| \\
\text{H}
\end{array}
$$

condensation polymerization methods typical of carbon-based monomers. Rather, some unique reactions are used to form silicon-based polymers. All of these reactions result in the formation of siloxane bonds which are alternating Si and O as shown in Figure 9.7. These polymers can be called *polysiloxanes* or, simply, *silicones.*

Each silicon atom in a silicone polymer would be bonded to two oxygen atoms and would also be bonded to two pendent groups. In the most common commercial silicones, these pendent groups are common carbon-based entities, such as methyl groups or *phenyl* groups. (Phenyl is a name given to an attached benzene ring.) A typical silicone polymer repeating unit is shown in Figure 9.7 where the methyl groups are the pendent groups. This polymer, dimethylsiloxane, is the most common silicone. The pendent groups can be changed to give specific characteristics in addition to the normal silicone properties. Methyl groups impart low surface tension and water repellency to the polymer whereas phenyl groups impart thermal stability, organic compatibility and water repellency. Aliphatic groups with three or more carbons are used when compatibility with organic compounds or paintability are desired. If the aliphatic group has a high fluorocarbon character, solvent resistance is enhanced.

Silicones can be polymerized into three common forms—oils, elastomers, and molding compounds. The oils are low-molecular-weight thermoplastic materials that are usually liquids at room temperature. They are used as mold releases, coolants, foam suppressants, lubricants, hydraulic fluids, processing aids, surfactants, and water repellents.

Silicone elastomers can be made by increasing the molecular weight and crosslinking the chains. For elastomers to have good mechanical properties, they must be filled, usually with silica. (This is similar to the need to fill carbon-based elastomers, often with carbon black.)

If the molecular weights are increased even further, silicone molding resins can be made. These are also crosslinked but usually do not contain fillers.

The advantages of silicone elastomers and molding resins are similar and include the following: low surface tension, nonionic/nonpolar characteristics, unique solubility characteristics, hydrophobicity (water repellency), thermal stability, oxidation resistance, good dielectric properties, low freezing and pour points, low volatility at high molecular weight, high volatility at low molecular weight, minimum viscosity/temperature slope, shear stability, high gas transmission rates, low flammability, relatively inert, nontoxic, and environmentally safe. A few specific performance details are that silicones will retain properties up to 400°F (200°C) and as low as −535°F (−315°C); and are almost immune to UV degradation because silicones are almost transparent to UV radiation. The mechanical properties of silicone materials are usually somewhat lower than carbon-based materials.

The applications for silicone elastomers and molding resins include: sealants, caulks, elastomeric coatings for waterproofing, flexible molding materials, nonstructural plastic parts, encapsulating gels, adhesives, and insulation materials. Silicones are also used for prosthetics (artificial ears, etc.) and have been used for breast implants and for other internal medical applications. Some claims of toxicity or carcinogenic behavior have limited the use of silicones in these applications. Studies on the effects of silicones in these applications are ongoing.

Silicone elastomers and molding compounds must be crosslinked to achieve the properties that are desired for most of these applications. Silicones can be cured at room temperature (called room temperature vulcanized, or *RTV*) or at high temperature *(HTV)*. Convenient one- and two-component systems are available. Some one-component systems use the absorption of water to react with chemically active sites along the polymer chain to form crosslinks. Other one-component systems cure with the absorption of CO_2. Hence, all that is required to crosslink these materials is to allow normally moist air to diffuse into the polymer. Many of these vulcanization reactions release acetic acid or ammonia which can be mildly corrosive. In some cases, the RTV materials are molded into specific shapes. The vulcanization can be done during the forming operation or after the part has been shaped.

Two-component cure systems are much like epoxy systems with a hardener. Two components are mixed in the proper ratio and allowed to react. The advantage of the two-component systems is that reaction is less likely to occur by accident, as it might if the container of the one-component system is broken or otherwise compromised.

HTV compounds are cured by heat. These usually involve a crosslinking based upon the action of a peroxide. Therefore, a heat-activated peroxide is added to the silicone along with other additives that might be needed for the proper crosslinking to occur. These are then heated as they are molded. The heating causes the peroxide to activate and crosslinking to then occur.

PROCESSING OF ELASTOMERS

Most traditional plastic processing methods are discussed in the later chapters of this book. In the case of elastomers, however, some aspects of the processing methods are quite different from other plastic processing methods because of the uniqueness of elastomers. Therefore, these unique processing features are outlined in this chapter on elastomers so that they can be considered along with the discussions on the natures of the elastomeric materials.

Compounding

This name is given to mixing when that mixing is an intensive, high-shear process as is required to disperse additives in elastomers. The high viscosity of the elastomers dictates the use of high-shear mixing techniques. The most common machines used to achieve this type of intensive mixing in rubber compounding are called *Banbury mixers*. A diagram of a Banbury mixer is given in Figure 9.8.

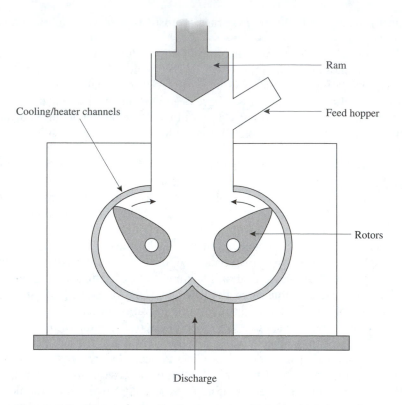

Figure 9.8 Diagram of a Banbury mixer, which is a high-intensive mixer often used in rubber compounding.

In the Banbury mixer the material enters the side of the upper chute and is compressed into the mixing chamber by the ram. The mixing chamber contains two counter-rotating rotors which are turning at different speeds, thus creating a high shearing action. The mixing chamber is jacketed so that it can be heated or cooled, depending upon the requirements of the particular mix and the stage of mixing that is being done. The mixed material is discharged through a port in the bottom of the mixing chamber.

To prevent premature crosslinking (called *scorching*), the mixing of additives into the rubber material is usually done in two steps. In the first step the rubber is mixed with the filler (usually carbon black) and with oils and other minor constituents that are useful in processing and prevention of oxidation or UV degradation. During this phase, the rubber material is broken down into a less viscous, extended-chain polymer mass. This is the state that allows powders and other minor constituents to be mixed in.

The crosslinking or vulcanizing agent (usually sulphur for traditional rubbers) is not added during stage one. The temperature during stage one mixing will often reach high levels (typically 300°F, 150°C) that would cause the material to crosslink. These high temperatures are reached because of the high intensity of mixing that is required to get good dispersion. The forces during mixing are very high and Banbury mixers are very sturdy machines with powerful motors capable of high loads.

The crosslinking agent is added in stage two. This step is usually less intensive and temperatures can be kept under 212°F (100°C). The elastomer is already pliable in stage two and some processing aids further improve the mixing of the crosslinker with the elastomer blend.

The blending of small amounts of material into an elastomer and the finish blending to achieve good homogeneity of the mixture can also be done using a two-roll mill. In this operation the elastomeric mass is placed into the nip between two massive rolls, which are counterrotating at slightly different speeds. As the elastomeric mass is placed into the nip, it is flattened and begins to follow one of the rolls. With each turn, the material is further squeezed and flattened as it passes through the nip. The high shear created between these rolls gives the mixing action desired. Two-roll mills mix much more slowly than Banbury mills, but excessive heat creation is less of a problem.

Preforming

The thoroughly mixed elastomer batch is discharged from the mixer onto a two-roll mill or, perhaps, into a short extruder. The two-roll mill converts the material from a formless mass into flat sheets that can be cut into strips for molding. The extruder converts the mass into a rope that can also be used in the molding process. Care must be taken in these preforming processes that the temperatures do not increase appreciably, otherwise scorching could occur.

If the elastomer is not discharged directly into the mill or extruder but, rather, is kept in storage, the elastomer will again take its highly viscous form. When this occurs, the material must be elongated before being preformed. This secondary elongation is most often done on a two-roll mill.

Molding

Molding is usually done in conventional molding equipment used for thermosetting resins. The most common molding systems of this type are compression and transfer molding. Some elastomers can be extruded and injection molded, although this is less common. When large sheets are desired, the elastomers can be calendered. All these molding methods are discussed in detail in later chapters. The vulcanization occurs during the molding phase so the molding is usually done at elevated temperatures and sufficient time is required in the molding to ensure that crosslinking is complete.

Dipping

One important rubber-processing method that is different from all the others just discussed is dipping or dip coating. This method begins with either a latex of the rubber or a solution of the rubber. This latex or solution is quite pure and would not be made simply by using the latex rubber as it comes from the rubber tree. A form is dipped into the purified latex or solution, sometimes repeatedly to build up thickness, and then set aside or in an oven to dry. When the water or solvent has evaporated, the rubber material or rubber-coated product is removed from the mold. This is the most common method for making thin rubber products such as surgical gloves.

CASE STUDY 9.1

Elastomeric Lining for a Pump

A major cause of acid rain and other corrosive pollutants has been the presence of sulphur oxides in the exhaust (flue) gases of utilities and other coal-burning facilities. Coal often contains large quantities of sulphur that are converted to sulphur oxides when the coal is burned and then become corrosive acids when the sulphur oxide gas combines with water in the atmosphere. The efficient reduction of these sulphur oxides from the flue gases is critical to control of this major source of air pollution.

A process which gives 95% elimination of the sulphur oxides from flue gases has been developed and has proven to be cost effective and reliable, although some key developments were required to achieve the level of reliability required for commercial application. This method involves the spraying of a *slurry* of limestone and water into the flue gases. (A slurry is a mixture of fine particles of a solid in a liquid.) The limestone, which is alkaline, reacts with the acid in the flue gas to form gypsum, a mineral that can be sold for applications such as the making of wall board. The conversion of the limestone to gypsum is very efficient, usually approaching 100%. Therefore, not only are the sulphur oxide pollutants removed from the flue gases but a commercially viable by-product is produced.

When first implemented, the slurry process had some serious problems, especially in the reliability of the pump that was needed to move the slurry and generate the pressure for the spraying. The slurry pumps wore out quickly, both from the corrosive nature of the slurry and the mechanical impinging of the slurry on the inside of the pump. A specification for the pump was established. The pump had to have the capacity for pumping large quantities of slurry (up to 300,000 cubic feet/hour, 10,000 m^3/h), be corrosion resistant (24,000 hours or about 3 years of continuous duty without maintenance), and have the capability of quick maintenance when it was required. The strength and capacity were met with a cast iron pump, which could be made in sizes ranging from inlet sizes of 14 to 28 inches (355 to 710 mm) and casing heights of 60 to 120 inches (1600 to 3000 mm). However, the cast iron pump would not meet the other specifications.

Rather than change from cast iron to a corrosion-resistant metal, which would be prohibitively expensive, the suggestion was made to line the pump with elastomeric material. Several types of elastomers (natural rubber, isoprene, SBR, and neoprene) were used and found to be successful in combating the corrosion problem and reducing the mechanical wear. The rubber materials had the following advantages over the bare metal: abrasion resistance, natural corrosion resistance, hydraulic noise dampening, and low cost versus noncorrosive metal pumps or liners. The impeller was also coated with elastomers to reduce wear. After operating the elastomeric-coated pumps, the coated impellers were found to retain their ability to build pressure for a longer time than noncoated impellers, thus extending the life of the impeller as well as the pump casing.

The major disadvantage to using elastomers to line the pump was the difficulty in changing the lining when maintenance was required. The elastomer coating was bonded to the cast iron case and, therefore, was difficult to remove. This problem was solved by bonding the elastomer to a fiberglass-reinforced shell that was precast in the shape of the inside of the pump. (A patent for this invention was granted to Baker Hughes Incorporated, patent #5,219,461.) When maintenance was required, the old fiberglass shell was simply popped out and the new one with the liner already attached was installed.

The presence of the fiberglass shell also improved the ability to install the elastomer. If the elastomer were installed in the pump, the elastomer sheets would collapse and hit the installer on the head, or require several people to support the elastomer sheets while they were being bonded to the pump. With the shell, the elastomer sheets could be laid into the shell (by inverting it) and thus eliminate this annoying problem.

Other improvements with the shell included the reduction of flutter (pump chatter) due to improper flow conditions that might occasionally arise (turbulent flow instead of laminar flow). Also, the shell held the elastomer in place so that the normal shrinkage that occurred with cooling and curing of the elastomer was not present and better tolerances could be achieved.

The combination of an elastomer coating and a fiberglass-reinforced plastic shell has proven to be a major factor in solving the pumping problems of the wet limestone process for scrubbing flue gases. Field experience has confirmed the superiority of the method. The elastomer lining has proven to be effective in reducing corrosion and mechanical abrasion. The type of elastomer chosen can be varied to meet the particular conditions of this and many other corrosive and problem pump applications. New methods for installing the elastomeric coating are now being tested.

SUMMARY

Elastomers are polymeric materials that can stretch to over twice their normal length and then, when the stretching force is relieved, return immediately to their original shape. This behavior is called elastic elongation. Most elastomers will also compress to one-half their original size and return again to their original shape after the compressing force is removed. This property is called resiliency. The stress-strain curve for most elastomers shows a low initial slope (low modulus) and a long elongation with very little force required.

If elastomeric materials are crosslinked, they continue to exhibit this elastomeric behavior provided that the number of crosslinks is not too high. The crosslinking or curing of elastomers is called vulcanization. Typically there is one crosslink for every 100 or so atoms along the polymer backbone. This amount of crosslinking is much less dense than in the thermoset plastic molding resins that were discussed in the chapter on thermosets. The amount of crosslinking in elastomers is sometimes called *lightly crosslinked*. The crosslinks have the beneficial effect of giving an ultimate limit to the amount of stretch which the molecules can have. This stretch limit gives a limit to the amount of creep, thus solving one of the most difficult problems confronting the early users of elastomeric materials, the softening and creeping that occurred at high temperatures.

Another advantage of crosslinks is that the limit to stretching gives a limit to the elastic range, thus defining the maximum amount of stretch that can be obtained without permanent deformation.

Elastomeric materials are usually highly random, aliphatic molecules that have no crystallinity in their relaxed state. These molecules can be pulled into more oriented configurations, which is what occurs during elastic stretching. The most common materials that possess these properties are built on a four-carbon backbone unit with a carbon-carbon bond between the middle two carbons. Some molecules have one or more pendent methyl groups which add intramolecular and intermolecular interferences that result in higher stiffness, higher strength, and higher hardness. In some applications these interactions are desirable, whereas in other applications they are not. Some elastomers have added bulky comonomers (like styrene) to achieve similar interactions. Chlorine can be added as a pendent group to the basic four-carbon repeating unit to give intermolecular interactions and, in addition, oil resistance.

Thermoplastic elastomers can be made by copolymerizing ethylene and propylene. As copolymers these materials are noncrystalline and have elongations over 200%, thus qualifying as elastomers. The advantage of these materials over crosslinked elastomers is that they can be processed in traditional thermoplastic processes, such as extrusion and injection molding. If a monomer having two carbon-carbon double bonds is added to the ethylene and propylene monomers when they are reacted, a crosslinkable elastomer from ethylene and propylene can be formed. This material is designated EPDM.

Fluorocarbon polymers can be made that have 200% elongation and are therefore elastomers. These fluorocarbon elastomers have excellent resistance to solvents, oils, acids, alkalis, and most other solvents. This resistance extends over the entire use range, which is much broader than for other elastomeric materials. The fluorocarbon elastomers also have excellent resistance to oxidation and UV radiation.

Silicones are polymers based on the silicon atom rather than on the carbon atom. When polymerized, the silicon monomers form repeating silicon-oxygen bonds which are called siloxane bonds. Therefore, silicon polymers are polysiloxanes or, more normally, silicones. Silicones can be oils, elastomers or molding compounds, although the elastomers are the most important. When these elastomers are cured at room temperature (usually by the absorption of either water or carbon dioxide), the curing system is called room temperature vulcanization (RTV). When cured at high temperatures, usually with peroxides, the system is called high-temperature vulcanization (HTV).

To achieve good durability, as would be needed in automobile tires, elastomers need to have a filler added. The most common filler is carbon black which also imparts resistance to UV radiation. This carbon black must be well dispersed in the elastomeric batch. To accomplish this task, high-intensity and high-shear mixers have been developed. The most common type of mixer for this purpose is called a Banbury mixer. Two mixing steps are normally required when using a Banbury mixer. The filler and other minor constituents are added in the first mixing step which is usually done at high temperatures. A second step to mix in the curing agent is done at lower temperature so that the mixture will not prematurely crosslink. After mixing is completed, the elastomeric batch is preformed, usually on a two-roll mill, and is then molded and cured.

GLOSSARY

Balatta   The trans form of polyisoprene, also called gutta percha.

Banbury mixer   A high-shear, intensive mixer often used in the blending of compounds into a rubber mixture.

Buna-N   A polymer, sometimes called nitrile rubber, that is the combination of butadiene and acrylonitrile.

Buna-S   A polymer, sometimes called butadiene rubber, that is the combination of styrene and butadiene, also called styrene rubber.

Butadiene rubber   A synthetic substitute for natural rubber that has now gained application in a wide variety of polymeric systems.

Butyl rubber   A synthetic elastomer, also called polybutylene, has less elongation than butadiene and is not crosslinked but has excellent low permeability and UV resistance.

Chlorine rubber (chloroprene)   A polymer in which a pendant chlorine is attached to a butadiene backbone, also called chloroprene rubber or neoprene rubber.

Cis   A prefix used in chemistry to describe the situation where the two pendant groups associated with a carbon-carbon double bond are on the same side of the bond.

Crosslink density   The total number of crosslinks in a thermoset or crosslinked elastomeric system.

Elastic elongation   The elongation of any material when that material is at its yield point or, in other words, the elongation of a material at the point where the strain is at the limit of total recovery.

Elastomer   A polymer with more than 200% elastic elongation.

Emulsion   A suspension of a nonsoluble component in a solvent.

Entropy   A thermodynamic measure of the randomness of a system.

EPDM   Ethylene propylene diene monomer rubber, a principally thermoplastic elastomer with a small amount of diene added to give a very light crosslinking.

Gutta percha   The trans form of polyisoprene, also called balatta.

Hard rubber   Natural rubber that has been so highly crosslinked that the elongation is less than 200%.

Hevea rubber   Natural rubber.

High-temperature vulcanization (HTV)   Curing of a thermoset, usually a silicone or an epoxy, using heat in the process.

Isoprene   An alternative name for polyisoprene, the polymer basis of natural rubber and gutta percha.

Latex   An emulsion or suspension of a nonsoluble component (such as a polymer) in water.

Latex rubber   Natural rubber.

Lightly crosslinked   A polymer system in which only a few crosslinks exist.

Neoprene rubber   Another name for chloroprene.

Nitrile rubber (NBR)   A copolymer of butadiene and acrylonitrile, also called buna-N.

Phenyl   The name applied to a benzene ring that is attached to another chemical entity (such as when benzene is pendant to a polymer chain).

Resilience   The ability of a material to recover elastically from a compressive force.

Room temperature vulcanization (RTV)   Curing of a thermoset, usually a silicone or an epoxy, without the need for heating.

Rubber   A term that applies to natural rubber, a derivative of the sap of the rubber tree *(Hevea braziliennsis)*, and also to elastomers in general because of their properties that are similar to those of natural rubber.

Scission   Breaking of the chains in a polymer, usually by action of oxygen, ozone, or UV light.

Scorching   Premature crosslinking of a polymer, especially an elastomer.

Silane   Silicon-based monomers used to make silicone polymers.

Silicone   Polymers based on silicon and oxygen bonds down the backbone with pendant carbon-based groups; also called siloxanes.

Siloxanes   Polymers with silicon and oxygen bonds forming the backbone.

Slurry   A mixture of fine particles of a solid in a liquid.

Styrene rubber (SBR)   A copolymer of styrene and butadiene, sometimes called buna-S or GSR (government styrene rubber).

Terpolymer   A polymer made by the copolymerization of three monomers.

Trans   A prefix used in chemistry to describe the situation where the two pendant groups associated with a carbon-carbon double bond are on opposite sides of the bond.

Vulcanization   Originally a name given to the curing of natural rubber with heat and sulphur but now applied to any curing (crosslinking) reaction, especially of elastomeric systems.

QUESTIONS

1. What are the key polymeric structural features common to most elastomeric materials?

2. What is the mechanism and thermodynamic explanation that explains why elastomers recover to their original shape after being elastically stretched?

3. Why is the cis form of polyisoprene softer than the trans form?

4. Explain why polybutadiene is softer than polyisoprene.

5. Explain why silicones have higher gas permeability than do carbon-based molecules with equivalent pendent groups.

6. Describe the changes in properties that are likely to occur when the temperature is lowered below the T_g for an elastomer. What does this indicate about the usable lower temperature range for elastomers?

7. What is the effect of crosslinks on an elastomer and what is the structural explanation for their effect? What would happen to the amount of stretch, hardness, strength, and creep in an elastomer if the crosslink density were increased?

8. Explain why copolymers of ethylene and propylene can be thermoplastic elastomers when the pure polyethylene and pure polypropylene are not elastomeric. Discuss whether you would expect a blend (instead of a copolymer) of polyethylene and polypropylene to be elastomeric?

9. Explain what may happen to the properties of a rubber material if the carbon black filler is poorly mixed into the batch.
10. Why would fluorocarbon elastomers be so effective in resisting solvents?
11. (a) Explain why stretching an elastomer causes an unstable state and, therefore, when the stretching force is removed, the elastomer recovers. (b) Explain why stretching of some polyethylene molecules results in "blushing" and this situation is not eliminated when the stretching force is removed.

REFERENCES

Chanda, Manas, and Salil K. Roy, *Plastics Technology Handbook* (2nd ed), Albany, NY: Delmar Publishers, Inc., 1989.

Charrier, Jean-Michel, *Polymer Materials and Processing,* Munich: Hanser Publishers, 1991.

Driver, Walter E., *Plastic Chemistry and Technology,* New York: Van Nostrand Reinhold Company, 1979.

DuPont Products Guide, E.I. DuPont de Nemours Company, Wilmington, DE, GS-11258, July 1985.

"Engineering Plastics," *Engineered Materials Handbook,* Vol.2, Metals Park, OH: ASM International, 1988.

Fatzinger, John, and Mark Attride, "Slurry pump designs for wet limestone scrubbing," *World Pumps,* Feb. 1994, 46-51.

Seymour, Raymond B., *Engineering Polymer Sourcebook,* New York: McGraw-Hill Publishing Company, 1990.

Seymour, Raymond B., and Charles E. Carraher, *Giant Molecules,* New York: John Wiley and Sons, Inc., 1990.

A Silicone Primer: An Overview of Silicon-Based Chemistry for the Plastics Industry, Midland, MI: Dow Corning Corporation, Form No. 24-934-91.

Ulrich, Henri, *Introduction to Industrial Polymers* (2nd ed), Munich: Hanser Publishers, 1993.

CHAPTER TEN

DESIGNING WITH PLASTICS

CHAPTER OVERVIEW

This chapter examines the following concepts:

- Design methodology
- Layout/drawing
- Constraints (mechanical (including FEA), appearance, environmental, testing, safety/codes/standards, manufacturing method/design rules, economic)
- Material choice (databases, selection methodology)
- Prototyping (traditional methods, rapid methods)

DESIGN METHODOLOGY

This chapter examines the methodology of designing a part out of plastic. The part-design process is part of a much larger product-realization process that is best understood if learned and executed formally. The product-realization process is simply the entire process to make or realize a new part, including design of the product, design of the process to make the product, actual production runs to make the product, and all the functions that support the making of the new product. Often the product-realization process is attempted informally, that is, by developing the product on an intuitive basis. By its nature, however, product realization involves many different disciplines that one person can rarely represent well. Therefore, the intuitive process often causes unnecessary readjustments and delays as oversights are corrected. These problems can be eliminated by making sure that all of the steps are done appropriately. This requires a formal approach. Thus, this chapter begins with a formal description of the steps involved in the product-realization process. The remainder of the chapter is devoted to part design, with the understanding that it is but one step in the overall process. It should be kept in mind that the overall process can be used for making a part and for making a system or any complex assembly, and that a system is just the sum of several parts.

The making of a part or a system involves a number of steps. These steps, from concept to realization, are as follows:

- Need recognition.
- Functional specification.
- Concept generation and evaluation.
- Part design.
- Process design and planning.
- Prototype creation and verification.
- Production implementation.

The flow of these steps is diagrammed in Figure 10.1. Notice that part or system design (often the responsibility of mechanical engineers) and process design (the responsibility of manufacturing engineers) are shown as occurring simultaneously, with interactions between the groups so that each group can respond to the design being created by the other. This simultaneous design of part and process emphasizes the iterative nature of these steps. When the part-design function is separated from the process-design function, or when they occur in sequence rather than at the same time, much time is lost and many errors occur. The linking of these two design stages is called *simultaneous engineering,* or *concurrent engineering,* and is an important part of modern new-product development.

The recognition of a need or the recognition of an opportunity generally is considered to be the beginning of the part-creation process. This need may be seen in the marketplace, where customers have a problem that is not being adequately addressed. The need may be recognized within a company, where an existing product is seen to be inadequate or where an improvement in a product is conceived. The need can come from many other sources, but the need is only an idea until the next step is taken.

The formal annotation of the need is done by making a *functional specification*. It is called functional because this specification details **what** must be done to satisfy the need, that is, what function the product must be able to accomplish to satisfy the need. No attempt is made at this point to say how that will be done. Marketing or direct customer input is essential at the functional specification step to insure that the part to be created will satisfy the need.

The next step in the creation process is called *concept generation and evaluation*. During this step, ideas on **how** to meet the functional specification are presented. This is generally a brainstorming session where many people, often led by technical personnel, give their ideas. Initially, as many ideas as possible are presented, without any attempt to evaluate them for feasibility or practicality. Eventually, however, the test of reason and practicality is applied to each of the concepts, and they are ranked. The best of these concepts is chosen as the initial basis for proceeding through the remaining steps. If the best option selection is found to have some problems as the ensuing steps in the product-realization process are taken, the next best concept is then chosen and the steps are begun again with the new concept as the basis.

The next steps are the two design steps: part or product design and process design. In brief, the part-design process involves simultaneous layout or drawing of the part, establishment of constraints on the part, detailed analysis of the layout (often mathematical),

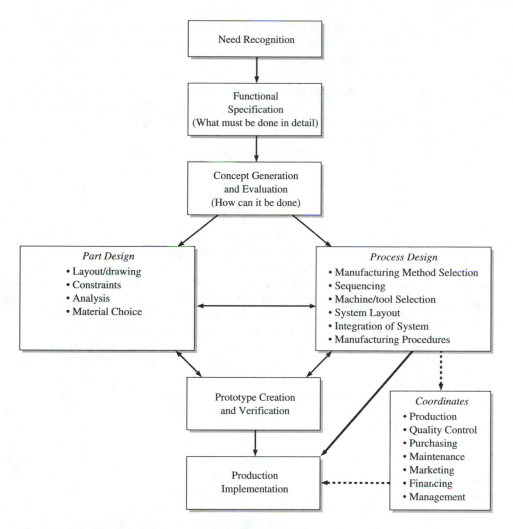

Figure 10.1   Steps in the process of realizing or making a new part.

and choice of materials. All of these steps within the part-design process are discussed in detail in this chapter. In addition, the chapters in this text that deal with plastic properties and specific materials assist in the part-design phase.

The process-design phase involves selection of the manufacturing method, sequencing of manufacturing steps, machine and tool selection or design, layout of the system, integration of the system, and establishment of manufacturing operation procedures. The majority of the chapters in this text that deal with plastics manufacturing methods can assist in the process-design phase.

After the part and process designs have been created, a prototype of the part is made. This prototype may be made by the eventual process envisioned, but is more likely to be made some other way since the envisioned process is really a method for making production quantities of the part rather than the limited quantity needed for a prototype. Remember that the purpose of the prototype is to confirm that the part, as designed, functions properly and meets the functional specifications. Of course, the making of the prototype may point out problems in the part design that will require the design process to be redone, at least in part.

After the prototype is verified as meeting the need, the design of the manufacturing process is then verified and implemented. Test runs will usually be done to confirm that the process design is correct. After these test runs, full production implementation can be started.

The product-realization process described herein applies to parts and systems made of any material. The process can easily be applied to parts made of plastics, which is the focus of this text. The use of plastics is increasing rapidly, and the methods that can be used to properly design a plastic part have become very important in the modern world.

As indicated, much of this chapter focuses on the part-design phase of the product-realization methodology. This discussion includes a number of assumptions, to wit: the need has been recognized and properly characterized in a functional specification, the concept generation and evaluation have been completed and a ranking of how the functional specification is to be met has been defined, and the implementation plan has defined a specific part that must be designed and manufactured. Although the discussion concentrates on the part-design phase of the product-realization process, some references are given to the process-design phase, which must be done simultaneously. In fact, inputs from the process-design phase that must be taken into account before the part design can be completed are formally identified.

LAYOUT/DRAWING

The making of a detailed layout or drawing of a plastic part should be viewed as a process of making decisions and not just as a reduction of a sketch to a detailed representation. The engineer or engineering technologist who supervises the overall part design should, therefore, be intimately involved in the layout or drawing of the part.

In the past, the most common method of layout or drawing was with a blueprint created by manual drawing. Today, the drawing is usually done using a computer. The programs that allow these drawings to be made are called computer-aided design (CAD) programs. These are available from many different companies and for several different types of computers ranging from personal computers (PCs) to workstations to large mainframes. The reliance on CAD drawings is so complete in some companies that no actual paper print of the part is ever made. Quality control, engineering changes, shop floor needs, and all other functions that have used prints in the past now simply refer to the computer rendering of the part.

Some CAD systems are able to render the part in only two dimensions, much as is done with blueprints. Most of the newer and more powerful CAD systems, however, can provide representations that are more complete—three-dimensional pictures of the part. These three-dimensional pictures can be rotated, expanded, shrunk, and quickly modified (even while keeping certain dimensional constraints), so that the part can be well visualized and understood. Dimensions and tolerances can be listed on the screen.

The process of making the CAD representation forces decisions about size, shape, dimensions, and, if desired, tolerances and other process-related and material-related parameters. Since many of these parameters are not well defined at the initial stage of CAD drawing, the drawing process is certainly iterative with changes being made as the design process progresses. This ability to make rapid changes to the drawing is one of the most important advantages of CAD over traditional manual drawing. Changes based on processing requirements can be used as an important criterion for deciding which of several competing processes might be the best for the part. Hence, part design is an integral contributor to process design.

CONSTRAINTS

The identification of constraints on the design of the part is carried out simultaneously with the drawing of the part. Constraint identification is a way of forcing detailed decisions to be made in the part design, usually reflecting these constraints in the drawing. Several important categories of constraints are discussed.

Appearance

The shape and size of the part are two of the most important general constraints. The general shape and size are usually, but not always, dictated by the functional specification, since function is closely related to shape and size. However, that is not always the case, and some of the most innovative designs occur when shapes and sizes are left open to experimentation and are not defined until after other constraints have been set. For instance, bottles need not be round or even symmetrical, steering wheels need not be circular, and housings need not follow the shape of the parts inside. To obtain originality in part appearance, many companies employ industrial designers who have strong artistic backgrounds. These designers can also contribute important *ergonomic* (human factors) considerations. (Ergonomics is the study of how a design can be optimized for accommodating human constraints, capabilities, and preferences. For instance, a tool with a handle diameter too big to be gripped by the average person would be poorly designed from an ergonomic standpoint.)

Color, surface quality, streamlining, and other appearance parameters should be decided during the part-design process. Some of these qualities are functional but some are picked on nonfunctional bases, perhaps because of uniformity with other products made by the company, simple eye appeal, or cost.

Mechanical

The mechanical constraints on a part often are some of the most important in satisfying the functional specification. The mechanical constraints may involve such parameters as tensile stiffness and strength, elongation, flexural properties, torsional properties, abrasion resistance, impact toughness, burst pressure, and creep. In many parts, these parameters are important in satisfying both short-term and long-term functional requirements. Fatigue life often is strongly influenced by the mechanical properties of the part. Therefore, long-term performance should be a part of the functional specification of the part when long-term use is anticipated. (Long-term use can also be affected by environmental parameters, discussed in detail later.)

Many designers who have not had experience with plastics have difficulty with the proper specification of the mechanical properties of a plastic part. Unlike other materials such as metals and wood, the testing of mechanical properties of plastics is time-dependent. Hence, the data used to design the plastic part may not reflect the in-use, time-dependent mechanical conditions. This problem was highlighted in a study made by the auto industry in the design of a plastic bumper to replace a traditional steel bumper. The designs made using normal flexural properties were off by an average of 10% from actual experimental data. The deviation was caused by a difference in the conditions under which the standard tests were conducted and the conditions of actual part use. This deviation was within the normal safety factor for such designs, but calculations and designs can be much more accurate, resulting in more efficient material use, if the mechanical properties of the plastic are corrected for nonlinear behavior.

An important method for conducting analyses of the mechanical properties of a design for a part is *finite element analysis (FEA)*. Using a CAD rendering of the part, this analysis technique applies simulated stresses on the part and then measures the resulting properties. Because most of the parts studied are irregular in shape, the results of the applied stresses are complicated and difficult to interpret. Therefore, the FEA technique involves dividing the part into many very small, uniformly shaped elements. By distributing the stress over these elements the resultant strains and other properties resulting in each element can be determined. The total strain or other resultant property for the entire part is obtained by examining the results for the individual elements. Areas of high stress concentration can be detected because the small elements in those areas all have high stresses. Typically, these high values are shown in graphic form. Figure 10.2, for example, shows the CAD drawing, the division of the part into elements and the results of an analysis that uses shading to show areas of stress concentration.

Environmental

The environment in which a part is used always has an effect on the properties of the part. Sometimes these effects are small in comparison to other properties that are critical in the proper performance of the part. In many cases, however, the environment can be extremely important in part performance. For instance, the electrical resistivity of a part can be the overriding property when the part is to be used as an electrical insulator. Plastics used as storage tanks or conveying pipes should be insensitive to the liquids that are to be contained within the plastic parts. Thermal conductivity of the plastic is important for applications such as frying pan handles. Sunlight (UV) and oxidation resistance can be important for a part that is to be used outside for extended periods of time. There-

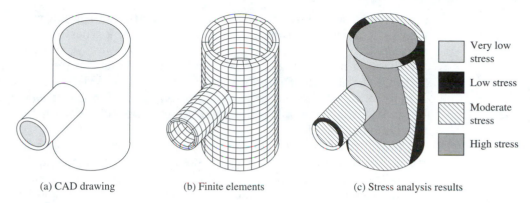

(a) CAD drawing   (b) Finite elements   (c) Stress analysis results

Very low
stress

Low stress

Moderate
stress

High stress

Figure 10.2   Representations of the finite element analysis (FEA) of a part showing the part CAD drawing, the elements, and the stress analysis.

fore, the constraints placed on the part by the environmental conditions of use should be carefully envisioned. Both short-term and long-term environmental effects should be considered.

One of the most important of all environmental factors is temperature. Plastics generally have lower temperature capabilities than metals or ceramics. Hence, operating temperatures should be considered carefully in the design of a plastic part. An example that shows the critical nature of temperature conditions is the design of automobile engines with plastic components. For instance, the rapid action required of some modern engine parts, such as valve lifters, favors the use of plastics over metals because of the lower density of the plastic materials. In general, though, the high thermal environment in an automobile engine has precluded the use of plastics in these applications. However, the need for plastics in some applications, such as the valve lifters, was sufficiently great that plastic parts were tested. Initially, a 20% glass-filled nylon (6/6) was used in the demonstration tests. After many hours in the automobiles, the plastic valve lifters showed some shape distortion. Because this distortion could result in premature failure, a change in the part design was required. A different material, 20% glass-filled nylon (4/6) was then tried and, after a normal useful life, no shape distortion was noted.

A key to the proper design of the part is anticipating and then stipulating the constraints that should be placed on the part from environmental forces over the entire useful life of the part. The chapters in this text that discuss the physical and chemical properties of plastics are useful in considering these environmental effects.

Testing of plastic products is used for both design and quality assurance. Numerous standard tests assist in correlating data from one site and one time to another. More product-specific tests have been developed to monitor the performance of a specific product or group of products, but these have not been officially adopted by one of the standard testing groups. Several of the standard tests are discussed briefly in this book to provide some background for the types of tests that are commonly performed on plastic products.

The product-specific tests are not discussed because they usually are proprietary to a company or so unique to a product that their inclusion in a general text of this type is not warranted.

The most important testing society in the United States is the American Society for Testing and Materials (ASTM). Emphasis is given to these ASTM standards, although similar standards have been developed by European standards organizations such as International Standards Organization (ISO) and Deutches Institut für Normung—German Institute for Standardization (DIN).

Several methods have been employed to measure properties of plastics that are difficult to measure without destroying or invading the plastic material. For instance, measuring the thickness of a film material while it is being extruded generally involves cutting the sample out of the sheet. One of the techniques used for nondestructive testing is ultrasonic testing. In this method an ultrasonic signal is passed through the sample. The time for the signal to be reflected off the opposite side of the sample is measured and the thickness calculated. Ultrasonics can also be used to investigate the presence of internal flaws because these flaws give reflections that notify the tester that something within the sample is causing the signal to behave in a nonstandard way. Another method of determining thickness is with a β-gauge (beta-gauge). This device measures the amount of beta rays that are absorbed when a beta source is directed through a sample. Several other NDT techniques, such as X rays, ultraviolet and infrared lights, and fluorescent dyes on the surface, can be used to measure various properties of plastic materials.

Safety, Codes, and Standards

Some constraints are imposed by safety considerations or by regulation. Regulations are usually also safety related (such as OSHA regulations in the United States), but may arise from other governmental or societal requirements such as protecting the environment or recognizing consumer testing agency approvals. Codes can also be industry dictated, such as the size of a particular part fitting being standard across an industry so that parts from all manufacturers can be used with the same connection device. Plastic PVC pipe, for example, is a regulated product—the diameter of the pipe is fixed by agreement so that standard PVC fittings can be used. The wall thickness (actually the burst pressure) is also fixed by standard because the pipe must meet safety and performance standards as set by a regulatory agency.

Manufacturing Method/Design Rules

The input from the manufacturing process design will place important constraints on the part design and vice versa. Therefore, a preliminary manufacturing method should be chosen early in the part-design process and then continually reconsidered as the part design progresses. Often, details of the part design will be affected by the ability of the manufacturing method. For instance, a hollow part with a narrow opening would be difficult to make using injection molding but would be easy to make with blow molding or rotomolding. However, if the wall tolerances are very tight, neither blow molding nor roto-

Table 10.1 Categories of Design Rules for Injection Molding

• Shrinkage	• Radii
• Tolerances	• Sinks
• Inserts	• Wall thicknesses
• Draft angles	• Flatness
• Undercuts	• Warpage
• Intersections	• Fasteners
• Ribs	• Surface characteristics
• Bosses	• Surface texture
• Fillets	

molding may be able to make the part within the tolerances required. Therefore, the tolerances may need to be changed or, if the tight tolerances are indeed required, a combination of injection molding of the parison and then blow molding of the part may be required. Alternately, casting, or perhaps injection molding with a very complex mold, may be an option.

Each manufacturing process has certain suggestions or rules, called *design rules,* that should be followed for optimal performance in that particular process. For instance, the rules suggested for control of constant part thickness are much more important in injection molding than in casting. Similarly, rules that govern the placement of inserts are more important in injection molding and transfer molding than in thermoforming. Because of the diversity and complexity of design rules for each manufacturing process, no attempt is made here to list the design rules for all of the processes. Instead, Table 10.1 provides a brief list of the types of categories covered by design rules for one process—injection molding.

This list shows the broad scope of rules that must be decided in the design of the part. Each of these rules has a specific value or range of values that is dictated by both the process and the particular material or, in some cases, by the particular grade of material that is chosen.

Economics

The cost of a part is a consideration in any commercial enterprise. Often, the cost places constraints on the design that affect choices such as what material or processing method can be used. Cost considerations also sometimes disallow the inclusion of desired properties that are available with a higher-priced design. If the costs are too high even to meet the minimum performance specification, then the specification and/or the costs will need to be changed. This may require having customer input in the design process. Customer input can efficiently be obtained using *focus groups*—small (5–6 people) groups of potential customers that meet to discuss the advantages and disadvantages of a new product design.

Economics is often a function of production volume. While increasing the volume does produce more favorable economics, the reality of the product sales may not follow the increase in volume. Therefore, caution must be exercised in estimating the sales volume for a particular part.

The economics of the part involves many other factors. Some of the most important economic factors are material costs, production labor costs, production machine burden (that is, time on the molding machine), production overhead costs, and corporate general costs.

MATERIAL CHOICE

This text considers only those materials that are used for solid products. The first consideration in material choice is whether plastics or some other solid material such as metal, ceramic, or natural polymer such as wood will be used. All these materials have advantages in various applications. Table 10.2 shows the areas of advantage for plastics, which are linked to plastic's unique properties.

Databases

After the choice has been made to use a plastic material, the choice of a specific plastic must be made. The wide variety of plastics and grades within each plastic type make this choice interesting and challenging. Several databases of plastics properties are available to assist in comparing the properties of each plastic type. These databases include on-line computer services (such as the GE database), encyclopedia sections (such as the *Modern Plastics Encyclopedia* or the *Plastics Technology Manufacturing Handbook and Buyer's Guide*), tables and charts published in plastics textbooks and computer databases available on disc and CDs (such as from IDES). A searchable database is especially valuable as it allows rapid access to data for many products and sorting of plastics based on multiple selection criteria.

Table 10.2   Advantages and Disadvantages of Plastic Materials Versus Other Common Materials Such as Metals, Ceramics, and Wood

Advantages of Plastics
Ease of fabrication at relatively low temperatures
Low density
Relatively low material cost
Little or no corrosion
High strength and stiffness when compared to weight
Low thermal conductivity
High electrical resistance

Disadvantages of Plastics
Low compression, shear, and bearing strength
Long-term elevated-temperature limitations
Aging and weathering under certain environments
Fatigue under stress conditions

Although valuable, databases must be considered only as general guides. Several factors contribute to the need for caution in using the data contained within the databases. The following is a description of some of these factors.

Multiple Grades.   Each plastic is offered for sale in a number of grades. These grades, which are largely functions of density (crystallinity), melt index (molecular weight), and molecular weight distribution, can be very significant in determining the properties of the plastic. The values in the databases often are averages for the several grades.

Multiple Additives.   Some resin manufacturers offer plastic resins that are formulated for special properties. These resins can be considered different grades of a particular resin type. For instance, fire-retardant grades, impact grades (to which tougheners may have been added), filled grades (with inorganic fillers), reinforced grades (with chopped fiberglass), and many others can be purchased in most resin types.

Different Blends.   The ability to produce resins with a wide range of properties by mixing two or more polymers has led to a wide variety of new resin types. These blends may be standard products for a resin manufacturer, but materials made on demand by compounders also are available. Even within a well-known resin type, such as ABS, the ability to copolymerize, graft polymers together, and blend and mix polymers leads to innumerable possible combinations of the basic components and, therefore, many different property combinations.

Improvements.   New polymers are constantly being developed, and existing resins are constantly being improved. These improvements, either for cost or property enhancement, inevitably change the properties. Because improvements are made on a continuous basis, databases are always outdated, no matter how often they are reprinted or changed.

Additives.   The database data generally omit the effects of additives. In many cases the presence of major additives, such as an inorganic filler, can significantly change the properties.

Selection Methodology

The wide variety of plastics and the wide range of properties available among the plastic materials suggest that some organized method be used to help in the process of selecting a particular resin for a particular application. The following multistep process is one selection methodology that can be used.

Step 1: Define the Criteria for Material Selection.   The overall product-realization process provides the basic criteria that will serve as a basis for the material selection. These criteria involve the functional specification (which defines what the part must do and the environment in which it must operate), the appearance characteristics, the constraints on the part, the consequences of part failure, the codes and standards under which it might be required to operate, and the costs. The criteria should be well defined before the material selection process is begun.

Although no list can cover all the properties that must be considered for all applications, the list in Table 10.3 covers properties that should be considered in almost all applications.

Step 2: Determine the Advantages and Disadvantages of Each Polymer Type. This step is a preliminary screening of the entire breadth of polymers to determine which of them can logically satisfy the need. If the properties of certain plastics being considered are unknown, then those plastics should be retained for consideration at this phase of the selection process. Generally, however, the properties of plastic types are understood sufficiently to eliminate some plastic types and thus simplify the selection process.

Some key properties or constraints of the particular application are usually sufficient to make this preliminary screening. For example, if the application is a structural one, low-density polyethylene can usually be excluded. Likewise, if the application requires high elongation, acrylic and polystyrene can be excluded. The exclusion of polymers that do not meet performance specifications will result in a shortened list of candidate polymers.

Table 10.3   Factors (Constraints) to Consider in the Design of a Part

Appearance
- Shape
- Size
- Ergonomics
- Color
- Clarity
- Paintability or ability to be decorated
- Surface quality

Mechanical
- Tensile strength
- Tensile modulus (stiffness)
- Elongation
- Flexural strength
- Flexural modulus
- Torsional properties
- Abrasion resistance
- Lubricity
- Impact toughness
- Burst pressure
- Creep
- Fatigue life
- Ability to be bonded to or to be sealed

Environmental
- Flame retardance
- Electrical resistivity
- Solvent resistance
- Thermal conductivity
- UV resistance
- Oxidation resistance
- Thermal use limit
- Water resistance
- Permeability
- ESCR

Safety, codes and standards. (Often application or industry specific)

Design rules. (Depend on each manufacturing process chosen)

Economics
- Material cost
- Labor cost
- Machine burden cost
- Overhead cost

Step 3: Select Polymers for Special Properties. This step involves considering the polymers in terms of special properties, especially those associated with the environment in which the polymer must operate. This step eliminates many polymers, further shortening the list of acceptable ones. These special properties may be somewhat more subtle than those properties considered in previous steps. The performance of the plastics in these properties may not be as clear-cut, thus requiring a simultaneous evaluation of the importance of the special property in the actual use of the material along with the relative excellence of the performance of the plastic. For example, an application for a toy may be chiefly indoors, thus limiting the amount of UV exposure that will be encountered. Under this condition, a material that is UV sensitive could be acceptable for the use, although a UV-tolerant polymer could be rated slightly higher because of its additional versatility, that is, the part could be used outdoors without difficulty.

Those polymers that are acceptable in a particular property might have different degrees of acceptability and therefore might be ranked according to their performance in the various environments, although some may simply be "acceptable" with no ranking possible. For instance, nylon is acceptable for most moderately high-temperature environments, but it is not as good as phenolic. The following are some environmental conditions that might be considered:

- Water solubility.
- Oil solubility.
- Solvent resistance (to a particular solvent or group of solvents).
- Flame retardance.
- Corrosion resistance.
- Oxidation resistance.
- UV resistance.
- Heat resistance.
- Electrical resistance or some other electrical property.
- Permeation resistance (barrier properties).

Other performance criteria should be considered at this stage as well. For instance, the ability of the material to be painted is important in some applications such as automotive exterior panels. The ability of the material to be bonded is important if the part is a component in an assembly that must be bonded together. The functional specifications should be carefully considered for any special performance criteria that are not among the more common environmental considerations mentioned above.

Step 4: Rank the Resins and Make a Preliminary Selection. The short list of acceptable resins produced by following the previous steps can now be ranked according to the full range of performance criteria. Some criteria and constraints, such as size, shape, and mechanical properties, still have not been applied. These criteria are interrelated and so must be considered concurrently. For instance, if the constraint on size is quite broad, even materials normally considered to be flexible or weak can be strengthened by making them thicker or by making them in some structural form, such as an I-beam. Hence, the application of a constraint and the consideration of mechanical or other properties should be done iteratively.

Ranking the resins is done by considering the most important property and then giving a numerical rank to the resins in the list based on that property. (The **highest rank number** should be given to the resin with the **highest performance.** In other words, if four resins are being ranked, the highest-ranked resin would be a "4.") As an example of how this might be done, suppose that stiffness is the most important property. A database of plastic properties would be consulted for all of the resins on the short list and the stiffest resin would receive the highest ranking on that property, with others ranked in descending order based on their stiffnesses. A multiplier, or *weighting factor (WF)*, is then assigned to the property to reflect the importance of this property relative to the other properties. For instance, stiffness might be given a weighting factor of 10, which would be the highest weighting factor because stiffness is the most important property. All of the stiffness ranks given to all potential resins would then be multiplied by this weighting factor to obtain a stiffness score for each resin for the stiffness property.

All the candidate resins would then be ranked according to the next most important property, which might be solvent resistance. These ranks would be multiplied by a weighting factor (7, perhaps) that represents the importance of solvent resistance to the final performance of the part. Scores for solvent resistance would thus equal the ranking multiplied by 7.

This procedure would be followed for all of the properties having significance. When the resins have been ranked and scored for all significant properties, the scores are totaled for each resin. The resin with the highest total score would then be the first choice. This process is illustrated in Table 10.4.

Considerable ambiguity is likely at this stage of the material-selection process, especially as the various specific grades of resin are considered. Still, by considering the ranking that was obtained in the previous steps and the general capability of each resin and grade in meeting all of the performance criteria, an overall ranking of resins can be achieved. The highest-scoring resin is, then, the first choice for making the prototype or working model.

Table 10.4 Methodology for Ranking and Scoring Resins According to the Importance of the Properties

	Significant Properties						Score
	Stiffness		Solvent Resistance		Low Cost		
Resin	Rank	Score (WF=10)	Rank	Score (WF=7)	Rank	Score (WF=3)	Total
PE	1	10	4	28	4	12	50
PP	2	20	3	21	3	9	50
PS	4	40	1	7	2	6	53
Nylon	3	30	2	14	1	3	47

Computer-aided Selection

An alternate to the selection methodology suggested above is a computer-aided selection process. These semiautomated systems usually employ some formalized decision-making process that asks the designer to answer a series of questions and thereby narrow the selection process down to a few material candidates.

All these computer-aided selection systems require that the preliminary steps in the design process be completed. The functional specification must be prepared, a sketch of the part made, the constraints identified, the processing rules established, mechanical considerations made with respect to the part size and shape, and cost considerations applied. These functional specifications and constraints are input to the computer by establishing the primary and secondary functions. For instance, the primary function of a case for a hand calculator might be to "enclose components." Secondary functions might be to "resist impact" or to "mask scratches." Using this approach, the designer is required only to specify the functions to be performed and the conditions under which a part is expected to operate. In satisfying the functional requirements of the part, the designer may modify the basic geometry or some special environmental need that will modify the material choice.

All the standard choices are already embedded in the software so that the input for a particular part is simply a process of answering questions. The answers lead through a tree structure that eventually ends up with a complete or nearly complete function description of the part within the computer memory.

The material selection is done by the computer based on the major and minor functional specifications. The computer database has properties for all of the major plastic materials. As the functional specifications are input, the computer eliminates the inappropriate materials from candidacy. This elimination process sometimes requires decisions in addition to the functional specification. For instance, the computer may ask what manufacturing process is to be used. Eventually, the computer narrows down the choices to a single material.

PROTOTYPING

Traditional

The traditional method of making a prototype is by machining, joining, simple molding, or some other manual-intensive method. These methods are almost always slow and are not used for large production quantities, but are acceptable to make a few items for prove-out. Ideally these prototypes are made from the same material that will be used for the final product, but that is not always possible. When it is not possible, the major purpose of the prototype is to verify the part dimensional characteristics. When the prototype can be made from the same material as the final product, a full range of tests can be carried out to verify all the part-design parameters. The standard against which the prototype should be compared is the functional specification.

The process design must be compatible with the part design for the prototype to have any real meaning. If the part design does not satisfy all of the criteria or design rules for the manufacturing process that is envisioned, neither the part-design dimensions nor the performance characteristics can be verified.

Rapid Prototyping

Several systems have been developed to make prototypes much more quickly than could be done using traditional prototyping methods. These systems use the CAD drawing of the part and then, by various methods, manufacture a part that duplicates the CAD drawing. The prototype may be of the same material or may be of some different material, depending on the rapid prototyping system.

Rapid prototyping systems are very valuable in shortening the time required to create a prototype, sometimes reducing the time by a factor of 10. Such time saving allows several different prototypes to be compared and evaluated, thus improving the efficiency of the entire product-realization process.

The value of rapid prototyping has spurred the development of several competing systems. The most important of these systems are stereolithography, laminated object manufacturing, selective laser sintering, fused deposition modeling, and ballistic particle manufacturing.

Stereolithography.   The term *stereolithography* means "three-dimensional printing." In this process the three-dimensional CAD representation of the product is loaded onto the computer that controls the stereolithography machine. This computer has a program that divides the CAD representation into many thin cross-sectional slices (about 0.004 inch to 0.02 inch or 0.1 mm to 0.5 mm thick), thus creating a sectioned CAD representation. Each slice becomes a path for a laser to trace on the top of a table that is located inside a vat of *photopolymer*, a polymer that polymerizes when a light is shined on it. The first slice is polymerized on the thin film of photopolymer that covers the platform. After this slice has polymerized (it takes only a few seconds), the platform is lowered the thickness of a slice. This lowering allows a thin film of photopolymer to cover the polymerized slice already on the platform. The laser then follows the path of the second slice as dictated by the sectioned CAD drawing, thus causing the resin covering the platform to be polymerized in the shape of the second slice. The process is repeated for all of the slices of the sectioned CAD drawing. The finished prototype rests on the platform and is surrounded by the photopolymer. The prototype is extracted simply by raising the platform and allowing the photopolymer to run off.

Because the prototype is made by building up many slices of cured polymer, the prototype generally is not strong. However, considerable strength can be achieved by postcuring it under a UV light. Because the prototype is not constructed out of the same material as the finished part, only dimensional and conceptual proveout are possible. The prototype can, however, be used as a model for shell casting so that a cast metal or cast plastic part could be produced in a production mode. The stereolithography system is depicted in Figure 10.3.

Laminated Object Manufacturing (LOM).   The LOM process also uses a CAD representation of the part that is sliced to create a multilayered object. In the LOM process a

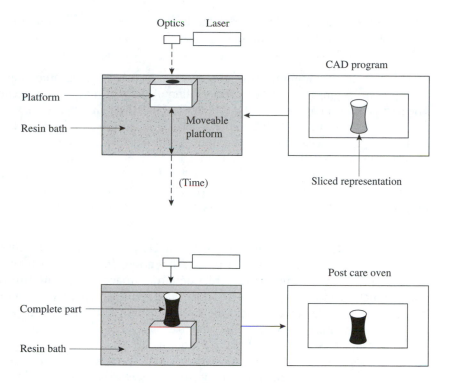

Figure 10.3   Diagram of the stereolithography system.

sheet of paper is placed on the top of a platform and the slice pattern is cut into the paper by a laser. Areas of the paper that are to be removed are cut with cross-hatching so that they will easily separate from the final prototype. A second layer of adhesive-coated paper is then placed on top of the first and bonded to it. The laser then cuts the next slice pattern into the paper. This system of bonding a layer of paper to those already on the platform and cutting subsequent slices in each layer continues until the entire structure has been created. The prototype is then removed from the platform and the cross-hatched areas are removed, leaving just the prototype, which is made of laminated paper. As with stereolithography, the prototype is not made from the same material as the final part so it can be used only for concept, shape, and size considerations.

Selective Laser Sintering (SLS).   The sliced CAD representation method used in other processes also is used in SLS. In SLS a laser is used to fuse a thermoplastic powder that is spread across the top of the platform. After the fusion of each layer, the platform is lowered and a new thin layer of thermoplastic powder is spread across the top. The laser then fuses the powder in a new pattern and, by fusing slightly deeper than one layer, fuses the top layer to the one below it. Thus, if the final part is to be made of thermoplastic, the prototype can be of the same material. Mechanical properties are not quite comparable to the molded part, but are becoming very close.

Fused Deposition Modeling (FDM).   As with the other rapid prototyping methods, the FDM process begins with a CAD representation of the part that is cross-sectionally sliced. The part is built up by depositing a thin layer of hot thermoplastic resin, which is applied by using an application head. The thermoplastic is fed into the head as a ribbon that is heated inside the head and then laid onto the surface of the platform to form each slice pattern. The platform is then lowered and, after the previous slice has cooled and solidified, the next slice pattern is laid down. The FDM process is capable of making a prototype out of the same material as the final part, although some postforming consolidation (by heating) may be necessary to obtain properties that are comparable to the molded part.

Ballistic Particle Manufacturing (BPM).   The BPM process also relies on a sliced CAD representation as the point of process initiation. In BPM each layer is applied by spraying a very fine thermoplastic powder onto the surface of the platform. This is a technology much like that used in ink-jet systems for printing. Each layer is sprayed onto the previous layer with a binder to hold the layers together. Although highly porous, the prototype can sometimes be made of the same material as the final part. Sintering is required to obtain properties similar to those of a molded part.

All the rapid prototyping methods can be used to make models for soft molds. Many can also be used in investment casting in the same way that wax is traditionally used. In fact, some of the processes allow wax to be used as the prototype material so that traditional investment casting methods can be used.

===================================== CASE STUDY 10.1 =====================================

Design of Plastic Stakes for Concrete Tilt-up Walls

An inventor's idea can sometimes be an excellent example of the design process. Such an idea is a plastic stake used to hold a slab of polystyrene foam onto the surface of a concrete wall. These stakes are placed through the polystyrene foam slab and the slab is laid on top of the uncured concrete. The concrete cures and secures the stakes, which also secures the polystyrene to the concrete. After the concrete is completely cured, the stakes are used to tilt the concrete slab upright so that it forms a wall. The foam improves the thermal insulation of the wall. The wall is connected to other walls to form a structure. When the walls are firmly in place, a foamed concrete is sprayed over the top of the stakes to give a finished, decorative surface (see Figure 10.4).

Need Recognition

As noted previously in this chapter, the first step in the design process is to identify the need. The need for the stake occurs because existing stakes used to hold the insulation to

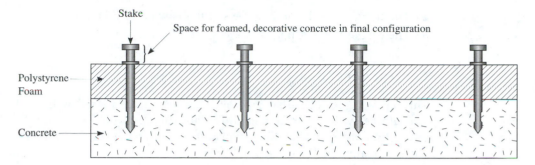

Figure 10.4   The use of plastic stakes for holding a polystyrene slab against a concrete wall.

the concrete have heretofore been made out of metal. The metal stakes perform well mechanically, but are a serious problem with thermal insulation. The metal stakes conduct heat readily, thus reducing the value of the polystyrene foam slab as an insulator. To make matters worse, when the temperature of the outside concrete is very low, the metal stakes, which are also at this low temperature, can cause condensation to form on the inside wall at the stake heads. The inside wall is therefore spotted. The metal stakes are relatively high priced, thus giving the plastic stake a market opening.

Functional Specification

The functional design of the stakes is as follows:

- The stakes must be long enough to go through a slab of 2-inch (5-cm) polystyrene foam and protrude into the uncured concrete just less than 2 inches (5 cm). The length of the stake from the top of the foam to the tip of the stake cannot exceed 4 inches (10 cm). The stake should allow for 1.5 inches (4 cm) of foamed decorative concrete on top of the foam slab. Overall length is, therefore, about 5.5 inches (14 cm).

- The stakes can be of any diameter but a head with a 0.75-inch (2-cm) diameter is common in the industry. The smaller diameter of the stake should be approximately 0.5-inch (1.3-cm) diameter. These sizes ensure that equipment used with existing metal stakes when lifting the wall to the vertical would be compatible.

- The stakes must be securely bonded into the concrete. The pullout strength needs to exceed the tensile strength of the stake. The weight supported needs to be at least 1100 pounds (500 kg). This gives a 10× safety factor on the typical load.

- The stake should not easily deform. Therefore, the tensile modulus must exceed 300,000 psi (2.0 MPa).

- The stake should not be easily knocked off if a lifting device might happen to hit it. Therefore, the stake must have an Izod impact strength of 10 ft-lb/in. (21 J/m) or more.

- The material must not corrode in concrete.

- The thermal insulation must be substantially lower than for steel.
- The cost of the part must not exceed $0.25 each.

Concept Generation

The key requirement is that the stake have a high tensile strength and that the pullout strength be even higher. To meet this requirement, the stake would probably have some change in cross section in the region of the stake that would be within the concrete. This change in cross section would allow the concrete to form around the stake and then, when cured, give an interference to pull out. The requirements of noncorrosion and good thermal insulation can be satisfied by many resins. However, only the engineering resins are likely to be able to meet the strength, stiffness, and toughness requirements. Even with some of the engineering resins, fiberglass reinforcement may be necessary. The high volume of production required suggests that injection molding be the method of manufacture.

Part Design and Process Design

These two steps were conducted simultaneously. The general design of the part, drawn with a CAD machine, is shown in Figure 10.5a.

The input from the process design suggested that the part may have a faster molding cycle if the thickness of the part could be reduced. This reduction was accomplished by

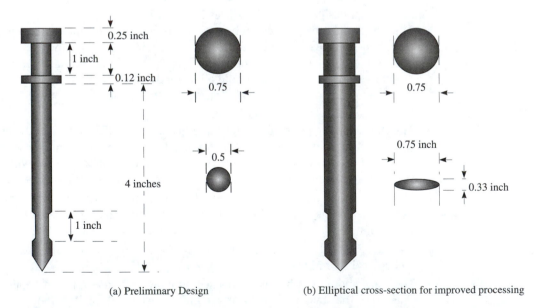

(a) Preliminary Design   (b) Elliptical cross-section for improved processing

Figure 10.5 Design of a plastic stake for binding foam to a concrete wall.

changing the shape of the part from a circular cross section to an elliptical cross section. The overall cross-sectional area was to remain the same and the maximum diameter was not to exceed 0.75 inches (2 cm). Therefore, the minor radius was calculated to be 0.33 inches (0.8 cm). The elliptical design, shown in Figure 10.5b, should also reduce the tendency of the material to form voids or sinks.

The material choices were narrowed to polycarbonate and 30% glass-filled nylon, which both met the major criteria. The polycarbonate was found to have a better Izod impact strength and was therefore chosen as the material. Table 10.5 shows how the two materials were ranked and scored.

Prototype Creation

The CAD drawing was input directly to a numerical-control milling machine using a computer-aided manufacturing program. This program allowed the intended tool path to be traced so that interference with the machine or with the fixtures could be examined prior to actual cutting. When that was verified, a cavity for an injection mold was cut. The cavity was made of aluminum. The mold was assembled and placed into a small injection molding machine. Several parts were molded and tested to verify that the specifications had been met. They were successful.

Production Implementation

The part was shown to potential customers, who were pleased and placed orders for purchase. The small one-cavity mold allowed these initial orders to be filled. Field tests on the parts confirmed that they were acceptable.

Orders began for large quantities, so a multicavity production-quality mold was made to meet the demand. The part is now in production and all specifications, including price, have been met.

Table 10.5   Table of Materials Ranks and Scores for Stake

| | Properties | | | | | | | | Score |
| | Tensile Strength | | Izod Impact | | Flexural Strength | | Termal Conductivity | | |
Resin	Rank	Score (WF=10)	Rank	Score (WF=8)	Rank	Score (WF=5)	Rank	Score (WF=5)	Total
PC	1 (9400)	10	2 (15)	16	1 (350)	5	2 (4.7)	10	41
Nylon	2 (25000)	20	1 (4)	0*	2 (1200)	10	1 (5.1)	5	35

*Doesn't meet specification

SUMMARY

A design process conducted without any particular plan or scheme can be successful, but it can also lead to many design changes that might have been avoided if a more formalized design process were followed. This chapter outlines a process in which each step logically follows the former and in which steps that should be conducted simultaneously are clearly identified.

The realization of a need usually is the first step in the overall process of product realization. Then the need is formalized in a functional (performance) specification. This step is very important so that the "target" at which the designer can "shoot" is well-defined. Either the customer or marketing personnel, who may represent the customer, should be involved in the writing of the functional specification.

After defining what the part must do, the design team then discusses how the functional specification can be met. The purpose of this brainstorming session is to identify many possible solutions. Creativity is the key at this stage of the design process. After sufficient ideas on how to accomplish the function have been given, these ideas are then screened for reasonableness and feasibility.

The formal design steps are then begun. Two designs steps are carried out simultaneously: product design and process design. These must be done concurrently because input from one affects the other.

The product- or part-design phase includes a rendering of the part, usually done today using a CAD system. The design also includes an identification of the constraints under which the design must be done. These constraints typically arise from the functional specification, but can also be dictated by process-design rules, safety codes, and the environment in which the part must operate. The choice of material is also made during the part-design phase. Material choice is a function of the needs that must be filled by the part and the specifications that are developed to fill those needs. A system is introduced for rating the most important performance parameters and then arriving at a suggestion for a material that is the best overall choice. Analyses may also be done to insure that the specification and the material are appropriate for the need.

After the product and process have both been defined, a prototype of the part is made. This can be made by traditional machining, simple molding, or by one of the new rapid prototyping methods. The prototype is verified to ensure that the specifications are met.

When the prototype is certified, the actual production system is assembled and production is begun. This step relies on the proper design of the process and should be checked for the normal process parameters of quality, efficiency, and overall productivity.

GLOSSARY

Concept generation and evaluation   The step in the part-creation process that discusses how to meet the need that has been identified.

Concurrent engineering   Designing both the part and the process at the same time, through an iterative process; also called simultaneous engineering.

Constraint   Restrictions on the design of a part or process that are dictated by the use of the part or process.

Design rules   A set of suggestions or rules for part design that depend upon the manufacturing process chosen for making the part.

Ergonomic   Human factor considerations.

Finite element analysis   A method, usually employing computers, in which a complex part is divided into very small sections that can individually be analyzed under simulated stresses so that, in the sum, the entire part is analyzed.

Focus group   A small group of potential customers who meet to evaluate a product.

Functional specification   A statement of what must be done (functions to be satisfied) to meet the requirements for a product to satisfy a particular need.

Photopolymer   A polymer that cures by the action of light (usually UV light).

Simultaneous engineering   Designing both the part and the process at the same time, through an iterative process; also called concurrent engineering.

Stereolithography   A rapid prototyping system in which a photocurable polymer is laid onto a surface and then cured by a laser.

Weighting factor   A number assigned to a property based upon the importance of that property in the performance of a part, used in the design process for choosing a material.

QUESTIONS

1. Define simultaneous engineering, and point out a problem that might arise if it is not done.
2. Performance tests of valve lifters in an automobile engine showed distortion when the material used was 20% glass-reinforced nylon (6/6). Explain why a shift to 20% glass-reinforced nylon (4/6) was suggested.
3. Indicate why design rules must be done after a preliminary choice is made on both process and material.
4. Give an example of why the functional specification must also include some idea of the production volume.
5. Discuss the advantages of rapid prototyping methods versus conventional prototyping.

REFERENCES

Allen, Dell K., and Allen L. Findley, "Computer-aided Plastics Selection," *Design Engineering Conference of American Society of Mechanical Engineers,* Chicago, IL, Mar 28–31, 1983.

MacDermott, Charles P., *Selecting Thermoplastics for Engineering Applications,* New York: Marcel Dekker, Inc., 1984.

Maniscalco, Michelle A., "Automotive Plastics Impact Design," *Machine Design,* Aug 8, 1994, 65–72.

Modern Plastics Encyclopedia, Hightstown, NJ: McGraw Hill, Inc., Yearly Publication.

Prospector: Plastics Materials Database, version 2.06a, Lanamore, NY: Integrated Engineering Systems (IDES).

Seymour, Raymond B., and Charles E. Carraher, *Structure-Property Relationships in Polymers,* New York: Plenum Publishing Corporation, 1984.

Todd, Robert H., Carl D. Sorensen, and Spencer P. Magleby, "Designing a Senior Capstone Course to Satisfy Industrial Customers," *Journal of Engineering Education,* Vol. 82, No. 2, April 1993, 92–100.

"Valox® Design Guide," GE Plastics, VAL-50E(6/89)RTB.

CHAPTER ELEVEN

EXTRUSION PROCESS

CHAPTER OVERVIEW

This chapter examines the following concepts:

- Purpose, advantages, disadvantages, and cost elements
- Equipment (extruder, die, cooling, puller, removal, special equipment, plant concepts such as layout and controllers, and capacity)
- Normal operation and control of the process (including startup, part dimensional control, critical operational parameters and techniques, and maintenance and safety)
- Extrusion problems and troubleshooting
- Material and product considerations (including material differences, general shape and design considerations, and discussions of extrusions of profiles, pipes, coating of wire and cable, sheets, blown films, coating of paper and fabric, calendering, and synthetic fibers)
- Postextrusion forming
- Coextrusion

INTRODUCTION

The word *extrusion* comes from Greek roots that mean to "push out." This phrase correctly describes the extrusion process, which is, in essence, a pump that supplies a continuous stream of material to a shaping tool or to some other subsequent shaping process. Plastic materials can be extruded, and are the major subject material for this chapter. However, metals can also be extruded, such as aluminum window frames. In the broadest sense, squeezing a tube of toothpaste is an extrusion process. A caulking gun is a familiar form of extrusion, as are grease guns and cake decorating tools. The forming of spaghetti or noodles by mechanically pushing the dough through holes of the proper shape is another form of extrusion. Hamburger is made using a grinder in which the chunks of meat are placed into the top and then fall onto a screw auger that both chops the meat and pushes it forward until it exits the machine as a continuous mass. A hot-melt glue gun is especially illustrative of extrusion because it also involves heating the glue, a feature also contained in the extrusion of plastic materials. Many of the elements of plastics extrusion are similar to these other common extrusion processes.

In addition to the shaping of parts by the extrusion process, extrusion is the most efficient and widely used process for melting plastic resin as part of the process of adding or mixing fillers, colorants, and other additives into the molten plastic. Extrusion can be used to shape the part directly after this mixing or an extruder can be used as the melting device that is coupled with other shaping processes. When used for direct shaping, a shaping device or tool is placed directly on the end of the extrusion machine (extruder) and the process is called *extrusion molding* or, more simply, *extrusion*. The making of pasta would be such an example as would the production of plastic drinking straws. If an extruder is used to prepare the material that will be shaped in some secondary process, such as when the hamburger from the meat grinder is conveyed directly to a press that forms it into patties, then the extruder machine is considered to be a captive and integral part of the subsequent shaping process and the term extruder is not generally separately applied. Examples of the use of extruders as integral parts of other plastic-forming operations would include injection molding, blow molding, and some foam making, all of which are discussed in later chapters.

A schematic view of the extrusion process is given in Figure 11.1. In normal plastics extrusion, plastics granules or pellets and any other materials to be mixed with them are fed into a hopper attached to the extrusion machine. From the hopper the material falls through a hole in the top of the extruder (feed throat) onto the extrusion screw. This screw, which turns inside the extruder barrel, conveys the plastic forward into a heated region of the barrel where the combination of external heating and heating from friction melts the plastic. The screw moves the molten plastic forward until it exits through a hole in the end of the extruder barrel to which a tool (die) has been attached. The die imparts a shape to the molten plastic stream which is immediately cooled (in a water tank) to solidify the plastic, thus retaining the shape (with only minor changes) created by the die.

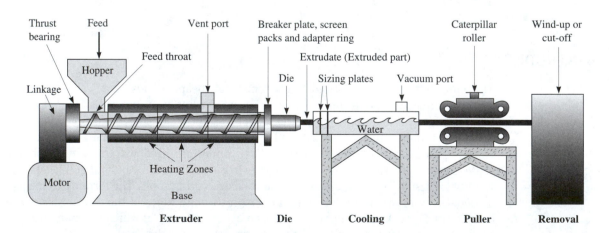

Figure 11.1   Typical extrusion line showing major equipment.

The output of the extruder is, therefore, a continuous part *(extrudate)*. Auxiliary equipment is used to pull the part away from the extruder at the proper rate. Other auxiliary equipment cuts the part to the proper length and packages it for shipment, perhaps by coiling or by stacking.

During the extrusion process, volatiles such as solvents, water, or trapped air can easily be removed. Therefore, extrusion is used extensively by resin manufacturers to eliminate volatile contaminants that might be residues from the polymerization process and to add materials such as antioxidants and processing aids that the manufacturer wants to include in the resin before it is shipped to the part manufacturer or molder. The volatiles can be captured at the vent in the extruder for proper recovery. The normal form of the product from these extrusion operations is thin rods that are chopped into the pellets that are so familiar in the plastics industry. Hence, even when the pellets are to be used in some other shaping process, the pelletization step itself is an extrusion operation.

In addition to the long, thin rods that are chopped into pellets, other common extruded shapes include: pipe, sheets, fibers, coatings on wire and cable, coatings on paper and other products, and various parts of many different shapes which are collectively called profile extrusions. These parts can be small (such as medical catheters) or very large (such as industrial pipes). Photo 11.1 shows some typical extruded products.

Thermoplastics are the most common plastics that are extruded. However, some thermosets (especially rubber) can also be extruded, provided the temperature of the extruder

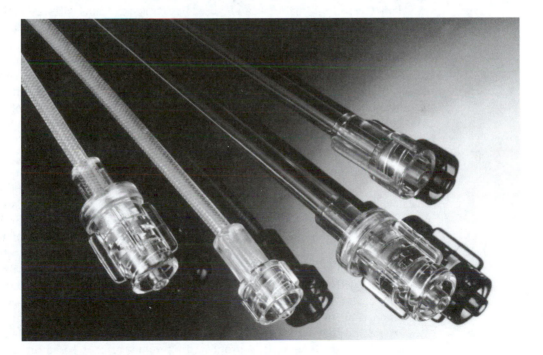

Photo 11.1   Extruded medical tubing with injection molded end-fittings. (Courtesy: Merit Medical Co.)

is kept below the temperature that initiates the cure. Extruders that are frequently used for thermosets will have easily removable ends so that cleanout can be done simply to avoid buildup of crosslinked material inside the extruder.

Detailed comparisons between the various processes used for forming plastic parts are difficult because of the different types of products and uses that are associated with each process. However, some broad comparisons can be made that point out basic advantages and disadvantages of extrusion versus the other processes, especially in determining when extrusion can be especially valuable as a production method and for understanding its basic nature.

Extrusion's advantageous continuous nature leads to high production volumes, thus making it the least expensive forming method. A limitation of extrusion is the limited complexity of parts that can be made. Extrusion dies cannot be as complex as injection molding molds, for instance. A further limitation on part shape is that parts must have a uniform cross-sectional shape over the length of the part. This constant cross-section limitation is imposed because the die has no moving parts and cannot, therefore, create parts with any changing cross section. (Some postextrusion forming can be done that allows minor changes in cross-sectional shape, but this postextrusion forming is less common.)

Another advantage of extrusion is the efficient manner in which the plastic is melted because extrusion uses both external heat and internal frictional heat. These two heating methods can be used together to increase production rate or, if desired, can be used separately to optimize heat profiles for particular resins or processing conditions.

Many types of raw material can be used in extrusion. Resins with residual solvents and other volatiles can be used. Other typical raw materials include resin pellets, resin granules, resin flake, resin regrind, which is usually chopped to a flake form, fillers that are often powders, and small amounts of liquids when added with dry materials. These materials can be mixed together and extruded even though they often have widely differing densities and shapes. The result is a well-mixed, uniform pellet or part. These mixing capabilities have given rise to the use of extruders as the preferred method of mixing resins with colorants, fillers, processing aids, and with other resins to create blends. This mixing of materials with resins is called *compounding*. Companies that specialize in color matching and other compounding tasks are called *compounders*.

The costs of extrusion depend upon several key elements. These elements include setup and die change time, heating time, extrusion time and rate, cooling time, overhead costs, which include amortization of the equipment and dies, and the standard costs of any product such as raw material costs and plant overheads. When parts of similar shape are made at high volume, such as with PVC pipe, the major cost is the raw material (as much as 60%) with only minor additional costs for the actual extrusion process. In cases such as this, the product is often sold on a weight basis, which implies that extrusion is a standard process with little special expertise from one extrusion company to another. In some cases the customer for the extruded part will design the part and simply contract an extrusion operator to convert the raw resin into the part. These types of extruders are called *converters,* and the process of extruding for someone else's product is called *tolling*.

In other products, especially profiles, the product price is based on each part and a large premium is usually charged. This premium price reflects the difficulty and costs associated with special shapes and shorter production runs. It can also reflect the costs of

Table 11.1 Comparison of Extrusion to Other Plastics Molding Processes

Advantages	Disadvantages
Continuous	Limited complexity of parts
High production volumes	Uniform cross-sectional shapes only
Low cost per pound	
Efficient melting	
Many types of raw materials	
Good mixing (compounding)	

designing the part if the extrusion company is also the part designer. The advantages and disadvantages of extrusion are summarized in Table 11.1.

EQUIPMENT

A typical extruder and the auxiliary equipment usually used in an extrusion line are shown in Figure 11.1. The five principal parts of the extrusion line are: extruder, die or tool, cooling, puller, and removal. Each of these parts is discussed here.

Extruder

The extruder sits on a base so that the extruder and its output part are at working height, about 3 feet (1 m) high. The base is massive to insure that the extruder does not move relative to the downstream equipment and to minimize vibrations. The base also provides a convenient anchor for the drive motor.

The **drive motor** turns the screw and therefore provides the power for the operation of the extruder to push out the plastic material. The drive must be capable of variable speeds and DC motors are the most common drive motor although some adjustable-frequency AC motors are also used. DC motors cost less initially but have lower operating efficiencies and have more speed drift because of mechanical resistances within the extruder. The speed of DC motors can be conveniently changed simply by changing the input voltage. Most other features of the DC and AC motors are similar in performance. A key specification for an extruder motor is the power capability (horsepower or kilowatts). The required extruder power is increased (1) as output increases, (2) as the barrel diameter increases, (3) as the screw length increases, and (4) if high outputs are required at high temperatures. Power requirement is also a function of resin type and mold design. With all of these performance factors affecting power requirements, an analysis by an expert in extrusion machine construction and an understanding of the probable uses of the extruder are important to a cost-effective purchase.

A key monitoring parameter of extruder operation is the motor current usage, which is measured with an ammeter. High amps indicate that the motor is working hard to melt and pump the resin. Extruder operation can be monitored for proper heating of the material and proper screw speed by noting the amperage and adjusting the heating and screw speed for optimum performance. A high amp value could also indicate that some foreign material (such as metal or wood) has entered the extruder, is wedged inside, and is partially restricting the material flow. If the current (amperage) is too high, the motor could be burned up; therefore, a high-amp shutdown is provided as a protection on most systems.

All the motors run much faster than typical screw rotation speeds so a gear-reduction linkage between the motor and the screw is required to reduce the motor speed (1750–2000 rpm) to a workable screw rotation speed (15–200 rpm). This reduction is accomplished with gears or with a combination of gears and pulleys. Small extruders are usually run at low revolutions because the barrels are so short that if run at high speeds the material would not have enough residence time to be fully melted. A typical speed for a 1-inch or 2.5-cm (1.0-inch inside-diameter barrel) extruder would be 50 rpm. Very large extruders would also be run at low speeds because of so much frictional heating potential that the resin will exceed its decomposition temperature, at least in some local areas, and will scorch (referred to as *extruder burn*). Therefore, a 12-inch (30-cm) machine (that is, a barrel with a 12-inch inside diameter) would also be run at about 50 rpm. The maximum speed would be for 3- to 5-inch (7.5- to 12.5-cm) extruders and would typically be in the 100- to 200-rpm range.

A large **thrust bearing** is mounted on the screw near where the screw attaches to the linkage arrangement. This thrust bearing prevents the screw from moving backwards and absorbs the thrust of the screw as it turns against the resistance of the resin. The thrust bearing will wear, especially at high back pressures that occur when the screw turns at high rates and high throughputs and when very viscous materials are processed. The thrust bearing should be part of a routine maintenance program in which it is inspected and lubricated regularly.

The *barrel* of the extruder provides the rigid surface against which the thrust bearing seals and is the chamber in which the screw turns and the resin flows. The barrel would typically be made of hardened steel that is lined with wear-resistant and corrosion-resistant metal. The inside diameter of the barrel is an important parameter in specifying the size and capacity of the extruder. Production machines would typically be (in English units) 2, 2.5, 3, 3.5, 4.5, 6, 8, 10, and 12 inches, or (in metric) 45, 50, 60, 75, 80, 100, 130, and 165 mm. The metric machines may not have any direct correspondence to English-unit machines, because each set of machines was developed independently. For instance, 100-mm extruders are common in metric but the nearest English size, 4 inches, is a size that is uncommon among English-unit machines. Small extruders for laboratory use or for small-output applications would be commonly made in English units at sizes of 0.75-, 1-, and 1.5-inch diameters.

The outside of the barrel is jacketed with electrical heating elements to aid in the melting of the resin. These heating elements are normally divided into independently controlled zones so that the heating profile along the extruder can be varied to optimize the melting and viscosity control of the resin. Thermocouples are placed on the outside of the barrel wall to monitor the temperature in each zone and provide feedback for automated temperature control.

An opening in the top of the barrel just beyond the thrust bearing is called the *feed throat* and is the inlet for the resin. Often the resin feeds into the extruder and onto the screw by gravity from a standard hopper that is mounted over the feed throat. The hopper can be filled manually or by an automatic or semiautomatic pneumatic feeding system. Hoppers can be equipped with augers, weighing devices and other measuring equipment to give constant feed by weight or volume of the entire batch or of various components that are separately metered into the feed throat. Hoppers can be fitted with mechanisms that dry and feed *hygroscopic* (moisture-absorbing) materials.

The **extruder screw** is attached to the drive linkage through the thrust bearing and rotates inside the barrel. The screw is machined out of a solid rod and, when finished, is like a shaft on which a helical screw has been built. Each turn of the helix is called a *flight* and is like the thread of a screw, especially an Archimedean screw that is used for conveying materials. The clearance between the tips of the flights (that is, the outside diameter of the screw) and the barrel wall is approximately 0.004 inches (0.10 mm) and is constant over the length of the screw. This clearance increases as the screw wears. After considerable wear has occurred, the screw can be rebuilt but this should not be done until problems, such as severe loss of output, are clearly identified. Rebuilding a screw is expensive, as much as 75% of the cost of a new screw, and will not necessarily improve production. Screws are made of steel and chrome plated to resist corrosion. The tips of the flights can be coated with wear-resistant material.

The extruder screw has several important functions to perform such as (1) conveying the resin through the extruder, (2) imparting mechanical energy as part of the melting process, (3) mixing ingredients together, and (4) building pressure in the extruder so that the resin will be pushed through the die. Screw design is the subject of much study and design effort to maximize the performance of the screw in accomplishing these functions. One major development in screw design is the creation of twin-screw extruders in which two screws operate together to achieve unique performance characteristics. However, over 90% of extruders are of the single-screw type, which are described in this section. Twin-screw extruders and other special extruder modifications are discussed later.

An important parameter associated with the screw is the ratio of the length of the flighted portion of the screw to the inside diameter of the barrel or the outside diameter (tip-to-tip distance) of the screw flights (which is almost the same as the ID of the barrel). This ratio is called the *L/D (length/diameter)* of the screw. The L/D is a measure of the capability of the screw to mix materials and of the ability of the screw to melt certain hard-to-melt materials. It is also a measure of the amount of energy needed to run the extruder. High L/D ratios indicate good mixing and good melting capabilities but also high energy requirements. Typical L/D ratios are 16:1 to 32:1, with a tendency in newer machines toward higher ratios. Elastomeric materials such as thermoset rubbers would have smaller L/D ratios.

Some of the terminology and concepts of extruder screws are shown in detail in Figure 11.2, which is a portion or segment of a typical barrel and screw and will assist in the discussion of screw characteristics. The barrel diameter is constant over the entire length of the extruder. The *root* is the measure of the diameter of the shaft of the screw. The root diameter can vary along the length of the screw. The *flights* rise above the shaft creating a *flight depth,* which is the difference between the top of the flight and the root diameter. The outside diameter of the flights is constant along the entire length of the screw so that the clearance between the top of the flights and the inside of the barrel is constant. As the

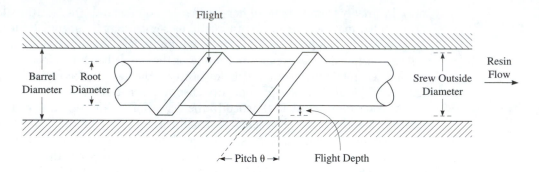

Figure 11.2   Details of a typical barrel and screw.

root diameter changes, the flight depths will correspondingly change. Therefore, if the root diameter is small, the flight depths are large and vice versa. The flights travel around the shaft at an angle identified as the *pitch*. The most common pitch angle is 17.5°, and it is usually constant over the entire length of the screw.

Diagrams of two types of single screws that have been designed to optimize extrusion performance of two different materials, polyethylene and nylon, are shown in Figure 11.3.

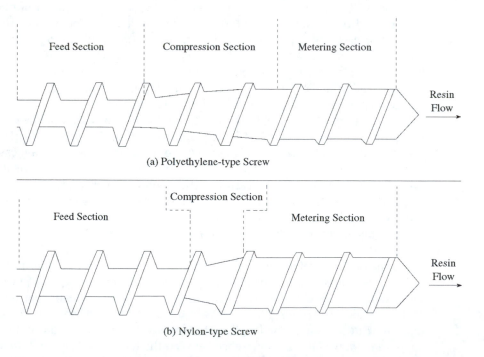

Figure 11.3   Screw designs showing differences in design for (a) polyethylene-type and (b) nylon-type screws.

Each of these screws has three major sections that are present in almost all standard extruder screws. These sections are: feed, compression, and metering. Each of these screw sections is discussed in detail with comparisons of the two screws, polyethylene-type and nylon-type, to give insight into the functions of each of the screw sections.

The *feed section* is located directly under the feed throat and is the portion of the screw where resin first enters the extruder, usually about the first three to ten turns of the flight. The purpose of the feed section is to convey the solid material away from the feed throat into the forward sections of the extruder where the material is melted. The feed section has a small, constant root diameter that results in large, constant-depth flights to accommodate the bulky dry solid resins and other additives that might be included in the material to be extruded. The deep flights also allow for material expansion as the resin is heated in the downstream portions of the section. Care should be taken that the material is not heated too quickly. If premature heating occurs, the resin could stick to the screw in the feed section and simply rotate with the screw rather than being moved forward. This condition is called *bridging* and it will stop the movement of material through the screw and prevent the proper operation of the extruder. To ensure that bridging does not occur and to improve the efficiency of the transfer of material down the screw, some screws are cooled by circulating water through a hole that has been drilled through the center (along the center line) of the shaft extending from the back of the screw through the feed section area. Screws in which this hole has been drilled are called *cored screws*. Bridging is worse with large-diameter screws, high screw speeds, and resins which soften at low temperatures.

The next section of the screw is called the *compression section,* the *melting section* or, less commonly, the *transition section,* and can be identified by the gradual increase in the diameter of the root along the length of the section. This root diameter increase means that the flight depth gradually decreases throughout the compression section, compressing the resin and forcing the air or other volatiles out of the resin melt. These volatiles usually escape by flowing backward through the gap between the screw and the barrel or through a vent port in the extruder barrel. The removal of these volatiles is important in making a pore-free (void-free) product.

The compression of the resin imparts shear on the resin which adds mechanical energy and heats the resin. This heating is sometimes called *adiabatic heating,* which implies that no external heat is added. With some resins, adiabatic heating is preferred over normal thermal heating because of the tendency of these resins to burn or degrade if overheated. Adiabatic heating is more efficient in disentangling the resins than is simple thermal heating because of the shear action imposed on the polymers. In most normal operations, however, *thermal heating* is contributed by the heating bands that surround the barrel. Most resins are best melted by a combination of adiabatic and thermal heating.

Compression of the resin can be increased by decreasing the pitch of the flights. This method is far less common than changing the flight depth but gives the screw designer another variable that might be useful in some situations.

A comparison of the polyethylene-type screw and the nylon-type screw in Figure 11.3a and b reveals the differences in the feed and compression sections in screws because of the differences in melting characteristics of resins. Polyethylene has a low melting point and a much broader melting range than does nylon. Nylon will melt much more sharply but at a higher temperature. Therefore, the feed section for polyethylene is short to ensure that the polyethylene does not prematurely melt and the compression section is long to allow for melting both low and high molecular weight material. Nylon, on the other hand, has a long feed zone that allows for the addition of much thermal energy to raise the resin temperature without melting and then a very short compression zone because when nylon is at the melt temperature, it melts quickly.

In many extrusion operations a single extruder can be used to make several different products. Changing from one product to another would often mean that the die would be changed and that the type of resin used would also change. However, the special screw designs that are illustrated in Figure 11.3 would probably not work well when several different resins are extruded. Since changing screws is a difficult task, it is not generally done when resins are changed except when very long runs with a particular resin are made. Instead, screw designs that are compromises of the optimal designs for each resin are used. The screw designs used in these multiuse extruders are called *general purpose screws*. The performance of these general purpose screws can be modified by changes in the operational settings of the extrusion system, such as temperatures, screw speed, and so on. Further discussion of these operational changes is given in a later section on extrusion control.

The section of a screw that follows the compression section is called the *metering section* and is characterized by a constant root diameter and very shallow flight depths. The resin should be completely melted by the time it reaches the metering section, but these shallow flight depths insure that high shear is added to the resin to accomplish any melting of residual solids. The high shear in the metering section also builds pressure on the melted resin so that it can be pushed out of the end of the extruder. The metering section also gives final mixing and ensures thermal uniformity. The metering section in the nylon-type screw is longer than in the polyethylene-type screw to ensure homogeneity of the material.

Another important extrusion parameter is the *compression ratio*. The compression ratio is the ratio of the flight depth in the feed section to the flight depth in the metering section. Values as low as 1.1:1 and as high as 5:1 are known with about 2.25:1 being typical. The compression ratio is a measure of the work that is expended on the resin.

The **head zone** portion of the extruder follows the end of the screw. The head zone and the die section that is attached to it are illustrated in Figure 11.4.

After leaving the end of the screw the plastic flows through the *screen pack* and then through the *breaker plate,* which are in the head zone. The screen pack is a collection of wire screens, usually of different meshes, that filters out any unmelted resin that may remain and any particular contaminants. The screen pack is backed by the breaker plate which is a disc of sturdy metal with many holes drilled through it and can be thought of as a very coarse screen. Screens of various sizes are often grouped together into a single pack. Screens with small openings provide filtration while screens with larger openings but stronger wires give support.

The screen pack will eventually become clogged with filtered material and must be changed. At this point the screen pack is said to be *blinded.* To accomplish the screen pack change, the extruder head can be unbolted and rotated open after the extruder is stopped, thus exposing the screen pack. Screen pack clogging is noted by an increase in the back

Figure 11.4 Head zone and typical die.

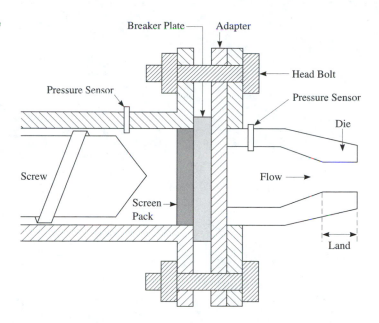

pressure in the extruder, which is usually sensed by measuring the pressure before and after the screen pack. If this pressure differential is high, the packs need to be changed. Screen pack changers are available that will change the screen pack without requiring that the extruder head be opened. One method works by sliding a new set of screens into the proper location when the pressure reaches a certain differential. The sudden pressure change when the clean screens are installed may cause some disruption in the material flow. Another method moves a strip of screen pack material continuously across the flow path. The continuous screen pack changers can be driven by the flowing action of the resin so no additional power source is needed.

By flowing through the screen pack and breaker plate, some additional homogenizing of the resin is accomplished. A natural twist orientation is imparted to the polymer molecules because of the rotation of the screw and this is reduced by the flow through the breaker plate and screen pack. A small pulsation in the flow of the resin is common because of small nonconcentricity between the screw and the barrel and other small rotational effects. The screen pack and breaker plate also help to smooth this pulsation.

Die

The die is the shaping tool that is mounted on the end of the extruder, usually onto a ring called the *adapter*. The purpose of the die is to give shape to the melt so that after leaving the die the melt can be cooled into the shape desired. The key die features can be seen in Figure 11.4, showing a die used for making a rod and in Figure 11.5, which is in more detail. In solid extruded parts, such as a rod, the inside of the die gives the desired shape.

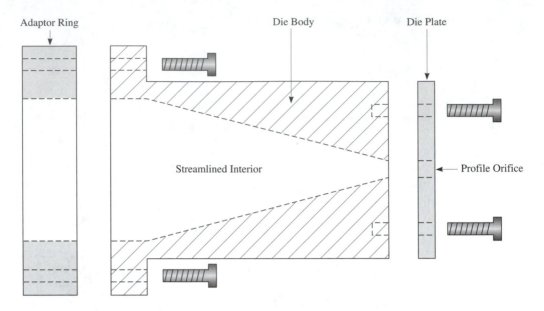

Figure 11.5 Extrusion die parts.

These are the simplest of the shapes commonly extruded. Some important shapes, including the die, other equipment, and any particular operating conditions unique to that shape, that will be discussed are pipes, profiles (nonregular solid and hollow shapes), wire and cable coating, paper and fabric coating, sheets, films, and fibers.

Die Parts. To a much greater extent than with molds (especially those for complex molding processes such as injection molding), dies used in extrusion are made by the plastic fabricator (in-house) rather than by a professional moldmaker. The ability and tendency to make the dies in-house comes from three factors: the relative simplicity of the extrusion die in comparison with a mold; the lower accuracy requirement of dies versus molds; and the availability of standard extrusion dies for making many standard parts that may need only minor modifications to make the specific part to be extruded. These modifications often can be made in the die face only, with no change required at all in the die body. These die parts are shown in Figure 11.5 for a simple profile part. The taper of the die body is somewhat difficult to machine, but the dimensions are not critical. The critical dimensions are in the profile orifice, and these are easily machined in the relatively thin plate that constitutes the die faceplate.

The shapes of dies are as varied as the shapes of the parts that are made. In all dies, however, some general considerations are constant.

The flow pattern inside the die is extremely important. The flow should be streamlined with no opportunities for stagnation, which could cause the material to sit for long periods and degrade. Streamlined flow is less likely to cause shear stresses in the melt which might cause unwanted deformations and defects in the final part.

The exit orifice of the die is usually proceeded by a short section where the walls are parallel. This section is called the *land* (see Figure 11.4). The purpose of the land is to build a uniform pressure in the melt so that the material is stabilized and flows evenly before exiting the die. The land is usually the zone of maximum pressure in the entire extruder and can have a strong influence on the back pressure on the screw. Therefore, if the land is too long, excessive back pressure could reduce extruder output and cause internal wear on the thrust bearing and other extruder parts. On the other hand, if the land is too short, the resin flow can be erratic and the part will be less uniform. Hence, some experimentation in determining the land distance is often required in building a new die. Experience in practice dictates making the land slightly longer than is thought necessary and then removing material from the face of the die, thereby shortening the land. More information on dies is given later in this chapter.

Die Materials. Most extrusion dies are made of stainless steel, although the use of tool steels such as P20 and H13 is not unusual. The dies and die faces can be chrome plated to give additional wear when the resin may decompose to form a corrosive by-product, such as PVC decomposing to give HCl. The dies can also be coated with tungsten nitride to give a harder surface that will withstand the abrasiveness of resins containing fiberglass or some other brittle reinforcement or filler. Some extrusions, especially those for making fibers, are done from a solution. If the solvent is corrosive, the die (spinnerette) can be coated or, perhaps, made from a highly corrosion-resistant metal such as Hastaloy C or Inconel.

When only a small number of a particular part will ever be made, uncoated carbon steel or even aluminum dies can be used, but these wear quickly. Tolerances cannot be maintained over a high volume of the part because of the wear, especially at high temperatures.

Cooling

Upon exiting the die, the extrudate must be cooled to retain its shape. This cooling is most often done by introducing the extrudate into a cooling bath in which water is circulated or is sprayed onto the part. A simple trough, tray, or box is commonly used for this purpose. The water can be room temperature or cooled, depending upon the needs for cooling the particular part. When the cooling requirements are not as demanding, the extrudate can be cured with air.

To assist in defining the shape of the part during the cooling process, plates or rings with holes of the proper size and shape can be placed inside the cooling bath so that the extrudate passes through them. These are called *sizing plates*.

With hollow parts, such as tubing or pipe, a vacuum can be applied to the water tank that creates a lower pressure in the tank space surrounding the part than exists inside the part. This pressure differential forces the part outward against the sizing plates and gives better control of the outside diameter of the part.

Puller

After the part has been cooled sufficiently that it will retain its shape under moderate tension and radial compression forces, the part enters a puller. The purpose of the puller is to draw the material away from the extruder and push the material into the removal equipment. If the material is not pulled from the extruder, the extrudate would flow down the face of the die and onto the floor. Therefore, pulling is required to move the molten extrudate away from the extruder and through the cooling system. The puller cannot be placed before the cooling system because the hot extrudate would not have sufficient strength to resist the pulling and squeezing forces required to move it.

Adjustable-speed belt or caterpillar pullers are commonly used. These machines have continuous belts that move away from the extruder and grip the extrudate and move it through the puller. The gap between the belts can be adjusted so that the gripping force will not crush or deform the part. Belt speed can be adjusted to match the rate of extrusion or can be slightly higher or lower than the extrusion speed as required for proper part-size control.

Removal

After the puller, the material must be removed from the line and prepared for shipment. The continuous nature of the extrusion operation suggests that automated equipment be used for this purpose. The exact nature of that equipment depends upon the nature of the product and much variation can be found, even within similar product types.

Rigid parts are often cut off into standard lengths. This type of operation is common in rigid PVC pipe. The lengths can be changed from order to order but are often 20 feet, or 6 m. The cutoff device, a saw in most cases, has a controller that senses the length of pipe that has passed through between cuts. When the proper cutoff length is reached, the saw begins its cut. To ensure that the cut will be straight across the end and perpendicular to the long axis of the pipe, the saw moves with the pipe. When cutoff is complete, the pipe section is guided off the cutter table and into a stacking machine.

When the familiar bell end is to be placed on the pipe section, the cutoff pipe moves into a secondary operation in which the pipe end, approximately the final 5 inches (15 cm), slides into a small oven where it is heated to a temperature that will allow thermoplastic forming. When the pipe end is at the proper temperature, the pipe is removed from the oven and a short plug is pushed into the heated end. The outside diameter of this plug is the same as the desired inside diameter of the belled end. The plug is held in place for a few seconds so that the pipe will cool and retain its expanded shape. The plug is then removed and the pipe moves onto a stacking or some other machine that controls final packaging. The belling operation usually takes less time than the time required to extrude a section of pipe so only one belling device is needed. If the belling time is too long, then two devices are needed so that cutoff pipe will not back up on the line.

Flexible product is often rolled up onto reels or into coils. Again, automated equipment is used whenever possible. This equipment will also have a cutoff device, but the lengths of the packaged material are usually much longer than with rigid parts. With high extrusion rates, two reeling machines or stations may be required so that cutoff and final packaging of the reels can be done without interruption of the extrusion operation.

Special Equipment

Twin-screw extruders have already been mentioned as a type of extrusion equipment that is valuable with certain resins, especially those that are sensitive to heat such as PVC. The use of twin screws, which operate side by side in tandem, requires different linkages and barrels than would be used for single-screw operation. Therefore, an extruder cannot be converted from a single-screw to a twin-screw machine.

The screws in twin-screw extruders intermesh so that the relative motion of the flight of one screw inside the channel of the other acts as a paddle that pushes the material from screw to screw and from flight to flight as seen in Figure 11.6. This continuous and steady pumping pattern keeps the material moving forward in a positive pumping fashion.

Twin-screw extruders have a more positive pumping action than single-screw extruders and can therefore be used more effectively in high-output situations. This feature of twin-screw machines has led to their use in some applications in which volume is critically important even though the resin may not be particularly heat sensitive.

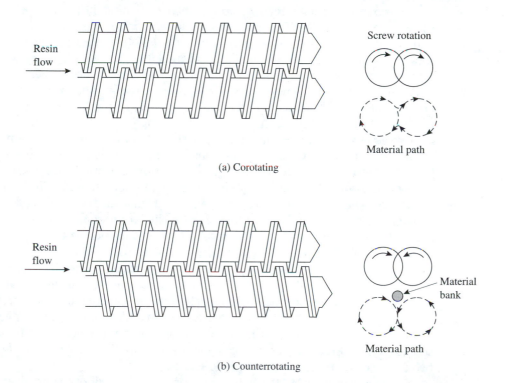

Figure 11.6 Twin-screw extruders showing the two screw arrangements: (a) corotating and (b) counterrotating.

Two different patterns for intermeshing twin-screw extruders are possible. In one pattern the screws rotate in the same direction, that is, both turn clockwise or both turn counterclockwise. This pattern is shown in Figure 11.6a and is called *corotating*. In the other intermeshing pattern the screws rotate counter to each other, that is, one rotates clockwise and the other rotates counterclockwise. This pattern is shown in Figure 11.6b and is called *counterrotating*.

In the corotating screws system the material is passed from one screw to another and follows a path over and under the screws. As shown in Figure 11.6a, the material moves alternately from the top of one screw to the top of the other, and then around the outside of second screw until it is on the bottom of the second screw, after which it then moves to the bottom of the first screw and then around the outside of that screw until it reaches the top again, but at a point farther along the screw length. This gives high contact with the extruder barrel, which improves the efficiency of the thermal heating. The path also ensures that most of the resin will be subjected to the same amount of shear as it passes between the screws and the barrel. The self-wiping nature of the corotating screws is much more complete than in the counterrotating system, thus in the corotating case there is less likelihood that material will become stagnant. Mixing is better in corotating systems than in either counterrotating or single-screw extruders. Corotating twin-screw extruders are, therefore, popular for compounding where good mixing is essential.

In a counterrotating screws system the material is brought to the junction of the two screws and builds up in what is called a *material bank* on the top of the junction. This buildup of material is conveyed along the length of the screw by the screw flights. As shown in Figure 11.6b, the material passes between the screws, high shear is created, but shear elsewhere is very low. Only a small amount of material passes between the screws. Therefore, total shear is lower than in single-screw extruders and in corotating twin-screw extruders. Most of the material in the bank along the junction of the screws is simply carried toward the end of the extruder. Hence, pumping (forward movement of the material) is more efficient in counterrotating screw systems than in corotating systems.

Another special extrusion system is one in which volatile materials are removed from the extruder. In some materials, such as those in which solvents are present or that are highly hygroscopic, venting beyond that occurring in normal extrusion (by flowing backwards over the flights) is necessary. This is accomplished by putting a venting port through the barrel of the extruder. To prevent melted resin from exiting through this port, the flight depth can be suddenly increased (the root diameter can be reduced) directly under the vent. This increase in flight depth reduces the amount of compression, which lowers the pressure in that region. The volatiles can then be extracted easily without loss of resin. The flight depth is decreased (root diameter increased) immediately after the venting section to restore the pumping action of the extruder and the pressure buildup necessary to push the plastic out the die. Screws that have this change in flight depth for venting are called *two-stage screws*. Two-stage screws must be longer than nonvented screws to give time and pumping to rebuild the pressure. Efficiency is, therefore, lower for two-stage screws than for normal single screws.

Additives can be added through the vent port in the wall of the extruder. A common additive inserted at this point is chopped fiberglass. This procedure minimizes breakage and wear on the fiberglass because it only passes through a short section of the extruder.

Other special screw designs include provisions for additional mixing. One type of special mixer has parallel interrupted flights with much different pitch angles. These interrupt the normal flow and cause the material to swirl as it passes through the section. Another device used to improve mixing is a fluted mixing section that forces the resin through a circuitous path as it moves along the screw. Pins can be added to the screw to break up the normal flow and induce turbulent mixing. Other barriers, such as rings and smaller flights at different angles, have also been used. Some of these mixing devices have special names such as Maddock section and barrier screw.

Another method of improving mixing is to place a mixing device at the end of the screw. Obviously, the screw must be considerably shorter than the barrel in order to accommodate these mixers which are typically about 8–12 inches (20–30 cm) long. Because these mixers do not rotate with the screw, they are called *static mixers*. They depend upon the flow of the melt through some tortuous path to accomplish their mixing.

In some operations, such as when the output of the extruder must be precisely metered into the die or into some other device that is placed between the extruder and the die, a gear pump is attached to the end of the extruder. Such a case might be for extrusion of very fine fibers or for extrusion of materials that are to be foamed out of a solvent and the solvent mixer must come between the extruder and the die. The gear pump is a positive displacement pump that precisely pumps the material at a steady output. This prevents surges and drifts in extruder output from heat cycles, pressure changes, resin changes, or other factors and gives close control over product thickness and output rate.

Gear pumps can also be useful in reducing the load on the extruder. The die is removed from direct contact with the extruder and this lowers the back pressure. The output of the extruder can then be increased. The temperature of the melt might also be lowered and the total work done by the extruder could be reduced. The lower pressure can also extend the life of the thrust bearing.

In lines where a gear pump is utilized, the gear pump becomes the master or controlling device and the extruder is simply the feed device to the pump's inlet. Therefore, the speed of the extruder can be controlled through a feedback loop governed by the pressure at the pump inlet.

When a gear pump is used in the extrusion line, some safety concerns should be met that are not necessarily required in standard extrusion operations. The major concern is that the gear pump forces its output almost without regard to downstream pressure. (This is called *positive pumping*.) Therefore, if the downstream line is plugged or the downstream flow is somehow stopped (say, at the die), the gear pump may continue to pump and therefore create enormous pressures upstream of the plug or stoppage. As a safety check, pressure relief valves are often installed so that when a set pressure is reached in the extruder and in the line after the gear pump, the valves open in a predictable way at a predictable point. A design for a pressure valve that is often used is simply a plug that has a thin metal cap over the end. This plug is screwed through the wall of the extruder at the points where pressure buildup is most likely. The thickness and strength of the cap are carefully controlled so that it will rupture at a precise pressure. Thus the danger of explosion from overpressurization is controlled by providing safety outlets.

Plant Concepts (Layout and Controllers)

Extrusion lines are long; therefore, adequate space must be provided to accommodate these lines. A typical extrusion line might be 45 feet (15 m) and would need additional space around the ends of the lines for product handling, screw removal, and product removal, all usually requiring forklifts. If the extrusion rate is high, the line may be even longer to provide for adequate cooling capability. The lines should be straight, with all the equipment in a line, so that no stresses are introduced into the product by trying to make it go around a curve.

If the hoppers are not fed automatically, a common practice is to place a mezzanine above the feed end of the extruders. This mezzanine provides a convenient platform on which to place the resin packages and facilitate adding the resin and other ingredients into the hopper. The difficulty with the use of the mezzanine is that the likelihood of foreign objects being dropped into the hopper is much greater than if feeding is done by climbing ladders up to the hopper. The best feeding method is pneumatic loading which is both more convenient and less likely to introduce contaminants. Pneumatic feeding is discussed in detail in the chapter on operations.

Note: The equations that allow extruder capacity to be calculated are presented in the following discussion (boxed text). This discussion can, however, be passed over by those not concerned with the model of extrusion output or capacity optimization.

Capacity   Note: Optional subject

The single most important parameter that determines extruder capacity is the size of the screw. A convenient formula for calculating the total flow of an extruder (that is, the total amount of material that passes through the extruder) is as follows:

$$\text{Total flow} = \text{drag flow} - \text{pressure flow} - \text{leakage flow} \qquad (11.1)$$

where the *drag flow* is a measure of the amount of material that is dragged through the extruder by the friction action of the barrel and the screw. The *pressure flow* is the flow that is caused by the back pressure inside the extruder. The pressure flow is counter to the drag flow and therefore has a negative sign in equation (11.1). The final component of the total flow is the *leakage flow* which is the amount of material that leaks past the screw in the small space between the screw and the barrel. The leak flow also has a negative sign because it diminishes the total flow.

The drag flow can be determined by a consideration of flow between parallel plates in a classical analysis of Newtonian fluid flows. The result of that derivation gives the following equation for the drag flow:

$$\text{Drag flow} = (1/2)\pi^2 D^2 NH \sin \theta \cos \theta \qquad (11.2)$$

where D is the diameter of the screw, N is the speed of the screw, H is the flight depth in the metering section, and θ is the pitch angle. This equation indicates that the flow through the extruder is increased by increasing the diameter of the screw,

---

increasing the speed of the screw, and increasing the flight depth. The optimum value of the pitch angle is somewhat more complicated and is found to depend strongly on the number of flights, the flight width, and the screw diameter. In almost all screws, the pitch angle is a constant at 17.5°.

The pressure flow component of the total flow can also be found using classical Newtonian flow analysis. By this analysis, the pressure flow is:

$$\text{Pressure flow} = \frac{\pi D H^3 P \sin^2 \theta}{12 \, \eta L} \tag{11.3}$$

where D is the diameter of the screw, H is the flight depth, P is the back pressure, θ is the pitch angle, η is the viscosity, and L is the length between flights.

The leak flow is small compared to drag and pressure flow and may usually be neglected in finding total flow.

The total flow is therefore given by the following equation:

$$\text{Total flow} = (1/2)\pi^2 D^2 N H \sin \theta \cos \theta - \left(\frac{\pi D H^3 P \sin^2 \theta}{12 \, \eta L} \right) \tag{11.4}$$

The use of this equation requires a detailed knowledge of the dimensions of the screw being used. These parameters are, of course, fixed for that screw and are usually represented as constants in the equation so that the effect of the operating variables can be more clearly seen. Therefore, if the screw dimensional parameters D, H, θ, L and the other constants are combined into two constants, α and β, the resulting operational equation for total flow is reduced to:

$$\text{Total flow} = \alpha N - \left(\frac{\beta P}{\eta} \right) \tag{11.5}$$

This equation, then, says that increasing the speed of the extruder *(N)* will increase the output of a particular screw, and that the output of that extruder will be decreased by an increase in the back pressure *(P)*. The back pressure will increase significantly as the screen packs become contaminated. Therefore, a good operational parameter to monitor is the pressure differential before and after the screen packs. If the pressure differential becomes too high, the flow can be interrupted, as it would when the screen packs are blinded.

If the viscosity (η) decreases, as it would when the temperature is increased, the second term of equation (11.5) will increase since the viscosity is in the denominator of the term. An increase in the second term will decrease the total output. Therefore, increasing the temperature or decreasing the viscosity in any way will decrease the total output. The decrease in output with a decrease in viscosity can also be envisioned by remembering that the back pressure will have a greater effect on low-viscosity materials and will tend to retard their advance.

In some cases the output of screws of various diameters is to be compared. A simple estimate that relates these different screws, assuming that the screws are operated at the same speed and that the screw parameters are comparable, is:

$$\text{Total flow (approximate)} = A D^{\delta} \tag{11.6}$$

where the constant A is a function of the units chosen. The empirically determined exponent for D is usually of the order of 2.0–2.2.

The positive pumping nature of twin-screw extruders vastly simplifies the output calculations for these extruders. The output of a positive pump such as a twin-screw extruder can be calculated as follows:

$$\text{Total flow} = CN \tag{11.7}$$

where C is a constant dependent on the geometry of the screws and N is the speed of the screw.

NORMAL OPERATION AND CONTROL OF THE PROCESS

The dynamic, continuous nature of extrusion suggests that the process will make good-quality product most efficiently if the process is stable and in a steady-state condition. Getting into this stable steady-state operation often takes considerable time. Therefore, disruptions in extrusion operations are highly undesirable and are to be avoided. The most serious disruptions include shutdown, contamination, resin changes, die changes, screen pack changes, and other similar events that cause the flow to be interrupted. After one of these disruptions has occurred, certain key parameters can be followed or controlled to move the process back into the steady-state condition. When the process is in steady state, these key parameters can be monitored to keep the process within acceptable operational limits (in proper control) so that good-quality products will continue to be made. These parameters will be considered first during the startup sequence and then during normal steady-state operation. They will then be examined as they are used for troubleshooting problems.

Feedback or automatic controllers and monitors are used extensively for monitoring portions of the extrusion operation, such as heating zones. Thermocouples are placed along the outside of the barrel to sense the temperatures and send their signals to the temperature controllers that turn the heat on or off as the situation may require. In this sense, these thermal controllers have full feedback control. Controllers of various sophistications are available to smooth the controlling of the on and off cycles and define the temperature range over which the temperature will be controlled.

Pressure in the melt is measured by a thin metal disc set flush with the wall, usually at the tip of the screw but still inside the barrel, where it will reflect the screen contamination, the load on the thrust bearing, and mixing conditions in the final turns of the screw. A transducer inside the pressure-sensing device converts the motion of the metal disc into an electrical signal that is transmitted to a display device. This pressure measurement can be compared with another pressure measurement from downstream to trigger, when appropriate, the automatic screen-changing mechanism.

Some successful attempts have been made to control the speed of the feed hopper auger by the turning speed of the extruder, or to control the speed of the extruder by the thickness of the part. The many other factors which control part thickness make this latter control system difficult. The objection to most control systems that attempt to control the entire line is that too many variables exist that can each give similar results in some area but have different effects in another. Nevertheless, the use of feedback controllers in extrusion is growing.

Start-up

The extruder should be preheated before attempting to turn the screw. This preheat should include the screw heating zones and the die. During this preheat, the die is often removed and sometimes the exit end of the extruder is opened to avoid buildup of pressure should a resin be present that decomposes into a gas. The opening of the end of the extruder also allows the screen pack to be changed. The heaters should be started so that the die end of the extruder is heated first. This will permit gases to escape through the die end. Head bolts should be tightened and the die should be securely fastened to the head after installing a new screen pack. The die heating should be turned on.

Care should be taken to ensure that the hopper has sufficient material and that any drying, blending, or other material treatments are ready before the screw rotation is activated. The extruder screw (rotation) is turned on after preheating and the material treatments have been completed. If throat cooling is used (as with a cored screw), that cooling should have been turned on before the rotation of the screw is started. The turning speed of the screw should initially be slow and gradually increased to the desired operational value as steady-state operation is reached.

When some resins are used in extrusion, especially those likely to decompose with prolonged heating, these resins are removed from the extruder by running another resin through the extruder just before shutdown. This process is called *purging*. The purging resin should be easy to melt, have sufficient density that it will sweep the prior resin from the extruder, and be known to present no startup problems. Polyethylene is a convenient purge material for many applications, although some specific purge resins are also commercially available. As part of the start-up, the hopper is filled with the regular resin, which then replaces the purge material as the extruder is brought into steady state.

Initially the extrudate is allowed to simply drool onto the floor. The initial material may not be all melted and the temperatures may have to be raised to above-normal levels for a period of time to obtain full melting. Care should be taken that no wires or other equipment are under the die. When temperatures, speeds, and pressures are near steady state conditions and the purged resin has been replaced by the regular resin, string up can be made. *String up* is the manual pulling of the extrudate through the cooling system and into the puller. Gloves should be worn during string up because the extrudate is hot.

To begin the string up process, the extrudate that is falling from the end of the extruder is cut off near the opening of the die so that a fresh end can be inserted through the opening of the cooling bath. A lead device (such as a clamp with a rope lead) can be used to facilitate moving the end of the extrudate through the various holes and other guides downstream from the extruder. With the top of the bath open, the extrudate is threaded through the sizing plates or rings and then through the remainder of the tank. The extrudate should be manually pulled away from the extruder at the same rate it is coming out of the extruder. This uniform and steady pulling rate will prevent sagging of the extrudate and maintain a reasonable thickness. The extrudate is manually inserted into the puller and the puller speed is matched to the speed at which the extrudate is being extruded. During the entire string up process, it may be useful to slow the speed of the extruder so that the speed of string up (manually pulling the material away from the extruder) is sufficiently slow to allow for some momentary difficulties and the operator doing the string up is not under a tight time constraint. When string up is complete, the speeds of both the extruder and the puller can then be increased.

After the puller and extruder have been synchronized, the cooling tanks can be closed (if a vacuum is used) and other conditions such as speed, temperatures, and pressures can be brought into their steady-state conditions. Final adjustments can be made in part size after the steady-state conditions are reached.

Part-Dimension Control

Some adjustments can be made in the dimensions of the part beyond those available by modifying the geometry of the die. The geometry of the die is the major influence on setting the part size and shape, but other factors can have a significant influence as well. These adjustments largely affect the conditions of the extrudate between its exit from the die and when it has become cool enough that its shape is fixed. The extrudate is dynamically changing throughout this distance. One of the most important phenomena that occurs in this region is the swelling of the size (cross section) of the extrudate as it exits the die. This swelling is called *die swell* and is illustrated in Figure 11.7. The die swell is measured as the ratio of the diameter of the extrudate to the die orifice diameter (D_x/D_d) after exiting the die.

Die swell is caused by the viscoelastic nature of the polymer melt. Because of viscoelasticity, the polymer melt dissipates forces slowly. The compressive forces that are

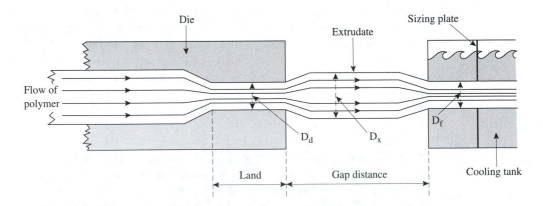

Figure 11.7   Illustration of die swell.

needed to push the polymer melt through the land and the small die orifice are not completely relieved by the time the polymer exits the die. The polymer therefore expands when it exits the die in response to the residual compressive forces, thus returning to an approximation of the shape the material had just before it entered the constrictive land portion of the die. This shape recovery appears as a swelling of the polymer after the die. This property of slow dissipation of forces by a viscoelastic material has been called *plastic memory* because it acts to restore the shape previously held, as if the material had a memory. High molecular weight increases plastic memory because the dissipation of the forces requires disentanglement of the molecules which is more difficult as the molecules get longer. Other material properties, such as intermolecular bonding (hydrogen bonding, etc.), can also increase plastic memory. Of course, if the material is a solid, then plastic memory is the same as elastic recovery. Therefore, plastic memory is a term that applies only to molten or semimolten material.

Die swell can be reduced by extending (lengthening) the land so that the polymer has sufficient time under the compressed conditions to dissipate the compression forces. Increasing temperature will also reduce die swell as it imparts the energy needed to disentangle the molecules. If nothing were done to reduce die swell or to make other changes in part dimensions after the die, the finished part would be larger than the die orifice by the amount of the die swell.

In normal extrusion practice, several other variables can be and often are changed that affect part dimensions, such as adjustments in the distance between the die face and the water tank. As the distance between the die and the water tank (called the *gap distance*) is shortened, the extrudate is drawn more quickly into the water tank through the entry hole that is, usually, smaller than the diameter of the swelled extrudate. Therefore, the part again becomes constricted. Further adjustments in part diameter made at the cooling bath can occur by the choice of hole size in the sizing plates that are located within the cooling bath. These phenomena are illustrated in Figure 11.7.

The relative speeds of the puller and the extruder also have critical effects on the extrudate while it is cooling. If the puller speed is slightly faster than the extruder, as is normally the case, the part is thinned slightly while it is cooling. A measure of the amount of reduction in size from the speed differential between the extruder and the puller is the *drawdown ratio* which is defined as the ratio of the maximum diameter of the swell to the final part diameter (D_z/D_f in Figure 11.7).

In summary, part dimensions are affected by the following: die orifice, die land, temperature, material properties such as molecular weight and hydrogen bonding, gap distance, sizing plate size and placement, and extruder speed as compared to puller speed. These many variables make proper sizing of a part complicated but also give many different combinations that can be used to achieve the desired part dimensions.

Many of these variables affect other part properties besides size. One of the most important variables having other effects is the drawdown ratio. An increase in the amount of drawdown, such as when the speed of the puller is increased relative to the speed of the extruder, results in the molecules' being stretched more and becoming more aligned. This alignment will increase the strength of the part in the machine or flow direction, with a corresponding decrease in strength of the part in the radial or cross direction. Generally, in pipe and tubing applications, this loss of strength in the radial direction will adversely affect the burst pressure. The orientation of the molecules in the machine direction will also increase the tendency of the material to stress crack. Therefore, large drawdown ratios are generally avoided except for those applications in which high speed is very important and other factors do not apply. Other machine variables that increase molecular orientation are longer land length, higher molecular weight, and decreasing part temperature.

Critical Operational Parameters and Techniques

In addition to those parameters used to control part dimensions and orientation, other operational parameters are important to the successful running of an extrusion line. One set of key parameters is the adjustment in the line for the type of plastic resin that is being extruded. As previously discussed, when changing from one resin to another, one important change could be a different screw that would optimize the melting characteristics of the material. In some extrusion operations, especially those where many different resins are extruded and changes from one resin to another must be done quickly, a different approach is taken. Instead of a specialized screw, which must be changed with each change in resin, a general-purpose screw is used that is a compromise between the screws for particular resins. The general-purpose screw would have a feed section that is longer than in a polyethylene-type screw but shorter than in a nylon-type screw and a compression zone that is shorter than in the polyethylene-type screw but longer than in the nylon-type screw. (See Figure 12.3 for illustrations of polyethylene-type and nylon-type screws.) To compensate for the nonoptimal performance of the general-purpose screw for a particular resin, other parameters, chiefly the heating zone temperatures are changed and have been found to compensate quite well.

To understand how the temperatures of the heating zones might be changed to compensate for a nonoptimal screw, two cases can be considered: extrusion of polyethylene and extrusion of nylon, both using the same general-purpose screw. (The term *ideal* will be used to indicate how conditions would be set if the material and the screw were matched.) In the polyethylene case, the general-purpose screw will have a longer feed zone than ideal, since most resins require more heat input than polyethylene, and therefore the material will remain in the feed zone for a longer period of time than it would in the ideal screw. This longer residence time means that more heat will be added in the feed zone assuming no change in heating zone temperatures was made. To compensate for this situation, the heating zones in the feed zone should be lowered from the normal flat profile that would be used with the ideal polyethylene-type screw.

The polyethylene material being extruded with a general-purpose screw will have a shorter compression zone than the ideal (that is, the compression zone will be shorter than in a screw made specifically for polyethylene). This shorter compression zone means that the mechanical heating that would normally occur will not occur to the same extent and the temperature of the polyethylene material as it leaves the compression zone will be lower than in the ideal case. The shorter compression zone may also leave some of the polyethylene unmelted. To compensate for the temperature difference from ideal and to ensure that complete melting will occur, the heating zones of the extruder near the end of the compression zone and in the metering zone are raised above the flat temperature

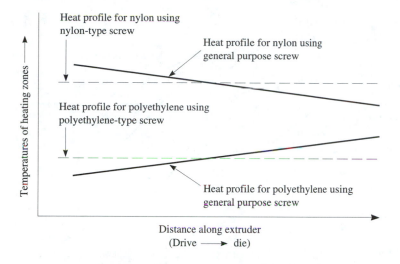

Figure 11.8 Temperature profiles for polyethylene and nylon when extruded with general-purpose screws versus resin-specific screws.

profile used with an ideal polyethylene-type screw. This temperature profile is shown in Figure 11.8.

The reverse case is true when nylon is extruded with a general-purpose screw. The general-purpose screw will have a shorter feed zone than will the ideal nylon-type screw and so the temperatures in the feed zone are raised to compensate. The compression zone will be longer than with the ideal screw and so the heating zones over the compression and metering sections are lowered to compensate.

The temperatures of the heating zones and of the die can be used to adjust the viscosity of the resin as it exits the extruder. If this viscosity is too low, the melt strength will also be low and the part shape will be difficult to maintain until it is cooled. This allows sizing and other dimensional controls to be made more easily. The importance of this temperature control can be seen from a plot of viscosities of various common resins as a function of temperature. This is presented in Figure 11.9.

The mixing of two or more resins is strongly dependent upon viscosities. Materials are mixed more efficiently when their viscosities are similar. Therefore, conditions of mechanical heating and thermal heating might be varied to give the resins as close to the same viscosity as possible. The proper extrusion temperature might be approximated for mixing and melting two different resins by considering Figure 11.9. For instance, the temperature for extrusion mixing of polyethylene with PMMA might be chosen as approximately 424°F (218°C), the temperature where the viscosities of the two resins approach the same value.

Changes in the heating zones can also affect the throughput of the resin. If the heating of the feed zone is too high, bridging occurs, as has been previously described. When bridging occurs, the resin has a greater adherence for the screw than for the barrel. Under these conditions, the resin simply rotates with the screw and is not advanced forward. This

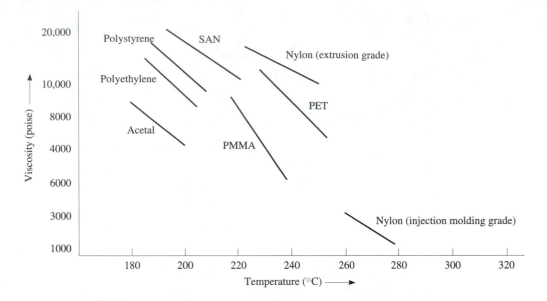

Figure 11.9 Plot of viscosities of common resins as a function of temperature.

situation must be avoided. Lowering the temperature of the feed zone is one way to prevent bridging.

The binding affinity (that is the tendency of the resin to stick to the metal) of most resins to metal reaches a maximum near the resin melt temperature. Therefore, the relative tendency to adhere to the screw versus the barrel can be controlled by properly balancing the heating of the resin by mechanical energy input or thermal input. In some resins, such as PMMA in which the bonding tendency is very high, the walls of the feed zone can be hot, to encourage bonding to the barrel rather than the screw, and resin transport is therefore improved. With PVC there is little tendency to adhere to any metal and so the heating of PVC using strong mechanical action of the screw will not cause appreciable bridging. Polyethylene has a moderate tendency to adhere to metals and so the heating by the screw should be balanced with the heating through the barrel.

The overall balance of mechanical versus thermal heating is also strongly dependent on the heat capacity of the resin. Simply stated, the combination of the two types of heating must supply sufficient energy to bring the temperature of the resin up to the melting point and then through the melting point transition. Heat capacity is the controlling resin parameter in determining what heating is required.

The die is generally separately heated and is important for final control of the viscosity and temperature of the part. With any resin, low-viscosity conditions (high die temperatures) require less energy to extrude. High-viscosity conditions (low die temperatures), however, give better part definition. Low die temperatures mean lower material temperatures which leads to a reduction in the amount of cooling that is required to set the shape of the part. When less cooling of the part is required, the speed of the extruder can often be increased whenever part cooling is the limiting factor in extrusion speed. Obviously, some compromise must be reached in selecting the proper temperature for the die.

Maintenance and Safety

Extruders can last many years with good care, including routine, preventive maintenance. Proper maintenance requires a familiarity with the owner's manual and the manufacturer's recommendations. Some general suggestions for maintenance for each of the major components of an extrusion line are as follows:

- **Base.** The extruder should be securely bolted to the base and the base immobilized in position on the floor. Proper alignment of the downstream equipment into a straight line from the base should then be insured.

- **Drive.** Most DC motors are air cooled with a fan to blow air out of the motor case. Be sure that the fan is turning in the proper direction or dust will be pulled into the motor. Some motors have an air filter and this should be routinely changed. The inside of the motor should be cleaned and the brushes replaced periodically when they show excessive wear. The gears should be well lubricated and the lubricant inspected often to insure that the oil contamination is not excessive. The lubricant and filters should be changed in accordance with the recommended schedule. The drive should also be inspected for unusual vibrations. The concentricity of the screw in the drive mechanism should be checked.

- **Thrust bearing.** Look inside the inspection space at the drive end of the extruder to check if any damage or excessive wear to the bearing can be seen. (Anthills of powdered material are one sign of excessive wear.) Check for leaking lubricant which may signal excessive wear.

- **Screw.** The screw should be removed occasionally and measured to determine the amount of wear. The flights should be inspected for severe gouges or chips and cracks. Clean the screw of any excessive buildup of carbonized or crosslinked material. Check the screw seat for rust or other deposit or corrosion and remove it. Apply an appropriate lubricant.

- **Barrel.** Look inside the barrel with a light and inspect for excessive wear or contamination.

- **Heating or cooling system.** Inspect contact surfaces of the barrel heaters and check electrical continuity of all heating units. Check for proper temperature capability of the thermocouples. Ensure that the temperature controllers are properly working. Replace wiring that has damaged insulation. Inspect all electrical connections for corrosion and tighten loose connections. Check water flow throughout the cooling system. Flush the cooling system and descale if necessary.

- **Head and die.** Look for leakage at joints, especially around the breaker plate. Clean off any carbonized or crosslinked material from the breaker plate and check that the breaker plate is flat. Renew screen packs and shear pins. Calibrate pressure sensors and thermocouples. Clean and adjust all dies. Test alarms and safety devices, if any.

■ **Other.** Replace water hoses that show excessive wear and inspect all connections for tight fit. Check for proper vacuum and seal. Calibrate the puller speed and lubricate properly. Sharpen or change blades in the cutoff system and lubricate any moving parts.

Safety is a major concern with large equipment like extruders. The most obvious problems are the heated surfaces and hot material. Gloves should be worn when any work is done on a heated extruder or when handling hot plastic. Care should be taken in handling large masses of plastic because they can appear to be cold but may be still quite hot.

The extruder has many moving parts. Loose clothing should be avoided. Safety guards, interlocks, and alarms should be provided and not removed, ignored or turned off while the machine is in operation.

The product may be heavy as are many dies and other pieces of equipment associated with the extruder. Care should be taken in handling these and in loading the hopper. Proper lifting techniques should always be followed.

The area around the extruder should be kept clean. Resin pellets on the floor are slippery and should be swept up if accidentally dumped. Masses of extrudate should be cleaned up and properly discarded.

EXTRUSION PROBLEMS AND TROUBLESHOOTING

Melt Fracture

When the extrudate has a rough surface, especially with short cracks or ridges that are oriented in the machine direction or helically around the extrudate, the phenomenon of *melt fracture* has likely occurred. This material defect occurs because the tensile forces on the extrudate exceed the critical shear stress and shear rate of the melt. Hence, the material experiences random fractures. The fracture of the material is generally throughout the extrudate, not just on the surface.

Turbulent flow in the die is one of the principal causes of melt fracture. This turbulence (flow that has eddies and convolutions rather than being smoothly layered) is most often present when the die is not properly streamlined. Examples of nonstreamlined and streamlined dies and the differences in extrudates are given in Figure 11.10.

Melt fracture can also occur with streamlined dies, especially when the diameter of the part is small. Low temperatures of the melt and high molecular weights also contribute to melt fracture.

Melt fracture can be reduced by the obvious steps of streamlining the die, raising the melt temperature, and selecting a resin with a lower molecular weight. A long land length will also help in reducing melt fracture because the long land forces the material into laminar flow (smoothly layered flow) and provides sufficient time to relieve stresses that might have been introduced in the melt. Low molecular weight fractions or other processing aids that facilitate melting and reduce shear can also help in reducing melt fracture. Lowering the extruder speed will help reduce melt fracture.

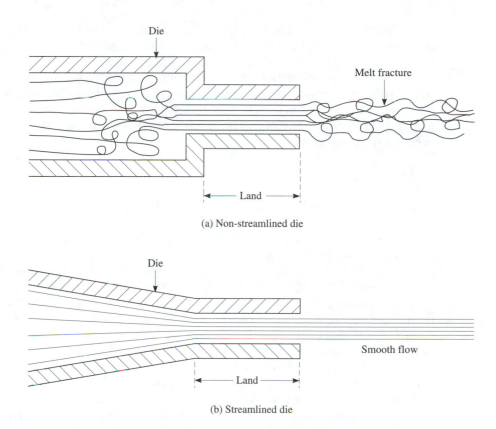

Die

Melt fracture

Land

(a) Non-streamlined die

Die

Land

Smooth flow

(b) Streamlined die

Figure 11.10   The effect of streamlining in a die to prevent melt fracture.

Sharkskin or Alligator Hide

If the surface of the extrudate is rough but with lines running perpendicular to the flow direction, the phenomenon is likely due to a tearing of the surface of the melt. This phenomenon is called *sharkskin* or *alligator hide*. This defect is caused by tensile stresses that are different from melt fracture stresses because the stresses that give rise to sharkskin arise from laminar flow. As the material is flowing, it will develop a flow profile with the center of the material flowing faster than the edges, because of the friction of the material against the edges of the die. The material in the center of the flow column slides against itself easier than it can slide against the die walls. As the material leaves the die, the outside material has to accelerate to the velocity at which the extrudate is leaving the die. This generates tensile stress and, if the stress exceeds the tensile strength, the surface ruptures causing the defect. In some cases, the outer surface will "snap back" to relieve the tensile stresses and the part will take on the appearance of a sectioned rod, somewhat like bamboo, and *bambooing* is a term that is applied to this severe state of the defect. When the differences between the applied stresses and the tensile strength are small, the material may simply form dimples on the surface. This phenomenon is called *orange peel*.

Sharkskin, alligator hide, bamboo, and orange peel are worsened if the resin is stiff (high modulus) and if the pressure in the extruder (high extrusion speed) is excessive. Lower temperatures in the die also exaggerate the problem.

The problem of these defects can be relieved by heating the resin, especially at the die, and by reducing pressure and the speed of the extruder. A broad molecular weight distribution will also reduce the defects. A larger gap in the die would reduce the stresses, but more drawdown would be required to maintain the same part size. Use processing aids to lower overall viscosity. Run slower.

Uneven Flow and Surging

A cyclical variation in the extrudate thickness with a cycle time between surges of from 3 seconds to 3 minutes is called uneven flow, or *surging*. The ammeter on the extruder will often reflect this surging pattern by cycling over a wide amperage range. The pressures could also reflect the same pattern.

Several different problems in the extruder could cause this uneven flow. One cause of the surging could be inadequate screw speed control. The motor could be undersized or the speed control inadequate for the size of the screw or the type of material. If surging is from this problem, the cure is to move to another machine with greater capacity or, perhaps, raise the temperatures so that the load on the extruder will be less.

A major contaminant, such as a piece of metal, could also cause surging as it is pushed against the barrel by the flights of the screw. If the contaminant is large, the screw may have to be pulled and cleaned to cure the problem.

Another possible cause is associated with a mismatch between the screw dimensions and the bulk density of the material being extruded. If the screw was designed for pellets and a less dense, fluffy material is being extruded, the depth of the flights in the feed section may not be sufficient to move enough material to keep the extruder flowing smoothly. The result could be a surging of the extrudate. The problem can be solved by changing to a deeper-flighted screw or by densifying the incoming fluffy material. Densification can be done by several methods including: (1) extruding the fluffy material into pellets using a machine with a more appropriate screw, (2) compressing the material in the hopper with an auger (called a *cram feeder*), (3) spinning the fluff at high speed so that it shrinks and balls up into little particles, or (4) if the fluff is made by chopping filaments or film, eliminating the chopping of the material by feeding the scrap directly into the extruder.

Another method for getting poorly flowing material into the extruder is *starve feeding*. In starve feeding the screw runs so fast that whatever material drops from the hopper is immediately carried away. Then by limiting the flow of material through the hopper, all material is transferred evenly and the flow is controlled.

Partial bridging could cause surging. In this case the resin could be clinging to the screw in the feed zone. Traditional bridging would be solved by lowering the heat in the feed zone of the extruder, but partial bridging could arise from on/off heating cycles that are too long. This could be detected by comparing the frequency of the surging with the heating cycle. If the heating cycle is the problem, changes in the heater control cycle will solve the problem.

Similarly, the feed from the hopper could be uneven, perhaps dropping as clumps into the feed throat and then moving in an uneven flow through the extruder. A feed auger could solve this problem. An investigation into the cause of the clumps should also be made.

Surging could also be caused by a locking up of the solid mass of resin in the decreasing flight depths of the compression zone and subsequent irregular melting. Sometimes raising the temperature of the feed zone or the compression to get better melting will alleviate this problem.

A downstream cause of surging could be slippage in the puller. This can be solved by tightening the grip on the part or by changing to different puller belts or a different type of puller.

If none of these solutions work, simply lowering the speed of the extruder may help, as could increasing the back pressure by using more or finer screens in the screen pack. If these do not work, the ultimate solution to surging is to install a gear pump.

Note that noncyclical thickness changes might be mistaken for surging. These changes are more likely due to gradual drifts in the extrusion line, as from slow plugging of the screen packs or from long-term temperature drifts, perhaps even day/night cycles.

Degradation

This defect is detected in the part as discolorations and lower physical or mechanical properties. A strong odor may also occur to indicate degradation.

If the degradation is general, that is, the entire extrudate is affected as shown by discoloration throughout, although darker streaks may also be present, the most likely cause is that the heat is too high for the speed of extrusion. The obvious solutions are to reduce the heat or to increase the extrusion speed. Some combination of these two is likely, since the speed of the extruder affects mechanical heating of the material.

Nonuniform degradation could be evidenced by specks of dark material in the extrudate. This type of degradation could come from material that is trapped or adhering to the surfaces inside the extruder and are therefore degraded by the long residence times at high temperatures. The remedy for this problem is to improve the flow pattern of the material through the extruder. These changes may be difficult because they require equipment modifications. However, the most likely region for problems is in the die, which can be modified more easily than other parts of the extruder. Problems upstream from the screen pack will usually show on the screens and probably will not show up in the extrudate.

If resin changes are frequently made, the degraded material may be a past resin that is not fully purged. In this case a better purging material or procedure is indicated.

Poor Mixing

Streaks or particles in the extrudate could also result from poor mixing. This improper mixing usually results from running the extruder faster than it can mix the materials. Slowing the extruder speed is the most obvious remedy. Increasing the back pressure will also improve mixing and may be advantageous because output will not be reduced as much.

The back pressure could be increased by using more or finer screens and by cooling the metering zone and die. Heating farther back in the extruder could also improve mixing.

Some mixing problems are so difficult that a separate step in a special mixer should be done before the material is introduced into the extruder. Another method is to extrude the materials separately and then add these premixed materials as a pellet that can be mixed with the resin more effectively. Such materials are called *concentrates*. Materials that are often concentrated in this fashion are carbon black, antioxidant, and thermal stabilizers.

Changing the screw or adding special mixing devices inside the extruder barrel are other methods of improving mixing. Some examples include increasing the L/D, using a static mixer at the end of the screw, and using a screw that is optimized for the particular resin and extrusion conditions, but these are difficult and costly changes to make.

Contamination

Contamination could be yet another cause of streaks or spots in the extrudate. If these spots are small dimples or discolorations on an otherwise smooth surface, they are sometimes called *fish eyes*. The contaminants can be distinguished by examining the part under microscope, with solvents, or by some chemical analysis technique. Examination of the screen pack will often confirm the existence of contaminants. Contamination may be difficult to distinguish from incomplete mixing as might be the case with incompletely mixed carbon black or other pigments.

Contamination is a common problem, especially when the extruder is used with many resins. One source of contamination is, therefore, material from a previous run that is not fully purged.

The most common location for contamination to enter the extruder is through an open hopper. Contaminants could include dust, other resins that might have been in the conveying lines, incompatible materials introduced with regrind, paper or other resin packaging materials, or simply materials that fall into the hopper or other parts of the resin conveying system.

The solutions include keeping the hopper covered, putting filters in the conveying system, inspecting incoming material, and isolating regrind operations by resin type. Decreasing the opening size of the screens can also stop contamination, although this will result in faster plugging of the screens, less even flow, and more frequent screen pack changes.

Bubbles in the Extrudate

Excessive moisture or volatiles can be absorbed by the resin and then vaporize when the melt exits the die and the external pressure is reduced, resulting in bubbles within the extrudate. Surface defects can also result. Some resins, such as PET, nylon, and polycarbonate degrade severely when heated in the presence of moisture. Moisture levels of less than 0.1% are recommended for these resins.

Several methods are available to reduce the amount of moisture and volatiles. The most effective in reducing moisture is to pass all of the resin through a dehumidifying dryer and to pneumatically convey the resin through the drying system and into the hopper. Other drying methods include blowing dry air though the resin while it is in the hopper, using the resin directly from the bag (if bagged resin is used), and storing the resin in a low-humidity location, usually at or below the extruder ambient temperature. Venting the extruder will also remove moisture and volatiles.

Air entrapment will also cause bubbles in the extrudate. These bubbles tend to be less regular and less numerous than the bubbles from moisture and volatiles. The air entrapment is usually the result of improper match between resin and screw. The usual remedial actions are to increase the back pressure by using more or finer screens and to lower the barrel heats.

Because of the screw-dependent nature of air entrapment, another remedy is to lower the screw speed which will allow the air to flow backwards in the extruder and will permit the screw to act more effectively.

Troubleshooting

The problems in extrusion are very complex and any attempt to reduce them all to one table or guide and to present some possible solutions to those problems is a gross oversimplification. However, some value can be obtained from a presentation of some of the most typical problems so that the thought process involved in solving the problems might be examined through the perspective of seeing many different solution steps for a single problem. Therefore, some typical problems and some suggested solutions are given in Table 11.2.

Table 11.2 Troubleshooting Guide for Extrusion

Problem	Possible Cause
High drive motor amperage	• **Resin not heated enough.** Possible bad heater or heater temperatures too low. Raise the temperatures and check the electrical output of the heaters. • **Resin not correct.** Resin molecular weight may be too high. The resin may be crosslinking or degrading. • **Plugged (blinded) screens.** Change the screens. • **Motor.** Maintenance needed. Motor speed too high. • **Contamination.** A large contaminant such as a soda pop can or a pencil could have dropped into the extruder. You may have to pull the screw to check for this. Make all other investigations first.
Interrupted resin output	• **Hopper.** Clumping in the hopper (resin density too low or some contaminant). Temperature of the feed zone may be moving into the hopper and causing the resin to stick together, especially with a low molecular weight resin. Lower the feed-zone temperature. • **Bridging.** Temperatures in the feed zone too high; lower the heater. Density of the resin too low; use a cram feeder of extrude the material into pellets in a separate operation. • **Clogging.** Check the screen pack. Look for degraded or crosslinked resin.

(continued)

Table 11.2   (Continued)

Problem	Possible Cause
Uneven flow (surging)	• **Temperatures.** Raise the temperatures in the heating zones especially if the extrudate has a high viscosity. If an internal mixer is used, it may have to be removed. If partial bridging is suspected, lower the feed-zone temperature, which may require that the metering and, perhaps, the compression-zone temperatures are raised.
	• **Contamination/plugging.** Check the pressure across the screen pack and if high, change the screen pack; if found that they are not blinded, increase the screen openings. Check for plugging in the hopper. If no other solution is found, pull the screw and check for a large contaminant.
	• **Equipment.** Extruder too long for the barrel or improperly placed in the thrust bearing, thus causing the screw to drag on the bottom of the barrel. This requires pulling the screw. Motor not functioning properly due to a need for maintenance or because it is undersized. The puller could be slipping; monitor its speed and the speed of the material going through it. If irregular, increase the pulling pressure on the part.
	• **Resin.** Density of resin could be too low, thus requiring a cram feeder, starve feeding, or pelletization.
Unmelted particles in the extrudate	• **Screen pack.** Hole in the screen pack. Replace screens.
	• **Temperatures.** Raise the temperature, especially in the compression and metering zones. Check for bad heater.
	• **Contamination.** Crosslinked or degraded polymer, especially in the die. Lower the die temperature if the material seems off-color or if the particles won't melt if put on a hot plate. If the particles will melt, raise the die temperature to melt them. Streamline the die better.
Discolored extrudate	• **Degraded polymer.** Temperatures too high.
	• **Poor mixing.** Pigments or dyes not well mixed; increase temperatures and/or screw speed and/or add an internal mixer.
	• **Die design.** Die not streamlined.
	• **Output control.** Screw speed too high, especially if resin is subject to degradation from adiabatic heating. Extruder too large for the output.
Pressure drop across die too high	• **Plugging.** Plugged screen pack. Screen pack has openings that are too small or has too many screens in the pack.
	• **Melting.** Temperatures too low.
Extrudate viscosity too low	• **Resin.** Melt temperatures too high. Resin melt index too high; change resins. Resin molecular weight distribution too narrow.
	• **Temperatures.** Die temperature too high.

Table 11.2   (Continued)

Problem	Possible Cause
Part diameter too small or walls too thin	• **Equipment coordination.** Puller speed too high or extruder speed too low. Gap too large, so move the cooling tank closer to the die face. Melt temperature too low. • **Equipment design.** Die opening too small or land too long. These require modifying the die. Sizing plates in cooling tank too small requires changing the plates. • **Resin.** Molecular weight too high.
Part diameter too large or walls too thick	See diameter too small and adjust in the opposite manner.
Surface of part is rough—lines running in the machine direction or helically around the extrudate (melt fracture)	• **Die.** Not streamlined properly. Land too short. Die temperature too low. These require modifying the die. • **Resin.** Melt temperature too low. Molecular weight too high or molecular weight distribution too narrow, which both require changing the resin.
Surface of part is rough—lines running perpendicular to the flow direction (alligator hide or sharkskin or bambooing or orange peel)	• **Die.** Die temperature too low. Land too long. Resin gap in the die might be too small. • **Resin.** Modulus of resin too high or too narrow molecular weight distribution, which require a change in resin. • **Operation.** Extruder speed too high; lower the revolutions per minute. Back pressure could be too high; change screen packs. Raise melt temperature by increasing heater temperatures.
Spots on the part surface (fish eyes)	• **Contamination.** The contaminant could be from degradation or from material introduced at the hopper. Check the screen pack for discolored material, which would indicate a hopper origin. Water could be the contaminant; this would require drying the resin. • **Degradation.** Temperatures are probably too high, especially in the die.
Bubbles in the part	• **Water.** Dry the resin. • **Degradation.** Check for an odor and if present, lower the temperatures of the melt.
Warped part	• **Die.** Spider mandrel not concentric in the die opening and therefore needs adjusting. Land is not even on all sides. Entry angle of the die is not uniform on all sides. • **Cooling tank.** Part is being twisted or turned as it enters the cooling tank. Align the tanks to be parallel with the extruder outlet. • **Part design.** Look for nonsymmetries in the part, which may induce internal stresses.
Sheet not uniform in thickness	• **Die.** The coat-hanger die is not properly adjusted. Check by injecting a colored material just upstream of the die and seeing if it comes out uniformly. Die temperatures not uniform. • **Downstream equipment.** The rollers could be nonparallel. Puller could be unaligned with extruder.

(continued)

Table 11.2   (Continued)

Problem	Possible Cause
Blown film not uniform	• **Die.** Mandrel not concentric. Air inlet off center. Flow channel in the die could be partially plugged. Temperatures in the die may not be uniform. • **Downstream equipment.** Puller not aligned with the die. Sizing rollers/cage may not be aligned.
Coextruded product	• **Extruders.** May not have matching speeds. • **Resins.** May not be compatible. • **Die.** Flows may not be matched or a channel may be partially plugged.
Postextrusion molding	• **Downstream equipment.** May not be properly aligned with the extruder. May not be adjusted for correct speed. • **Resin.** Temperature may not be correct. May not have correct molecular weight or distribution.

MATERIAL AND PRODUCT CONSIDERATIONS

Materials

Most thermoplastic and some thermoset materials can be extruded. The thermoplastic materials that cannot be extruded are materials such as PTFE that have such high melting points that they begin to decompose if melting is attempted. (PTFE is compressed into a desired shape in a process that is sometimes called ram extrusion and then sintered. This process will be discussed in detail in the chapter on Compression and Transfer Molding.)

Rubber materials are the most commonly extruded thermoset materials. The extruders for these thermosets are usually very short (low L/D ratios) and have provisions for easy cleanout should the extrusion be stopped for some reason and the thermoset begin to gel in the extruder. Thermosets are extruded below their cure temperatures to help prevent premature curing. Thermoset materials that are extruded may also be inhibited; that is, materials are added that retard the curing reaction provided temperatures are below certain levels or times are less than certain values. Care must be taken during start-up of extruders in which thermosets have been extruded so that the materials will not set up during initial heating. To prevent this from occurring, thermosets are purged from the extruder at shutdown. Typical products made by thermoset extrusion include wire and cable coating, weatherstripping, and extruded strand that is later used for feeding molding operations. Even though these thermoset extrusions may be quite common, extrusion is still considered to be principally a process for thermoplastic resins.

Not all thermoplastic resins are extruded with equal ease and not all resin grades are appropriate for extrusion. It is, in fact, quite common when specifying a resin to specify that the resin be "extrusion grade" or some other grade that is optimized for a particular process. Extrusion-grade resins have some characteristic properties that enhance the ability of the material to be extruded. The most important of these are high molecular weight and broad molecular weight distribution.

High molecular weight is desired for good extrusion. As discussed in the chapter on micro structures in polymers, molecular weight is usually measured by the melt index test. A low melt index indicates a high molecular weight. Extrusion resins would typically have low melt index numbers, occasionally less than one (1). These very low melt indices are called *fractional melt materials*. High molecular weight is important in extrusion because the molten material must be able to retain its shape for a brief period of time, that is, after leaving the die and before it is cooled. The material must also be able to be pulled in tension during this same brief time. Materials that possess this ability are said to have *melt strength*. Water, for example, has little or no melt strength. Honey has melt strength. Materials with high molecular weight have been found to have high melt strengths. These materials also have high viscosities (that is, are very thick). Most thermoplastic materials have this property, provided the temperature is not too high.

A broad molecular weight distribution is important in extrusion because the melting process is facilitated by having low molecular weight materials that can first melt and then help to lubricate the action of the screw on the higher molecular weight materials. Hence, broad molecular weight distributions (MWD) are also favorable for extrusion-grade resins.

Resins that are particularly heat sensitive, such as PVC, can be extruded with less degradation if low molecular weight materials are added to the PVC. The most common of these additives are low molecular weight acids, such as stearic acid. These additives are sometimes called *lubricants*.

Although materials of widely differing bulk densities are often extruded together, such as resin and filler, powder and pellets, and virgin and regrind, some problems in feeding and mixing these different materials can arise. Often, the problems center around the poor flow characteristics of the low-density material. When such problems arise, *force feeding* of the low-density material (sometimes called *cram feeding*) can make feed more uniform and predictable.

Some materials, especially those that might be mechanically damaged by extrusion are sometimes fed into the extruder through vent ports that are located downstream from the normal feed zone. Chopped fiberglass and other fibrous reinforcements are commonly added to resin by this method.

The rigidity of a resin contributes to the ability of that material to be held in tight tolerance in the extrusion process. Rigid PVC and polystyrene can have tolerances of 5.0% in very thin parts and 1.3% in thick parts, whereas flexible vinyl and polyethylene can be held to only 9.6% in the same dimensional thin part and 2.1% in an equivalent thick part.

General Shape and Design Considerations

Extruded parts are characterized as having a constant cross section and being continuously produced. These criteria apply to many products such as pellets (chopped extruded strands) that are the output of compounders and other resin manufacturers and modifiers, pipes, tubing, rods, fibers, coatings of other products (such as wire, cable, and paper) and various profile parts that have cross-sectional shapes that are not just very simple geometrical shapes. Products with other shapes that are extruded and then require post extrusion shaping are discussed later in this chapter or, if the shaping is from blowing to expand against a mold, in the chapter on blow molding.

The element of the extrusion process that is critical in determining the part shape is the die. A simple die for making rods has already been discussed. When more complex shapes are made, innovations in die design become very important. Some considerations for improving die design and dimensional shape control and thereby increasing output include the following:

- Whenever possible parts having a constant material thickness in all sections of the part are superior to parts in which the thickness varies within the part.

- Completely closed sections are to be replaced by partially closed sections unless the completely closed section is required for functionality.

- Warpage is often caused by uneven flows in the die. Consistent warpage to one side may be caused, for instance, by one side of the land being longer or having a rougher surface than the other side of the land.

- Part shrinkage is not a major consideration in extrusion as it would be in other plastic-molding processes, but die shape must allow for changes of the part cross section because of flow differences between restricted and nonrestricted parts of the die. This point is discussed more fully in the section on profiles.

- Surface finish, dimensional control, and tolerance control cannot be as fine as in some other plastic-shaping processes because of the dynamic nature of extrusion. Typical surface finishes are 4–40×10^{-5} inches (1–10 μm). Angles can be controlled to 4 or 5 degrees and dimensional tolerances to 8 to 10%. Thickness variation can be controlled to within 0.0004 inches (0.01 mm) in relatively thin sections.

These factors are also dependent upon resin type as well as part design. Therefore, the effect of material properties on the extrusion process should also be considered.

The extrusion of various shapes is so different that extrusion processes are subdivided by the type of shape that is being extruded. Therefore, each of the major shapes will be discussed separately so that the unique characteristics of each can be understood and appreciated more fully.

Profiles

Solid parts that have shapes other than simple round rods or flat sheets are called *profile shapes* or, simply, *profiles*. If a profile part is to be made, the inside diameter of the die gradually converges to the shaped orifice that determines the shape in the final part. The resin fills the exit orifice of the die and the part is then cooled.

Parts that are hollow are also called profiles provided that the outer shape of the part is something other than cylindrical. Hollow cylindrical parts are called pipes or tubing.

The shape of the exit orifice may be slightly different from the shape of the final part because of the tendency of molten plastic to flow more in the larger parts of the die (such as the center), to swell as it exits the die, and to shrink when cooled. Profile dies must allow for the fact that extruded parts change shape—they swell when they leave the die and encounter reduced pressure and then they contract when their temperature decreases. The swelling process, called *die swell,* distorts the shape of the extrudate.

Most generally, this distorted shape is the precursor to the final shape since significant further shaping is not likely to be done and the only additional changes in shape from the time the material exits the die to the final part are shrinkage from thermal cooling and drawdown from the pulling mechanism. Therefore, the shape of the die opening must compensate for the die swell and the distortion it causes.

The general rule in distortion compensation is that the part flows and swells most in the areas of lowest restriction. Therefore, if a die has a sharp corner, the flow will be less in that location and the sharpness of the corner will be rounded. A further rule is that the polymer material will attempt to pull into itself and make round holes. This is largely due to the melt strength of the polymer. Therefore, if a sharp angle is desired in the final part, the die should have a slight excess of material in the region between the points of the sharp angle. This principle is illustrated in Figure 11.11. More complicated shapes that illustrate the same principle are shown in Figure 11.12.

The extrusion of synthetic fibers is similar to conventional profile parts except that the fibers are much smaller. However, except for the small diameter, the concepts for extrusion of fibers, the detail needed, and the allowances for swelling and contraction are like those for normal profiles.

Even through fibers are very small, some of them can be extruded into quite complicated shapes. For instance, the oval part with four holes in it shown in Figure 11.12 is a fiber cross section. The die interrupts the flow of the plastic material as it exits the die and causes an opening in the melt. The plastic attempts to heal this opening by closing again, but the first step in this closing is to minimize the energy associated with having a large new surface formed. This minimization of surface energy causes the interruption to take a circular shape. Then, before the circle can close, the fiber is cooled, thus trapping the circular shape inside the fiber.

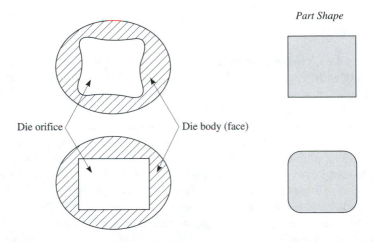

Figure 11.11   Changes in shape that occur between the die orifice and the finished part.

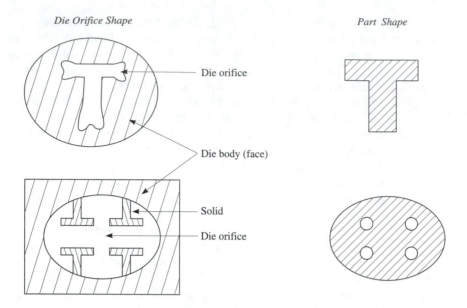

Figure 11.12 Die-swell compensation by changes in die orifice design in complicated extruded profile parts.

Fibers with circular holes running along their length, like macaroni, have the ability to break up light that impinges on them. This diffraction of light assists in hiding dirt that might be present on the surface of the fibers. Therefore, hollow fibers are used for carpets. The hollow tubes also allow air to enter the fibers. This air provides insulation, thus giving *hollow-filled fibers* some of their benefits as packing material for sleeping bags and blankets.

Pipe and Tubing

These are seamless, cylindrical hollow parts of constant cross section that are differentiated by rigidity with pipe being rigid and tubing being flexible. Hose is another cylindrical part, but it is often a complex material with many layers with an extruded outer layer. Hoses are, therefore, discussed separately.

If a hollow extruded part is to be made, some method must be provided to force the molten material into a cylindrical shape. This is done by mounting a solid metal obstruction in the middle of the stream of flow. This obstruction can be small in diameter (like a pin) or somewhat larger (like a rod). In all cases, the obstruction is called a *mandrel*. The upstream end of the mandrel is usually streamlined so that the resin will not stagnate as it flows around the mandrel. A die with a mandrel is illustrated in Figure 11.13.

The mandrel is supported as far back in the die as possible and cantilevered forward. This arrangement minimizes disruptions in the flow that are caused by the support pins because the resin will have time to be smoothed before the resin exits the die. The cross-sectional shape of the pins is usually streamlined to minimize disruption of the flow. When

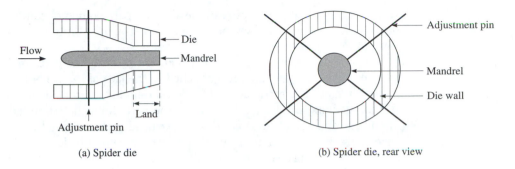

Flow

Die

Mandrel

Land

Adjustment pin

Adjustment pin

Mandrel

Die wall

(a) Spider die

(b) Spider die, rear view

Figure 11.13   Spider die with mandrel as seen from (a) side and (b) rear.

viewed from the rear, the die, supporting pins, and mandrel look somewhat like a spider thus giving the name *spider die* to this type of die, as shown in Figure 11.13. The pins not only support the mandrel, but they can also be used to adjust the position of the mandrel so that it is centered in the exit opening, thus giving uniform wall thickness to the part.

Mandrels can also be supported on the face of the breaker plate. This eliminates the need for adjusting pins but also eliminates the capability of moving the mandrel within the die for centering. Furthermore, the upstream end of the mandrel must be attached to the breaker plate, which implies considerable surface mating, and such a large occluded surface may lead to stagnation on the upstream side of the breaker plate. Mandrels can also be extended backward if an offset tube connecting the die with the extruder is used. The offset die is discussed in detail in the section on wire and cable coating.

The outside diameter of the downstream end of the mandrel is smaller than the inside diameter of the die at the exit point, thus creating a space surrounding the mandrel. The difference between the mandrel and die diameters dictates the thickness of the wall of the cylindrical part. Some die swell occurs to thicken the walls of the part (die swell occurs both internally and externally in the cylindrical part), but the swell is uniformly reduced with drawdown. Wall thickness, outside diameter, and inside diameter are all important variables for pipes and tubing, although not all can be controlled simultaneously. The interaction of these dimensions and the various variables that control them were previously discussed.

As discussed previously, the outside diameter of a hollow part can be sized by applying a vacuum to the part in the cooling tank which pulls the part out against the sizing plates. This same effect can be achieved by pressurizing the inside of a hollow part rather than by applying a vacuum to the outside. The inside pressure can be added by directing a pressure line down the middle of the mandrel around which the part is extruded. The connection of this line to the outside of the die should be done as far from the die exit as possible so that flow disruptions can mend themselves, just as would be done with the mandrel adjusting screws. Some operators have used crosshead dies (discussed in the following section) as a convenient method to introduce this internal pressure by simply putting the pressure line in place of the wire or cable. The use of internal air can also cool the part if additional cooling is required. Such cooling may be necessary if some internal feature is being extruded. Any feature inside a hollow part would not come in contact with the cooling water and would, therefore, cool slowly. The use of internal cooling air would be beneficial in such cases.

If the inside diameter is the critical dimension of a hollow part, an *internal sizing mandrel* can be used. This mandrel is attached to the face of the die but extends into the cooling bath region. The extrudate simply flows over this internal sizing mandrel as it is cooled, thus defining the inside diameter of the part. Note that simultaneous sizing of both the inside and outside diameters of an extruded part is difficult as that requires simultaneous pressure against an inside sizing mandrel and outside sizing plates. The best that can usually be expected is to use either an inside or outside sizing mechanism and then to carefully control the wall thickness of the part by methods that are discussed in the section on extrusion control.

Plastics parts are often measured for dimensional tolerances using standard dimensional devices such as micrometers, calipers, and straightedge rulers. When rapid measurements are desired, such as for quality control procedures, special gauges often are fabricated. These gauges are made so that parts that have the proper dimensions fit within the gauges and parts that are either too big or too small do not fit. Such devices are commonly called *go/no-go gauges*. An example of a go/no-go gauge for plastic tubing is shown in Figure 11.14. The measurement is made by inserting the gauge into the end of the tubing and

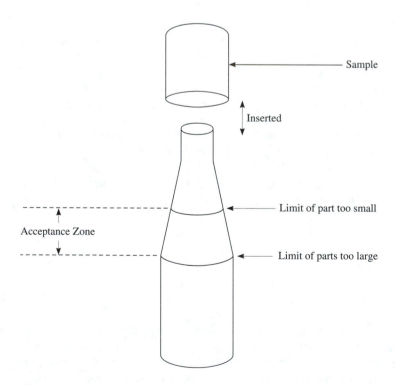

Figure 11.14   A go/no-go gauge for plastic tubing.

carefully moving the tubing along the gauge until a tight fit is obtained, making sure not to stretch the tubing. If the end of the tubing lies between the two lines on the gauge that mark the limits for tubes that are either too small or too large, then the tubing is accepted.

When automatic dimensioning or dimensioning of large parts is done, coordinate axis measuring machines can be used. The part to be measured is precisely located and then anchored onto the table of the machine and a touch-ball, mounted on a movable gantry over the part, is used to touch various locations. The machine indicates the relative positions that are touched.

Optical and video comparators are also used to measure part dimensions. These machines display a magnified image of the part and thereby allow close tolerances to be determined.

Coating of Wire and Cable

The key to coating wire and cable and most other extrusion coated products is the *crosshead* or *offset die*, which is illustrated for the wire and cable case in Figure 11.15. In this process the flow of resin is directed to the side of the extruder through an offset channel which is just a tube attached to the end of the extruder with a right angle turn to move

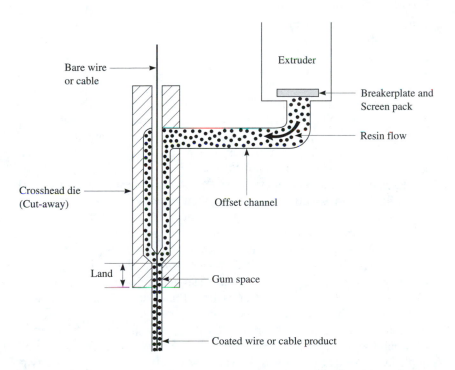

Figure 11.15   Offset or crosshead die for coating wire and cable (top view).

the molten resin to the side of the extruder. This offset channel then merges into the crosshead die, making another right angle turn and then streamlining to force the molten plastic onto the wire or cable. The wire or cable is brought into the back of the crosshead die through a cavity in the die that tapers toward the exit end and eventually closes to a hole just large enough for the wire or cable to enter through. At this point the wire or cable enters the flow of the molten resin and is coated by the resin. The diameter of the die exit orifice is slightly larger than the wire. This difference in diameter is called the *gum space*. A land just before the exit of the die ensures even flow and gives final definition to the diameter of the coating.

The extrusion rates for wire and cable coating can be very high because the amount of material extruded over the wire or cable is small, the extruders can be very large, and the process is inherently simple. Speeds of 700 feet (210 m) per minute have been reported. At these high speeds, screw designs are generally modified to insure proper mixing and temperature control but the major difficulties in lines of this speed is the smoothing of the wire or cable before it enters the die, cooling, and part removal equipment. Large amounts of wire and cable are crosslinked by radiation for improved physical performance. This process is discussed in the chapter entitled Radiation Processes.

A coating for hoses can be applied in much the same manner as wire and cables. Hoses are typically complex parts made with several layers of material including braiding, wrapping or fabric. One common manufacturing method is to form a central tube that is cooled, dried, and then wrapped with reinforcement fibers. The part then enters an offset die and is coated with a layer of plastic. This process of coating and wrapping can be continued until the proper number of layers have been builtup. All of the processes can be done in-line so that part removal need only be done once.

Sheets and Flat Film

Sheets and flat film are made in essentially the same way, the difference between them being thickness. Sheets have been defined as flat materials having a thickness of at least 0.004 inches (0.10 mm). Flat films have thicknesses less than 0.004 inches (0.10 mm). Another method of making film, blown film, differs significantly from the sheet and flat film method and is discussed later in this chapter.

Extruding flat sheets and film presents a special problem because of the need to move the melt from an essentially compact cylindrical shape at the entrance to the die to a wide, flat shape within the relatively short distance of the die. If this is poorly done, shear will be created within the melt and the resulting part may have nonuniformities across its width. The nonuniformities could be internal stress regions that might cause unwanted curl or areas of uneven mechanical properties in the sheet. Hence, the construction of the die is critical to the success of the flat sheet and film extrusion process.

The most efficient method of making this shape transition is through the use of a *clothes hanger die*. In this die, which is illustrated in Figure 11.16, the molten polymer enters the die near its center (from left to right) and is directed to both left and right through channels that distribute the melt across the width of the die. The distribution channels are angled toward the front of the die. The physical arrangement of an entry point and two angled distribution channels forms the shape of a clothes hanger and hence the name.

Figure 11.16 Clothes hanger die for extruding sheet (top view).

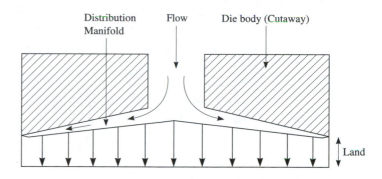

After being distributed across the die, the melt flows through gaps between two metal plates and moves toward the front of the die. These plates and gaps are called the land and serve the same purpose as the lands of other dies, that is, flow regulation, pressure buildup, surge suppression, and setting final part thickness. The result is, therefore, a sheet of plastic material exiting from the downstream end of the gap.

The construction of the die in this shape restricts the flow in some areas so that all portions of the melt have approximately the same equivalent travel distance and therefore the same amount of shear. Note that the melt material that enters near the middle of the die must flow through a longer land than material that is distributed to the edges. Therefore, because the flow through the land is more restricted than the flow through the distribution channel, the total flow of the melt can be made constant for all parts of the die. To achieve uniformity of flow, the diameters of the channels and the width of the gap are adjusted so that the exit paths through the die are all equivalent. When perfectly done, the molecules that flow into the die together in a mass will exit the die at the same time, although now they will be spread into a sheet form. This phenomenon is called *plug flow* and indicates that the material would move as a plug from entrance to exit, with no internal shear being created. In this way, material that flows straight through the land becoming the center of the sheet will exit at the same time as material that arrived at the entrance to the distribution manifold at the same time but was directed to the edge or any other region of the sheet. Sheets up to 10 feet (3 m) wide have been successfully extruded through clothes hanger dies.

Other die designs can be used to achieve these same flow uniformities, but the clothes hanger die is the most common and seems to be the most effective.

The cooling and other downstream equipment is slightly different for sheet and flat film than for the other extruder parts that have previously been considered. Instead of entering a cooling bath, the sheet material is wrapped around a series of cooling and finishing rolls. These rolls, which are usually filled with chilled water, cool the sheet or flat film and give the surface a smooth finish. Farther downstream are the puller rolls and either the cutter or the reel-up, the two most common removal devices for sheet and flat film, respectively. A typical arrangement for the sheet-making line is given in Figure 11.17.

The thickness of the sheet and film is set by the combination of extruder speed, die gap and the relative speeds of the rollers and pullers. Increasing the speed of the

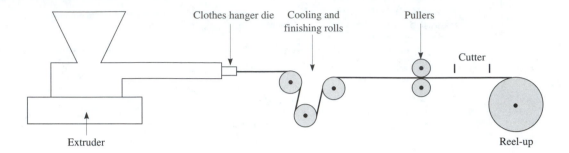

Figure 11.17   Sheet- and flat film–making equipment.

downstream rollers relative to the upstream rollers will thin the material. This process of reducing the thickness of the film by differential speeds of downstream rolls is called *drawing*. Drawing orients the molecules of the sheet or film into the machine (drawn) direction. Hence, sheets have more strength in that direction than in the cross direction. In some sheets and films, this orientation is undesired. Therefore, these sheets and films can also be mechanically grasped along their edges and pulled sideways to induce cross direction orientation. The side-pulling devices are called *tenterhooks*. The side pull is usually done at the same time as the machine direction pull. Sheets and films that have been drawn in two directions are called *biaxially oriented*.

These biaxially oriented films have strengths that can be twice as high as nonoriented films and sheets. Biaxially oriented films not only have strength in two directions, they also possess superior clarity to single-direction-oriented films. This clarity comes from the equalization of internal stresses when the part is drawn in two directions. Hence, film that is used for video cassettes, for movies, and for transparencies would generally be biaxially oriented. The increased strength from this drawing process is also important, especially because these materials are quite thin in many of their applications. Just as films and sheets can be drawn in order to increase their strength, fibers can also be drawn. In fact, almost all fibers are drawn, thus achieving many times more tensile strength than they would have without drawing.

Blown Films

This is the process that is used to make thin films like garbage bags and other high-volume, thin-walled film materials.

In this process the extruded material flows through a *tubular die,* which is like a combination of an offset die and a pipe die in that the material turns as it flows into the die, thus providing access to the back of the die and allowing the extrudate to travel upwards. The upwards motion is preferred because it makes gravity uniform over the entire part. Downward extrusion would also accomplish this purpose but most extrusion operations have much more room overhead than below for the additional downstream equipment. The melt flows around a mandrel and exits the die as a tube. A cooling ring is placed at the exit of the die to give the tube some dimensional stability since the material is air-cooled rather than cooled in a water tank.

The treatment of the tube after it has exited the die is the critical part of the blown film process. Air is introduced through the back of the die and flows upward inside the middle of the tube of material. This air flow pushes the tube outward. The tube, which is sometimes called the bubble, continues to expand, cool, and crystallize until the radial (tensile) strength of the plastic equals the pressure of the air inside or until some outer mechanical limit is reached. The cooling and crystallization increase the radial strength so that the bubble size is limited. The bubble continues upward with approximately the same shape (sometimes through a basket of rollers or an external air chamber to ensure that the shape is constant). The bubble is then forced into a flat sheet by the *collapsing guides* (also called the *tent frame*) and moved into the nip rolls. These rolls pinch the material together so that the bubble is maintained. The nip rolls also pull on the plastic in the same way that a puller system works with a traditional extrusion line. After the nip rollers, the material travels down over some rollers and enters the takeoff equipment. The material can be perforated and sealed (for bags) between the nip rollers and the takeoff equipment. The cut is usually perforated so that windup is still easy to accomplish. The high speed and continuous nature of this process often dictates that a two-station windup system be used for takeoff. A depiction of a blown film line is given in Figure 11.18.

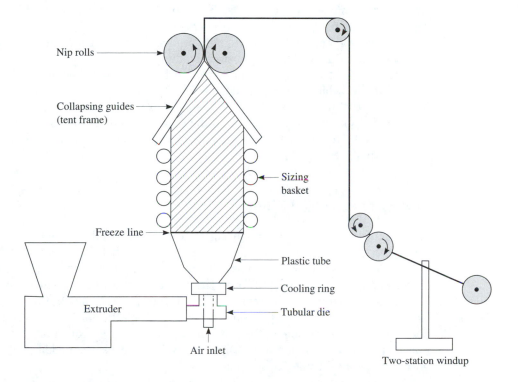

Figure 11.18   A blown film line.

The stretching and cooling of the tube causes the molecules to be oriented. The push of the internal air orients the molecules in the radial direction while the simultaneous pulling by the nip rolls orients the molecules in the machine direction. Blown films are, therefore, biaxially oriented. This orientation often causes some crystallization in the film. The amount of expansion of the bubble is important in controlling the process and predicting mechanical properties. Therefore, a parameter that measures the amount of expansion of the tube has been developed. This parameter is the *blow-up ratio* and is the ratio of the final tube diameter after blow-up to the diameter of the orifice of the die. Blow-up ratios of 3:1 are common. Most mechanical properties are increased as the blow-up ratio increases because orientation increases.

The cooling of the tube can be followed by the change in crystallinity that occurs with the cooling and stretching. The material exiting the die is amorphous and clear. As it cools and stretches the molecules become more oriented and develop a closer-packed configuration, similar to crystalline structures. These close-packed structures reflect light and therefore become less transparent. The line on the bubble that marks the onset of this close-packed molecular arrangement can be identified by a loss of clarity (haziness). Because it looks like frost or freezing has occurred at that point, the line is called the *frost line* or the *freeze line*.

To allow films of different thicknesses to be made using the same line, many tubular dies have a mechanism that allows the mandrel to move (by screwing) either closer to or farther away from a bushing. The gap between the mandrel and the bushing sets the exit gap for the die and, therefore, the thickness of the tube.

Some customers would prefer to have product that is not tubular. These products are made by slitting the material just before windup.

The crystallization rate of some resins, such as polypropylene, is so slow that it cannot be effectively blown as just described. For these types of materials, the system is modified so that the material is cold-water quenched as it exits the die. This creates a rubbery, amorphous tube that is then heated to the temperature that will cause crystallization to be most rapid. The tube is then blown at this temperature, cooled and passed to nip rollers in the normal way.

Coating of Paper and Fabric

Paper and fabric are also commonly coated with plastic. In these products the extruder is fitted with a wide die, such as a clothes hanger die, so that the extrudate is spread over a wide path. Just as with sheet and flat film production, the extrudate is passed over a series of rollers that cool and finish the part. The difference in the equipment when paper or fabric is to be coated is that the paper or fabric material is mated with the hot plastic at the first chill roll and is pressed into the surface of the plastic sheet or film. The pressing is done by the use of tension on the paper or fabric as it moves from a tensioning roll to the first chill roll. The passing of the sheet/paper material between additional rolls ensures good bonding of the plastic/paper laminate.

A system for paying off the paper or fabric and for guiding and tensioning the paper or fabric is important for proper alignment and mating of the paper or fabric with the plastic.

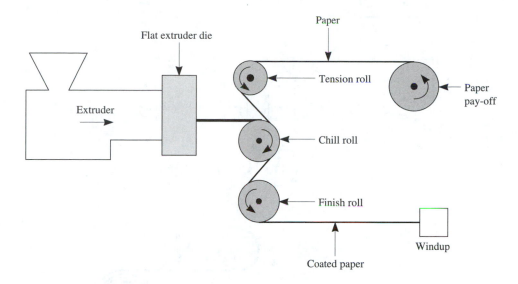

Figure 11.19 Extrusion coating of paper.

Likewise, proper takeup equipment is required to ensure that the material is not wrinkled or unnecessarily squeezed when rolled up. This process is illustrated in Figure 11.19.

Other materials besides paper or fabric that can be formed into a sheet form can also be coated by extruded film or sheet. The processes for these materials would be similar to those described for paper and fabric. It is also possible to bond two sheets together without the use of an extruder. This process is called laminating and is discussed in the chapter on finishing and assembly.

Calendering

Calendering is an alternate method for making sheets or flat films that does not use a die to spread the extrudate into a wide, flat film. Calendaring is not really a separate form of extrusion but is included here because it uses extrusion and does not warrant a separate chapter. In the calendering process the extrudate is extruded directly into the nip area between two rolls. The gap between these two rolls is small and therefore the plastic material builds a small pool on top of the nip area. This pool of material is called the *bank*, or *extrudate bank*. The rolls have a small gap between them and the plastic is forced through this gap by the counterrotation of the rolls. The gap sets the initial thickness of the sheet material that exits from between the first two rolls onto a second and, perhaps, a third roll. The stack of rolls further defines the thickness of the sheet or film by adjusting gaps and relative rotational speeds. The gaps between subsequent rolls get smaller and the temperatures of the rolls lower to finally set the dimensions of the part. Finally, the material moves to a takeoff station, usually a roll up. The calendering system is illustrated in Figure 11.20.

The advantage of calendering over direct sheet or flat film extrusion is that a complicated and expensive die, such as a clothes hanger die, is not needed. Also, some additional mixing can be done as part of the calendering process, which allows materials that might not be

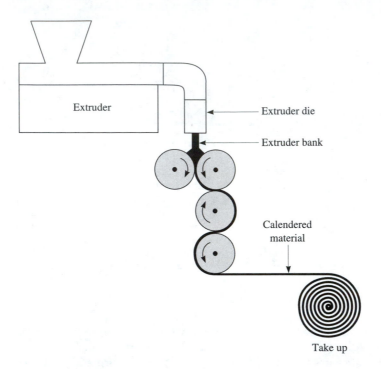

Figure 11.20
Calendaring system.

easily extruded to be added after the material has exited the extruder. The calendering process also provides a convenient method of texturing the surface of the sheet or film by simply texturing the outer surface of the final roll. Paper or fabric can be added in the calendering process much like they are added in direct film and sheet extrusion and coating.

The calendering process subjects the plastic to less heat than direct sheet extrusion and is therefore a convenient method for PVC and thermoset and elastomeric materials. Rubber membranes are made by this method as are fabric-reinforced rubber belts.

Examples of applications for calendered plastics include vinyl films for baby pants, inflatable toys, shower curtains, table cloths, pool liners, wall covering, and veneered panels. Credit cards, blister packaging, floor tiles, and floor covering (linoleum) are also made by calendering.

Synthetic Fibers

Although several methods for making synthetic fibers are employed, the method that involves extrusion (melt spinning) involves most of the general principles and so the fiber-making processes are discussed in this chapter. The processes are called: melt spinning, wet spinning, dry spinning, and gel spinning.

Many of the terms associated with making synthetic fibers are retained from the much older processes of converting natural fibers into long threads. One of the most obvious of these holdover terms is *spinning*. Natural fibers such as wool or cotton occur as short lengths (only a few inches or centimeters long) that must be made into a long fiber. These short fibers (called *staple*) are simultaneously overlapped and twisted (spun) so that a long fiber results. Hence, the method for creating a natural fiber was called spinning. The method for making a synthetic fiber is also called spinning, although nothing gets spun.

In melt spinning the thermoplastic resin is melted, much like any other shape would be melted except that the viscosity of the melt must be very low so that it can pass through the very small holes that are needed to form the fibers. Instead of moving directly from the end of the extruder to the die, the melt is conveyed to a manifold that distributes the melt to a series of dies. These dies which are called *spinnerets*, are like normal extrusion dies for making rods except that the exit orifices are very small and many orifices are cut into a single orifice plate. The holes are of the order of 0.005 inches (0.12 mm) and are streamlined to avoid melt stagnation and the associated problems like melt fracture and sharkskin. Some spinnerets can have hundreds of these holes in a single plate. The melt passes through the holes as it would through an extrusion die except that the normal orientation of the spinneret is to have the extrudate leave vertically downward. The extrudate is a thin rod or filament that is called a fiber. The cooling is done by blowing cool air across the fibers just after they exit from the spinneret. This process is shown in Figure 11.21.

The efficiency of the cooling is increased if the fibers from each spinneret are kept inside a small containment box and cooling air introduced into each box. This arrangement also prevents interactions between the fibers from different spinnerets since they may be lined up in close proximity.

The fibers leave the containment box and are gathered together by a roller. The group or bundle of fibers is called a *tow*. This roller directs the fibers into a drawing apparatus where the fibers are wrapped around rollers of different diameters or different rotational speeds that stretch or draw (elongate) the fibers. The stretching of the fibers aligns the molecules along the long axis of the fibers, thus increasing their tensile strength. This is the same effect as the drawing process that was previously discussed with films. The fibers are then wound onto bobbins. To help keep the group of fibers together, the tow can be twisted slightly and it then is called a *yarn*.

Some uses for fibers require that the fibers be chopped into short lengths (staple) after drawing and before windup. The fibers would be chopped and then baled for shipment instead of wound onto a bobbin. Staple fibers are especially useful if natural and synthetic fibers are to be blended. The blending is usually done when the natural fibers are spun into a yarn in a traditional fiber-spinning operation.

The gradual clogging of the screen pack can cause disruptions in the flow of the fibers. This problem is especially severe in fibers because of their small size and extreme sensitivity to pressure variations. To prevent this from occurring, melt pumps can be added to the melt line.

The shapes of the spinneret holes need not be round. Fibers with trilobal cross sections, square cross sections, and even holes down the middle like macaroni (sometimes with up to four holes) have been made. The different shapes are largely intended to

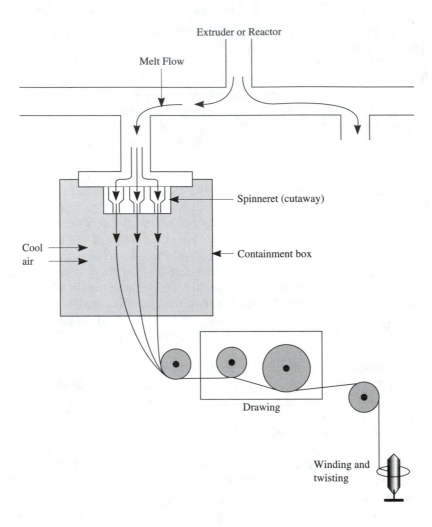

Figure 11.21   Melt spinning showing a synthetic fiber spinneret.

assist in soil-hiding properties and insulation. Dies to make these shapes have already been discussed.

Melt spinning is used to make nylon, polyester, olefin, and several other fibers.

Wet spinning is the method that was used to make the earliest synthetic fibers and is still used today for cellulosic-derived and some totally synthetic fibers. These fibers are made by dissolving cellulose and then forcing the solvated cellulose through spinnerets that are similar to those already discussed. The fibers are, at this point, streams of very viscous liquids. Instead of moving into an air gap after exiting the spinneret, the filaments pass directly into a chemical bath where they are solidified (also called *regenerated* because a solid material is regenerated). Because of the bath on the exit side of the spinneret, the process is called wet spinning. This process is used to make rayon (cellulosic), acrylic, aramid, modacrylic and spandex fibers.

In dry spinning the polymeric material is also dissolved in a solvent. In this case the fibers enter air when they exit from the spinneret and solidify because the solvent escapes to a reclamation process. This process is used to make acrylic, spandex, triacetate, and vinyon fibers.

In gel spinning, filaments coming from the spinneret form a rubbery solid on cooling. The process can also be described as dry-wet spinning since the filaments first pass through air and then are cooled further in a liquid bath. Some high-strength polyethylene fibers are produced by gel spinning.

POSTEXTRUSION FORMING

Postextrusion forming is any process that changes the shape of an extruded part while it is still on the extrusion line. Therefore, off-line processes such as adhesive bonding, cutting and shaping of discrete parts, and finishing by traditional methods are not part of postextrusion forming. However, some methods that have already been discussed are postextrusion forming. For instance, the shaping of a bell on the end of an extruded PVC pipe is done on-line and therefore qualifies. The drawing of sheet, film and fibers also qualifies.

A postextrusion forming process not already discussed is the making of corrugated plastic sheet. This sheet is extruded flat and then, on line, reheated slightly and passed between shaped rollers that deform the part and cause the familiar ripple pattern. Similarly, hot pipe can be passed between shaped rollers so that a wavy corrugated pattern is created in the pipe. The corrugation gives additional strength to the pipe and is therefore used with thin-walled pipe to improve crush strength. Some agricultural drainage pipe is corrugated by this method. Flexible drinking straws are similarly molded to create the familiar rippled, or corrugated, shape.

Some parts are simply heated on line and then bent into new shapes. These are most often done when the shape desired would be very difficult to create directly from the die, or when only some sections of the part are to be shaped.

Postextrusion forming is used to create expanded (porous-wall) medical tubing. This tubing is made by normal extrusion using an offset die through which a long hollow air conduit has been passed. This air conduit serves as the mandrel but it extends well beyond the end of the die and into the cooling tank. Air is blown through the air conduit at a rate that will cause the thin walls of the tubing to break apart slightly and become porous. The resulting product is used for passing materials through the tube and into the body, such as in kidney dialysis, and oxygenation directly into the bloodstream.

COEXTRUSION

This is a process that joins two or more streams of molten plastic into a single extrudate stream in such a way that the materials bond together but do not mix. Generally a separate extruder is needed for each stream and then the extruders are linked so that the extrudates can flow together in a way that is appropriate for the particular application. Three general locations for joining the various extruders are common. The first is to combine the melts just upstream of the die using a special adaptor. Laminar flows keep the materials from mixing. This method is the least expensive system because the adaptor is generally not expensive and the existing die can still be used. This method does not work well if the polymers are not well matched so that they bond easily. A typical example of this technology would be attaching a small extruder just upstream of the die to pump a different-colored resin into a larger extruder so that the final part will be striped. Drinking straws are a good example of this technology.

A second location for combining the streams is inside the die. This is the most common method and seems to insure the most reliable control over the mating of the materials since the length over which the separate laminar flows must be maintained is short. The disadvantage is the cost of a new die, which could be very complex, depending on the number of layers to be joined.

A die that embodies many of these features is the tubular die, which is used to make coextruded blown film. A mandrel is located in the center of the die to force the resin into

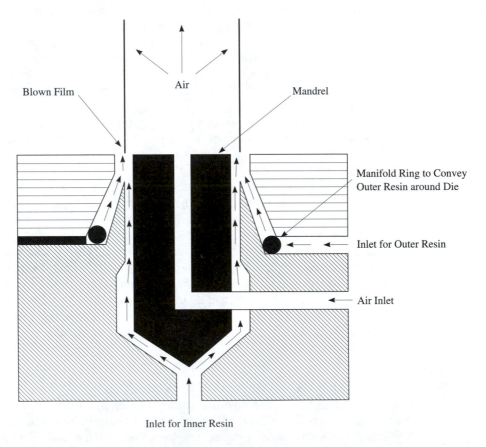

Figure 11.22 Coextrusion die for blown film.

a hollow circular form. Then, around the perimeter of the circle another die opening provides a path for a second resin to enter and surround the first resin. This makes a plastic tube within another plastic tube. The two resin tubes are brought into contact at the die opening and wed together, hopefully. (The resins should be chosen to have the compatibility to bond when they are brought together.) This type of die is shown in Figure 11.22. Two resin inlets are shown, one from each of two separate extruders. The air inlet in the middle of the mandrel is necessary because the blown film must have air pressure inside to accomplish the blowing of the film. The air pressure or volume is adjusted to give the correct bubble size.

The third location for combining the streams is at the lips of the dies. This method may include rollers that press the streams together. The advantage of this method is good control over the thickness of each layer. The disadvantage is the complexity of the machinery in the critical gap zone between the die exit and the cooling system. Also, bonding is likely to not be as good.

Some applications of coextrusion include the following:

- A special resin could be coextruded with the main resin so that special bonding, printing, or some physical or mechanical property might be introduced into a portion of the main part. An example of this would be the extrusion of a soft vinyl strip along one edge of a PVC vinyl window so that the window would have an integrally bonded weatherstripping.

- Wire and cable coatings are sometimes done in several steps with several different plastics being used as the coatings, one over the top of the other. This can be done in one extrusion operation by putting multiple extruders on the line and simply passing the coated wire from one die to another in a sequence.

- A coextrusion method that has proven to be very cost effective is to extrude a thin layer of high-value and costly resin over the top of an inexpensive resin. The inexpensive resin might be *regrind* (that is, material which has been extruded but did not meet specifications and has been ground or chopped into small pieces for easy processing) or simply a cheaper grade. This technique has been used in polyethylene pipe where the high-performance outer material is needed for stress crack resistance.

- Perhaps the most important commercial application of coextrusion is for making multiple layered films. In this case several extruders would feed together into a complex die that would merge each of the resin streams into a sandwich material. These multiresin films are valuable in obtaining good barrier properties for packaging films. No single plastic can give good barrier properties to all gases so by combining several plastic films together, the desired properties are obtained. In some cases additional layers are added to these barrier films to enhance some other property besides permeability. For instance, the ability of the material to be heat sealed or the flexibility of the material can be greatly improved by coextruding a plastic with these specific properties, or, a government-approved material may need to be the layer contacting the food. A similar application is the coextrusion of trash bags where a strong but expensive layer is extruded between weaker layers.

- In some cases, a mutually compatible layer must be extruded between two other materials because the outer materials are incompatible with each other and will not bond. The compatibility of all coextruded resins should be verified as bonding problems occur frequently.

■ Another use of coextrusion is to coat a foamed material, such as a sheet that will be later formed into an egg carton, with a layer that can be easily decorated and provides an attractive outer surface.

■ Other applications include coextruding a material with one desired property with a second material that has another desired property. For example, extruding ABS (for toughness) with polystyrene (for low cost) for use as refrigerator door liners and margarine tubs; or PMMA (for weather resistance) with polystyrene (for economy) for outdoor products; or polystyrene (for low cost) with polysulfone (for high-temperature stability) in appliance applications near the motor. Polycarbonate could be extruded on PMMA to obtain a tougher, more crack-resistant surface. Polyethylene or polypropylene could be extruded inside ABS so that chemical resistance can be imparted to a rigid, easily joined, and tough pipe.

The possible combinations of materials and applications are limitless. Certainly coextrusion has become a major manufacturing method and seems to have potential for many more applications. An alternate method to obtain some of these properties is through lamination, which is discussed in the chapter on Finishing and Assembly.

===== CASE STUDY 11.1 =====

Extrusion of Irrigation Tubing

Drip irrigation is an industry that utilizes plastics extensively for both the lines that carry the water and for the fittings and devices that are used to place the water near the plants and trees. The water is brought to the field through PVC pipe and is filtered so that the drip system (that part of the water system that is actually located in the field) will not be clogged by sediment in the water. PVC pipes also distribute the water to the heads of the major sections of the field or orchard so that the drip irrigation system can be connected to the PVC. The drip irrigation system then distributes the water directly to the plants and trees.

Two general types of drip irrigation systems exist and they are known as permanent and annual. The permanent system is used for permanent crops such as orchards, vineyards and other crops that remain in place for multiple years. This type of system has semirigid polyethylene tubes (called hoses) that run along the line of trees or vines. The tubes carry the water throughout the field, generally by running off the PVC header at the top of the orchard or vineyard. Sometimes the hoses will circle the trees or have smaller branch hoses that circle the trees. Small devices are attached to the hoses by penetrating the hose wall. These devices provide a water exit path that reduces the pressure of the water inside the hose so that it exits under very low pressure, generally as a trickle or drip. Therefore each device is a point source for water emission. These devices are called drip irrigation *emitters* and are built specifically to give a precise and preset amount of flow. Several emitters could be placed around a tree as the water needs of the tree require.

Although the hose is extruded, and is therefore of some interest in this chapter, the extrusion is rather common and will not be considered further in this case study. The emitters and fittings are injection molded. The second type of drip irrigation system is much more interesting in extrusion.

The second type of drip irrigation system is called annual, implying that the drip irrigation system is used only for the life of a single crop. The crops in this case are annual crops, such as tomatoes, strawberries, or sugarcane, that must be plowed and replanted each year or every two years. When the crop is finished and the field is plowed in preparation for the new crop, the drip irrigation system is removed. Some farmers attempt to save and reuse this tubing, but most simply throw the tubing away because it costs less to buy new tubing than to store and refurbish old tubing. In either case the tubing is removed from the field after each crop.

An annual system consists of flexible, thin-walled tubing that is attached to the PVC header in much the same way that the permanent hose is attached. The difference is that the tubing is much thinner and, instead of emitters that are inserted into the hose, holes are predrilled in the flexible tubing to emit the water. The flexible tubing runs along each of the plant rows and emits water through the predrilled holes. The spacing of the holes is usually from 12 inches (30 cm) to 40 inches (1 m) depending upon the spacing of the plants along the row. No attempt is made to line up the holes with the plants. The natural capability of the soil to distribute the water is usually sufficient to insure that the root system of all the plants is wetted.

Several designs of this thin-walled tubing have been made and their differences illustrate well some of the characteristics and innovations in design that can be made using extrusion as the principal manufacturing method. The designs of the most common, both historically and commercially, are illustrated in Figure 11.23. The object of all designs is the same—to provide a low-cost tubing that will distribute water evenly along the entire length of the row.

The first design of tubing (illustrated in Figure 11.23a) (called monotube) is simply an extruded tube with wall thickness of about 0.16 inches (0.4 mm). The holes that emit the water were punched or drilled on the extrusion line after the tubing was made using lasers that could be focused to ensure that no damage would occur to the inside of the tubing opposite the emitter hole. (Attempts to punch or drill the holes mechanically invariably caused damage to this wall opposite the hole.) The cost of this tube was very low; however, the emission of the water was not even from head to tail along the row. The problem was that when the water entered the tube it was at a higher pressure (typically 10 psi, 70 KPa) than when it reached the end of the row. This decrease in pressure was due to the natural frictional pressure loss experienced along all tubes and pipes. The typical length of a crop row was 300 feet (100 m), which gave pressure differences of 10 to 20%, head to tail along the row. Variations in water supply of this magnitude made significant and unacceptable differences in plant growth and yield. Therefore, even though monotubes were inexpensive to make, the field performance was not acceptable and these tubes are no longer made commercially.

A second type of tube was invented that consisted of two chambers, one to carry the high-pressure water and the second to distribute and emit the water. The high-pressure chamber would have pressures of about 10 psi (70 KPa), much like the monotube case,

Figure 11.23
Several designs of
drip irrigation tubing.

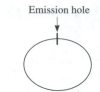

(a) Monotube, extruded and then punched

(b) Twin-Wall® tubing made from flat sheet that is punched and seamed

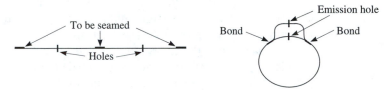

(c) Twin-Wall II® tubing that is punched and double-seamed

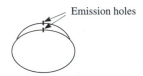

(d) Bi-Wall® extruded profile then drilled with no bonding

and would carry that water to the end of the row. However, because this high-pressure tube had only a few holes (perhaps every 6 to 10 feet, 2 to 3 m) that allowed water to flow, the pressure drop would be much less than with the monotube, about 3 to 7%, head to tail. These differences would be acceptable for plant growth uniformity.

However, with only a few emission holes in the high-pressure chamber, the water would not be emitted frequently enough to wet the roots of all plants. Therefore, the second (low-pressure) chamber was attached to distribute the water more evenly. The second chamber would have holes to the outside about every 12 to 36 inches (30 to 100 cm), perfectly acceptable for good plant wetting. The pressure in the second chamber would be about one-tenth the pressure in the high-pressure chamber, hence the water would trickle or flow from the outer emission holes. The problem with uniformity down the plant row would not be present with the double-chamber tube because the length of the distribution tubes would effectively be only the distance between the holes from the high pressure chamber. Furthermore, the pressure in the distribution chamber would be so low that small variations in pressure would be insignificant in terms of actual water flow.

A tube named "Twin-Wall™" was manufactured in this double-chamber design and a patent obtained. It was fabricated from two extruded sheets of about 0.1 inches (0.3 mm) thickness and 1.3 inches (3.5 cm) and 2 inches (5.0 cm) wide that were punched to give the proper holes (size and spacing). These two strips of flat sheet were then bonded together with an ultrasonic sealer to form the double-chambered tube shown in Figure 11.23b. The extrusion was simple (flat sheet) and the punching was easily done online. The cost of the Twin-Wall tube was higher than the monotube. Even though the extrusion rate for the flat sheets was much faster than for the monotube and each Twin-Wall sheet was thinner than the monotube, the total amount of material in Twin-Wall was significantly more because of the twin-chamber design. A cost reduction was achieved when wide sheets 6 feet (2 m) were extruded and then slit to the proper widths. The remaining major difficulty was the seaming which was slow and could break out from the pressure in the high-pressure tube. However, this product sold well for many years.

The next-generation product was also made by the same company that made Twin-Wall and was called Twin-Wall II. The main difference was the amount of material that was used and the seaming locations. The product was made from a single flat sheet (cut from a larger sheet) that was punched with both the inside and outside emitter holes. This sheet is illustrated in Figure 11.23c, where the locations of the holes are shown. The flat sheet was then wrapped around two mandrels to form the high-pressure chamber and the smaller distribution chamber and two seams were made to form the two chambers. The two seams raised the chance for failure, but much less material was needed than the original Twin-Wall design. Hence, the cost of this product was much less than the original Twin-Wall.

While much of the development of Twin-Wall was occurring, a competitive product was being developed. This product was also two-chambered but had no seams. Both chambers were completely formed by extrusion. This product was called "Bi-Wall™" and is illustrated in Figure 11.23d. Initially the Bi-Wall tubing was much thicker than Twin-Wall because of the problems associated with extruding the difficult 2-chambered profile. The problem was in keeping the sizes of both the large and small chambers simultaneously constant and in keeping constant wall thicknesses, especially the wall separating the chambers (called the web). Keeping the walls constant was important because the holes were made with lasers. These lasers were mounted on the extrusion line (just after cooling) and were time triggered. The inside holes were made by laser drilling through both the outer and inner walls by changing the focal length of the laser and by changing the time of the laser pulse. If either the outer wall or the web was too thick or thin, the size of the hole would be less or greater than specification. Hence, drawdown, mandrel sizes, the gap between the extruder and the cooling tank, the sizing plates in the cooling tank, and the relative speeds of the extruder and the puller had to be carefully monitored and controlled.

The cost advantage of Bi-Wall came from the ability to make the product all in one operation. Twin-Wall was first extruded into sheet (by one manufacturer) and then punched and then slit and then folded and then seamed, whereas Bi-Wall was made in one pass on one line. Further cost advantages could be obtained if Bi-Wall could be extruded faster and with thinner walls. Therefore, great efforts were made to increase the speed of the Bi-Wall extrusion operation while simultaneously reducing the wall thickness.

The first attempts were to simply reduce the wall thickness. Because the fittings that connected the Bi-Wall to the PVC header pipes depended upon having a constant inside diameter for Bi-Wall, any reduction in wall thickness would have to be made by keeping the inside diameter constant. This was done by maintaining the diameters of the mandrels and by reducing the gaps between the mandrels and the die walls and the mandrels from each other. A reduction in wall thickness from 0.02 inches (0.6 mm) to 0.01 inches (0.3 mm) was made by this method over a period of about four years. The long times involved in making this change were associated with the need to retime the lasers to get the proper hole size, insure that the change in thickness did not adversely affect field performance, and ensure that the thinner small chamber would not be crushed during the windup so that it would not perform properly in the field. Eventually the product wall thickness was reduced to 0.006 inches (0.15 mm).

The initial attempt to increase line speed was simply to increase the speed of the extruder and the puller. This resulted in increased drawdown and, disastrously, increased susceptibility to environmental stress cracking. After many thousand meters of Bi-Wall stress cracked in the field, the amount of drawdown was returned to the original level and attempts for increased speed were made elsewhere.

The next attempt to increase line speed was to improve cooling because it was found that a major problem in increasing speed was cooling the web. This part of the tube never touched the cooling water and therefore depended upon heat conduction and convection through the air for cooling. If the web was not cool by the time the holes in it were drilled, the holes would change size with cooling and the water emission rate would change. The initial improvement in cooling was simply to extend the length of the cooling system and to decrease the temperature of the water bath. Both of these helped, but further cooling was required.

The major breakthrough in increasing speed was made by changing the type of die that was being used. Initially Bi-Wall was made on a standard spider die with a second, smaller spider die for the upper chamber. The change was to use an offset die, much like a wire and cable coating die. Instead of using the wire or cable as the mandrel, hollow tubes were used. Cool air was blown through these hollow tubes to provide direct cooling on the web. In combination with the improved water cooling system and other mechanical improvements in pulling and windup, the speed of the line was increased from an initial rate of 300 feet (100 m) per minute to an amazing 1600 feet (500 m) per minute.

The net result of these changes in wall thickness and speed was a dramatic reduction in the cost of the Bi-Wall product from approximately $0.25 per foot ($0.80 per meter) to $0.06 per foot ($0.20 per meter). The market share of Bi-Wall reached over 90% in some markets and it became the most widely used drip irrigation tubing in the world.

SUMMARY

The extrusion process is the most fundamental method for processing thermoplastics because it efficiently melts, mixes, and pumps viscoplastic materials. Since almost all thermoplastic materials must have these processes done to them at some time, extrusion is used more than any other plastic-forming method. Sometimes extrusion is a captive process within another plastics-manufacturing method, as in most injection molding machines.

In some cases, other processes could be used to make the same parts. For instance, some sheets can be made by laminating or by calendering as well as by extrusion. It is the role of the plastics-manufacturing engineer to choose the process that is most appropriate for each circumstance. The purpose of this text is to acquaint the reader with as many of the processes as possible so that a well-informed choice can be made.

The machinery elements of extrusion have been discussed in detail. The extruder itself is complex, having many components that all must work in concert for the part to be made most efficiently. The extruder base, motor, gear linkage, thrust bearing, and length of barrel are critical equipment parts to be considered when buying an extruder, but are of lesser interest during normal operations. Other equipment such as the screw, heating sections, die, cooling chamber, puller, and part removal are considered much more frequently. The operating parameters of the extruder are, of course, considered continuously and the settings of these parameters dictate the operation of the extruder within the larger constraints of the equipment type.

This chapter has discussed all of the equipment and key operating parameters both from the viewpoint of required function and operational effect. One important operational concept for extrusion that should be reemphasized is that extruders run best when in a steady-state condition. The operating parameters should be carefully monitored so that steady state can be quickly reached and then easily maintained.

A list of the most common problems in extruder operations was presented. Some procedures were outlined for solving each of these deviations from the normal steady state. The procedures had many options in most cases and so judgment and experience should be used in deciding what remedial path to follow. Although not specifically mentioned, a frequent suggestion is that changes in machine settings should be bold if the problem is severe. Many hours have been wasted by making timid changes and waiting for some evidence of improvement. Extruders will self-compensate against many changes making incremental changes difficult to effect.

The major types of products that are extruded were discussed. The simplest parts are solid rods that are extruded out of a simple, streamlined orifice. When pellets are made by resin manufacturers or by compounders, the pellets are made in dies of this type. Of course, many other rods are made which are, themselves, the finished or near-finished part.

Extrusion dies are usually made by the extrusion operator or can be purchased for typical, standard parts (such as PVC pipe). The tolerances allowed in the body of the die are much less demanding than would typically be required in an injection molding die. Only the face of the die must have careful dimensional control, and even that is less stringent than the typical injection molding cavity. Extrusion has more latitude to change the dimensions of the part during the process than can be done in injection molding.

If the rod is shaped into any other form than cylindrical, then the part is called a profile. Profiles can have simple cross sections such as squares, triangles, or angles stock, or they may be very complex, with holes, sections of various widths and lengths. All of these parts must be made from a die that is cut appropriately for the particular shape desired.

As the polymer may swell more in some portions of the die than in others, the cutting of profile dies can be an art since actual running of the die is so difficult to model.

Pipes and tubing dies introduced the concept of a mandrel, as in a spider die. This important feature allows hollow parts to be extruded. In some cases, the die is offset from the end of the extruder and the rear of the die is open which allows entry of a wire or some other device that moves through the die and becomes part of the finished product, as with coated wire and cable. The offset die was developed to allow access to the back of the die so that the wire or cable could be moved through the die.

Sheet extrusion introduced the concept of a wide die (clothes hanger die) that allows even or plug flow through the die so that internal stresses will not be created. This type of die is also used in coating wide materials such as paper and fabric. A wide die can also be used to feed material to a calendering operation in which rolls are used to change the shape of the molten plastic material.

Blown film is made with a tubular die that has air blown up the center so that the polymer extrudate will be expanded into a bubble. The size of the bubble can be controlled and that will dictate the amount of orientation that the molecules obtain. The molecules are oriented in both the machine and the cross direction by the blowing of the film to give strength in both directions.

Synthetic fibers are made by forcing a polymer melt or solution through tiny holes that set the size of the fiber. The plate through which the holes are drilled is called a spinneret and it serves the same purpose as the die in normal extrusion. The fibers can enter either a solvent or dry space after leaving the spinneret, depending upon the nature of the polymer and the process that is used to make it into a fiber. The process of making fibers is called spinning.

Although extruded parts have a constant cross section, that cross section can be changed online to give some special products. Corrugated panels and expanded tubing are two such products. The process of changing the shape of the part while it is still on the extrusion line is called postextrusion forming.

Coextrusion is a process in which two or more resin streams are brought together and formed into a single part. Generally the streams are from separate extruders that are mated either just before, at, or just after the die. The laminar flow of the polymer in these regions of the extruder usually results in a laminar product.

The concepts involved in extrusion are useful in understanding many other processes. Therefore, extrusion is the first of the plastics processes to be considered and was considered in great detail.

GLOSSARY

Adapter   A ring on the end of the extruder to which the die is mounted.

Adiabatic heating   Mechanical heating induced by the turning of the screw.

Bambooing   A defect in the skin of an extruded part that resembles bamboo; a severe form of sharkskin.

Bank   A mass of resin that builds up during certain extrusion processes, including twin-screw extrusion and calendering.

Barrel   The cylindrical chamber in which the extruder screw turns.

Bi-axial orientation   The process by which a film or sheet is stretched both in the machine direction and in the cross direction, thus increasing the orientation of the molecules in both directions and, therefore, increasing the strength in both directions.

Blinded   When the screen pack is clogged.

Blow-up ratio   The ratio of the size of the bubble in a blown film to the size of the die opening.

Breaker plate   A metal plate with small holes in it that supports the screen pack.

Bridging   Partial melting of the polymer in the feed zone and subsequent sticking of the polymer to the screw and cessation of the conveying of material along the barrel.

Clothes hanger die   Named for its shape, a die used to spread the molten resin from the compact cylinder shape of the flow as it enters the die into a wide, flat shape needed to make some films and sheets.

Collapsing guides   A set of guides that force a blown film bubble into a flat shape for shipping.

Compounder   A manufacturer who does compounding.

Compounding   The mixing of additives into a plastic, a process often done with extrusion.

Compression ratio   The ratio of the flight depth in the feed section to the flight depth in the metering section.

Compression zone   The region of the screw in which melting occurs; the melting section or the transition section.

Concentrates   Premixed and, usually, preextruded (pelletized) materials that are to be blended into the regular resin; the use of concentrates facilitates the mixing in of these materials.

Converter   A manufacturer who performs extrusion for another company, where the customer company has designed the product, often provides the tooling, and may even supply the resin.

Corotating screws   Twin screws both rotating in the same direction.

Cored screws   Extruder screws in which water can be circulated, chiefly to prevent bridging.

Counterrotating screws   Twin screws rotating in opposite directions.

Cram feeder   A type of auger that is mounted to the hopper to compress the material (usually a fluffy or powdery form) that is being fed to the extruder.

Crosshead die   Used to coat wire and cable and other applications where the back of the die must be accessible; also called an offset die.

Die swell   The distortion in the shape of the extrudate because of expansion when exiting the die.

Drag flow   The measure of the amount of material that is dragged through the extruder by the friction of the barrel and the screw.

Drawdown ratio   The ratio of the maximum diameter of the swell of an extrudate to the final size of the part.

Drawing   The process of pulling or orienting a sheet, film, or fiber and causing the molecules to line up in the direction of pull, thus increasing strength in that direction.

Drive motor   A motor that provides power for turning the extruder screw.

Emitters   Devices that reduce the water pressure at the delivery points to the crops in drip irrigation systems.

Extrudate   The material that is extruded (the plastic part in an extrusion operation).

Extruder   An extrusion machine.

Extruder burn   Local decomposition of the plastic inside an extruder, caused by excessive mechanical heating.

Extrusion   A shaping/molding process used in plastics processing in which the material is melted and then pushed out the end of the machine, usually through a forming die.

Feed section   The first portion of the screw where the principal function is to convey material forward.

Feed throat   The entry port for the resin into the extruder barrel.

Fish eyes   Defects in a molded plastic part that resemble fish eyes and are often caused from contamination of the resin.

Flight   The rises in the screw pattern on the surface of the screw root; sometimes also a single turn in the helix or screw pattern along the screw.

Flight depth   The distance from the tip of the flight to the surface of the root.

Force feeding   Compacting a low-density resin in the hopper or feed throat so that it can be extruded more easily; see also *cram feeder*.

Fractional melt materials   Resins with a melt index lower than one (1)—that is, resins with high molecular weight.

Freeze line or frost line   The line on a blown film bubble that marks the onset of crystallization. It is detected as the line where the plastic changes from clear (amorphous) to translucent (crystallized).

Gap distance   The distance between the end of the die and the cooling tank.

General-purpose screw   A screw with compromises in the lengths of the various sections so that many different resins can be run, usually with the assistance of temperature variations in the heating zones.

Go/no-go gauge   A quality control device that allows rapid determination of whether a plastic part is of the correct size by inserting the part into the gauge or inserting the gauge into the part and noting whether the engagement line of the part and gauge falls within an acceptable region.

Gum space   The small space between the wire or cable that is coated in an offset die and the interior diameter of the die orifice; this space determines the thickness of the coating.

Head zone   The region at the end of the screw and before the die.

Hygroscopic   Materials that absorb moisture.

Inside sizing mandrel   A mandrel used inside a hollow part when the part is inside the cooling tank, thus controlling its internal diameter.

Land   The outlet portion of an extrusion die; characterized by a flat zone in which the walls of the zone are parallel, thus imparting a straight, linear flow to the exiting melt.

Leakage flow   The amount of material that leaks past the screw in the small space between the screw and the barrel.

Length/diameter (L/D) ratio   The fraction representing the length of the flighted portion of the screw divided by the overall diameter of the screw.

Lubricants   Materials added to a resin that facilitate the extrusion of the resin, usually from acting to coat the resin or the extruder but perhaps also as plasticizers.

Mandrel   A rod that is placed inside a die to force the molten resin to flow around it, thus creating an opening; used to make tubes, pipes, and other hollow extruded parts.

Material bank   A mass of plastic material that builds up on the top of the twin screws in a counterrotating system.

Melt fracture   Short cracks or ridges that appear in the extrudate; usually caused by poor flow of the molten polymer.

Melt strength   The property of a molten resin that allows it to be pulled in tension.

Metering zone   The region of the screw in which the principal purpose is pumping the resin forward and ensuring that all the material is melted.

Offset die   A die in which a tube connects the extruder output into the side of a die, thus allowing entry into the back of the die for some other material, such as wire or cable; also called a crosshead die.

Orange peel   A defect in the surface of an extruded part in which small dimples are formed; related to sharkskin.

Pitch   The angle of advance of the screw flights.

Plastic memory   The tendency of a molten or semimolten material to return to its original shape after a deforming stress has been removed.

Plug flow   Flow in which all parts of the molten material enter and exit the die at the same time, regardless of their geometrical distribution across the face of the die.

Positive pumping   When a pump, usually mounted at the exit of the extruder, forces its output, regardless of opposing pressure.

Pressure flow   The measure of the amount of material that tends to flow backward in the extruder barrel because of the back pressure created by the extruder screw's pumping action.

Profile shapes   Extruded parts that have shapes other than very simple rods, sheets, tubes, and pipes.

Postextrusion forming   The process of giving a different shape to an extruded part by molding the part after it has exited the extrusion die but while it is still on the extrusion line.

Puller   A machine that pulls on the extrudate after it has been cooled.

Purging   Running the extruder just before shutdown or when changing resins to eliminate the presence of a particular resin; usually done by adding another resin to the hopper and running the extruder. Sometimes the purging resin is specifically tailored for purging.

Regenerating   The process of resolidifying cellulosic fibers or other fibers made by wet spinning at the time of spinning into synthetic fibers.

Regrind   Resin that has been extruded or molded but, because of some defect, is ground into flakes to be used again in a molding or extrusion operation.

Root   The shaft of the screw.

Screen pack   A group of screens that act as a filter; located at the end of the screw in the head zone.

Screw   The long, solid shaft that turns inside the extruder barrel and that has an Archimedes screw pattern machined on the outer surface so that the screw will move material along.

Sharkskin (alligator hide)   A defect in an extruder part in which the outer surface of the part is rough with lines running perpendicular to the flow direction; a tearing of the outer surface; usually associated with stresses in the extrudate from sticking to the die walls.

Sizing plates   Plates with holes in them or rings located in the cooling bath that give the extrudate its final shape.

Spider die   A die used for forming hollow parts in which the mandrel is supported by several pins inserted through the die walls; when viewed from the back, the pins resemble the legs of a spider.

Spinnerets   The dies through which synthetic fibers are made (spun).

Spinning   The process of forming a fiber, usually a simple extrusion through a die when the fiber is synthetic (plastic).

Staple   Short fibers.

Starve feeding   A technique for running the feed of an extruder in which the speed of the extruder screw is so fast that the material that drops onto the screw is immediately carried away, thus assisting in control of the feeding of difficult (such as powdery) materials.

Static mixers   A short device that causes the molten resin in the barrel to flow in a mixing pattern; the device is placed inside the barrel in the space between the end of the screw and the breaker plate.

String up   The manual pulling of the extrudate through the cooling system and into the puller during startup of an extruder.

Surging   Regular variations in the speed of the extruder motor or in the extruder flow; often associated with the presence of a solid contaminant inside the barrel, a misaligned screw, partial bridging, or some other improper flow/melting pattern for the melt.

Tent frame   A set of collapsing guides that force a blown film into a flat shape for shipping.

Tenterhooks   Devices that grip the edges of a film to allow the film to be drawn sideways.

Thermal heating   Heating of the plastic due to the heat from electrical heaters surrounding the barrel.

Thrust bearing   A bearing that supports the screw inside the extruder barrel and prevents plastic material from moving into the gear mechanism.

Tolling   The process of converting resin to shapes when the resin, die, and design are supplied by another company; in essence, the customer company rents time on the extruder.

Tow   A group or bundle of fibers.

Tubular die   A die that is used to make blown film and therefore forms a tubular shape but also allows access to the back of the die for air for blowing the film to be introduced.

Twin-screw extruders   Extrusion machines that have two rotating screws operated in tandem and mounted so that the flights of one screw mesh with the flights of the other.

Two-stage screws   Extruder screws that have a sudden reduction in the root diameter to cause a drop in pressure, usually in the vicinity of the vent port.

Yarn   A twisted tow, thus assisting in keeping the fibers in a bundle together.

QUESTIONS

1. What are the differences between adiabatic heating and thermal heating as applied to extruders and why are both important?

2. Discuss why the compression zone of a nylon-type screw is so much shorter than the compression zone in a polyethylene-type screw.

3. Discuss the temperature profile shapes normally used for nylon and polyethylene when using a general-purpose screw. Which is higher in actual value? What are the slopes of each?

4. Describe a spider die. How are adjustments made?

5. Discuss three operational steps that can be taken to reduce the thickness of an extruded part.

6. Explain the effect of molecular weight distribution on extrusion.

7. Discuss the relationship between the melt temperature and the location of the frost line in blown film manufacturing.

8. Discuss why melt pumps are rarely used with twin-screw extruders.

9. During the operation of an extruder in making thin-walled tubing, several defects have been encountered that appear to be small globules of resin. Investigation of these globules reveals that they will soften but not melt when heated. Explain what these globules might be and suggest two possible remedies for eliminating them.

10. Explain why outside diameter dimensional control is easier than inside diameter dimensional control for extruded pipe.

11. Why do the cross-sectional shapes of the die orifice and the extruded part sometimes differ?

REFERENCES

Brydson, J. A., *Handbook for Plastics Processors,* Oxford, UK: Heinemann Newnes, 1990.

Chandra, Manas, and Salil K. Roy, *Plastics Technology Handbook,* New York: McGraw-Hill Book Company, 1973.

Charrier, Jean-Michel, *Polymeric Materials and Processing,* Munich: Hanser Publishers, 1991.

Frados, Joel (ed.), *Plastics Engineering Handbook* (5th ed.), Florence, KY: International Thomson Publishing, 1994.

Griff, Allan L., *Plastics Extrusion Operating Manual,* Bethesda, MD: Edison Technical Services, 1990.

Manufactured Fiber Fact Book, Washington: American Fiber Manufacturers Association, Inc., 1988.

Milby, Robert V., *Plastics Technology,* New York: McGraw-Hill Book Company, 1973.

Morton-Jones, D. H., *Polymer Processing,* London: Chapman and Hall Ltd, 1989.

Richardson, Terry L., *Industrial Plastics: Theory and Application* (2nd ed.), Albany, NY: Delmar Publications, Inc., 1989.

Saechtling, H., *International Plastics Handbook* (2nd ed), Munich: Hanser Publishing, 1987.

CHAPTER TWELVE

INJECTION MOLDING PROCESS

CHAPTER OVERVIEW

This chapter examines the following concepts:
- Equipment (injection unit, mold, clamping unit, plant concepts, safety)
- Material and product considerations (materials, shapes, part design)
- Operation and control of the process (critical parameters, troubleshooting and quality improvement, and maintenance)
- Specialized injection molding processes

INTRODUCTION

In contrast to the extrusion process, which makes continuous parts of constant cross section (as discussed in the chapter on the extrusion process), injection molding makes discrete parts that can have complex and variable cross sections as well as a range of surface textures and characteristics. The wide variety in the types of parts that can be made by injection molding is a key reason that more injection molding machines are used for plastics processing than any other type of molding equipment. Almost all thermoplastics and some thermosets can be injection molded, thus adding to the flexibility of the process. Another important reason for the popularity of injection molding is that parts are highly repeatable, that is, the parts can be made with very little part-to-part variation. Furthermore, injection molding is highly automatable with little post-molding finishing required in most parts and high output rates, so that labor costs can be a small portion of the part cost. For some complex parts, such as those shown in Photo 12.1, injection molding is the only practical manufacturing process.

The difficulties of injection molding are the high cost of the machines and the molds, the relatively high amount of scrap (process residue) that must be reprocessed, the high level of competition because of the large number of molders capable of using the process, and the need for development and proper use of automated feedback and control systems.

The injection molding process is conceptually simple. In the process a plastic is melted and then forced into the cavity of a closed mold which gives shape to the plastic.

419

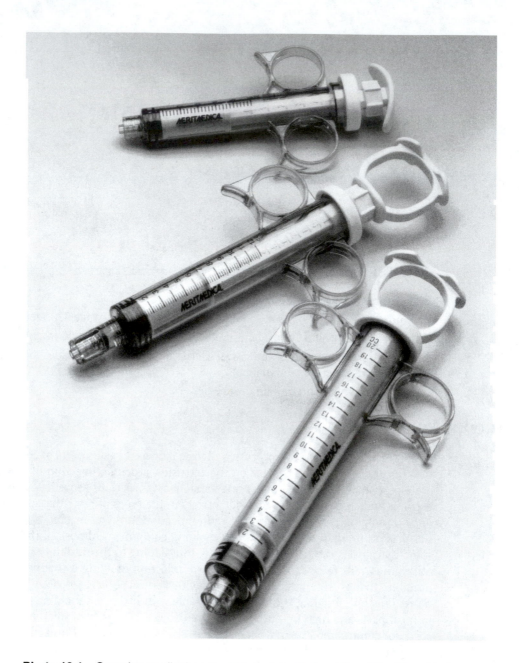

Photo 12.1   Complex medical parts made by injection molding. (Courtesy: Merit Medical Systems, Inc.)

After sufficient time for the plastic part to solidify (usually by cooling), the mold opens and the part is removed.

The keys to success in injection molding are to have (1) the proper machine for good melting and injecting of the resin, (2) the proper resin for appropriate part performance, (3) a good mold for part definition and removal, and (4) proper operation for efficient molding cycles. Each of these will be discussed in detail in the sections that follow.

The major factors that determine costs are material type and mold cycle. The material type is usually set by the part performance requirements, therefore, most efforts at cost reduction are aimed at reducing mold cycle. The mold cycle is strongly dependent upon the design of the mold and the parameters of the manufacturing operation. Mold cycle costs also include direct labor and overhead costs such as mold and machine amortization. The equipment and its proper operation are, therefore, keys to the cost and operation of the process.

EQUIPMENT

The equipment for injection molding is divided into three main functional units: (1) injection, (2) mold, and (3) clamping. These sections are each shown in Figure 12.1 along with the major components of each unit. Each of these units or functions will be discussed separately. Photos of an injection molding machine and its controller are given in Photo 12.2.

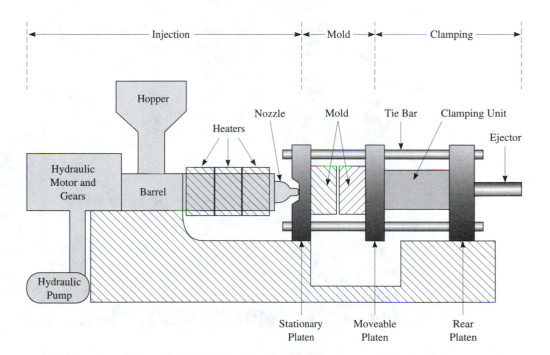

Figure 12.1   Injection molding machine showing three major functional units (injection, mold, and clamping) along with major components of each unit.

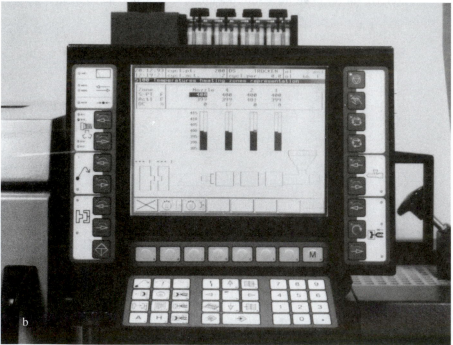

Photo 12.2 (a) Commercial injection molding machine and (b) controller. (Courtesy: Boy Machines, Inc.)

Injection Unit

The purposes of the injection unit are to liquify the plastic material and then to inject the liquid into the mold. Several component parts are required to accomplish these tasks. The components work together but are considered separately.

The resin is usually introduced into the injection molding machine through a hopper. The hopper can be fed manually or pneumatically (either semi-automatically or fully automated). Several hoppers can be clustered on one machine so that fillers, colorants, or other additives can be fed simultaneously. In these cases the injection molding machine is also a mixer. However, because of the limited size (barrel length) of most injection molding machines, mixing capability is poor and so most fillers and other resins are blended in a separate extrusion operation. Colorants are commonly added as preextruded concentrates rather than directly in the injection molding machine. Some hoppers have volumetric or gravimetric units attached to ensure precise feed rates for all materials being added. Drying is very common on injection molding hoppers because of the hygroscopic nature of several important injection molding resins such as nylon, polycarbonate, PET, and ABS.

The resin enters the barrel of the injection molding machine by gravity from the hopper through a hole in the top of the barrel called the feed throat. The barrel is a heavy steel cylinder built to withstand the pressures and temperatures involved in melting the resin.

Two systems have been used in injection molding machines to melt and inject the resin. The most commonly used type uses a *reciprocating screw,* which has many similarities to an extruder screw but with a unique reciprocating (back-and-forth) action. The other type of injector system is the *ram injector,* which is discussed later. The action of the reciprocating screw machine can best be understood by referring to Figure 12.2.

The design of the screw is similar to an extrusion screw. Both have three sections: feed (to advance the resin), compression (to melt the resin), and metering (to homogenize the resin and pump it forward), just as for extrusion screws. Because of the wide variety of resins that are typically run on a single injection molding machine, most screws used in injection molding machines are general-purpose screws, not specialized to the melting characteristics of only one resin. Hence, compromises in obtaining ideal molding have been made, again reducing mixing efficiency in injection molding when compared with extrusion.

Injection molding machines and screws are much shorter than extruders and therefore the L/D ratio for injection molding screws is lower than for extruders. Typical values for injection molding machines are 12:1 to 20:1. These low L/D ratios suggest that mixing and blending are less efficient in injection molding machines than in extruders. The compression ratios for injection molding screws (ratio of the diameter of the root in the feed zone to the diameter of the root in the metering zone) are often in the range 2:1 to 5:1, which is somewhat lower than for extrusion. This lower compression ratio simply means that less mechanical action is added during the melting process and more thermal energy is needed.

The first step in the cycle of a reciprocating screw injection molding machine is the turning of the screw to melt the resin. This is the step that is like the operation of an extruder and is depicted in Figure 12.2a. The resin is melted by both mechanical shearing

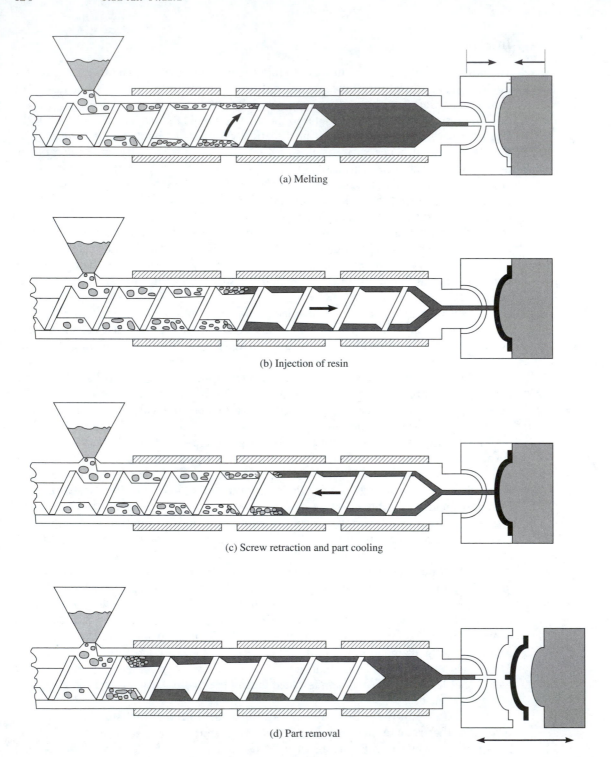

(a) Melting

(b) Injection of resin

(c) Screw retraction and part cooling

(d) Part removal

Figure 12.2   Injection cycle for a reciprocating screw injection molding machine.

action and by thermal energy from the heaters that surround the barrel. The molten resin is conveyed to a space at the end of the screw where it collects in a pool. During this step, the mold is closed.

In the second stage of the cycle (see Figure 12.2b), the entire screw moves forward (usually driven by a hydraulic mechanism at the drive end of the machine) and pushes the molten resin out through the end of the barrel. To ensure that the resin does not flow backward, a *check valve* or nonreturn valve is attached to the end of the screw. The action of the check valve is shown in Figure 12.3. Normally the screw will stay in the forward position until the resin begins to harden in the mold. This creates pressure in the mold, thus ensuring that the mold fills completely. When the screw retreats in a later step in the cycle, the valve opens and allows the resin to move forward to re-form the resin pool.

Perhaps the most important measure of the size of an injection molding machine is the weight of resin that can be injected. This weight is called the *shot size*. Typical shot sizes for injection molding machines range from 0.7 ounces (20 g) to 700 ounces, about 44 pounds (20 kg). Because the shot size is dependent upon the density of the plastic being injected, polystyrene has been chosen as the standard for rating of machines. A

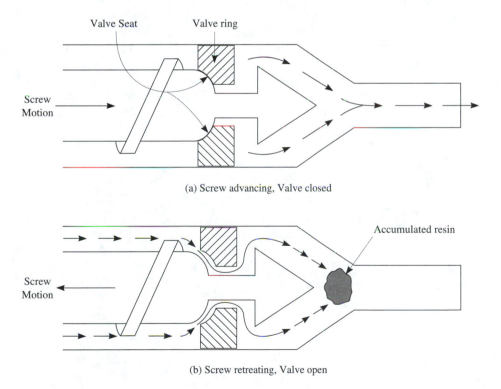

(a) Screw advancing, Valve closed

(b) Screw retreating, Valve open

Figure 12.3 Check valve (or nonreturn valve) operation, which prevents resin flow when the screw is advanced during resin injection.

slightly different rating system has been suggested that would give the shot size as the volume of material that can be injected against some standard pressure, such as 15,000 psi (100 MPa).

During the injection cycle, the resin moves forward through the nozzle, the open end of the barrel that is shaped to fit into a bushing on the back of the mold. The nozzle was moved into place against this bushing during the start-up of the operation of the injection molding machine and it remains in contact with the bushing during all the normal operational cycles.

The third and fourth stages of the cycle (see Figure 12.2c and d) are the retraction of the screw to again create the space at the end of the screw and the continued cooling of the part in the mold until it is cool enough to be removed. During stage four, while the part is cooling, the screw turns and melts additional resin. The turning of the screw pumps the resin forward through the normal pumping action of the Archimedes screw design, which the screw possesses, thus filling the pool at the end of the barrel in anticipation of the next cycle. (This pumping is not the same as the reciprocating motion of the screw associated with injection of the resin into the mold.) The mold is then opened and the part removed. The mold is closed again and the cycle is restarted.

The second type of injection system is called ram injection. An injection molding machine that uses *ram injection* is depicted in Figure 12.4. This plunger-type machine was used prior to the invention of the reciprocating screw machines. In the ram injection machine the resin is fed from a hopper into the barrel and heated through the input of thermal energy from the heaters around the barrel. The molten resin collects in a pool in the barrel called the injection chamber. The molten resin is then pushed forward by the action of a plunger (ram or piston) driven by a hydraulic system at the head of the machine. To facilitate the melting of any residual solid material and to give better mixing of the melt, the molten resin is pushed past a *torpedo* or *spreader* that, along with a back pressure

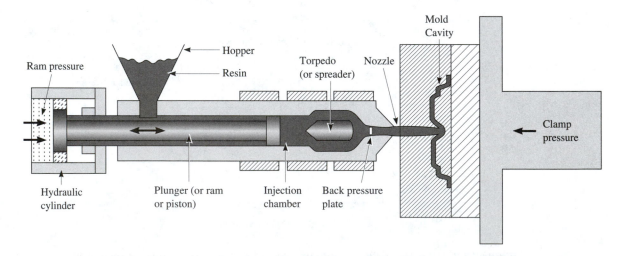

Figure 12.4   Ram injection in an injection molding machine.

plate, imparts shear to the melt. Then, as with the reciprocating screw machine, the resin flows through a nozzle into the mold.

Even though the ram injector machine was developed before the reciprocating screw machine and is generally less expensive to build, some important advantages of the reciprocating screw machine have given it a preeminent position in the marketplace. These reciprocating screw advantages include: (1) more uniform melting, (2) improved mixing of additives and dispersion throughout the resin, (3) lower injection pressures, (4) larger permissible part area, (5) fewer stresses in parts, and (6) faster total cycle.

One further advantage is that with screw machines the shot size can be a much smaller percentage of the total shot capability than in ram machines. A rule of thumb is that for reciprocating screws, the minimum shot size should be at least 1/20 of the total volume of the resin pool cavity. For ram injection machines, the minimum should be at least 1/5 of the total shot size. For example, if a reciprocating screw machine with a 70-ounce (2000-g) shot capability is being used, the minimum part that could be made would be 3.5 ounces (100 g) (1/20), whereas if a ram injection machine with a 70-ounce (2000-g) shot capability is being used, the minimum part that could be made would be 14 ounces (400 g) (1/5). Hence, the reciprocating screw machine is more capable of making small parts from large machines. However, both machines operate better if the part size is between 1/2 and 3/4 of the total shot capacity.

Both the reciprocating screw and the ram injection system combine the melting of the resin and the injection of the resin into a single chamber. When extremely fast cycles are desired, the melting of the resin can be separated from the injection function. This is done by attaching the barrel in which the melting is done (either by a screw or a ram) to a second barrel that contains a plunger and an injection chamber. The melting barrel is called the *preplasticizing cylinder* and the entire arrangement is called a *preplasticizing machine*. The discharge of the preplasticizing chamber is into the injection chamber of the second barrel. This discharge can be continuous, hence the faster cycle capability. When called for in the cycle, the ram or plunger in the second barrel simply moves forward and pushes the resin into the mold. However, if the melting of the resin is not the limiting factor in the cycle, preplasticizing machines have little value, unless, of course, a second mold is attached in some alternating injection sequence. Preplasticizing machines are not widely used.

Molds

Designing and fabricating molds for injection molding is far more complicated than making extrusion dies. This section presents an overview of the process involved and discusses some of the most important aspects. Several entire books have been written (see references at the end of the chapter) on the details of making injection molding dies, and even these point out the need for extensive training and experience in this highly specialized field.

Mold Parts. As shown in Figure 12.1, the mold is between the *stationary platen* and the *moveable platen* of the injection molding machine. The mold has two main sections called the *stationary plate* and the *moveable plate* which are mounted, respectively, onto the stationary and moveable platens of the injection molding machine. Hence, the moveable plate of the mold moves with the moveable platen of the injection molding machine. It is this movement that opens the mold and allows the part to be removed. The mechanism that moves the platen is part of the clamping unit and will be described later.

The connection from the injection unit to the mold is through the nozzle on the end of the injection unit that inserts through a hole in the stationary platen and touches against a bushing, called the *sprue bushing,* that is on the outside of the mold. A locating ring on the face of the mold assists in the alignment of the nozzle with the mold. This and other key features of the mold are illustrated in Figure 12.5, which gives a general overview of a typical mold.

The nozzle is heated so that the resin remains liquid within it. The sprue bushing is part of the mold and is cooled so the resin solidifies in the bushing. Therefore, when the part is removed, the break point, which is at the solid/liquid junction, is between the nozzle and the bushing.

The channel that runs through the stationary plate of the mold is called the *sprue channel* and the material that is in the channel is called the *sprue.* Because the sprue is within the mold (where it is cold enough to freeze), it is solid and is therefore removed with the part. (Molds in which the sprue and other parts of the mold are heated can be made and will be discussed later in this chapter.) The solid sprue is removed from the finished part assembly after the part is ejected from the mold. The sprue can generally be reground and run with virgin resin in subsequently produced normal parts.

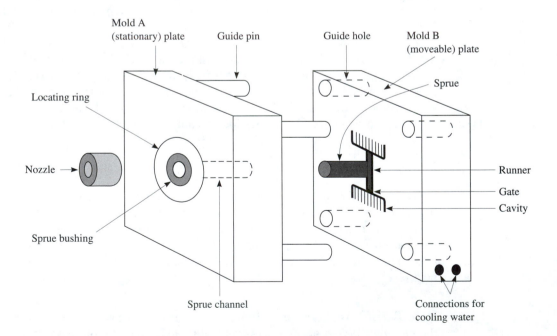

Figure 12.5 Mold diagram showing various key elements.

The resin flows from the sprue through the *runners* or connecting channels to the *mold cavities,* which are carefully shaped, recessed areas on the face of the mold in which the resin solidifies to form the part. The runners are, therefore, part of the waste material that is reground with the sprue. Although molds can be made with just one cavity, the more common practice, especially for small parts and large quantities, is to have several cavities in one mold. Such molds are called *multicavity molds.* Multicavity molds are discussed later in this chapter.

Several arrangements are possible for these various mold parts, but one of the most common arrangements (which is a general example of the other types) is shown in Figure 12.6. The entire assembly is called the *mold base.* A photo of a mold base is shown in Photo 12.3.

Mold Bases. Mold bases can be purchased as entire units and then the cavities are cut into the front- and rear-cavity plates (A and B) for the specific part. The size of the mold base cannot be larger in surface area (plate dimensions) than the opening of the tie bars on the machine to which the mold base is mounted. The machine must also be able to open sufficiently to allow proper operation, including mounting, opening, and ejection of the mold part. Other considerations for the size of the mold base are the size of the part, the number of cavities, and the pattern of the cavities.

The parts are made by the resin filling the cavities cut into plates A and B. When the mold opens, the parts are retained on the B side. The ejection system then activates to push the parts out of the cavities. The retention on the B side can be ensured by using a sprue puller pin (discussed later in this section) or by adjusting the draft angle of the cavities or the core placement so that shrinkage against the mold cavities is slightly greater on the B side than on the A side. Other methods for ensuring that the parts are on the B side also exist. The opening line or dividing line between the cavity parts on the A and B sides is called the *parting line.* The mold base shown in Figure 12.6 has only one opening between the plates and hence only one parting line.

Mold bases can also be designed so that more than one mold opening occurs, that is, there is more than one parting line. These types of molds, called *stacked molds,* are used when relatively thin parts are to be made in a large machine with high injection capacity. Rather than create many parts on one mold plate, the parts are created on two mold plates. Hence, stacked molds are like having two (or more) molds running simultaneously in the machine.

Some mold bases combine the top clamp plate with plate A and combine the support plate with plate B. These simpler mold bases are most commonly used when the parts are very simple and small and need (usually) only one or two cavities to be cut. The danger in combining the plates is that the cavities will be cut so deep that the plate width is insufficient to give proper resistance to the injection pressure, possibly distorting the cavities.

The mold base is mounted to the platens of the injection molding machine by mold clamps or screws through the top clamping plate and the ejection housing. The mounting must position the sprue bushing to align with the nozzle of the injection molding machine. The locating ring assists in this alignment because it contains a conical-shaped hole lined with a soft metal that allows some minor adjustments to ensure exact alignment. When properly aligned, the nozzle tip slides through the locator ring and fits

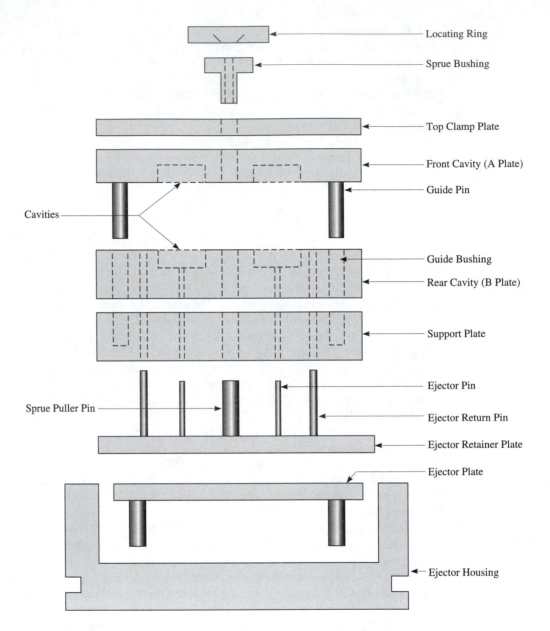

Figure 12.6 Typical mold base showing the various components.

exactly against the top of the sprue bushing. The size of the nozzle should be slightly smaller than the opening of the sprue bushing to prevent leakage. Often, the top of the sprue bushing is concave and the end of the nozzle is convex to facilitate their mating.

The nozzle is heated and the mold is usually cold, making the logical separation point in the resin stream the line that separates the molten resin in the nozzle and the

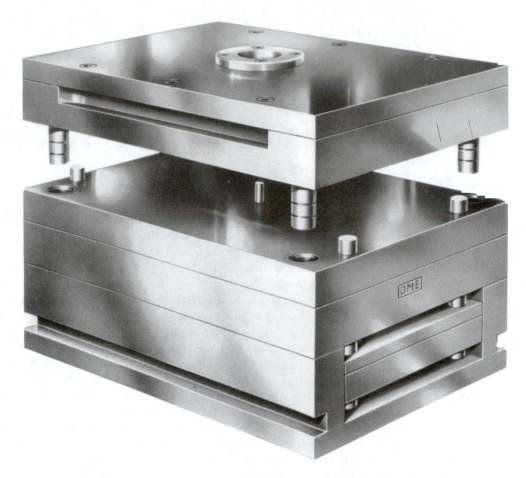

Photo 12.3   Typical mold base (Courtesy: D-M-E Company)

solid sprue. Therefore, the sprue is usually withdrawn with the part and runner system and is then removed from the part by clipping or some other mechanical cutting operation. The resin flows down the sprue cavity and reaches a pin that is inserted into the B plate when the mold is closed. This pin, called the *sprue puller pin,* forms the bottom of the sprue cavity when the mold is closed. The pin is mounted onto the ejector plate. When the resin flows against the sprue puller, the shape of the end of the sprue puller pin is designed to give some form of attachment to the sprue. Common attachments are a lock hook (or z-puller), a reverse taper, and a groove, which all give an undercut to the sprue and hence secure the sprue to the pin. These various sprue end types are shown in Figure 12.7.

When the mold is opened, the sprue puller pin ensures that the runner system and the parts stay on the B side of the mold. The ejector system is then activated and the parts are pushed out of the cavities and off the end of the sprue puller pin.

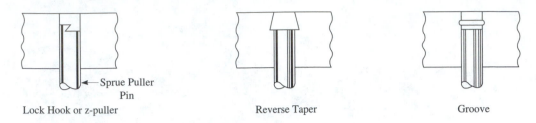

Figure 12.7 Sprue puller pin end configurations that hold onto the sprue and allow withdrawal.

The ejection system is driven by a power system separate from the one that opens and closes the mold and is described later in this chapter.

The mold base shown in Figure 12.6 is for a relatively simple part. Complications arise when the part has an undercut or a threaded section. When these occur, the mold base must allow for moving or sliding sections, which move out of the way so that the mold can open. The force of the machine hydraulics in opening the mold usually drives these slides or screws and occurs simultaneously with the mold opening.

The mold base is usually cored for cooling. This coring is done by boring channels in the cavity plates or in the support plates adjacent to the cavity plates. Water or some other coolant liquid is run through the cooling channels. If cores are used inside deep, hollow parts, the cores often are channeled for cooling. The layout of the cooling channels can be very important in determining the precise dimensions and properties of the parts. For instance, if the part is generally circular and the cooling channels are arranged in a square pattern, some portions of the part will cool much quicker than other part sections. This differential cooling can cause differences in thermal contraction and crystallization throughout the part. Symmetry of the cooling system and symmetry of the parts are important and should be matched.

Runners.   *Runners* are the distribution system for the resin from the sprue to the cavities. The runner system can be important in determining the flow of the resin through the mold base and can have quite a significant effect on the properties of the part. Cooling takes place along the runner walls, so runners must be large enough to ensure that the proper amount of resin flows through the runner system and fills the cavities.

Therefore flow characteristics (viscosity, melt strength, etc.), the temperature, and other factors such as the need for dimensional preciseness and part uniformity are all important factors in determining the diameter and length of the runners. The diameters and pattern of the runner system have traditionally been developed by mold designers based upon their experience. Recently, several computer programs have been developed that assist the mold designer to properly construct the runners. These computer systems use resin properties and flows to calculate the suggested runner diameter and length. If the diameter of the runner system is too small or the length of the runner too long, the resin can freeze in the runners before the mold is completely full. Runners that are too small can also cause excessive molecular orientation. If the runner system is too large, then excess material would be injected and too much regrind created.

When using resins that have a very high viscosity, the sizing of the runner channels is especially important to prevent premature freezing of the part and inadequate cavity filling. For instance, polycarbonate, which has a high viscosity, should have larger runners than nylon, which has a comparatively low viscosity.

The shear forces imparted by the runner system are also important. All resins are viscoelastic and hence the flow is shear dependent, but some resins are especially shear sensitive. For instance, acetal can decompose if subjected to excessive shear forces, especially if the resin temperature is high. In general, shear should be minimized.

Other considerations in the sizing of the runners are the volume of the parts, the size of the parts, and the thickness of the parts. Each of these part parameters has a slightly different effect on the runner-system sizing.

The optimum flow of the resin through the runner system depends on the shape of the channels as well as the diameter. Round channels give the best flow characteristics but are difficult to machine. Machining costs can be reduced by machining only one side of the mold plates. This is less desirable but can be made acceptable if certain rules are observed. Merely cutting a semicircular channel is unacceptable. A better shape is shown in Figure 12.8, where the depth of the channel is at least two-thirds the size of the width and the sides are tapered between 2° and 5°.

The pattern used for the runner system is also important in the proper performance of the injection molding system. Generally, the length of the runner system should be minimized. This reduces the need for the resin to flow and also reduces the amount of resin that must be reground as part of the runner and sprue system.

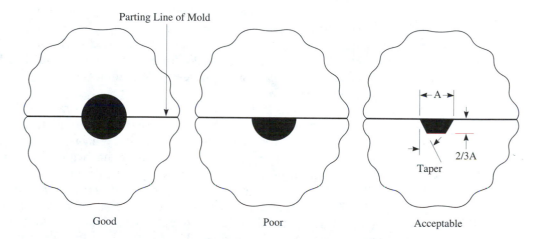

Figure 12.8   Runner channel shapes showing the preferred shape (circular cross section), an unacceptable shape (semicircular), and an acceptable shape.

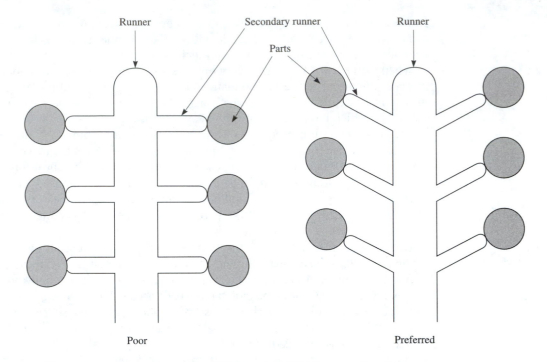

Figure 12.9   Secondary runners showing nonstreamlined and streamlined designs.

When secondary runner channels are used, as is the case with multicavity molds, the flow into the secondary channels should be streamlined, that is, angled in the flow direction (see Figure 12.9). This streamlining minimizes shear on the resin.

Accuracy in part dimensions and uniformity from part to part can best be obtained with symmetrical runner systems that also have few bends. Examples of nonsymmetrical and symmetrical runner systems are illustrated in Figure 12.10. Note that the nonsymmetrical system also has several right-angle turns that cause unwanted turbulence and poor resin flow and should therefore be avoided in any runner design. Symmetrical molds are also called *balanced molds*.

The accuracy of the parts in the symmetrical runner system is improved because each of the cavities receives the same injection pressure since the runners are the same length and size. In the nonsymmetrical case the most distant cavities would receive less injection pressure and, therefore, would have minor size variances from the cavities closer to the sprue.

Note the small extensions of the runner system when right-angle turns are made. These small channels beyond the runner are called *cold-well extensions*. When the molten plastic flows down a runner, the leading edge is against cold air and tends to cool faster than other parts of the flow. After a long distance of flow in a long runner system, the cold end can go into the gate and possibly clog the gate. To prevent this from occurring, the cold-well extension at the point where the flow changes directions takes the cold end and captures it. After the well is full, the flow is diverted down the new runner di-

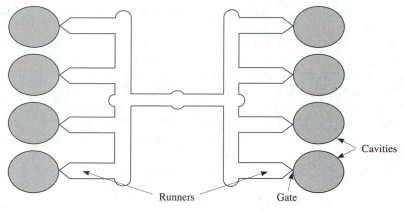

(a) Nonsymmetrical runner system

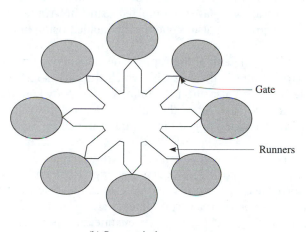

(b) Symmetrical runner system

Figure 12.10 Runner systems illustrating (a) nonsymmetrical and (b) symmetrical (balanced) runner systems.

rection with a fresh (hot) end. This hot end passes through the gate with much less likelihood of plugging (see Figure 12.10a).

When the cavities are not producing the same part, the runner system can be very complicated. Some formulas have been adopted and some computer programs developed that allow runner systems to be sized for shape and length according to the type of resin and other factors that affect the runners. These programs give an approximation of the proper runner system. Then, when the mold is run, the runners are adjusted slightly to optimize the flows and part properties. The adjustments in the runner system are made by machining additional material out of the runners and enlarging the diameters. Another method of adjusting the runner system for different-sized parts is to place a restriction in the runner going to a smaller cavity. This restriction acts like a valve and reduces the flow down that particular runner.

Hot-runner molds are gaining popularity because of their higher production rates and because they eliminate the need to trim the runner and sprue from the parts. In hot-runner molds, the resin is conveyed from the sprue to the cavities through a manifold system that is kept hot, usually by cartridge heaters. Thus, the resin is prevented from solidifying. The separation point for the liquid resin and the solid part is at the gate. This means that the runner system is kept full and is immediately ready for the next shot as soon as the previous shot is ejected. Filling time is therefore reduced. The size of the runner system must be matched closely to the shot size so that excessive residence time is avoided.

The heating of the runner system and sprue is usually done with electric cartridge heaters embedded in the face of the mold around the runners. This added heating significantly increases the cost of the hot-runner molds (by about 30%). The higher cost of the mold often can be recovered in improved operational costs, especially for high-volume parts. Small parts in which the weight of the runner system is high compared to the weight of the part are also favorably molded using hot-runner molds. Medical products that must avoid regrind are also economically molded using hot-runner molds.

Gates. The *gate* is the end of the runner and the entry path into the cavity. The shape of the gate strongly affects the ease with which the part is removed from the runner system. The gate shape can also affect the filling of the cavity and, therefore, the dimensions and properties of the part. Since the gate is the most restricted point in the injection molding system, it is the point of highest interference for reinforcements and, perhaps, fillers.

Several gate designs are in common use. These are pictured in Figure 12.11.

The *edge gate* is a small rectangular opening at the end of the runner channel that connects to the edge of the cavity. The edge gate can be below the parting line if the channel and part are also below the parting line, as shown in the figure, or it can be symmetrical about the parting line if the runner channel and part are also on both sides of the parting line. The edge gate should be cut off after the part is molded. The major advantage of the edge gate is the low cost of making the gate.

 The *submarine gate* starts from the edge of the runner and goes into the cavity edge at an angle. It narrows to a point as it moves from the runner to the cavity. The submarine gate is used to shear off the runners when the part is ejected. This shearing occurs at the point and generally leaves a small blemish on the part at that point. The advantage to the submarine gate is that separation of parts and runner systems is automatic. A disadvantage is that the gate cannot be used for some resins because of high shear.

The *tab gate* is formed by connecting the runner directly into the cavity with no reduction in runner cross-section. The tab gate is used for very large parts where a reduction in flow would disturb the resin's flow pattern and might result in inadequate flow into the cavity.

The *fan gate* is made by reducing only the thickness and not the diameter of the runner channel as it goes into the cavity. This gate is used for parts of intermediate size and when reinforcements in the resin cannot flow through the edge gate. Removal of the fan gate is approximately the same as with the edge gate.

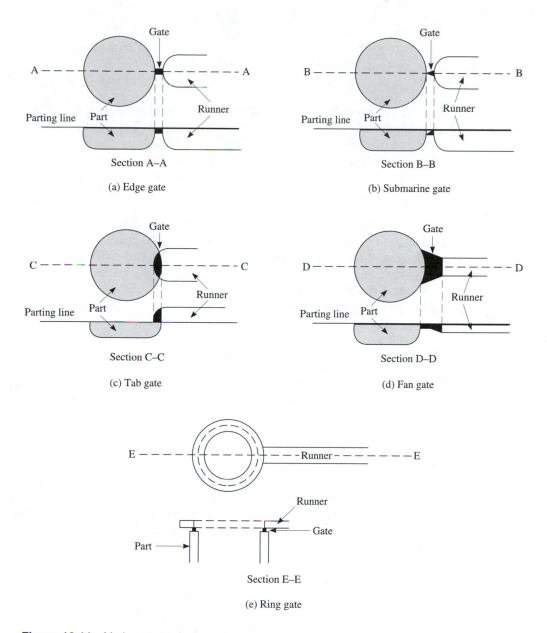

Figure 12.11 Various gate designs.

The *ring gate* is used to make hollow cylinder parts. The ring gate covers the entire top of the cylindrical part so that the resin flow is downward into the walls of the part. The ring gate can also be used for hollow parts and other parts where the walls cannot have weld lines.

Cavities. The *cavities* are the actual molding locations. Resin enters the cavities through the gates, fills the cavities, and then cools to form the solid, molded plastic part. The parts are then ejected and finished or otherwise prepared for actual use. The cavities are the heart of the molding process and must be precisely as desired for the part to be properly made.

The shape of the cavities determines the shape of the part, since the plastic will replicate the surface of the cavity. Great care must be taken in making the cavity so that it will produce the part desired. Great care must also be taken in protecting the cavity surface during operation of the injection molding machine. For instance, only tools made of soft materials such as brass, aluminum, or wood should be used to clean the cavities or to pry loose a part that might be stuck in a cavity. In order to allow for easy removal of the part from the cavity, the walls are machined with a slight angle, called the *draft angle*, so that the top of the cavity is slightly larger than the bottom. All features of the part, that is, all features in the mold cavity should be drafted so that the part will be removed easily. Some exceptions are discussed in the section of this chapter dealing with part shapes and materials.

The cavities are usually placed so that part of the cavity is on the stationary plate of the mold and part is on the moveable plate. When this arrangement is used, the split in the two parts of the mold can often be detected in the part. This split is seen as a small line, called the *parting line*. Ideally this parting line would be nearly invisible. However, if the molds are poorly aligned or worn or if the clamping pressure is not sufficient for the particular part, the parting line can become quite prominent.

When resin is injected into the mold cavities, the cavities are full of air. This air can be compressed and cause disruptions in the even flow of the resin in filling the mold. Therefore, a *vent* is provided for each cavity that allows the air to escape from the cavity as the resin enters. These vents are usually small, thin channels that lead from the side of the cavity opposite the gate to the outside of the mold. The vent diameters are large enough to allow air to pass freely but small enough that resin will not enter them. Typical dimensions are 0.0005 to 0.003 inch (0.0015 to 0.080 mm). The tops of ribs and other places where air might get trapped should also be vented.

The essential functions of a mold are to direct the flow of the resin into the cavities where the resin is shaped and cooled so that a solid part can be removed. Many molds have water circulated through channels bored in the mold plates to facilitate this cooling. The water temperature is selected to give the proper cooling rate for the particular plastic being run. In some cases, the water is considerably hotter (200°F, 93°C) than room temperature, especially when cooling the plastic too rapidly could cause internal stresses or not give time for crystallization. The cooling channels flow near the cavities so that heat transfer is rapid. For the maximum in part dimensional control, the cooling channels should be arranged so that each cavity receives the same amount of symmetrical cooling.

Multicavity Molds. Careful consideration should be given to the number of cavities in a mold. The costs of making the mold must be balanced against the production requirements for the part. Good marketing data that can predict the sales volume for the part often are the key to successfully determining the proper number of cavities. Single-cavity molds most often are used for limited production runs, such as when a part is in development and a limited quantity is needed for marketing and product testing. Single-cavity molds are also important when the part is very large, so that the size requirement of the injection molding machine does not become excessive.

When more than one part is made in the same mold, the mold is called a *multi-cavity mold*. Typical nomenclature reserves the name multicavity for molds in which all of the parts are the same.

When multi-cavity molds are indicated, a mold that allows the use of *mold inserts* might prove valuable. The inserts are metal plugs inserted into holes that are machined in the A and B plates of the mold. The plugs are retained in the plates by a small flange on the back of the insert and then pressed in place by the clamp plate and the support plate. The plug inserts have the cavities cut into them. Therefore, when a part needs to be changed, the inserts for one part can be removed and the inserts for the second part put in their place. This allows one mold base to be used for many different parts. The only real restriction is that the size of the part must fit within the dimensions of the insert. The insert system, illustrated in Figure 12.12, is especially useful for multicavity molds.

Some molds are built so that the cavities are different from each other. These molds, called *family molds*, are usually constructed so that all of the parts of a particular assembly can be molded at the same time (like model airplanes, etc.). The obvious advantage of family molds is that only one mold needs to be used to make all of the parts in the assembly. However, some disadvantages also occur with family molds. If parts of

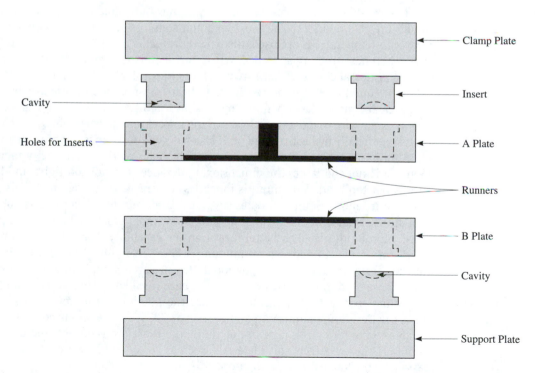

Figure 12.12   Placement of inserts into the A and B plates of a mold so that the mold base can be used for several different parts.

different shapes are to be made in one mold, the runner system should be sized so that the flow down each leg of the runner system is the same, thus filling all of the molds simultaneously. The sizing of runner systems for nonuniform parts is very complicated and is usually only approximately successful. That results in a slight decrease in part accuracy in molds where the cavities are different. Also, if one of the cavities is damaged so that it cannot be used, then a separate mold must be constructed for just that part. When this happens, many of the advantages of family molds disappear. Experience has shown that family molds are most useful when small quantities of an assembly are to be made or if the mold is just for prototyping and separate, multicavity molds making identical parts are to be constructed after the part has been exactly defined and the market established.

Mold Materials.   The type of material used to make the mold, especially the plate containing the cavities, depends on many different factors. For instance, if the total production expected is high, harder and stronger materials would normally be used. If the part has reinforcements, harder and stronger materials would be chosen, perhaps with special heat treating. Materials that decompose to form corrosive materials require stainless steels or some other corrosion-resistant metal. Thermoset materials and some high-performance thermoplastics are usually run at higher mold temperatures and therefore require a high-temperature alloy. Parts that demand very close tolerances will usually require very hard steels.

The most common material for making injection mold cavities is tool steel. Typical grades are P20 and P21. The most common stainless steel for molds is type 420. When very high temperatures are used, the type H13 steel is preferred. The O1 and O2 steels are used when heat treating is needed. When mirror finishes are needed, the A2 and A6 steels are the principal choices. There are many other acceptable alloys in each of the categories.

Some mold cavities require very high heat transfer. Often these are long parts. Such molds can be made of a beryllium-copper alloy that has very high heat transfer. Beryllium-copper is superior to most steels in absorbing impacts and in being able to be hardened more than superficially by heat treatment. It also has good resistance to abrasion and fatigue.

Aluminum can be an effective material for mold cavities when the number of parts to be made is not too large, the dimensional tolerances are not too tight, and the temperature is not too high. Aluminum is much less expensive than steel materials and is much easier to machine. Some estimates suggest that aluminum machining is 40% faster than tool steels. Lead time for making an aluminum mold is about half that required of tool steel. Aluminum also has better heat-transfer capability than steel, making possible a short cure cycle. In addition, aluminum is lower in weight than steel, which assists in moving, installing, and storing the mold. Recent developments in aluminum alloys, especially those used in the aerospace industry, have resulted in some very strong and durable aluminum materials that further enhance the potential of aluminum as a material for mold cavities. Some aluminum alloys, such as 7075, can be anodized to a hardness comparable to the tool steels. The use of aluminum therefore reduces mold costs by as much as 50%. Other relatively soft materials, such as epoxy and brass, have also been used occasionally for low-volume production.

Table 12.1 compares materials for making mold cavities by the properties that are of most interest in mold making and mold performance. The ideal mold material would have

Table 12.1 Comparison of Mold Materials

	Material				
Property	Tool Steel (P20)	Stainless Steel (420)	Hotwork Die Steel (H13)	Beryllium-Copper	Aluminum
Machinability[1]	0.65–0.80	0.45	0.43	0.60	1.2
Coefficient of Expansion[2]	7.1	5.7	6.1	9.34	12.8
Specific Heat Capacity[3]	0.11	0.11	0.11	0.10	0.23
Thermal Conductivity[4]	20	14.4	16.3	68	80
Density[5]	0.28	0.28	0.28	0.30	0.10
Hardness[6]	28–37	52	52–54	36–40	about 10
Yield Strength[7]	130–135	215	228	140–175	68–78
Weldability	Good	Moderate-difficult	Difficult	Moderate	Moderate

[1]Ratings from Tool and Manufacturing Engineering Handbook, SME, Vol. 1. High numbers are better. Standard soft steel is basis at 1.0.

[2]Dimensions are inch/inch °F. Higher numbers indicate increased expansion.

[3]Dimensions are Btu/(pound °F). Higher numbers indicate greater amount of heat required to heat the mold.

[4]Dimensions are Btu/[(h ft 2°F)/ft]. Higher numbers indicate more thermal conductivity.

[5]Dimensions are pound/cu. inch. Lower numbers indicate lower weight.

[6]Readings are in Rc. The actual values for aluminum are Brinnel hardness of 163–167.

[7]Dimensions are (psi) × 1000. Higher numbers are stronger.

high machinability, low coefficient of expansion, low specific heat capacity, high thermal conductivity, low density, high hardness, high yield strength, and good weldability.

Parts of the mold base other than the cavities and the cavity plates use many different materials because of the varied requirements for these parts. The support plates are typically inexpensive steels such as 1020 or P20 tool steel where additional strength and stiffness are needed. Most slides are made of P20 tool steel so that they have good wear and low-friction sliding surfaces. The core pins, ejector pins, and other auxiliary parts that need strength are made from H13 or a shock-resistant steel.

Construction of the Mold.   The mold base is usually purchased as a premade unit, with only the cavities to be made for each part. Therefore, the focus of this section is on the making of the cavities.

The principal method for making the cavities is traditional metal machining, such as milling, turning, drilling, or grinding. The machining operation finishes the cavities so that with only a small amount of finish work, usually hand polishing, hardening, or lapping, the cavities are completed. Modern machine shops use computer-aided design (CAD) systems to design the runners, gates, and cavities. The runner system and gates can be checked for size using one of the mold-flow programs now available. The design output of the CAD system can be used to create a machine code for numerical control (NC) machines to precisely perform these machining operations. Machining may cause some internal stresses in the material, so annealing to remove these stresses often is done as a part of the machining operation.

The second most common method of making the cavities is *electrical discharge machining (EDM)*. In this process a graphite electrode, which has been carefully cut into an exact mirror image of the cavity, is pressed against the face of the cavity plate and used to cut the cavity. The cutting is done by a spark between the graphite electrode and the work piece. The spark causes the work piece to erode in the exact pattern defined by the geometry of the electrode. The erosion process is quite slow, but when one cavity is finished, the same electrode can be used to make other cavities that have the same shape as the first. Because the electrode plunges gradually into the face of the plate to create the cavity, this process is also called *plunge EDM*. The graphite electrode can be shaped by any convenient cutting process. One popular process, an EDM process called *wire EDM,* uses the spark between a wire and the graphite to cut the graphite into the desired shape.

Another method for making cavities is *hobbing,* which was much more popular before EDM became well known. In the hobbing process a hardened tool, which is in the mirror image of the cavity shape, is mechanically pushed into the metal such that the cavity is created. The metal plate must be relatively soft so that the tool can press into the surface and create the cavity. As a result, some plastic deformation of the metal surrounding the cavity usually occurs. The stresses from this deformation are relieved by annealing. Changes in the plate surface near the cavity are machined away, and the plate often is hardened.

Casting the metal around a model of the part to make the cavity can be done, but it is uncommon. The problem with casting is that tool steels and other high-strength materials are not easily cast. When casting is done, aluminum or some other soft metal is more likely used as the cavity material. Cast molds generally have a lower life expectancy (in terms of the number of cycles expected) than do the harder steel molds.

Surface finishes are important for determining the surface quality of the plastic part. Therefore, much care should be spent on this part of the mold-making process. After the cavity has been cut, the surface is usually smoothed and polished and then, possibly, heat-treated. Coating may also be done in some molds.

The surface may also be textured by machining or by other processes. For instance, chemical etching of the surface can be done by masking those parts of the surface that are not to be attacked by the chemical. After masking, the chemical etching compound is placed on the surface and allowed to chemically dissolve those unmasked portions to whatever depth is desired. When the required depth is reached, the chemical is removed. The mask is then also removed and the textured surface is completed.

Ejector System.   To facilitate part removal, most molds have some mechanism to knock out the parts. The most common ejector system uses *knock-out pins,* which are mounted on an *ejector plate* attached to a piston called the *ejector rod* as shown in Figure 12.13.

When the mold is open and the part is to be removed, the ejector rod advances, thus moving the ejector plate and ejector pins toward the mold. The pins move through holes cut in the moveable platen and moveable plate and enter the back of each of the cavities. The pins simply push the parts out of the cavities. Some care must be taken to ensure that the part has been cooled and is sufficiently solid that the pushing of the ejector pins will not damage the part.

An alternative ejection system does not use an ejector rod, but simply has the ejector pins and plate positioned so that when the mold opens by the sliding back of the moveable

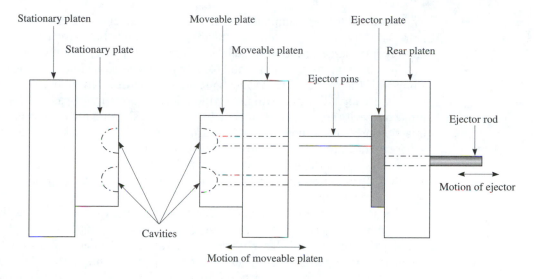

Figure 12.13 Ejector system showing pins, plate, and ejector rod.

platen, the pins slide through the appropriate holes in the moveable platen and enter the back of the mold. Therefore, no additional ejection mechanism is required other than the opening of the mold. The difficulty with this system is that if the parts are stuck and an additional ejection movement is desired, it must be done by moving the entire moveable platen rather than by just advancing the ejector rod.

In order to use the ejector system, the part must remain with the moveable plate rather than with the stationary plate when the mold opens. Strategies to accomplish this include use of sprue pullers, differential cooling of the moveable and fixed sides of the mold, and adjusting the draft angles of the cavities so that the part is held more tightly on the moveable side. These are discussed elsewhere in this chapter.

Clamping Unit

The purpose of the clamping unit, the third major section of an injection molding machine (Figure 12.1), is to hold the molds together while the resin is being injected and the part cooled. The clamping unit also provides the force to open the mold and, as has been indicated, may also provide the force for ejecting the part from the mold.

The amount of clamping force *(F)* needed is related to the injection pressure *(P)* and to the surface area of the cavities *(A)* by the following formula:

$$F = P \times A$$

or, more particularly,

$$\text{Clamping Force} = (\text{Injection Pressure})(\text{Total Cavity Projected Area}) \qquad (12.1)$$

This formula shows that increases in the injection pressure or in the total cavity *projected area* (the area as projected into a single plane, that is, the widest area of the part) will require a greater clamping force to keep the mold closed. Hence, large parts that require large injection pressures to fill the cavities and that have large cavity projected areas require large injection molding machines with large clamping forces.

The clamping force is a method of rating the size of an injection molding machine, along with the shot size, as has been previously discussed. In metric units, the clamping force is usually given in newtons (N), the injection pressure in pascals (Pa) and the total projected cavity area in square meters (m^2). In English units the force is given in tons, the injection pressure in pounds per square inch *(psi)* and the total projected cavity area in square inches (in.^{2}). Tons of force or clamping force is the total force, in pounds, divided by 2000. For instance, if the total force were 350,000 pounds, the clamping force would be 175 tons. Typical sizes of injection molding machines would range from 20 tons to 10,000 tons in English units.

The force required to inject resin depends upon many factors, such as the viscosity of the resin being injected, temperature, size of the runners, length of the runners, gate type and size, whether the mold is a hot-runner system, and the depth of the cavities. However, for most purposes a general guideline has proven to be quite useful. The guideline states that 350 pounds of force are needed for each square inch of cross-sectional area (when that area is measured parallel to the mold face).

Two types of clamping system are in common use. One type uses hydraulic pressure to close the mold and apply the clamping pressure. This hydraulic action is provided by a

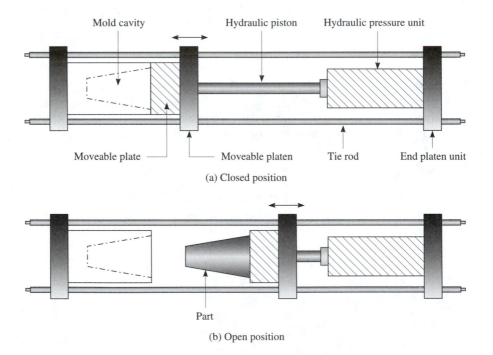

(a) Closed position

(b) Open position

Figure 12.14　Hydraulic clamping unit showing (a) closed position and (b) open position.

piston that is attached to the moveable platen from a hydraulic pressure unit mounted on the end platen. These hydraulic machines often use one hydraulic motor to power the clamping unit, the reciprocating motion of the screw, and the ejector unit. An example of a simple hydraulic clamping system is shown in Figure 12.14.

The other type of clamping system is a mechanical clamp that usually employs a toggle system. A toggle is a pair of mechanical arms that are rotated about a pivot point so that they can be open or locked into a position to hold pressure. A simple toggle system is depicted in Figure 12.15. Some toggle systems utilize several pairs of toggle arms rather than the single pair of arms depicted in Figure 12.15. The actuator rod for the toggle system can be either mechanical or hydraulic. When it is hydraulic, the system is called a mixed or hydraulic/mechanical clamping system.

Each of the two types of clamping systems has some advantages. The hydraulic system is capable of applying varying amounts of pressure, up to the maximum available, whereas the toggle system has much less control and generally provides either no pressure or full

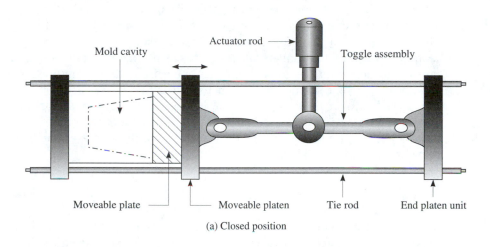

(a) Closed position

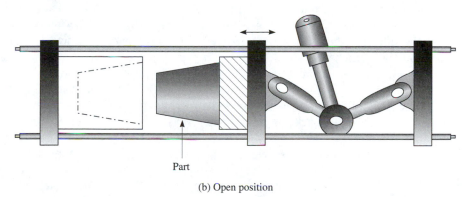

(b) Open position

Figure 12.15 Toggle clamping system showing (a) closed position and (b) open position.

pressure. The pressure of the toggle system is, therefore, difficult to monitor and the setting of the opening distance requires careful adjustment. The toggle system is generally less expensive but it wears more quickly than the hydraulic. Toggle systems are faster than hydraulic.

Some have criticized hydraulic systems because so much hydraulic fluid must be pumped to achieve both closing and clamping. However, some new two-stage hydraulic systems use only a small flow of fluid for the closing step when little pressure is needed, and then increase the amount of fluid pumped for the clamping part of the cycle.

Plant Concepts

Injection molding machines are generally self-contained, except for service lines such as cooling water, air, power, or vacuum. Hence, many injection molding machines can be located in a relatively small area and several machines can be serviced by a common source of service lines and one operator, whereas extruders are long and occupy considerable space, generally requiring one operator per machine. The principal duties of an operator for an injection molding machine are to ensure that good parts are being made, to separate the part from the runners and sprue, to ensure that the resin is adequately being fed to the hopper, and to monitor operations to ensure that temperatures, pressures, times, and other machine settings and parameters are within proper operating ranges. If inserts are placed in the mold, the operator may be required to place them, although that task can often be automated with a robot. Robots are also useful in removal of parts, either to increase automation or because the parts are so large that operator handling of the part is unwise.

Safety

The safety aspects of injection molding are chiefly concerned with ensuring that the machine's built-in safety devices are properly utilized. For instance, injection molding machines have a guard with an interlock system (usually mechanical and electrical) that prevents the mold from closing while the guard is open. The interlock should never be overridden. Other safety concerns are the high temperatures near the barrel and nozzle and the high temperature of the resin before being cooled in the mold. Another safety problem focuses around the weight of the molds, which can often be well over 100 pounds (50 kg). When changing molds, operators may try to remove the old mold or install the new mold without the benefit of some mechanical lifting device. This is dangerous because of potential back strains and other muscle injuries as well as the possibility of dropping the mold.

Some resins present special problems which should be considered. For instance, acetal decomposes readily when left at high temperatures. Therefore, whenever acetal is molded, the machine should be purged before shutdown and should be started up on another resin and then switched into acetal or, if started on acetal, should be started under low-temperature molding conditions. The mold open or dead time should be minimized since resin is not moving and could be overheated in some locations within the barrel. Other times should also be kept as short as possible.

Some indications of possible degradation are: frothy nozzle drool, spitting nozzle, pronounced odor, discolored resin, parts with whitish deposits, and evidence of the screw being pushed back from internal pressure. If any of these danger signs are noted the operator should avoid looking directly into the hopper or nozzle. The operator should note whether the nozzle is plugged (a common cause of degradation) and free the plug by heating the nozzle by raising the nozzle temperature in roughly 10°F (6°C) increments or, if that fails, by cooling the barrel. Purging with polyethylene or some other purge resin may be required. Proper ventilation of the room should be provided and the machine should be run manually until it is again running smoothly. Other resins that may decompose include PVC and various resins that are moisture sensitive such as PET, polycarbonate and nylon.

MATERIAL AND PRODUCT CONSIDERATIONS

Materials

Almost all thermoplastics can be injection molded. The resin manufacturers should be consulted for initial settings for operating conditions for their resins. These settings should then be adjusted for each specific mold to achieve the minimum molding cycle with the optimum part performance and quality. As discussed in the chapter on extrusion, the grades of resin for injection molding and extrusion are usually quite different. Whereas in extrusion a resin with high melt strength is desired so that the part will retain its shape while cooling, the opposite is true with injection molding. Resins with low melt viscosity are desired so that the flow through the runner system and gate and into the mold will be done easily with minimum injection pressure. Injection molding grades would generally have low molecular weights and narrow molecular weight distributions.

Some thermosets can be injection molded provided that temperatures in the barrel are kept low, the L/D for the screw is low, and cycles are quite rapid. Provision should be made for quick removal of the nozzle and for purging, should something happen that interrupts the continuous flow of material through the machine.

Shapes

One of the advantages of injection molding is that parts with very complex shapes can be made. For instance, solid parts with many structural features, such as pins, ribs, bosses, etc., can be molded simply by cutting these features into the mold cavity. The molten resin will flow into all parts of the cavity, thus forming the features desired.

Hollow parts can be created by allowing the moveable plate to protrude into the cavity of the stationary plate as shown in Figure 12.16.

Although somewhat more complicated to mold, threads can also be placed on the inside of a part. This is done by the use of a *core pin* that is inserted into the cavity where

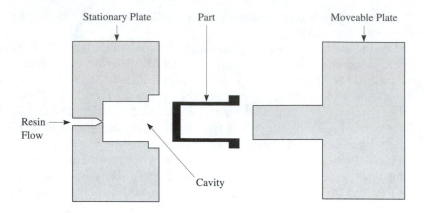

Figure 12.16   Hollow part made by a protrusion of the moveable plate into the cavity of the stationary plate.

the threads are desired. The pin has thread grooves cut into it so that when resin flows around the core, threads are created on the inside of the part. Such an arrangement is shown in Figure 12.17.

When threads are molded into plastic parts, the largest size and coarsest class of thread should be used. These are the easiest to mold and also the strongest in operation. When very small or tight-tolerance threads are needed for part operation, metal inserts should be considered.

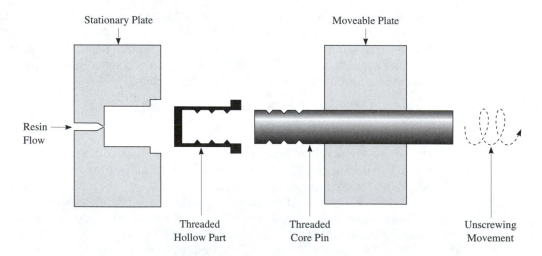

Figure 12.17   Threaded part made with the use of a threaded core pin and unscrewing retraction.

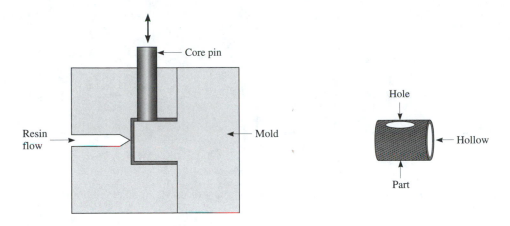

Figure 12.18 Hollow part with hole in the side illustrating the use of a core pin.

A hollow part with a hole on the side is even more complicated. This type of part is illustrated in Figure 12.18, where the use of a core pin is shown. The core pin slides into position after the mold is closed. The core pin seals against the surface of the moveable mold, in this case, and prevents the flow of resin into the area occupied by the core pin. The result is a hollow part with a hole on the side.

The insertion and withdrawal of core pins can become very complicated but with careful planning and proper design, parts of amazing complexity can be injection molded. The force for the actuation of the pins is usually from the mechanical action of the opening and closing of the mold by using slides that draw the core pins in and out.

Part Design

The underlying principles behind part design, other than functionality, are cooling and heat transfer from various sections, and thermal shrinkage of the plastic parts. Functionality must, of course, be preserved in all part designs. However, strength, stiffness, and other mechanical and physical properties can be achieved in several ways, some of which are better from a heat transfer and thermal shrinkage standpoint.

Heat transfer is best when the parts are of the same thickness. Remember that the inside of part sections cool more slowly than the surfaces. Therefore, because the part will shrink as it cools, the center of thick sections will shrink more than the surrounding areas. This is sometimes noted as a *dimple,* or *sink mark,* in the part. In other parts, the shrinkage can occur internally and voids will be created. To prevent this from occurring, all sections of the part should be the same size and should be as thin as possible. One possible design modification for a thick part is shown in Figure 12.19. In this part ribs are added in place of the thick section. Experience has found that ribs of this type can be just as strong as thick sections under loads in the same direction as the ribs.

Ribs also have the advantages of saving weight (economy of resin usage), of faster cooling than thick sections (shorter cycles), and of uniform cooling in all directions.

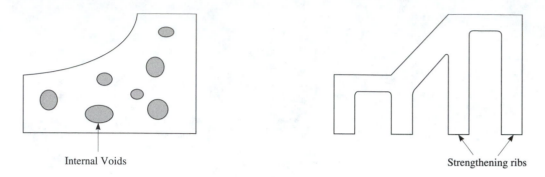

Internal Voids   Strengthening ribs

Figure 12.19 Ribs shown as a design modification to eliminate internal voids in a thick part.

Sharp bends and corners are generally to be avoided in cavities. These are difficult for the viscous resin to fill and can be the origin of stresses in the part. Figure 12.20 shows the elimination of sharp corners and changes to make thicknesses more uniform.

A minimum thickness for a part usually depends on the type of resin used. High-viscosity resins would have higher minimum thicknesses than would low-viscosity resins. These same rules suggest the fineness of the threads that could be made with a particular resin, although the strength of the plastic would also affect the thread fineness. Rules for each resin can usually be obtained from the resin manufacturer.

The finish on the surface of the cavities will be reflected in the part surface. The cavity finish can be specified using the standard system of either the Society of the Plastics Industry (SPI) or the Society of Plastics Engineers (SPE). Some textured surfaces are so rough that the resin becomes mechanically entrapped in the texture and causes the part to stick in the cavity. These situations should be avoided whenever possible.

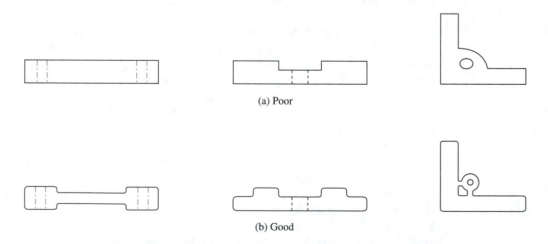

(a) Poor

(b) Good

Figure 12.20 Elimination of sharp corners and redesign to make thicknesses more uniform in injection molded parts.

Mold shrinkage is the difference in the part dimensions and the dimensions of the mold, that is, the change in size of the part in the mold due to thermal and other contractions. Mold shrinkage occurs in all plastic parts. This is a consideration in designing the mold so that the finished part will have the correct dimensions. Therefore, a calculation of mold shrinkage is an important part of plastic part and mold design. The amount of shrinkage is dependent upon several factors including (1) the temperature of the resin, (2) the type of resin (each has its characteristic shrinkage that depends upon the nature of the microstructure of the molecule), (3) the flow of the resin in the mold cavity, (4) operational factors such as injection pressure and the length of time that pressure is applied to the mold, and (5) additives to the resin.

Even though these factors are complicated, the most important factor is the characteristic shrinkage of the plastic and so an estimate of the change in part size due to this shrinkage factor can be made. Characteristic shrinkages for several common polymers are given in Table 12.2. These values are used by calculating the dimensions of the mold cavity from the following equation:

$$\text{Mold dimension} = \text{Part dimension} (1 + \text{Shrinkage value}) \qquad (12.2)$$

Notice the strong dependence of shrinkage on the presence of a reinforcement such as fiberglass. Plastics without reinforcement have a substantially greater shrinkage than when reinforced with fiberglass (compare polycarbonate with and without reinforcement in the table). The resin group with the lowest shrinkage, polyester BMC and SMC is reinforced with fiberglass. This group also has a high filler content. Both the reinforcements and, to a lesser degree the fillers, hold the resin in place during cooling and, therefore, the shrinkage is less. This effect is most important in the direction of orientation of the reinforcement. Across the orientation direction of the fiber reinforcement, little change in the shrinkage from the characteristic value of the nonreinforced plastic is seen. The fillers simply occupy space among the resin molecules and restrict the movement of the molecules from getting closer together.

Table 12.2 Characteristic Shrinkages for Various Plastics

Shrinkage (mm/mm)	Type of Material
0–0.002	Polyester (thermoset) BMC, SMC
0.001–0.004	Polycarbonate, 20% fiberglass
0.002–0.008	Acrylic
0.002–0.003	PVC
0.004–0.007	ABS
0.004–0.006	Polystyrene
0.005–0.007	Polycarbonate
0.005–0.008	Polyphenylene oxide
0.008–0.015	Nylon (6/6)
0.010–0.020	Polypropylene
0.018–0.023	Acetal
0.007–0.025	LDPE
0.020–0.040	HDPE

In general, resins that are crystalline tend to have higher characteristic shrinkages than do amorphous resins. The amorphous polymers (acrylic, PVC, ABS, and polystyrene) all have lower shrinkages than do the more crystalline polymers (nylon, acetal, polypropylene and polyethylene). Within the polyethylene group, the more amorphous LDPE has a lower shrinkage than does the crystalline HDPE. The greater bonding and order in the crystalline polymers draws the molecules together and results in this greater shrinkage.

Although not as important a factor as characteristic shrinkage, orientation of the resin molecules because of mold design (placement of gates, etc.) can have an effect on shrinkage. The shrinkage will be less in the flow direction because the oriented molecules tend to hold the plastic in place, much as a reinforcement would restrict the movement of the resin during shrinkage. This effect can be seen in Figure 12.21.

Shrinkage in a part is three dimensional and it is in the direction toward the center of mass. The amount of this shrinkage will be dependent upon orientation of fibers or molecules, but in most parts these orientation effects are small. Therefore, as a first approximation, shrinkage can be considered to be *isotropic,* that is, the same in all directions. Nonuniform section thicknesses will result in distortions of the part because of differential cooling resulting in nonuniform stresses. This is another reason to strive for uniform thicknesses throughout the part.

Additional changes in part dimensions can also occur after the part has been removed from the mold. One of the most common of these deformations is *postmold shrinkage.* This is caused by continued internal changes in the part because of additional cooling or, perhaps, because of crystallization. Some resins require considerable time to fully crystallize and for practical purposes cannot be kept in the mold until the crystallization is complete. In parts with complex structures, this additional crystallization can cause warping or other dimensional changes that are unacceptable. One method for preventing this warpage is to cool the part under pressure or in a holding fixture. The dimensional changes are normally handled through mold modifications.

Another characteristic of some plastic parts is *springback,* which occurs when a part is molded with a particular angle between two sections of the part and the angle changes

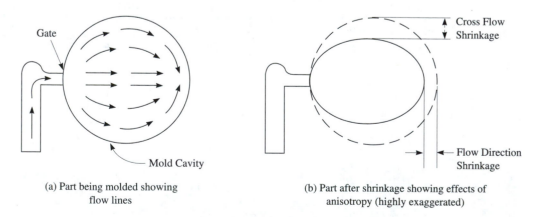

(a) Part being molded showing
flow lines

(b) Part after shrinkage showing effects of
anisotropy (highly exaggerated)

Figure 12.21   Shrinkage due to resin orientation from flow in the mold cavity.

after molding. This phenomenon occurs because of residual stresses in the part either from flow orientation, cooling, or design. Springback is most often handled by cooling the part in a holding fixture that sets the desired angle or contour. Some care must be taken when using this technique as it does not eliminate the internal stresses, it merely freezes the part in place. Therefore, if the part is heated sufficiently to relieve the stresses, the part dimensions may change, that is, the springback may reoccur.

The placement of the gate is an important consideration that can often affect shrinkage, molding efficiency, and part performance. In general, the gate should be placed at the thickest section of the part. This placement will mean that the resin will not have to flow through a thin section to fill a thick section, with the accompanying difficulty of freezing off in the thin section prematurely.

A further advantage to placing the gate at the thick section is that this placement allows additional resin to be injected into the mold even after much of the resin has solidified. When the solidification occurs, the plastic shrinks, both from crystallization and from thermal contractions. This shrinkage occurs most in the thickest regions. Hence, if additional material is added to the thickest region, the effects of shrinkage are minimized. This process of injecting additional resin during cooling is called *packing* the mold.

The vast majority of cavities have only one opening (gate). However, some large parts or complicated parts in which the flow of the resin in the cavity would not properly fill the cavity may have more than one opening. These are called *multi-injection points*. Whenever a cavity has multi-injection points, a major concern is the proper joining or *knitting together* of the resin flows. This knitting together of the flows is complicated because of the high viscosity of many resins, especially as the resins are cooling inside the mold. Whenever multipoint injection is used, mechanical weakness is possible at the place where the resin flows meet, called the *knit line* (see Figures 12.22 and 12.23).

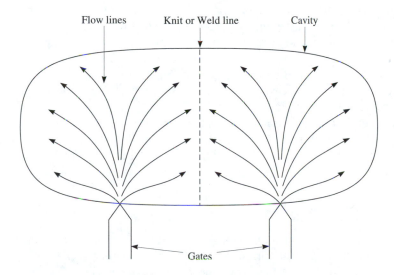

Figure 12.22 Knit line in a multi-injection-point cavity.

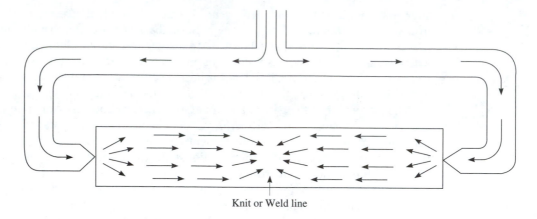

Knit or Weld line

Figure 12.23   Multiple injection points causing a weld line.

When multiple injection points cannot be avoided, some operational changes can be made to minimize the weld line. The temperature of the resin can be increased, the injection speed increased, the injection pressure increased, and the cycle time increased. Other changes that would reduce resin viscosity, such as raising the resin melt index, would also help. All of these factors will give greater mobility and time for the molecules in the two flow patterns to diffuse together.

To facilitate the removal of the part from the cavity, the cavity walls are cut with a slight angle so that the opening is slightly larger than the base. This angle, called the *draft angle,* is usually between 2° and 5° depending on the type of resin and the thickness of the part. Ribs and other vertical sections are also tapered inward to facilitate removal of the part.

Any portion of the cavity walls that is narrower than the portion of the walls directly below it is called an *undercut,* or a *negative draft.* Undercuts make removal of the part more difficult and should be avoided unless the design of the part specifically calls for an undercut. An example of a designed undercut is a latch clasp. If the plastic material is flexible enough, the part can be deformed sufficiently to push the lower portion of the part past the undercut in the cavity and simply eject the part in the normal way. This process is called *stripping* the part.

When undercuts are large or the part is stiff, stripping may not be possible. In these cases, the cavity walls must move out of the way to allow the part to be removed from the cavity. This is done by using slides and other mechanical mechanisms in the mold that open the cavity when the mold opens and before ejection is activated.

OPERATIONS AND CONTROL

A visual representation of the molding cycle of an injection molding machine may assist in the understanding of the various key control parameters and their interactions. This diagram is given in Figure 12.24. The cycle begins with the closing of the mold. The screw or plunger then moves forward to inject the resin that has collected in the injection pool.

Figure 12.24 Diagram of the total mold cycle of an injection molding machine.

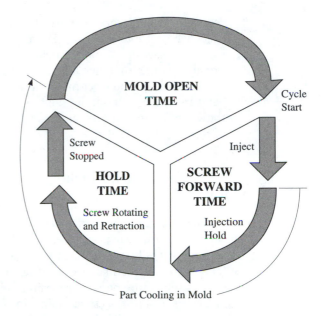

This continues until the mold is filled and held for the appropriate time to set the dimensions. The screw then retracts while rotating and melts additional material which is pumped forward into the injection pool. When sufficient material has been accumulated to make the next shot, the screw rotation stops and the machine waits until the part is cooled. The mold then opens and the part is ejected. After ejection, the mold is closed to complete the cycle.

The key parameters that are used to control the injection molding process are: (1) temperature of the melt, (2) temperature of the mold, (3) pressure of injection and hold, and (4) various times. The most important times are: injection time, dwell time, freeze time, and dead time. Each of these key parameters are considered separately and then considered jointly as part of the discussion of troubleshooting.

Temperature of the Melt

Initial settings for the heaters in the injection system (along the barrel and the nozzle) should be those recommended by the resin manufacturer. These temperatures are generally below the melting point of the resin, which allows for some mechanical heating but does not overheat the resin that would lengthen the cycle time and could cause some degradation in more sensitive resins. It is important to match (1) the heating with the type of resin, (2) the characteristics of the molding machine, and (3) the shot size being molded.

As with extrusion, the temperatures of the heating zones along the barrel can be adjusted independently to fine-tune the melting in the barrel and the viscosity of the melt. For instance, nylon will normally be run with a high temperature in the feed zone and decreasing temperature profile as the resin moves through the barrel. This adds heat to reduce the high torques required to melt and to assist in the molding and yet doesn't put excessive heat into the melt. (However, too much heat in the feed zone can cause bridging.) The viscosity of molten nylon is quite low, so flow into the mold is easy and additional thinning from increased temperatures is not needed.

The nozzle should act merely as a melt-conveying pipe and should not affect the temperature of the melt. Ideally, melt temperatures entering and leaving the nozzle should be the same. The nozzle temperature setting will depend on the design of the nozzle, the type of resin, and the shot size.

Temperature of the Mold

The mold temperature is controlled with water that can be circulated through the mold, although it should be remembered that the temperature of the water is not the same as the mold surface temperature. Both temperatures should be monitored. In some cases cooling or chilling is required while in others the water may be heated to control crystallization and thermal stresses. Cooler molds give faster molding cycles so the coldest mold temperature possible is desired. Limitations on the mold temperature are freezing of the resin in thin sections and resin cooling properties (crystallization, etc.). Therefore, a compromise must be struck between the flow of the resin and the mold temperature. Whenever possible, the critical freezing location should be at the gate. The size of the gate can then be adjusted to achieve the best flow in the mold at the lowest temperature possible.

Under some conditions, the temperatures of the fixed and moveable sides of the mold may be different. One such condition is when the temperature is to be used to assist in retaining the part on the moveable side of the mold. If cores are used in the mold, the largest core is generally on the moveable side which gives additional surface area to the moveable side and encourages the part to be retained there. If there is still a problem retaining the part on the moveable side even with a large core, the moveable side can be cooled relative to the fixed side and thus shrink the part onto the core.

If no core is present in the mold and the part sticks to the fixed side, differential cooling of the fixed side will cause the part to shrink slightly more on the fixed side, thus releasing the part slightly more than on the moveable side. Hence, the part goes with the moveable side. However, be cautious to not make the temperature differential too great between the two mold halves as this could result in some pin misalignment. If a large temperature differential is needed to retain the part on the fixed side, other techniques (such as screw pullers and draft angles which are discussed elsewhere in this chapter) should be used.

Some factors that affect the temperatures of the system, both the melt temperature and the mold temperature are:

- Shot size—larger shots take more heat.
- Injection rate—faster filling creates higher melt temperatures because of shearing.
- Size of runner system—long runners require higher temperatures.
- Part thickness—thick parts require more cooling time and are molded at lower temperatures.

Pressure of Injection and Hold

In general, injection and holding should be done at the lowest practical pressures possible without resulting in *short shots* (that is, injection shots that do not fill the mold completely). The pressures should also be high enough and maintained long enough to minimize shrinkage by adding more resin into the mold during the time that the part is shrinking. This addition of resin during shrinkage is called *packing* the mold. The limitations on injection pressure are, of course, the capacities of the machine to inject and clamp as well as the potential in some parts for stresses from overpacking. The injection pressure may also be limited by the mold construction.

Two-stage pressurization may be desired for some operations. In the first stage, high pressure is used to inject and fill the part. Then a lower pressure is used for the holding part of the cycle.

Injection or Fill Rate and Time

The injection time is the time to force the molten resin into the cavity. This time is controlled by the fill rate. In most cases, the resin should be injected as quickly as possible although the optimal fill rate depends upon the geometry of the part, the size of the gate, and the melt temperature. It should be remembered that the temperature of the mold is below the freezing point of the resin, so long injection times will increase the likelihood of premature freezing. With thick parts or with very small gates, the injection rate is sometimes lowered so that a continual passage of hot resin through the gate will prevent freezing and allow the part to fill completely. Parts with only thin sections are filled at high injection rates to prevent freezing before filling.

Recent efforts at controlling precisely the movement of the screw during resin injection have allowed shot size to be standardized and thus have improved part dimensional control. Although the shot size variations are small in most cases, for precise repeatability, the shot must be the same each time and that is done by controlling the screw movement and ensuring proper closing of the check valve at the end of the screw.

Dwell Time

The dwell time is the time that force or pressure is applied to the cavity after the cavity is filled. This force helps with packing and ensures that the dimensions of the part are set.

Freeze Time

The freeze time is the interval after the pressure is relieved and before the mold is opened. This time is governed by the ability of the part to harden sufficiently to be ejected. This time is often shortened by cooling the mold.

Ejection (Dead) Time

The dead time counts the time to open the mold, eject the part, and close the mold. Some care must be taken that the ejection of the part is not done too quickly, otherwise the part may be damaged or strained. The efficient mechanical operation of the machine is obviously a critical factor in determining dead time. Dead time is also affected by the ability to remove the part easily from the mold. To insure that the mold leaves the mold cavity easily, *mold release* is occasionally coated onto the mold cavity. Mold release is any substance that will lubricate the removal of the part from the cavity. Typical mold releases contain silicone, fluorocarbons, and other materials that do not readily stick to the mold. In some resins, especially thermosets, the mold release can be included in the resin. When this is done, the term *internal mold release* is used.

If metallic *inserts* are used in the part, these are inserted during the dead time. Inserts placed by hand can cause substantial delays because of the need to carefully place the inserts in the proper location. For this reason, automated insert placement is preferred, although the problem with automated systems is the potential for mold damage should the mold close on a poorly placed insert.

Most modern injection molding machines use timers or computers to control the various times. These times are set after some experimentation to determine the optimized cycle. These times would then be saved for use on the part when it is next molded.

Some new machines have reduced the number of input parameters greatly by storing resin flow information and other factors that affect the operation of the machine. One machine requires only the following inputs: mold shut height (which sets the closing distance for clamping), opening distance (which determines the amount of distance the machine must open to eject the part), part thickness, part weight, and material type. These latter parameters determine the cycle times, temperatures, and pressures. After these parameters are input, the machine can be operated and the cycle optimized by fine-tuning the parameters. The adjusted settings are then stored for use when the part is next molded.

Merely storing data about a run does not guarantee that a future run can be made without operator intervention and, perhaps, changes in the machine settings. This fact reflects the tremendous complexity of the injection molding process and the number of variables, some quite obscure, that must be controlled.

Therefore, the approach that is most often taken in control of injection molding is statistical. Control charts and other statistical methods are now available as part of the computer-control system of newer injection molding machines. They are also available as add-on equipment for older machines. With this assistance, operator intervention can be minimized, although all these statistical methods assume that some basic process parameters (such as the resin characteristics and often the molding environment) are constant from run to run. Some graphical presentations are also available to help interpret the data.

Another approach that is now being developed is the use of advanced neural networks (ANN) to control the system. Like statistical control, a database of experience must be built up for the process to work. Unlike statistical methods, however, the ANN has the capability of recommending process settings, even for conditions that might never have occurred before in running that machine. This capability arises from the artificial intelligence system that is part of the design of the ANN. Recent work has shown that ANNs, purchased off the shelf and trained during normal molding, can be quite accurate in predicting the quality of an injection molded part while the part is still in the mold. Furthermore, some early work even suggests that ANNs might be effective in suggesting molding conditions for a new part, identified to the machine only on the basis of geometry and resin characteristics.

Although optimization is usually directed toward the shortest cycle, some care should be exercised to ensure part reliability and dimensional stability as well. It is possible that the ideal cycle will not be the shortest because of these dimensional characteristics of the part.

Troubleshooting and Quality Improvement

The injection process is very complex and the interactions are numerous. When a problem has been noted, several parts should be made to ensure that the process is stable and the defect is not transitory. No specific procedures exist for the correcting of injection molding problems. But some defects are characteristic of certain operational problems and these, at least, suggest possible solutions for correcting the defects. Table 12.3 outlines the most common defects and some of the suggested corrections.

Maintenance

As with extrusion, the motor and hydraulic system should have routine preventive lubrication, inspection, and fluid changing. Filters should be routinely cleaned. Many electrical components can be checked while in operation, but that should be done on a scheduled basis.

The care of molds is especially important. These should be cleaned carefully and coated lightly with a rust preventative. They should be stored under a cover to prevent dirt accumulation and should be inspected for proper operation and cleaned before each use.

SPECIAL INJECTION MOLDING PROCESSES

Coinjection Molding

Just as multiple resins can be coextruded by attaching several extruders to a single die, a process has been developed that allows multiple resins to be injected into one mold to make a single part. This process is called *coinjection molding*. It too requires a separate machine for each resin to be injected. The separate machines are all connected to a single mold, usually through different channels in a single nozzle. The materials are injected either at or near the same time so that the part contains both materials. Normally the materials are injected so that one material forms a skin and the other material forms a core inside it. This is done by injecting some of the skin material into the mold and then injecting the core material as additional skin material continues to be injected. The core material pushes the skin material against the cavity walls, when the geometry of the injection nozzles and operating conditions are correct.

Table 12.3 Troubleshooting Guide for Injection Molding

Problem	Possible Cause
Burning or scorching of molded parts (black specks, spots, or streaks in the part)	• **Melt temperature.** Heaters may be overheating material in barrel. Reduce temperatures or time in the barrel. One of the heater bands may be delivering more heat than the others causing a hot spot. Frictional heat may be causing burning of pellets which would require increase in feed-zone temperature or use of resins with internal lubricant. • **Venting of mold.** Heated air trapped in a poorly vented mold may produce local burning. • **Design of barrel or nozzle.** Machine may permit resin to be trapped and begin to degrade. Check for smooth flow lines. Purging may be needed on routine basis. • **Hold-up time.** Cycle may be too long causing resin in barrel to overheat. Decrease temperatures and determine cause of extended cycle.
Poor weld and flow lines (weak welds, obvious poor knitting)	• **Melt temperature.** Low temperatures can prevent resin from knitting together across a weld line. Temperatures that are too high can cause degrading and gassing which can also prevent good knitting. • **Mold surface temperature.** If too low, material flowing into cavity will be cooled excessively and prevent knitting. • **Pressure on melt.** Too low injection pressure will not force molten material together at the weld line. Runners and gates may be too small resulting in low injection pressure. • **Choice of material.** Materials which flow more easily (higher MI) may be needed if part performance allows. • **Mold lubricants.** Mold lubricant may be pushed into weld by advancing molten polymer. • **Venting.** Vents may be too small or improperly positioned causing gases to be trapped at weld line.
Warpage of part (lack of dimensional stability)	• **Temperature of molded piece when ejected.** Part may be too hot so that ejector pins are warping it. Internal stresses or residual crystallization may be causing the warpage which can be relieved by cooling further in the mold or in holding fixtures. Uneven cooling may cause problem. Parts dropped into a box may warp from accumulated heat of parts in box. • **Mold surface temperature.** Uneven surface temperature may cause warpage. Check surface temperatures. Complex-shaped parts may cause uneven cooling, especially in molds that are not chilled. • **Knockout pins.** May not be working or may be improperly spaced. • **Undercuts in mold.** May be too deep so that removal of part is difficult. • **Part geometry.** Nonuniform wall thicknesses may cause residual stresses.

Table 12.3 *Continued*

Problem	Possible Cause
Unmelted particles in molding	• **Melt temperature.** If too low, some molding powder may pass through barrel and into cavity without being melted. Increase barrel temperatures, increase back pressure, or decrease screw speed. Screw may need to be changed to a different design. • **Temperature distribution in barrel.** A burned-out heater band may be causing poor melting. Check controller and thermocouple. • **Capacity of heating.** Barrel may not be long enough to melt properly. • **Screw speed.** Use slowest rotation possible for cycle to get more thermal heating, or if the plastic is highly susceptible to adiabatic heating, use the highest rotation possible.
Surface imperfections (bubbles, splay marks, smears, sinks, pits, orange peel, wrinkles, jetting, blush, frost)	• **Melt temperature.** If too high, material could decompose forming gasses that could cause splay marks or smears. • **Moisture.** Excessive moisture could cause splay marks. Dry the resin. • **Pressure on the melt.** If too low, incomplete mold filling could cause wrinkled surface or pits. • **Mold lubricant.** May be incompatible with resin. • **Mold construction.** Metals with low thermal conductivity, such as stainless steel, will improve surface quality by slower cooling. Gate position may need to be changed to eliminate wrinkles. • **Filling rate.** Faster filling will improve surface gloss because mold is filled before resin hardens, but filling too fast may cause other imperfections.
Flashing	• **Melt temperature.** If material is too hot, viscosity can be too low and resin will leak between mold plates. Look for nozzle drool. • **Pressure on melt.** Machine may not be able to clamp sufficiently. • **Fill rate.** Reduce, especially at point when cavities are nearly full. • **Mold construction.** Check alignment of mold cavities. Be sure of adequate venting. Check on warpage of plates. • **Machine settings.** Check on clamp force.
Sink marks or shrinkage voids	• **Injection time.** If the part weight is increasing as the injection time is increased, the gate is not sealing before the pressure is reduced. Lengthen injection time. • **Pressure on melt.** Increase pressure to pack more. • **Melt temperature.** Part may be too hot when ejected and sink marks could occur postmolding. • **Gate design.** Too small a gate will prevent adequate packing. When molding parts with large areas and thin walls, it may be necessary to use large or multiple gates.

(continued)

Table 12.3 *Continued*

Problem	Possible Cause
Internal voids	• **Melt temperature.** If too high material could decompose and form gasses. • **Mold surface temperature.** If mold is too cold, outside of part could set up before interior and cause voids. Raise mold temperature and extend cycle. • **Moisture.** May cause bubbles. • **Pressure on the melt.** Too low can cause insufficient packing. • **Runners and gates.** If too small, packing may be incomplete. • **Injection time.** If too short, the gate may not have frozen and voids could be created by shrinkage.
Incomplete filling of cavity (short shots)	• **Melt temperature.** Should be high enough to ensure complete melting and proper viscosity. • **Pressure on the melt.** Should be as high as possible without causing problems such as flash, etc. • **Back flow (nonreturn) valve.** Valve may be damaged resulting in insufficient shot size. • **Venting.** Pressure in mold can prevent proper filling. • **Runners.** If too small, resin may freeze before filling is complete. • **Gates.** Gate may be too small. Gate land may be too long. • **Mold temperature.** Increase if possible. • **Shot weight.** Machine may not have sufficient capacity or injection stroke may be improperly set. If multicavity mold is being used, some cavities may need to be blanked off. • **Injection speed.** If too slow, resin may freeze before mold is filled. • **Filling pattern.** If the same cavities consistently give incompletely filled parts, the cooling of the mold may be nonuniform, perhaps partially plugged cooling lines. Also, gates or runners may be nonuniform. If incomplete filling is random among the cavities, look for unmelted resin or contaminants.
Ejectability	• **Condition of mold.** If mold has undercuts, burrs or damaged surface, defect should be corrected to aid ejectability. If surface of cavity or core pins is highly polished, or if taper of core and cavity is insufficient, sticking will be problem. Mold release or rework may be necessary. • **Alignment of cores and cavities.** If not properly aligned, mold may be difficult to open and part will not eject easily.

Table 12.3 *Continued*

Problem	Possible Cause
	• **Length of cycle.** If too short, material may be too soft. Larger knockout area may be required. If too long, overpacking may have occurred causing too tight a fit on cavity walls or cores. • **Mold temperature.** If too high, part may not be sufficiently cool or rigid to eject properly. • **Melt temperature.** If too hot, resin may have degraded and degraded resin is difficult to eject. • **Sprue.** Check for damage to sprue which could cause entire part to be ejected poorly. If sprue is too large, material may be building up on the outside of the bushing. Enlarge bushing or reduce diameter of sprue. Taper of sprue may not be adequate. Sprue puller may not be adequate. • **Ejector pins.** If worn, flash may be filling in cracks and preventing proper action. Pins may be bent.
Brittleness of molded parts	• **Rate of filling cavities.** If too high, flow may not be uniform and gasses may be trapped. • **Sprue, runners and gates.** If too small, flows may not be stable. • **Melt temperature.** Should be high enough to insure complete melting of material. • **Contamination.** May have noncompatible resins. • **Part design.** Sharp angles and abrupt changes in thickness of wall sections may cause erratic flows and internal stresses. • **Mold temperature.** Generally cold molds benefit toughness in thin parts while thick parts suffer. Cold molds should be avoided where stress or solvent cracking may be a problem.
Contamination	• **Nozzle or barrel.** If contamination appears as a continuous line through the sprue, problem may be degraded material breaking loose from barrel wall or from cracked nozzle, bushing, or barrel. • **Mold lubricant.** Lubricating material from ejectors or mold release may be contaminant. • **Material fed to hopper.** Check for paper, incompatible resins, or dirt. Hopper may need to be covered. Pneumatic lines may be open. Filter in line may be contaminated. If material is dried in a drying unit, check for contamination from previous drying of noncompatible resin. • **Reground material.** May have become contaminated from noncompatible resin. May have started to degrade from previous heat history. May have been contaminated through in-plant handling.

Some users of coinjection molding have focused on the use of an inexpensive resin as the core with a superior resin as the skin. The inexpensive or poor-performing resin could be recycled material and the outer skin virgin, which is required for some performance criterion such as food packaging, UV light protection, or stress crack resistance. Others have suggested a rigid core with a resilient, perhaps elastomeric, outer skin for applications such as handles, appliance panels and housings, normally foamed and painted parts that can have the colored layer injected as the skin, and applications in which the skin and core have different properties, such as electrical or heat conductivity.

Coinjection machines have been available commercially for many years, especially in Europe. The costs are obviously higher than traditional injection molding machines, but significant material savings or performance advantages are available for some applications which justify the cost.

One drawback to the process is the need for compatibility between the resins. If not compatible, separation between the skin and core can occur, with accompanying decreases in product performance. In general, however, the difficulties with the process are rapidly being solved and coinjection molding is growing rapidly.

Flow-Controlled Injection Processes

Some injection processes use various methods to control the injection process and thereby improve properties in the product. One of these processes is called Scorim and uses multiple live feeds. In this process the injection stream is split by a special injection head into several streams, each with its own entry into the mold. The streams are independently pressurized. By alternating the injection from the several streams the material can be induced to flow back and forth in the mold, thus inducing excellent knitting and eliminating weld lines. When operated in tandem, the multiple injection system achieves more uniform and higher-density packing than conventional injection molding. Special orientation effects can also be achieved by selective pressurization of the streams.

A related process is called push-pull injection molding. It uses two injection systems that are mounted so that each is connected to a separate inlet to the mold. By alternating injection from the two injectors, the material inside the mold can be caused to flow back and forth as it freezes, thus creating layers within the part having different orientations. When used with short-fiber-reinforced resins, the orientations of the fibers are very high, thus leading to parts with enhanced physical properties.

Another flow-controlled injection process is called lamellar injection molding. In this process streams from two machines (coinjection) are split in the nozzle into several layers with the different resins alternating in the layered flow. This layered flow is then injected into the mold to create a laminar part. The advantages are that resins of quite different properties can be combined in this way to give parts with properties that can be optimized between the properties of the resins. For instance, improvements in gas barrier properties, physical properties, optical properties, heat resistance, and environmental stress crack resistance have all been demonstrated.

Low-pressure molding is another technique to improve physical properties and operating conditions in injection molding. In this technique, the injection pressure is lowered with accompanying careful control of other operating parameters such as temperature, runner size, injection rate, and back pressure. When all of these are correctly set, parts can be made at much lower pressures, thus making larger parts on smaller machines. These low-pressure parts have less internal stress and thus improved physical properties.

Gas Injection Molding

Injecting a gas into a partially filled injection molding cavity solves many problems with injection molding of foamed products and with some conventional solid products. Gas injection will significantly reduce the incidence of sink marks, will improve product definition, and will improve surface quality for many parts.

The gas may be injected either through the nozzle or through the runner system. In both processes, the cavity is partially filled and then an inert gas (usually nitrogen) is introduced into the cavity either simultaneously or alternately with additional material. The gas pushes on the low-viscosity material in the center and pushes this material into all parts of the cavity, pressing it against the cavity surface. Hence, voids are created inside thick sections and excellent surface definition is made against the cavity. The voids can be filled with subsequent resin injection.

The advantages of gas injection molding over conventional injection molding are:

- Improved conformation of the surface to the mold cavity surface resulting in superior surface definition and finish.
- Minimization of pits and sinks over thick sections, ribs, and bosses.
- Possible elimination of runner system if gas injection is through the nozzle.
- Lower cost tooling and machines because the pressure is lower than the injection pressure of a normal injection molding process.
- Increased part design flexibility allowing parts with very thick sections and sharp radii.
- Reduction in built-in stresses because the pressure on the part during cooling would be lower.

The obvious problem with gas injection molding is the newness of the process with the accompanying development of machines and techniques that is required. Some considerations that need to be addressed include the optimization of the gas injection location, the timing and method of addition of material after injection has commenced, design rules for parts made by this method, various details of tooling modifications to handle the special aspects of the process, and operational variable control for process optimization.

Injection-Compression Molding

In this process material is injected into a cavity but not so that the cavity is completely filled. Then the back of the cavity moves on a pressure cycle to reduce the size of the cavity so that the material is squeezed into the smaller volume. This technique forces the material against the cavity walls to obtain improved surface quality and forces the material into all parts of the mold to obtain better part dimensional control and shape definition.

Injection-compression molding has the advantages of injection molding so far as injecting into a closed mold and injecting a liquid material. It also has the advantages of the compression process because part definition is improved. The complication of the process is the complexity of the mold and the lack of design rules for parts and tools.

MODELING AND COMPUTER-AIDED MOLD-FLOW ANALYSIS

After a mold has been made, tests are run to verify a number of mold features. In properly made molds, the filling pattern completely fills the mold and does so in the minimum time, the shrinkage estimated for each region in the part is correct, the part does not warp, the part does not have excessive internal stresses, the gates and runners are sized correctly, and the orientation of the molecules and/or reinforcements are as expected. Tests to verify these features require many different runs, often with very carefully controlled conditions to ensure that the desired investigation is done correctly. The complexity of injection molds (and to a lesser extent, other molds) increases the number and complexity of these tests. Such tests can, of course, lead to changes in the mold, which can lead to further testing after the changes have been made.

Much of this complexity can be eliminated by using computers to simulate the injection molding process. These simulation programs are available from several software companies and seem to have some basic similarities and features that are very valuable for the adjustment and prove-out of an injection mold.

These programs require some important information as input data. These data include the following: fill time, plastic melt temperature, packing pressure, freeze temperature, mold temperature, plastic ejection temperature (the temperature when the part will be ejected), part ejection time (as set on the machine), coolant circuit temperatures and pressures, mold exterior boundary conditions, and insert boundary conditions. Most of these parameters can be approximated from prior experience or determined quickly from a few simple measurements on the mold or on a mold similar to the one to be built.

The software program then builds a model of the mold and the part to be made. The model divides the part into many small sections, each of which is treated as a small, discrete unit on which forces can be imposed. These small units are chosen small enough that they are regular in shape, so calculations for each of the small units are straightforward. This system of dividing the large part into very small units is called *finite element analysis (FEA)*. The FEA methods used within the programs to establish a model of the part can also be used to calculate traditional mechanical properties such as strength, stiffness, and elongation.

The system usually has stored in memory pertinent data on the specific plastic material to be molded. Thus the key molding parameters and results can determined for the specific case in question.

One of the most important outputs of these programs is the filling pattern. This output shows the amount of the mold that is filled at particular times. These data allow problems such as air entrapment, poor knitting, and premature freeze-off to be seen. Figure 12.25 shows the simulation of an experiment wherein several short shots are made in a

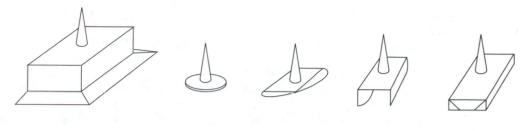

(a) Steps (short shots) in Injection Molding a Box

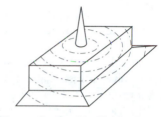

(b) Flows as Predicted by Molding Program

Figure 12.25 Mold-flow experiment using short shots compared to a computer simulation of a mold fill showing lines of different fill times.

mold to see where the filling has progressed. The figure compares the series of short shots with the mold-filling output from a typical simulation program.

The simulation programs are very valuable for predicting shrinkage, which, in complex parts, is otherwise very difficult to predict. Shrinkage looks at four factors: volumetric shrinkage, crystallinity, mold restraint, and orientation. Shrinkage is calculated by examining these four factors in each specific area within the part. This calculation can also predict warpage (caused by nonsymmetrical shrinkage) because the program can be used to compare shrinkage in different regions of the part. Sink marks, which are shrinkage in thick sections, can also be predicted.

Many of the computer programs can calculate stresses caused by packing the mold. These stresses result from shrinkage or warpage that is not allowed to occur because of restraints on the part or within the part. Hence, the stresses are built into the part by the processing.

Some of the programs also assist in the sizing and length determination of runners. These factors are determined by the time required to fill the mold and the nature of the plastic material to be molded. Typically, the program identifies which of several suggested runner patterns is acceptable. The program also indicates whether a particular runner diameter is too small to accommodate the required flow to fill the mold.

Some recently developed programs are able to show the effects of gas injection in injection molding. These programs can also show the effect of orientation of fiber reinforcements in the part and how those orientations change from one region of the part to another.

CASE STUDY 12.1

Estimating the Cost of an Injection Molded Pocket Knife

Many methods are used to determine the cost of an injection molded part. Some methods are based upon rough estimates and rules of thumb while others are extremely detailed with allocated costs for numerous plant functions and overheads. The method chosen in this case study is a compromise between these two extremes. The major cost elements are identified explicitly for a particular part, a pocket knife with a plastic handle. The knife analyzed in this case study is made with a nylon handle and is illustrated in Figure 12.26. It is made in a two-cavity mold, somewhat like the mold illustrated in Figure 12.5 except that the knife handle requires cores and is therefore a more complicated mold.

The cost elements are calculated in such a way that the same procedure can be applied to other parts with relative ease. The method shown has proven to be effective in determining a price that will allow sales prices to be set to assure profitability.

The method involves a systematic examination and calculation of each of the major cost elements. A form is created on which the key information can be gathered and which will serve as a historical document so that pricing estimates can be reviewed from time to time. Such a form is given in Table 12.4. A blank form is given in the appendix.

Introduction Section

It is assumed the form will be filled out by the company doing the molding of the part (the molder). This company may not be the same as the customer who will use the part in some other assembly or otherwise sell it. The molder and the customer can be different entities, although, of course, they could also be the same. Mold making is another function that may be done by the molder or by a different company. The mold could be secured by the customer or through the molder. Both cases are explained in the detailed instructions on using the form.

The name of the part, part number, part print number and revision are self-explanatory. The customer should be identified and some reference given for the customer's request for quote so that any specifications or other details can be referred to. The date, name of the estimator, and name of the approver should be noted for historical reference purposes.

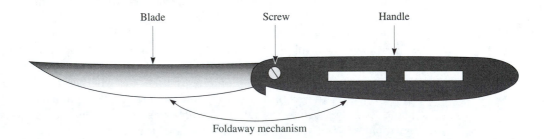

Figure 12.26   Pocket knife with foldaway blade and injection molded handle.

Table 12.4   Cost Estimating Form for Injection Molding

Part name: *Knife handle*	Customer: *ABC Company*		Date: *8 July 1994*
Part number: *1234*	Part print/revision: *A-1*		Estimator: *ABS*
Request for quote reference: *Letter from ABC on 17 June 94*			Approved: MMS

Resin and Additive Costs	Resin and Grade	Source	Price	Additives	Additives Costs	Additive Content per Part	Total Material Cost per Part
	nylon (6/6)-Zytel 101	*DuPont*	*$1.36/lb in pillar packs*	*Colorant (black)*	*$3.00*	*1.0%*	*$1.38/lb*

Part Cost	Part weight (grams)	Waste Factor	Adjusted Weight	Material Costs per Part		Part Cost Contribution per 1,000 Parts	
	3.8	*1.05*	*4.0*	*$0.0122*		*$12.20*	

Tooling Costs	Tool Description	External Cost	Internal Cost	Total Cost	Expected Quantity	Tooling Cost Contribution per 1,000 Parts	
	2 cavity, steel	*$5,000*	*$15,000*	*$20,000*	*4,000,000*	*$5.00*	

Machine Costs	Cavities	Cycle Time (s)		Parts/h	Auto/ Operator	Machine Rate	Machine Cost Contribution per 1,000 Parts
	2	*30*		*240*	*Operator*	*$25.00*	*$104.17*

Secondary Operation Costs	Operation	Cycle Time	Rate	Operation	Cycle Time	Rate	Secondary Cost Contribution per 1,000 Parts
	Attach.	*10 s*	*$7.00/h*	—	—	—	*$19.44*

Purchased Items Costs	Item	Source	Cost/ 1000 Parts	Item	Source	Cost/ 1000 Parts	Purchased Cost Contribution per 1,000 Parts
	Blade	*XYZ*	*$1,250.00*	*Screw*	*A and L*	*$2.00*	*$1,252.00*

Packaging and Shipping Costs	Material	Source	Cost/ 1000 Parts	Material	Source	Cost/ 1000 Parts	Packaging Cost Contribution per 1,000 Parts
	Bubble	*ACME*	*$50.00*	*Box*	*Consolid.*	*$0.70*	*$50.70*

Total Factory Cost per 1,000 Parts	**$1443.51**
General and Administrative Costs per 1,000 Parts	$144.35
Marketing Costs per 1,000 Parts	$288.70
Total Cost per 1,000 Parts	**$1,876.56**

Resin and Additive Costs

The type of resin and particular grade should be very clear. The source is the resin manufacturer or wholesaler from whom the resin is obtained. The cost of the resin should be stated in a typical order quantity. Many resins are sold at quantity discounts. The packaging of the resin might also be important to note since shipments by rail car are often less than shipments by bulk trucks. If shipped in *pillar packs* or *gaylords* (palletized boxes), the cost is higher than when the resin is shipped in bulk. The highest-priced packaging of all is bags. In the example of the knife, the resin is to be nylon (6/6), Zytel 101, which is to be obtained from DuPont. The cost is $1.36 per pound and the material is to be bought in pillar packs at order quantities of 40,000 pounds.

If an additive is included in the resin, the type of additive and its cost are noted. Several additives could, of course, be added, and each of these would be shown on the form. In this case, the only additive is a color concentrate (black) which has a price of $3.00 per pound.

The concentrations of all the additives would be shown separately in the next column of the form. Since each additive may have a separate price, it is important to reflect the concentration and price of each. In this case, the concentration of colorant is 1.0%.

The total material cost is determined by Equation (12.3):

$$\text{Total Matl Cost} = (\text{Resin Cost})(\text{Resin Fraction}) + \sum_i [(\text{Cost of Add.}_i)(\text{Fraction of Add.}_i)] \quad \textbf{(12.3)}$$

The summation is over all the different additives. If more than one additive is included, several terms could be included. The only requirement is that the sum of all the fractions be unity. In the knife handle case, the total material cost calculation is given by Equation (12.4).

$$1.38 = (1.36)(0.99) + (3.00)(0.01) \quad \textbf{(12.4)}$$

Part Costs

The part cost is a function of the actual amount of material in the part (part weight) and the amount of product that is waste. The part weight is determined by simply weighing several parts and averaging their weight. In this case the weight for each part is 3.8 g.

The waste factor is not determined from the weight of material such as runners and sprues or even bad product, that will be reground and then reused in the process. The waste factor is found by dividing the total weight of resin purchased, less the total weight of resin currently in the plant, by the total weight of the resin (parts) that is shipped. This waste factor will therefore account for all resin losses. Other methods for determining the amount of unrecoverable waste would also be acceptable. In the present case, a waste factor of 1.05 (5% total waste) was used.

The adjusted weight is the actual weight times the waste factor. The adjusted weight of the knife handle is 4.0 g.

The part material cost is calculated by multiplying the adjusted weight of the part by the total material cost which includes resin and additives. Because the total material cost is usually given in $/pound (in the United States), a conversion to similar weight units is required if the weight is in grams. In the case of the knife blade, the part material cost is $0.003 per part.

The cost contribution per 1,000 parts is determined for each of the major cost components. The cost contribution for part costs is calculated by multiplying the part material cost by 1,000. The quantity of 1,000 parts was chosen because of the small size of many injection molded parts. The cost contributions are then in convenient price figures. The part cost contribution for the knife is $12.20.

Tooling Costs

Since many parts may be made on similar tooling and the cost of the part is dependent upon the tooling used, a description of the tooling should be given so that the particular tool can be exactly identified. The knife blade is made on a two-cavity steel mold.

The external cost is a key element in determining the cost of the mold and represents the cost quoted by the mold maker. When the mold is made by the molding company, the external costs are zero unless some special work has been done externally. The internal costs are those costs associated with handling, modifying, and otherwise preparing the mold to run properly and make parts to specification. Even when the mold is made externally, some minor adjustments are usually needed and can most reasonably be made by the molding company. When the mold is made internally, the internal costs represent all costs of the mold. In the knife blade case, the mold is made internally but some special machining is done externally, hence the external costs are $5,000 and the internal costs are $15,000.

The total cost of the mold is the sum of the external and internal costs. If the tooling is to be sold to an outside customer, the molder places a markup on the total price and sells it to the customer. The markup is justified because the molder accepted some responsibility when the molder agreed to obtain the mold. The customer price is the sum of total cost and markup.

Another circumstance could be that the customer owns the mold but has requested that the molder amortize the mold by adding the cost of the mold to the cost of the part. This is also the normal practice when the customer and the molder are the same entity and is the case used in the knife handle example. When the cost of the mold is to be amortized by adding to the cost of the parts, the number of parts to be made on the mold must be estimated. In the knife handle case, the number of parts is 4,000,000. The cost contribution from tooling costs is the total cost of the mold divided by the number of parts to be made multiplied by 1,000 to get the cost per 1,000 parts. For the knife handle the tooling cost contribution is $5.00.

One of the major drawbacks to the use of injection molding is the relatively high cost of the tooling. Hence, systems that allow tooling costs to be spread over the parts to be made are very inviting to new companies that are trying to get started with a new product. Someone, however, must take the risk that the projected number of parts will actually be produced. This risk is rarely borne by the custom molder. Hence, even when the molder agrees to amortize the cost of the mold over the cost of the parts, some payoff clause is usually inserted in the contract to ensure that the required number of parts are made within the agreed-upon time.

Another circumstance is possible for determining the tooling cost. Since the mold is normally owned by the customer, the customer may supply the mold to a custom molder with no cost to the molder. In this case, no mold cost is included in the price of the part.

Machine Costs

The cost of the machine is dependent on the time the machine is in use to make the parts and whether the machine is attended by an operator or not. Normally each machine will have two rates (operator or automatic). These rates are determined by the original cost of the machine, the ongoing costs to operate that particular machine (maintenance, lubricants, etc.), and any special equipment that might be added to the machine (such as special control equipment) that might allow a premium to be charged for the machine. The rate charged for a particular machine may also be dependent upon the demand for the machine.

In the case of the knife handle, the mold has two cavities and the mold cycle is 30 seconds. This results in 240 parts per hour. The cost of the machine is $25.00 with an operator and $18.00 without an operator. In this case the operator is needed to cut the parts off the sprue and runner system.

The machine cost contribution is determined by dividing the machine rate per hour by the number of parts made per hour and then multiplying by 1000 to get the contribution per 1000 parts. The machine cost contribution for the knife blade is $104.17.

Secondary Operation Costs

Many injection molded parts are subjected to other operations after molding. For instance, the part might be put into an assembly (as is the case with the knife handle), or it might be drilled, machined, or bonded to another part. The costs of these operations are normally figured at some rate which depends upon the nature of the equipment involved and whether an operator is required. Therefore, the secondary costs are identified with a cycle time and a rate. The secondary cost contribution from secondary tasks is the sum of all the costs of all secondary operations required. In the case of the knife blade, the only secondary operation is securing the blade into the handle. This is done by screwing together the blade and the handle. The rate for this operation is $7.00 per hour and each operation takes 10 seconds.

Most plastic parts must also have the runner and sprue system removed. Normally the machine operator does this task and it is not considered a secondary operation. The machine operator's time has already been included in the cost as part of the machine cost. If, however, the operator does not remove the runners, then the cost should be added as a secondary operation.

Purchased Items Costs

These costs are usually just the cost of the item that is included in the assembly of the product. In the case of the knife, the blade and the screw to attach the blade to the handle are purchased items. The purchased cost contribution is the sum of these purchased items. For the knife the costs are $1250 per 1,000 for the blades and $2.00 per 1,000 for the screws.

Packaging and Shipping Costs

The costs of boxes, bags, blister packs and any other packaging material are important costs of the product. The pocket knives are packaged in a blister pack ($50.00 per 1,000) and then in a cardboard box ($0.70 for a box that holds 1,000 parts). The cost of transportation can be added if that is included in the quoted price. However, goods are usually quoted with the price at the factory. This is called f.o.b. factory. (The abbreviation f.o.b. stands for "freight on board" and indicates that the costs are for goods loaded on a truck or train but not moved from the location stated, in this case the factory.)

Total Factory Cost per 1000 Parts

This amount is simply the sum of the various components of factory cost. For the knife this cost is $1,443.51.

General and Administrative Costs per 1000 Parts

This cost is associated with the general running of the company. It includes all costs that cannot be directly allocated to a product or service. It includes such items as general management, accounting, legal services, rent, utilities, insurance, and other overheads, which can include services such as quality control, warehousing, and maintenance. The G and A costs are normally a percentage of the total factory cost. This percentage is determined by experience, although some allocation methods can be used. In the case of the knife, a 10% G and A cost is used. This value is typical of the plastics industry.

Marketing Costs

The marketing costs are usually an estimate of the costs associated with marketing which include the salaries of inside sales staff, commissions to sales people and agents, and advertising costs. The 20% value used here is typical. In some companies, the G and A costs are figured after marketing costs. The reverse has been done here because sales of this type of product are often done by outside commissioned sales representatives and their commissions would be added after G and A.

Total Cost per 1,000 Parts

The total cost is the sum of the total factory cost plus the G and A and the marketing costs. The knife cost is $1876.16 per 1000 parts or $1.88 per knife. The sales price of the knife would only partially be based on cost. Other factors that affect the sales price are demand and reputation. Hence, the sales price is quite variable and dependent upon specific market conditions. However, the cost should be somewhat stable and can serve as one of the key components in determining sales price.

CASE STUDY 12.2

Mold Costs and Selection

Traditionally, two methods are used for estimating mold cost. The first method is an analytical approach, in which a mold is outlined piece by piece on paper, and cost estimates for each piece are made. Cost estimates are based on the amount of time to grind, bore, drill, mill, and, in general, work the metal into the desired shape. The total number of hours is calculated and multiplied by a labor rate to get a cost estimate. The second method attempts to estimate the cost through a complex combination of past experience, mold design, economic considerations, and familiarity with the vendor. The estimator tends to evaluate these factors subconsciously, relying on a feel for the present mold-buying situation.

The problems with the first method are that it is time-consuming and requires many skills that plastic-part manufacturers may not have. Also, estimates may be required long before a moldmaker is selected and the methods of mold construction determined, thus complicating the cost for time and moldmaker profit.

The second method is used much more often and lends itself to the expertise that plastic-part manufacturers have or can develop through a judicial survey of past experience and some basic assumptions on cost estimating. This method is the basis of the example shown in this case study.

Selecting the right size of mold for a product involves more than simply calculating the quantity of pieces to be produced by a certain number of cavities over a period of time. There are many other considerations to be evaluated before the optimum mold size is determined. Marketing has the responsibility of determining the number of parts that can be sold (produced) in a given period of time. The timing should be phased for weekly, monthly, and yearly increments with high-demand periods clearly shown. The limitations on storage and inventory must also be determined. If these market and inventory data are unknown, the emphasis should be on the lowest-cost production possible.

Since the number of working days available and the working time each day are part of the calculation that will be made, a convenient number of working hours per day is 20 and the number of minutes in each working hour is 50. These underestimates will give a margin of safety to the calculations and will allow for 80% efficiency.

Determining the costs associated with a particular part begins with a design of the part and the limitations imposed on the design for tolerances, both of the part and of other parts that might be in the same assembly, since the tolerances in these other parts may dictate the tolerances allowable in the part to be molded. The precision required will affect the number of cavities that can be used in the production. Normally, the more cavities there are, the higher the variation between cavities and, therefore, the higher the total variation in the part.

Another restriction that strongly affects the number of cavities that can be used is the cosmetic effect of the gate. If the gate location is not in an apparent spot or if the nature of the part is such that the gate is not cosmetically important, then a high number of cavities can be used. As the gate appearance gains importance, the location of the gate becomes more important and the number of cavities is reduced simply by the physical arrangements possible.

An additional limitation is imposed if the system must be a hot-runner mold. The cost of the hot-runner system significantly increases the cost of the mold in a nonlinear way with an increase in cavities. For instance, in a cold-runner system, a second cavity would cost approximately 60% more than a single cavity. If a hot-runner system is needed, the cost of the second cavity would be 100% more than a single-cavity system. This differential continues with each increase in the number of cavities.

These are special considerations that affect the cost of the mold. The primary considerations are sales volume, part weight and size, mold cost, and cycle time. Making some reasonable assumptions, several cases can be drawn to examine the total costs of these factors. Assume a part that is 10 inches (25 cm) in diameter and 0.1 inches (0.25 cm) thick with a weight of 8 ounces (0.2 kg). Assume that the annual quantity is 200,000 parts. Other assumptions regarding the making of the mold are evident from Table 12.5, which summarizes the case.

Note that the table considers four cases for total number of parts produced. The lowest cost for 200,000 total parts is for the two-cavity mold ($0.504/part). As the volume increases to 400,000 total parts, the most favorable number of parts changes to four cavities ($0.484/part); it changes to eight cavities at 800,000 parts ($0.459/part). Note that the cost per part decreases as the number of total parts increases. However, as stated previously, other considerations that may be dictated by the part itself can have an overriding effect such that fewer cavities can be tolerated and still achieve part performance.

Table 12.5 Economic Factors in Choosing Mold Size

Variable	One-Cavity Mold	Two-Cavity Mold	Four-Cavity Mold	Eight-Cavity Mold
Size of part (diameter, in inches)	10	10	10	10
Weight (ounces)	8	16	32	64
Thickness (inches)	0.1	0.1	0.1	0.1
Mold cost	$5,000	$8,000	$12,000	$20,000
Delivery time (days)	90	120	150	180
Annual quantity sold	200,000	200,000	200,000	200,000
Cycle time (seconds)	60	65	68	70
Production rate (parts/hour)	60	110	210	410
Hourly rate or machine burden ($/hour)	$12	18	32	33
Material cost/part	$0.30	$0.30	$0.30	$0.30
Yearly Amortization ($/part)	$0.025	$0.040	$0.065	$0.100
Total cost/part (200,000 parts)	$0.525	$0.504	$0.517	$0.534
Total cost/part (400,000 parts)	$0.512	$0.485	$0.484	$0.484
Total cost/part (600,000 parts)	$0.508	$0.477	$0.474	$0.467
Total cost/unit (800,000 parts)	$0.505	$0.474	$0.468	$0.459

The figures used in the example must be updated periodically to reflect current costs and other conditions. However, the basic reasoning of the system is sound. The principles illustrated are important to understand. The costs are dependent on multiple factors that can change with changes in the size of a part, the costs and complexity of increasing the number of cavities, the annual volume, the cost of equipment, and the total volume of parts.

SUMMARY

Injection molding is an extremely important molding process for plastics. Injection molding machines are used for plastics processing more than any other type of molding equipment. This high popularity is due to the wide variety of shapes that are possible with this process, the high degree of repeatability that can be attained, and the high level of automation that is possible. Almost all thermoplastic and many thermoset resins can be injection molded.

The major drawbacks to injection molding are the relatively high cost of the equipment and the high cost of the molds. The high cost of the equipment suggests that as many parts as possible be made in very high volumes so that equipment cost can be spread over many parts and not represent too high a portion of the total cost of each part. A key element in the cost of injection molding is the cycle time, that is, the time the machine takes to complete the part and become ready to begin the next part. Cycle time should be minimized.

Injection molding machines are self-contained units except for service lines. The resin is introduced into a hopper, either manually or automatically. Some resins need to be dried to remove the moisture that is absorbed naturally. This is often done by blowing hot (low-humidity) air through the resin, either in the hopper or in a separate drying step prior to introduction. Colorants and other additives can be added in the hopper and/or also in a previous mixing step. The resin feeds from the hopper into the feed throat by gravity.

Two types of injection molding machines have been in common use. These are the ram and the reciprocating screw machine. Today, the reciprocating screw machine dominates the market. Reciprocating screw machines have better melting characteristics, are easier to control, require less thermal heat because they also induce mechanical heating, have a more uniform melt, and can inject smaller quantities from the same machine. In reciprocating screw machines, a screw with three zones is used to convey the material forward (feed zone), to melt the material (compression zone), and to add pressure and insure that the resin is completely melted (metering zone).

After passing along the screw, the molten material collects in a pool at the end of the screw in a region called the injection zone. At the appropriate part of the cycle, the screw stops turning and advances forward, thus pushing the resin in the pool out through the end of the barrel. A sliding check valve prevents the resin from flowing backward along the screw. This step is called injection. The resin flows through a nozzle that is mated to the surface of a mold. It then flows through a passageway in the mold (called a sprue) and into a channel system (runners) that connect the sprue to the mold cavities. Just as the resin enters the cavity it passes through a narrow opening called the gate. The resin then sits in the cavity until it has cooled sufficiently to be solid.

Injection molds are much more complicated than extrusion dies. The injection molds are made from several plates that are assembled together. Some of the plates give support so that the assembly can withstand the pressure of the injection, some contain the cavities, and some operate the ejection system. The entire assembly is the mold base.

Generally, the plates of an injection mold are made of hard tool steel. Other materials such as beryllium-copper, stainless steel, and aluminum can also be used and are even preferred in some applications.

The runners, gates, and cavities are cut into the mold so that the plastic material flows smoothly and does not freeze off until the mold is completely filled. The patterns and styles of these entities can be of several designs, depending on the number of parts, the need for close tolerances, the need to have the part automatically removed from the runner system, and the appearance of the part.

The cavities can be made by several methods. Machining is the oldest method and is still the most common. However, electrical discharge machining (EDM) has some major advantages and is now emerging as a major method for cutting cavities. Hobbing is an old method that is now in decline, especially with the growth of EDM.

After solidification, the mold opens, which activates some ejector pins that assist in removing the part. In most cases the runners and sprue are also ejected with the part. These are removed, usually by clipping them off, and are reground to make additional parts.

In most injection molding operations, the ejected part is nearly the final shape and little additional shaping is needed. However, several types of secondary operations may be done to get the finished assembly. For instance, the part may have some additional part attached (as in the case study in this chapter), or may be welded or attached into an assembly. Many of these secondary operations can be done automatically or semiautomatically to reduce the costs of labor involved in the total operation.

The control of the injection molding process is complex because so many of the variables are interrelated. In general, the material should be injected into the mold at the fastest rate and with the highest pressure practical. The limits on these parameters are the clamping capacity of the machine, the physical limitations of the nozzle, runners, gates, cavities, and the characteristics of the resin. The temperature should be chosen to have a resin with a relatively low viscosity so that it will flow easily and fill the mold, but not be so hot that the time to cool is excessively long. Other variables, such as cycle time, temperature profile, time the pressure is held against the filled mold, and treatments to the resin such as drying or use of additives are all secondary parameters but can be important in obtaining good-quality parts.

To assist in the building of molds, computer simulation programs have been developed. These programs show the filling of the mold and predict key parameters associated with the molding process such as shrinkage, warpage, venting requirements, orientations, and runner and gate system sizes.

The costs of an injection molded part are dependent upon the costs of (1) the resin, (2) the tooling, (3) the machine, (4) the secondary labor, and (5) shipping and packaging. All of these costs are outlined in the first case study.

GLOSSARY

Cavities   The shaped, recessed areas on the face of the mold into which the resin flows and becomes solid, thus creating the part.

Check valve   A valve that allows material to pass in only one direction; such a valve is mounted near the end of the screw in an injection molding machine to ensure that during the injection motion of the screw, the resin does not flow backward.

Coinjection molding   A process in which more than one resin is injected into a single mold.

Cold-well extensions   Small channels that are part of the runner system and extend beyond the normal path of the resin at points where the resin turns a corner; these are used to trap cold material that is leading the resin flow.

Core pins   Inserts placed into the mold cavity that force the molten resin to flow around them, thus creating a hollow area in the part.

Draft angle   The slight angle in the walls of the cavity, sloping so that the opening of the cavity is the widest part, thus allowing the parts to be removed easily from the cavity.

Edge gate   A small rectangular opening at the end of the runner that connects to the cavity; can be below the parting line or symmetrical with the parting line.

EDM (electric discharge machining)   A method of metal removal using a spark; often used in the creation of the cavities in injection molds; plunge EDM and wire EDM are two methods for doing this type of material removal.

Ejector pins   See *knock-out pins*

Ejector plate   A plate that is activated by the ejector rod and that pushes on the knock-out pins during part removal from the cavities.

Ejector rod   A rod that activates the motion of the ejector plate, which, in turn, pushes the ejector pins or knock-out pins into the cavities.

Fan gate   A small channel from the runner to the cavity in which the thickness but not the diameter of the runner is reduced; often used with reinforcements.

Gas injection molding   A process in which a gas is injected into the mold while the mold is still filling with resin.

Gate   The entry portion of the cavity.

Gaylord   A resin box of roughly 4 foot × 4 foot × 4 foot dimension, almost always shipped on a pallet; also called a pillar pack because of the support pillars often placed in the corners of the box.

Hobbing   A method for making cavities in a mold, involving mechanically pushing a shaped die into the metal to create a cavity.

Hot-runner molds   Molds in which the runner system (and the sprue) are heated, thereby eliminating these as part of the regrind; the separation in these systems is at the gate.

Inserts   Components of the injected part that are placed into the mold cavity before the resin is injected; often metal.

Internal mold release   A mold release added to the resin.

Isotropic   When a property is the same in all directions.

Knit line   The place where two resin streams meet; also called the weld line.

Knitting of the resin   When two streams of resin flow together.

Knock-out pins   Long pins that push into the cavities during the part-removal phase of the injection molding cycle, thus pushing the parts out of the cavities; also called ejector pins.

Mold inserts   Cavities made in small metal plugs, which are then inserted into holes in the plates of the mold; this allows changing cavities while keeping the rest of the mold set.

Mold release   A material applied to the surface of the mold or added to the resin that facilitates removal of the part from the cavity, usually by coating the cavity walls with a slick film.

Mold shrinkage   The amount of size reduction that a plastic part undergoes while in the mold.

Moveable plate   The side of the mold that is attached to the moveable platen; the B side of the mold.

Moveable platen   The large metal plate (platen) that is attached to the moving part of the injection molding machine (the side of the machine where the sliding hydraulic or mechanical mechanism is located) and to which one side of the mold is attached.

Multicavity mold   A mold in which more than one cavity is created on the mold face.

Multi-injection points   When two or more gates enter into the same cavity.

Nozzle   The end of the barrel through which the resin is injected into the mold; it is shaped to fit into the sprue bushing.

Packing   The process of continuing to press resin into a mold even after the resin in the mold has begun to solidify.

Parting line   The line along which the mold opens and, therefore, the line that splits the cavities between the A and B sides.

Pillar pack   A resin shipping box with rough dimensions of 4 foot $\times$ 4 foot $\times$ 4 foot with support pillars in the corner, almost always shipped on a pallet; also called a gaylord.

Postmold shrinkage   Shrinkage that occurs in a part either because of an annealing process or naturally because of the movement of the molecules, usually to relieve internal stresses.

Preplastizing machine   A type of injection molding machine that has a chamber in which the resin is melted and separate injection chamber into which the molten resin flows prior to being injected into the mold.

Projected area   The surface area of a cavity or of all the cavities against which the pressure of the molding machine must push; the area at the widest part of the cavities.

Ram injector   A type of injection molding machine that is characterized by a melting of the resin only by external heating and then an injection of the resin by a simple piston motion.

Reciprocating screw machine   A type of injection molding machine that is characterized by a screw that melts the resin in combination with external heating, and then a reciprocating action that injects the resin.

Ring gate   A small channel connecting the runner to the cavity that covers the entire top of a cylindrical part so that the resin flows downward into the cylindrical cavity.

Runner   The channels through which the resin moves across the face of the mold as it travels from the sprue channel to the cavities; also the name of the material that solidified in the runner channel.

Short shot   When the amount of material injected is not sufficient to completely fill the cavity.

Shot size   The amount of resin that can be injected by a particular injection molding machine.

Shrinkage   The amount of size reduction that a plastic part undergoes as it cools or cures in the mold.

Sink mark   A surface imperfection in injection molding that often results from uneven shrinkage of the part, especially in sections that are quite thick.

Spreader   A device in the end of a ram injection molding machine that forces the resin against the walls of the barrel and adds shear energy to the melt.

Springback   The postmolding change in an angle between two sections of a part, usually due to movement of the molecules to relieve internal stresses or from shrinkage.

Sprue   The material that is solidified inside the sprue channel during the cooling stage of the molding operation; it is connected to the runners.

Sprue bushing   The connection on the back of the mold that mates with the injection nozzle.

Sprue channel   The channel or hole in the mold through which the resin flows as it passes from the back of the mold to the side where the mold cavities are located.

Sprue puller pin   A pin that resides at the downstream end of the sprue (located in the B side of the mold) that has a mechanical undercut to which the resin bonds, thus ensuring that when the B side of the mold opens, the sprue and the rest of the solid resin in the runners and cavities will move with the B side and not stay on the A side.

Stacked molds   Mold bases having more than two plates in which cavities are cut, thus increasing dramatically the number of parts that can be made during one molding cycle.

Stationary plate   The side of the mold that is attached to the stationary platen; the A side of the mold.

Stationary platen   The large metal plate (platen) that is attached to the fixed part of the injection molding machine (the side of the machine where the screw is located) and to which one side of the mold is attached.

Stripping   The process of ejecting a part out of a cavity even though the part has a section that is wider than the opening through which it is moved.

Submarine gate   A small channel that connects the end of the runner to the cavity in which the runner tapers to a point and is located below the parting line.

Tab gate   A small channel from the end of the runner to the cavity that is not reduced in size from the runner itself; used for very large parts.

Torpedo   Another name for a spreader device.

Undercut   A cavity in which some dimension in the cavity is greater than the opening.

Vent   A small hole or channel that allows air to escape from the cavity as it is filled with resin.

Weld line   The place where two resin streams join; also called the knit line.

QUESTIONS

1. Why are injection molding machines not as effective for mixing additives or other resins as are traditional extrusion machines?
2. Where is the normal separation point between the material that is removed with each cycle and the material that is in the machine and used in the next cycle?
3. What feature in a mold will allow a hollow, cylindrical part to be made?
4. What two measures are used in rating an injection molding machine?
5. What is packing the mold and why is it important in obtaining good injection molded parts?
6. What is a vent in the mold, what problems are prevented by the presence of a vent, and what parameters control its size?
7. How does high crystallinity in a resin affect the way the resin is injection molded, including any postmolding considerations which might be made?
8. Explain the purpose of a preplasticizing unit.
9. Why is it important to have the sections of the molded part as uniform in thickness as possible?
10. If nonmelted resin particles are noted in the molded part, what changes might be made to correct the process?
11. Assume that you are assigned to determine the minimum clamping force for a part to be molded out of polystyrene. The part cross-sectional area is 10×14 inches. What is the clamping force required if, as a general rule, 2.5 tons of force are needed for each square inch of cross-sectional area?
12. Why is good weldability desirable in a mold cavity material?
13. Why is low specific heat capacity desired in a mold cavity material for some applications and a high specific heat capacity desired in others?
14. Why are ejector pins made of hard yet shock-resistant materials?
15. Identify four advantages and two disadvantages of using aluminum rather than tool steel to make cavities for injection molds.
16. Identify one major advantage or use of the following gates types: submarine, edge, and ring.
17. What is the purpose of cold-well extensions?
18. Identify three ways to ensure that an injection molded part stays on the B side of the mold when the mold opens.
19. Identify two advantages of using inserts in injection molds.

REFERENCES

Ashby, Steven, "Molding Stronger Plastic Parts," *Mechanical Engineering,* Nov. 1993, pp. 56–59.

Beall, Glen L., "Plastic Part Design for Economical Injection Molding," Seminar notes prepared for Storage Technology Corporation, Littleton, CO, 1993.

Brydson, J. A., *Handbook for Plastics Processors,* Oxford, UK: Heinemann Newnes, 1990.

DeGrospari, John, "Low-Pressure Alternatives for Molding Large Automotive Parts," *Plastics Technology,* Sept 1993, pp. 60–65.

Frados, Joel, (ed.) *Plastics Engineering Handbook* (5th ed.), Florence, KY: International Thomson Publishing, 1994.

"Gas Injection Molding of an Automotive Structural Part," *Plastics Engineering,* Oct. 1991, pp. 21–26.

I-DEAS Master Series, *Student Guide* (2nd ed.), Milford, OH: Structural Dynamics Research Corporation, 1994.

"Injection method may end troubles with anisotropy," *Modern Plastics,* Jan. 1990, pp. 12–14.

Kamal, M. R., and Thanasis D. Papathanasiou, "Filling of a Complex-Shaped Mold with a Viscoelastic Polymer. Part II: Comparison with Experimental Data," *Polymer Engineering and Science,* Mid-April 1993, pp. 410–417.

Menges, Georg and Paul Mohren, *How To Make Injection Molds* (2nd ed.), Munich: Hanser Publishers, 1993.

Milby, Robert V., *Plastics Technology,* New York: McGraw-Hill Book Company, 1973.

"Mold Analysis Software Developing at a Rapid Pace," *Plastics Technology,* January 1993, pp. 21–25.

"Moldmaking: A Statistical/Heuristic Approach to Estimating Mold Costs," *Plastics Engineering,* June 1989, pp. 51–53.

Morton-Jones, D. H., *Polymer Processing,* London: Chapman and Hall Ltd, 1989.

"Multinational Producibility: Plastic Molded Parts," Xerox, MN2-104.1 Revision 6, Sept. 1987.

"New Mold-Analysis Software for Beginners and Sophisticated Users," *Plastics Technology,* August 1993, pp. 17–21.

Papathanasiou, Thanasis D. and M. R. Kamal, "Filling of a Complex-Shaped Mold with a Viscoelastic Polymer. Part I: The Mathematical Model," *Polymer Engineering and Science,* Mid-April 1993, pp. 400–409.

Richards, Peter and Ed Galli, "Estimating: Putting Method in the Madness," *Plastics Machinery & Equipment,* Feb. 1982, pp. 15–18.

Richardson, Terry L., *Industrial Plastics: Theory and Application* (2nd ed.), Albany, NY: Delmar Publications, Inc., 1989.

Rosato, Donald V., David P. DiMattia, and Dominick V. Rosato, *Designing with Plastics and Composites: A Handbook,* New York: Van Nostrand Reinhold, 1991.

Spier, I. Martin, "The Economics of Mold Selection," *PM&E Mold and Die Corner Collection.*

Stoeckhert, Klaus (ed.), *Mold-Making Handbook,* Munich: Hanser Publishers, 1983.

"Troubleshooting guide for injection molding," DuPont, E-67417.

CHAPTER THIRTEEN

BLOW MOLDING

CHAPTER OVERVIEW

This chapter examines the following concepts:

- Extrusion blow molding (continuous and intermittent)
- Injection blow molding (hot and cold parisons)
- Molds and dies (programmable parison formation, sliding/compression dies)
- Plant concepts (equipment layout, capacity)
- Product considerations (materials, coblow molding, shapes and part design)
- Operation and control of the process

INTRODUCTION

Blow molding is a plastic-forming process that is especially well suited for the manufacture of bottles and other simple, hollow-shaped parts. The essence of the process is the formation of hollow parts from a preformed plastic tube. While blow molding is competitive with other processes that can make hollow parts, specifically injection molding and rotational molding, blow molding has some advantages in forming some parts that make it the process of choice for the majority of medium-sized hollow containers. Blow molding can create parts with much lower mold costs than injection molding and can create parts with narrow openings and wide bodies whereas parts with those characteristics would be very difficult to form by injection molding. Rotational molding can also create hollow parts but the cycles for rotational molding are much longer and therefore, rotational molding is used more extensively for very large hollow parts or for parts with shorter runs, especially where exact part definition is not critical, where both injection molding and blow molding are less suitable. Blow molding can, however, be economically used for parts ranging from a few ounces or milliliters to over 120 gallons (450 L) in volume. The process is suitable for high production rates and, in these cases, is usually widely integrated with automatic resin feeding and part removal. Such a system is seen in Photo 13.1.

Photo 13.1   Integrated blow molding machine (Courtesy: Battenfeld Blowmolding Machines, Boontown, NJ)

The blow molding process consists of the following general steps:
1. Melting the resin. This is done in an extruder that is captive to the blow molding machine. Typical extrusion screws, heaters, and other extrusion equipment are used.
2. Form the molten resin into a cylinder or tube. This cylinder or tube is called a *parison*. The parison can be formed by two methods and these constitute the two main types of blow molding—extrusion blow molding and injection blow molding. (Each of these methods is considered in detail, but a brief overview assists in understanding their differences.) In extrusion blow molding an extrusion die is utilized to form the parison. The parison formation can be either continuous or intermittent. In either case, some method must be provided to close the ends of the parison so that it can be inflated. The most common method of closing the ends is simply to pinch them off by the closing of a two-part mold. In injection blow molding the parison is formed by injecting the resin over a core pin. The injection molded parison is molded with a closed bottom and the top is closed by the mold closure.
3. After the parison has been formed, either by extrusion or injection molding, the parison is placed inside a mold (usually with a sealing of the top and/or the bottom of the parison) and inflated so that the plastic is pushed outward against the cavity walls. The inflation is done through a pin that is usually inserted through the opening in the bottle. This is the heart of the blow molding process and the step that truly distinguishes blow molding from other plastic molding-methods. This step is illustrated in Figure 13.1.

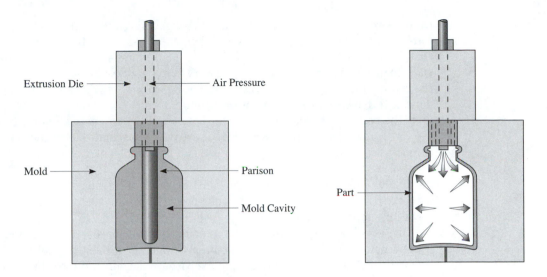

Figure 13.1   Blowing step in the blow molding process.

4. The part is allowed to cool in the mold and is then ejected.
5. The part is trimmed. When the end of the parison is closed by pinching off, flash is formed. This flash must be removed.

Blow molding was practiced anciently by glass blowers to make bottles and was used in the early part of the 20th century to make small bottles, but has only recently been adopted widely as a major manufacturing process. Some of the reasons for this surge in popularity are the increased use of plastics for containers and the development of some highly integrated blow molding machines that give very high production rates and excellent repeatability of quality parts.

The blow molding process has the advantage of producing hollow parts with small or large openings in a rapid and efficient way. Cycles are usually very short, so the costs of the equipment can be spread over many parts. Therefore, the key cost elements are material costs including the amount of material in the part.

Blow molding is divided into two main divisions based on the method of formation of the parison. The parison can be formed using an extrusion die and an extrusion machine in one method or by injection molding using an injection molding machine in the other method. These two methods will be discussed separately and then some comparisons will be made.

Extrusion Blow Molding

In *extrusion blow molding* the parison is formed from an extrusion die that is similar to the type that would be used for the forming of blown film. (Blown film extrusion is discussed in the chapter on extrusion.) Just as with blown film, the molten material would flow from the extruder through an adaptor that would change the direction of extrusion from horizontal to vertical (downward for blow molding), thus allowing gravity to act uniformly on the part. The material would then enter the die and flow around a mandrel so that the extrudate would be cylindrical. The die for blow molding would produce thicker walls than would be produced for blown film. In many cases the die would have a hole down the center of the mandrel so that air could be blown into the cylinder of material, just as is done with blown film. In some blow molding operations, however, the air is introduced from the bottom through an inlet created as part of the pinching off of the parison.

In the manufacture of blown film, the extruder runs continuously and the film is continuous, usually with a provision for rapid and automatic changeover of the takeoff equipment. Blow molded parts are not continuous but discrete, that is, each part is molded individually. Therefore, some provision must be made to manufacture these discrete parts with extrusion, an inherently continuous process forming the parison but requiring discrete time to mold the parts.

One possible situation is that the molding cycle is shorter than the time required to make a new parison. Under these conditions the parison must be captured by the mold, the parison blown, the part cooled, and the part ejected, all within the time required to extrude a new parison. This must all take place without interfering with the formation of the next parison. In almost every practical case, the mold cycle is longer than the time required to extrude the parison. Therefore, some provision must be made to match the output of the extruder and the molding cycle. Two alternatives can be adopted. The first is called *continuous extrusion blow molding* and the second is called *intermittent extrusion blow molding*. Continuous extrusion blow molding is accomplished by using multiple molds to match the mold cycle to the extrusion speed while intermittent extrusion blow molding uses either a reciprocating screw or an accumulator system.

In continuous extrusion blow molding the extruder is run continuously and the output of the extruder is matched by having multiple molds which seal and blow the parison and then move away from the extruder to cool and eject. For example, if the mold cycle is no more than twice as long as the time to create a parison, a two-mold system can be used. This system uses moving molds and is illustrated in Figure 13.2. In this system one mold captures the parison (shown in the figure as mold A) while a part is cooling in the other mold (B). Mold A then withdraws the parison from the extruder while the part is ejected from mold B. In the next step the part is cooling in mold A, which has moved out of the path of mold B, while mold B is moving into position around the parison. The cycle is completed by further cooling of the part in mold A with eventual ejection, and capture of a new parison in mold B followed by blowing of the part into the mold cavity. This method is sometimes called the *rising mold system*. Systems are also in use in which more than two molds are used to alternately capture and withdraw the parison.

In a system that is similar to the rising mold system, the parison is cut from the die by a knife and transferred by a mechanical arm to a mold where it is blown, cooled and ejected. While this is occurring, other parisons are cut and transferred to other molds. This system, called the *parison transfer system,* is used rather than the rising mold system when the parts are quite large and the moving of large molds becomes mechanically difficult. This system is illustrated in Figure 13.3.

Another continuous extrusion blow molding method uses a *multiple-mold system* to match the output of the extruder with the molding cycle, where the molds are mounted

Figure 13.2 Two-mold system used in continuous extrusion blow molding.

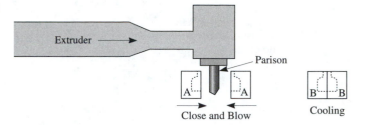

(a) Capture of Parison (A) and Cooling (B)

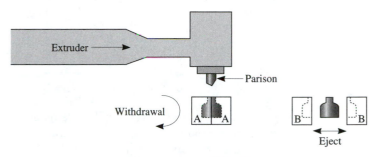

(b) Withdraw (A) and Eject (B)

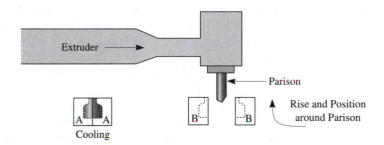

(c) Cooling (A) and Positioning (B)

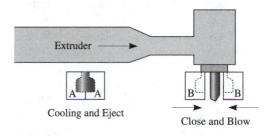

(d) Cooling (A) and Capture (B)

Figure 13.3 Parison transfer system for continuous extrusion blow molding.

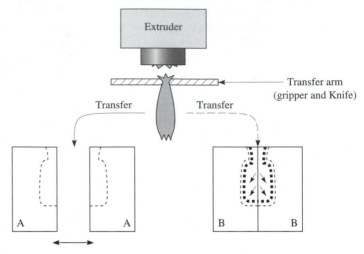

on a rotating wheel. Several molds are required in this method, as shown in Figure 13.4. While one mold is closing to capture the parison, the mold ahead is in position for blowing of the part, and other molds are closed while the part is cooling. Still farther around the wheel the mold opens for ejection of the part and in preparation for capturing another parison. In this system the rotational speed of the wheel and the number of molds mounted to the wheel are matched to the extruder output speed. Equivalent systems have been developed in which the wheel is mounted horizontally instead of vertically.

The major disadvantage of the rotating wheel versus the moveable mold system is that the rotating wheel is much more complicated mechanically and the cost of the system with all of the additional molds is quite high. The advantage of the rotating wheel versus the moving molds is that mold cycles that are much longer than the extrusion rate can be accommodated.

The continuous extrusion blow molding methods are best suited to high production of small to medium-size parts (up to about 8 gallons, 30 L). With small parts, the time to form the parison is short and the mold cycle is usually also short. Therefore, multiple molds can effectively be used to capture the parison and move it from the die area so that another mold can move into place. The limit of the system is the mechanical movement of the molds.

If the parts are large, the time to form the parison by continuous extrusion is relatively long. This long extrusion time can result in excessive cooling of the parison such that it cannot be effectively blown and excessive sagging of the parison under its own weight. Sagging can cause necking or stretching of the part with accompanying poor control of final part dimensions or breaking off of the part at the die. Therefore, a method of forming the parison that is faster than the extrusion speed of the extruder must be used. This rapid forming has been accomplished by two methods—the *reciprocating screw system* and the *accumulator system*. In both these systems the parison is formed intermittently rather than continuously. The intermittent extrusion blow molding methods use only one mold.

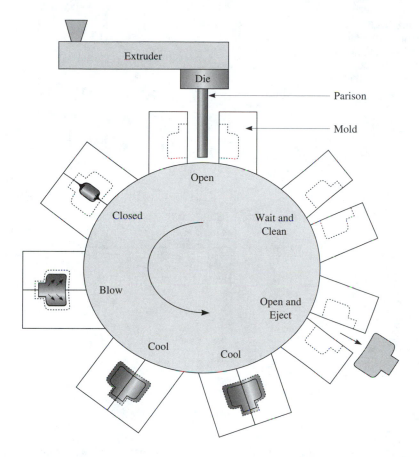

Figure 13.4 Rotating mold system used in continuous extrusion blow molding.

In the reciprocating screw method a specific amount of molten resin is injected through the extrusion die to form the parison. In this manner, a parison can be formed in one or two seconds that would take much more time to form by continuous extrusion. During the subsequent blowing and cooling, the extrusion screw is retracting and accumulating another charge. No parison is being formed during this part of the cycle so the parison formation is intermittent. The reciprocating screw system is similar to traditional injection molding.

If the blowing and cooling times are long, the screw may actually stop when sufficient material has accumulated for the next parison shot. This stoppage may cause some disruptions in the extruder so another method, which uses an accumulator and allows continuous running of the extruder to solve this problem, has been developed.

The accumulator system also forms the parison intermittently. In the accumulator system the extrudate flows from the extruder into an external chamber or accumulator. At the appropriate moment in the cycle, a ram in the chamber advances and injects the resin

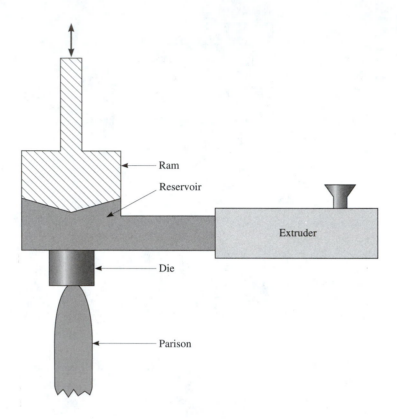

Figure 13.5
Accumulator system used in intermittent extrusion blow molding.

through a die that is mounted on the outlet of the accumulator to form the parison. Very large parts (up to 120 gallons, 450 L) can be made using the accumulator system because the volume of the accumulator can be several times larger than the injection volume possible with a reciprocating screw machine and injection using the ram can be very fast. Some systems even use several extruders to feed the accumulator thus further increasing the size of the parts that can be made. In most accumulators, the ram is lifted by the rising of the resin, thus eliminating air pockets in the accumulator that could lead to inaccuracies in the injection of the resin. The accumulator is heated to maintain the proper temperature of the resin and so only resins with good heat stability should be used since the time that the resin may be at a high temperature could be quite long. The accumulator system is depicted in Figure 13.5.

Injection Blow Molding

In injection blow molding the parison is formed by the injection of molten resin into a mold cavity and around a core pin. The parison is, therefore, discretely formed as a part in an injection molding process. This contrasts with the formation of the parison through an extrusion die as is done with extrusion blow molding. The similarity of the injection blow molding process to traditional injection molding is obvious. The difference is that in in-

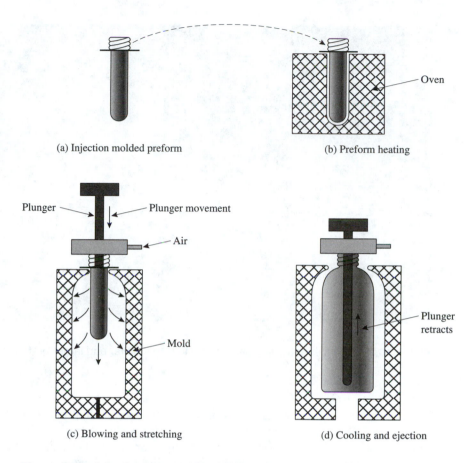

(a) Injection molded preform   (b) Preform heating

(c) Blowing and stretching   (d) Cooling and ejection

Figure 13.6   Injection blow molding process.

jection blow molding the parison is not a finished part but is subjected to a subsequent step that forms the final shape. That second step is, of course, blowing of the intermediate part in a second mold to form a traditional blow molded shape. Because of the distinct separation of the two steps, the parison made by injection molding is sometimes called a *preform*. The injection blow molding process is shown schematically in Figure 13.6 and a photo showing preforms and bottles for two different soda bottles is shown in Photo 13.2.

During the injection cycle a traditional injection molding machine is used to create the parison. The mold is closed with the core pin in place. The resin is then injected to form a cylindrical part around the core pin. The threads, if any, are also formed at this stage. The mold is then opened, the core pin is removed and preform is ejected. Either the preform is transferred to a blowing station while it is still hot or it can be reheated. After the second mold has closed and mated with the top of the preform, air is injected into the heated preform, usually through a hole in the core pin. This blows the preform against the inside walls of the mold to create the part. The mold opens and the part is ejected.

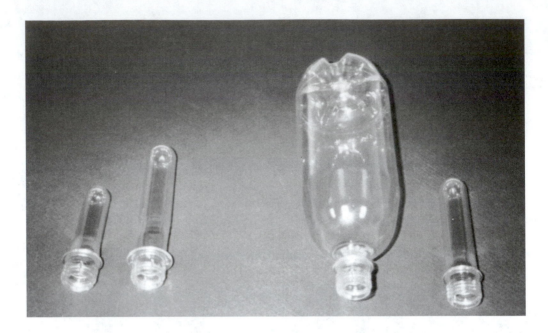

Photo 13.2 One-liter bottle and preform along with smaller and larger preforms for other bottle sizes.

The injecting and blowing cycles of injection blow molding need not be done at the same time or even at the same location. The parisons can be made by injection and then either stored until the finished blow molded parts are needed or shipped to a satellite location where they can be blown. When the blowing is not done immediately after the injection, the parison must be reheated. The flexibility of separating the two cycles has proven to be valuable in the manufacture of soda pop bottles. The parisons are made in a central location on large, multicavity injection molding machines, which gives economy of scale and close engineering control over the injection molding step, the more critical of the two steps. The parisons are then moved to the blow molding machines (which can be at separate sites), reheated and blown into the familiar soda bottles. The blow molder does not need to have an expensive injection molding machine and injection mold but needs only an oven and a blowing station.

Injection blow molding allows the formation of a parison that can have a nonconstant cross section. This flexibility in shape provides a method for blowing bottles with better wall thickness uniformity than can be made with parisons from extrusion blow molding. The blowing process naturally causes thin sections where the parison has to be blown farther, such as in corners. By carefully designing the shape of the core pin and mold cavity in injection blow molding, the parison can be thicker in just the correct areas to compensate for this greater stretching.

Injection blow molding offers the possibility of another innovation in blow molding technology. This innovation is called *stretch blow molding* and is a technique that uses

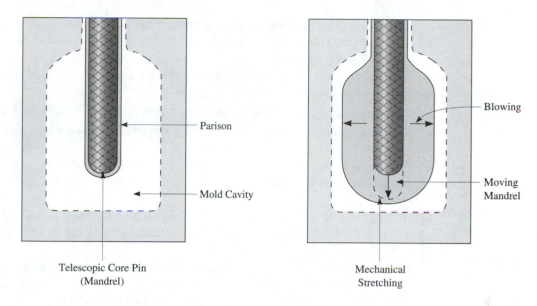

Figure 13.7   Stretch blow molding.

mechanical assistance to stretch the part in the longitudinal direction at the same time the blowing of the part stretches it in the hoop or radial direction. This biaxial stretching occurs when the parison or preform is blown into the desired shape in the blow mold. The longitudal stretching of the parison is accomplished by using a telescoping mandrel or core pin that extends to push on the bottom of the preform at the same time that the air is being injected to push against the walls to stretch the material radially. For this process to have its maximum benefit, the temperature of the preform is carefully adjusted to the correct temperature for orientation. This process is shown in Figure 13.6 and 13.7.

The major advantage of stretch blow molding is improvement of mechanical properties. Polymers are stronger in the direction of orientation of the molecules. Therefore, by stretching in two directions, the parts can be made higher in both burst strength (dependent on hoop orientation) and impact strength (improved by longitudinal orientation). PET soda pop bottles are a prime example of the use of this technique. If not stretched, the PET resin will remain amorphous and will not meet the drop-breakage test that is expected of a soda bottle. (Dropped full from 6 feet without breakage.) The crystallization forced by stretching also improves the resistance of the PET to gas permeation. These and other advantages are discussed in the case study at the conclusion of this chapter.

Comparison of Extrusion and Injection Blow Molding

The similarities and differences between extrusion blow molding and injection blow molding are important in the choice of the method for forming the part. A brief review of the characteristics of each of the methods is therefore given.

Extrusion blow molding is characterized by:

- It is best suited for bottles over 1/2 pound (200 g) in weight, shorter runs, and quick tool changeover.
- Machine costs are comparable to injection blow molding.
- Tooling costs are 50% to 75% less than injection blow molding.
- It requires sprue and head trimming, which generates 20% to 30% scrap.
- It requires additional equipment to grind scrap and reintroduce into molding machine.
- Total cycle is shorter than with injection blow molding, since parison formation and blowing can be done on same machine with no need to transfer to another mold.
- Wider choice of resins is possible with extrusion under present technologies because resins that do not liquify sufficiently for injection molding can be extruded into a parison and blown.
- Final part design flexibility can be greater because asymmetrical openings can be made by extrusion blow molding as can additional openings and blown handles. These design features are difficult with injection blow molding.

The characteristics of injection blow molding are:

- Best suited for long runs and smaller bottles.
- No trim scrap.
- Higher accuracy in the final part, that is, the precisely formed parison is more likely to make consistent and accurate parts.
- Uniform wall thicknesses.
- No seam lines or pinch marks; more-stable base designs are possible with injection blow molding.
- Better transparencies with injection blow molding because the crystallization can be better controlled and the blowing can be more stress free.
- Can lead to improved mechanical properties from improved parison design and from the possibility of stretch blow molding.

MOLDS AND DIES

General Die and Preform Mold Considerations

Extrusion dies used to form parisons are usually made from tool steel to give long wear without significant dimensional changes. The material and construction methods are similar to those used for extruded pipe and blown film dies.

Molds for making the preforms used in injection blow molding are made of the same general materials (usually tool steel) as other injection molding molds. Practice has been to make the tools for applications such as soda bottles with 64 or more cavities so that the economies of scale will benefit the production of the bottles.

Programmed Parison Formation

A method of improving the part uniformity of extrusion blow molded parts is through programmed parison formation. This method employs an extrusion die that has a mandrel with a conical-shaped end. The flow path inside the die is also conical-shaped. A die of this type is shown in Figure 13.8. When the mandrel moves within the bushing or the die body (whichever the actual geometry of the die requires), the gap between the mandrel and the bushing/die body changes. This feature is used to make a parison that is thicker at the bottom than at the top, thus compensating for the natural thickness variation in blow molded parts because of stretching. The variation in thickness is accomplished by timing the movement of die/mandrel with the extrusion of the parison. A cam or piston is usually used to move either the die body or the mandrel.

Parison programming is not effective for very small parts. There is just not enough time to move the mandrel while the parison is being formed. The programming of a parison can affect the optical characteristics of the final part. Some parts may have a wavy appearance because of the changes in wall thickness in the parison which are carried over into the part.

General Mold Considerations (for the Molds in Which the Part Is Blown)

Many plastic-molding processes require less pressure in the molding step than injection molding. Lower-pressure molding methods include blow molding, rotomolding, thermoforming, casting, and most processes making composite parts. The lower-pressure requirement means that molds can be made of lower-strength, less expensive materials.

Figure 13.8 Die for automatic parison programming.

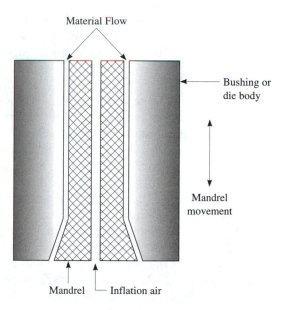

Generally, therefore, the predominant materials of injection molds (tool steels, stainless steels, beryllium-copper, and soft steels) are not needed for the lower-pressure processes. Instead, the dominant materials for the low-pressure applications are aluminum and various plastic and composite materials. Hard metals may be used as inserts in some particular locations where wear is expected to be high, such as pinch-off points.

Aluminum is by far the most common material for blow molding molds. This process has only modest pressures, and the resins used in these processes are generally not highly abrasive. The durability of aluminum is, therefore, sufficient for most production molds. The high heat transfer of aluminum is a further advantage in these applications.

The aluminum molds for blow molding can be either machined or cast, with casting as the preferred method when the molds are large. The cast aluminum molds generally are hardened after casting and final machining. The surface quality of aluminum molds can be polished to give a smooth, but not a mirror, finish.

Machined aluminum molds for blow molding are made of alloy aluminums that are inherently hard or that can be hardened after machining. A typical aluminum of this type is 7075, although many other aluminum alloys are commonly used.

Aluminum molds can also be used for rotomolding, thermoforming, casting, composites molding, and other low-pressure molding processes, although the very low pressures and often low surface-quality restrictions of these other processes allow the use of many other materials.

The major disadvantage of aluminum molds is the excessive wear they tend to exhibit, especially in thin areas such as the pinch-off. Steel inserts can be used in these areas of high wear to extend the mold's life. Also, aluminum molds distort somewhat after prolonged use. When very long production runs are anticipated, steel molds can be used. When very fast molding cycles are needed and higher heat transfer is desired so that the part will cool more quickly, beryllium copper molds can be used. If only a few parts are to be produced, such as for prototype production, cast resins (such as epoxy or epoxy with aluminum filler) can be used for the molds.

Blow molds are usually made in two symmetrical halves. This simplifies mold making and replacement. It also helps in the monitoring of wear. Hence, a parting line is common for blow molded parts. The symmetry also helps in allowing for shrinkage.

Blow molds are usually provided with some arrangement of cooling passages through which the cooling medium (usually water) is circulated. Best results are achieved when uniform temperatures are maintained throughout the mold.

Proper venting of the mold is important to achieve smooth and even expansion of the part against the mold. If air is entrapped, the part may be misshaped and the cycle may be lengthened. Surface blemishes may also be created. Most blow molds can be vented conveniently through the parting line but other vents can be created if air entrapment is noted in a particular area.

The surface of the mold cavity is generally not polished or chrome plated. These treatments do not substantially improve the mold surface and may cause sticking of the part in the mold. The most common mold surface is slightly roughened, usually by grit blasting. This gives a means for improving the removal of air and still seems to give acceptable part surfaces. In fact, the surface quality of the part is often improved because trapped air will often create small pockmarks in the surface of the part (called orange peel) that are eliminated by the slightly roughened surface.

The engraving of logos and other identification marks on the mold cavity is common. These will, of course, show up on the surface of the final part in mirror image and relief or indentation. Release from these engravings is usually not a problem because the blowing of the semisolid parison against the engraving is not as binding as would occur when a liquid is injected into the mold.

The blowing point is usually through the hole that forms the opening of the part. This can be either at the top or the bottom of the mold, depending on the orientation of the part in the mold.

Ejection of the part can be either by gravity or by mechanical assist. The mechanical assist can be linked to ejection pins, as with injection molding, to air jets, or to grabbers that use the flash in the pinch-off to withdraw the part. Highly automated systems, such as for the making of milk bottles, usually use the mechanical assist because that allows the part to be positively withdrawn, thus giving a greater measure of control to the handling of the part.

Sliding/Compression Blow Molds

The molding of a recessed ring or lip onto a blow molded part can be done through the use of sliding molds that apply compression to certain areas of the part. The use of this technique to make a flowerpot is seen in Figure 13.9.

In this process the mold closes around the parison and the parison is blown in the normal method for blow molding. As can be seen by careful inspection of Figure 13.9, the parison flows around a recess in the side of the mold at what will become the top of each flowerpot. Before the parison is cool, the upper and lower sections of the mold slide toward the center portion, which is fixed. This sliding motion compresses the material that is in the gaps between the sliding parts of the mold and the fixed parts. The reinforcement ring of the flowerpot is thus created.

After the part is compressed, it is allowed to cool under this compression pressure. When it is cold, the slides move away from the fixed part of the mold and the mold opens to eject the part. Two flowerpots are made in this fashion by the mold design shown, and they are separated by cutting at the ring.

Sliding parts of the mold can also be used to create undercuts in parts. These may or may not require compression. When an undercut is to be made, the parison is blown around the slide to create the undercut. After the part is cooled, the slide moves out and frees the part as the mold opens.

PLANT CONCEPTS

Blow molding machines are generally self-contained and do not need extensive cooling or part-removal equipment, as would be the case with extrusion operations. Therefore, plant facility requirements are rather easy to satisfy.

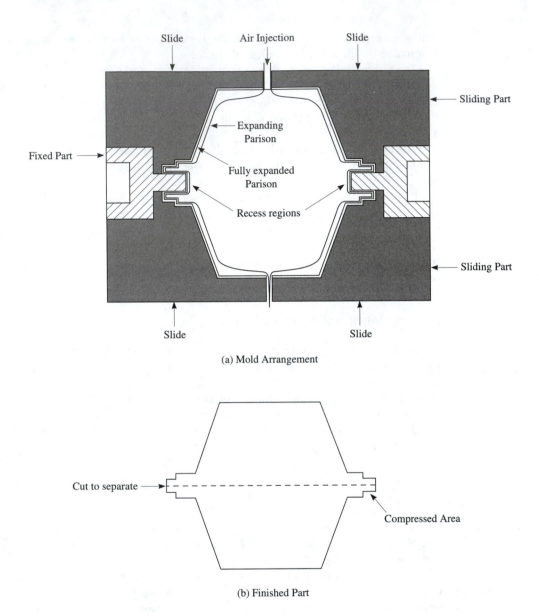

(a) Mold Arrangement

(b) Finished Part

Figure 13.9 Use of sliding/compression and blow molding to make a flowerpot.

The removal and handling of scrap (trimming and deflashing) are important for many blow molding operations. A scrap removal station is usually provided so that the pinch-offs can be removed (often by automated or semiautomated cutting). The scrap is then chopped and blended into the virgin material for reprocessing. Special reduction of the reground material to pellets is usually not necessary. As with regrind from all processes, care should be taken to not contaminate the regrind with other resins or with material that may be somehow contaminated. This is especially true for food and beverage applications. Blow molded operations have a general goal of using no more than 2% regrind with virgin material, but can use up to 50%, although in some operations more regrind than 20% causes significant operational problems.

Bottles are usually pressurized on-line to ensure integrity of the walls. The high-volume production of large quantities often requires online filling and/or labelling that is coordinated with the blowing operation. This is usually done automatically in a line that is adjacent to the blow molding line. The filling line can be integrated with the automated removal equipment on the blow molding line.

PRODUCT CONSIDERATIONS

Materials

Even though most thermoplastic resins can theoretically be blow molded, some materials dominate the blow molding industry. One critical requirement is that the polymer must have good melt strength. If the polymer melt strength is too low, the parison will rip when it is blown. The most common plastics and their principal applications are: HDPE (stiff bottles, toys, cases, drums), LDPE (flexible bottles), polypropylene (higher-temperature bottles and cases such as those used for sterilization), PVC (clear bottles, oil-resistant containers), PET (soda pop bottles), polycarbonate (housings), nylon (automotive coolant bottles, power steering reservoirs) and FEP (chemical-resistant bottles).

The polyolefins (HDPE, LDPE, and PP) are easy to process and give the advantage of good electrical properties, moisture resistance, and low cost. They are, however, somewhat sensitive to oils and, especially HDPE and PP, can have stress crack problems.

PVC is very temperature sensitive. That must be considered in the processing of this resin. For instance, accumulators are rarely used with PVC because the resin would be kept at a high temperature for a relatively long time. Injection blow molding is, likewise, rarely used for PVC because the process requires two heat cycles.

Some of the problems of these resins can be solved by a process called *coblow molding*. This process uses coextruded materials which are extruded into a multilayered parison and then blown in the normal way. (This extrusion process is similar to the extrusion of multilayered blown film discussed in the chapter on Extrusion.) These coextruded materials also offer advantages in permeation control, flexibility, toughness, and appearance. Each layer can contribute some beneficial property to the part. Cost advantages can also be achieved when one of the layers is a regrind material or, perhaps, a low-cost resin that serves as a filler layer.

An advantage of blow molding over injection molding is the forgiving nature of the process toward the use of postconsumer regrind. The flow requirements of the resin are less stringent in blow molding than in injection molding, thus allowing for some mismatch between the characteristics of the main resin (usually the virgin resin) and regrind. Therefore, postconsumer regrind can often be used in blow molding at quite high concentrations without a significant deterioration of part performance.

Shapes

Blow molded parts must be designed to take into account the specific physical phenomena involved in and resulting from the blow molding process. By their nature, blow molded parts are basically hollow. Most traditional parts have tended to be generally cubical or cylindrical, although many newer applications utilize broad, flat sections. The hollow interior can be injected with foam or some other substance for improved insulation or stiffness. The interior can also be used as a form of inner ducting or conduit for wiring. Sometimes the molded object (part) can be cut in two to yield two parts from the same production cycle.

Some parts which are made by blow molding include bottles of all types (especially for disposable packaging), barrels, automotive tanks, automotive ducts, trash carts, flowerpots, double wall cases for tools and insulated storage, and toys. Wall thickness is usually limited to about 1/2 inch (1 cm) or less.

As in most plastic-forming processes, the ideal is a uniform wall thickness in the part. Wall thickness in a blown part must be expected to vary due to the nature of the process. The part will typically be thicker on the bottom surface and thin towards the top due to parison sagging. The corners will be thinner still. Points of extreme stretching or of too deep a draw should be avoided. The ideal blow molded part would be completely symmetrical. This would allow uniform stretching and as even a wall thickness as possible.

The stiffness of the final part is largely dependent on the wall thickness, so control of wall thickness throughout the part is critical to good performance in applications where stiffness is important. Many bottle products require good stiffness to stand straight and to withstand the pressures of filling and labelling.

Care must be taken when designing the part so that the corners and edges are adequately rounded. Rounded corners lessen the effect of the thinning because the material need not stretch as far to completely fill the mold. Another consideration is that rounding the corners reduces the sensitivity of the plastic to forming cracks that are initiated by small, sharp corners (stress riser locations) and thereby improves mechanical properties.

The shape of the bottom of the container is also important in part design. Normally a concave (push-up) shape is preferred over a flat bottom. The concave shape is thicker in the middle (because the plastic did not travel as far) and is therefore stronger but also shrinks more. The concave shape allows for this shrinkage and other dimensional changes without causing the bottle to have an unstable (rocking) bottom as can often happen if the bottom is molded flat.

The opening through which the part is blown is often utilized in the final part as a neck and a closure location, such as for a threaded cap or lip. The threads for such a closure can be easily molded when the part is blown. However, the threads should be coarse and must allow for considerable variation in dimension since blow molding is not a particularly precise molding operation. Injection blow molded parts can be made with much tighter tolerances than can extrusion blow molded parts. Threads can be made as part of the injection process in injection blow molding and then retained during the blowing process by carefully screening the thread area from heat when the preforms are reheated prior to being blown.

The volume of the container can be adjusted slightly by using inserts in the mold. These inserts are usually disks that are attached inside the cavity and therefore protrude into the space (volume) that would otherwise be taken up by the plastic. In other words, the disks reduce the volume of the plastic part. When the volume of the container is slightly too large, a disk is put into the cavity. Then when the part is molded, the part is blown around the disk, thus reducing the volume of the cavity and, therefore, the volume of the bottle. The disk position is seen in the side of the bottle as a round indentation. This indentation decreases the volume. Milk jugs often utilize this volume adjustment system.

Handles can be molded into the part by blowing the part past a pinch point. When this occurs, the pinch point should be open so that further blowing of the part and development of the handle shape can be completed. Most milk jugs utilize this type of handle.

Handles and inserts can also be attached to the outside of the part during the blowing sequence. This is done by placing the insert into a recess in the mold cavity before the part is blown. Then when the part is blown, the interior pressure forces the part to mate with the insert. The mating surfaces of the insert and the part should be compatible materials and compatible shapes. The temperature of the parison and the handle should be appropriate for good bonding.

An important parameter for blow molding is the *blow ratio,* which is defined by Equation (13.1).

$$\text{Blow Ratio} = \frac{\text{Mold Diameter}}{\text{Parison Diameter}} \qquad (13.1)$$

This ratio assumes a generally cylindrical part. Blow ratios between 1.5 and 3 are common, but ratios up to 7 are possible with some materials and some part shapes. The ability of the plastic to expand is often measured by the blow ratio for the plastic which serves as a convenient guide in the ordering of plastic grades for various specific parts.

OPERATION AND CONTROL

In many blow molding operations the control of the process is not as tightly dictated as would be the case in injection molding or even extrusion. There is, therefore, more art in the operation of blow molding and greater operator skill is required. This flexibility in the process increases as the size of the part increases. Some points of variability include the following: (1) stretch (sag) of the parison, (2) temperature of the parison and temperature of the surrounding space, (3) melt flow characteristics of the resin, (4) speed of parison formation, (5) crystalline nature of the polymer, and (6) cooling capability of the mold. These parameters are critical to the proper operation of the blow molding process. Therefore, every attempt to control and standardize these parameters will improve the repeatability of the process.

Troubleshooting

The problems in blow molding can be conveniently divided into two groups—(1) problems in forming the parison and (2) problems in molding. The most typical problems and some suggested solutions are given in Table 13.1.

Table 13.1   Troubleshooting Guide for Blow Molding

Problem	Possible Cause
Parison Faults	
Parison curl (parison is not straight as it leaves the die)	• Align the die • Temperature variations in the die region, especially the die ring relative to the outer part of the die • One side of the die, especially the land, longer than the other • Blow-up ratio too high
Parison lengthens excessively (sag or stretch)	• Melt index of the material too high • Processing temperature too high • Extrusion rate too low • Screw speed too slow
Parison has rough surface	• Processing temperature too low • Extrusion die too cold • Parison extruded too rapidly (melt fracture)
Parison has marks running lengthwise	• Die ring damaged • Die ring needs cleaning (probably degraded material) • Mandrel support spider not streamlined properly or misaligned
Parison exhibits local discolorations	• Material contaminated • Proportion of reclaimed material too high • Degraded material in the extruder
Parison exhibits brown stripes or discolorations	• Melt temperature too high • Residence time of material too long • Defective connection between extruder and die • Mandrel support not properly holding
Molded Part Problems	
Part ruptures as it is blown	• Parison wall not uniform • Parison not grasped evenly by the mold halves • Resin melt index too high • Too much regrind • Contaminated resin • Parison not at the proper temperature • Blow-up air enters too slowly or too rapidly
Problems with insertion of blowing needle	• Insertion rate too slow or fast • Needle stroke too short • Blow-up air flowing prematurely from the needle • Bore diameter of needle too small
Molding not fully inflated	• Blow-up air pressure too low • Blow-up time too short
Molding sticks to mold	• Mold too hot • Mold cycle too short

Table 13.1 *Continued*

Problem	Possible Cause
Molded Part Problems	
Wall of molded part is too thin	• Processing temperature too high • Parison not correctly dimensioned • Parison sagging excessively
Wall thickness not uniform or changes	• Parison sagging • Parison not formed properly (see parison faults) • Programming of parison not working properly
Molding surface has uneven appearance	• Mold cavity not sufficiently vented • Mold cavity too smooth
Molding surface contains many small round or lens-shaped inclusions (gels)	• Material needs to be dried • Pellets need more heating in extruder
Molding tear after demolding	• Blow-up ratio not appropriate • Temperature of plasticization unit or extruder too high • Mold cycle too short
Excessive shrinkage	• Walls too thick • Mold not cool enough • Melt temperature too high • Blow pressure too low
Variable part weight	• Part sagging

CASE STUDY 13.1

Making Soda Pop Bottles

The rapid expansion of the use of plastic containers for soda pop bottles was the result of several events and developments which all had a major impact on this particular market and product. One of these events was the oil embargo of the mid-1970s. The major impact of the embargo was to call attention to the strong dependence of the United States on imported oil, especially as a fuel. Many efforts were directed toward reducing the amount of fuel that was consumed. More fuel-efficient engines were developed, mass transportation was emphasized, and lighter vehicles were created. The much greater weight of glass versus plastic bottles was a major impetus to change to the plastic bottles because transportation costs would be reduced.

At the same time that the weight savings advantages of plastic bottles over glass became important, plastic resins with good barrier properties were developed. Many programs to develop resins with high barrier properties for carbon dioxide were under way throughout the plastics industry. Some were based upon traditionally good barrier materials such as styrene acrylonitrile (SAN), PVC, and polyvinylidene chloride. While these plastics were good, some taste tests seemed to indicate that residual chemicals from the resin-making process could be detected in the soda pop. The resin that proved to have the correct combination of properties was polyethylene terephthalate (PET). A typical set of performance requirements were that after 120 days at room temperature, the bottles would have <15% loss of CO_2, no off-taste, no change of shape (such as swelling), no fall in liquid level, and be able to survive a drop test in which the full bottle was dropped from 6 feet (2 m) height. PET was found to have excellent barrier properties. PVC has a gas permeability of 2 times that of PET. The permeability of HDPE is 52 times greater, PP is 57 times greater, and LDPE is 114 times greater than oriented (crystallized) PET.

Another development that coincided with the others to make the use of plastic soda pop bottles possible was the development of stretch blow molding. PET needed to be oriented to achieve the crystallinity for low permeability and for mechanical toughness and strength required for the soda bottle application. The orientation could conveniently be done with stretch blow molding. (Stretch blow molding is illustrated in Figure 13.7.) With stretching, the PET was tough enough to withstand being dropped without breaking, and strong enough to allow the bottles to be filled with a pressurized solution and not burst or deform. Other advantages of stretch blow molding include:

- Greater rigidity so that the pouring of the bottle will be easier.

- Reduced permeation over nonoriented bottles because the orientation moves the molecules closer together, thus making the movement of the gas through the molecular layers more difficult.

- Greater precision in dimensional control.

- Ability to give special shapes to the bottom. This capability is used for some bottles that are fluted at the bottom both to give stable surface and to increase strength.

- Elimination of bottom beads, which are present when the bottom is pinched off rather than molded.

- Ability to match the wall dimensions of the preform to the expansion of the bottle so that wall thicknesses are more uniform.

- Increased burst pressure.

The soda pop bottles are small-pressure vessels and so the optimum shape for pressure containment is a sphere. A reasonable compromise is the familiar cylindrical shape with hemispherical ends. To allow the bottle to stand upright, a base cap of high-density polyethylene was bonded onto the bottom. Recently, shaped (fluted) bottoms have been developed that eliminate the need for the base cap.

The economics of soda bottles suggested that the industry needed to have the capability of shipping preforms to the many bottlers rather than requiring that the bottles run a blow molding machine. With stretch blow molding, the preforms can be made at a central location using well-controlled injection molding conditions. These preforms can then be shipped to other locations where a far less complicated machine is needed to reheat and blow them. The blowing sites would not need to handle resin pellets or worry about the problems associated with a full injection molding operation. Molds for the blow molding at the blowing locations would be low-cost blow molds rather than the much more expensive injection molds.

Some processing considerations are important, however. In order to achieve the proper crystallization of the PET during the blowing operation, three separate processing stages must be achieved.

1. The PET preform is injection molded at a hot temperature (480–540°F, 250–280°C) and quenched so that it will be amorphous. (If the injection molded preform is slowly cooled, crystals will form giving a preform that is much more difficult to blow.) The injection pressures are usually quite low and, to achieve the quench, the mold temperatures are very low. Because of the tendency of PET to degrade when hot and wet, the resin must be carefully dried and should be of a very high purity.

2. The preforms can be stored or shipped to a separate location for final forming. When ready to be blow molded, the material is heated to 200–212°F (95–100°C), a temperature that is about 60°F (30°C) above T_g but not nearly as hot as would be required to injection mold the material.

3. When the preform is uniformly hot, it is then stretch blow molded. Small crystals are induced into the material. (This process is called *stress-induced crystallization*). These crystals give the required physical and mechanical properties but are so small that they do not appreciably refract light. Therefore, the crystallized material is transparent. If the preform is further heated (to about 300°F, 150°C), the crystallization is enhanced and the mechanical properties are further improved. (PET film is subjected to a similar heating and stretching process to develop improved physical and mechanical properties.) The resulting bottle is *biaxially oriented,* that is, molecules are oriented in both the longitudinal and radial directions.

The bottles can be capped with an HDPE base or simply use the base that is created during the blowing operation. The threads are created as part of the molded preform and are maintained through the subsequent heating and blowing steps.

Initially PET soda pop bottles were only economical in large (2-L) bottles. Recently, the development of improved processing techniques has resulted in acceptable economics for the creation of smaller and larger bottles. The future for plastic barrier bottles for sodas and other drinks appears to be very favorable.

SUMMARY

Blow molding is the principal method used to make bottles and other hollow shapes in which the mouth is smaller than the body of the container. Blow molding is a two-step process. In the first step a parison or preform is created. Then, in the second step, that parison is expanded into a mold. The expansion is done by injecting air inside the hot parison which forces it against the walls of the mold cavity.

Two principal methods are used to form the parison and are called extrusion blow molding and injection blow molding. In extrusion blow molding the parison is formed by forcing the plastic through an extrusion die. This process can be either continuous, in which case multiple molds are required, or intermittent. Intermittent extrusion blow molding uses an accumulator to build up the charge of material so that it can be pushed rapidly through the die to form the parison. Intermittent blow molding is used for large parts. Alternately, the extruder would be stopped.

The second principal method for forming the parison is injection molding. In this method a standard injection molding machine is used and the parison is simply injected around a core pin. The parison is then transferred to another machine where it is placed into a mold, heated, and blown into shape. A modification of the injection blow molding system uses a mechanical device to stretch the parison in the longitudinal direction so that biaxial orientation can be achieved. This process is called stretch blow molding and is the method that PET soda pop bottles are produced.

The process of blow molding can be somewhat complicated with problems related to both the formation of the parison and the later formation of the finished part. A troubleshooting guide should be consulted to assist the operator in examining all of the possible causes of a particular defect.

GLOSSARY

Accumulator system   A system in which the extruder runs continuously and feeds an accumulator to accommodate the times required for the molding cycle.

Biaxial orientation   When the molecules in a part are oriented in two directions.

Blow ratio   The ratio of the mold diameter to the parison diameter.

Coblow molding   A process in which two or more resins are blown simultaneously into a bottle.

Continuous extrusion blow molding   During this process the extruder runs continuously, thus making a continuous parison.

Extrusion blow molding   The blow molding process in which the parison is formed by an extrusion process.

Injection blow molding   A process in which a preform is made by injection molding and then the preform is heated and blown, using blow molding, into the final shape.

Intermittent extrusion blow molding   During this process the extruder is stopped during the time that the molding occurs.

Multiple-mold system   A continuous blow molding system that uses multiple molds to match the output of the extruder when forming the parison.

Parison   The cylindrical tube that is trapped within the mold and then blown to fill the mold in a blow molding process.

Parison transfer system   A blow molding system in which the parison is removed from the die and transferred to a mold, thus giving space for another parison to be formed while the first part is cooling.

Preform   An injection molded part that is to later be heated and then formed into a shape in blow molding.

Reciprocating screw system   A blow molding system in which the turning of the screw of the extruder is stopped in order to accommodate the times required to mold the part; usually the extruder advances intermittently to push some resin forward and form the parison.

Rising mold system   A system in which two (or more) molds are used to mold parts from one extruder during continuous extrusion blow molding.

Stretch blow molding   Blow molding in which the parison (or preform) is expanded in both the radial direction and, simultaneously, in the longitudinal direction (usually by using a mechanical plunger).

Stress-induced crystallization   Crystal structure formed when a plastic is stretched.

QUESTIONS

1. What is the principal problem in forming a bottle using injection molding?
2. Why is continuous extrusion blow molding not recommended for large blow molded parts?
3. List three advantages of the moveable mold system versus the rotary system for continuous extrusion blow molding.
4. In a blow molded part, where are the thinnest sections likely to occur?
5. Why can blow molding molds be made out of aluminum, whereas injection molding molds are usually made out of tool steel?
6. What is the advantage of a programmable parison device?
7. What method is used in injection blow molding to achieve the type of part wall thickness control that can be obtained in programmable parisons?
8. Describe sliding/compression blow molding and indicate its advantage.
9. What factors are likely to determine the maximum size part that can be blow molded?
10. What key processing considerations must be met in order to use PET to make soda pop bottles?

REFERENCES

Blow Molding of Thermoplastics, Hoechst Plastics, KV 202e - 7076 (8355).

Brydson, J. A., *Handbook for Plastics Processors,* Oxford, UK: Heinemann Newnes, 1990.

Frados, Joel (ed.), *Plastics Engineering Handbook* (4th ed.), Florence, KY: International Thomson Publishing, 1994.

John, Frederick W., "Blow-molding grown up," *Machine Design,* Apr 2, 1970, pp 89–95.

Lee, Norman C., *Blow Molding Design Guide,* Munich: Hanser Publishers, 1998.

Lee, Norman C. (ed.), *Plastic Blow Molding Handbook,* New York: Van Nostrand Reinhold, 1990.

Morton-Jones, D. H., *Polymer Processing,* London: Chapman and Hall Ltd, 1989.

Schwartz, Seymour S., and Sidney H. Gooding, *Plastic Materials and Processes,* New York: Van Nostrand Reinhold Company, 1982.

CHAPTER FOURTEEN

THERMOFORMING PROCESS

CHAPTER OVERVIEW

This chapter examines the following concepts:

- Forming processes (fundamental vacuum, pressure, plug-assist, reverse draw, free, drape, snap-back, matched die, and mechanical)
- Equipment (machines, molds, plant considerations)
- Product considerations (materials, shapes and part design)
- Operation and control (critical parameters, troubleshooting, maintenance and safety)

INTRODUCTION

Thermoforming is a process used to shape thermoplastic sheets and films into discrete parts. (In general, both sheets and films can be thermoformed but, for simplicity, the term sheets will be used in this chapter to represent both types of materials, unless specifically stated otherwise.) The basic principles of the thermoforming process are to heat a thermoplastic sheet until it softens and then force the hot and pliable material against the contours of a mold by using either mechanical, air, or vacuum pressure. When held against the mold and allowed to cool, the plastic retains the shape and detail of the mold. The cooling step is usually short.

The thermoforming process is significantly different from other plastic processing methods considered thus far because in thermoforming the material is not melted. The material is heated only enough to soften it. Because cooling can be much faster than when melted, lower molding times are possible. Note that because this process uses heat with subsequent forming, thermoset materials, which would cure under these conditions, cannot be used.

Another major difference between thermoforming and other processes considered thus far is the lower pressures that are required to thermoform. Both the mechanical and pneumatic pressures used in thermoforming are just slightly greater than atmospheric pressures. Hence, the forming equipment and the molds can be made of less sturdy materials than are required for high-pressure plastic-forming processes such as extrusion, injection molding, and even blow molding. Large parts can therefore be made using thermoforming without the high capital costs of large molds and pressurizing machines.

A significant disadvantage of thermoforming versus the other processes is the much greater amount of scrap it generates. Because the parts are made from portions of a sheet, each part must be trimmed and the excess material recycled. Furthermore, the cost of the sheet material is raised because a separate sheet-forming step must be done to create the starting material of the thermoforming process.

The designs of parts made by thermoforming are more limited than in injection molding because the plastic does not melt and flow into intricate shapes. Parts made by thermoforming are generally open structures with the diameters of the openings greater than the diameter of the body. Parts with sharp bends and corners are difficult and parts with thick and thin walls, such as bosses and solid ribs, are generally not possible. Wall-thickness control is also difficult and some areas are inherently thinner than others because of uneven stretching of the material as it is pressed into the mold.

When using thermoforming to make certain parts, it should be remembered that the process inherently results in internal stresses, and some applications may result in considerable stress on the part.

For many applications, the advantages of thermoforming outweigh the disadvantages. These are summarized in Table 14.1.

Table 14.1   Advantages and Disadvantages of Thermoforming Versus Other Plastics Molding Processes

Advantages	Disadvantages
Low machine cost	High cost of raw materials (sheets)
Low temperature requirement	High scrap
Low mold cost	Limited part shapes
Low pressure requirement	Only one side of part defined by mold
Large parts easily formed	Inherent wall thickness variation
Fast mold cycles	Internal stresses common

The number and variety of applications for this fast-cycle and low-cost method of forming plastics are increasing rapidly, among them: signs, light fixtures, luggage, trays, housings, cases, covers, drawers, tubs, snowmobile bodies, small boat hulls, boat windshields, refrigerator door liners, ice cube trays, egg cartons, and both blister packaging and skin packaging.

FORMING PROCESSES

The wide variety of parts made by the thermoforming process has led to the development of several modifications of the basic technique to optimize the making of particular shapes and to improve upon some of the inherent problems associated with thermoforming.

These process modifications reflect changes in the type of mold and the method of forcing the plastic material into the mold. The techniques can be grouped into several major types which will each be considered separately.

Fundamental (Straight) Vacuum Forming

This is the simplest thermoforming technique and the one most commonly envisioned when thermoforming is discussed. All thermoforming techniques were once referred to as *vacuum forming.* (The terms *straight* and *fundamental* are used merely to emphasize that this is the simple method of vacuum forming and not one of the several variations.)

In this processing technique the material is clamped into a frame, which holds the material around its periphery. The material is then heated and when it begins to sag (the *sag point*) it is transferred so that it seals against the mold. In this process the mold is a *female mold* or *cavity mold,* although in other, related processes male molds are used. A vacuum is immediately applied to the back of the mold through an air space. The vacuum connects with the mold cavity through small vent holes that are drilled through the mold walls. When the vacuum is applied, the plastic material is drawn against the mold cavity walls. (Actually, the outside pressure of the air pushes the plastic against the mold because of the partial vacuum created between the plastic and the mold. Hence, the highest pressure that can be achieved with the vacuum technique is approximately 1 atmosphere [14.5 psi, 100 kPa].) The vacuum is continued while the material cools and takes the shape of the mold cavity. The vacuum forming technique is illustrated in Figure 14.1.

When the sagging occurs, the center of the material moves downward and stretching of the material occurs near the clamp and results in a thinning in that region (the periphery). As the material then moves against the cavity walls when the vacuum is applied, stretching occurs uniformly over the sheet until the sheet touches the mold. Wherever the sheet touches the mold the thickness at that location becomes fixed. Further stretching must occur in the areas that have not yet touched the mold, typically in the corners. Therefore, the thinnest areas are in the corners. Other thin areas are near the clamps and also in any other area where the material must travel longer distances. Hence, uneven wall thicknesses are inherent in this technique.

Fundamental vacuum forming is used as a technique when the outside of the part (the side against the mold) must have fine detail or close tolerances. To minimize scraps in processes where several molds are used simultaneously, the female molds can be placed close together. A photo of vacuum formed parts is given in Photo 14.1.

Pressure Forming

The *pressure forming* technique also forms the heated plastic material in a female mold. In this case, however, positive air pressure on the top of the plastic is used to force the material against the mold.

In pressure forming the material is clamped and heated just as with fundamental vacuum forming. The softened sheet is transferred to the molding area and a seal is made

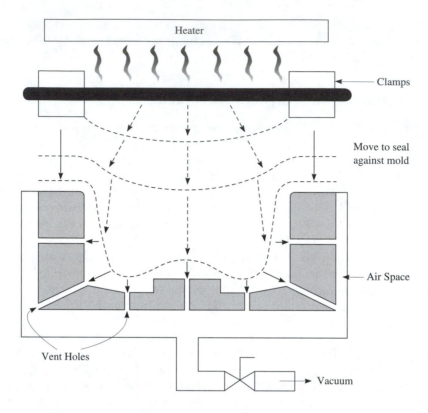

Figure 14.1 Fundamental, or straight, vacuum forming technique.

so that the upper chamber, above the plastic, is air tight. The material is also sealed against the mold as is done with vacuum forming. Air pressure is then introduced into the area above the softened plastic, usually through an air baffle to insure that the air is evenly distributed across the surface of the plastic. The air pressure forces the plastic against the mold. The vent holes are still used to ensure that no air is trapped between the plastic and the mold. This technique is shown in Figure 14.2 and is similar to blow molding.

The air pressure used in this technique is generally 14.5 to 300 psi (100 to 2000 kPa), and it should be applied as quickly as possible to prevent the sheet from cooling or from excessive sagging.

The advantage of the pressure method is that mold cycles can be faster than with fundamental vacuum forming, the sheet can be formed at a lower temperature because the forcing pressure is higher, and greater dimensional control and part definition can be obtained. Vacuum pressure is limited to about 1 atmosphere, whereas the positive pressure in pressure forming can be much higher. When a positive pressure source is used, the pressure can be created much more rapidly and generally at lower cost.

Photo 14.1   Standard vacuum forming machine (Courtesy: Brown Machine)

Plug-Assist Forming

Just as the softened material can be forced into the mold by positive air pressure as is done in pressure forming, it can be forced downward by mechanical pressure. In *plug-assist forming,* a plug is used to force the material into the mold. Generally the plug will not push the material completely into the mold but, rather, only partway to a positive seating. A vacuum is then applied to draw the material against the cavity walls and complete the forming operation. This technique is shown in Figure 14.3. Instead of a vacuum being used to force the part against the walls, air pressure could be used. Both techniques (vacuum or pressure) are called plug-assist forming when a plug is used.

The major advantage of plug-assist forming is better wall thickness uniformity than can be obtained with vacuum or pressure forming, especially for parts with deep draws such as cup or box shapes. The plug can be used to carry material towards the areas that would have been too thin if just fundamental vacuum or pressure forming were used. In plug-assist forming the initial sagging of the material is kept to a minimum. The plug is moved against the material and wherever the plug touches the material, the

Figure 14.2 Pressure forming.

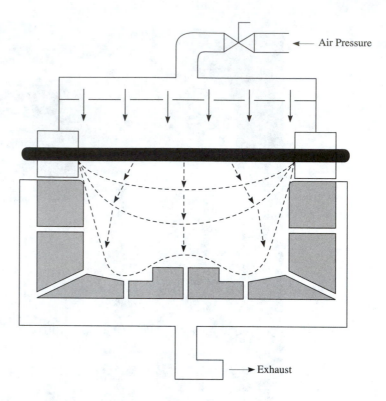

Air Pressure

Exhaust

thickness of the material is fixed at the contact points. Therefore, by shaping the plug appropriately, the plug can "carry" some of the material toward the corners and force the areas that would be thick in fundamental vacuum forming to be stretched. (The areas that would normally be stretched during the plug-forming stage would be the center of the bottom and the sides.) Then, when the vacuum is applied, the material moves outward off the plug and stretches uniformly until the material makes contact with the mold. As with fundamental vacuum forming, this contact first occurs at the center of the bottom and along the walls. The thickness then becomes fixed at those locations and stretching occurs in the corners and other locations not yet in contact with the mold. In the case of the plug-assist forming, however, the material in the corners is thicker than in the fundamental vacuum case and can stretch more without becoming too thin. Therefore, the plug has resulted in a retention of the thickness in the corners for a portion of the forming process such that the finished part is more even overall. Many fundamental vacuum forming and pressure forming machines have the capability of plug-assist forming.

Plugs can be made of metal, wood, or thermoset plastics. To prevent premature cooling of the material, especially if the plug is metal, the plug is usually heated to a few degrees less than the temperature of the plastic. The plug should be 10% to 20% smaller in length and width than the female cavity.

Figure 14.3 Plug-assist forming.

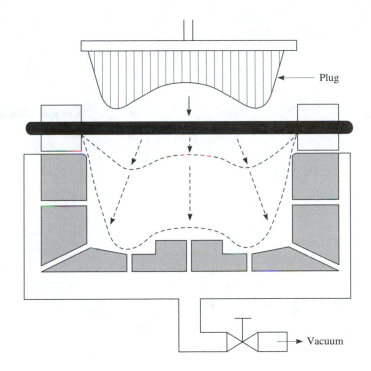

Reverse Draw Forming

The *reverse draw forming* technique (also called *invert forming, pillow forming,* or *billow forming*) is used when very deep draws are needed. In this technique the material is heated to the sag point and then blown away from the mold. The purpose of this blowing or reverse drawing is to thin the material in the center of the sheet that is the thickest area during fundamental vacuum and pressure forming. The size of the bubble is often controlled automatically by an electric eye. When the bubble reaches the correct size, the pressure is stopped and a plug presses on the material to force it back into the mold. As with normal plug-assist molding, the points of contact with the plug are fixed in thickness and all other areas are thinned by the movement of the plug as it further stretches the material. After the plug has moved the corner material close to the walls of the mold, a vacuum is activated through the mold to draw the plastic against the mold walls. The stretching occurs in the normal way with the material travelling the furthest being stretched the most. The molding is completed by cooling the material against the mold. The reverse draw forming method is shown in Figure 14.4.

This technique has the advantage over plug-assist molding in that an additional stretching step has been added (the reverse drawing step), which allows even better control over the part thicknesses. The penalty for this improved control is a longer mold cycle. Fine adjustments in the thickness of the part can be achieved by varying (1) the

Figure 14.4 Reverse draw forming or billow forming.

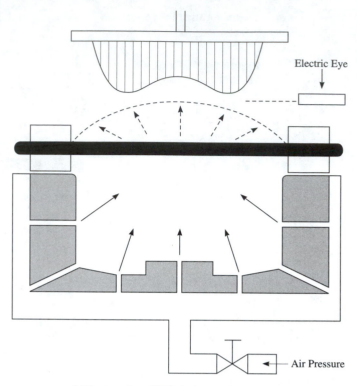

(a) Reverse draw (billowing) step to pre-stretch the material

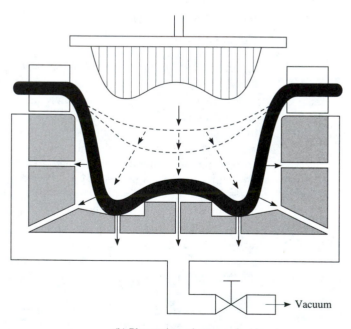

(b) Plug-assist and vacuum forming step

temperature at which the material is blown, (2) the size of the bubble, (3) the size of the plug, (4) the shape of the plug, (5) the speed of travel of the plug, and (6) the timing of the imposition of the vacuum or air pressure.

Free Forming

In some products, especially those where very high optical quality is required, the forming is done without a mold. (Touching the mold can result in undesirable changes in the surface quality of the part.) This technique is called *free forming* or *free blowing* and is illustrated in Figure 14.5.

In free forming the part is expanded with air pressure, much as it would be in the first step of reverse draw forming. The size of the bubble is often monitored by an electric eye. When the bubble reaches the desired size, the air pressure is reduced to a level that maintains the size of the bubble while the part cools.

The complexity of shapes of parts made by free forming is much more limited than could be made if a mold were employed. However, some shape control can be achieved by varying the shape of the clamping ring. For instance, if the clamping ring is a circle, the part will be a hemisphere. If the shape of the clamping ring is teardrop, the resulting part will be elongated and streamlined. This streamlined shape is often used for canopies for racing vehicles where the high optical clarity is needed. (Forming against a mold can cause some loss in optical clarity.)

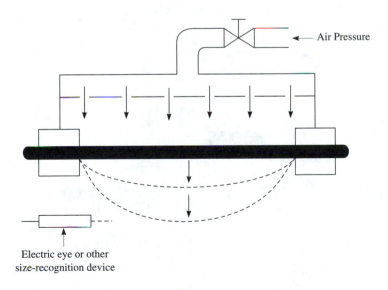

Figure 14.5 Free forming or free blowing.

Drape Forming

Drape forming is the term applied to thermoforming using a male (plug) mold. In this technique the material is clamped and heated to the sag point and then forced over the mold by moving the material onto the mold. Simultaneously, vacuum is applied through vent holes in the mold, thus drawing the part tightly against the outside surface of the mold. The part is cooled in contact with the mold. The drape forming technique is shown in Figure 14.6.

The movement of the mold and material together is much like the movement of a plug in plug-assist forming except that the plug is smaller than the size of the finished part and, of course, the plug withdraws from the part before final forming is done. Nevertheless, the actions of the two processes are similar.

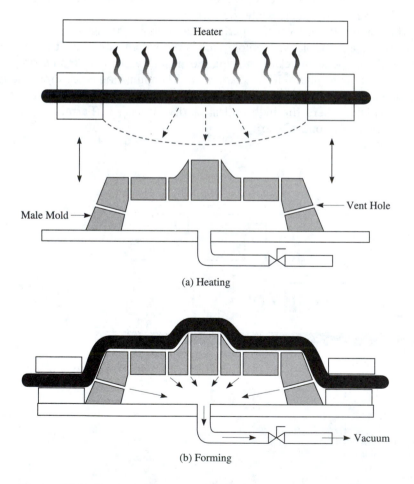

Figure 14.6   Drape forming on a male (plug) mold.

Drape forming can be combined with a lateral stretching technique to achieve a biaxially oriented part. When this technique is used, the heated material is stretched outward by the clamping mechanism before it is forced over the male mold. This gives orientation to the sheet. When the sheet is then forced over the male mold, further orientation occurs. The resulting part is, therefore, stronger in the oriented directions than would be otherwise possible.

The most obvious use of male molds is when the inside of the part needs to have good surface definition or detail and dimensional control. This is achieved because the material is forced against the outside of the mold, which becomes the inside of the part. Parts made by drape forming would include those for which a grain surface is required on the inside, bath tubs and spas, and other deeply drawn cup-shaped parts.

A major advantage of drape forming over forming in a cavity mold is the greater depths of draw that can be achieved without excessive thinning. The part is forced over the mold with relatively little thinning. Therefore, when the final forming takes place, the thickness uniformity is much better than could be achieved with a cavity mold.

Another advantage of drape forming is the lower cost of making a plug mold versus a cavity mold. It is easier to cut and polish on the outside of a surface than in a cavity. Plug molds are also more easily damaged so additional storage and use care is required.

Just as plug and pressure forming are faster than fundamental vacuum forming because the material is forced into contact with the mold by fairly high pressures, so too is drape forming faster than fundamental vacuum forming. Therefore, drape molding is favored for parts that are extremely sensitive to mold cycle.

A disadvantage of drape forming is that more space is required around the mold so the trim (scrap) is greater. This requirement arises from the need to clamp the material on the outside of the mold. Therefore, very high-production lines in which many molds are used to form similar parts (such as low cost cups) would use female molds rather than male molds so that the number of parts molded in a set area of sheet would be higher.

Snap-back Forming

Snap-back forming is a variation on drape forming. In snap-back forming the material is heated to the sag point and then drawn slightly into a vacuum box below the part. This prestretching thins the center of the part and is usually about one-half to two-thirds of the total draw that the part will receive. A second step is then activated to give more draw. The plug mold is pressed against the material to draw it farther. During this drawing step, the thickness of the center of the material is fixed by contact with the mold and thinning occurs near the edges. Finally the part is finish formed by applying a vacuum through the plug mold and causing the part to "snap-back" against the outside of the plug mold. The part cools against the plug mold to take its final shape. The snap-back forming process is depicted in Figure 14.7, where the prestretching, drawing and forming steps are shown.

If the part is oddly shaped, such as having a very long draw in just one area in the part, the vacuum box can be shaped to accommodate the unusual features.

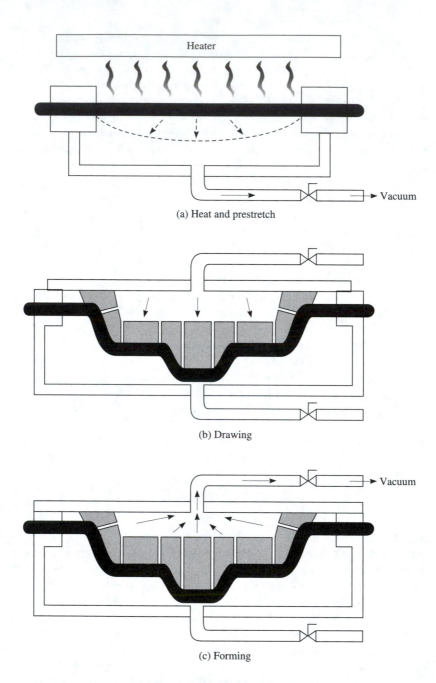

(a) Heat and prestretch

(b) Drawing

(c) Forming

Figure 14.7 Snap-back forming showing prestretching, drawing, and forming steps.

The uniformity of wall thickness is very well controlled by this snap-back forming process and its popularity is increasing in spite of the typically longer cycle times that are required.

Matched Die Forming

Some parts, especially those with very complicated shapes, can best be formed if both plug and cavity molds or dies are used together. These molds must be *mated* or *matched* so that they fit together without interference and leave only the space between them for the material. In matched die forming the clamped material is heated to the sag point and is then transferred to a molding station where the plug and cavity molds are immediately brought together to squeeze the material. Vent holes in both molds allow any air that might be trapped between the material and the molds to escape. No vacuum or air pressure is used during this forming process. The material is mechanically pressed into the shape defined by the matched molds and is allowed to cool while the molds continue to press against it. Matched die forming is depicted in Figure 14.8.

The pressures involved in matched die forming are somewhat higher than in other thermoforming methods but are not nearly as high as injection or blow molding nor as high as matched die compression molding, which is discussed in the chapter on

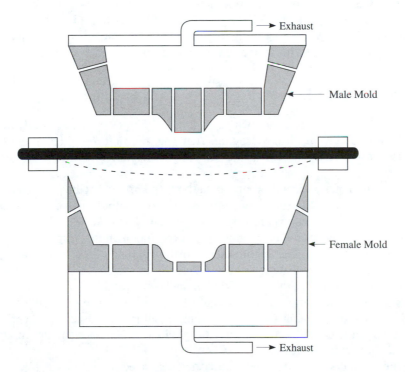

Figure 14.8 Matched die forming.

compression and transfer molding processes. The lower pressures result from the soft and pliable state of the plastic material when it is formed and the relatively small distances over which the material must be moved during the forming process.

Matched die forming is used most often for parts that do not have large draws. Very large parts can also be molded effectively using matched die forming. Matched die forming is also used for parts that must have excellent definition or dimensional control on both sides.

Mechanical Forming

Mechanical forming is, perhaps, the simplest of all the thermoforming techniques. In this method the heated sheet material is formed by mechanically pressing it using forming tools (not molds) to give the desired shape. For instance, if a sheet is to be attached inside a bracket, the bracket itself can be used as the shape-defining device. The plastic material would simply be cut to the appropriate size and then heated to the sag point. While still hot the material would be pressed against the inside of the bracket to create the desired shape. The pressing could be done by a roller, a block, or any other convenient instrument that will apply a relatively uniform pressure on the material as it takes the shape desired. The material is then cooled in place.

Another example of mechanical forming would be the wrapping of a heated plastic sheet around a rod. In this case the material is cut to the desired size and one end is inserted into a clamp attached to the rod. The material is heated to the sag point and then pressed onto the rod as the rod rotates. The material is again cooled in place.

The examples of forming in a bracket and around a rod are illustrated in Figure 14.9.

Blister Pack and Skin Pack

Blister pack and *skin pack* are two packaging techniques used extensively in retail sales of small items and in medical packaging. They use thermoforming techniques that are similar to those already discussed but still sufficiently different that they warrant separate consideration.

In both blister pack and skin pack methods the product to be packaged is captured between the plastic sheet (or, more commonly, plastic film) and a piece of cardboard. The plastic, product, and cardboard comprise the display unit.

Blister packaging is usually done by first thermoforming a thin sheet of relatively stiff, clear plastic material. The mold to do the forming would typically have many cavities that are shaped to accommodate the size and, sometimes, the shape of the product. The formed shape in the sheet is called a *blister*. Many blisters would be formed simultaneously. The forming technique can vary but, because the depth of draw is usually not large, either straight vacuum forming or drape forming is commonly used. The depth of the cavity would be sufficient to accommodate the product and allow a small amount of product movement within the cavity.

The products to be packaged would then be placed into the cavities while the sheet is still oriented with the cavities down. After the products are placed, a printed sheet of card-

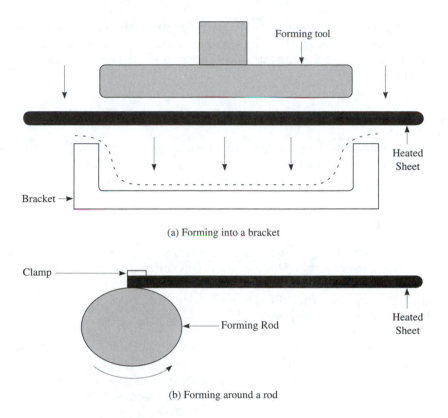

(a) Forming into a bracket

(b) Forming around a rod

Figure 14.9   Mechanical forming for a bracket and a rod.

board onto which some adhesive material has been applied would be laid over the sheet of plastic and the two would be pressed together, thus trapping the product. The sheet would then be cut and the individual packs would be ready for display.

The characteristic that is most noticeable in the blister pack method is the well-defined cavity in which the parts are located. This type of packaging is very desirable, although the cost is somewhat higher than skin packs.

Skin packs also retain the part between the plastic film and a piece of cardboard. In the case of skin packs, however, the plastic is not stiff and no cavity is formed. Rather, the plastic is pliable and the forming is done so that the part is encapsulated by the film against the cardboard. This is done by using sheets of cardboard that are printed and then perforated with many very small holes. The cardboard would then have an adhesive applied to the upper (printed) surface. The products are placed so that one product would be in each finished pack. The heated plastic material is then laid over the sheet of cardboard and a vacuum is applied under the cardboard so that the clear plastic sheet is drawn against the cardboard. The plastic is bonded to the cardboard by the adhesive and the part is retained between them. The sheet is then cut so that each product is separate.

The skin pack method pulls the plastic directly onto the product. Therefore, wrinkles in the plastic are inevitable and no product movement is possible. Most retailers find that this package is less pleasing than the blister pack but the costs of skin packaging are much lower.

EQUIPMENT

The thermoforming process is much simpler than the other molding processes that have been discussed heretofore and therefore the equipment (both the molding or forming machines and the molds) is also considerably simpler and less expensive.

Machines

Much less uniformity of design is seen in thermoforming machines than in machines for other molding processes. However, all thermoforming machines must provide for the following basic functions: heating of the sheet, clamping the sheet, moving the sheet and mold into the proper relationship for forming, a vacuum or pressure system, and appropriate controls and safety devices.

The simplest machines are single-station machines in which a single part is molded. A machine of this type is depicted in Figure 14.10 and in Photo 14.2. The pressures are not high and so the frame can be built of standard angle iron.

The plastic sheet is placed on a transport mechanism that will allow the material to be moved in and out of the oven. A clamping ring, secured by a clamping device, is placed around the perimeter of the plastic sheet. The clamping ring is usually a simple shape made of box steel and the clamping devices are toggles that are mounted to the transport mechanism.

The heater can be of any type that will heat over a wide area so that the plastic will be uniformly heated. Infrared heaters are probably the most common. The heaters can be placed both above and below the plastic if thick plastic sheets are to be formed. The heaters are usually enclosed in an insulated box (oven) so that heat losses are minimized.

The mold is placed on a table that is connected to a vacuum and, often, a pressure source through an appropriate valving arrangement. The vacuum table is connected to a vacuum pump but, because the vacuum should be applied very quickly when forming, a more common arrangement is to have a large surge tank between the continuously running vacuum pump and the vacuum table. Then, when vacuum is to be applied, a valve is opened connecting the evacuated surge tank and the vacuum table.

The vacuum/pressure table sits on the top of a hydraulic ram that can raise the mold so that it moves against the plastic sheet. A seal is made between the vacuum/pressure table and the clamping mechanism. Both female and male molds can be used on a machine of this type.

Many thermoforming machines also have a ram assist device that is also mounted to a hydraulic ram so that it can be lowered onto the top of the plastic. A cooling blower can also be included.

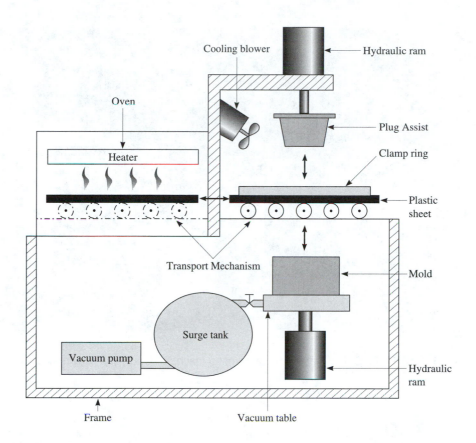

Figure 14.10   Single-station thermoforming machine.

A provision is usually provided for trimming the part. This can be done at the forming machine or on a separate machine.

Higher-speed production can be achieved by using multistation thermoforming machines. These are often laid out in a rotary fashion such that the plastic sheet is loaded at one station, heated at another, formed at another, and trimmed and removed at the original loading station. These machines can be highly automated or partially automated. Multistation machines are especially suited to long production runs, especially when some manual labor is involved in one or more of the steps.

Another arrangement for even higher-speed production is a continuous thermoforming machine. This type of machine is discussed in the case study at the end of this chapter and is shown in Photo 14.3.

Molds

The low pressures involved in thermoforming allow great latitude in the selection of mold materials. The most common material is aluminum, which can be either machined or

Photo 14.2   Single-station thermoforming machine (Courtesy: Brown Machine)

Photo 14.3   Continuous thermoforming machine (Courtesy: Brown Machine)

cast, but other materials in widespread use are hard woods, cast epoxy and other thermosets, water-cooled steel, and other cast or machined metals. If handled carefully, plaster molds, often reinforced with fibers, are also acceptable.

Thermoformed molds usually have a draft angle of from 2° to 7°. This amount of taper will give good removal from the mold. Cavity molds can have less taper than this amount because shrinkage is away from the mold.

After the mold has been properly shaped, the vent holes are created. Vent holes should be placed in all low parts of the mold cavity to provide an easy escape for any trapped air. Experience has shown that if the vent holes are backdrilled, as shown in Figure 14.11, the air can be evacuated much more quickly than if the holes are just straight.

The sizes and spacing of the holes is a function of the type of material. The hole dimensions shown in Figure 14.11 are suggested minimums for commonly thermoformed materials. If the material is very soft or very thin, smaller diameters might be needed to avoid having the material flow into the holes and become detectable on the molded part. The spacing of the holes is generally not critical. A sufficient number of holes should be made so that the vacuum can be quickly applied and the part formed without any gaps where the material was not drawn properly against the mold. Where very fast vacuum application is desired, slots rather than holes can be used.

Plant Considerations

Thermoforming machines can vary in size from about 1 square yard (1 m^2) to over 30 square yards (30 m^2), with the large machines being chiefly the automated, sheet-feed type. The machines have few special utility requirements except for the common electrical power and, perhaps, factory high-pressure air. The vacuum system is usually contained within the machine.

Figure 14.11 Vent holes in a thermoforming mold with backdrilled holes shown.

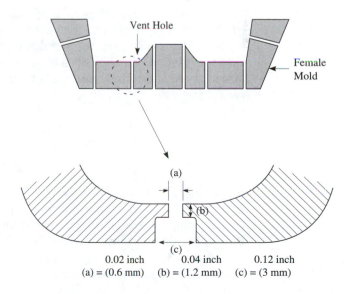

The high output rate of most machines requires that some provision for product movement be made so that the space around the machine can be kept uncluttered. Shearing and finishing space must also be provided unless it is integral to the thermoforming machine.

PRODUCT CONSIDERATIONS

Materials

Thermoforming has become such an important manufacturing method that for many plastics the principal output of extrusion for these plastics is thermoformable sheet. Although almost every thermoplastic has been successfully thermoformed commercially, the most commonly thermoformed plastics are ABS, PMMA, polyolefins, HIPS, and PVC. Common plastics that are not usually thermoformed include acetals and nylons, which melt sharply and therefore are difficult to control in the thermoforming operation. The easiest plastics to use in thermoforming are those with a wide melting point and a high melt strength, allowing the material to be heated and formed without tearing.

Although extrusion sheets are the most common source for thermoplastic starting materials, sheets made by calendering, laminating, casting, and blown films are also used. Reinforced sheets can be used, although thermoforming speeds and the maximum draw for these sheets are usually somewhat lower to keep from tearing the sheets during the forming operations.

PMMA is commonly used for backlit signs and for many other outdoor applications. When a clear and tough material is needed, the most commonly thermoformed plastics are PVC and polyvinylidene chloride.

Shapes and Part Design

Thermoforming makes relatively simple parts that are generally open-top or hollow structures with the opening wider than the rest of the part. An important exception to this rule is the ability to make parts with modest undercuts, that is, with some areas that are wider than the opening. The ability of a plastic material to deform slightly and then return to its former shape allows these undercuts to be made without resorting to sectioned molds. A typical part with such an undercut would be a tray for holding medical instruments. Each cavity in the tray would have a minor undercut so that the part would stay in the tray for shipping. This would be called a *snap fit*. Removal takes a minor force because the plastic must be deformed slightly to spread the top of the cavity apart. Such a design is shown in Figure 14.12.

When large undercuts are required, a sliding mold is used. This type of system is illustrated in Figure 14.13. The part of the mold that slides is called a *slug*. When the part is pulled onto the mold, the slug is in a position to form the undercut. After the part is cool, the slug is moved so that the undercut is removed and the part can be removed from the mold without any interference.

Ribs, bosses, and areas of varying thickness are generally not possible with standard thermoforming techniques. When reinforcing structures are required in a thermoformed

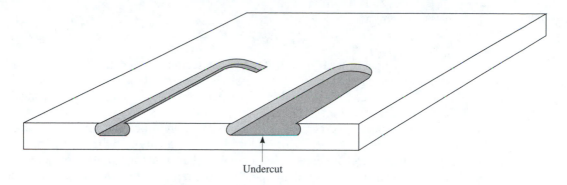

Figure 14.12 Undercut design for a shipping tray that can be made with a single-piece thermoforming mold.

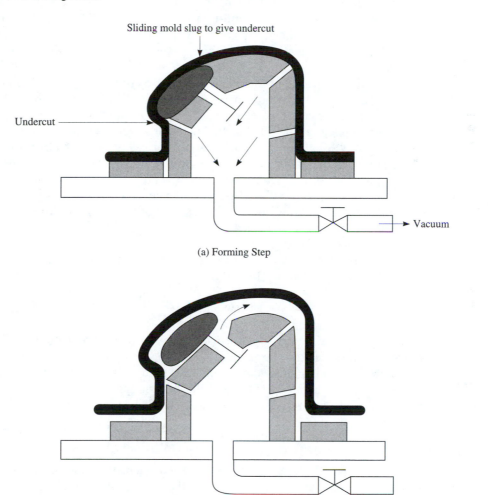

Figure 14.13 Undercut part made by a sliding mold system.

part, the normal procedure is to form a hat section so that the structure can be made without any variation in wall thickness. This general structural technique is frequently used in sheet metal stock used in construction materials such as "corrugated" roofing, siding, etc. A hat section is shown in Figure 14.14.

As with plastic parts made by most common processing methods, sharp angles should be avoided whenever possible in designing a part to be made by thermoforming. Simple shapes are preferred over complex shapes except that very smooth surfaces should be avoided because of the potential to trap air between the part and the surface of the mold. Very smooth surfaces also have a tendency to hold the plastic material, especially polyethylenes.

The practical thickness limit of thermoformed parts is one that can be uniformly heated and formed, which for most materials, is about 1/2 inch (1 cm). Most thermoformed sheets are, however, much thinner than this maximum with 0.08 to 0.16 inch (2 to 4 mm) being typical. Films as thin as 0.02 inch (0.5 mm) are routinely thermoformed.

Inserts can be placed in thermoformed parts by using the thermoforming material to capture the insert. Two arrangements to do this are shown in Figure 14.15. In the first arrangement the insert is placed on a pedestal in the mold. A small indexing pin can hold the insert in place on the mold as the material is thermoformed about it. In the second method of capturing an insert the material is placed on the outside of the mold and the material is then thermoformed over the insert. Care should be taken to locate vent holes so that the plastic is tightly formed around the insert.

When molding, especially with a female mold, care should be taken that the depth to which the part is drawn is not excessive in relationship to the width of the part opening. This relationship has been defined as the *draw ratio* and is an important parameter that is used to estimate the amount of wall thickness variation that might occur. High draw ratios are more likely to result in excessive thinning and wall nonuniformities. The draw ratio is defined in Equation (14.1).

$$\text{Draw Ratio} = \frac{\text{Depth of Part}}{\text{Width of Part}} \tag{14.1}$$

In general, draw ratios for parts made on cavity molds should not exceed 2:1. For plug molds, the draw ratio can be as high as 7:1.

The *area ratio* is another important factor used in designing thermoformed parts. This ratio is defined in Equation (14.2) and allows some approximation of the amount of thinning that will be experienced by the plastic sheet when it is thermoformed.

$$\text{Area Ratio} = \frac{\text{Area of Sheet Before Forming}}{\text{Area of Part After Forming}} \tag{14.2}$$

If a sheet is 30 square inches (200 cm^2) and will be thermoformed into a part that has a total area of 60 square inches (400 cm^2), the area ratio is 1:2. The overall average thickness of the part will therefore be one-half the original thickness. The area ratio is often used to calculate the size of the *blank* (unformed sheet) that must be used to make a particular part. This calculation is important and so a sample calculation is provided. (See Sample Problem 14.1.)

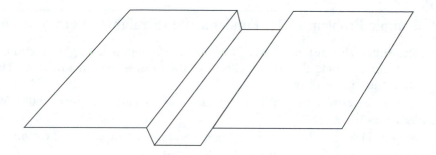

Figure 14.14 Forming a hat section to give structural reinforcement to a thermoformed part.

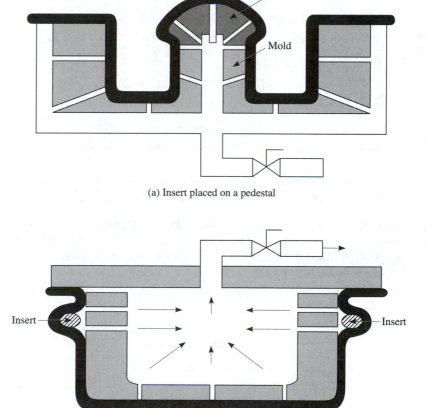

(a) Insert placed on a pedestal

(b) Insert placed around mold

Figure 14.15 Capturing an insert in a thermoformed part.

Sample Problem 14.1   Fundamental (Straight) Thermoforming

Determine dimensions of the blank for fundamental (straight) vacuum thermo-forming assuming that the thickness of the finished part is uniform. (The part is shown in Figure 14.16.)

Step 1. Determine the finished dimensions of the part to be made: $60 \times 50 \times 5$ cm by 2 mm thick.

Step 2. Make an allowance for clamping and mold clearance = 2 cm on each side.

Step 3. Dimensions of blank (with allowance) = 64×54.

Step 4. Calculate the surface area of the part.

$$50 \times 60 \times 1 \text{ side} = 3{,}000 \text{ cm}^2$$
$$50 \times 5 \times 2 \text{ sides} = 500 \text{ cm}^2$$
$$60 \times 5 \times 2 \text{ sides} = 600 \text{ cm}^2$$

$$\text{Total Area of Finished Part} = 4{,}100 \text{ cm}^2$$

Step 5. Calculate thermoformable area of the blank.

$$60 \times 50 = 3{,}000 \text{ cm}^2$$

Step 6. Calculate area ratio.

$$3{,}000 \text{ cm}^2/4{,}100 \text{ cm}^2 = 1{:}0.7317$$

Step 7. Calculate thickness of blank.

$$2 \text{ mm}/0.7317 = 2.73 \text{ mm}$$

Sample Problem 14.2   Reverse Draw Thermoforming

Determine the dimensions of a blank to thermoform a part $60 \times 50 \times 20$ cm and 2 mm thick. Assume that the part is to be made by reverse draw thermoforming and that 10% additional material is needed for clearance between the sheet clamping frame and the mold. Also allow 1.5 cm for clamping the sheet.

Step 1. Determine the dimensions of the blank by allowing for clearance and clamping.

$$(60)\,(0.10) + 3 = 69 \text{ cm wide}$$
$$(50)\,(.010) + 3 = 58 \text{ cm long}$$

Therefore, blank dimensions are 69×58.

Step 2. Calculate the area of the finished part.

$$(60)\,(50)\,(1 \text{ Side}) = 3000 \text{ cm}^2$$
$$(60)\,(20)\,(2 \text{ Sides}) = 2400 \text{ cm}^2$$
$$(50)\,(20)\,(2 \text{ Sides}) = 2000 \text{ cm}^2$$

$$\text{Total Area} = 7400 \text{ cm}^2$$

Step 3. Calculate the amount of area that is available for thickness reduction. Assume that when the part is reverse drawn, the thickness at the top of the bubble is the finished thickness (2 mm). Therefore, the entire area of the bottom is fixed and not available for thinning.

$$\text{Nonthinnable Area} = (60)\,(50) = 3000 \text{ cm}^2$$
$$\text{Thinnable Area} = 7400\ 2\ 3000 = 4400 \text{ cm}^2$$

Step 4. Calculate the area ratio.

$$4400/7400 = 1{:}0.59$$

Step 5. Calculate the required thickness of the blank.

$$2 \text{ mm}/0.59 = 3.36 \text{ mm}$$

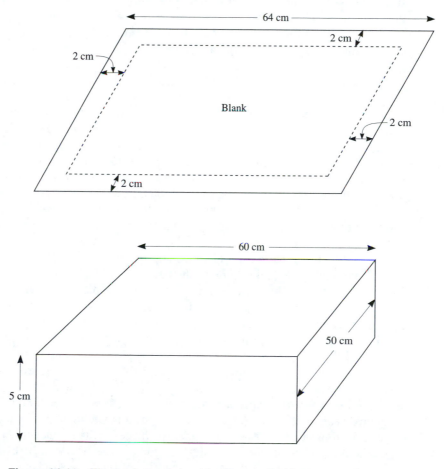

Figure 14.16 Thermoformed sheet for Sample Problem 14.1.

OPERATION AND CONTROL

Critical Operational Parameters

The inherent variability of the formation of the bubble and timing of the forming operation make thermoforming more variable than other plastic-molding operations. Successful control of the thermoforming operation can best be accomplished by standardizing the critical parameters associated with the process. These parameters include: sheet properties, heating conditions, and forming operation.

The most important sheet property to control and standardize is the thickness. Variations in thickness over the sheet should be kept under 5%. Heating of the sheet is normally kept at the minimum amount to shorten the cycle, so if large thickness variations occur in the sheet, some additional heating of the sheet may be required to ensure uniform temperature.

Uniformity from sheet to sheet is also desirable so that the cycle can be kept as constant as possible over an extended time. When sheets are made during the same production run, hopefully one after the other, sheet-to-sheet uniformity is more likely. If sheets are purchased from a distributor of plastic unfinished shapes, the sheets are more likely to come from different production runs or even different production facilities and may have greater variation.

A key property that should be controlled from sheet to sheet is the melt index. If one sheet has a lower melt index than another, the amount of heat to achieve the same formability will be higher in the sheet with the lower melt index. (The molecules in the sheet with the lower melt index are longer.) The melt index can be measured for each sheet, although typically it is measured only when sheets from different runs are likely.

Other variables that might change from sheet to sheet and could affect thermoforming cycles include: density, regrind content, and molecular orientation. These properties are generally not tested for unless some problems are encountered. The normal method of controlling them is to work with the supplier to maintain these parameters within certain limits or to receive a notification of any change in the normal processing conditions that may affect these properties.

Ideally, identical parts from identical sheets would be made in long runs of thermoformed parts. In general, it is more likely that runs will be short and the parts made will be vastly different, sometimes even from different types of plastic. Under these conditions, the best uniformity of operation can be obtained by keeping very close records on each run so that when a part is made again, some parameters for consistent operation will be available.

The key parameters during the forming operation are (1) the speed of vacuum application, (2) the temperature of the mold, (3) the size of the bubble in reverse draw forming, and (4) the size of the plug in plug-assist and reverse draw forming. Some guidance for each of these parameters is as follows:

- **Speed of vacuum application.** In general, the vacuum should be applied as quickly as possible. The valve in the vacuum line should be of the type that one turn will open or close the valve completely. A vacuum surge tank should be provided to allow a large evacuation over a short period of time. The vent holes in the mold should be large enough that the vacuum is not highly restricted.

- **Temperature of the mold.** The mold is normally at room temperature or at some temperature that is below the solidification point of the plastic. Whenever the mold temperature increases, the mold cycle lengthens and the shrinkage increases. Sometimes, especially during high production runs, the mold can be heated by repeated moldings. When this occurs, some cooling should be provided into the mold. This can be done with external fans or cored cooling within the mold.

- **Size of the bubble.** When reverse draft forming, the size of the bubble should not exceed about one-half to three-quarters of the shorter dimension of the clamped sheet. The clearance between the bubble and the female mold should be 2 to 4 inches (6 to 10 cm) to avoid premature touching of the bubble against the mold and subsequent fixing of the thickness at the touching point.

- **Plug size.** The plug size is normally no more than 70% to 85% of the mold cavity and the shape of the plug should generally mirror the shape of the cavity. The areas in the sheet that travel long distances should mate with a raised portion of the plug so that the plug will touch first at those points and fix the thickness. The plug travel speed is balanced against the vacuum force so that the initial movement of the material is controlled by the plug and then subsequent movement, especially close to the end of the plug travel, is controlled by the vacuum.

Troubleshooting

The correction of problems in thermoforming is largely done by inspection of the part with subsequent modifications in the process. The troubleshooting guide in Table 14.2 will assist in understanding the problems behind the changes.

Table 14.2   Troubleshooting Guide for Thermoforming

Problem	Possible Cause
Incomplete forming of part	• Sheet too cool
	• Not enough vacuum/pressure or not fast enough
Blisters or bubbles	• Sheet too hot
	• Excess moisture in sheet
	• Uneven heating
Webbing or bridging	• Mold corners too sharp
	• Not enough vent holes
	• Sheet too hot causing too much material in forming area
Mold release difficult	• Draft insufficient
	• Undercuts
	• Rough mold surface
	• Part has shrunk on mold
	• Part temperature too high

Continued

Table 14.2 *Continued*

Problem	Possible Cause
Warping	• Sheet (all or part) too cool when formed • Poor design • Mold temperature too low
Tearing	• Design exceeds maximum elongation • Sheet too hot • Not enough clearance between mold and plug or bubble • Sheet too cool (especially if very thin)
Excessive shrinkage of part	• Stresses in sheet because it was too cool when molded • Part not cooled enough in mold • Sheet molecular orientation incorrect so rotate sheet with respect to mold
Cracking	• Angles in mold too sharp • Part too cool when molded
Pinholes or mold mark-off	• Vent holes too large • Sheet temperature too high • Vacuum or pressure too high
Blushing	• Sheet too cool • Vacuum application too slow or not high enough vacuum
Sheet scorched	• Outer surface of sheet too hot so reduce heat and lengthen heating time • Heat both sides
Mottled surface	• Entrapped air • Moisture in sheet • Mold surface too shiny

CASE STUDY 14.1

Continuous Thermoforming

An application for thermoforming that is rapidly gaining widespread application in industry is the continuous, automated thermoforming station. A particular application for continuous thermoforming done in a single machine is called *form, fill, and seal.* These systems are characterized by the continuous feed of a roll of film or sheet which is then thermoformed by some high-speed thermoforming system, either vacuum or pressure. The parts are usually filled while on the line and then sealed with a top material. The packages are then moved into a cutting station where they are separated from the scrap (nonformed) part of the sheet. The loose packages are then boxed for delivery or warehousing. This process is illustrated in Figure 14.17.

In form, fill, and seal machines, the film unwind system feeds the sheet material into a gripper system so that movement though the machine will be positively controlled. The heating section, which is shown in Figure 14.17 as separate from the forming station to

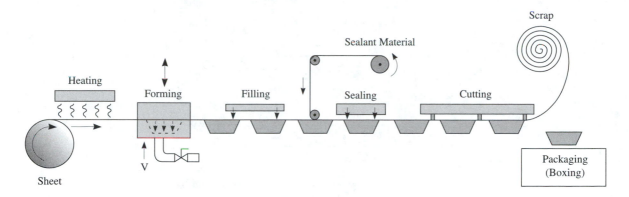

Figure 14.17 Form, fill, and seal automated thermoforming and packaging.

clearly indicate the two functions, can be located within the same enclosure as the forming station. Several heater configurations are possible. The heater can be over the film, under the film, both over and under, or in preheating stations. The placement depends on the nature of the film material, for instance, whether the film has a metallic face.

The thermoforming station is usually a device that sandwiches the sheet between two platens that move together to form the part and then separate to allow passage of the formed sheet. Forming is usually with a vacuum draw into a female mold with pressure assist on the upper side of the sheet. This allows faster cycles, better part definition and formation of more complicated shapes. When very difficult shapes are to be formed, a plug assist can be used. Forming on male molds can also be done.

The formed containers (pockets) are then loaded with the desired parts while moving along the machine. This filling can be done by robots, but is more often done manually. The speed of filling is often the limiting factor in the speed of the machine.

The containers are then sealed by bringing a paper, plastic, or other sealing material over the tops of the containers. This sealing material would have previously been treated with a pressure-activated adhesive. After contact with the filled container, a fluorocarbon-coated heated sealing bar applies both heat and pressure to the seal. In some machines the container can be evacuated and/or filled with a special gas prior to sealing. This is an advantage in applications where perishable materials are packaged.

The containers are separated from the sheet with traditional roll, blade or die cutter machines that trim along the edges. They are then moved into locations that label and box the products.

Other methods of continuous thermoforming are commercially available. In one of these systems, a wheel on which molds are located is used to form the parts after they have been heated. This configuration is shown in Figure 14.18a.

In another configuration, the sheet is extruded directly into the automated thermoforming machine. This method is perhaps the lowest-cost method, although coordination between the various parts of the system must be good. In the system shown in Figure 14.18b, two wheels with matched dies are used to quickly form the parts.

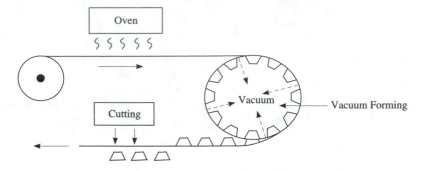

(a) Roll feed with continuous thermoforming

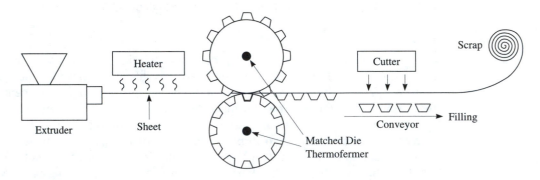

(b) Extruder feed with matched die thermoforming

Figure 14.18   Continuous thermoforming.

The configuration shown in Figure 14.18b has been used to make margarine tubs because of their extremely competitive cost requirement. The total cycle time for the margarine tubs is 2 seconds with a production rate of 75,000 tubs/hour.

SUMMARY

The thermoforming process uses sheet or film plastic as the starting material for the process. This sheet material is heated to a point where it is soft and pliable and is then formed into a shape by placing the material into a mold and then some pressure method to deform the softened sheet into the cavity.

The simplest forming method uses vacuum to draw the sheet into a female mold. This method is called vacuum forming. The vacuum pulls through small vent holes in the sides of the mold. These draw the material tightly against the mold cavity where it cools and takes the shape of the mold cavity. A similar method, called pressure forming, uses air pressure to force the material into a female mold. Wall uniformity is not good in these two methods.

To improve wall uniformity and to enable more difficult shapes to be formed, a mechanical device to force the softened plastic into the mold can be used. This mechanical device is called a plug and the forming system is called plug-assist forming.

Even better wall thickness uniformity can be achieved by forcing the plastic material away from the mold (using air pressure) and then using a plug and vacuum to complete the forming. Some parts are made by forcing (blowing) the material through a ring and allowing the material to cool in the blown position. This method gives excellent optical clarity and is called free forming.

Some parts require that the inside of the part be against the mold for part definition. In these cases a male (plug) mold is used and the forming method is called drape forming. Vacuum can be used to draw the softened plastic against the mold in much the same way as with vacuum forming. A major advantage to plug molds is that wall uniformity is usually better than with cavity molds.

Several other methods for molding have been introduced to meet specific needs. In one of these methods, called snap-back forming, the material is blown away from a male mold, the male mold is then used like a plug to draw and stretch the material, and then a vacuum is applied to "snap" the material against the mold for final shape definition. In another forming method matched plug and cavity dies are used to stamp the softened plastic sheet. This method is called matched die forming. Still another method simply mechanically presses the heated plastic around a mandrel, into a box, or into some other required shape without using a die. This is called mechanical forming.

Thermoforming machines are not as complicated or costly as are the plastics processing machines that melt the plastic material. The essential elements of the machine are a heating chamber or oven, a method for removing the material from the oven, and locating it so that the sheet and the mold can be brought into airtight contact. The vacuum or pressure must be applied and the part cooled against the mold. Machines to accomplish these functions can be single-mold batch systems or continuous, highly automated machines giving very high production throughputs.

The pressures in thermoforming are not high so molds can be made of a variety of materials. Aluminum molds are most common, but wooden, plaster, and steel (especially cored for cooling), are used widely.

Parts made by thermoforming are generally simple hollow shapes in which the opening is wider than the part body. Some undercuts and inserts are possible, but these require some special techniques to accomplish.

One of the most important applications for thermoforming is packaging. High-speed machines called form, fill, and seal machines allow all the critical packaging operations to be done on a single machine at very high production rates. Other high-speed automated machines are becoming more common where production runs are long and the cost of the thermoforming must be kept low. Thermoforming is certainly a rapidly growing plastic-manufacturing method.

GLOSSARY

Area ratio   The ratio of the area of the sheet before forming to the area of the sheet after forming.

Billow forming   Another name for reverse draw forming.

Blank   An unformed sheet.

Blister   The formed cavity in a blister pack.

Blister pack   A packaging style in which a thermoformed cavity is used to contain the part.

Cavity mold   A female mold.

Drape forming   Thermoforming using a male (plug) mold.

Draw ratio   The ratio of the depth of the part to the width of the part.

Female mold   A mold in which the part is pressed into a cavity.

Form, fill, seal machines   Thermoforming machines that comprise an entire product-filling production line in which the plastic sheet is thermoformed, filled with product, and then sealed (usually with a cardboard or paper backing).

Free forming   A thermoforming technique in which the forming is done only by air pressure, without a mold.

Fundamental vacuum forming   The simplest vacuum forming method in which only a vacuum is used to force the hot plastic against the mold surface.

Invert forming   Another name for reverse draw forming.

Matched die forming   A thermoforming technique in which two molds press against opposite sides of a heated sheet to form it.

Mated forming   Another name for matched die forming.

Mechanical forming   A technique in which a heated sheet is formed using only mechanical pressure.

Plug-assist forming   A thermoforming technique in which a plug or shaped pressure device is used to partially form the sheet prior to full pressure against the mold.

Pillow forming   Another name for reverse draw forming.

Pressure forming   A thermoforming technique in which positive pressure (rather than vacuum) is used to force the plastic against the mold walls.

Reverse draw forming   A thermoforming technique that uses a blowing step to stretch the heated plastic prior to its being forced against the mold.

Sag point   The condition (temperature, time, thickness, etc.) at which a sheet begins to sag inside the thermoforming oven.

Skin pack   A packaging style in which a sheet is thermoformed against a part resting on a perforated card through which the vacuum is supplied to the sheet, thus encapsulating the part against the cart.

Slug   A part of a thermoforming mold that can be moved to accommodate an undercut.

Snap-back forming   A thermoforming technique in which the hot sheet is first stretched and then partially formed with a plug and then finished formed by a vacuum against a male mold.

Snap fit   A thermoformed tray in which the thermoformed cavity has an undercut and thus retains parts that are inserted into the cavity.

Straight vacuum molding   Another name for fundamental vacuum forming or, simply, vacuum forming.

Thermoforming   A process used to shape thermoplastic sheets in which the sheet is heated and then forced into a mold.

Vacuum forming   A thermoforming process in which vacuum is used to force the plastic material against the walls of the mold.

QUESTIONS

1. Discuss the pressure capabilities of vacuum forming relative to pressure forming.
2. List three considerations that must be taken into account in the placement, number, and size of the vent holes in a female mold used for fundamental (straight) vacuum forming.
3. What is backdrilling of a vent hole and why is it done?
4. A thermoformed part has a total surface area of 1000 cm^2 and a required thickness of 1 mm. It is formed from a blank sheet with a total usable surface area of 500 cm^2 (not counting the part of the sheet used for clamping) and a thickness of 2 mm. When forming the part, the walls often tear. Indicate what might be the problem.
5. Discuss the differences between blister packaging and skin packaging.
6. Why is the thickness of a section fixed when it touches a solid object such as the walls of the cavity or the plug?
7. A deep cup (20 cm deep $\times$ 10 cm diameter) is to be formed. Compare thermoforming, blow molding, and injection molding as processes for making the cup. What technical and economic considerations should be considered in determining which method is best?
8. Give two possible causes for the thermoformed part to stick to the mold.
9. What is the purpose of a vacuum surge tank?
10. Calculate the size and thickness of a blank needed to make a part that is $100 \times 50 \times 10$ cm with a finished part thickness of 3 mm. Assume 2 cm per side are required for clamping.

REFERENCES

Baird, Ronald J., and David T. Baird, *Industrial Plastics*. South Holland, IL: The Goodheart-Willcox Company, Inc., 1986.

Frados, Joel, *Plastics Engineering Handbook* (4th ed.), Florence, KY: International Thomson Publishing, 1994.

Milby, Robert V., *Plastics Technology,* New York: McGraw-Hill Book Company, 1973.

Morton-Jones, D. H., *Polymer Processing,* London: Chapman and Hall Ltd, 1989.

Richardson, Terry L., *Industrial Plastics* (2nd ed.), Albany, NY: Delmar Publishers Inc., 1989

CHAPTER FIFTEEN

ROTATIONAL MOLDING PROCESS

CHAPTER OVERVIEW

This chapter examines the following concepts:

- Process overview
- Equipment (machines, molds, plant concepts)
- Product considerations (materials, shapes, design)
- Operation and control (critical parameters, troubleshooting, maintenance and safety)

INTRODUCTION

The *rotational molding* process, often called *rotomolding,* uses the rotation of a mold in a heated chamber to form the part. Rotational molding is ideally suited to the formation of very large, seamless, and stress-free hollow parts, such as 25,000-gallon (100,000-L) agricultural tanks, although the process has also been used to make parts as small as ping-pong balls. The process uses no pressure and so the molds can be simple and inexpensive. Most processes that use no pressure are called casting or cast molding. In that sense, rotational molding is a type of casting and is sometimes called rotational casting. Because of the growing importance of the rotational molding process, it has been separated from the casting chapter in this text. The temperatures used in rotational molding can be lower than in many of the plastics molding processes because the plastic is never fully melted. Rotational molding can be used to make parts that are functionally competitive with other processes, such as injection molding and blow molding, and offers some unique processing advantages that, in some cases, give capabilities not possible with the other processes. Likewise, rotational molding has some limitations that must be accounted for in making parts. These advantages and disadvantages will be discussed in detail after the rotational molding process itself is explained.

Rotational Molding Process

The rotational molding process has four principles steps. These are (1) loading, (2) heating, (3) cooling, and (4) unloading. Several different activities and phenomena occur during each of these steps. A common type of rotational molding machine employs a carousel to move the molds through the four steps of the rotational molding cycle. This situation is depicted in Figure 15.1.

The *loading step* begins with the careful weighing of a charge of the starting material. The starting material for the rotomolding process is usually a finely ground thermoplastic powder. Other starting materials, which are not as common, could include dispersions of thermoplastic powders in nonvolatile solvents and even some thermoset resins. The amount of the charge of material is the desired weight of the finished part.

The charge of starting material is loaded, usually manually, into an open, cold mold that should have been prepared by coating the inside with a mold release agent. (The mold release need not be reapplied after each molding cycle because some residual mold release from previous cycles may still be present.) After loading, the mold is closed, usually manually, by simply clamping the lid and body of the mold together. The mold is then moved into an oven where it can be heated.

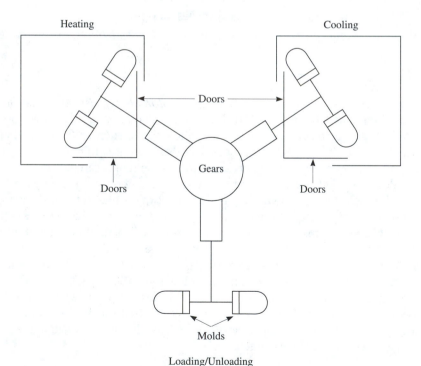

Figure 15.1   Rotational molding equipment in a carousel arrangement showing loading/unloading, heating, and cooling stations.

The *heating cycle* is done in an oven. The heating oven must be large enough that the entire mold assembly can be placed inside it and rotated freely. The mold assembly is rotated continuously throughout the heating and the cooling cycles.

The heating cycle for rotational molding is often quite long and so it is common to have multiple molds that are cycled together. Two molds are attached to each of the arms in the carousel machine depicted in Figure 15.1.

During the heating cycle the rotation of the mold assembly causes the plastic powder to tumble inside the mold. The plastic powder eventually becomes hot and tacky and begins to stick to the inside of the mold. With additional heating and tumbling, all the plastic eventually sticks to the walls of the mold, soon joining together into a continuous mass in a process called *fusion*. Fusion occurs because the plastic particles are in a lower energy state when the number of surfaces is reduced. Obviously, the smallest number of surfaces possible is when all the plastic particles have become one solid mass without interior surfaces. Therefore, the preferred energy state occurs when the particles join together or fuse. Note that fusion is a process that joins solid materials rather than materials that change to a liquid (molten) state and then resolidify. The particles in the rotational molding process are never actually melted. This process of coating the mold and then fusing the plastic particles is depicted in Figure 15.2.

The heating cycle is the longest part of the rotational molding process and is, therefore, the rate-determining step. The process of heating the resin and fusing the particles normally takes 7 to 15 minutes, but 30-minute cycles are not uncommon. Large

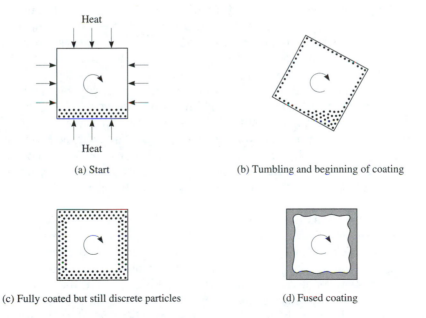

(a) Start

(b) Tumbling and beginning of coating

(c) Fully coated but still discrete particles

(d) Fused coating

Figure 15.2   The heat cycle of rotational molding showing tumbling inside a heated mold, coating of the mold walls, and fusion.

and especially thick parts may require even longer cycles. The cycle increases dramatically with thickness because the buildup of plastic on the walls of the mold acts as an insulating barrier separating the external heating from the resin that is still tumbling in the interior of the mold. Some resins are crosslinked during the fusion step extending the heating cycle even further.

To obtain an even coating of the walls of the mold, the mold must be rotated about more than one axis. Experience has shown that an even coating can be obtained if the rotation is *biaxial* that is, about two axes, provided that certain constraints are met. The first constraint is that the rotation about the two axes must be done simultaneously. The second constraint is that the rotational speeds must be different. For best uniformity of coating, the rotational speed about the major axis (the axis running parallel to the longest dimension of the part) should be faster than the rotation about the minor axis (perpendicular to the major axis and running parallel to the smaller dimension of the part). A third constraint is that these rotational speeds should not be whole-number multiples of each other. For instance, rotational speeds about the two axes of 22 rpm and 5 rpm would be acceptable, but speeds of 25 rpm and 5 rpm would be less desirable. This relationship of not being integer multiples of each other gives a randomness to the rotation of the mold which results in more even coating. A common ratio of rotational speeds is 3.75:1. A fourth constraint is that the rotational speeds are slow, usually no more than 60 rpm. At these slow rotational speeds, the particles tumble inside the mold without any centrifugal or spinning forces. The only external force desired is gravity. Biaxial mold rotation is shown in Figure 15.3.

The heating of the air inside the mold will cause the air to expand. To avoid any buildup of internal pressure that might distort the mold, a vent is usually provided. This vent can be as simple as a small pipe that runs from the inside of the mold cavity through the wall of the mold. To prevent resin from entering the vent as it rotates, a small amount of fiberglass can be placed in the end that is inside the mold.

The presence of this vent pipe means that the part will have a small hole in it where the pipe penetrates the mold wall. In many parts this small hole can be placed so that the function of the part is not affected. In other cases, the vent hole can be placed in a portion of the molded part that is removed before use. Such an example is shown in Figure 15.4. In this example the part is a trash container with a lid. A narrow region is molded between the lid and the container. As shown in the figure this narrow region, which contains the vent hole, is removed after molding. The removal of this region separates the lid from the body and, because of the molding shape, creates a lip on the container over which the lid can seal.

Many small parts and spherical parts need not be vented. In these parts the structure of the part resists the distortions that may be caused by the internal pressures.

The *cooling step* occurs after the plastic material has been fully fused. Without stopping the rotation, the mold is transferred to a cooling station where it is cooled as quickly as possible without causing the part to shrink away from the mold. If the part shrinks away from the mold, shape distortion will occur when the part has cooled. The cooling rate must, therefore, be controlled. This control is achieved by choosing the type of cooling medium. Fastest cooling is achieved by spraying cold water over the surface of the mold. Slightly slower cooling occurs when a mist of water is used and even slower cool-

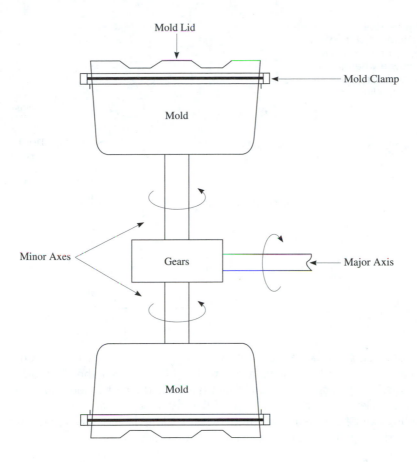

Figure 15.3 Biaxial rotation.

ing can be obtained by just blowing cool air on the mold. Sometimes the mold is cooled with air and then, later in the cycle, is completely cooled with a water spray.

Some molders have used a small internal pressure of 3 to 4 psi, (20–30 KPa) to help prevent the pulling away of the part from the wall of the mold. Before this technique is used, however, a calculation should be made to ensure that the mold can withstand this pressure without deformation.

After cooling, the part is transferred out of the cooling chamber to an unloading station. (Note that if a molding system such as the carousel is used, the mold cannot be transferred to the unloading station until the rate-determining step, normally the heat cycle, is complete.)

The *unloading* step is begun by removing the lid of the mold. The cold part is removed manually or, if needed, with a mechanical assist of some kind. The length of the heating portion of the cycle usually allows time for trimming the part, loading the resin charge for the next part, and closing the mold; all done by a single operator.

Figure 15.4
Placement of the vent hole so that it can be removed when the molded part is trimmed to form a lid.

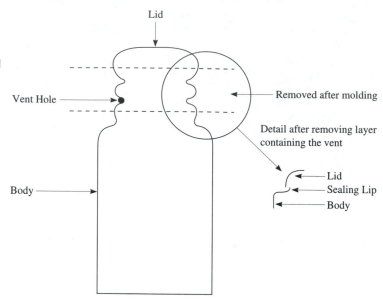

Lid

Vent Hole

Removed after molding

Detail after removing layer containing the vent

Lid
Sealing Lip
Body

Body

Unloading can be very difficult with some parts. Mold release is usually sufficient to allow removal, but occasionally the parts will stick even though mold release is used. When sticking happens, air can be blown around the part to loosen it. Another technique used to release a stuck part is to cool the inside of the part and cause it to shrink. This cooling can be done with cold air or, when sticking is very severe, dry ice can be put into the part or applied to the area where sticking is occurring.

Advantages and Disadvantages of Rotational Molding

The advantages and disadvantages of rotational molding, compared to other plastic-molding processes, are summarized in Table 15.1. One major advantage of rotational molding when compared to other plastic-molding processes is the ability to make very large parts with relatively low-cost equipment. This advantage arises from the low pressures involved in rotational molding and the relatively low heat required because the material is not fully melted. The size of parts made by rotational molding is limited only by the size of the oven and cooling chambers and the mechanical capability of rotating the molds.

The shapes allowable in rotomolding are more limited than those permitted for injection molding but are of about the same complexity as the shapes that can be made with blow molding or thermoforming. Blow molding and thermoforming are especially competitive with rotomolding for smaller parts with the advantage of faster cycles. A unique characteristic of rotomolding is that the resin tends to be slightly thicker in the outside corners of a rotomolded part rather than thinner as with blow molding and thermoforming.

The ability to make stress-free parts is also an advantage of rotomolding. Because of the low stresses, brittleness is reduced and impact toughness can be improved. However,

Table 15.1   Advantages and Disadvantages of the Rotational Molding Process Versus Other Plastics Molding Processes

Advantages	Disadvantages
Low pressures	Simple shapes only
Low-cost molds	Poor dimensional tolerance control
Thicker corners	Generally thicker overall walls
Very large parts possible	Slow molding cycles
Low equipment cost	Low part mechanical properties
Stress-free parts	Limited set of usable resins
Easy color and resin changes	
Easy mold changes	

many factors in rotomolding can eliminate these advantages, especially where impact toughness is concerned. For instance, if the plastic resin is not fully fused, the impact toughness decreases rapidly. Also, the ability of a resin to fuse easily is enhanced by using resins with low molecular weights but low molecular weights decrease impact toughness. Therefore, impact toughness can be either improved or decreased by rotomolding in comparison with other molding processes.

If parts of equivalent size are to be made, the investment costs of rotational molding are somewhat less than injection molding or blow molding and about the same as thermoforming. The low cost of the equipment and the ease of making the mold enable the use of rotomolding as a method of manufacturing prototypes, provided, of course, that the part is hollow and fits the other criteria for rotomolding.

Rotomolding allows color changes with every part because the charge is made separately with each part and the mold is clean after each molding. Mold changes are likewise easy to make and can be done rapidly. These easy changes give rotomolding an advantage when production runs are short.

EQUIPMENT

Machine Types

The most common commercial equipment for rotational molding is the three-station carousel type that is shown in Figure 15.1. Provided that the times required for the three stations (heating, cooling, loading/unloading) are not too different, the carousel method can be very efficient in the use of equipment.

A system that is mechanically simpler than the carousel is the *batch rotational molding system*. In this system the mold is loaded in one location and then manually moved to the heating station and then, later, to the cooling station. The mechanical drive mechanisms for rotating the mold are fixed at each station and the mold is simply engaged into a mating system to achieve a mechanical linkage. To facilitate movement between stations, a track or conveyor system is often employed. The batch system is less expensive and less prone to mechanical breakdown than the carousel system. A system of the batch type is shown in Figure 15.5.

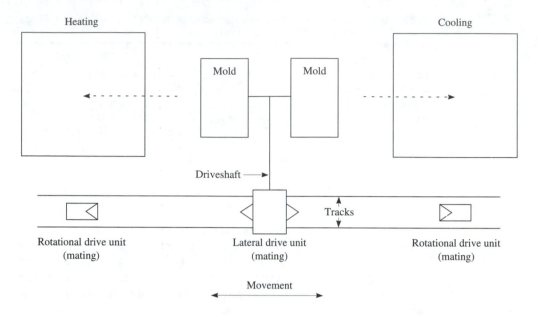

Figure 15.5   Batch molding system for rotational molding.

A variation on the batch system is the *straight-line rotational molding system,* which can be envisioned by analyzing Figure 15.5 and making a few changes. In the straight-line system the drive units at each station and the gear mating box are eliminated. In their place a single drive unit is attached to the drive shaft and mounted on a carriage that moves with the molds. The entire drive unit therefore shuttles back and forth along the tracks with the mold assembly. The straight-line system is especially advantageous for very large parts.

Rather than moving the mold assembly from station to station, another approach is to jacket the mold so that heating fluids and, alternatively, cooling fluids can be circulated around the molds. This system is called *jacketed rotational molding.* The heating fluid is usually oil, but some molten salts can be used when the improved heating capability of the salts is needed to shorten the heating cycle. The cooling fluid is usually water. The higher cost of the molds and the greater problems with safety and convenience in handling the molds generally limit the use of jacketed molding to resins that are very thermal-sensitive and need the greater precision in temperature control that can be obtained with jacketing. However, the improved mechanical reliability and smaller space occupied by a jacketed rotational molding machine are increasing its use. Single-station systems where a box serves for both heating and cooling are much lower in cost than other systems and are also gaining acceptance.

In the systems that use separate heating and cooling stations, the heating station is simply a large oven. The heat is usually supplied by a gas-fired, forced-air system. While this type of system is economical to run, the thermal uniformity within the oven is often poor. Thermal uniformity can be improved by using multiple heat inlet points and diffusers at the inlets to break up the air flow. Thermal uniformity can affect the uniformity of wall thickness in the part because the material will preferentially stick to any surface that is slightly hotter than the surrounding surfaces. Even though the molds are rotating in the oven, some preferred orientations are difficult to eliminate and thus thermal uniformity within the oven is important.

Molds

The most common mold material is cast aluminum. Other materials used are sheet metal, machined aluminum, steel (especially when many parts will be made and for jacketed molding), and copper or a copper alloy when high heat transfer is needed for very complex shapes.

Most molds are two-part molds with a body and a lid. The lid can simply be clamped in place because internal pressures are low. The surface of the part will replicate the inside surface of the mold so if a shiny outer part surface is desired, a polished inner mold surface is required. The inner surface of the mold is usually coated with a spray-on mold release, although permanent coatings for mold release (usually fluorocarbon plastics) can also be used.

Molds usually have a 4 to 6° draft angle to permit easy removal of the part. When the inside surface of the part is very smooth or shiny, removal is more difficult and a slightly larger draft angle may be required. The draft angle may also change depending upon the material.

Multiple molds are common on one shaft so that the number of parts per cycle can be increased. The molds are usually the same size, but that is not necessary provided that the cycles are not too different for the different parts. The principal limitation on the number of molds that can be used is the mechanical ability of the system to rotate with the increased weight and the obvious size requirement to fit within the heating and cooling chambers.

Plant Considerations

Rotational molding machines can be very large because of the size of the parts that are made. The plant may need to have a high ceiling and be quite spacious to accommodate large oven and cooling chambers. See Photo 15.1 for an example of a commercial rotational molding machine.

Special auxiliary services are not required except for the natural gas to feed the oven burner and the chilled water for cooling.

PRODUCT CONSIDERATIONS

Materials

Several factors limit the thermoplastic resins that can reasonably be rotomolded. Some of these factors are chemical, that is, related to the molecular structure of the resin, whereas others are more mechanical. The overwhelming preponderance of polyethylene as the

Photo 15.1   Production-type rotational molding machine (Courtesy: Ferry Industries, Inc.)

resin of choice in rotational molding (perhaps 90%) signals the difficulties that are encountered in finding usable resins that meet both performance and economic limitations. A good resin for rotomolding can be defined by the following parameters:

- **Thermal stability.** As has been previously discussed, PVC is highly heat sensitive and can be rotationally molded only with great care. The problem is not so much the high temperature as it is the long cycle time at a relatively high temperature. Another resin that is difficult to rotomold for this same reason is ABS. The butadiene component has a strong tendency to degrade at the long heat cycle times required in most rotomolding operations. However, recent work has shown that certain types of ABS polymerizations result in polymer particles in which the butadiene is less susceptible to degradation, and these resins can be successfully rotationally molded.

- **Excessively high fusion temperature.** The high fusion temperature of many plastics limits their use as rotomolding resins. Plastics such as the advanced engineering thermoplastics have glass transition temperatures (a reasonable estimate of the temperature at which fusion takes place) that are so high that effective fusion cannot occur within the temperature capabilities of most commercial rotomolding ovens.

- **Grindability.** The grindability of a material means that the resin can be ground to a fine powder. Resin grades that have very low melting points may not be easy to grind in the high-speed impact mills that have proven to give the most consistently good powder because they melt. In some cases a low melting resin can be ground under an atmosphere of liquid nitrogen or some other cooling method so that the material will not become too hot during the grinding operation.

- **Particle distribution.** The distribution of powder particle sizes should be narrow. The presence of very small particles with large particles means that the heat absorption of the particles is quite different and could lead to nonuniform coating of the mold.

- **Mesh size.** The *mesh size* is a measure of the size of the screen mesh through which 95% of the particles will pass. The meshes are defined by standards adopted by each country. In the United States, the meshes are called United States Standard Sieve Sizes. While particles are still characterized widely by mesh size, the current trend is to characterize them by the size of the opening in millimeters. A mesh size of 35 would have an opening of 0.5 mm. The common mesh sizes for rotomolding powders are 16 to 50, which correspond to hole openings of 1.19 to 0.297 mm, with the higher mesh number corresponding to the smaller hole size.

- **Pourability.** To tumble properly in the mold, the material plastic powder must flow easily without any external pressure (other than gravity). This property is called *pourability.* The pourability is measured by simply noting the time for a given weight of the powder to flow through a standard funnel. When the funnel that is defined by ASTM Test Procedure 1895–69 is used, a minimum flow rate of 185 grams/minute characterizes acceptable rotational molding powders.

- **Bulk density.** The *bulk density* is a measure of the density of the powder before it is heated or compacted. In other words, bulk density is the density of the material just after it is ground. For best rotational molding, the higher the bulk density the better because the particles are able to naturally pack tightly together. This tight packing promotes the fusion of the particles and minimizes the amount of air space between the particles. Some grinding systems result in long, hairlike tails on the particles. These tails are undesirable and lower the bulk density because the particles are prohibited from packing. The long tails also worsen the pourability of the powder.

- **Fusability.** The particles must fuse together easily during the heating cycle. If the molecular weight of the plastic is too high (melt index too low) the particles will require too much energy to move together, which occurs when they fuse, and high temperatures will be required. If the temperatures are too high, degradation could occur. If the molecular weight is too low, the material will melt rather than fuse. If melting occurs, the material will puddle in the mold and the walls will not be uniformly coated. Therefore, for each resin, a range of molecular weights must be maintained. This range is usually expressed as a range of melt index numbers. For HDPE the normal melt index range is 3 to 70. Within that range, resins with high melt index numbers are chosen when the part is very complex and good flow into complicated areas is required. Resins with low melt index numbers are chosen when improved stress crack resistance, impact toughness, or creep resistance is needed. From a physical performance viewpoint, the low–melt index resins would always be preferred, but they are difficult to mold and cannot be used for some parts.

Crosslinked Parts

A method for solving the dilemma of needing low melt index for properties and high melt index for processing is to mold with a high–melt index resin so that the part can be well defined and fusion occurs readily and then to increase the molecular weight by chemically crosslinking the resin. The resins normally used in rotational molding (such as polyethylene) are not normally crosslinkable by chemical methods, but under certain circumstances they can be crosslinked. These circumstances involve the addition of a peroxide initiator, a crosslinking agent, heat, and time. The crosslinking reactions are shown in Figure 15.6.

The peroxide initiator becomes activated at elevated temperatures. (Examples of such a peroxide would be benzoyl peroxide and lauroyl peroxide.) When these materials are

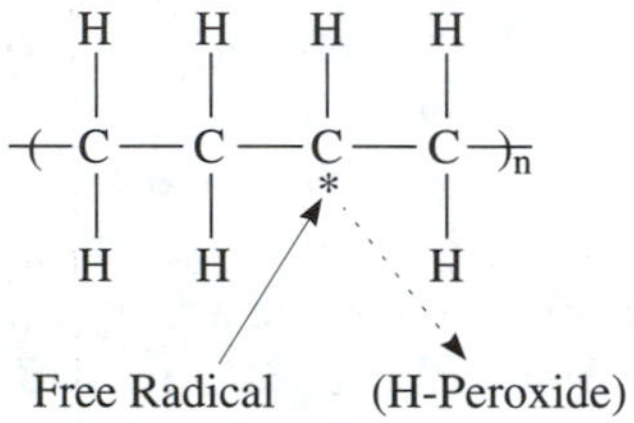

(a) Formation of a free radical by extraction of a hydrogen atom from Polyethylene (PE) by a peroxide

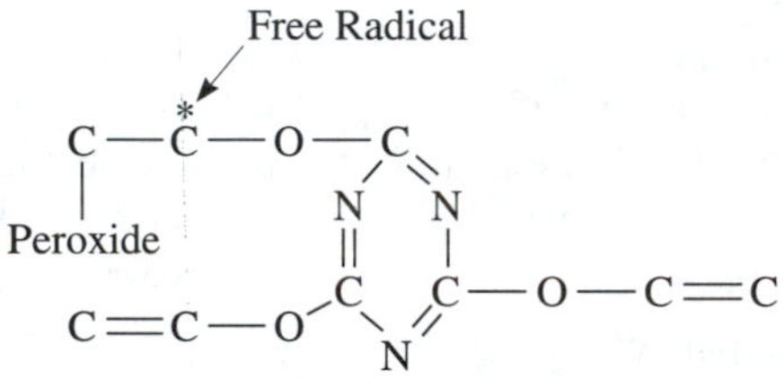

(b) Formation of a free radical by reaction of a peroxide with the carbon-carbon double bond in the tri-allycyanurate (TAC) crosslinking agent

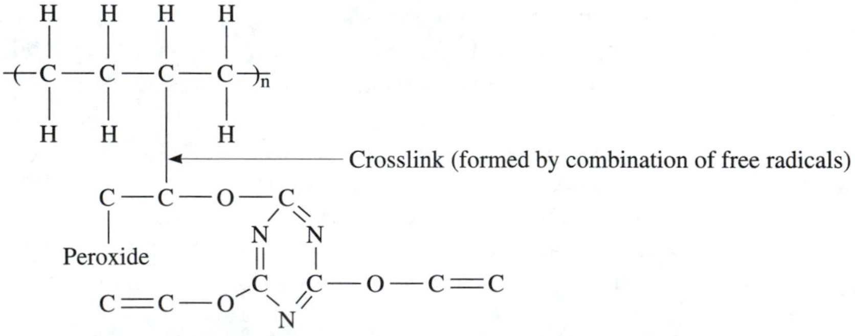

(c) Reaction of the free radicals on the PE and TAC to form a crosslink

Figure 15.6   Crosslinking of polyethylene using a peroxide and a crosslinking agent tri-allyl cyanurate (TAC).

heated, they break apart and form two free radical molecules. A free radical can extract a hydrogen atom off the polyethylene backbone. The hydrogen atoms have also become somewhat activated by the high temperature, thus facilitating their removal. When a hydrogen atom is extracted from the chain, a free radical is created on the carbon that was formerly bonded to the hydrogen. This free radical is very reactive and will combine readily with other free radicals in the vicinity. This step is shown in Figure 15.6a.

The peroxide initiator, when heated, also reacts with the crosslinking agent. The most common of these crosslinking agents used in rotational molding is tri-allyl cyanurate (TAC) which is characterized by three branches of atoms each of which has a carbon-carbon double bond. This step is shown in Figure 15.6b.

The free radical on the polyethylene and the free radical on the TAC will readily react to form a bond. This joins the TAC with the polyethylene. The TAC molecules can make similar reactions on the other carbon-carbon bonds with other polyethylene molecules that are similarly activated by the peroxide initiators. The TAC molecule then becomes the bridge (crosslink) that connects several polyethylene molecules. When these polyethylene molecules are linked together, the molecular weight is the sum of the molecular weights of the combined molecules. Hence, a rapid increase in molecular weight is achieved with a corresponding increase in physical properties, especially impact toughness, stress crack resistance, and creep resistance.

The time required to effect this hydrogen atom extraction and subsequent crosslinking reaction throughout the molecule is quite long. This extended time is necessary because the thermoplastic material is a soft solid. (The more rapidly crosslinked thermoset materials are usually liquids.) In a solid, the movement required to bring the free radicals into close proximity so that they can bond occurs slowly. The formation of a crosslinked thermoplastic material will increase the heat cycle time by roughly 50%. This added cycle time is the major disadvantage to using crosslinked thermoplastics made by rotomolding. In some applications, however, the added value justifies the cost.

Some molders have achieved improved properties by changing the type of polyethylene resin rather than by crosslinking. This change is usually to a linear low-density polyethylene (LLDPE) or, perhaps, a linear medium-density polyethylene (LMDPE), which is similar in properties. The toughness, stress crack resistance and, perhaps, the creep resistance are improved over most low- and medium-density rotomolding resins but are typically not as high as the fully crosslinked material. However, for many applications the improvement in properties obtained with the linear resins is sufficient for the needs of the application and the cost advantage of not crosslinking can be significant.

When crosslinked, the crystalline regions become disrupted and the molecules spread apart slightly (less crystallinity) and the density of the material is slightly less than it was before crosslinking. This decrease in density has only a minor effect on stiffness, solvent sensitivity, and other density-related properties.

The success of chemical crosslinking of thermoplastic resins during rotational molding has led some processors to investigate the use of powdered thermoset resins in rotational molding. They have reported some success and cite faster cure cycles as an advantage over crosslinking of thermoplastics. Another advantage for some thermosets could be a nonheated cure. At this time, few commercial applications have been reported using traditional thermoset resins in rotomolding.

Mixed Resins

A technique that is not widely used but has great promise for certain applications is the use of mixed resins. These resins must be compatible and bond tightly together. The resins are also characterized by two different melting points. The different melting points in the resins almost always mean that the two resins will have different *softening temperatures,* which are important in rotational molding since the softening temperatures are the temperatures at which the resins fuse and stick to the walls of the mold. The softening temperatures are lower than the melting points and usually higher than the glass transition temperature, but the offset between the melting points and the softening temperatures differ among resins depending upon several resin characteristics such as crystallinity, extent of interpolymeric interactions, molecular weight, and molecular weight distribution.

When mixed together and heated, the resin with the lower softening point will coat the inside of the mold first, forming a skin. The second resin will then soften as the temperature increases and will stick to the first material forming an inner layer. This technique allows parts to be made with a tough outer shell and a stiffer inner body. Special solvent or environmental resistance can also be achieved on either the inside or outside as the application may require.

Additives

Some additives have a strongly beneficial effect on the properties of rotomolded parts. For instance, antioxidants are routinely added to rotomolding resins to retard the degradation that inevitably occurs to some extent during the heating cycle. These antioxidants prevent this degradation and therefore maintain the molecular weight of the material. The effect of the antioxidant is highly dependent on the type of resin and the type of antioxidant. For some resins, the addition of one type of antioxidant can increase the impact toughness by over 10 times while another antioxidant may increase the impact toughness by only 20%. The best procedure is to investigate thoroughly the resin and antioxidant combinations before commercializing a product.

Other additives can significantly decrease the properties of a rotomolded part. An example of this would be a filler that is present in a high enough concentration to inhibit the fusion of the resin molecules. This inhibition is especially detrimental with fillers that are poorly dispersed in the resin or that tend to form aggregates. The concentrations at which these fillers become significant depends upon the nature of the filler and the resin, but some fillers, such as color pigments, can be detrimental at less than 5% concentration.

Shape and Design

While all rotomolded parts must be hollow, some very different shapes can be achieved. Typical shapes would include agricultural tanks, trash carts, buoys, and highway safety barriers. The ability to make large molds of reasonable complexity has allowed rotomolding to be used for many large toys, especially the climb-in or climb-on variety. Riding

Figure 15.7   A rotomolded infant seat showing the use of a double-wall construction.

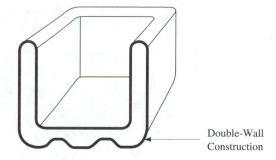

Double-Wall Construction

horses, slippery slides, animal-shaped toy boxes, and playhouses are just some common examples of these types of toys.

When relatively stiff resins such as polycarbonate or nylon are used, rotomolded parts can compete well with fiberglass-reinforced parts such as truck cabs and snowmobile housings. These applications are growing rapidly as the need for increased toughness that can be obtained with thermoplastic materials becomes more evident.

Rotomolding has become an important molding method for parts that are not as big as those just described and have much more complicated shapes. While still hollow, it is the ability to shape the outside of the part that is most important. A typical example of such an application is shown in Figure 15.7. The part shown is an infant seat with a double-wall construction. The seat is rotomolded in a mold that has just enough volume inside to allow the resin particles to tumble freely. Other common parts made by this method include truck-bed liners (Photo 15.2), instrument housings, and carrying cases.

While great flexibility of design is possible with rotational molding, some important limitations should be noted. Large, flat sections on parts should be avoided. When the part has large, flat sections, ribs or domes should be used to give structural reinforcement to those areas because of their tendency to warp after removal from the mold.

Wall thickness variation is usually better with rotationally molded parts than with blow molded or thermoformed, but not as good as with injection molded parts. With rotomolding, wall thickness variations of no more than ±10% are common, provided the process is in control. The major factors that would contribute to variations in wall thickness beyond these limits would be thermal variations in the oven and cooling chamber and problems with the rotational speeds about the two axes.

The thickness of a part is limited by the ability of the resin to transmit heat. Most rotationally molded parts are between 0.3 and 1.2 inches (7.5 to 30 mm) thick. Increases in thickness can have a major effect on the cycle. For instance, an increase of 0.3 inch (7.5 mm) in thickness for a part made of nylon (6) will require an increase in heat cycle time of 2 minutes. If the heating time becomes too long, the parts can begin to degrade. Therefore, the limit on thickness is dependent on the thermal sensitivity of the resin.

Parts with undercuts are difficult to rotomold because of the problem of removal from the mold. If an undercut must be made, a sliding mold is the normal method of achieving it. The obvious problems with such a mold are the added cost of the mold, the potential for mechanical problems with the sliding mechanism (especially since rotational molds are manually operated), and the potential for resin to bind the sliding mechanism.

Photo 15.2 Large rotationally molded part with double walls. (Courtesy: Pawnee Rotational Molding Company, L.P.)

Outside threads can be molded into rotationally molded parts but care should be taken to avoid threads with a sharp "V" shape. The resin particles do not readily flow into the bottom of the "V" and bridging is likely. A much better shape for the threads would be a coarse thread with a rounded bottom.

Metal inserts can be molded into a part with little difficulty. The insert is set into the wall of the mold so that it protrudes into the mold cavity. During the heat cycle, the resin flows around the metal insert and captures it. The only difficulty is removal of the part. Some provision must be made for the insert to slide out.

OPERATION AND CONTROL OF THE PROCESS

Several operational and control parameters have already been identified as the particular components of the process were discussed. For instance, thermal uniformity in the heating and cooling chambers is important for thickness control and part physical properties. The complexity of these variables may not, however, be appreciated. The heating cycle is dependent upon (1) the nature of the resin (thermal conductivity, particle size), (2) the characteristics of the part (shape, wall thickness), (3) the number and nature of the molds (material, wall thickness), (4) the oven characteristics (size, burner efficiency, air circulation), and (5) deviations from normal procedures such as changes in normal resin characteristics within accepted norms, environmental temperatures, and the weight of the charge.

Occasionally these variables can have a significant effect on the quality of the part. For instance, if the resin charge weight is less than it should be, the wall thickness will be reduced (which can affect the part physical properties). Furthermore, the reduction in wall thickness can mean that the heating cycle could be too long for the material and degradation could occur. Hence, control over process parameters is very important.

The cooling cycle is also complex. It is dependent upon many of the same variables as the heating cycle, although the problems associated with improper cooling are less severe on part performance than are the problems that arise from improper heating. Cooling affects shrinkage and crystallinity, which are relatively minor in importance for most rotomolded parts, although premature shrinkage of the part away from the mold can cause severe surface imperfections.

The major potential problem in rotational molding is inadequate fusing of the particles. The potential changes in properties with inadequate fusion for polyethylene are shown in Table 15.2. The changes in physical properties with inadequate fusion are most evident in elongation, impact strength, and flex life. The appearance characteristics are useful because they permit a visual assessment of whether the part has been fully fused. If the interior of the part is smooth and glossy, if few bubbles are apparent, if the exterior of the part is free of pin holes, and if the color is rich and fully developed, the chances are good that the part was fully fused. Underfused parts would, on the other hand, exhibit orange peeled appearance in the interior surface, many bubbles that could be seen, especially on the edges, and pin holes on the exterior surface.

Troubleshooting

The rotomolding process is highly sensitive to many variables that are difficult to control and even more difficult to change as the size and type of part being made are changed. For

Table 15.2 Physical Properties and Appearance Characteristics of Fully Fused and Underfused Rotomolded Polyethylene Parts

Physical Properties	Fully Fused	Underfused
Tensile strength (MPa)	18.4	17.7
Elongation (%)	60	10
Flexural modulus (MPa)	748	680
Impact strength @ $-30°C$ (J)	40	13
Flex life, F_{50}, cycles	1500	650
Appearance		
Interior surface	Smooth, glossy	Orange peeled
Bubbles	None	Many
Exterior surface	Smooth	Pin holes
Color	Fully developed	Poorly developed

instance, the movement of the air in the heating chamber or the uniformity of cooling are both set for particular conditions and circumstances and are generally not modified when the molds are changed. As a result, optimum conditions for a particular mold are not achieved.

Other variables can be controlled and modified whenever molds are changed. The weight of the charge is an obvious variable that must be changed whenever a different part is to be made. In most cases, however, the charge weight is manually weighed and little automatic control is exercised. As a result, weight is a characteristic that has a wide range of variability and can cause problems in getting adequately fused parts.

A method for proceeding to minimize the problems associated with these changes in the mold and other processing changes is to identify the best molding conditions possible and then repeat these conditions whenever possible.

The analysis and correction of part defects is an important method of identifying the proper conditions for each mold and part. A set of troubleshooting procedures to take corrective actions is given in the troubleshooting guide (Table 15.3).

Maintenance and Safety

Maintenance in rotomolding focuses on the proper lubrication and repair of the mechanical movement of the molds. The equipment is subjected to thermal cycles which can have a negative effect on lubricant life and on part meshing. Therefore, mechanical movement should be carefully monitored for wear.

The oven burner should be routinely examined to ensure that it is clean and functioning properly. The overall operation of the oven should be monitored by taking temperature readings in many different locations within the oven and comparing them with historical records.

The major problem with the cooling system is clogging of water sprays which should be visually inspected often. Treated water can be used to reduce the plugging.

The major hazards in a rotational molding operation are in the mechanical turning and movement of the molds. Care should be taken in moving around the equipment to prevent accidental impact or getting hands or clothing caught in the gears or other moving apparatus.

The hot molds and hot oven (which is usually kept on continuously during operation) are other potential problem areas in a rotomolding operation. Even if the mold may have been moved out of the oven, care should be exercised and, when in doubt, it should be considered to be hot.

The manual removal of parts will occasionally cause back strain or other muscular problems. These can best be avoided by exercising the proper techniques for lifting or other physical exertions. Mechanical removal devices or assistance from others should also be available.

Vent pipes should be carefully installed to ensure that pressures do not build up inside the mold. Likewise, pressure buildup from excessive peroxide can be a problem, especially if the peroxide is added to the powdered resin rather than melt-blended in an extruder.

Table 15.3 Troubleshooting Guide for Rotational Molding

Problem	Possible Cause
Uneven wall thickness	• Improper mold rotation (perhaps mechanical problem or rotation ratio is not correct) • Mold thickness variations that can cause uneven heating and therefore uneven sticking • Uneven heating from oven air flow
Incomplete fusion	• Heating cycle too short • Oven temperature too low • Inadequate air flow in oven • Excessively thick mold walls • Resin may have incorrect melt index
Bridging of powder in mold	• Rotate mold faster • Use powder with increased flow • Design recesses with larger radii • Vent mold at bottom of recesses
Mold release difficult	• Draft insufficient • Undercuts • Rough mold surface • Degradation of plastic • Lack of sufficient mold release
Warping	• Uneven cooling (may be clogged water nozzles) • Part pulled away from mold due to cooling too quickly • Poor venting has created a vacuum inside part
Low impact strength	• Insufficient fusion • Cooling too slow • Poor dispersion of additives • Not enough crosslinking
Surface pitting	• Contaminated mold surface • Polymer degraded

CASE STUDY 15.1

Trash Cart Manufacturing

The production and sale of large trash carts that can be wheeled by the homeowner to the curb for pickup by trash trucks is a growing industry worldwide and has become very competitive. In the United States the most common size of these carts is about 95 gallons (360 L), although 65-gallon (250-L) and 35-gallon (130-L) sizes are also available. The carts have two wheels that enable easy movement of the cart even when it is loaded. The maximum load expected is about 200 pounds (90 kg), although the carts will still function reasonably well with loads up to 350 pounds (160 kg).

The designs of these carts differ somewhat from company to company and from process to process but the cart depicted in Figure 15.8 reasonably represents the general features of all of the major brands currently used for this application.

The competition in the manufacture of these carts is not only between several manufacturers, but is also between three different manufacturing processes that can be used for making the carts. Each manufacturing process is touted as superior to the others and each process does give some advantages in part appearance and performance. The three processes are rotational molding, blow molding, and injection molding. These processes have all been described previously, and so the comparisons that are made between these processes assume that each is well known.

Rotational molding was the most common method for manufacturing trash carts for many years and is still widely used for this application. The size of the carts fits about midway in the normal size range for rotationally molded parts so making the carts in normal and readily available equipment is not a problem. The shapes of the carts are easily made using rotomolding. The lids for the carts are molded with the bodies and then separated by mechanical cutting after the part is removed from the mold. Brackets or holders for the axles and handles are molded into the part and then, after the part is removed from the mold and the lid has been cut off, holes are drilled in these brackets or holders and the handles and axles are inserted. The wheels are then installed onto the axle and secured in place. Occasionally other handle parts are also assembled at the time the part is finished.

To obtain the performance required of the carts (strength, impact toughness especially at low temperatures, ability to regain shape after squeezing), the carts have traditionally been crosslinked. Although crosslinking is still widely used, some manufacturers have introduced the use of the linear low- and medium-density resins without crosslinking. The linear low- and medium-density resins are inherently tougher and stronger resins than

Figure 15.8   Typical trash cart design.

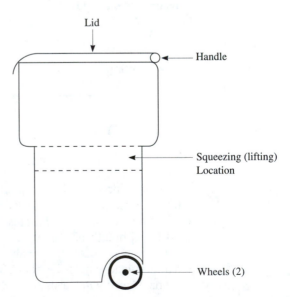

the base resins (noncrosslinked) that are used to make crosslinked carts. Many rotomolders, however, believe that the performance of the rotomolded noncrosslinked carts, even with the linear resin, is not acceptable in some of the critical tests and, therefore, prefer to continue the crosslinking of the carts. Even those supplying noncrosslinked carts generally agree that the crosslinked carts are superior in performance. The question is whether the higher performance is an "overkill," especially in light of the much longer molding cycles and, therefore, higher costs of the crosslinked carts.

The long processing time involved in making the rotomolded carts has invited competition from other manufacturing processes. One competitive process is blow molding. The initial difficulty with blow molding of these trash carts has been the size of the cart. Few blow molding operations can make a part as large as the 95-gallon cart. The problems with the size involve the size of the accumulator and the excessive sag that occurs while trying to drop such a large parison from the accumulator. The introduction of new and larger machines has solved these problems and allowed blow molded carts to be very competitive, especially because of the much shorter molding cycle that is possible. In fact, a machine that is capable of molding two 95-gallon carts simultaneously is now in production.

Initial tests of the blow molded carts indicated less than satisfactory impact toughness. This problem has been corrected by use of a resin with a much lower melt index. This low–melt index resin can be used successfully in blow molding because the only limitation would be its melting. Other processing features, such as melt strength and resistance to creep, would be improved in blow molding with the low–melt index material. This is not the case, however, when trying to rotomold with a low–melt index resin. The use of low–melt index resins would generally not be possible in rotational molding because of the difficulty of fusing these resins. Hence, the gain in properties from low melt index resins which would be expected in most manufacturing processes would not be possible in rotomolding. (This is another reason that crosslinking has been so highly successful in rotational molding.)

Another process that has become competitive in making these large trash carts is injection molding. Just as with blow molding, the size of the carts is greater than the shot size that can normally be injection molded. Very large injection molding machines have, however, been constructed. These machines, perhaps the largest injection molding machines commercially available, have proven to be reliable and practical for the production of these trash carts.

To make the mold cycles reasonable, the molds are made of beryllium copper, at least in the areas that need to have high heat transfer, but they are very expensive. The costs of the injection molding process for these carts are also raised by the use of automated equipment to remove the parts from the injection molding machine and move them around the plant. Although the costs are higher than most other injection molding operations, the large number of carts and the relative high price per cart seem to justify the high capital costs.

The performance of injection molded carts has been quite good although the impact toughness has been an occasional problem. Changing to a low–melt index resin is more difficult in injection molding than in blow molding. In injection molding the slow flow of the low–melt index resin in the mold can substantially lengthen the molding time and may also result in poorly filled molds. Hence, a practical lower limit of melt index exists for this application. Other resin properties, such as density and molecular weight distribution can, however, be adjusted to compensate.

Injection molding has a major advantage over both rotational molding and blow molding—the ability to have much more complicated shapes in the parts. Therefore, areas of the cart that are weak can be reinforced with a rib or some other structural feature. This capability also allows injection molded parts to be substantially thinner than rotationally molded carts. Blow molded carts can also be thin because of the superior resins that can be used.

A recent requirement of many municipalities is that the carts sold to that city must contain some postconsumer recycled material. This postconsumer recycled material would be obtained from recycling milk jugs or other typical polyethylene products. After chopping, cleaning, and, perhaps reextruding and regrinding of the resin, the postconsumer regrind is mixed with the virgin material in the required proportions. Experience seems to indicate that the blow molding process is able to use higher levels of this regrind material than are the other two processes. This seems to confirm that blow molding is more forgiving in the type of resin that can be used than are injection molding and rotational molding. However, the other processes are developing methods to use this regrind and, ultimately, all of the processes may be able to compete equally in this regard.

Standardized tests have confirmed that if the carts are made properly, carts made by all three of the processes can be expected to last for 10 years of normal use. Some differences in performance are, of course, seen. The relative importance of these differences is a judgment that should be exercised by the purchaser (usually a municipality) based upon experience and local conditions. (For instance, a city that has a very cold winter may want to be especially careful to buy carts with good low-temperature properties.) All three processes are close enough in cost to be competitive, with the major differences in prices arising from factors such as resin costs, producer efficiencies, producer overhead costs, market strategy, and distribution costs rather than process differences.

SUMMARY

Rotational molding is capable of molding much larger parts than any of the other common processes for thermoplastic resins.

In rotational molding, also called rotomolding, a cold mold is filled with powdered resin, the mold is closed (usually manually), and is then heated while being rotated about two axes (at different speeds). This rotation causes the powdered plastic to tumble inside the mold and, as the mold walls become hot, to stick to the walls. Eventually all the plastic sticks to the walls; if the mold rotation has been done appropriately, the walls will be uniformly coated.

The mold is then moved into a cooling chamber where either cool air or water is sprayed on the mold while it continues to be rotated. When the mold and contents are cold, the mold is moved to an unloading location where the top of the mold is removed and the part extracted.

During the molding cycle, the plastic material is never fully melted. Rather, the plastic powder softens until it becomes sticky and then, with further heating, the small plastic particles fuse together to give a solid mass. If, however, this fusion is incomplete, the parts will be porous and several mechanical properties will be adversely affected.

The equipment requirements for rotational molding are not as capital intensive as other plastic-molding processes such as injection molding or blow molding. The heating is less because the plastic resin is not melted. The pressure requirements are negligible so the machines need not be heavy duty for withstanding injection pressures. The equipment should be large enough to accommodate the parts that will be made and it should function well mechanically.

The rotational molding method is less automated than many other plastic-molding methods and so some judgment needs to be exercised to ensure that the parts are made appropriately. This judgment is not too difficult to acquire and several clues are available, such as examining the part for bubbles, shiny surfaces, and pinholes to determine whether the part has been molded correctly.

Rotational molding can compete with blow molding and injection molding in making many parts. Generally, rotational molding becomes more cost effective as the size of the part increases. Parts such as 95-gallon trash carts are near the upper limit for blow molding and injection molding but are mid-range for rotational molding.

GLOSSARY

Batch rotomolding system   A simple mechanical machine for use in rotomolding.

Biaxial   Rotation about two axes.

Bulk density   A measure of the density of the powder before it is heated or compacted.

Cooling step   The time in the rotomolding cycle in which the part and mold are cooled.

Fusion   The joining together of separate resin particles to make a unified structure.

Heating step   The time in the rotomolding cycle during which the mold and its contents are heated and during which the resin particles fuse to a unified part.

Jacketed rotomolding system   A rotomolding system in which the molds are integrally heated through channels in the mold wall.

Loading step   The time in the rotomolding cycle in which powdered resin is placed inside the mold; the mold is then closed and secured.

Mesh size   A measure of the size of the screen mesh through which 95% of the particles pass, thus allowing the grind of the particles to be classified according to the size of the resulting powder.

Pourability   The ability of a material to slow under gravity and, therefore, make uniform parts by the rotomolding process.

Rotational molding   A process in which plastic powder is inside a heated, enclosed mold and, through turning of the mold, the plastic powder sticks to the walls of the mold and then to itself to form a hollow part that has the shape of the inside of the mold cavity.

Rotomolding   Another name for rotational molding.

Softening temperature   The temperature at which a resin will stick to itself or some other resin and fuse.

Straight-line rotomolding   A mechanically simple rotomolding system that is related to the batch system.

Unloading step   The time in the rotomolding cycle during which the part is removed from the mold.

QUESTIONS

1. What constraints are necessary for biaxial rotation to evenly coat the mold in a rotational molding process?
2. Discuss three ways to alleviate the sticking of a part to the mold.
3. Explain how polyethylene is crosslinked in a rotomolding operation.
4. Describe a technique to form a part with an outer coating and a different inner material using rotomolding.
5. Why is there a limit on the thickness of parts that can be reasonably made by rotomolding?
6. Describe why fillers and other additives may increase or, in some cases, decrease rotomolded part mechanical properties.
7. What is the major limitation on the distance between walls in a double-walled carrying case made by rotational molding?
8. Give five part characteristics that indicate a poorly fused part.
9. Describe why the weight of the resin charge is so important in part performance.

REFERENCES

Crawford, R., "Sintering," *Developments in Plastics Technology—3,* eds. A. Whelen and J. L. Craft, New York: Elsevier Science Publishing Company, Inc., 1982.

"The Engineer's Guide to Designing Rotationally Molded Plastic Parts," Association of Rotational Molders, 1982, Reprinted 12/85.

Frados, Joel, *Plastics Engineering Handbook* (5th ed.), Florence, KY: International Thomson Publishing, 1994.

Phillips Chemical Company, "Rotational Molding," Bartlesville, OK.

CHAPTER SIXTEEN

CASTING PROCESSES

This chapter examines the following concepts:

- General concepts, advantages, and disadvantages
- Processes (mold casting, embedding, encapsulation, dip casting, fluidized bed coating, painting, slush casting, static power casting, cell casting, continuous casting, and film casting)
- Equipment (machines, molds, plant concepts)
- Product considerations (materials, shapes, design)
- Operation and process control

INTRODUCTION

Casting processes are characterized by the use of a liquid or powder starting material that is shaped without the application of significant pressure. The absence of pressure is such an important characteristic of casting processes that all processes for forming plastics that do not use pressure might be considered casting processes. For instance, some authorities classify rotational molding as a casting process. Rotational molding has become such a specialized process, however, that it is considered separately from casting in this text.

Pressure can be used in some casting processes provided that it is minimal. Some casting processes, for instance, use the weight of a top plate in a mold to define the surface quality of the cast part. Other casting processes use a vacuum (which creates the pressure differential on the part) to insure that the material flows properly into the cavity in which the liquid material is to harden. Such uses of pressure are considered insignificant and do not remove these processes from the general casting category.

The absence of significant pressure means that the molds and support equipment used in casting need not be as strong as would be required for a high-pressure molding process such as injection molding. Molds for casting can be made of wood, plaster, plastic, aluminum, rubber, and other materials that may fit specific applications. Some molds can be flexible, thus allowing for removal of parts that have undercuts, even without sliding sections as would be the case with hard molds. The absence of pressure allows large parts to be made because the molds need not be massive, as they would with pressurized processes. Because there is no size limitation inherent in the process, large plastic parts (such as large nylon gears) that cannot be made by any other process can be made by casting.

Heat is sometimes added to hasten hardening, although this is not required in all cases. Some resins have heat applied prior to forming and then are cooled to harden. The application of heat depends on the type of resin being cast. Materials can achieve the solid state by chemical reaction, by external heating followed by cooling, or by solvent evaporation.

The materials used in casting and the various techniques used to cast parts are closely related. In some cases the processing technique was developed to use a particular material more efficiently. Therefore, this text discusses casting materials and casting techniques together. For organization and clarity, the discussion begins with a brief introduction to casting materials, moves to a presentation of each of the major casting techniques, and ends with an in-depth discussion of the materials.

The most common types of materials used in casting are *liquid resins,* namely, *monomers, syrups,* or low molecular weight thermosets. These are materials that harden through some chemical process, usually polymerization or crosslinking. Liquid resins can therefore be monomers or short-chain polymers that form either thermoplastics or thermosets when hardened. Typical examples, all of which are discussed in some detail later in this chapter, are nylon (6) monomer (caprolactam), acrylic syrup (which contains acrylic polymer dissolved in acrylic monomer), polyester resin (which is a liquid at low molecular weights and before crosslinking), and phenolic resole (which is a B-staged thermosetting resin). Most casting techniques can use liquid resins as starting materials.

Another type of casting resin is the *hot-melt plastics.* These are fully polymerized thermoplastic materials that have been liquified by heating them above their melting point. These materials harden by cooling. The casting processes that can use hot-melt plastics are more limited than those that use liquid resins because hot-melt plastics are generally quite viscous. These materials may also require some heating of the mold to insure that it fills completely and then cooling of the mold to harden the resin. This requirement complicates the process over what would be required with liquid resins. Another complication is the frequent need to add heat stabilizers to prevent resin decomposition during the extended time that the resin must be held as a liquid.

A third general type of casting materials is the group consisting of *plastisols* and *organisols.* Both of these are plastic particles that are suspended in a plasticizing solvent. They harden either by evaporation of the solvent or by reaction of the solvent with the plastic. In plastisols the concentration of solids is greater than 90%; in organisols the concentration is generally between 50% and 90%. Plastisols generally have no volatile solvent, whereas organisols generally do. Adding a volatile organic solvent to a plastisol converts it into an organisol. Plastisols are generally thick, viscous liquids and organisols are somewhat less viscous.

The fourth type of casting material is simply *dissolved plastic.* This type of material is made by dissolving plastic particles (usually a powder so that dissolution is easy) into a volatile solvent. This material is hardened by evaporating the solvent.

A fifth type of casting material is a powder. The powder must be free-flowing, that is, able to flow and fill a cavity without the need for pressure. The powder material is solidified by heating to fuse the particles.

CASTING PROCESSES

Mold Casting

The processing technique most commonly associated with casting is *mold casting*. In this technique the casting material is poured into an open mold. Two minor modifications of this technique are shown in Figure 16.1. In the first technique, called *solid casting,* the

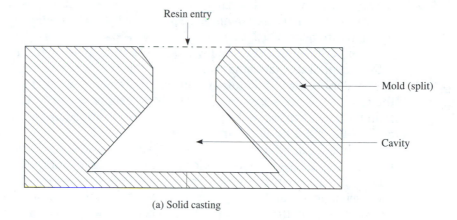

(a) Solid casting

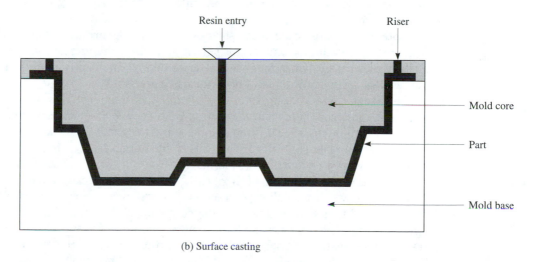

(b) Surface casting

Figure 16.1 Mold casting showing two techniques: (a) open mold casting and (b) surface or gated casting.

resin is simply poured into the open mold until the mold is filled. The resin hardens to form the part. Because it is difficult to remove solvents from a solid part, especially if it is thick, casting materials that rely on solvent evaporation to harden are rarely used for solid casting.

In Figure 16.1a, the part has an undercut and so the mold is split to remove the part. Parts without undercuts need not be split because the parts can be withdrawn through the opening used for resin entry.

The molds used in mold casting are typically made from a prototype of the part itself. This process can be quite complicated, depending on the materials to be used in the final mold and the degree of precision that is needed in the final part. In general, however, the mold-making process is begun by making a prototype that has the same shape as the final part to be made by casting. The prototype is often made by machining but, in many cases today, is made by one of the rapid prototype processes described in the chapter on design. The type of materials used to make the prototype is usually unimportant as long as the dimensions are maintained.

After verifying that the prototype has the correct dimensions, the prototype is coated with a material that takes the exact shape of the prototype and then hardens. Some common materials used to do this are silicone elastomer, thermoset epoxies, and plaster. In the case of silicone elastomer, it hardens to a semihard state and then can be removed from the part. Removal of the part can be by direct extraction through the opening, especially if there are no undercuts or if the undercuts are so small that the silicone elastomeric mass can be stripped away from the prototype. The silicone can then be used as the casting mold, with the realization that the forces and problems of removal of the prototype will be encountered with each cast part. If these removal problems are too difficult, the silicone mass can be split, thus providing for easier removal but then requiring a jig or fixture to hold the two silicone parts together when actual casting is done. The thermoset epoxies and the plaster are much more likely to require splitting of the mold in order to remove the part.

Note that in parts made by the solid casting method, shrinkage is common, especially when the part is thick. Some of this shrinkage is from the increase in density that normally occurs with polymerization and/or crosslinking. The inclusion of fillers in the casting resin can reduce the amount of shrinkage appreciably.

Shrinkage can also occur because of thermal contractions. This type of shrinkage is, of course, most common when the casting resin used is a hot-melt thermoplastic. Little can be done to reduce the shrinkage from cooling thermoplastics except to add additional material as the resin pool shrinks, but this is difficult and can cause a poor weld line between the two resin masses. Some thermal shrinkage occurs because of the exotherm that accompanies the crosslinking of thermosets. This shrinkage can be reduced significantly by slowing down the crosslinking reaction. For instance, rather than trying to cure the resin overnight, as is often the case, the cure components are adjusted to give a much longer curing cycle, thus never allowing the exotherm to rise and, therefore, increase the reaction rate in the material, which in turn further increases the exotherm. By slowing the reaction, the exotherm can be readily dissipated and no secondary rise from thermally induced reactions will occur. This technique was used once in the casting of a large telescope mirror (approximately 30 feet (10 m) in diameter by 4 inches (10 cm) thick, which was cured over the course of several months to minimize the shrinkage and also any optical aberrations that might arise from thermal stresses associated with this shrinkage.

The second modification of the mold casting technique uses a gate to fill the mold. Risers, which are placed at the far ends of the mold, help determine when the mold is full and also ensure that the material flows to the upper portions of the mold. The mold core defines the thickness and upper surface of the part. This process, called *surface casting* or *gated casting,* is similar to traditional metal casting in sand molds.

The mold sections for the gated casting process are usually made from blocks of resin that have been cast about a prototype of the part or have been machined. The two parts of the mold (the mold base and the mold core) must mate in such a way that a cavity is formed between them, that cavity being the space occupied by the cast part. This mating suggests that the mold base and mold core must be kept apart so that the cavity remains. Several methods can be utilized to ensure that the mold parts are correctly positioned. One method employs standoff inserts placed in the mold cavity. Another method simply uses pins through the side of a frame that supports the mold core above the mold base.

A vacuum is sometimes applied to the risers to facilitate filling and ensure that the entire mold cavity is filled with resin.

The most common materials used in mold casting are liquid resins. These materials do not have any evaporating solvents that would be difficult to remove from thick parts typically made by the mold casting technique.

Embedding and Potting

A process that is closely related to mold casting is *embedding.* In this process an article to be encased with resin, such as an electrical part, is placed in the mold so that the resin that is poured in completely submerges the article. (In the case of electrical components, the electrical leads may protrude outside the mold.) The part can be retained in its position in the mold by several means. The most obvious in the case of electrical parts is to use the electrical leads to hold the part in place. If the electrical leads are too thin and flimsy to hold the part, a small rod can be attached to the part that parallels the wires and holds the part in its correct position inside the mold. When casting is completed, the rod can be trimmed to the surface of the part. Another method of retaining the part in the mold is to use support inserts in the mold, which position the part or pins through the mold walls.

A vacuum sometimes is used to ensure that the resin completely fills the mold and surrounds the article. The embedded article is removed from the mold when the resin has hardened, and the mold is reused.

The most common resins used in embedding are liquid resins and hot-melt resins. Solvent-based resins are rarely used because it is difficult to remove solvents from the part. Solvents may also attack the item to be embedded.

The shape of the mold (the shape of the resin defined by the inside surfaces of the mold) is usually simple. In some cases, however, the shape of the mold defines some assembly cavity into which the encapsulated part must fit. The embedding process is illustrated in Figure 16.2.

Figure 16.2 The casting processes of embedding and potting. In embedding the embedded part is removed from the mold; in potting the mold remains with the embedded part.

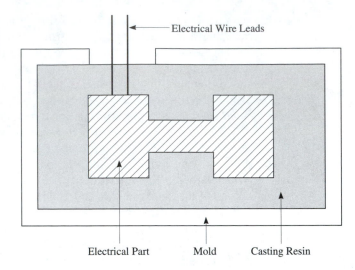

Electrical Wire Leads

Electrical Part   Mold   Casting Resin

Common examples of articles that are embedded include a variety of electrical components and encapulated metal parts. Urethane wheels for roller skates, skateboards, and in-line skates are made by embedding the roller bearing in a urethane casting resin. Other embedded parts include scientific specimens that are embedded in a clear resin for protection during examination and for long-term preservation.

Figure 16.2 also depicts the *potting process*. Potting differs from embedding in that the potting mold remains attached to the embedded part after hardening has occurred and thus is part of the product. Potting molds often are simple cups that define the shape of the resin but are expendable and therefore are shipped as part of the embedded part. The molds may even serve as a protective cover for the embedded assembly. As already suggested, potting is used widely in the electronics industry.

Encapsulation, Dip Casting, Fluidized Bed Coating, and Painting

In some cases an article must be encased in a plastic but the shape of the final part must conform to the shape of the article inside. When the shape of the article defines the shape of the final product, the process is called *encapsulation*. The encapsulation process produces a thinner layer of plastic coating than does embedding or potting (see Figure 16.3).

In some cases, the material must be coated repeatedly to build up the resin to the desired thickness. This repeated application can easily be done by dipping the article to be coated in the resin, a process called *dip casting*.

Almost all casting resin types can be used for encapsulation. The article can be submerged in a liquid resin, hot-melt resin, plastisol, organisol, or a solvent-based casting resin. The evaporation of the solvent poses no problem because the resin layers are thin and no mold is used.

The article to be encapsulated can be at room temperature or it can be heated. This choice depends, to some extent, on the type of resin used for the encapsulation. Plastisols

Figure 16.3 The encapsulation casting process.

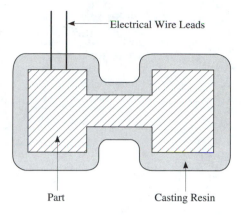

and organisols often work best when the part is heated slightly, as heating increases the ability of some resins to stick to the surface, especially if repeated dippings are made.

Heating the article to be encapsulated also permits the use of a finely ground powdered resin instead of just liquid materials. When powdered resins are used, the heated article is submerged in a bed of the resin powder that is fluidized by having air blown through it. The powder coats the part and then cools to form an encapsulation. In some cases the powder-coated article is heated further so that the resin fuses to a continuous coating, a process called *fluidized bed coating*.

The simplest of all encapsulation processes is the application of a solvent-based casting resin with a brush or as a spray. *Painting* is such a process. The term painting is normally reserved for those encapsulation processes where the layer of coated material is very thin and the casting resin is solvent based.

Slush Casting

A casting process that is closely related to rotational molding is *slush molding*. In this process the mold is filled with a *plastisol*. The mold would be similar to the type used for mold casting or for rotational molding. After closing the mold, it is rotated in an oven, similar to rotational molding except that the time for rotation is limited to only the time necessary to congeal the desired thickness of material onto the surface of the mold. (The plastisol congeals or solidifies by becoming hot and sticking to the mold.) After the desired amount of material has solidified, the mold is removed from the oven, the lid is taken off and the liquid material that still remains in the center of the mold is poured out. The lid is then replaced, the mold is again put into the oven and heating is continued until the solid material has been properly fused.

After heating, the mold is moved to a cooling chamber and then, when fully cooled, to an unloading location where the lid is removed and the part extracted from the mold. A representation of the mold and plastisol material at the stage where both solid and liquid material is still in the mold is given in Figure 16.4.

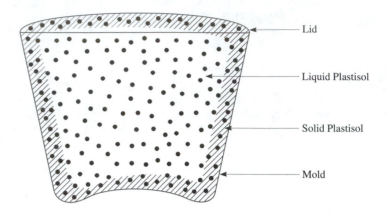

Figure 16.4 Slush molding showing the point in the process where some of the plastisol has solidified (congealed) and some remains liquid but has not yet been poured out.

Lid

Liquid Plastisol

Solid Plastisol

Mold

Because the slush molding operation can be easily automated, it is frequently used for small items such as vinyl boots and gloves. A conveyor system carries the molds through an oven for the time required to form the desired solid thickness. The molds exit the oven, are automatically turned and the liquid emptied, and then the molds go into another oven for fusing. The conveyor system then moves the molds to a cooling section and finally to an unloading area where the parts are automatically removed from the mold.

A slight modification of slush molding is called the *multiple-pour method*. In this method the plastisol is added to the mold and the mold is heated in the normal way for slush molding except that the heating time is very short. Normally only enough material is solidified to form a thin layer against the mold. The mold is then removed and poured. At this point a new plastisol material can be added and the mold closed, and reheated until the desired thickness of the new plastisol is formed against the mold walls. The mold is then reopened and the liquid material is poured out. The process then continues as would be normal for slush molding.

The advantage of multiple-pour slush molding is that a coating can be formed with a different material than the bulk of the part. This allows parts to be made with special tough or colored surface materials.

Thermoset materials can be rotomolded using the slush molding process. The thermoset resin, usually a viscous liquid or a low-melting solid, is poured into the mold as would be done with a plastisol. The mold is then placed into an oven where the outer material becomes hot faster than the material in the bulk of the liquid. The initiator, which would be heat activated, would react first with the outer material. When enough of this material had reacted, the remaining material would be poured out. The part would then be cured to complete the formation of the part.

Static Powder Casting

If a powder is used instead of the plastisol in slush molding, the process is called *static powder molding*. A mold is filled with the powdered material and heated in an oven for a short time. The heating causes the powder to fuse, but the time is chosen so that only a thin layer or coating forms on the inside of the mold. When the desired thickness is built, the mold is inverted and the excess, nonfused material, is dumped out (much like in slush casting). The mold is then returned to the oven to complete the fusing of the plastic and is then cooled. After cooling, the part is removed from the mold. The principal material used in this process is polyethylene.

If pressure were used in this process to compact the resin into the mold prior to heating, the static powder molding method would be very similar to sintering, which has been described previously when discussing PTFE resins. Because of their high melting point and the difficulty of melting without decomposing them, sintering is used to make solid parts that are then machined or otherwise shaped into final form.

Cell Casting

Another casting technique, called *cell casting,* uses a mold that is defined by two parallel plates (usually polished glass plates) separated by a gasket (which retains the material and sets the thickness of the part). (See Figure 16.5.) The resin is simply poured between the sheets and allowed to harden. The process is a batch process with each cast sheet made individually.

The principal material used in cell casting is acrylic syrup, which is a liquid casting resin. Because the process is confined, cell casting rarely involves solvent-based materials. The difficulty of pouring a hot material between two sheets that may be separated only by a small distance also restricts the use of hot-melt resins.

Continuous Casting

Another method for making clear plastic sheets is *continuous casting.* In this technique the liquid resin material, usually acrylic syrup, is poured between two continuous belts separated by a gasket. The gasket retains the liquid resin and defines the thickness of the

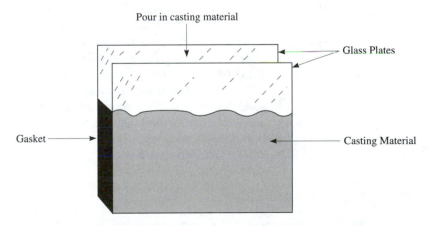

Figure 16.5   Cell casting for the production of a clear plastic sheet.

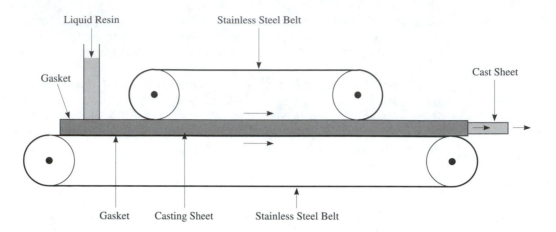

Figure 16.6   Continuous casting method.

sheet. Continuous casting is shown in Figure 16.6. Both continuous casting and cell casting make cast sheets that are largely free of distortions. The major advantage to continuous casting over cell casting is volume production. Also, continuous cast sheets often can be processed (such as thermoforming) somewhat more easily than can cell cast sheets because the casting is done horizontally and the effect of gravity is uniform over the entire surface, thus creating minimal molecular orientations within the sheet. This property is especially noticeable in cross-linked continuous cast sheets, which, when heated for later processing, have uniform shrinkage in all directions.

Cell casting allows more freedom in the use of resins than does continuous casting. With cell casting, very high molecular weight resins can be used because the pouring to fill the cell can be done slowly, thus allowing the casting resin to level itself and fill all parts of the cell. The casting resin may be quite thick when high molecular weight resins are used. Other advantages of cell casting over continuous casting include close control over the product thickness and surface and, some believe, slightly better optical quality.

Cell cast and continuous cast sheets have some major advantages over extruded sheets. Cast sheets can be crosslinked, whereas extruded sheets cannot. Cast sheets can use resins with higher molecular weights than can extruded sheets, giving cast sheets greater strength and impact toughness. Cast sheets can be made distortion free, whereas extruded sheets have areas of inherent stress due to the orientation of the molecules when they are pushed through the die. Perhaps the most important advantage of cast sheets is that their optical quality is superior to that of extruded sheets.

Cast sheets, both cell and continuous, are used to make many products. These include advertising signs, skylights, display cases, and windows.

Film Casting or Solvent Casting

In *film casting* or *solvent casting* the starting material is a liquid consisting of a resin (usually powdered) dissolved in a volatile solvent. (Note that the term film casting is also used for the method of making a film by extruding the film onto a set of rollers that smooth and stretch it. In this author's opinion, the extruded processing method is incorrectly named and thus is not called film casting in this text. Only the true casting process, without pressure, is called film casting.) The resin solution is sprayed or poured onto a stainless steel belt that conveys the material through an oven where the solvent is evaporated. The film of the resin that remains on the belt is then stripped off and wound onto a reel. A solvent recovery system is an important part of this process because it helps to prevent air pollution and allows economical reuse of the solvent. This system is depicted in Figure 16.7.

Films made by solvent casting have some distinct advantages over films made by extrusion. With solvent casting no heat stabilizer is needed in the resin because the resin is never melted. The heat stabilizer may adversely affect properties such as clarity and raises the cost of extruded film. The solvent cast film can be made free of distortions, unlike the extruded films. Again, clarity is improved when the film is distortion free. One other major advantage of a cast film over an extruded film is the uniform thickness that can be achieved with cast films. When the solution is sprayed or poured onto the belt, the solution flows to cover the belt uniformly (assuming the belt is level). The thickness uniformity of extruded films depends on the die gap uniformity, which is often difficult to maintain. Another advantage of cast films is that extremely wide films can be made simply by using a very wide belt and applying the solution somewhat uniformly. In extruded films, if very wide extruded films are needed, very wide dies must be created, assuming that the

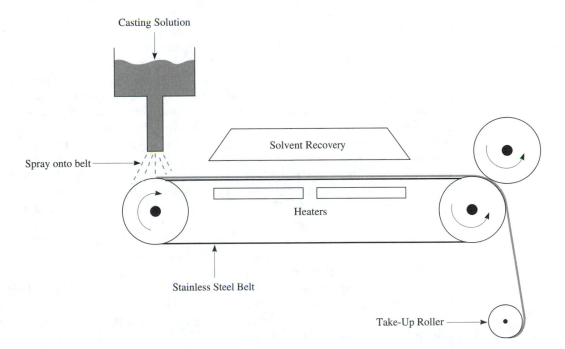

Figure 16.7   Film casting or solvent casting system.

extruder has the output to feed such wide dies. These wide dies are expensive to make and difficult to maintain. Finally the cost of an extruder is much higher than the cost of a film casting unit and is more costly to operate. The major advantages of extruded films are high production rates and elimination of the need to recover the solvent.

EQUIPMENT

Machine Types

The casting processes give greater freedom in making processing machines than any other plastic process method. This arises from the low pressures and, often, low temperatures inherent in casting processes. No particular equipment type is required for the casting processes. As already indicated, the size of casting equipment can be very large, thus allowing the fabrication of large parts that would be difficult if not impossible to make by other methods.

Molds

Molds can be made of almost any solid material. The materials most commonly used commercially are aluminum (both cast and machined), wood, plaster, rubber, silicone, epoxy and other thermoset plastics, glass, and steel. The major considerations for choice of material are durability, ease of fabrication, porosity, and convenience.

In noncommercial applications or for prototyping, the softer mold materials are especially useful. Molds made of silicone rubber, for instance, are very common for producing limited quantities of parts. These molds have the advantage that they can be cast around a model, separated into halves to remove the model, and then used for making a small number of parts that duplicate the shape of the model. With low-viscosity silicone the model's details can be replicated precisely. Undercuts in the model can also be duplicated and the replicate parts removed from the mold because the mold is flexible and can be expanded and twisted to move it around the undercuts. The nature of silicone also provides inherent mold release properties to these molds, further simplifying the molding.

Plastic, either rigid or elastomeric, and reinforced plastic molds are common, especially when the forces on the molds are very low and temperatures are not too high. Plastic and composite molds are inexpensive because they are easy to make by molding (usually cast) against a model. In this process a model of the part is made. The model can be made of any convenient material that allows the plastic to be molded around it. The model is placed in a box so that the resin flows around the part and forms a mold. The part may need to be supported in the mold to allow the entire part to be put into the cavity. After the resin is cast around the part, the mold must be cut in half to remove the part and provide a way to extract the part from the mold when actual molding is done. Alternately, the part can be placed on the top of a box so that only half of the part is covered. Then, when that part of the mold is cured, the part is inverted and the opposing mold is cast. This is similar to traditional metal casting.

Plastic and composite tools have a unique advantage in replicating highly detailed and complex patterns in parts. The light weight of the plastic and composite molds and the easy mold-making methods mean that very large molds can be built and used to make parts by casting or other low-pressure processes. For instance, ship screw propellers, which are over three meters in diameter with highly convoluted surfaces, can be cast in with plastic molds but would be too ponderous for metal molds.

Plastic molds also have a unique advantage in replicating highly detailed and complex patterns in parts. This is especially true of the elastomeric mold materials. These soft mold materials are cast as liquids that readily flow into the detail areas and fill them. The materials then cure into a semirigid elastomer that can be stripped from the master and then used as a mold to make parts. The detail of the master is usually reproduced very accurately in the mold and, consequently, in the parts. The flexibility of the elastomer allows the mold to be stripped away, even when undercuts and other complicated surface patterns may not permit the removal of other mold materials.

Another method of making plastic and composite molds is to machine the molds directly. This is less common because of the difficulty of machining versus the ease of casting the mold, but its simplicity makes it a method of choice for parts that are not too complicated.

The most common plastic materials used for casting molds are epoxies, silicones, and urethanes. The epoxies are rigid, and the silicones and urethanes are elastomeric. Because epoxy molds are too brittle for some applications, they often are filled or reinforced. Metals can be used as filling materials to give additional strength to the mold and to increase the heat-transfer capability. These molds are called *mass cast molds*. Aluminum powder is commonly used as one of the metal filler materials. The mass cast molds can be hard enough to be used for relatively low-volume production of blow molded and thermoformed parts.

Products that later will be manufactured in metal molds sometimes are made by other methods for the purposes of market testing and analysis. The most common methods to make these parts are by machining and by casting in soft tooling. Hence, prototypes are predominantly made by these two methods, even when the part will later be made in a hard mold for injection molding or compression molding.

Plant Concepts

The wide variation in processing equipment and molds for the various casting processes means that plant considerations, such as layout and auxiliary equipment, must be considered only in general terms. Plant layout is a function of the size of the parts to be made and the automation that may be employed. Some parts are made singly, in batch mode, while others are made in an assembly-line-type operation. For instance, bathroom sinks can be made along a production line in which many molds are conveyed under a large mixing tank, where the casting resin (an acrylic syrup or a polyester liquid that has high filler content) is pumped into the open molds. The parts are formed with the sink bowl downward. This arrangement results in a smooth bowl that is against the mold surface and a rough side opposite the bowl that is the top of the part as molded. The filled molds then move through an oven, where the resin is cured. Then, the solidified parts are removed from the molds and finished by drilling appropriate holes for the spigots and drain. (Some molds have inserts and pins so that the holes are created as part of the molding process.) The entire operation is done in a continuous conveying system that takes up considerable space within a factory, especially when the storage and raw material conveying systems and mixing systems are included.

Casting is a slower process than many of the other plastic-molding operations. Hence, some additional space may be needed to allow the same production rates as in pressurized processes. This additional space permits many molds to be used while the resin is curing.

PRODUCT CONSIDERATIONS

Materials

Almost all resins can be used for casting, either as liquid resins, hot-melt resins, plastisols, organisols, solvent-based resins, or powders. Some of the most commonly used resins are listed below.

Acrylic syrup. Acrylic syrup is made by mixing acrylic polymer (PMMA) with acrylic monomer. The viscosity can be adjusted by varying the concentrations of these two components and is usually like thin honey. When fillers are added, as is commonly done for applications such as imitation marble, the viscosity may need to be modified to ensure that the fillers remain in suspension. The syrup also has a peroxide added (usually of the heat-activated type) so that the acrylic monomer in the syrup will polymerize when heated. Other additives are inhibitors to give the material longer shelf life and accelerators or promoters that shorten the time to cure and can be added just prior to use. Crosslinking agents, usually of the type that has several carbon-carbon double bonds on a single molecule, can also be added so that a crosslinked acrylic product will result. A crosslinked product has greater strength and stiffness than a noncrosslinked product. One problem with acrylic sheets is poor surface abrasion resistance, which can be corrected by coating the sheets with an abrasion-resistant material, often a thin layer of silicate (glass). A paper covering is also added to protect the sheets during shipping.

Other syrups include polystyrene and caprolactam, the monomer for nylon (6). Caprolactam is the preferred monomer for use in nylon liquid resins because it polymerizes without the formation of water. The caprolactam is a ring molecule that opens to create the active sites for polymerization. Other advantages of caprolactam are that it has low shrinkage, good toughness and strength, and a low exotherm so heat dissipation is not a problem. Some typical applications for cast nylon (6) are large parts such as gears, bearings, and tanks, shutters, furniture, and buckets.

Polyester (thermoset) resins. Low molecular weight unsaturated polyesters can be used as liquid casting resins. A peroxide initiator is added so that the material will crosslink when heat is added. As discussed in the chapter on thermoset materials, a crosslinking agent such as styrene monomer may also be added to facilitate crosslink formation. The styrene can also serve as a solvent to adjust the viscosity of the casting resin. As with acrylic syrups, inhibitors are often added to prolong the shelf life of the casting resin and then, just prior to use, accelerators can be added to increase the speed of the curing reaction. The crosslinking reaction is strongly exothermic so care should be taken not to allow excessive heat buildup in the part. This is best done by limiting the thickness of the part to be cured, by minimizing the amount of heat applied to initiate the peroxide reaction, and by reducing as much as possible the amount of accelerator used. The technique of making successive pours of the casting resin can also be done if the part is extremely thick.

Like syrups, polyesters often have fillers. Reinforcements, such as fiberglass, can also be added to polyester casting resins to achieve higher strengths, stiffnesses, and other enhanced mechanical properties. The specific processes for molding reinforced polyesters are discussed in the chapter on composite materials and processes. Heat-dissipating fillers can also be used to reduce the effect of the exotherm.

Phenolics. Resoles are phenolic resins that are B-staged and then later cured by heating. Although water is condensed out of these resins when the polymerization/crosslinking reaction occurs, heating often is sufficient to evaporate the water and form relatively bubble-free parts. Almost all phenolics are filled to improve physical properties and reduce costs. Glass microspheres are commonly used as a filler to reduce the density of the casting. Phenolic resins are especially important for embedding electrical devices and for producing very hard materials such as piano keys and billiard balls.

Epoxy resins. Epoxies are widely used for casting resins and for tooling made by casting. The normal method of use is to combine the resin and the hardener just prior to use. Accelerators can also be added to reduce the curing time. The reaction is highly exothermic so the same procedures used with polyesters to control the effects of the exotherm should be used with epoxies. Epoxies are used for electrical embedding, potting, and encapsulation, for sealants and adhesives, and for rigid molds that will be used for making a small number of cast parts.

Polyurethanes. These resins cure into either thermoplastics or thermosets without the evolution of any condensate and are therefore well suited for casting. The major drawback of the urethanes is that once the two components (the polyol and the isocyanate) are mixed, reaction begins immediately and the resin must be cast at once. The flexibility and toughness of urethanes have led to many applications including wheels for skateboards and in-line skates, gaskets, and impact barriers around delicate instruments. Rigid urethanes can also be cast and some of the most important products include furniture and panels.

Silicones. Silicone casting materials are used for mold casting wherein solid silicone parts are made, for embedding, potting, and encapsulating, for sealing and adhesives, and for making flexible molds for other casting resins. Silicones are especially valuable for mold making because they have inherent mold release properties for many of the common casting resins. If the silicone is cured at room temperature, the material is called RTV (room temperature vulcanization). If cured by adding external heat, the silicone material is called HTV (high-temperature vulcanization).

Vinyls. Most plastisols and organisols are based on PVC as the principal resin. These materials are made of PVC resin (usually a powder) combined with a plasticizer to form a paste or thick liquid material (plastisol). Plastisols usually have over 90% solids. If a solvent is added, the material becomes an organisol with a solids content of 50% to 90%. The most common processing method for these materials is dip coating, although they are also used for embedding and encapsulation. Nylon is another resin that can be used in plastisols and organisols.

Hot-melt casting resins. Hot-melt casting resins are like candle wax, in that they are solids that are easily melted, cast into a shape, and then resolidified. Although most thermoplastics can be used as hot-melt casting resins, some resins, such as cellulose acetate, ethyl cellulose, polyethylene, and polyamides (nylons), are used more often than others. Some resins are not widely used because their melts are so viscous that they do not flow well without pressure or because they decompose easily when heated for long periods, as is often the case in casting. The most important of these poor casting resins are acetals, polycarbonate, and polypropylene. Powders are made from most thermoplastic resins, although polyethylene is the most widely used. The casting processes are most often used with these materials when stress-free or very large parts are made.

Shapes and Part Design

Many parts have already been discussed in the section on materials. The shapes and designs of these parts represent the wide variety of products that can be made with casting processes. Other important applications for cast parts include jewelry, optical lenses, tool handles, knobs, countertops, sinks, and buttons.

The design of cast parts has fewer restrictions than for most other plastic manufacturing operations. This design freedom is a result of the fluid nature of the starting material, low pressures, and low temperatures that lead to the possibility of using flexible molds. Features that would be very difficult in other processes can be done more easily with casting. These features include severe undercuts, complicated surfaces and details, and widely varying thicknesses in the part.

OPERATION AND CONTROL OF THE CASTING PROCESS

Far fewer operational parameters are critical in casting than would be critical in most other plastic processing methods. The most important parameters with casting concern ensuring that the starting material is of the correct viscosity for molding. Most liquid plastics will change viscosities over time because of slow curing, temperature changes, or solvent evaporation. A good procedure for handling cast resins is to monitor the viscosity of batches prior to use so that processing consistency can be maintained from batch to batch. Proper viscosity control is also helpful in eliminating bubbles from the casting resin. If the bubbles do not rapidly leave the mix, a vacuum can be applied to speed the process.

Another important part of the casting process that can result in problems is removal of the part from the mold. The complexity of the mold cavities (the degree of detail often present in a cast part) and the presence of undercuts or other unusual features can cause the part to stick to the mold. Therefore, a mold release (also called a *parting agent*) is usually required. However, too much parting agent can interfere with the replication of the mold surface by filling in some of the surface indentations present on the part to be molded. Therefore, caution should be exercised in applying the parting agent.

Control of the exotherm can be an important parameter to control in those casting processes where polymerization or crosslinking occurs. If the exotherm gets too high, the part can expand excessively and then crack when cooling. Also, the part may develop internal stresses that later result in weaknesses or visual distortions. Bubbles can also form from hot solvent or resin. As suggested earlier, the exotherm can be controlled by design of the part (making all of the sections thin), including heat dissipating fillers or attaching a heat dissipating mass, slowing the cure so that heat can naturally be conducted away, pouring the plastic in stages to minimize the thickness of the material curing at any one time, and cooling the part as it is being cured.

Troubleshooting

The problems in casting are summarized in the troubleshooting guide given in Table 16.1. As with the other processes, only the most common problems are addressed.

Table 16.1 Troubleshooting Guide for Casting

Problem	Possible Cause
Resin not filling the mold or poor part definition	• **Resin not heated enough if a hot-melt.** Temperature may not be high enough to allow for the normal cooling that occurs during mold filling. Heating device may be faulty. • **Plugged runners.** The runners may be undersized for the viscosity of the resin, especially if the resin has fillers or reinforcement added. The runner could have residual material from a previous run. • **Air entrapment.** The mold may need to be vented to allow the air to escape. • **Part complexity.** A vacuum may be needed because of the complexity of the part. The viscosity of the resin may be so high that it cannot go into all of the small areas of the part, especially if the resin is filled or reinforced.
Cracking of the part	• **Exotherm.** The exotherm may be so high that the shrinkage of the part causes it to crack when it cools. The exotherm can be reduced by slowing the crosslinking, adding filler, cooling the mold, or making the part thinner. • **Part complexity.** Cracking could be associated with an undercut in the part that cannot be easily stripped out of the mold. Also, very fine details on the surface of the part can result in excessive sticking. • **Mold release.** Not enough mold release in the mold or part.
Bubbles in the part	• **Solvent.** Residual solvent might be causing the part to partially foam. The part cure or solidification should be slowed. Heat and/or a vacuum might be applied to the part to accelerate solvent evaporation.

(continued)

Table 16.1 *Continued*

Problem	Possible Cause
	• **Air.** Air can easily be entrapped in the casting resin during mixing of the resin or even during pouring into the mold. This can be removed by applying a vacuum to the mold or, if that is difficult, to the container in which mixing is done. • **Condensate.** Water or other condensates result from some of the crosslinking or polymerization reactions of cast resins. This water is most often removed by applying a vacuum to the mold as the cure is progressing. The condensate can also be removed more easily from thin parts so a redesign may be necessary. The part might also be heated to vaporize the condensate. • **Degradation.** Hot-melt resins could be degrading from extended heating prior to casting, thus creating decomposition gases which result in bubbles.
Nonuniform part density	• **Incomplete mixing.** A casting resin with filler or reinforcement may need extensive mixing to evenly distribute the filler or reinforcement. This mixing may need to be continued to the moment of pouring so that the materials stay in suspension. • **Viscosity.** If the casting resin viscosity is either too high or too low, the filler and reinforcement may not flow uniformly during casting. Adjustments in resin viscosity can be made by adding or subtracting monomer or solvent, advancing or retarding the stage of cure at which casting is done (for B-staged resins), or adding a viscosity adjustment additive.
Surface imperfections	• **Mold release.** Either too much or too little mold release can cause the surface of the part to be mottled or splotchy. • **Improper filling.** The comments already given for poor mold filling could apply with special attention to casting resin viscosity and temperature control. • **Degradation.** The part may be degrading during heating and casting.
Embedded part not properly placed	• **Locating pins.** The casting of the resin may have moved the locating pins. These can be anchored more firmly, perhaps even by a mechanical holding device or by bonding.

===== CASE STUDY 16.1 =====

Casting a Polyester Thermoset Part in a Silicone Mold

The process of making a silicone mold and then using the mold to cast a polyester part has wide applications in laboratories for education, in commercial research and development for making prototypes, and in commercial applications where the number of parts to be made is small (roughly up to 200).

The fabrication of the silicone mold is done using a casting process. The required steps are carefully outlined below.

1. The making of a silicone mold begins with the construction of a model or pattern, which can be of virtually any solid material. Wood, metal, and plastic are commonly used because they are such easy materials to machine. A recent innovation is to make the pattern using one of the rapid prototyping methods outlined in the chapter on design. This part can then be used as the model.

2. If the model or pattern has a flat side, the pattern is fixed to the bottom of a box that will serve to limit the size of the silicone mold. If the part is to be completely molded, it is mounted on the mold bottom with some standoff pins so that resin can flow under the part and create the desired thickness of mold material. After securing the model in the box, a mold release is applied to the model and to the inside of the box.

3. The type of silicone chosen is determined by many factors. One factor is the amount of working time needed before the material becomes hard. Other factors include the complexity of the model and the detail of the model surface that must be replicated. By lowering the viscosity of the silicone casting resin, very high detail can be replicated from the model to the mold.

4. The silicone casting resin is prepared by carefully measuring or weighing the resin and the initiator in the proper amounts, as indicated by the silicone resin manufacturer. These materials are then mixed thoroughly. Any bubbles mixed into the casting resin are removed by placing the mixture under a vacuum.

5. The casting resin is carefully poured into the mold box that contains the model. Care should be taken to ensure that no air is trapped and that the model is completely covered with resin. The thickness of the silicone layer in the mold box is determined by the strength needed, the amount of use required of the mold, and the complexity of the part, since very complex parts may require more twisting of the mold to extract the part.

6. The silicone is allowed to gel. Gelation can take from a few hours to one day, depending on the thickness of the part and the type of silicone chosen.

7. After cure, the model is removed from the silicone mold. If the part had a flat side and was attached to the bottom of the box, removal is accomplished by removing the bottom of the box and *stripping* the pattern away from the silicone. (Stripping is the process of removing the part by deforming the mold or the part to allow them to separate.) If the part was totally encased by the silicone, the bottom of the mold box is removed and the pins detached from the mold box. The silicone is then slit to precisely the depth of the silicone over the part and to approximately the same length as the part. This slit is then used to extract the model from the silicone mold. This removal is also called stripping the part from the mold. Some care should be taken to ensure that the silicone does not tear during this removal process. If desired, the silicone mold can be completely removed from the mold box.

8. The mold should be prepared for the casting resin that will make the part. Most silicone molds do not need additional mold release, although with some resins and after several runs, the application of a mold release is advisable. The mold is now ready for casting.

9. The thermoset polyester casting resin is prepared by mixing the resin with the amount of initiator (catalyst) suggested by the resin manufacturer. Normally, this amount is 1% to 2% of the weight of the resin. In some cases, the resin will already be initiated and this should be determined carefully. If a fast cure is desired, an accelerator can be added. Other additives such as colorant, filler, and antioxidant can also be added at this stage. All the materials are then mixed thoroughly, and any bubbles are removed by vacuum.

10. The casting resin (thermoset polyester and additives) is poured slowly into the mold so that air is not trapped in the mold. The casting resin should fill the cavity in the mold that was left when the model was removed.

11. If the part has a flat side, that side is defined by placing a lid on the resin. Make sure that the lid has been properly prepared with mold release. If the part is totally encased, the material is poured into the mold until the mold is full. The open part of the mold can be closed by clamping lightly if needed.

12. The resin is allowed to cure. This can be done at room temperature or at an elevated temperature, depending on the choice of resin and initiator. If done at elevated temperature, some care should have been taken to ensure that the silicone mold can withstand the curing temperature without degradation. Curing normally takes 3 to 8 hours.

13. The product is removed from the mold just as the pattern or model was removed when the silicone mold was made. Some trimming of small amounts of flash may be necessary. The part may also need finishing such as buffing or painting.

With only minor modifications casting of embedded materials and inserts can be done. The procedure outlined is general in nature and can be used for other types of mold materials and for other types of casting resins.

SUMMARY

Casting is a process that can be used equally well for the production of large quantities of materials (such as cast acrylic sheets) or for prototypes where only one or two of the parts will be produced. The simplicity of the process and, especially, the lack of externally applied pressure create this wide applicability.

The starting materials for the casting processes are polymeric fluids. These fluids fall into five main classes: liquid resins (including monomers, monomers with polymer dissolved in them, and low molecular weight polymers that will crosslink); hot-melt resins (thermoplastics that are melted and then cast); plastisols and organisols (resin particles with small amounts of plasticizer added); plastics dissolved in volatile solvents; and fine powders.

A number of casting techniques offer varied opportunities for the use of these casting resins. The most important of these casting processes are solid mold casting, embedding or potting, encapsulation, dip casting, fluidized bed coating, painting, slush casting, static powder casting, cell casting, continuous casting, and film casting.

The molds used in the various casting processes can be made from many materials because of the low pressure requirements and often low temperature requirements of casting. Materials commonly used for molds include wood, metal, plastic, plaster, and rubber.

Casting can be done quite simply or can be an involved semiautomated process. Comparatively few parameters are required to be held closely because of the simplicity of the process. Casting is highly forgiving of minor deviations from the optimum process conditions and is therefore valuable at many levels of processing sophistication.

GLOSSARY

Cell casting A process in which a casting liquid is poured between two plates, usually glass, which have a gasket between them to form a cell to contain the casting liquid; then the resin solidifies, usually through polymerization or crosslinking.

Continuous casting A process of casting resins in which the casting resin is introduced between two moving sheets separated by the thickness desired in the final part; then when the casting resin hardens, a continuous or near continuous part is formed.

Dip casting Repeated dipping of an item in a coating resin to build up the thickness of the coating.

Dissolved plastic A solution of a plastic resin in a volatile solvent.

Embedding A casting technique in which a part is encased with a resin by placing the part in a mold and then casting the resin around the part.

Encapsulation Using a resin to coat a resin in such a way that the shape of the finished part is the shape of the item that was coated.

Film casting A process in which a casting resin is sprayed or otherwise coated onto a moving belt, which is then passed under heaters to remove the solvent and/or cure the resin.

Fluidized bed coating A coating technique in which a powdered resin is used as the coating and a part, usually heated, is dipped into the bed of powder, which is made fluid by blowing air upward through the powder.

Gated casting A casting technique in which a casting resin is poured into a gate that conducts the resin into a cavity between the mold and the mold core, which thus determine the shape of the cast part; usually there are risers on the distal ends of the mold to ensure that the mold has been completely filled.

Hot-melt plastics Thermoplastic resins, usually of moderate melting point, that are melted and then cast and cooled to form the part.

Liquid resins Either monomers or syrups, both of which contain materials that can undergo polymerization or crosslinking to harden.

Mold casting A process in which the molding resin is poured into an open mold.

Mold release A substance applied to the mold surface or, occasionally, to the molding resin that facilitates the removal of the part from the mold by reducing the tendency of the part to stick to the mold.

Multiple-pour method A slush molding process in which the casting liquid is removed and a second, different, type of casting resin is introduced, thus creating layers of different materials on the inside of the mold.

Organisol A mixture of resin particles suspended in a plasticizing liquid in which the concentration of the resin is between 50% and 90%.

Painting A resin casting process in which the resin, often in a solvent base, is sprayed or brushed onto the item to be coated.

Parting agent Another name for a mold-release material, a substance added to assist in removing the part from the mold.

Plastisol A mixture of resin particles suspended in a plasticizing liquid in which the concentration of the resin is greater than 90%.

Potting   A casting process in which an item is encased in a resin and the mold that defines the shape of the resin remains with the finished part.

Slush casting   A process in which a casting resin is placed inside a mold, which is usually heated to create a thin layer of solid material against the mold; then the remainder of the casting resin, which is still liquid, is poured out, thus making a hollow part.

Solvent casting   Another name for film casting.

Static powder casting   A slush molding process in which a powder is used instead of a liquid casting resin.

Surface casting   Another name for gated casting.

Syrup   A solution of a polymer in its monomer, thus creating a liquid with moderate viscosity that will polymerize into a single polymer type.

QUESTIONS

1. Briefly describe five different casting techniques.
2. Identify three major advantages of a cast film over an extruded film. Identify one major disadvantage of the cast film.
3. What are plastisols and organisols?
4. Identify three advantages of cell casting over continuous casting. Identify one advantage of continuous casting over cell casting.
5. Discuss the advantages of casting over other plastic molding operations in equipment size and cost.
6. Discuss what attributes might make a resin inappropriate for use as a hot-melt casting resin.
7. Why does casting have more design flexibility than other plastic manufacturing processes?
8. What is a parting agent, and why is it important in casting?
9. Identify three methods for controlling an exotherm in casting.
10. Identify four major types of polymeric liquids used for casting and give one specific example of each type.
11. Define and describe slush molding.

REFERENCES

Baird, Ronald J., and David T. Baird, *Industrial Plastics,* South Holland, IL: The Goodheart-Willcox Co., Inc., 1986.

Carley, James F. (ed.), *Wittington's Dictionary of Plastics,* Lancaster, PA: Technomic Publishing Co., Inc., 1993.

Frados, Joel (ed.), *Plastics Engineering Handbook* (5th ed.), Florence, KY: International Thomson Publisher, 1994.

Milby, Robert V., *Plastics Technology,* New York: McGraw-Hill Book Co., 1973.

"Plexiglas Acrylic Sheet," Philadelphia: Rohm and Haas, PLA-22, February 1992.

Richardson, Terry L., *Industrial Plastics* (2nd ed.), Albany, NY: Delmar Publishers, Inc., 1983.

CHAPTER SEVENTEEN

FOAMING PROCESSES

This chapter examines the following concepts:

- General concepts, advantages, and disadvantages
- Processes to create foams (mechanical, chemical, physical, hollow spheres)
- Processes to shape foams (molding, casting, extrusion, skiving)
- Rebond materials
- Product considerations (materials, shapes, design)
- Operation and process control

INTRODUCTION

Foaming processes are characterized by techniques that cause tiny bubbles to form within the plastic material such that when the plastic solidifies the bubbles, or at least the holes created by the bubbles, remain. The solidified bubble-containing material can be thought of as a cellular structure. The products made by these processes are referred to as *foams* or *cellular plastics.*

Two types of cells occur in cellular plastics. The first cell type is a *closed cell* structure, wherein each of the cells within the plastic is a separate, discrete entity. These closed cells can be compared to tiny balloons or pockets. The walls have no holes in them. If the walls are appropriately impermeable, each cell can hold a gas. The second cell type is an *open cell* structure, wherein the cells are interconnected (each cell is connected to other cells through holes in its walls). The cells cannot hold gas. Rather, gases move easily within and throughout the entire cellular structure. This type of structure is like a sponge.

Plastic foams can also be classified on the basis of wall stiffness. If the walls are stiff, the foam is called a *rigid foam*. If the walls collapse when pressed, the foam is called a *flexible foam*. Both open and closed cell foams can have either flexible walls or rigid walls.

Plastic foams have some physical characteristics that are valuable for several important applications. The cellular structure means that much of the space in the plastic foam is filled with air or some other gas. The low thermal conductivity of gases means that these foams are very good thermal insulators. Some applications that utilize this insulating property are hot and cold cups, building insulation slabs, and pipe insulation.

The open structure of the plastic means that the material is inherently lightweight. This light weight coupled with the nonabsorbing nature of closed cell foams has led to widespread use of closed cell foams for flotation devices such as life jackets, buoys, and pontoons for boats and planes. The use of foams for flotation applications is further enhanced because the closed cell foams do not deflate when punctured since only the few damaged cells collapse.

Another characteristic property of foam structures is the high energy dissipation of impacts on the foam. This occurs because the cell walls, especially if they are flexible, can collapse when the foam is impacted. This collapsing causes the walls to flex and therefore absorb some of the energy of the impact. Applications that take advantage of this energy absorbing property include delicate instrument packaging, cushioning for furniture and other seating, carpet padding, shock absorbers, and crash pads for automobiles and other vehicles. Many of these applications are also enhanced by the light weight of foams.

A somewhat surprising property of some rigid foams is their high ratio of load-bearing strength to weight. The walls of the cells in these foams act like many tiny columns and support quite heavy loads, especially when the weight of the foam is taken into account. Hence, foams of this type can be used as structural parts and cores for structural parts, especially where weight is important. Furniture frames, airplane wings, space structures, and some auto parts, for example, use this load-bearing capability of rigid foams. Many other applications require that the foam material occupy space to give a desired shape and resist moderate impacts, such as would occur in automobile dashboards.

The advantages and disadvantages of foams are outlined in Table 17.1.

Most thermoplastic and thermoset resins can be foamed to make cellular plastics. In each case two phenomena must occur. First, the resin must be foamed. Second, the foamed resin must then be shaped and solidified. Some resins are foamed as molten polymers and then cooled. Other resins are liquids that are foamed and then reacted to form solids. The two stages of foam processing (foaming and forming/solidifying) are discussed separately.

Table 17.1 Advantages and Disadvantages of Foamed Versus Nonfoamed Plastics

Advantages	Disadvantages
Light weight	Slowness of most processes
Low thermal conductivity	Evolution of gases in some processes
High support per unit weight	Variable density with some parts
Low-cost molds	Loss of some mechanical properties
Many methods available to create foams	
Many molding methods	
Excellent impact protection	

PROCESSES TO CREATE FOAMS IN RESINS

The wide variety of resins capable of being foamed suggests that many processes or techniques have been developed for foaming the resins. This text examines four major methods: mechanical foaming, chemical foaming, physical foaming, and addition of hollow glass spheres.

Mechanical Foaming

A liquid resin or resin solution can be mechanically beaten or whipped to disperse air throughout the material. The frothy liquid can then be shaped and hardened. This mechanical foaming process is similar to whipping cream or making a milk shake. Resins most commonly foamed by the mechanical method are plastisols, vinyl esters, urea formaldehydes, phenolics, and polyesters. Two important applications for mechanically foamed plastics are as a foamed backing on carpets and as a part of thick linoleum flooring.

Chemical Foaming

Chemical foaming results from the formation of a gas through the breakdown of special chemicals called *foaming agents* or *blowing agents*. When these foaming agents are added to liquid or molten resins, foaming occurs. The breakdown is usually triggered by heat. Hence, the foaming agents are added to the liquid resin, the mixture is heated, and foaming occurs. Shaping and solidifying can then be done.

One of the most common examples of this type of foaming is the rising of bread dough. The chemical reaction that forms the gas (carbon dioxide) is the action of yeast. Several organic molecules that evolve gases when heated are used commercially to foam plastics. Common resins foamed by this method are polyurethanes, most elastomers, urea formaldehydes, and silicones.

A related process is foaming by the flash (rapid) evaporation of water formed as part of the resin polymerization or crosslinking reaction. While not specifically considered a foaming agent, water vapor is certainly the material that produces the foam. Because the water is formed by a chemical reaction, this foaming technique fits in the chemical foaming category. Polyurethanes often are foamed by water evolution and volatilization.

Physical Foaming

If a gas is forced into a liquid or molten resin and then the pressure is reduced, the gas is liberated quickly and a foam is created. This is called physical foaming because no chemical bonds are broken in the process of evolving the gas. An example of this method is the foaming that occurs with shaving cream as it exits an aerosol can.

Another type of physical foaming occurs when a volatile solvent is dissolved in a resin and the solution heated. The heating causes the solvent to volatilize and a foam is created. An example of this method is the foaming of polystyrene with pentane. Other plastics foamed by the dissolved solvent method include SAN, PP, PE, and PVC.

Hollow Glass Spheres

Hollow glass spheres (microballoons) can be mixed into a liquid or molten resin as a filler. The spheres have a very low density and so the resultant material has the characteristics of a foam. These materials are called *syntactic foams*. These foams are closed cell and have found many applications in flotation devices, noise alleviation, thermal insulation, and high-compression-strength devices. These materials are also used widely to make low-cost molds for low-pressure molding processes.

PROCESSES TO SHAPE AND SOLIDIFY FOAMS

After the foams have been created, several processing methods can be used to shape and solidify the foams into useful products. In all these processes, the pressures normally involved in shaping the foams are relatively low since high pressures would likely cause the foam bubbles to collapse and destroy the foam. The low pressures allow the use of low-cost molds or other shaping devices, as is the case with other low-pressure plastic-molding methods. Some foams are shaped and solidified without the need to externally heat the foam. This low-temperature foaming capability gives additional latitude in the choice of mold material and construction. Foam molds can, therefore, be made from a wide variety of materials including metals, wood, rigid plastics, and elastomers.

As previously noted, foams can be either flexible or rigid. The flexibility of foams also allows great latitude in molding and shaping since undercuts and other fairly complex and difficult-to-mold shapes can be extracted (stripped) from the molds without the need of complicated slides or removable cores.

Because all foaming methods cause the volume of the fluid material to expand, all such methods can be called *expansion processes*. In a similar view, all plastic foams can be called *expanded plastics*.

Each of the major shaping and solidifying processes can be used with any of the foam-creating processes previously described in this chapter. These foam-creating processes can be used both with thermoplastics that are melted, foamed, and then cooled to solidify, and with thermosets that are liquids that are foamed and then crosslinked to harden. In discussing these shaping processes, the term resin will be used to include the resin itself and any filler, solvent, initiator, or other additive that might determine the specific properties of the product. The presence of foaming or blowing agent will be specifically noted since the time for adding that agent may be determined by the shaping process.

The processing methods to shape foams differ fundamentally in the manner in which the foamed materials are *introduced into the molds* or other shaping devices. They also differ in the manner in which the *foaming process is controlled*.

Molding

Foamed liquid materials can be injected or poured directly into molds that define the shape of the product after solidification. Two somewhat different methods are used to control the manner in which the foam expands within the mold. In the first method, called *low-pressure foam molding,* a metered volume of liquid resin containing the foaming agent is introduced into the mold. The volume of this as yet unfoamed or partially foamed material is much less than the volume of the mold, but it is soon allowed to expand to fill the mold. The trigger for initiation of the foaming action can be a reduction of pressure when the material is introduced into the open mold, much like the foaming of materials that occurs when the materials are released from a pressurized can (such as shaving creams, whipped cream, and tire stop leak), or it can be an increase in temperature if the mold is heated. The type of triggering action depends on the type of foaming agent used.

Expansion of the material takes place because the foaming agent creates bubbles within the liquid resin that cause the volume of the resin to expand (see Figure 17.1). This expansion is limited by the size of the mold, and is called low-pressure foam molding because the expansion is done against no external pressure. The expansion process itself may create a slight pressure, but this pressure is trivial.

The other method of controlling the expansion of the foaming process in the mold involves expansion of the foam against an external pressure. Hence, this method is called *high-pressure foam molding.* It is important to note that even though the process is called high-pressure molding, the pressures are high only in comparison with the essentially zero pressure of the low-pressure foam molding method and are not high when compared to plastic-molding methods such as extrusion and injection molding.

In the high-pressure foam molding method the liquid resin with foaming agent is introduced into a mold until the mold cavity is filled completely. Then, the mold cavity itself

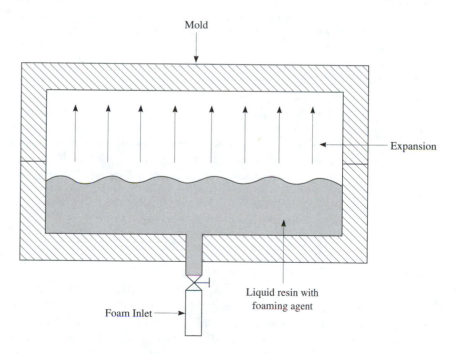

Figure 17.1 Low-pressure foam molding.

is mechanically expanded at a set rate as the foaming occurs. The expanding cavity controls the amount of foaming that occurs and therefore determines the density (volume and amount of foaming) of the foam. This process is shown in Figure 17.2.

The types of products made by the two foam molding processes can be quite different. The low-pressure process tends to create open cell foams while the high-pressure process creates predominately closed cell foams. Under certain conditions, however, closed cell foams can be made by the low-pressure method and open cell foams can be made by the high-pressure method. For instance, if the expansion of the mold during high-pressure foam molding process is quite rapid, the cell structure may convert from closed cells to open cells. This event is known as "blowing the cells open" and can be a problem when closed cell foams are desired. The variance in molding conditions that will change a closed cell foam into an open cell foam can be very slight, perhaps only a few degrees difference in the temperature of the liquid or a slightly faster expansion speed in the mold.

In both molding processes the cells that touch the walls of the mold and some of the cells just inside these outer cells collapse and form a nonfoamed plastic coating, or *integral skin*. The thickness of this skin is determined by the amount of material introduced into the mold, the temperature during molding, the nature of the mold surface, the pressure of mold release, and the pressure during the expansion process. The type of foaming agent used can also affect the thickness. Integral skins also form around inserts placed within the molds.

Integral skins affect several important properties of foams. For instance, the permeation of a liquid or a gas through the skin is much lower than would occur through the remainder of the foam, especially in open cel foams. The integral skin increases the strength of the foam. This feature of integral-skinned foams is used to limit the indentation of springs into the foam when the foam is used as cushioning over furniture springs, for example.

An important process for foam molding is called *reaction injection molding (RIM)*. In RIM two reactive components are metered together and mixed in-line so that they begin to react in either a polymerization or crosslinking reaction. (The RIM process, which is also used to make nonfoamed parts, is described in more detail in the chapter on compression and transfer molding.) To make a foam by the RIM process, a foaming agent is added to the resin stream and mixed in the in-line mixer. This mixed resin material is then injected into a closed mold. The liquid resin containing the foaming agent begins to expand inside the mold. By balancing the rate of resin injection and the rate of foam formation the mold will fill with a material of the desired density. This process is illustrated in Figure 17.3.

The reactive nature of the resin components is the origin of the "reaction" part of the RIM name. The most commonly used resin for RIM is polyurethane, where the polyol is part A and the isocyanate is part B. In some cases, the foaming can occur by the reaction of the isocyanate with water to create carbon dioxide. Hence, the foaming agent can simply be water, although other foaming agents can also be used.

In contrast to the low-pressure foam molding method, in RIM the injection of material into the mold continues as the foaming inside the mold also continues. Therefore, the expansion is against a constant pressure (the injection pressure) rather than a free expansion. Another difference is that the RIM process can be used to encapsulate an insert or a core, whereas the free expansion of low-pressure foam molding would not effectively surround such a barrier to expansion within the mold. When the core is a reinforcement such

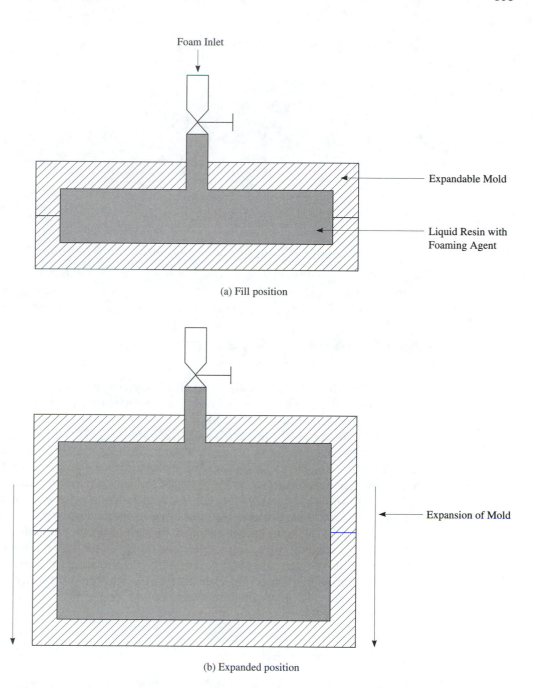

Figure 17.2   High-pressure foam molding.

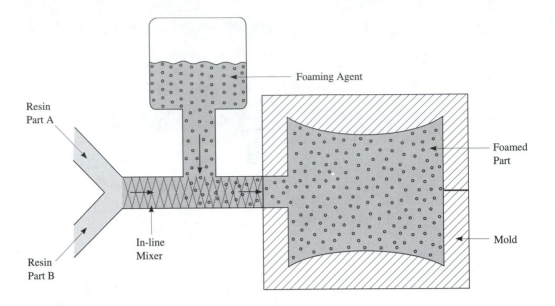

Figure 17.3   Reaction injection molding (RIM) foaming method.

as fiberglass, the process is called structural reaction injection molding (SRIM). This process and a similar process called resin transfer molding (RTM) are discussed in detail in the chapter on composite materials.

Another foam molding process is called *foam-in-place*. This process is similar to RIM in that the foam is injected into a cavity at a rate that balances the foaming of the material in the cavity and the injection rate. This balance gives the proper foam density. The difference between foam-in-place and RIM is that in foam-in-place the cavity is between the walls of the product and the foam is not removed. Foam-in-place is done, for instance, to insulate between the walls of a refrigerator by injecting the foam between the joined inner and outer walls. The foam simply solidifies in place. This method allows sequential assembly steps, including the wiring of lights and other devices within the refrigerator walls, then joining of the inner and outer walls, and finally the injection of the thermal insulation foam. The ability to inject the foam after all the other steps have been done simplifies the assembly process. To facilitate the foaming into long, deep cavities, a foam-injection nozzle places the foam inside the cavity extremities (see Figure 17.4).

Extrusion Foaming

The process of *extrusion foaming* is the same as normal extrusion except that a foaming agent is added in the extruder so that when the resin exits the extrusion die, the resin begins to expand. This expansion is controlled by some mechanical system such as conveyor belts that are spaced apart to give the desired foam density and final thickness. Paper or some other sheet material often is used as a conveying medium so that the foam will move

Figure 17.4 Foam-in-place technology used to insulate between the walls of a refrigerator.

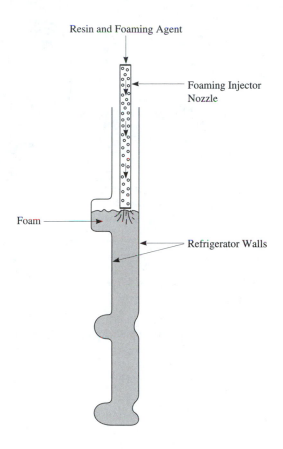

easily between the conveyor belts. Paper can be used to limit the expansion on top and bottom and on the sides. Therefore the expansion occurs inside a moving box. A extrusion foaming system is illustrated in Figure 17.5.

Extrusion foaming is principally used to produce flat foam slabs, such as the type used widely for building insulation panels. These slabs are cut off by a saw that is mounted on the conveyor line and that moves with the part so that the cutoff is square. The paper or other coating material can be removed on line or shipped as a protective covering.

Thin extrusion-foamed sheets have found a major application in the packaging industry, where they are thermoformed to create clamshell containers used for packaging eggs and fast foods. Egg cartons utilize the foam's cushioning and shock-absorbing nature; fast-food containers use the foam's thermal insulative capability.

Casting Foams

The process of *casting foams* differs from the other foaming processes because in casting no mechanical limit such as a mold or conveying system is placed on the foam's expansion. In the most common technique for casting foams, *pan casting,* a predetermined

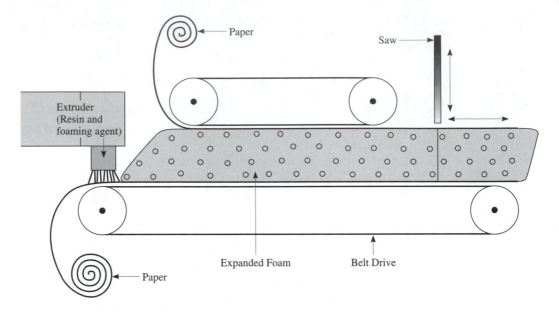

Figure 17.5   Extrusion foaming.

amount of resin and foaming agent are combined and then poured into an open pan. The foaming agent expands the resin, causing it to rise until the foaming agent has been spent. (This process is similar to the rising of bread dough in a bread pan where the dough is the resin and the yeast is the foaming agent that chemically produces the gas.) The pan casting method is illustrated in Figure 17.6.

The resulting foam material is called a *bun* and has an integral skin that is denser than the rest of the foam. This integral skin is like the crust on a loaf of bread and is formed in both cases from the collapse of the cells on the surface of the foam when they touch a material that is different from the surrounding foam material. In this case the external material is the air. When the foam is cured by heating, the crust thickens because the heating causes cells just below the surface also to collapse.

The foamed buns are removed from the pans and then sliced or cut into the desired shapes. The most common shapes are slabs. This slicing or cutting is given a special name, *skiving*. Skiving of foams is done with knives or saws as is done with other plastic materials, but it also is done conveniently with a hot wire. Typically, the wire is a high-resistance metal heated by simply running an electric current through it. The wire is held by a clamping mechanism that can be adjusted vertically to obtain various slab thicknesses. The bun is then moved against the wire slowly so that the wire melts the foam. As the bun is moved past the wire, a foam slab is separated from the rest of the bun. This process is illustrated in Figure 17.7.

Another foam casting technique is *foam spraying*. In this method the resin and foaming agent are mixed together and then held under pressure in a pressurized bot-

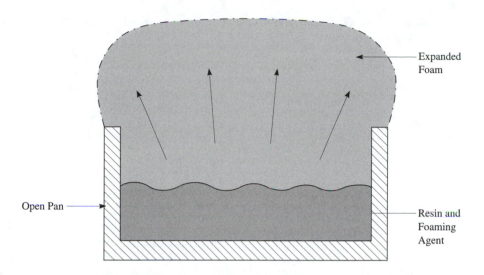

Figure 17.6 Foam casting in a pan.

tle. A propellant for carrying the resin and foaming agent is also included in the bottle. The material is sprayed onto a surface to be coated with the foam by simply depressing the valve on the pressurized bottle. The resin, foaming agent, and propellant are sprayed onto the surface of the item where the foaming agent causes the resin to

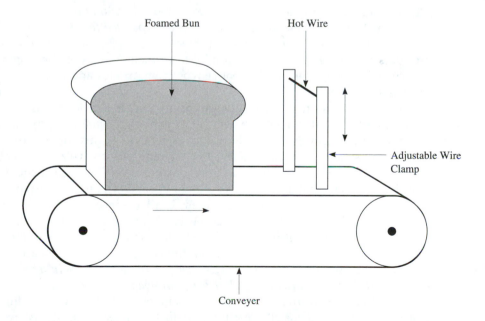

Figure 17.7 Skiving a foamed bun using a hot wire.

Figure 17.8 Foam spraying to coat an object with foam.

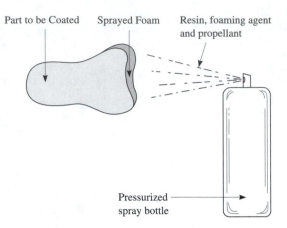

expand freely. The propellant escapes into the atmosphere. This system is illustrated in Figure 17.8.

Spraying is a convenient method for applying foam insulation to roofs, pipes, and walls where slabs might be difficult to use. Spraying also has a distinct advantage when the part to be covered with foam is irregularly shaped.

Expanded Foam Molding

Although all foaming processes can be considered expansion processes because resin volume expands, one process in particular is called *expanded foam molding,* or *expandable bead molding,* and has some unique characteristics not present in the other foam-shaping processes already discussed. The expanded foam molding process (see Figure 17.9) is of particular importance in making polystyrene foam articles and is illustrated using that resin as an example.

The basic starter material for the expansion process is made when polystyrene is simultaneously foamed and polymerized. This is done by polymerizing styrene in a solution of water and an organic volatile solvent such as pentane. In this environment, the polymer forms small spheres or beads of foamed polystyrene with the volatile solvent trapped inside the small, closed cells. The beads are approximately 0.02 to 0.08 inches (0.003 to 0.02 mm) in diameter, roughly the size of a grain of sand. The beads are removed from the water and are dried. In this form, the beads can be shipped to the molders and can be stored, provided they are stored at relatively low temperatures (below room temperature). This temperature will largely prevent the volatile gas from escaping out of the beads. However, some volatile gas is always present, so storage containers should be as airtight as possible and the storage facility should be well ventilated. *(Caution: The storage facility should contain only nonsparking electrical equipment.)*

The molder begins the actual expanded foam molding process by heating the beads so that they expand to about 20 times their original size, a step called *preexpansion.* Preexpansion is done by conveying the tiny beads into a chamber (expansion chamber) where

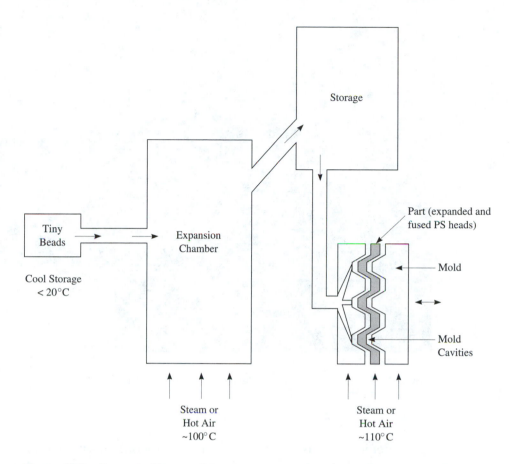

Figure 17.9   Expanded foam molding process.

hot air or steam is blown through them. The increase in temperature to about 212°F (100°C) causes the styrene to soften and volatilizes the trapped solvent. The gaseous solvent expands because of the heating and pushes out on the softened polymer, thus enlarging the spheres (beads). The size of the expanded beads is determined by the temperature and the amount of time the beads are left at the elevated temperature. A photo of the preexpanded and expanded beads is shown in Photo 17.1.

By preexpanding the beads before molding, the molding is much more uniform and predictable than it would be if the beads were molded and expanded simultaneously. Attempts to mold and expand the beads result in some sections of the molded parts having poorly expanded beads because heating is often less uniform within the confines of a mold than in the relatively open environment of an expansion chamber.

The expanded beads, roughly the size of apple seeds, are stored for about 8 hours to allow the internal pressure inside the beads to equilibrate with atmospheric pressure. Some of the volatile gas will inevitably escape during this storage period. Since a small

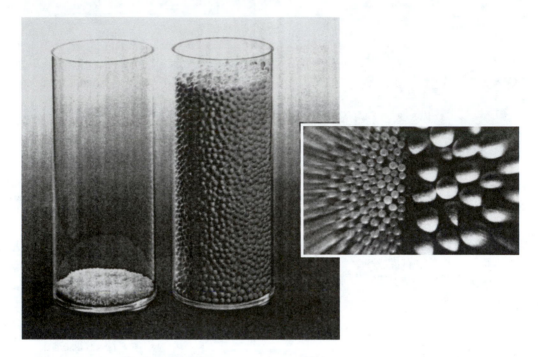

Photo 17.1   Expanded polystyrene beads as shaped and then after expansion (Courtesy: Marko Foam Products, Inc.)

amount of internal gas retention is required for the subsequent molding operation, storage time should not be so long that all of the volatiles escape. Storage at room temperature for up to a week has proven to be acceptable. Expanded beads are used for other applications besides molding, one example of which is as filling material in beanbag chairs.

Molding of the expanded beads is done by conveying the beads to the mold area and blowing them into the cavities of a closed mold. After the cavities are filled, hot air or high-pressure steam is blown through the cavities such that the temperature of the beads reaches about 212° to 220°F (100° to 110°C). At these temperatures the styrene again softens and the residual volatile solvent expands, further enlarging the beads. Because the beads are confined in the cavities, the softened particles press together and eventually fuse to become a single mass of connected beads. If done properly, the fusion of the beads is such that all holes (pores) between them are sealed by the softened resin. The parts are then cooled by blowing cool air or water through the mold or by running cool water through cooling pipes within the mold body. The mold is then opened and the parts removed. (The machine that holds, opens, and closes the molds is much like the hydraulic system and mold handling system of an injection molding machine.)

The expanded foam molding process is used extensively for making drinking cups, ice chests, and packaging. Careful examination of the products reveals their bead-nature.

REBOND

Rebond foam differs from the other foam materials discussed in that its starting material is not a resin with a foaming agent. Rather, the starting material is scrap flexible foam. This scrap foam is chopped into small pieces (about the size of peas), blended with a binder resin (usually the same resin as the bulk of the flexible foam), and then mixed until the binder has coated the chopped particles. Next, the blended mixture is placed into a mold and heated to cure the binder. The resulting part could be in the finished shape or a block of material that is then skived (slit into thin sheets) to make sheets of the foamed material.

Some consideration should be made regarding the nature of the binder. As already indicated, the binder should be the same material as the scrap to ensure successful bonding. The density of the binder is also a consideration. If the binder is too stiff, the rebond may not have the required resilience. Therefore, binders often are foamed, just as the rest of the scrap material. The only problem with this is that a foam binder has less tensile strength than a nonfoamed equivalent binder, which may reduce the rebond material's shear strength.

Heating the material in the mold often is done by blowing steam through the mold. This causes the binder throughout the material to cure more evenly than could be achieved by heating external to the mold, like in an oven. The problem with trying to use only external heating is that the foam is such a good thermal insulator that the inside of the bun may not reach the appropriate temperature until the outer material has begun to decompose.

In addition to the obvious cost advantage of using scrap material, this process allows the density of the final product to be controlled by the size of the mold. The scrap and binder mixture can be placed into a movable mold and then compressed so that the cured material has a higher density than the foam scrap from which it is made. This higher density is especially important for applications such as carpet padding where impact protection is an important property.

Rebond foam often is less costly than virgin foams, so it is useful for high-volume, low-value applications. In addition to carpet padding, these applications include low-cost molded packaging, low-cost comfort cushioning, and filling for pads where covers will be used to prevent shear forces on the foam itself.

One of the major problems with rebond is that the binder may not hold the chopped particles together well. Shearing forces are particularly destructive for rebond foams. Another problem is the speckled appearance that arises from the use of multicolored scrap.

PRODUCT CONSIDERATIONS

Materials

Most thermoplastic and thermoset materials can be foamed by one or more of the foam-forming processes described in this chapter. The most common resins used to make foams include polyurethane, PVC, polystyrene, polyethylene, polypropylene, epoxy, phenolic,

ABS, ureaformaldehydes, silicones, ionomers, and cellulose acetates. Fillers in moderate concentrations can be added to most of these resins to obtain desired physical, chemical, or mechanical properties, such as flame retardance. Short-length reinforcements can be added to obtain additional compression strength since the reinforcements tend to orient along the cell walls and therefore give additional strength when the walls are collapsing under a compression force.

One key filler for some resins is microballoons that are used to create the foam structure. These materials, called *syntactic foams,* are made by mixing small glass spheres (approximately 0.0012-inch, or 0.03-mm, diameter) with the liquid resin. The foam is, naturally, closed cell. Major applications for syntactic foams include low-cost tooling, noise damping, thermal insulation, and flotation.

Foams can also be made from resins blended with rubbers to achieve a natural resilience. Closed cell PVC foams with nitrile rubber are especially well known (Ensolite®). PVC can also be plasticized to obtain soft and resilient foams.

Polystryene foams, especially closed cell foams such as those made for the expanded foam molding process, are almost completely nonabsorbing of water. Other foams that are essentially nonabsorbing of water include closed cell polyethylene, polypropylene, and polystyrene.

Some resins, especially low-density polyethylene, can be crosslinked as part of the foaming and forming processes. Crosslinking strengthens the cell walls and reportedly improves several physical and mechanical properties over noncrosslinked polyethylene foams.

Applications

The major applications for foamed plastics, previously introduced in this chapter, are thermal insulation, lightweight, energy absorption, and load bearing with good strength to weight. To best meet the many requirements needed in these applications, foams can be made over a wide range of densities. The lowest-density foams are about 0.12 pound/cubic foot (0.002 g/cc). These foams are much less than the density of normal plastics, which range from 5.4 to 15 pounds/cubic foot (0.9 to 2.5 g/cc). The highest densities for foams would be just slightly lower than the normal density of the resin from which the foam is made. In other words, foam densities can range in a continuum from very low densities to almost no foaming of the resin at all. For comparison with foams that are reasonably well-known, foams used for cushioning are typically 2 to 5 pounds/cubic foot (0.03 to 0.08 g/cc) density.

The basic nature of the cell type has some relationship to the applications of the foam. Structural applications require that the integral skin be stiff and solid and can be either closed or open cell foams. Closed cell foams are usually superior to open cell foams in thermal insulation applications and flotation applications, whereas open cell foams are better for soft cushioning. Energy dissipation can be favored by either type of structure depending on the nature of the shock or vibration to be absorbed.

Several tests are used to measure the energy dissipation capability of materials. One of these tests is the traditional stress-strain test that determines the strength of materials. For important cushioning applications, the compression mode is the important type of stress-strain test. Because the foam materials are often very soft, an indentor with a large

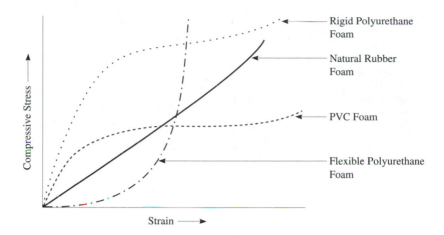

Figure 17.10 Compression stress-strain curves for soft foam materials.

area (typically a flat plate about 4 inches (10 cm) in diameter) is used. The cushion is placed under the indentor and the compression strength measured. Curves typical of some well-known types of foams are given in Figure 17.10.

Note that the curve for flexible polyurethane initially gives a high deflection (high strain) with only a minor increase in stress or force applied. This indicates that the material is quite soft. With continued applied force the curve suddenly rises, indicating that the material has "bottomed out" and that it has become stiff. This rapid change in the curve is the point where the cells have all collapsed and the material is behaving like a non-foamed plastic. Until this bottoming out point is reached, the foam is very soft and comfortable and would make an excellent cushion or packaging material where the object to be cushioned is not too heavy or the type of impact anticipated is not too severe.

The other materials also eventually bottom out but at widely differing stresses. Some of the other foams are initially stiff and then have a crushing of the cells at a fairly high stress level (such as the rigid polyurethane). The PVC foam also has an initial stiffness, although not as much as the rigid polyurethane, and then a long flat region where the cells are being crushed. The natural rubber foam has a nearly constant increase in stress with increasing strain indicating a resilience in the cell walls.

Another test that measures the ability of a material to absorb energy is conducted by placing an accelerometer in a device and then placing the device on top of a cushioning material. The package of device and cushioning material is then subjected to a force through the cushioning material, with the accelerometer recording the amount of transferred force. A high accelerometer reading means that the cushioning material did not absorb much of the energy. A low accelerometer reading means that the cushioning material dissipated the energy of the force well.

Figure 17.11 shows the curves for a flexible polyurethane foam, an expanded polystyrene foam, and a polyethylene foam. An idealized breaking point line is also depicted. Any acceleration (Gs) greater than the idealized line will result in damage to the part to be protected. At low stress levels, polyurethane gives the best protection. This indicates

Figure 17.11 Force transmitted through a cushioning material in terms of the stress of the initial force and the accelerometer reading (in Gs) of the response sensor.

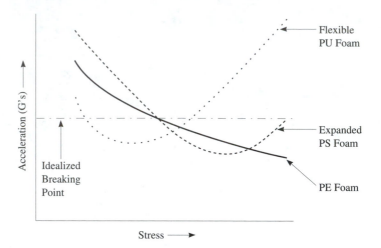

that the flexible polyurethane material is able to absorb and dissipate the applied stress. At higher stress levels, the flexible polyurethane is less effective, presumably because it is nearly bottomed out by the applied stress and therefore cannot absorb and dissipate the energy as well as it could at lower stress levels. The expanded polystyrene is better at higher stress levels than at lower stress levels. This stiffer material evidently transmits energy at low stress levels because there is little deflection of the cell walls and the material behaves almost as if it were not foamed. At higher stress levels, however, the walls are crushed by the stress (impact) and a large amount of the energy is absorbed. The polyethylene foam is similar in behavior to the expanded polystyrene.

Protection from continuous vibrations is another important aspect of packaging materials that can be measured. This measurement is done by placing an accelerometer on a part that is resting on the packaging material and then subjecting the package to vibrations of various frequencies. The accelerometer measures the amount of vibrational energy transmitted through the packaging material. Ideally, none of the energy will pass through over the entire frequency range. However, all materials have some frequencies where the material does not absorb energy well but rather, vibrates in resonance. For most materials, these frequencies of nonabsorption (resonance) correspond to the natural vibrational frequencies of the material. A vibrational frequency curve for a typical foam material is given in Figure 17.12.

These curves are used by comparing the amount of transmitted acceleration and the natural frequency of the material with the expected amount of acceleration that the part would be able to withstand over extended periods and the expected vibrational frequencies typical of the environment anticipated for the package. For instance, if the package is to be transported by rail, the most common vibrational frequencies that will be encountered are 2–7 Hz and 50–70 Hz. These frequencies arise from the frequencies of the joints in the rails and the swaying and vibrating of the cars. If the material has a natural frequency in the ranges for rail transportation, and if the amount of vibrational force transmitted through the material approaches the amount that will cause damage to the part, then the

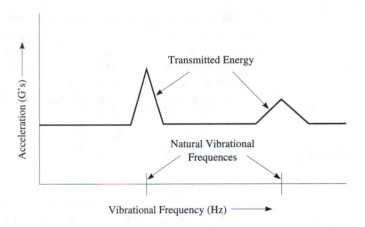

Figure 17.12 Vibrational frequency and transmission curve for a typical foam material.

material should be avoided. Typical vibrational frequencies for various transportational modes are as follows: train—2–7 Hz and 50–70 Hz; truck—4–6 Hz and 12–14 Hz; ship—2–20 Hz; and airplane—1–5 Hz and 25–200 Hz.

Another important characteristic of foams is the *compression set* taken by foams after repeated cycles of compression. Compression set is the amount of permanent indentation or loss of thickness recovery that occurs after repeated impacts or long-term compression. Rigid foams have more compression set than flexible foams, but all foams have some set. The amount of set increases with the number of impact cycles and with the length of time and amount of long-term load. Curves showing the compression set of a typical foam material are given in Figure 17.13.

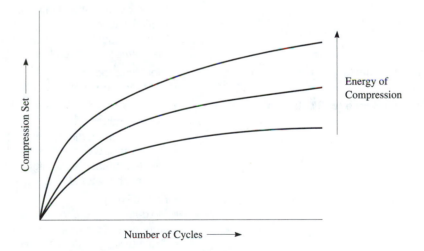

Figure 17.13 Compression set curves for a typical foam material showing the effects of the number of cycles and the amount of compression energy.

All of the curves initially show a significant amount of set. This is typical of most foams and reflects the inability of the cell walls to return to their original height after being crushed. After repeated cycles, however, cell wall recovery is more constant, probably reflecting the natural resiliency of the plastic material. Note that the amount of set is greater as the energy of the compression impact increases. This trend is as expected because the deflection of the walls is greater with more impact energy, so recovery is, on average, less.

CONTROL AND OPERATION

The most important process parameters in forming the foam are the amount of the foaming agent and the conditions under which the foaming occurs (temperature, pressure, and expansion mixing conditions). The composition of the resin component (amount of resin,

Table 17.2   Troubleshooting Guidelines for Foaming

Problem	Possible Cause
Incomplete filling of the mold	• Insufficient material • Mold too cold • Incomplete mixing of foaming agent • Loss of blowing agent • Trapped gas in mold or poor venting
Internal voids	• Incorrect pour • Large amounts of trapped air • Excessive foaming agent
Pits or holes in surface	• Too much mold release • Not enough material in mold • Mold too cold • Water present in mold • Melt temperature too low
Distorted parts	• Mold time too short • Poor mold design
Part too dense (cell collapse)	• Poor mixing • Contamination • Not enough foaming agent • Preexpanded beads too old • Injection pressure too high • Too much material
Part density varies	• Poor mixing • Mold temperature too low • Cure time too short • Too much catalyst (initiator)
Poor mold release	• Not enough mold release • Poor mold design • Cold mold

solvent, filler, and other additives) can also be important in determining foam characteristics. In general, the foam can be described by the number of cells and the cell size. The optimal number and size of cells depend on the application of the foam and cannot, in general, be stipulated.

The critical parameters for molding and shaping can vary greatly from one process to another. However, the amount of resin and the pressure under which it is expanded are generally key parameters to control.

Despite the diversity of foam-making processes, there are some general troubleshooting guidelines. These are presented in Table 17.2.

Maintenance and Safety

The most important safety considerations concern the flammability of gases used for foaming agents. One particularly dangerous gas is the pentane used in expanded polystyrene foams. Pentane is heavier than air and will burn readily. Therefore, rooms in which pentane might be present should be ventilated from floor level.

Another danger in foam manufacturing is exposure to isocyanates present in polyurethane formulations. Isocyanates react readily with water and can be hazardous if allowed to come into contact with the mucous membranes of the body (such as the eyes or nose).

CASE STUDY 17.1

Foam Insulation

The use of foamed plastic for consumer and industrial goods is a major industry and one of the most important segments of that industry is insulation. The high volume of material required for insulation places it at the top of uses for foamed plastics. The key property for these applications is, of course, the ability of the foamed plastic material to resist the transmission of thermal energy or, in other words, to thermally insulate.

Have you ever wondered why you can barely feel the heat from the hot chocolate in a foamed cup or why the ice in a picnic cooler melts at such a slow rate and can be used to keep food cool for long periods even on hot days? In both of these examples a foamed plastic is probably responsible for insulating the hot or cold material from the surrounding environment.

Plastic foams can be formulated to have excellent thermal insulative capabilities. Some foams are better than others and so a wide range of properties can be chosen with a balance between insulative ability and other required properties such as strength, chemical resistance, or cushioning capability. In some cases the plastic foam is in the form of slabs that are used as insulation in walls and other locations where the ease of application of a rigid material is of high value. In other cases the plastic foam is sprayed and it foams in place, thus conforming to the shape of the container or conforming to the shapes of the walls between which it is sprayed.

Thermal insulative capability can be measured quantitatively, either by *thermal resistivity (R)* or by *thermal conductivity (K)*, where $R = 1/K$. Thermal resistivity is a measure (using English units) of the increase in temperature (°F) that occurs in a specific length of time (1 hour) from the application of a specific amount of thermal energy (1 Btu) through a sample of the material that is of a standard area (1 foot^2) and thickness (1 inch). Some typical *R*-values for common insulative materials (and a few other materials for comparison) are given in Table 17.3.

R-values higher than those in the table can be obtained by increasing the thickness of the materials beyond the 1-inch standard. For instance, the *R*-value for fiberglass can be increased from 4.2 to 11 by increasing the thickness to 4 inches. At 6 inches the *R*-value for fiberglass is 20. Another method of increasing the *R*-value is to coat the materials with foil or some other reflective material. These reflective coatings reduce the amount of thermal energy that moves through the insulating material by one or more of the thermal conductive mechanisms.

A brief review of the mechanisms for heat transfer may be useful in understanding these *R*-values. Heat is conducted by three mechanisms: radiation, conduction, and convection. In *radiation* the heat travels in waves (similar to light waves). No medium is necessary for radiative heat transport. The existence of the heat source and a capability of that heat source to lose energy through emission are the only requirements to have the hot body, which radiates the energy, gradually lose that energy. Radiation can move through a vacuum.

Heat transfer by *conduction* requires contact between the hot body and the body to be warmed. This contact allows some of the heat energy from the hot body to move directly to the cold body. This heat energy is seen by an increase in the vibration of the atoms in the cold body (and a corresponding loss in the vibration of the atoms in the hot body).

Convection is a transfer of heat energy from a hot body to a cold body by some medium or material that possesses mass. Convection differs, therefore, from radiation,

Table 17.3   *R*-values for Some Typical Insulating Materials

Material	*R*-value (°F–h–ft^2/Btu–in.)
Foamed polyethylene	7.1
Foamed polystyrene	5.0
Fiberglass	4.2
Mineral fiber	4.2
Perlite	3.3
Cellular glass	2.6
Wood	2.5
Calcium silicate	2.5
Airspace	2.3
Concrete block	2.0
Concrete	1.0
Steel	0.0032

which needs no medium, but it is similar to radiation in that the hot and cold bodies can be separated. This mass transfer medium carries the heat with it. The most common form of convection uses air as the heat-conveying medium. Therefore, a hot body (perhaps a heating element) heats the air that is blown past it and then the air takes the heat to another location where it heats a second, colder body. This type of heating is typical of a convection oven.

Foams are excellent thermal insulators because they reduce significantly all three types of heat transfer. Radiative heat transfer is blocked by the presence of the foam, which has little response to radiation heating.

Foams physically separate the hot and cold bodies, thereby reducing conductive heat transfer. This separation effectively prevents the hot body from transferring its energy directly to the cold body by contact.

Heat transfer by convection is reduced by foams because of the interference between the air passing through the foam (that carries the heat) and the foam structure itself. This interference causes some of the energy being carried by the air to be transferred to the foam, thereby heating the foam rather than the cold body being insulated by the foam. In closed cell foams, this interference from the cell structure is even more pronounced than in open cells foams where the air can move through the structure. Additional insulative capability can be achieved in closed cell foams if the gas inside the cells is not air. For instance, the R-values of hydrogen gas (1.26), nitrogen gas (0.18) and chlorofluorocarbons (0.052) are all lower than the value of air (2.3). Therefore, the insulating capability of closed cell foams can be changed by the type of gas that is trapped in the foam. The trapped gas is generally the same as the gas that was used to produce the foam (the foaming agent).

The use of chlorofluorocarbons (CFCs), which are the best insulating gases for foams, has created problems. These CFCs are released into the atmosphere by the foaming process and when the cells of the foam rupture. CFCs are suspected of harming the earth's ozone layer by reacting with the ozone molecules. Although not proven, computer simulations indicate that this harmful effect could eventually reduce the amount of ozone and thereby cause an increase in the amount of harmful ultraviolet sun rays that reach the earth's surface. As a result of these suspicions, CFCs have been banned from some applications and a concerted effort is under way to eliminate their use completely. Other gases for foaming are being sought. Several alternatives have been found with reasonable, although somewhat inferior, insulating capabilities.

SUMMARY

Plastic foams are made by foaming a liquid plastic, either a melt, a liquid resin, or a solution, and then forming or shaping the foamed material. The foaming is done by mixing a foaming agent with the resin. Several types of foaming systems are used. These include mechanical foams (where air is beaten into the resin like air in whipped cream), chemical foams (where a chemical reaction produces the gas like yeast produces carbon dioxide in bread making), physical foams (where the solvent is caused to foam by some physical process, heating or reducing pressure, as is done with aerosol shaving cream), and hollow spheres (where the premade glass microballoons are mixed into the resin).

Shaping and forming the foams can be done using several techniques. One of the most common is molding. The mold is partially filled and then the resin containing the foaming agent simply expands to fill the mold. This system is called low-pressure foam molding. In another molding method, high-pressure foam molding, the mold is completely filled with the unfoamed resin. Then, the mold is slowly increased in size as the resin foams. This controlled foaming often produces a closed cell foam, whereas the low-pressure foam molding often produces an open celled foam.

Foam-in-place is a molding process where the mold is the walls of the container to be insulated. For instance, foam is sprayed between the inner and outer walls of a refrigerator where it then solidifies.

Another technique for foaming plastics is casting. In this technique the expansion of the foam is not contained. The resin with foaming agent is poured into a container shaped like a large bread pan. The foaming occurs and the resin rises to whatever height is dictated by the amount of resin, the amount of foaming agent, and the foaming and molding conditions. The resulting material is then usually sliced (skived) into slabs.

Extruded foams are formed when the foaming agent is mixed into the resin while it is still in the extruder. Then, when the resin exits the extruder, the foaming agent is activated (usually by the decrease in pressure) and the foam is created. The size of the extruded foam is generally limited by moving the foam between parallel conveyor belts. The solidified foam is usually made to the desired thickness, but it can be skived if desired.

Expanded foam is very popular with polystyrene. In this method the polystyrene is foamed as it is polymerized. When done properly, the product has the form of small polystyrene granules, about the size of grains of sand, and some of the foaming agent (a volatile liquid) is trapped inside the closed cells within the granules. These granules are then shipped to the molder. The molder expands the granules by blowing steam or hot air through them. At this stage the material is called preexpanded beads. These preexpanded beads are then conveyed into the molds where they are further expanded and fused together by the addition of heat. The resulting product is a closed cell, molded product that still retains some of the visual characteristics of the beads but has no holes between the beads when properly fused.

Almost all thermoplastic and thermoset materials can be foamed. Therefore, a wide variety of physical, mechanical, and chemical properties are possible. Furthermore, foam density can vary widely, thus giving further variety in the choice of application material.

Applications for foams include comfort cushioning, shock resistance, and insulation. Each of these applications has a set of tests that characterize the foam and match the foam's mechanical behavior to the requirements of the particular application.

GLOSSARY

Blowing agents   Another term for foaming agents.
Bun   A foamed part made by pan molding.
Casting foams   Foaming without the limitation of a fully enclosed mold.
Cellular plastics   Foamed plastic materials.

Chemical foaming   The creation of a foam through a chemical reaction, which produces a gas that causes the foam to form.

Closed cell foams   Foamed materials in which the cells are discrete, complete entities.

Compression set   The loss of resilience in a material because of prolonged compression.

Conduction   Transport of heat by contact between a cold and hot body.

Convection   Transport of heat by some medium that connects the hot and cold bodies.

Expanded bead molding   Another name for expanded foam molding.

Expanded foam molding   A process in which a foamed bead is expanded using a trapped volatile gas and then heated to fuse the expanded particles; also called expanded beam molding.

Expanded plastics   Foamed plastics.

Expansion process   A process in which a foam is shaped and solidified.

Extrusion foaming   Similar to normal extrusion except that the extrudate contains a foaming agent; therefore, the final part is foamed.

Flexible foam   A foamed structure in which the walls of the cells collapse readily when pressed.

Foaming agents   A chemical that releases a gas and thereby causes a foam to form.

Foaming processes   Techniques that cause tiny bubbles to be formed within the plastic material such that the solid plastic part retains these bubbles.

Foam-in-place   Injection of a resin with foaming agent into a space created by the walls of a part and then solidification of the foam within that space.

Foams   Foamed materials; in the case of interest here, foamed plastics.

Foam spraying   Use of a foamed resin, held under pressure, and then released as it is sprayed onto a surface.

High-pressure foam molding   A method of making a foam part in which the foamed resin expands inside the mold against a moderate pressure, which controls the rate of expansion.

Integral skin   A nonfoamed layer on the outside of the foam caused by the collapse of some of the cells.

Low-pressure foam molding   A method of making a foam part in which the foamed resin expands freely in the mold, that is, without the application of any external pressure.

Mechanical foaming   The whipping or beating of a resin to disperse air throughout and create a foam.

Open cell foams   Foamed materials in which the cells are interconnected, thus, allowing air to pas between the cells.

Pan foaming   Using an open pan as the foaming mold.

Physical foaming   Creation of a foam by forcing a gas into the resin.

Preexpansion   A step in the expanded foam molding process in which tiny, foamed beads are expanded so that they can later be molded.

Radiation   Heat transport via waves.

Reaction injection molding (RIM)   A plastic-molding process that can be used to both foam and shape the resin; two reactive resins and a foaming agent are introduced into a closed mold, where foaming and solidification occur.

Rebond   A material made by binding together chopped pieces of recycled foam, usually flexible polyurethane foam.

Rigid foam   A foamed structure in which the walls of the cells are stiff and do not collapse readily.

Skiving   Slitting into thin sheets.

Syntactic foams   A foam created by the addition of hollow spheres to a resin.

Thermal conductivity   The measure of a material's thermal conductive capability (reciprocal of the thermal resistivity).

Thermal resistivity   The measure of a material's thermal insulative capability.

QUESTIONS

1. Describe the four basic methods of creating a foam.
2. Explain why open cell structures are created when the mold expansion in high-pressure foam molding is quite rapid and why closed cell structures occur when the molding is slower.
3. What is meant by "blowing the cells open," and how can this event be controlled?
4. Describe the principal similarities and differences between RIM and low-pressure foam molding.
5. What is an integral skin, and how is it controlled?
6. Describe how foamed plastic egg cartons are made.
7. Explain how foam-in-place and RIM differ.
8. In an expansion molding operation, why does fusion occur in the mold and not in the preexpansion step?
9. Predict how the acceleration curves versus stress curves would appear for different thicknesses of the same material. Draw curves for 2 cm, 4 cm, and 6 cm thicknesses.
10. Explain how to determine if a packaging material is appropriate for a particular delicate instrument.

REFERENCES

Baird, Ronald J., and David T. Baird, *Industrial Plastics,* South Holland, IL: The Goodheart-Willcox Co., Inc., 1986.

Birley, A. W., R. J. Heath, and M. J. Scott, *Plastics Materials* (2nd ed.), New York: Chapman and Hall, 1988.

Chanda, Manas, and Salil K. Roy, *Plastics Technology Handbook,* New York: Marcel Dekker, Inc., 1987.

"Expanded Polystyrene Package Design," ARCO Chemical Co., ACC-P37-8810.

Milby, Robert V., *Plastics Technology,* New York: McGraw-Hill Book Co., 1973.

Richardson, Terry L., *Industrial Plastics* (2nd ed.), Albany, NY: Delmar Publishers, Inc., 1983.

"Technical Manual Dylite Expandable Polystyrene," Koppers, Bulletin C-9-273.

CHAPTER EIGHTEEN

COMPRESSION AND TRANSFER MOLDING PROCESSES

This chapter examines the following concepts:

- Compression molding (concept, machines and molds, comparisons with other molding processes)
- Transfer molding (concept, machines and molds, comparisons with other molding processes)
- Product considerations (materials, shapes, designs)
- Operation and process control (key process parameters, troubleshooting, maintenance and safety)
- Reaction injection molding (RIM)
- Cold forming, sintering, and ram extrusion

COMPRESSION MOLDING

Compression molding, also called *matched die molding,* is a molding process used almost exclusively for molding thermoset materials. The other processes discussed thus far are used chiefly for thermoplastic materials (extrusion, injection molding, blow molding, thermoforming, and rotational molding) or for both thermoplastics and thermosets (casting and foaming). When thermoplastics use molding methods similar to compression molding, these molding methods are usually called by separate names such as cold forming, sintering, or ram extrusion (methods discussed later in this chapter).

Concept

The principles of compression molding are quite simple. A charge of thermosetting resin is placed in the cavity of a matched mold that is in the open position. The mold is closed by bringing the male and female halves together, and pressure is exerted to squeeze the resin so that it uniformly fills the mold cavity. While under pressure, the material is heated, which causes it to crosslink (cure) and to harden. When the material is hard, the mold is opened and the part is removed. This process can be compared to waffle making.

Machines and Molds

Figure 18.1 is a diagram of a compression molding machine. This diagram shows the most important elements of the compression molding process. A compression molding machine is shown in Photo 18.1.

The molding machine itself (sometimes called the *compression press*) consists of a heavy metal base onto which slide rods are attached and a compression unit that slides up and down on the slide rods. These slide rods guide the movement of the compression assembly from its open to closed positions. The movement of the compression assembly and the force to clamp the compression assembly against the base are supplied by a hydraulic unit mounted above the compression assembly. This action can be fully manual (with the hydraulic pressure supplied by a manual pump), semiautomatic (where a pressure valve is activated by an operator-controlled switch), or fully automatic (where time is the triggering factor). Note that the unit shown in Figure 18.1 closes downward (called a *downstroke machine*), but could be reversed so that the compression unit is on the bottom and therefore closes upward *(upstroke machine)*. Compression machines are normally very rugged and massive, often lasting for several decades.

The physical opening between the base and the compression assembly is called the *daylight opening* and is an important factor in choosing the proper compression molding machine. The daylight opening should be sufficient to accommodate the platens and the molds and leave space for loading and unloading the mold. This can be a problem in some cases where the part is deep and the vertical height of the molds is great.

Compression molds are subjected to very high pressures, perhaps greater than those in any other type of plastic-molding process. Therefore, compression molds tend to be built on rugged, massive plates that can support the mold and withstand the pressures of the mating of the molds. A typical mold set is shown in Figure 18.2.

The mold base plates (platens) are attached to the base and the compression unit of the compression press with large anchoring bolts. These platens are large enough that they are cored for insertion of cartridge heaters or some other convenient heating system such as oil or, in some cases, steam. Care should be taken in choosing a particular heating system to ensure that the platens' heating capacity is sufficient to heat the molds to the proper temperature and to maintain that temperature during the molding cycle.

The molds are attached directly to the platens. Sometimes the platens have cutout zones for the insertion of the upper and lower interchangeable mold cavities, but in other cases the molds are simply bolted onto the surfaces of the platens.

The parts are normally ejected from the mold by using *knockout pins* or *ejector pins*. These can be activated by a small hydraulic cylinder or, if the knockout mechanism is placed within the compression assembly, the movement of the compression assembly away from the base can activate the knockout mechanism. To further facilitate part removal, the mold is coated with mold release (often bees' wax) and then, if sticking occurs, additional mold release agent is sprayed into the mold to assist with part removal.

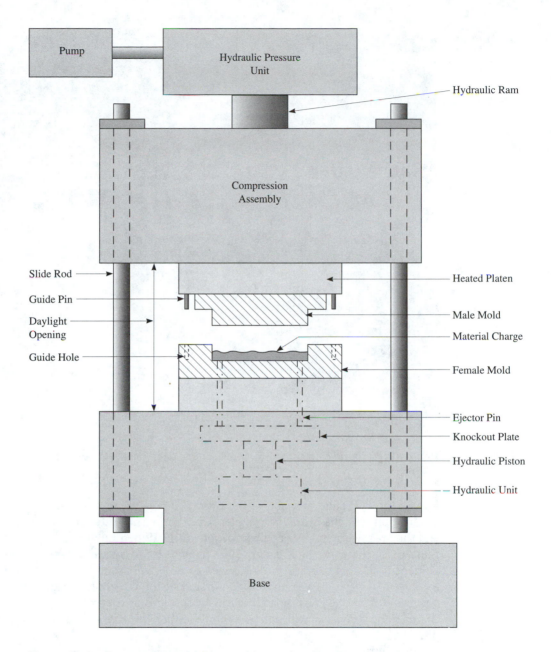

Figure 18.1 Compression molding machine and major process elements.

The maximum force required for a particular molding operation influences the choice of a compression press. The total force that can be exerted by the machine is called the *press capacity, machine rating,* or *machine size* and is stated in force units: newtons (N) in the SI system or tons in the English system. This machine capacity is a function of the

Photo 18.1   Large commercial compression molding machine (Courtesy: Wabash Hydraulic Presses)

area of the hydraulic ram and of the hydraulic pressure in the line going from the hydraulic pump to the hydraulic cylinder. The relationship between these hydraulic factors and the press capacity is given by Equation (18.1). (When English units are used in Equation (18.1), the conversion factor [2000 (pounds/ton)] is used because the size of a machine is typically expressed in tons. When SI units are used, the press capacity is usually expressed in newtons (N) and no conversion factor is needed because pascals (Pa) are equal to N/m^2.) The press capacity is usually specified by the equipment manufacturer and the operator generally will not be aware of either the area of the ram or the hydraulic pressure.

In English units:

$$\text{Press Capacity (tons)} = \frac{\text{Area of Ram (inches}^2) \times \text{Hydraulic Pressure (pounds/inch}^2)}{2000 (\text{pounds/ton})} \quad (18.1)$$

In metric units:

$$\text{Press Capacity } (N) = \text{Area of Ram (m}^2) \times \text{Hydraulic Pressure (Pa)}$$

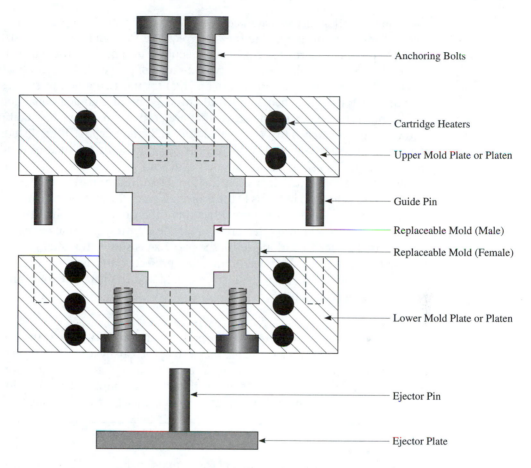

Figure 18.2 Compression molding mold set.

The *force required* to mold a particular part can be determined and then compared with the *press capacity*. If the force required is **less** than the press capacity, then the press can be used for that particular part. When selecting a machine for a particular molding job, a safety factor (approximately 1.3 times the calculated force required) is usually employed.

The force required depends on material characteristics and part geometry. Sufficient force must be exerted in the material to consolidate it and move it into all regions of the mold. Low-viscosity materials require less pressure than do those with high viscosities. Because of the complex nature of this material-dependency, it is difficult to develop a precise mathematical expression for determining the pressure required to mold a part. Sophisticated models have been created that give approximations of the pressure required to mold the part as functions of the material parameters (such as viscosity) but these are used mainly for research purposes rather than for production development.

Approximation factors have been developed experimentally that permit some simple calculations to be made relating the force required for molding and the size of the part. These factors give approximate pressures required to move a particular material laterally across the area of a mold cavity. The typical range of pressures for consolidation is 1.5 ksi (1 ksi = 1,000 psi) to 8 ksi (10 MPa to 55 MPa) for most common plastic resins.

Parts that are more than 1.5 inch (3 cm) deep require an additional amount of pressure to push the material into regions that are deep and narrow. This additional pressure factor is approximately 3 to 4 MPa per additional 3 cm of depth. (If English units are used, the added pressure requirement for depths beyond the first 1 inch of depth is approximately 500 to 750 psi per additional inch of depth.)

As the size of the part increases, the force required to mold that part also increases. The part size is defined in terms of the *projected area,* the area of the cavity using its maximum width and length. If the mold has more than one cavity, then the total projected area is the sum of the projected areas of all the cavities. For shallow parts (usually less than about 1.5 inch [3 cm] in depth), the part size is simply the total projected area of all the cavities. The method for determining the projected area is illustrated in Figure 18.3.

The relationship between the force required and the part size for a given material is represented by Equation (18.2),

$$F = (A)[(P_A) + (\rho d_e)] \tag{18.2}$$

where F is the force required to mold the part, A is the projected area of the part, P_A is the required cavity pressure for a particular material (P_A is usually developed experimentally), ρ is an additional pressure factor (developed experimentally) that accounts for the effect of depth, and d_e is the depth in excess of the minimum amount. Calculations for determining the size of an actual machine to be used for molding a part can be illustrated by Problem 18.1.

Problem 18.1

Calculation of the capacity (force) of an actual molding machine to be used for a particular molding job.

The mold has two cavities, each with a maximum width of 8 cm and a length of 20 cm. The depth of each cavity is 6 cm. The material factor for area (P_A) for a thin part (<3 cm) is 20 MPa. The depth factor is 1.5 MPa/cm for each 1 cm of depth beyond the first 3 cm.

Determine the minimum machine capacity needed to mold these parts.

Solution:

Total projected area:

$$(2 \text{ cavities})(8 \text{ cm})(20 \text{ cm}) = 320 \text{ cm}^2$$

Force required for molding (from Equation 18.2):

$$\text{Force} = (320 \text{ cm}^2)[(20 \text{ MPa}) + (1.5 \text{ MPa/cm})(3 \text{ cm})]$$
$$= 0.73 \text{ MN or } 730 \text{ kN}$$

Safety factor (1.3) applied to give 955 kN or 1 MN nominal size.

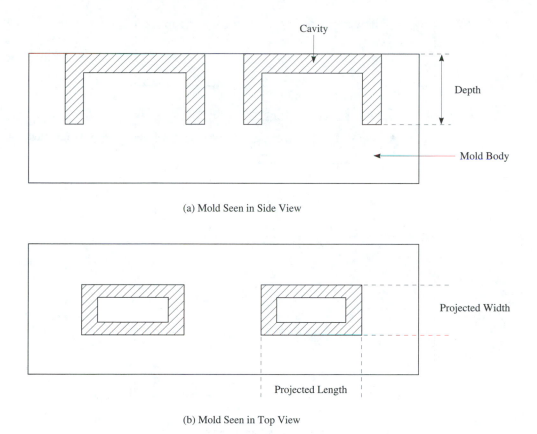

(a) Mold Seen in Side View

(b) Mold Seen in Top View

Figure 18.3 Projected area and depth for a multicavity compression mold.

Mold Closure Types

Three types of mold closures are typically used for compression molding. These are called the **flash type,** the **positive type,** and the **semipositive type.** These mold closures differ in the way the two mold halves mate, illustrated in Figure 18.4.

The *flash type mold closure* is the simplest of the three types and the least expensive to build. In this closure type the cavity is simply overfilled slightly and the excess material is squeezed out into the gap between the male and female mold halves. When cooled, this excess material is called *flash* and must be removed from the part. If insufficient material is charged into the mold, the part will be too small. The flash type closure works best for shallow parts and parts where dimensional control and property control are not critical.

The *positive type mold closure* differs from the flash type because of its long shearing surface between the male and female halves. This long shearing surface is made when the walls of the plunger and the walls of the cavity are very close together for a considerable distance. The gap between them is kept small so that material will not leak out. Hence, no

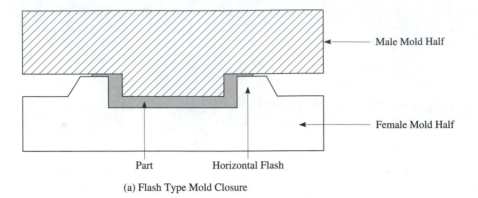

Male Mold Half

Female Mold Half

Part   Horizontal Flash

(a) Flash Type Mold Closure

Male Mold Half

Small Gap, No Flash

Female Mold Half

Part

(b) Positive Type Mold Closure

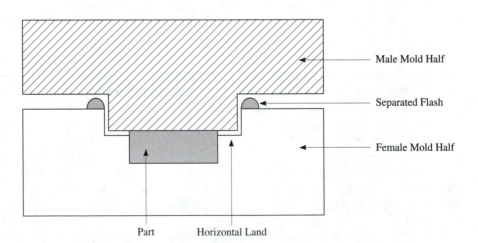

Male Mold Half

Separated Flash

Female Mold Half

Part   Horizontal Land

(c) Semipositive Type Mold Closure

Figure 18.4 Three mold closure types (flash, positive, and semipositive).

flash is created. The amount of material charged into the cavity must be carefully controlled so that the mold is filled but not to excess. If excess material is charged into the mold, the mold will not close. If too little material is charged, a short shot will result.

The *semipositive type mold closure* conveniently combines the features of the other closure types. In the semipositive system a horizontal land creates a gap between the plunger and the cavity, a gap considerably larger than that in a positive closure mold. If any excess material is in the cavity, this excess moves along the horizontal land and up the vertical land, which also has a greater gap than exists in the positive mold. The result is a pushing of the excess material past the gap regions into a horizontal area where it can easily be removed from the mold. Thus the part is essentially flash-free.

The most common materials for compression molds are the tool steels and stainless steels. Because of the high temperatures normally used, H13 steel is very common. Also common are the hardenable steels because hobbing continues to be an important method for making cavities and is best done in a steel that later can be hardened.

Although hobbing still is used for making the cavities for compression molds, machining continues to be the most important method. EDM is gaining in popularity, but is not as common in compression molds as it is in injection molds, perhaps because of the often larger size of compression molded parts.

Chrome plating is commonly used in compression molds to give parts a mirror finish. This method also reduces the cavity abrasion that results from the reinforcements commonly found in thermoset materials molded in compression molding.

Comparisons of Compression Molding with Other Molding Processes

The most obvious comparison for compression molding is to injection molding. Both processes produce discrete parts of widely differing geometries. Compression molding uses thermoset resins as starting materials, whereas injection molding uses thermoplastics. This difference dictates very different mold conditions. In compression molding the mold is heated so that the thermoset material will cure. In injection molding the mold is cold so that the melted polymer will freeze. Since curing usually takes longer than freezing, the molding cycle for compression molding is longer than for injection molding. Small compression molded parts might have a molding cycle from 1 to 2 minutes, whereas injection molding cycles for the same size part might be from 20 to 60 seconds. The injection molding process can be highly automated, virtually working without the need of operator intervention. Although compression molding cannot reach this same level of automation, it can be semiautomated, with many of the steps carried out with only minor operator involvement.

Another consequence of making thermoset plastic parts by compression molding is that reject parts cannot be reprocessed. Once thermosetting has occurred, the part cannot be remelted. Recycling of these parts usually involves grinding the part and use of the regrind as a filler.

The complexity of the parts that can be molded also differs. Compression molding is limited to shapes that can be made by charging into an open mold. Parts that have hollow sections and other features that require cores and slides are therefore very difficult to

Table 18.1 Comparison of Compression Molding, Transfer Molding and Injection Molding

Property	Compression Molding	Transfer Molding	Injection Molding
Material	Thermoset	Thermoset	Thermoplastic
Mold heating	Yes	Yes	No
Typical cycle	Minutes	Minutes	Seconds
Recycle parts	No	No	Yes
Mold open/closed	Open	Closed	Closed
Types of parts	Discrete	Discrete	Discrete
Part complexity	Simple	Complex	Complex
Hollow areas	Difficult	Easy with cores	Easy with cores
Inserts	Moderate	Easy	Moderate to difficult
Undercuts	No	Yes with slides	Yes with slides
Part size	Can be very large	Size is limited	Size is limited
Flash	Yes	No	No
Sprue/runners	No	Yes	Yes
Resin viscosity	Moderate to high	Low to moderate	Typically low
Reinforcements	Yes, can be long	Some, cannot be long	Some, cannot be long
Physical properties	Maintained	Minor orientation	Some orientation

make. Undercuts cannot be made by compression molding. However, the simplicity of the compression molds leads to much lower mold fabrication costs. Extensive analysis for sprue, runner, and gate designs are nonexistent.

Charging into an open mold does have some advantages, however. There are no sprues or runners that would be present in injection molding. The part can be used directly with, perhaps, only a flash-removal step. Another advantage of open molds is that the distance that the resin moves during the molding cycle is much shorter than in closed molds. This shorter path means less orientation of the resin molecules and, therefore, fewer problems with changes in physical and mechanical properties because of the molding process. The short path also allows high contents of filler and reinforcement and longer reinforcements than could be used in a closed mold process like injection molding. These comparisons are summarized in Table 18.1.

TRANSFER MOLDING

Concept

Transfer molding is a method of molding thermoset materials in a closed mold. In this process the thermoset resin is placed into a *transfer chamber* where the resin is heated until liquid. A plunger then forces the liquid resin through a sprue and runner system into the mold cavities. The heated mold is closed during the transfer process to allow the part to cure and solidify. After an appropriate time, the mold is opened and the part is ejected. The part is connected to a runner system and the transfer chamber, which must be trimmed off and discarded. Standard practice in transfer molding requires that a small excess of charge be placed into the transfer chamber to insure that the cavities are completely filled with each shot. This excess material, called *cull,* also cures because the transfer chamber is heated and is discarded as part of the runner system. The major components of this process are illustrated in Figure 18.5.

Equipment and Molds

The pressure requirements for transfer molding are somewhat less than those for compression molding because the material is liquid and therefore flows more easily than do some types of compression molding charges. The injection pressures are similar to those used in injection molding. The mold base for a transfer mold is similar to that used for injection molds, with the same type of plate supports, ejector system, and sprue, runners, and gates. Materials are also similar, with the exception that transfer molds more often tend to use the high-temperature tool steels (H13) because the molds are heated to cure the thermoset resin. A transfer mold base is depicted in Figure 18.6.

The low viscosity of thermoset materials allows the runners in a transfer molding system to be smaller and longer than in injection molding. Gates are designed much like those used in injection molding.

Transfer molds can be more complicated than compression molds and are, therefore, more costly. This greater design flexibility with transfer molding has proven to be especially useful for the encapsulation of parts. Electrical parts such as plugs on the ends of cords, integrated circuits, and devices with wire leads are frequently transfer molded. Transfer molding is also used to make handles for hardware tools. The bare tools can be loaded into the mold cavities, the mold closed, and then the resin transferred into the mold and around the insert. This process is shown in Figure 18.7, where cores or parts to be encapsulated are also represented.

Figure 18.7 also illustrates a slightly different sprue and runner system from that shown in Figure 18.5. The transfer chamber is separated from the runners and so a sprue is required. After the part is molded, that sprue and cull must be separated from the runners. This separation allows the parts and runner to be removed as the mold is opened and then, in a secondary step, the sprue and cull are removed from the machine before the transfer chamber is again filled prior to the next injection cycle. This is accomplished by creating a *break point* on the sprue (where it is thinnest) and pulling on the sprue by using an undercut on the bottom of the plunger. This undercut system is called a *sprue puller.*

Typically a compression molding press is used for transfer molding with the appropriate modifications made in the mold design to provide a heated transfer chamber with a plunger.

Comparisons of Transfer Molding with Other Molding Processes

The logical comparisons for transfer molding are with compression molding and injection molding. These comparisons are summarized in Table 18.1. When compared to compression molding, transfer molding has a slightly faster mold cycle because the material transferred in has already been heated. (Compression molding sometimes preheats the charge as well, thus eliminating this advantage for transfer molding).

Figure 18.5 Transfer molding process: (a) charge heating, (b) injection (transfer), and (c) part removal.

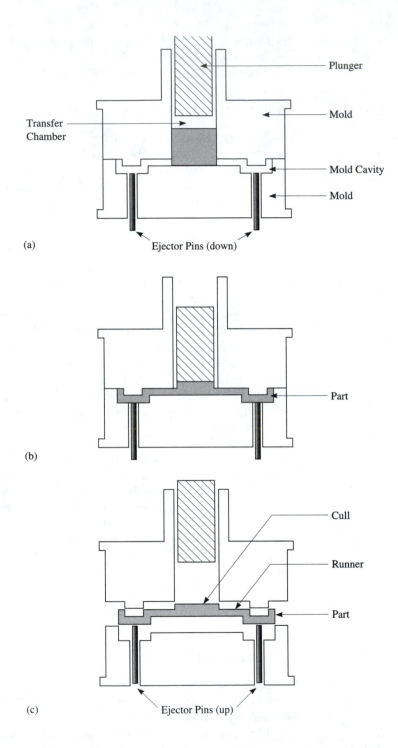

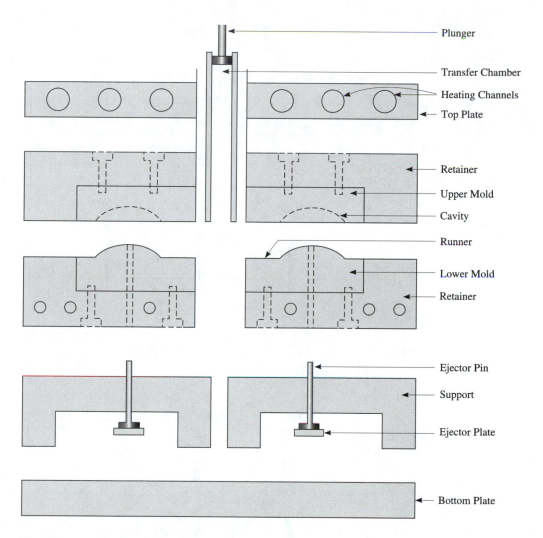

Figure 18.6 Transfer mold base (exploded view).

Transfer molding makes parts that are much more complicated than those made by compression molding. Part dimensions are more accurate with transfer molding than with compression molding and the parts typically do not have flash, although they do have runners, sprues, and culls. Inserts, especially delicate inserts, and encapsulated parts can be molded with transfer molding but are difficult to make with compression molding. Mold costs are higher for transfer molding than for compression molding.

The relatively long flow path of the material in transfer molding effectively eliminates the use of resins containing fiber reinforcements. Resin orientation can also be a problem with transfer molding, as it is with injection molding.

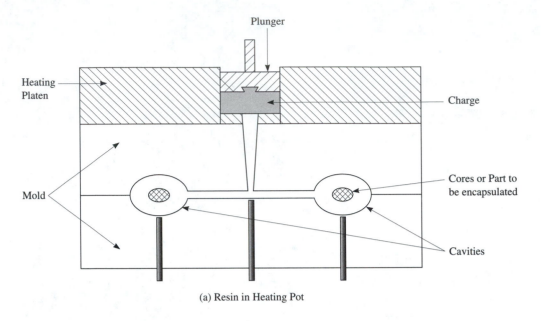

(a) Resin in Heating Pot

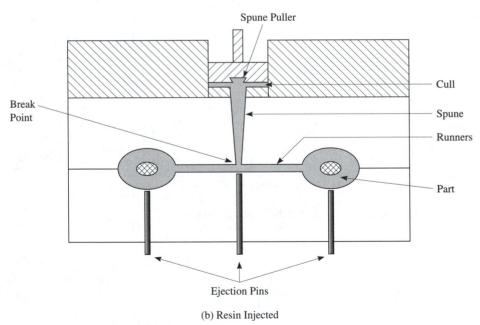

(b) Resin Injected

Figure 18.7   Transfer molding of cored parts or parts to be encapsulated.

Part complexity is about the same for transfer molding and injection molding and so molds are approximately the same cost. The molding times for transfer molding are generally much longer than for injection molding, reflecting the natural differences in thermoset and thermoplastic molding.

PRODUCT CONSIDERATIONS

Practically all thermoset resins can be compression molded. The starting materials can be liquids, pastes, doughs, granules, preforms, or doughs. *Pastes* are usually made by mixing a liquid thermoset resin with a filler. *Doughs* are pastes or, occasionally, pastes with reinforcement added. The reinforced material, also called *molding compound,* is discussed in the chapter on composite material processes. Granules are B-staged thermoset resins, such as phenolics, that liquify when heated and then cure. *Preforms* are thermoset granules pressed together into a particular shape for easy handling. Reinforced materials are generally not molded in transfer molding.

Viscosity control is an important factor in compression molding. The speed and pressure of the mold closure can cause excessive flash if the viscosity of the charged material is too low. Low viscosity may also be a problem when a fiber reinforcement is present because the mold may close on the fibers and not press on the resin that simply runs off the fibers. The secret in molding reinforced resins is to have the viscosity of the resin high enough that when the mold closes on the mixture, all the materials move together.

An interesting comparison of viscosities can be made between the resins used in transfer molding and those used in injection molding. At the moment of injection into the mold, thermoset materials are usually lower in viscosity than are thermoplastic resins. This is because of the lower molecular weight typical of a noncured thermoset resin compared to a thermoplastic. Therefore, flow into a complicated mold is actually easier for the thermoset, although the possibility of incidental flash from worn or mismatched molds is also greater. Hence, viscosity control for transfer molding is less of a concern than for either injection molding or compression molding.

The flexibility of design and restrictions of shapes have already been discussed in comparing compression molding, transfer molding, and injection molding. Those discussions indicated that the allowed complexity of compression molded parts would generally be less than for either transfer molded or injection molded parts.

CONTROL AND OPERATION

Critical Process Parameters

The most important process parameters in compression molding and transfer molding are the amount of charge, the molding pressure and closing speed, the temperature of the mold, and the molding cycle time. The criticality of the amount of charge has already been discussed in some detail. In general, if the charge is too large, flash will occur or, perhaps, the mold will not close. If the charge is too small, the part will be poorly formed and not well compressed, thus potentially giving rise to voids and poor mechanical properties.

Flash, when made, is usually removed by some automated or semiautomated mechanical process. A typical flash removal process tumbles the molded parts together in a rotating tub, somewhat like a clothes dryer. Removal is enhanced by freezing the parts, thus making the resin more brittle. Because flash is so thin, it becomes even more subject to mechanical removal when cold. Tumbling the parts with frozen rubber cylinders further improves flash removal. Processes that employ frozen parts are called *cryogenic flash removal*. Standard metal files can be used at the press to remove small amounts of flash.

The molding pressure and closing speed are important because they determine whether the material moves properly to fill the mold and then whether it is properly consolidated. If the pressure is too low, the part may be poorly formed or have internal voids. If the pressure is too high in transfer molding, the mold may spread apart and flash may occur.

Controlling the speed of pressure application is also important. In compression molding, the pressure should be applied evenly and smoothly after the initial contact with the resin charge has been made. If applied too quickly, the shear on the resin may disturb the viscosity balance and result in separation of filler, reinforcement, and resin. If applied too slowly, the resin may begin to cure and harden before the necessary movement within the mold is complete. Of course, a fast closing of the mold is desired to keep the mold cycle time to a minimum. One effective method is to have a very fast mold closure until the mold touches the resin charge. Then, the mold closure speed slows so that the material moves uniformly throughout the mold. This latter speed is determined by trial and error. Using two closing speeds is called *two-stage clamping*.

Pressure control and mold closure control are also required to eliminate a problem associated with thermoset resins that crosslink by a condensation reaction. (Phenolics are an example of such a resin.) The condensate often is water that evaporates at the molding temperatures and forms a gas within the mold. If not allowed to escape, this gas causes surface blisters and porosity. One method that allows the gas to escape is cracking open the mold slightly during molding, a process called *degassing* or *bumping the mold*. The molds are opened only slightly (typically 0.06–0.6 inches, 1.5–15 mm) and only for a moment. An alternative to degassing involves using a vented mold (a mold with a small channel that allows the gas to escape).

The optimum temperature of the mold depends on the resin being cured. Each curing reaction is unique for each resin and, to some extent, is also affected by the amounts of fillers and reinforcements that might be present. The shape and size of the part can also affect the optimal temperature, as can any preheating that must be done for transfer molding and is often done in compression molding to shorten the mold cycle.

If the mold temperature is too hot, degradation can occur. Degradation can be detected by scorching of the surface, discoloration, and by a decrease in mechanical and physical properties. The temperature of the mold also affects resin viscosity. Hence, trial and error may be required to establish the proper temperature for a particular part. A good starting place for temperature can usually be obtained by referring to literature from the resin manufacturer or to general literature (handbooks and encyclopedias) describing thermoset materials processing.

Molding cycle time should be kept as low as possible to achieve good process economics. The optimal time for opening the mold occurs when the part is rigid enough to be removed or, alternately, when the curing has progressed far enough to give the desired physical and mechanical properties. This is usually determined by experience.

To shorten the curing time, the temperature of the mold can sometimes be cycled (hot and cold). This requires that the mold be cored with both hot and cold tubing that can be shut off quickly so that they are not on at the same time. When this temperature cycling is done, the cold is usually applied only long enough to allow the part to stiffen so that it can be removed.

Troubleshooting

The relative simplicity of the compression molding and transfer molding processes allows the assembly of a reasonable troubleshooting guide. This guide is given in Table 18.2.

Table 18.2 Troubleshooting Guide for Compression Molding and Transfer Molding

Problem	Possible Cause
External blisters	• Mold temperature too high • Moisture • Insufficient cure time • Poor degassing or venting • Insufficient pressure
Marks from ejector pins	• Insufficient cure • Part not cold enough • Excessive ejection pressure
Internal voids	• Insufficient material load • Improper location of material • Mold too hot • Incorrect degassing • Insufficient cure time • Insufficient pressure
Incomplete parts	• Insufficient material load • Improper location of load • Press closing too slowly • Material preheated too long • Mold too hot • Viscosity too low • Material too old (partially cured)
Poor gloss	• Slightly insufficient charge • Excessive or incorrect release agent • Poor mold finish • Press closing too slowly • Material too old • Degassing too long • Insufficient pressure • Use dielectric preheater

(continued)

Table 18.2 *Continued*

Problem	Possible Cause
Orange peel surface	• Press closing too slowly • Mold temperature too high • Fines in molding granules
Sticking	• Insufficient or incorrect release agent • Nonuniform or low temperature • Lack of mold polish or scratched mold • Ejection pins not working • Insufficient cure time
Brittle parts	• Press opening or closing too slowly • Mold temperatures too high • Cure too long • Old material
Cracking on ejection	• Sticking in one area • Ejection not working • Undercure or overcure
Oversized vertical dimensions	• Flash remaining • Not enough pressure • Insufficient cure time • Excessive charge or poor charge location
Undersized vertical dimensions	• Slight undercharge • Distorted mold closure surfaces
Oversized horizontal dimensions	• Worn mold • Poor shrinkage
Warpage	• Nonuniform mold temperatures • Insufficient mold times • Male and female cavities different temperatures • Sticking on ejection • Insufficient pressure
Excessive flash	• Improper location of charge • Press closing too fast • Uneven mold temperature • Viscosity too low • Excessive material • Platens misaligned

Maintenance and Safety

The most important safety considerations concern the imminent danger of having a finger, hand, or some other object caught in a closing mold. This danger is so prevalent that special procedures should be enacted to prevent its occurrence. One preventive measure that can be taken is to require that two switches (usually buttons) be closed simultaneously in order to activate the mold closing. The buttons should, of course, be separated sufficiently so that two hands are required to activate them. Another safety measure involves locking out the machine activation system whenever working with the mold.

Another danger comes from the hot molds, which are generally unprotected. This danger is best avoided by wearing proper clothing and gloves.

REACTION INJECTION MOLDING (RIM)

The *reaction injection molding* process is similar in many ways to transfer molding. Both processes inject a liquid thermosetting material into a closed mold where the material cures and hardens. The major difference is that the type of resin used in RIM is quite different than the types used in transfer molding. In RIM the resin is a two-part reactive system. That is, parts A and B are mixed and, on mixing, they begin to react to form a plastic. The most common of these systems is polyurethane, where part A is a polyol and part B is an isocyanate. A RIM system is illustrated in Figure 18.8.

The reactive components (parts A and B) are liquids that are injected into the mold before any significant curing occurs. Therefore, the transfer chamber used in transfer molding is replaced by the pumping, mixing, and injection system used in RIM.

The parts typically made using the RIM system are quite large, including automotive bumpers and fenders. Many furniture components are also made by the RIM process. The relatively low pressures allow these large parts to be made without the need to use extremely large presses. No capacity limitation is dictated by the process itself.

The reaction times are comparatively short because of the fast-reacting nature of the common reactive components. For example, polyurethane reactions take typically less than a minute to achieve a cured part.

Foamed RIM products can be made by adding a foaming agent to the reactive mixture prior to its entry into the mold, as was discussed in the chapter on foaming. Fiber reinforced products can be made using RIM techniques by placing the reinforcement in

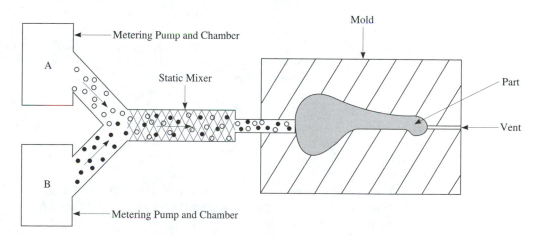

Figure 18.8   Reaction injection molding (RIM).

the mold and then closing the mold and injecting the resin. These processes (RTM and SRIM) are discussed in more detail in the chapter on composite processes.

COLD FORMING, SINTERING, AND RAM EXTRUSION

Cold forming, sintering, and ram extrusion have in common the fact that they use open molds to shape granular thermoplastic resins. They are discussed in this chapter because their shaping processes are similar to compression molding (a large press exerts pressure on the materials while they are in the mold).

Cold forming is a process in which granular thermoplastic resin is poured into an open mold, the mold is closed, and then high pressure is exerted to squeeze the material and form it into a solid part. Resin particles should be evenly distributed and level in the mold during filling, and the molds are vibrated or tapped during filling to ensure this. The most common pressing equipment for this process is matched die molds and compression presses, but metalworking machines such as forges, brake-presses, rollers, stamping machines, and coining machines can also be used.

During cold forming the particles are pressed into intimate contact and the air between them largely removed. The key parameter for the process is, of course, that the particles be compressed together tightly. *Compression ratios* (the ratio of the density of the *bulk material* or uncompressed material to density of the compressed material) are typically from 3:1 to 6:1. The pressures required to accomplish this compression are typically 3000 to 5000 psi (20 to 35 MPa).

If the molds are heated during the molding process, the operation is called *solid-phase forming*. In solid-phase forming, the temperatures of the mold are much lower then the melt temperatures of the resins (typically temperatures are 200°F, 100°C). At these low temperatures little corrosion takes place and mild carbon steels can be used for the molds.

The most common resins for cold forming and solid-phase forming are ABS, polycarbonate, polyolefins, PTFE, and rigid PVC. Brittle materials such as acrylic and polystyrene are difficult to form by these methods.

Parts made by cold forming and solid-phase forming are not strong. An analogous situation would be in ceramic production, where the unfired parts (called earthenware) can be used but have some strength and physical property deficiencies when compared with fired ceramic pieces. Applications that allow direct use of cold formed parts include thermoplastic seals and packing for valves. Parts made by cold forming and solid-phase forming are almost always simple shapes such as solid cylinders or blocks (called *billets*), hollow cylinders, and thick sheets. In the vast majority of cases, the cold forming and solid-phase forming methods are used to make parts that will be subsequently heated in a process called *sintering*.

The sintering process consists of three steps: preforming, fusing, and finishing. (Note that in compression molding of thermosets, preforming and fusing are done simultaneously during the molding operation.) The first step in sintering, preforming, is usually done by cold forming. The resultant preform is sometimes called a *green compact,* which implies that a subsequent heating step will be used to complete the shaping of the part.

(This terminology is taken from the ceramic-molding industry where "green compact" is the name given to a ceramic part before it is fired.) Preforms should ideally be compacted so that less than 10% shrinkage occurs during the fusing step, resulting in better particle fusion and fewer pores.

The fusion step can be done on the preform while it is still in the forming mold. If done under pressure, the mold is left in a press and heating is accomplished by heat jackets or heated molds. If done without external pressure, called *free fusion*, the parts can still be left in the mold, but the mold is moved to an oven where the entire assembly is heated. After fusion, the parts can be repressurized (usually to a lower pressure than was used for preforming) so that good compaction and part definition are assured. If the part is taken out of the forming mold, heated, and then transferred to another mold where it is pressed and cooled, the process is called *coining*. Complicated shapes can be done by the coining method.

Fusion of the particles in the preform is done by heating the preform to a temperature just below melt temperature or decomposition temperature, whichever is lower. At the fusion temperatures the resin particles fuse or coalesce into a solid mass. The resin reaches a gel stage but still has enough solid strength to support the pressure applied to it. When done properly, the fusion process causes the molecules from one particle to move across the surface boundary to entangle with the molecules of a neighboring particle. This results in a strengthening of the part. If compaction is good, voids are largely eliminated during the fusion step.

If the temperature during fusion is too low, the fusion time too short, or the compaction inadequate, the part will have poor fusion or a high void content. If the temperature during fusion is too high or the fusion time too long, degradation occurs. This degradation results in lower physical and mechanical properties. Therefore, narrow temperature and time windows exist for proper fusion. The control of temperature is quite important and ovens that can control the temperature to within 4°F (2°C) are commonly stipulated.

The cooling rate also affects properties. In crystalline polymers the amount of crystallinity is determined by the cooling rate. Faster cooling results in lower crystallinity because the crystals do not have time to form. High crystallinity usually increases part stiffness, creep resistance, and impermeability. Low crystallinity favors part flexibility, flex life, and compressibility. If cooling is too rapid, parts crack from the thermal stresses.

Thermal and mechanical stresses often are built into the parts by the sintering process. These stresses can be relieved by annealing, which involves removing the part from the mold and then reheating it, but to a temperature lower than the forming temperature. At this low temperature the molecules can reorient themselves to relieve the stress without affecting the overall shape of the part.

Finishing of the fused part often entails machining of the billet or sheet to some finished part shape. Typical metalworking equipment, such as mills and lathes, can be used.

Sintering is an especially important process for thermoplastic resins that do not have a well-defined melt phase. The most common of these are the fluoropolymers and the polyimides. In both cases the final shapes can be made either by machining out of stock material or by pressing the resin into the shape as a preform and then maintaining the part shape during the fusing step.

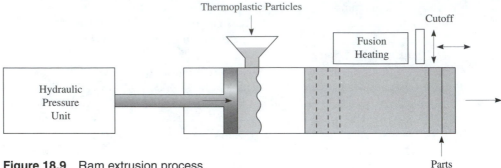

Figure 18.9 Ram extrusion process.

Ram extrusion is a process of forming powdered thermoplastic particles into continuous, long moldings. In this process small portions of powder are placed into a tube where each portion is compacted against previously compacted material by a reciprocating hydraulic ram. This previously compacted material is moved through a heating zone where it is fused. A high-quality finish on the inside of the tube results in a good finish on the billet and lower forces to push the billet through the machine. Eventually the fused solid mass exits the cylinder where it is cut to size and removed for additional finishing. The ram extrusion process is depicted in Figure 18.9.

The high thermal expansion and continued compression of the material inside the heating section give a high degree of compaction and low void content. This pressure also ensures that good fusion occurs if the temperature is correct. When the temperature is incorrect, the bonding between successive compacted portions is not good and the solid mass splits between the portions, a defect called *poker chipping*.

Hollow shapes can be made by ram extrusion. These shapes are used for pipe liners, seals, and gaskets. Rods can also be formed and are used as feedstock for automatic screw machines.

CASE STUDY 18.1

Manufacture of Automobile Body Panels

The automotive industry has become one of the principal users of compression molding. Easily molded plastic parts were used originally to replace metal parts that had complicated shapes and therefore required several metal-forming processes to make (such as heating ducts, trim, and rounded or nose sections on the ends of body panels). The lighter weight of plastic parts compared to metal parts of the same size has given further impetus to the shift from metals to plastics in automobiles. Plastics are also being used for assemblies of metal parts because one molded plastic part can sometimes take the place of several interconnected metal parts. This capability may hold the greatest potential for the use of plastics in automobiles because of the improved manufacturing time and cost obtained by reducing the number of parts and the number of required manufacturing steps.

This discussion focuses on exterior body parts. Simple matched die compression molding is the most important plastic-molding process for most of these parts. (Other plastic-molding processes such as rotomolding and blowmolding for tanks and injection molding for small actuator parts are ignored in this discussion.) To ensure quality parts, manufacturers must deal with potential problems inherent in the process.

Most important for the manufacturer of body panels is to have an exceptional surface finish capable of being painted to a class A standard. (This is a rating system used by automakers to classify the degree of finishing required to give a smooth, defect-free surface. Class A is the highest rating, requiring the least surface preparation.)

One requirement to obtain a class A surface directly out of the molding operation is that the part be made without surface waviness. Three variables are controlled to ensure that waviness does not occur.

1. The tool (matched die molds) is checked during manufacture for absence of waves. This is done optically with an instrument called a *defractoid*, which measures light-diffraction patterns off the tool surface.

2. Different combinations of additives are used with different resins, fillers, and reinforcements to give uniform distributions of the materials throughout the mold. The additives may adjust viscosity so that resin, filler, and reinforcement move together. The additives may also expand slightly during the curing phase so that the natural shrinkage of the resin can be offset. The emphasis here is to prevent warpage from local and nonuniform shrinkage.

3. The critical processing variables (charge amount, temperature, and pressure) are varied to fine-tune the process. Depending on the shape and geometry of the part, different charge amounts of the resin mixture are used. Pressure rate and temperature are also varied to provide a proper cure for the part shape and composition. These variables are not independent and an understanding of their interrelationship is key to part quality.

Another defect of concern to manufacturers of car body panels is porosity. The problem of porosity is handled in two ways. First the chemical formulation used is optimized to reduce voids and trapped air. Second, a vacuum is incorporated into the molding system so that trapped volatiles are removed. This eliminates the need for degassing.

Edge defects also concern manufacturers of plastic body panels. Edge defects are controlled by good mold design and careful processing. The mold must be designed to produce a thin and constant-thickness flash around the edges of the part. After the part is removed from the mold, flash removal results in smooth edges. The finishing process of flash removal must be gentle so that the part finish is undamaged.

Another area of concern for all part manufacturers is the reduction of molding costs. Shortening mold cycles is a key element in achieving lower costs. One way to do this is to adjust or manipulate the chemistry of the resin curing to give a shorter cure time. Resin manufacturers are constantly trying to accomplish faster cures. A caution here, though, is that changes in curing chemistry also affect other properties of the product. Another important variable in shortening cycle time is part geometry. A smaller, simpler part yields a shorter cycle time than does a thick, complicated part. Hence, overall part design should begin with this manufacturing process in mind. Other ways to shorten the mold cycle include moving the part out of the mold prior to full curing and then moving it through an oven to achieve the desired cure. Also, preheating the molding material shortens the cycle because the material reaches the cure temperature faster. Applying a vacuum to the mold also shortens the cycle by eliminating the need to degas the mold, by assisting in the rapid filling of the mold through elimination of any trapped air, and by pulling on the resin as it flows in the mold. Removing trapped air or moisture also improves the part's surface quality and void content.

The production of class A parts from compression molding is well established. Improvements can still be made, but the use of this molding system for production of plastic body panels and for other automotive parts will continue to grow as the advantages of plastic molding over metal parts are further recognized.

SUMMARY

Compression molding and transfer molding are processes used to mold thermoset materials. These are the most important processes for thermosets. In both processes the material is introduced into a heated mold where the material cures and hardens. The part is formed by the shape of the mold. After removal from the mold, the part may need to have some finishing to remove flash or an attached sprue and runner system.

In compression molding the mold is open when the material is introduced. This open mold arrangement has the advantage that molds can be simple and charging done either manually or automatically, as desired. Open mold charging means that the flow of the resin is short, normally just that required to fill the mold. Hence, resins mixed with fillers and long fiber reinforcements can be used in compression molding.

Transfer molding differs from compression molding in that the mold is closed when it is charged in transfer molding. The resin is preheated in a transfer chamber to the proper viscosity before being forced by a plunger into the mold cavities. Transfer molding uses a runner and sprue system to move the material from the transfer chamber to the cavities. An excess of material is usually charged into the transfer chamber to insure that the mold is completely filled. This excess material, which is cured because the transfer chamber is heated, is called the cull. Transfer molding gives better resin control than compression molding because the resin is usually lower in viscosity during the mold-filling stage. This high resin control and the closed-mold arrangement lead to the extensive use of transfer molding for the encapsulating of parts (many of them electrical) and the molding of handles.

The molds used in compression molding and transfer molding consist of two halves that mate to form the part between them (like a waffle iron). Molds must withstand considerable pressure and so are made from tool steel. Surfaces are usually polished and may be chrome plated.

The force for the molding is supplied by presses. These presses can be massive, depending on the size of the part to be molded. The important size considerations for determining the required molding force are the projected area and the depth of the part. Other considerations include the resin type and the nature of the fillers and reinforcements that may be present.

Most thermoset materials can and are molded by compression molding and transfer molding. Fillers and reinforcements are often added to lower the cost or to modify the mechanical and physical properties.

Reaction injection molding (RIM) is similar to transfer molding in that a thermoset liquid is injected into a heated mold. The difference is in the nature of the thermoset liquid. In RIM the liquid is a mixture of two materials that react on mixing and form a solid material. Polyurethanes are the most common example.

Sintering is a process for forming powdered thermoplastic resins and involves three steps: preforming, fusing, and finishing. Preforming can be done on presses similar to those used in compression molding and transfer molding. The preforming step compresses the bulk resin into a preform that is 90% of the density of the finished part. (If additional processing is not required, the preforming step is called cold forming or solid-phase forming, depending on whether the forming mold is heated.)

The fusion step in sintering is accomplished at temperatures below the melting or decomposition points of the resin. At these temperatures, the resin particles soften and coalesce, thus forming a solid mass that has a low void content (hopefully). After careful cooling, which sets the crystallinity of the part, the part is removed from the mold and machined to finish the shaping operation.

GLOSSARY

Billet   A solid cylinder or block used in cold forming or solid-phase forming.

Break point   The point in the sprue and runner system where break is most likely to occur.

Bumping the mold   Another name for degassing.

Coining   Fusion of a green compact done under pressure.

Cold forming   A molding process for thermoplastic granules in which the granules are pressed in a cold compression press.

Compression molding   A molding process in which a thermoset resin is placed between matched dies and then squeezed, under heat, until cured.

Compression press   The machine in which compression molding is done.

Compression ratio   The ratio of the density of the bulk material to the density of the finished part.

Cryogenic flash removal   Removal method in which the parts are frozen to facilitate the removal of the flash, usually by tumbling the frozen parts.

Cull   The excess material that is left in the transfer chamber after the mold is filled in a transfer molding operation.

Daylight opening   The space between the platens of a compression press when the platens are at their farthest distance apart.

Defractoid   A device that measures surface waviness.

Degassing   A procedure in which the compression mold is opened slightly for a brief period to allow the vaporized condensate to escape; also called bumping the mold.

Dough   Pastes made from thermoset resins (typically unsaturated polyesters), filler, and chopped reinforcements.

Downstroke machine   A compression press in which the platen moves downward to close the mold.

Ejector pins   Another name for knockout pins.

Flash   Excess plastic material that forms around the parting line of a part and is removed after molding.

Flash type mold closure   The type of mold in which flash is assumed because the mold is loaded with excess material and then that excess is squeezed out of the cavity and forms flash.

Force required   The amount of force needed to mold a particular part.

Free fusion   Fusion of a plastic part done without pressure.

Green compact   A cold-formed or solid-phase-formed part before it is sintered.

Knockout pins   Thin metal pins that are moved into the cavity of the mold and eject the part; also called ejector pins.

Machine rating   Another name for press capacity.

Machine size   Another name for press capacity.

Matched die molding   Another name for compression molding.

Molding compound   B-staged reinforced and filled thermoset resins, which are usually compression molded.

Parting line   The line that marks the division of the two mold halves; this can often be detected as a thin line on the surface of the molded part.

Poker chipping   A defect in parts made by ram extrusion, in which the parts have a tendency to separate into disks because of inadequate joining of the separate charges during the pressurizing step.

Positive type mold closure   A mold-closure system that seeks to eliminate flash by creating a shearing surface between the mold halves; requires close control of the charge amount.

Preform   Thermoset granules pressed together into a particular shape for easy handling.

Press capacity   The maximum force that can be exerted by a compression press.

Ram extrusion   A forming method in which a thermoplastic material is pressed into a shape using a continuous pressing machine, thus forming a long part of constant cross section.

Reaction injection molding (RIM)   A process, similar to transfer molding, in which a thermoset liquid is injected into a closed mold; different from transfer molding in that the liquid is composed of two components, which react during the injection cycle.

Semipositive type mold closure   A mold-closure system in which a small amount of flash is allowed to form but the flash is separated from the part by some shearing between the molds.

Sintering   Heating of a formed plastic part made from granules so as to join or fuse the granules.

Solid-phase forming   A process similar to cold forming but with heated molds so that some thermal sticking of the granules occurs.

Sprue puller   An undercut on the sprue that assists in retaining the sprue on one side of the mold, thus facilitating removal of the cull and sprue.

Transfer chamber   A chamber in a transfer molding machine where the thermoset resin is held (and usually heated to insure that the material is highly fluid) before being forced into the cavities of the mold.

Transfer molding   A thermoset molding process in which the thermoset material is forced from a transfer chamber into a closed mold and then allowed to cure.

Two-stage clamping   A technique in which the compression mold is closed to about 90% very quickly and then the remaining distance is closed more slowly, thus shortening the closing time and yet still allowing the material to move properly within the mold.

Upstroke machine   A compression press in which the platen moves upward to close the mold.

QUESTIONS

1. Calculate the size of the press required to mold a part using a multicavity transfer mold in which the part size is 2.5 inches by 2 inches. The mold contains 4 cavities. There are 4 runners, each of which has a diameter of 0.375 inches and a length of 1.5 inches. The sprue has a projected area of 0.25 square inches. Assume that a transfer pressure 8,000 psi is needed.
2. Discuss the differences in resin orientation that would be expected in a part made by transfer molding and a part made by injection molding.
3. Explain how both undercuring and overcuring can cause part cracking on removal from the compression molding process.
4. Explain why coining can be used to make more complicated shapes than can conventional sintering.
5. Describe poker chipping, and explain how it can be prevented.
6. Why not eliminate the cull and save this extra material when doing transfer molding?
7. Why can't the cull, sprue, and runners from a transfer molding operation be reprocessed as regrind like runners in an injection molding operation?
8. Describe the defect that will most likely occur if the press used for transfer molding is undersized in clamping pressure.
9. What is the major advantage of a flash type mold closure system in compression molding? What are the disadvantages?
10. What is meant by cryogenic flash removal?

REFERENCES

Carley, James F. (ed.), *Wittington's Dictionary of Plastics* (3rd ed.), Lancaster, PA: Technomic Publishing Co., Inc., 1993.

Chanda, Manas, and Salil K. Roy, *Plastics Technology Handbook,* New York: Marcel Dekker, Inc., 1987.

Frados, Joel (ed.), *Plastics Engineering Handbook of the Society of the Plastics Industry, Inc.* (4th ed.), New York: Van Nostrand Reinhold, 1960.

Gongal, S. V., "Polytetrafluoroethylene," E. I. DuPont de Nemours & Co., Inc.

Milby, Robert V., *Plastics Technology,* New York: McGraw-Hill Book Co., 1973.

Whelan, A., and J. L. Craft (eds.), *Developments in Plastics Technology—3,* New York: Elsevier Science Publishing Co., Inc., 1986.

CHAPTER NINETEEN

POLYMERIC COMPOSITE MATERIALS AND PROCESSES

CHAPTER OVERVIEW

This chapter examines the following concepts:

- Basic concepts of composite materials and processes, including advanced and engineering composites
- Matrix materials
- Reinforcements (fiberglass, carbon/graphite, organic fibers, reinforcement forms)
- Manufacturing methods for composite parts (processes for very short-fiber thermoplastics, matched die/compression molding, RTM, spray-up, hand lay-up for wet and prepreg materials, filament winding and fiber placement, pultrusion, processes for long and continuous fibers with thermoplastics)
- Plant concepts (layout, safety, material handling)

INTRODUCTION

In a broad sense, the term *composite materials* refers to all solid materials composed of more than one substance. A more narrow definition, and the one used in this book, is that composite materials are those solid materials composed of a *binder* or *matrix* that surrounds and binds together fibrous reinforcements. The binders of most interest in this text are plastic resins. Other matrix materials are metals and ceramics.

Polymeric (plastic) composite materials represent about 90% of all composites. These materials are made of fibrous reinforcements (usually fiberglass or carbon fibers) which are coated or surrounded by a plastic resin. The material is placed in a mold and solidified, either by thermoplastic or thermoset molding methods. The fibers give strength and toughness to the plastic. The plastic matrix allows the fibers to be shaped.

Nonpolymeric composites have either metal or ceramic as the binder material around the fibrous reinforcement. These nonpolymeric composites are used when temperature, strength, or some other desired property or operating condition prohibits the use of a polymeric composite.

Polymeric composites are divided into two groups. These differ principally in the type and length of the fiber reinforcement and in the type of resin used. One group of composites is called *advanced composites,* characterized by very long and very high-performance reinforcements. The resin types are also very high performance, meaning that the resins' thermal and mechanical properties are very high (superior). Advanced composites are typically used for aerospace applications (like rocket motor cases and airplane parts), but are also used for high-performance sporting goods (golf clubs, tennis rackets, bows and arrows) as well as an increasing number of parts that require the superb properties (primarily high strength-to-weight ratios) obtained from these materials.

Some of the products made using advanced composites are shown in Photos 19.1, 19.2, and 19.3. One application of advanced composites that holds tremendous potential is for wrapping concrete highway columns for protection against corrosion and for improvement in the performance of the columns during earthquakes. Photo 19.4 shows a column being wrapped with carbon fiber and epoxy composite material, which will then be cured to give a hard, strong overwrap on the column.

The second group of composites are the *engineering composites,* which are characterized by shorter fibers and fibers with lower mechanical properties. The resins are also lower in performance. Products made from engineering composites include boat hulls, canoes, tubs, shower stalls, spas, fuel storage tanks and other parts that contain short reinforcing fibers to give increased mechanical properties over parts made from conventional plastic resins. These latter parts are normally molded by the same processes as the nonreinforced plastic analogues. Since most of these processes have already been discussed (injection molding, extrusion, and so on), the section in this chapter that discusses these processes will focus only on the changes that must be made in the processes to allow for the presence of a fiber reinforcement.

Photo 19.1   B–2 Stealth bomber (Courtesy: Northrup Grumman Corporation)

Photo 19.2 Composite oars
(Courtesy: Advanced
Composites Manufacturing)

Some products made using engineering composites are shown in Photos 19.5, 19.6, and 19.7. The structures shown illustrate the capability of composite materials to give good strength and good durability.

Composite materials are part of our lives every day. We ride in cars and light-rail systems with composite panels, we fly in planes with composite parts, our homes have

Photo 19.3 Composite golf club shaft (Courtesy: Carbon Fiber Products)

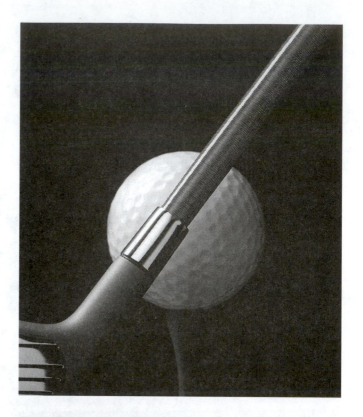

Photo 19.4 Wrapping highway columns using carbon fiber-epoxy composite material (Courtesy: TCR Composites)

Photo 19.5   Light-rail car made of fiberglass-polyester composite (Courtesy: Aero Trans, Inc.)

composite showers and numerous composite parts used in hidden places (such as in washers and dryers), and we enjoy recreation with composite tennis racquets, golf clubs, and boats.

Comparisons of Composites to Nonreinforced Materials

The presence of a reinforcement material can have major effects on the properties and processing of a plastic resin. As the lengths of the reinforcing fibers increase and the concentration of the reinforcement increases, these changes become more significant until, at fiber lengths over about 4 inches (100 mm) and concentrations over 40%, the fiber properties, rather than the resin properties, dominate the mechanical properties and processing characteristics of the composite. At the point where the fiber properties begin to dominate, the function of the plastic resin changes from the main source of mechanical and physical properties to a supporting role: holding the fibers together, giving shape to the part, and protecting the fibers from external forces and environmental conditions.

Photo 19.6   Composite structural roof (Courtesy: Delta Fiberglass Structures, Inc.)

Most mechanical properties of high-performance parts are, therefore, dominated by the fiber properties. For instance, a part having 60% long carbon fibers that are all aligned in one direction in an epoxy matrix typically has a tensile strength of 130,000 psi (900 MPa) and a modulus of 43×10^6 psi (300 GPa), whereas the part without the fibers could have a tensile strength of 8,000 psi (60 MPa) and a modulus of 43,000 psi (3 GPa).

The composite parts just described, with all of the fibers aligned in one direction, have comparatively little strength in the direction perpendicular to the fibers. In the perpendicular direction, the strength is approximately the same as the nonreinforced matrix material. For some applications, having the strength predominately in one direction is desirable. For other applications, however, strength in more than one direction is desired. An advantage to composite materials is that by using an appropriate manufacturing method, the fiber direction can be carefully controlled so that the part's mechanical properties are adjusted to optimum values in whatever direction or combination of directions is desired. Parts that have mechanical properties that are the same in all directions are called *isotropic*. Parts that have mechanical properties that are directional are called *nonisotropic* or *anisotropic*. Often a part is needed that has equivalent properties in a plane (such as the *x-y* plane) but different properties out of the plane (such as in the *z* direction). This type of part is called *orthotropic*. Orthotropic materials result when layers of materials are built up to make the part, as is the case with many composites. Composite

Photo 19.7 Composite tank (Courtesy: Delta Fiberglass Structures, Inc.)

materials made from several layers of materials are called *laminates*. The control of fiber direction is so important, especially when the fibers are very long, that most manufacturing processes for composites are characterized by different methods of controlling the fiber orientation and placement. The design of some composite parts carefully calculates and stipulates the fiber directions in the several layers of composites, a type of design called *laminate design theory*.

The ability to selectively control the direction of the composites' mechanical properties coupled with the inherent light weight of plastics means that composites often are much lighter in weight than competitive materials having the same mechanical properties. For instance, a carbon-fiber reinforced epoxy part weighs only one-ninth as much as a steel part having the same mechanical properties. The carbon fiber/epoxy part weighs approximately one-third as much as an equivalent-strength and -stiffness aluminum part. To facilitate these comparisons, two mechanical properties—specific strength and specific stiffness—are defined. The *specific strength* is the strength of the material divided by the specific gravity. The *specific stiffness* is the modulus of the material divided by the specific gravity. Therefore, materials with low specific gravities, such as the low-weight composites, have high specific strengths and high specific stiffnesses. The high specific strengths and high specific stiffnesses of composites have led to their extensive and increasing use in the aerospace industry. A plot of specific strength versus specific stiffness for various materials is given in Figure 19.1 and illustrates the comparative performance of composite materials and metals.

Another major advantage of composites, perhaps the one that will prove in the long run to be most important, is that they have mechanical properties that equal or exceed those of metals but with the ease of forming characteristic of plastics. (The forming is complicated somewhat by the presence of the fibers, but it is still easier than forming metals and usually has far less excess or wasted materials.) This combination of mechanical strength and formability means that assemblies that are made from several separate metal parts can be made (molded) in a single composite part, resulting in less assembly labor, fewer parts to fail, and usually less expensive production.

Composites also have some disadvantages when compared to other materials. Composite material costs are generally higher than metals. The processes for making composites are less well defined and optimized than are conventional metalmaking processes or plastic-molding processes. The design data for composites is less well known and more

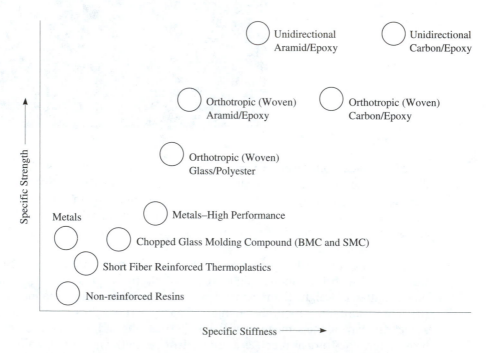

Figure 19.1 Specific strength versus specific stiffness for several materials.

complicated than design data for either metals or plastics. Compared to metals, high-performance composites tend to be easily damaged by low-velocity impacts.

MATRIX MATERIALS

Although both thermoplastic and thermoset materials can be reinforced and, therefore, can be made into composite materials, composites with very short fibers tend to have thermoplastic matrices. As the fibers get longer, the tendency increases to have a thermoset matrix. The discussion of matrix materials therefore is organized according to the length of the fiber reinforcement.

Very Short-Fiber Composites

Several factors contribute to the trend that links the type of matrix with the fiber length. First, the very short-fiber composites (with fiber lengths less than 0.2 inches, 5 mm) are usually processed by conventional plastic-processing methods. (These very short fibers are sometimes called *whiskers*.) When being extruded or injection molded, the fibers must be able to go through small clearances, such as the gap between the extruder screw and the extruder wall or the gate that connects the mold cavity with the runner system in both injection molding and transfer molding. Thermoplastics dominate these conventional processes and, therefore, dominate the composites made by these processes.

Another reason for the predominance of thermoplastics as matrices for very short-fiber composites is that thermoplastics often need the additional strength or additional stiffness gained from reinforcing with short fibers. Thermosets, because of their higher molecular weight from crosslinking, do not benefit as much from the very short fibers. The significant increase in properties obtained by adding short-fiber reinforcements to thermoplastics can be seen in Table 19.1. All of these composites containing very short fibers are part of the engineering composites group.

While most thermoplastic resins can be reinforced to give improved mechanical properties, engineering resins are the thermoplastics most often reinforced. These resins compete in applications where mechanical properties may be the critical factor in choosing the plastic. The most commonly reinforced resins are nylon, PET, polycarbonate, acetal, ABS, PPO, PPS, and other high-performance or high-temperature thermoplastics.

Intermediate-Length Fiber Composites

As the fibers get longer, the thermoset resins tend to dominate as matrix materials. There are several reasons for this. One reason is the greater difficulty encountered in coating the longer fibers with resin, a process called *wet-out*. The lower viscosity of thermoset resins makes wet-out easier and more complete. The fibers must be completely wetted with the resin to obtain the maximum in composite properties. The low viscosities of thermoset resins also help wet the fillers often used in high concentration in composites where the fibers are an intermediate length of .4 to 4 inches (10 to 100 mm).

Table 19.1   Effects of Fiberglass Content on Tensile Strength and Elongation
in Various Resins

Resin	Tensile Strength in Psi (MPa) for 0% and 30% Fiber Contents	
	Fiber Content 0%	Fiber Content 30%
ABS	4,300 (30)	14,500 (100)
Nylon	13,000 (90)	24,500 (170)
Polycarbonate	10,100 (70)	20,300 (140)
PET	7,200 (50)	22,000 (150)
PP	5,000 (35)	13,000 (90)
Sheet molding cmpd (SMC)	—	24,500 (170)

Resin	Elongation (%) for 0% and 30% Fiber Contents	
	Fiber Content 0%	Fiber Content 30%
ABS	30	2
Nylon	50	2
Polycarbonate	120	6
PET	200	5
PP	100	3
Sheet molding cmpd (SMC)	—	1

Fibers of this length often are called *chopped fibers* because of the way they are cut. Composites with chopped fibers are usually classified as engineering composites. Therefore, the engineering composites group consists of very short-fiber materials, usually with thermoplastic matrices, and intermediate or chopped fibers, generally with thermoset matrices and often with high filler contents.

By far the most common thermoset resins used in composites having intermediate-length fibers are the crosslinkable polyesters. The low cost of polyesters and the ability to use simple curing technologies at both room and elevated temperatures are important reasons that contribute to this wide use. The polyester and fiberglass composites are commonly called *fiberglass-reinforced plastics (FRP)*. Other thermoset resins used include phenolics, vinyl esters, and epoxies. New processes continue to be developed that allow conventional engineering thermoplastics to be used with chopped fibers and to compete with polyesters, although acceptance of these processes and products is slow.

Fillers are commonly added to many of the resins used with chopped fibers. This further reduces the cost and gives certain properties that depend on the presence of fillers.

Curing (crosslinking) of the thermoset materials is usually done as it would be for nonreinforced materials. These curing methods are described in the chapter on thermosets.

Very Long- and Continuous-Fiber Composites

These products have fibers that sometimes reach across the entire dimension of the part and are therefore called *continuous fibers*. As with chopped fibers, the wet-out of very long (greater than 4 inches, 100 mm) and continuous fibers is improved by the low viscosity of thermoset resins. Continuous and very long fibers are used to make advanced composites.

Another reason to use thermoset resins for advanced composites is the improved properties thermosets offer when compared with conventional thermoplastic resins. The critical properties in these applications are modulus (stiffness) and heat resistance, both of which are improved by crosslinking. In fact, the choice of a matrix material is often dictated by these properties.

Epoxies are the most widely used resins in composites having very long or continuous fibers. Epoxy resins generally have higher temperature and stiffness characteristics than do the crosslinkable polyesters, thus making epoxies preferred in these long-fiber composites where performance is valued more highly than cost or ease of production. Other thermoset resins used include phenolics (when flame retardance is important) and polyimides (when higher-temperature performance is important).

Some thermoplastic resins with exceptional properties are occasionally used for matrices in advanced composites. These resins, called *high-performance thermoplastics* or *advanced thermoplastics,* are chosen because they have some unique characteristics that fit a particular need. Some, for instance, combine high toughness, high thermal stability, and excellent solvent resistance, a combination especially useful for leading edges of airplane wings. Still, the problems with fiber wet-out and thermal stability must be solved for widespread acceptance of the high-performance thermoplastics in advanced composites.

REINFORCEMENTS

Three types of fibers dominate the reinforcement market and all three share some basic characteristics. They are all very strong, very stiff, and can be made in continuous lengths. These three fibers are fiberglass, carbon or graphite fibers, and organic fibers (the most important of which are generically called *aramids*).

The choice of which fiber will be used in a particular application depends largely on cost and performance. As can be seen in Figure 19.2, fiberglass is the least expensive fiber. Carbon fibers are the stiffest of the three major reinforcements and aramid fibers (Dupont's Kevlar® is the most widely used) are the toughest (highest strength and elongation). The data given show a range of values because each of the fiber types can be made in several different grades and is available from several different manufacturers. The numbers given on each figure are meant to show a representative value for each of the properties. Some applications combine fiber types so that properties of one type enhance the properties of another. For instance, aramid fibers are excellent in tension but poor in compression, whereas glass fibers are poor in tension but excellent in compression. Parts in which two or more fiber types are combined are called *hybrid composite parts.*

Fiberglass

Fiberglass is made by spinning molten glass in much the same way that a synthetic fiber is spun from a melt. The thin strands of fiberglass are cooled and joined into bundles of fibers called *tows*. These tows are then wrapped around spools and shipped as the raw material for making composites. By including certain chemicals in the molten glass and using some special processing techniques, different types of fiberglass can be made. These

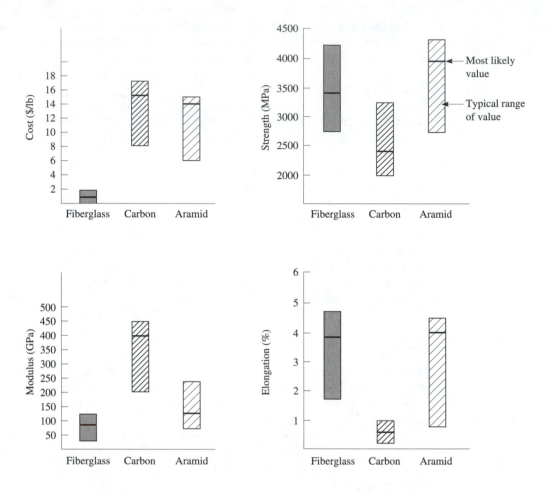

Figure 19.2   Comparisons of cost, tensile strength, modulus and elongation for fiberglass, carbon, and aramid fibers.

differ chiefly in their physical characteristics. Fiberglass that has improved electrical resistance is called *E-glass,* fiberglass with improved chemical resistance is called *C-glass,* and fiberglass with high strength is called *S-glass.* Some advanced composites use S-glass to achieve special mechanical properties, but the most commonly used fiberglass for composites is E-glass, because it is the least expensive type. The greatest amount of fiberglass for composite reinforcement is in engineering composites.

After the fiberglass is formed, it often is treated with a chemical to give improved abrasion resistance to the fibers. (Fiberglass is very brittle and susceptible to developing tiny cracks that can lead to premature failure.) Materials that are added to coat the fibers are called *sizings* or, sometimes, *sizes.* Many sizings are water-soluble polymers that can be removed easily before the fibers are coated with the resin to make a composite. However, this is rarely done with composites because the sizing is usually compatible with the resin.

Compatibility between the fiberglass and the resin is very important in achieving the optimum properties in the composite. If the resin does not bond well to the reinforcement, the fiber can slip inside the composite when a load is placed on it. This slippage will decrease the stiffness and strength of the composite. To assist in improving the bond between the fiber and the matrix, the fibers can be coated with chemical agents called *coupling agents*. These coupling agents are molecules that have one end chemically similar to the resin and the other end chemically like the glass. The glasslike end bonds to the fiberglass and then, when the resin is added to make the composite, the resin-similar end bonds with the resin, thus coupling the resin and fiberglass together. For simplicity, coupling agents often are added to sizing so that the fiberglass can be coated in one step. Both sizing and coupling agents are very important with fiberglass (more so than with the other kinds of reinforcements).

Carbon or Graphite Fibers

In the days when carbon fiber was first being produced, some distinction was made between standard carbon fibers and those that were subjected to a higher-temperature final-processing step and were therefore more purely graphite fibers. That distinction has now largely disappeared and so the terms *carbon fibers* and *graphite fibers* now refer to the same material.

Both carbon fibers and graphite fibers are commercially made from three different starting materials—polyacrylonitrile (PAN) fibers, pitch (a natural residue from coal or oil processing), and rayon fibers. The processes currently used for making carbon fibers require that fiber with high carbon content be used as the starting material. For PAN and rayon, that requirement is met automatically. For pitch-based carbon fibers, the pitch is heated and extruded through a spinnerette to make a fiber that is then used as the starting material for the process to make carbon fibers. These PAN, rayon, and pitch fibers are subjected to various stages of heat treatment while under tension and in a nonburning atmosphere. Under these conditions the fibers lose their noncarbon atoms (such as hydrogen and nitrogen). The general fiber shape is maintained throughout the process (except that the fiber diameter decreases), but the molecules within the fiber rearrange into the flat, stiff structure characteristic of graphite. The PAN, rayon, and pitch fibers are converted into fibers with almost all (over 90%) carbon content. These are carbon or graphite fibers.

After the final heating step, the carbon fibers are subjected to an electrolytic bath that cleans the surface of the fiber. The fibers are then coated with a sizing resin (usually epoxy), gathered into tows, and packaged on spools for shipping and later processing. The epoxy material can serve as both a sizing and a coupling agent for the carbon fibers, especially because epoxy resins are the most common matrix materials for composites containing carbon fibers. If some other matrix is used, then that matrix material can be substituted for the epoxy as the sizing.

The major advantage of carbon fibers over all other fibers is their very high modulus. Carbon fibers are among the stiffest of all known materials, especially when compared on an equal-weight basis. Since many applications for composites require very high stiffness at low weight, this combination of properties is highly valued.

The high cost of carbon fibers precludes their use in most engineering composite applications. Rather, carbon fibers are used in advanced composites, chiefly in aerospace industries but also in sports equipment and other high-value applications such as medical devices.

Organic Reinforcement Fibers

Most of the traditional organic fibers used for apparel (such as nylon or polyester) lack the high strength and high modulus required for reinforcing composites. The *aramids* (resins based on highly aromatic nylons) are a class of resins that can be made into fibers possessing these high mechanical properties. The most important fiber of this group is Kevlar®, which is made by the DuPont Company. Other manufacturers are now producing competitive aramid fibers but Kevlar® still maintains the predominant share of the market.

Aramid fibers are spun much as traditional synthetic fibers are spun. After being spun they are grouped into tows and wound onto spools. Coupling agents and sizes are generally not required.

Aramid fibers are strong and stiff but have their greatest use in applications where toughness is required, especially impact toughness. Impact toughness is difficult to measure on fibers alone (without the matrix or without making the fibers into some two-dimensional structure), so no specific toughness category is listed in Figure 19.2. However, some measure of toughness can be gained by looking at the strength and the elongation simultaneously.

Aramid fibers are used in applications where strength and toughness are both required. An example is the leading edges of airplane wings where strikes from birds or debris, especially during landing and takeoff, could damage the wing. Aramids are also used in armor applications where resistance to high-velocity impacts is critical. Body armor (bulletproof vests) is usually made from aramid fibers but without resin so that the armor can flex as a cloth material.

Another organic resin that has found some applications in composites is an ultrahigh molecular weight polyethylene (UHMWPE) called Spectra® in the United States. This material is made by Allied Signal. Like aramids this polymer fiber is used for applications where high-impact toughness is required. Such applications include bulletproof vests and those aerospace products where the material's extreme light weight and toughness are a strong advantage.

Comparison of Reinforcement Fibers

The different fiber types could be compared in a variety of ways, but a comparison of modulus and cost seems appropriate since modulus is most often the key design parameter. Cost, of course, is also important in the choice of reinforcement material and it becomes even more important as the number of applications for composite materials increases. Table 19.2 compares modulus values and costs for fiberglass, aramid fibers, and carbon (graphite) fibers.

Table 19.2 Modulus Values and Costs for the Major Types of Reinforcement Fibers

	Glass	Aramid	Carbon
Modulus psi (GPa)	10×10^6 (72)	17×10^6 (117)	33×10^6 (227)
Cost, $/lb ($/kg)	1.00 (2.20)	16.00 (35.00)	10.00 (22.00)

Reinforcement Forms

The fibers are all packaged on spools from the manufacturers but often are converted into some other physical form before being used to make composites. One of the simplest form conversions is to chop the fibers into shorter lengths. As already discussed, very short length fibers are used as reinforcements for thermoplastics that are to be processed in normal processing methods such as injection molding and extrusion. These very short fibers are usually blended into the resin by the resin manufacturer who chops the reinforcements. When the chopped fibers are longer, as they are for polyester and fiberglass molding compounds, the chopping is done by the manufacturer of the molding compound, often as part of a continuous process. When the chopped fibers are sprayed with the resin into a mold, the chopping is usually done at the time of the spraying. In this case the fibers are fed from the spools into a chopping mechanism, usually attached to the spray gun, and chopped as they are picked up by the spray stream. In general, the maximum length for any of these chopped fibers is approximately 4 inches (100 mm).

A somewhat more convenient form for the fibers, if a flat part is to be made or if the fibers are to be arranged in several layers, is a *mat*. Mats are made by spreading either chopped or continuous fibers onto a moving belt and adding a small amount of binder resin. The purpose of the binder is just to hold the mat together; the binder is not the composite matrix. The mats are then rolled up for transport to the composite manufacturing facility where they are unrolled and cut to size as they are used. Mats are made in several thicknesses.

Fibers can also be woven into cloth using several different weaving patterns. The choice of weaving pattern and the weight and number of threads per square area determine the handling properties of the cloth, such as pliability. Some cloths are more pliable than others and are used for composites that are highly shaped. Knitted fabrics can also be made from all of the reinforcement fibers for applications where the shape flexibility of knits is important. Cloth and knit fabrics are used for both engineering and advanced composites.

In some composite structures, especially advanced composites, the fibers need to be oriented at angles other than those that can conveniently be obtained from cloths. The most common way to obtain these other orientations is by using materials in which all the fibers go in one direction. These materials are called *unidirectional*. Different orientations are obtained by using several layers of unidirectional material. To keep the unidirectional fibers in place, and to ensure that the proper amount of resin is used, the fibers are precoated with resin in a separate step. Sheets of unidirectional fibers precoated with resin are called *prepregs*. Most prepregs are made of carbon fibers, although aramid and fiberglass prepregs are also available. Prepregs can also be made from cloths, although this is not as common as unidirectional fibers. The prepregs are usually sold as rolls in widths of 4 inches (100 mm) up to 40 inches (1000 mm).

MANUFACTURING METHODS FOR COMPOSITE PARTS

The key to manufacturing methods for composites is control of the direction, overlap, and other placement parameters of the fiber reinforcements. In some manufacturing methods this control is very precise, while in others the control is more relaxed and, as a result, the fiber directions are more random. Some methods are best suited to parts having a particular shape characteristic (such as a constant cross section), while other methods are used because they have great flexibility in the way the fibers can be arranged.

The nature of the molds used for shaping composites depends on the pressure and temperatures involved in the process. In all cases, however, some provision must be made to assist in removing the part from the mold. Generally, this involves applying a mold release to the mold prior to any material being placed into it.

Thermoplastic Processes Using Very Short Fibers

These processes are standard thermoplastic processes such as injection molding and extrusion. When used to make composite parts with very short fibers, only minor changes need to be made in the way the processes are done. The most important change is that all gaps in the flow path of the resin and fibers should be large enough to allow the fibers to pass. In injection molding, the critical points in the path are the gate at the entry to the mold cavities, the runners, and the orifice at the end of the injection nozzle. Similar close tolerances could exist in an extrusion die.

When these gaps are opened or enlarged to ensure good fiber passage, some decrease in back pressure can occur. This lowered back pressure may change the way the plastic material flows in the mold or die and can require some adjustments in temperature or pressure.

Another change in processing besides enlarging the gaps can be a change in the viscosity of the resin. The manufacturer of the resin may have intentionally changed the resin viscosity to give more uniform flow of the resin with the fibers. Therefore, the temperatures of the molding system may need to be adjusted to get optimum molding conditions.

The close tolerances between the screw and the barrel in the melting portion of an injection molding machine or an extruder can cause fiber damage. Whenever possible, the amount of melting from mechanical action should be minimized, usually by increasing the temperatures in the extruder.

The use of very short fibers in thermoplastic resins limits the use of these composite materials to engineering composite applications. These composite materials should be considered for the same applications as their nonreinforced analogues.

Matched Die or Compression Molding

One of the major advantages of compression molding over injection molding is the short flow path of the resin. In compression molding, the resin has to travel only the distance needed to fill the mold. This short path means that much longer fibers can be molded into composite parts by compression molding than can be molded by injection molding. Also, compression molding largely eliminates the problem of plugging the system with fibers too long to go through narrow openings. Maximum fiber length is that length that allows free movement of both resins and fiber within the mold. Fibers from 0.4 to 4 inches (10 to 100 mm) work well.

If, however, the fibers are not to move at all, as would be the case if a preform of an entire part were placed in the compression mold, continuous fibers and mats can be used effectively. When the preform method is used, the preform is placed in the mold and the resin is poured on top of it. The pressure from the mold closing then distributes the resin. Some fiber redistribution also occurs during this process, but is usually minor.

Compression molding with chopped fibers is much more important than molding with preforms. In the standard compression molding method, the fibers have been previously combined with resin (usually crosslinkable polyester) and a filler (such as calcium carbonate) into a molding compound. This molding compound is then used as the charge material in compression molding, by weighing the molding compound just prior to placing it into the open compression mold. Weighing and charging can be done either automatically or manually.

Matched die (compression) molding has the advantage over most other composite manufacturing methods in that it produces parts with both sides defined by the mold. Reasonably complex shapes (such as ribs and bosses) are possible in compression molding provided that the shapes are not so sharp or narrow that the fibers do not bend easily to fill them. Little trimming is required in matched die molding. The products usually have low void contents, and the reject rate is usually low.

Compression molding does have disadvantages. The molding compounds require refrigeration for storage, molds are usually metal so they cost more than molds used in lower-pressure processes, and the molding of large parts often requires large and costly presses.

Two types of molding compound are in common commercial use: *bulk molding compound (BMC)* and *sheet molding compound (SMC)*. These compounds differ in the way they are made, the way they are charged into the compression mold, and the length of the fibers commonly used.

Bulk molding compound is made by combining the resin, initiator, filler, and fibers in a low-intensity mixer (a typical ratio of resin/filler/reinforcement is 30/55/15). After thorough mixing, the consistency of the mixture is like a paste or dough. (BMC is also called *dough molding compound.*) This material is usually shaped into a log for easy transport and stored at low temperature until needed. Low-temperature storage is required to prevent premature curing of the BMC since the initiator is already included in the mix. When the BMC is molded, a portion of BMC is cut or broken off the log, weighed to insure proper charge weight, and then simply placed in the center of the compression mold, assuming that the mold is rather small. If the mold is large, the BMC can be placed in

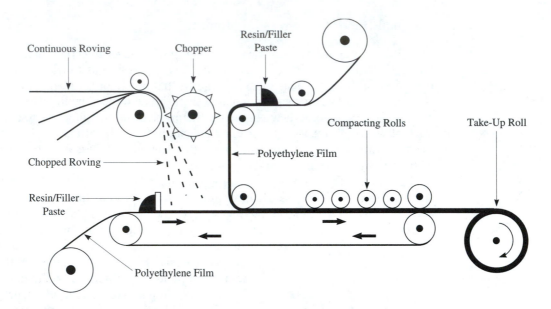

Figure 19.3   Process for making sheet molding compound (SMC).

several locations within the mold, although this may result in weak areas where flows meet. Closing the mold distributes the BMC throughout the mold by pressing on these lumps. The mold is then heated to cure the compound.

BMC is used extensively for molding small- to medium-sized parts where the movement of the BMC within the mold is not too great. This movement in the mold and the method of mixing the BMC components in a low-intensity mixer tend to limit the length of the fibers that can be included in BMC to about 3 inches (80 mm). Limiting fiber length also limits the mechanical properties obtained from BMC materials.

SMC is made by sprinkling chopped fibers onto a moving sheet of film that contains a layer of premixed resin, initiator, and filler (normally about 0.2 inches, 5 mm thick). Another sheet of film with the same type of resinous mixture is placed on top of the chopped fibers to form a sandwich of film, resin with filler and initiator, and chopped fibers. Next, this sandwich of materials is compressed by rollers to thoroughly mix the materials together. The sandwich, which is in the form of a continuous sheet, is then rolled up for storage and transport. This process is depicted in Figure 19.3. The fibers and the resinous mixture are mixed together in place, which allows longer fibers than can be used in BMC. SMC fibers up to 4 inches (100 mm) long are common.

When the SMC material is molded, it is cut from the roll in sheets and strips that are placed into the mold (after removing the film sheets). Little movement of material occurs within the mold because of this placement throughout the mold area. This limited movement also allows the use of longer fibers than can be used in BMC.

SMC is used to make large parts, such as the body of the Corvette automobile and bodies for truck cabs, as seen in Photo 19.8. It is also used to make parts that have high strength requirements (within the scope of engineering composites).

Photo 19.8   Composite truck cab body (Courtesy: Kenworth Truck Company)

Resin Transfer Molding (RTM)

Although *resin transfer molding (RTM)* is similar to both traditional transfer molding and reaction injection molding (RIM), it differs from them in that a reinforcement is molded with the resin. The physical arrangement and type of equipment used in RTM are like RIM; in fact, the same equipment can be used for both processes. In RTM, however, a fiber preform is placed in the mold and then the mold is closed. A fiber preform is made by forming noncoated fibers into the shape of the final part and then spraying them with a small amount of binders to hold them together. The fiber preforms used in RTM can be made from either chopped fibers, mat, or continuous fibers that are knitted or woven into the desired shape. Sometimes, the preform is stitched with reinforcing fiber to allow for transport and loading in the mold without significant preform damage. The resin is injected into the closed mold and wets out the fibers in the preform as the resin fills the mold.

Although polyurethane or other reactive resins can be used to make RTM parts, the more likely materials are polyesters and epoxies. The two pumping chambers contain polyester resin and initiator or epoxy resin and hardener. Resin injection is similar to that in RIM except that care must be taken to inject slowly enough that the fibers in the preform are not moved significantly as the resin fills the mold. Proper wet-out of the fibers is a major concern. A vacuum is often applied to the part of the mold farthest from the inlet to assist in resin movement inside the mold.

The capability of RTM to use both long and short fibers of any of the three major types and to use polyester or epoxy resins means that RTM can be used to make both engineering and advanced composites. This process is rapidly gaining popularity and seems to solve many of the problems inherent in other composite-making processes, such as how to make a composite part in an automated or semiautomated fashion, how to use fiber weaving and knitting technology to allow for more flexibility in the lay-up of fibers, and how to reduce pollution by confining the molding operation to a closed mold.

Spray-up

This method of manufacturing composite parts involves using a spray gun to which a fiber-chopping apparatus has been attached. As the resin and initiator are mixed in the gun and then sprayed out of the nozzle, the stream picks up fibers that have been chopped so that they drop into the stream. These fibers are then wetted by the resin and, simultaneously, blown forward. The operator directs the stream onto the surface of a mold. This process is depicted in Figure 19.4.

The need to transport the chopped fibers through the air in a spray limits the length of the fibers to about 3 inches (80 mm). The spray-up method depends on the skill of the operator to achieve an even coating of resin and fiber over the entire surface of the part. These two factors combine to dictate that the spray-up method be used only for engineering composites. Therefore, the most common materials used are polyester resin and fiberglass. Cures are done at room temperature, although modest heating is also common.

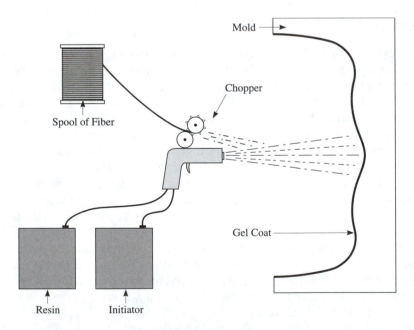

Figure 19.4   The spray-up process for making composite parts.

The spray-up method is well suited to making large parts such as boats and spas. No pressure is exerted by the process and so the molds can be made of almost any rigid material able to withstand the temperature of the particular cure. Materials such as wood, aluminum, and polyester or epoxy with fiberglass are common.

The spray-up method creates a rough and uneven surface. Therefore, a thin coat of resin is sprayed into the mold and allowed to partially cure (to a gel state) before the resin and fiberglass are sprayed on top. This precoat, called the *gel coat,* gives the part a smooth outer surface. It can be colored to give permanent color to the part and to assist in hiding the fiber pattern underneath. Furthermore, the gel coat serves as a physical (cured resin) barrier to prevent fibers from working their way to the surface of the part and thereby causing surface blemishes (a process called *fiber blooming*). Where the part is used in a wet environment, the gel coat also prevents wicking of the moisture along fibers that bloom to the surface. Wicking can cause weakness in the part. A very thin sheet of fiberglass (called a *surface veil*) can be placed against the gel coat to further prevent fibers from blooming. The veil also hides the fiber pattern and improves surface appearance.

Hand Lay-up (Wet and Prepreg)

Laying sheets of fibers into a mold by hand is obviously the simplest method of arranging fibers, and is one of the most important. Some parts are so complicated that only hand placement of the fibers is possible. Also, hand placement is popular, especially when only a few parts are to be made, because little equipment is needed and machine costs are minimized. The disadvantages to hand lay-up include the precision required of manual workers, the high costs associated with this labor, and the long cure times (often several hours).

The hand lay-up method for wet fibers is used for engineering composites. Polyester and fiberglass are the dominant materials. In this method a layer of fiber mat or fabric is placed in the mold, usually after a gel coat has been applied, and then the resin is poured on top of the fibers. The resin is then distributed over the entire fiber surface and into the fibers by using hand rollers. These rollers also compress the fibers to remove any air that might be trapped. If additional thickness is desired, another layer of fiber mat or cloth is placed in the mold and more resin is added and distributed. Sometimes, the layers of resin and fiber are allowed to partially cure before additional layers are applied.

As with spray-up, the molds used can be very large. Because no pressure is applied, mold materials can be almost anything able to withstand the curing temperatures.

Hand lay-up using prepreg materials is used for advanced composites. The prepreg materials are carefully placed in the mold so that the fiber orientation meets the design requirement and is smoothed flat so that no wrinkles occur in the prepreg sheet. Additional layers can be placed directly onto the previous layers. Prepregs are generally slightly tacky so the layers stick to each other.

After the required number of layers have been properly placed, the prepregs are subjected to a vacuum process called *debulking,* which removes any trapped air. The stack of prepreg materials on the mold can then be heated to cure the resin. Debulking and curing can be done simultaneously using a vacuum bag assembly like that presented in

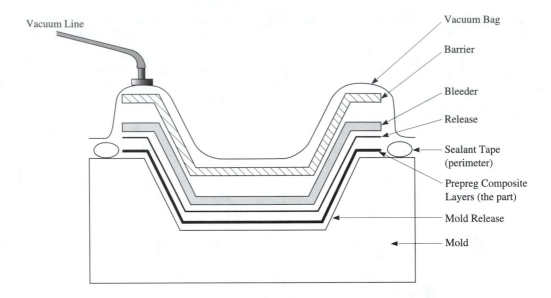

Figure 19.5   Vacuum bag assembly for debulking and curing prepreg materials.

Figure 19.5. The vacuum bag assembly also removes excess resin so that the fiber-to-resin ratio is high, thus maximizing the strength and stiffness to weight of the composite part.

In the vacuum bag assembly a layer of *release cloth* is placed on top of the prepreg layers. This cloth allows all the other materials in the vacuum bag assembly to be removed from the part after the part is cured. The release cloth is usually a fiberglass cloth coated with a release material, like a fluorocarbon plastic.

The next layer is an absorbent material called a *bleeder cloth*. The amount of bleeder cloth determines the amount of resin that is removed. Some experience is required in determining the amount of bleeder to ensure that just the right amount of resin is removed to give the optimal weight, but not so much that the fibers are undercoated with resin.

Additional layers of material can be placed on top of the bleeder to ensure that the resin stays within the bleeder layer. These layers are called the *barrier*. Finally, the entire assembly is covered with a *vacuum bag* (usually made from nylon film or silicone elastomer). The vacuum bag is adhered to the mold face and has a vacuum valve attached so that the entire assembly can be placed under a vacuum and compacted against the face of the mold. A vacuum is applied and the excess air is removed from the entire assembly.

After applying the vacuum, the mold and vacuum bag assembly are placed in an oven or autoclave to cure. The *autoclave,* which applies both heat and pressure, is used if very low void contents are required. After the required time at the proper temperature, the mold and assembly are removed and the vacuum bag assembly is taken off the part. The part is removed from the mold and, if required, finished by machining or other finishing methods.

Sometimes a very thick composite part is desired but weight constraints do not allow a solid composite to be made. If the compressive forces are not too high, a rigid, light-

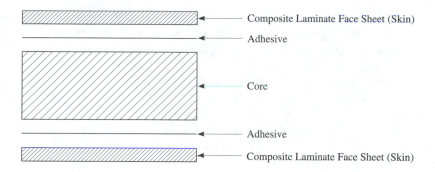

Figure 19.6 A sandwich composite.

weight material can be placed between composite laminate plates (called *faces* or *skins*) to give the thickness required without the weight. This layered construction is called a *sandwich composite*. The material in the middle of the sandwich is called the *sandwich core material*. The most commonly used sandwich core materials are rigid plastic foam, honeycomb, and balsa wood. The composite face plates are attached to the core material with adhesives. The sandwich construction is depicted in Figure 19.6.

In addition to thickness with little increase in weight, sandwich construction gives increased stiffness to bending forces applied against the face of the composite. The stiffness depends on the thickness of the core and increases geometrically with the thickness.

A composite manufacturing technique related to hand lay-up is *roll wrapping*. This technique is used for making generally cylindrical composite parts from prepreg materials. In roll wrapping a *mandrel* is used instead of a mold. A mandrel is a core (usually metal) on which the composite material is placed and then cured. In roll wrapping, the prepreg materials are cut so that they wrap around the mandrel an even number of times. The composite strips are then wrapped around the mandrel until the proper number of layers have been placed. (This wrapping can be semiautomated if desired.) The entire assembly is then overwrapped with a layer of cellophane tape that shrinks when heated to give the compaction necessary to bond the layers together and squeeze out the air. The assembly is then cured in an oven. After curing, the cellophane is removed and the mandrel is extracted from the composite part, which is then finished by machining.

Filament Winding and Fiber Placement

These two processes both use mandrels as the method of shaping the fibers and resins prior to curing, but they differ in how the fibers are placed against the mandrel. In one case (filament winding) the fibers are placed against the mandrel under tension. In the other case (fiber placement) the fibers are pressed against the mandrel by a compaction head. Both are highly automated processes as compared to roll wrapping, which is mostly manual.

Filament winding is more widely used than fiber placement. In filament winding the fibers are attached to the mandrel and then the mandrel is turned to draw the fibers off the spools. The fibers from the spools are gathered together into a ring or some other guide called the *payoff*. The payoff acts as a guide for the fibers and is mounted on a carriage or some other transport system that moves laterally along the long axis of the mandrel as the fibers are being drawn off. This payoff motion is synchronized with the turning of the mandrel to produce a pattern of fibers on the mandrel. The angles of the fiber pattern are determined by the relative motion and speeds of the mandrel and the payoff system.

Between the spools and the mandrel the fibers pass through a resin bath so that they are soaked with resin when placed on the mandrel. The most common resins used are polyester (when glass is the fiber) and epoxy (when carbon or aramid is the fibers). In both cases the resin in the bath is fully activated with initiator or hardener so that the only heat required is to cure the resin. The filament winding system is depicted in Figure 19.7 and Photo 19.9.

The use of a mandrel to shape the fibers and resin dictates that the general shape of parts made by filament winding is generally cylindrical. Pipes, pressure vessels, and similar products are most commonly made by filament winding. However, by allowing the payoff head to move vertically or to twist and rotate, quite complicated shapes can be made. All, however, must have an axis of rotation.

After winding is complete, the mandrel with the fibers and resin is placed into an oven or autoclave for curing. The curing is done at the temperatures appropriate for the par-

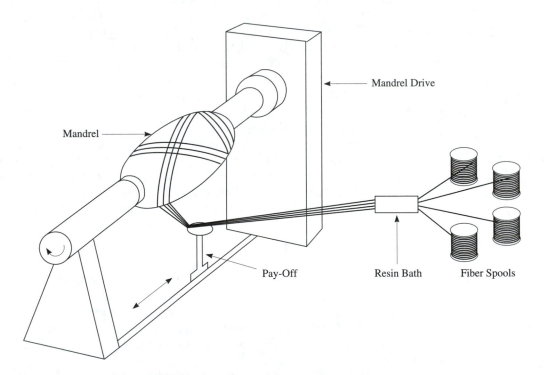

Figure 19.7   Filament winding system for making composite parts.

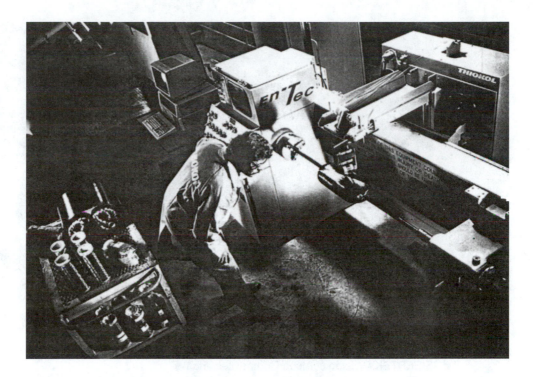

Photo 19.9   Filament winding machine (Courtesy: EnTec Corp.)

ticular resin used. The autoclave is used when high compaction (low void content) is required. Other innovative systems to obtain compaction include winding on an inflatable mandrel that is pressurized during cure to press the fibers and resin against a *clamshell mold* that has been placed over the mandrel assembly. (A clamshell mold is made in two halves that can be clamped together for molding and then separated for part removal afterwards.) Continuous filament winding on a moving mandrel with in-line curing is used to make pipe and other continuous products. Filament winding is used extensively for both engineering and advanced composites, with appropriate resins and fibers. Large parts made by filament winding are shown in Photo 19.10.

Fiber placement compresses the fibers against the mandrel by using a compaction head to payoff the fibers rather than using the turning of the mandrel to draw the fibers off the spools under tension. Normally, the fibers used are thin strips of prepreg rather than dry fibers tows wetted during the winding process. When these prepreg fiber strips are pressed against a turning mandrel, greater precision in locating the fibers is achieved. Furthermore, compaction against the mandrel allows shapes to be made using fiber placement that cannot be made with filament winding. For instance, with fiber placement the mandrel can be concave, whereas with filament winding a concave mandrel results in fibers bridging across the face of the mandrel rather than touching it.

The major drawbacks of fiber placement are the high cost of the machine and the relatively slow production rate. The high cost arises in part because of the complexity of the

Photo 19.10   Filament-wound fuselage for the Beech Starship aircraft (Courtesy: FIBERTEK Division of Alcoa Composites, Inc.)

compaction/payoff head. The payoff head must be able to rotate and move in several directions to accommodate the varied shapes of the mandrels that might be used. The high cost of making parts with fiber placement usually dictates that only advanced composite parts be made by this method.

Pultrusion

The *pultrusion* process makes parts that have a constant cross section. In this process, the fibers are drawn off the spools, through a resin bath, and then through a curing die. The force for drawing is supplied by a puller placed after the die. Therefore, as shown in Figure 19.8, the process continuously makes composite parts that are shaped by the die. A cutoff system after the puller cuts the parts to the desired length.

Parts made in pultrusion can be reinforced in directions other than the machine direction by pulling strips of mat or cloth into the die with the fibers. These can be pulled through the resin bath, but normally the fibers carry sufficient excess resin that wet-out of the mat or fabric occurs within the die. Compaction of these materials with the fibers is also accomplished within the die.

Pultrusion is a low-cost production method best suited to engineering composites because of the difficulty in precisely controlling the fiber placement. The pultrusion process is used principally with fiberglass and polyester. Some common composite parts made by pultrusion are moldings, sides of ladders, and tubes. Several pultruded parts are shown in Photo 19.11.

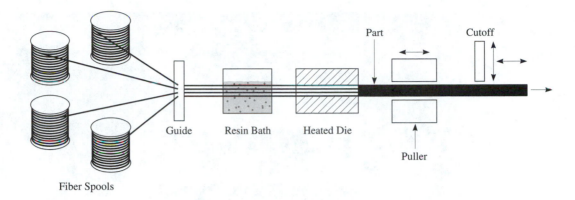

Figure 19.8 Pultrusion system for making composite parts.

Processes for Long and Continuous Fibers Using Thermoplastics

Some high-performance thermoplastic resins can be used as matrix materials for very long- and continuous-fiber composites. These resins are usually high molecular weight, highly crystalline thermoplastics that have a high degree of aromatic character. This combination of characteristics results in thermoplastics that have high stiffness, high strength, high toughness, and can withstand high temperatures and most solvents. The properties are comparable to high-performance epoxies with the added advantage of toughness (epoxies are brittle). Costs of these high-performance thermoplastics are above those of epoxies so the value of the thermoplastics is in applications where the toughness property or some other property is critical. With the development of new manufacturing methods particularly suited to the advanced thermoplastics, manufacturing cost advantages may also be realized over epoxy composites.

The major drawback in the processing of these high-performance thermoplastics to make advanced composites is the difficulty encountered in wetting out the fibers. The ability to wet-out fibers is complicated by the very high molecular weight of the thermoplastics, which is required to achieve the physical and mechanical properties. Some innovative methods do accomplish this wet-out, however. One method is to melt the resin and then apply the melt to the fibers while the resin is subjected to shear forces. The non-Newtonian nature of the resins means that resin viscosity is lowered by shearing. This property is called *shear thinning* and is used to make the resins thin enough to coat the fibers. Another technique is to powder the resin and then coat the fibers with the powder. During cure, the powder melts and covers the fibers. Still another method is to make fibers out of the thermoplastic resins and then coweave or comingle the thermoplastic fibers with the reinforcing fibers. Then, when the fibers are heated, the thermoplastic fibers melt and coat the reinforcements. One other method is to place a thin sheet of thermoplastic resin over a mat or fabric of dry fibers and then heat the thermoplastic until it is soft and press the resin onto the fibers. Some success has been achieved with each of these methods, but all continue to have significant problems.

By using the wet-out methods discussed, a significant amount of composites using high-performance thermoplastic resins and continuous or very long fibers have been

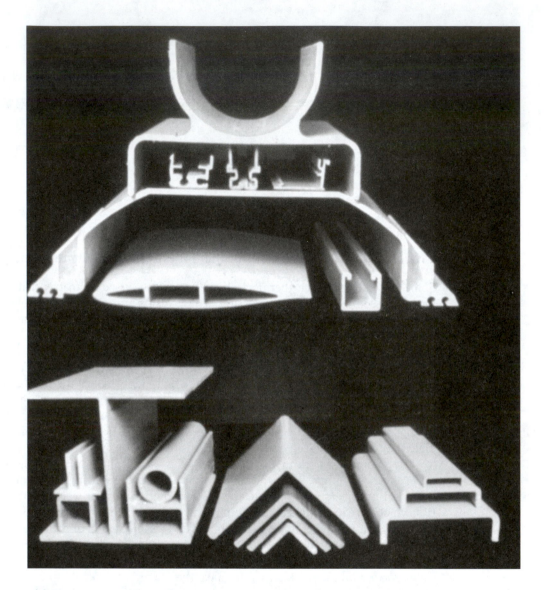

Photo 19.11   Standard pultruded structural and custom composite parts (Courtesy: Creative Pultrusions, Inc., Alum Bank, PA)

made. The most common manufacturing methods are those that are used for thermoset resins: hand lay-up, filament winding, and pultrusion.

One unique method for making thermoplastic composites has proven to be very economical and viable. Called *press thermoforming,* this method is similar to matched die thermoforming. In this method, the thermoplastic prepreg materials are stacked in the sequence, orientation, and thicknesses needed to meet the design. The materials are then

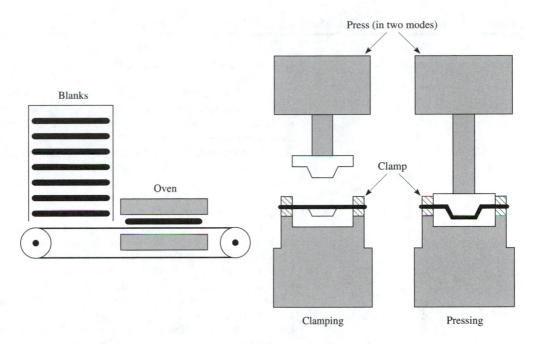

Figure 19.9 Press thermoforming of reinforced high-performance sheets to create advanced thermoplastic composites.

heated to the point where the thermoplastic resins begin to sag, placed into a matched die mold (which is cold), and pressed to give them the shape of the mold. The pressing cycle is very short (often less than 10 seconds) because the only thermal requirement during the pressing cycle is that the materials cool enough to solidify. This method holds great promise for making advanced composites out of thermoplastic resins. The method is illustrated in Figure 19.9, where both the open and closed modes of the press are shown.

Troubleshooting

The problems in composite manufacturing depend upon the process being used. The diversity of these processes complicates the design of a single troubleshooting guide, but, nevertheless, some common problems exist. These are summarized in a troubleshooting guide, which is given in Table 19.1.

PLANT CONCEPTS

Many composite manufacturing methods, such as spray-up, filament winding, and pultrusion, require considerable space, especially when making large parts. The autoclaves and ovens required to cure these parts can also be very large (and expensive). Therefore, facilities in which composites are made tend to be large. These facilities need to allow space for moving the large parts from one stage of the production sequence to another and for

Table 19.1   Troubleshooting Guide for Composite Materials Manufacturing

Problem	Possible Cause
Fibers not fully wetted	• **Resin viscosity too high.** In either manual or automated processes, the initial wetting of the fibers depends on the resin viscosity. The resin may have to be changed or slightly heated to lower the viscosity. Only rarely would a solvent be added to the resin, because solvents are difficult to remove and usually require extensive recovery systems. • **Residence time in the bath too short.** In automated processes such as filament winding and pultrusion, the speed of the line can affect the wet-out. Therefore, the speed may have to be reduced to give the resin time to soak into the fiber bundles. • **Not enough working in of the resins.** In manual processes such as wet lay-up and spray-up, the proper wetting of the fibers depends upon manual working in of the resin, usually with rollers or brushes. The solution if this is the problem is to simply spend more time doing the wet-out. Fortunately, when the fibers are properly wet out, there is a slight change in appearance of the fibers. • **Improper mix ratios.** The amount of filler might not be correct for the amount of resin. • **Improper amount of viscosity modifier.** Resin mixes may contain a thixotropic agent, which adjusts viscosity.
Cracking or warping of the part	• **Exotherm.** The exotherm may be so high that the shrinkage of the part causes it to crack when it cools. The exotherm can be reduced by slowing the cure, adding filler, cooling the mold, or making the part thinner. • **Part complexity.** Cracking could be associated with an undercut in the part that cannot be easily stripped out of the mold. Also, very fine details on the surface of the part can result in excessive sticking. • **Mold release.** Not enough mold release exists in the mold or part. • **Faulty laminar sequence.** The lay-up sequence must be both balanced and symmetrical; that is, the directions of the fibers in the layers must be the same on either side of the centerline in thickness of the part and on both sides of the centerline in length of part. Otherwise warpage can occur when curing because of the differences in contraction in the fiber direction and the cross direction. • **Excessive mandrel expansion.** The most common problem is that an inflatable mandrel is expanded too much and cracks the part as it is being cured. Also, if a mandrel has some ability to compress on the part as it cools, cracking from the pressure of the mandrel could result.
Poor or inadequate cure	• **Inadequate or improper initiator or hardener.** The type of initiator should be suited to the resin and to the cure conditions. Check labels and check viability of the initiator with a known good resin by running a gel time. Excessive initiator can cause premature exotherm and can also compete with the crosslinking reaction to reduce the crosslink density.

Table 19.1 *Continued*

Problem	Possible Cause
Poor or inadequate cure *(continued)*	• **Improper temperature.** The temperature must be reached and controlled. Air can easily be entrapped in the casting resin during mixing of the resin or even during pouring into the mold. This can be removed by applying a vacuum to the mold or, if that is difficult, to the container in which mixing is done. • **Improper accelerator content.** In many cure systems an accelerator is added to chemically begin the initiation of the reaction. It the accelerator concentration is wrong or the wrong accelerator is used, problems with the cure can result. Check with the manufacturer of the resin and/or the initiator to verify the compatibility of the acceleration system. • **Inhibitor.** Inhibitor is often added to extend the shelf life of a resin. The amount of initiator must be increased to compensate for the presence of the inhibitor. Hence, too much inhibitor could result in too little active initiator.
Part delaminates	• **Poor resin movement.** The layers of uncured composite must bond well together, and that process generally requires that the resin from one layer must mix with the resin in the adjacent layers. This mixing is improved when the viscosities of the resins are low and when the resins are compatible (ideally the same). Raising temperature will often help mixing, as will increasing the pressure during cure. Adding an absorbent layer to take up excess resin promotes resin movement, especially if the absorbent material is placed at the ends of the fibers as resin flows most easily in the fiber direction. • **Excessive cure of each layer.** If a layer is cured and then another is to be bonded to the first, that bonding is best if the molecules still have some ability to move and, therefore, intermingle across the boundary of the two layers. Therefore, stopping the cure of the first layer prior to full crosslinking can improve the bonding when another layer is applied to the first.
Part not adhering to an adhesive	• **Mold release.** Some mold releases have been known to inhibit adhesion of layers. Clean the layers of the parts with solvent if this problem is suspected. • **Adhesive is overly cured.** Just as the resin might have exceeded its shelf life, so too could the adhesive. Check for ease of bending and tackiness. • **Surface contamination.** The surface of the part could be contaminated. Usually a solvent wipe will solve the problem.
Properties not up to standard	• **Improper fiber laydown.** Most of the properties depend upon the fiber direction and content. Check that each layer is laid down in the right direction. Check for the correct number of layers.

storing the molds, which tend to be large as well. Lay-out needs to be well thought out, especially because some parts of the process, such as curing ovens or autoclaves, are used in many of the manufacturing operations and tend to take several hours to complete. Thus, these facilities are often shared by many parts at one time. Large freezers are also required for resin and prepreg storage. Normally, these materials must be kept at less than 0°F (-20°C) after the resin has been initiated to retard curing (crosslinking) and therefore extend the storage life of the material.

Controlling vapors from volatile solvents and resins must be carefully considered. Most polyesters have styrene, which is both harmful to breathe and flammable. Spray booths often are vented for this reason, as are curing ovens. General venting of the production facility is also advisable. Production personnel should wear masks when using most of the resins.

The finishing of composites can affect air quality. These materials create a considerable amount of fine dust when they are machined. This dust should be collected by a vacuum system, especially when the composites contain carbon fibers. These fibers conduct electricity and can cause electrical shorts when the dust settles on electrical equipment.

Composite parts and molds, though massive in many cases, are fragile and can be easily damaged from casual impacts. Therefore, moving and handling of composite parts and molds should be done with the greatest of care.

===== CASE STUDY 19.1 =====

FILAMENT WINDING OF THE BEECH STARSHIP AIRPLANE FUSELAGE

One of the great achievements in the history of composite manufacturing is the making of the fuselage of the 10-passenger Beech Starship airplane (see Figure 19.10) using filament winding. This was done at Fiber Technology Corporation in 1985 by a team of composite manufacturing and design specialists headed by Larry Ashton, one of the developers of the filament winding technique. Although not used in the current production of the

Figure 19.10 Drawing of the Beech Starship aircraft, which is largely made of composite materials.

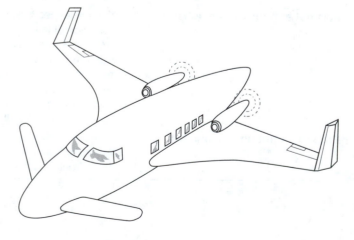

Starship, the demonstration of this method for manufacturing aircraft bodies may yet prove to be one of the most important developments in the fabrication of large composite parts such as airplane bodies and submarines hulls. The process holds the promise of low-cost manufacture with careful and precise control over fiber direction and placement and with little postmolding finish work.

The process began with the fabrication of the mandrel for the body. This mandrel was made of rigid foam (later, other more rigid materials were used) and was approximately 30 feet (9 meters) long and 8 feet (2.5 m) in diameter. The foam was cast around a steel mandrel that served as the means for turning the mandrel and moving it from one process step to another. The foam was machined to the tapered shape of the body of the airplane, including the rise for the cockpit area.

The mandrel was covered with a thin elastomeric material (about 0.2 inches, 5 mm thick), which would later be inflated as a means of pressing the uncured resin and fibers outward against a mold so that the outer surface of the body would be smooth and well defined by the mold. Therefore, this elastomeric covering was carefully joined together so that it would be airtight.

After machining, the mandrel was placed in the filament winding machine by locking the steel rod into turning chucks that were driven by the synchronized motors of the winder. Care was taken that the mandrel freely rotated and that no significant sagging of the mandrel occurred, as this could distort the shape of the final part.

After covering the elastomeric coating with a release material to ensure that it did not stick to the aircraft body after curing, the winding of the body was done. The filament winding machine had a movable carriage on which the spools of carbon fiber were mounted. The resin bath for fiber wet-out was also mounted on the carriage, as were the controls for the unit and a seat for the machine operator. The initial pattern was a wind at 30° off the horizontal, which gave both radial and longitudinal strength to the body. Other layers were wound at different angles as specified by the design to achieve maximum load carrying capability. The winding continued until the specified number of wraps and thickness of the aircraft skin were achieved.

After winding the inside skin, a layer of adhesive was applied manually to cover the entire surface of the airplane body. A layer of precut Nomex® honeycomb was placed onto the adhesive. Special cutouts and reinforcement sections for the side windows, cockpit, and doors were also added. Another layer of adhesive was then placed over the honeycomb. (The honeycomb and adhesives were used to give the body additional thickness and stiffness without significantly increasing the weight of the airplane.)

The outer skin was then wound onto the top of the adhesive and honeycomb. This outer layer was a mirror reflection of the inner skin. The total winding, including the application of the honeycomb and adhesives, was completed in 6 hours. This part is shown in Photo 19.10.

After winding, the entire assembly was placed inside a carbon fiber and epoxy clamshell mold that had been treated with mold release material. (The matching of materials in the mold and the body of the airplane ensured uniform thermal expansion of the body and the mold to minimize built-in stresses.) The interior of this mold was highly polished and made to match the desired dimensions, shape, and finish of the outer surface of the Starship. After clamping the mold closed, the mold and mandrel were moved into the oven for curing. At this point the elastomeric bag over the mandrel was inflated, which pressed the composite materials outward against the clamshell mold. The part was cured in this configuration.

After curing, the pressure on the elastomeric bag was relieved, the mold and contents were removed from the oven, and the mold was opened and removed from the part. The part was removed from the mandrel by cutting off the forward part of the airplane body along a line just behind the cockpit and at the maximum-diameter point of the body. Making this cut eased removal from the mandrel because the part was larger than the mandrel as a consequence of being inflated during curing. Having two separate body sections allowed easy access to both the cockpit and cabin during the remainder of the manufacturing process. The two parts of the body could then be rejoined using either mechanical fasteners or adhesives at any time in the aircraft manufacturing sequence. The windows and doors were simply cut open at designated· locations leaving the special molded-in reinforcements as templates.

Final assembly was done at Beech, including the final painting and finishing. The interior hardware, wings, engines, and other attachments were joined to the body.

This manufacturing method proved to be highly successful. It was, however, too advanced for the current inspection and certification capabilities required for aircraft. Therefore, the actual production of the Starship was done by hand lay-up. The time required for hand lay-up of the body was more than 10 times that required for filament winding.

SUMMARY

When fibers are added to plastics, the resulting materials are called composites. The properties of these composites vary widely depending on the nature of the reinforcing fiber (length, concentration, and type of fiber) and on the type of resin used in the matrix. At low-fiber concentrations and when the fibers are short, the properties of the composite are dominated by the matrix. These composites are like the nonreinforced plastics with the modification of fibers for a different grade having improved properties. When the fiber content is high and the fiber lengths are long, the reverse is true. The properties of the composite are dominated by the fibers.

Composites in which the properties are dominated by the matrix material (resin) are called engineering composites. Those composites that have properties dominated by the fibers are called advanced composites. Engineering composites cost much less than advanced composites.

Most thermoplastic resins can be reinforced with very short fibers and processed in the normal thermoplastic processes such as injection molding. When the fibers are intermediate in length, as they are when chopped, only thermoset resins are widely used as reinforcements. This restriction to thermosets results chiefly from the problems encountered in trying to wet-out the fibers with the very viscous thermoplastic resins. Many thermosets can be used for the intermediate-length fibers but the most common are polyesters, epoxies, and phenolics.

The matrices most often used with very long fibers are epoxies. Other matrices used commercially are polyesters, phenolics, and polyimides. Thermoplastic resins are rarely used because of the difficulty of wetting out the fibers, but some special wet-out methods allow high-performance thermoplastics to be used in a few commercial applications.

Three types of fibers dominate the market for reinforcing fibers: fiberglass, carbon (or graphite) fibers, and aramids. Fiberglass costs much less than the other two types and is used almost exclusively for engineering composites. Fiberglass composites often have a sizing material added to coat the fibers and protect them from damage during later processing, especially when the fibers are to be woven or knitted. The glass fibers also have a coupling agent added to improve the bond between the matrix and the fibers. Carbon fibers are coated with an epoxy that serves as both a sizing and a coupling agent. Aramid fibers usually are not coated.

Manufacturing processes for engineering composites are concerned with molding the resin without interference from the fibers. Manufacturing processes for advanced composites are concerned with placing the fibers accurately and ensuring that the matrix is evenly distributed over the fibers.

The processes used to make engineering composites include injection molding, extrusion, and, to a lesser extent, the other common processes for thermoplastic resins—matched die or compression molding, spray-up, hand lay-up (wet), and RTM. The processes used to make advanced composites include RTM, hand lay-up (prepreg), filament winding, fiber placement, pultrusion, and special processes for long fibers and thermoplastic resins.

GLOSSARY

Advanced composites Composite materials that are characterized by high performance, usually in strength and/or in temperature; typically employ long fibers as the reinforcement material and high-performance thermoset or thermoplastic resins as the matrix.

Advanced thermoplastics Another name for high-performance thermoplastics.

Anisotropic Another name for nonisotropic materials.

Aramid A thermoplastic reinforcement fiber that is a highly aromatic polyamide.

Autoclave A pressure vessel that can be heated so that composites (and other materials) can be cured in it.

Barrier material A sheet of material that allows the movement of air throughout the assembly of bagging materials.

Binder Another name for matrix.

Bleeder cloth A sheet of absorbent material that is used in a vacuum bagging molding to absorb excess resin.

Bulk molding compound (BMC) Composite material made from fiberglass, resin, and filler that is mixed together in a blender and then stored and B-staged until ready to use.

C-glass Fiberglass that is resistant to chemical attack.

Carbon fibers Fibers made from carbon-rich materials (polymers like PAN or from pitch), which are converted into graphitelike structures; also called graphite fibers.

Chopped fibers Reinforcement fibers of intermediate length (typically 0.4 to 4 inches, 10 to 100 mm).

Clamshell mold   A mold made in two halves that can be placed around a composite part during cure to give definition to the outer surface of the part, which is pressed against the inside of the clamshell mold, usually by applying internal pressure to the inside of the part and expanding it.

Composite materials   Solid materials with at least two substances that can, at least in theory, be physically separated; generally consist of a resinous matrix and a fibrous reinforcement.

Continuous fibers   Fiber reinforcements that extend for the entire length of a part.

Coupling agents   Chemicals added to the sizing and bond to the fiber and also to the matrix, thus increasing the bond between the fiber and matrix.

Debulking   Compacting layers of composite materials immediately after laydown.

Dough molding compound   Another name for bulk molding compound.

E-glass   Fiberglass that has excellent electrical properties; it is low cost and is the most commonly used type of fiberglass in composites.

Engineering composite   Composite materials that compete with many metals but do not have the very high properties of advanced composites; the fibers are usually short and the resins are typically inexpensive.

Faces   The composite parts comprising the outer portions of a sandwich construction; also called the skins.

Fiber blooming   The migration of fibrous reinforcements to the surface of a part.

Fiber placement   A process used for composite manufacturing in which the fibers are placed and pressed onto a mandrel rather than put onto the mandrel under tension, as is done in filament winding.

Fiberglass-reinforced plastic (FRP)   Material composed of moderate-length fiberglass, a low-cost thermoset resin such as unsaturated polyester, and, often, a filler.

Filament winding   A composite manufacturing process in which fibers are wetted and then wound onto a mandrel, usually in an automated process.

Gel coat   A layer of resin that is applied to the surface of a female mold prior to the application of composite materials into the mold, therefore the gel coat serves as the outer surface of the part.

Graphite fibers   Originally a very pure form of carbon fibers, although now that distinction has largely disappeared.

High-performance thermoplastics   Thermoplastic resins with high temperature and high mechanical properties.

Hybrid composite parts   Parts made from two or more types of reinforcement fibers.

Isotropic   Materials having their properties the same in all directions.

Laminate design theory   The mathematical theory that is used to calculate the strength, stiffness, and other properties of composite laminates.

Laminates   Parts made by joining layers together.

Mandrel   A solid, usually metal, piece that serves as a tool or mold on which composite materials are laid up and cured; the mandrel is usually removed after curing.

Matrix   The continuous phase of a composite material that surrounds and binds together fibrous reinforcements.

Nonisotropic   Materials that do not have uniform properties in all directions; also called anisotropic.

Orthotropic   Materials having properties the same in two directions but not in the third.

Payoff   The ring or other mechanical device through which fibers are moved just prior to their being laid down onto a mandrel, especially in filament winding.

Prepreg   Sheets of unidirectional fibers that have been coated with a resin and then B-staged so that the resin will stay on the fibers.

Press thermoforming   A manufacturing method for thermoplastic composites in which a prepreg sheet is heated and then pressed into the desired shape using cold matched dies.

Pultrusion   A composite manufacturing method in which fibers are drawn through a resin bath and then through a forming/curing die to create a constant-cross-section part.

Release cloth   A sheet of material that is applied on top of the noncured composite materials and that will allow the composite bagging materials which are subsequently applied to be separated from the molded part during a vacuum bagging cure.

Resin transfer molding (RTM)   A composite manufacturing process in which a fibrous preform is put into the mold and then, after closing the mold, the resin is injected to wet-out the fibers and then cured.

Roll wrapping   A process of making composite parts in which prepreg materials are wrapped around a mandrel and then cured.

S-glass   Fiberglass that is especially formulated for high strength.

Sandwich composite   A structure consisting of composite skins bonded to either side of a core material.

Sandwich core   The material that forms the center portion of a sandwich composite; typically these are honeycomb, light wood, or rigid foam materials.

Shear thinning   The reduction in the viscosity of a resin when the resin is sheared.

Sheet molding compound (SMC)   A molding material made from resin, chopped fiberglass, and filler that is mixed in a sheet form and then B-staged and rolled up until used.

Sizing   A coating put on fibers, especially on fiberglass, to protect them during mechanical handling and to help them bond to the matrix.

Skins   The outer materials in a composite sandwich construction.

Specific stiffness   The modulus of a material divided by the specific gravity or weight of the material.

Specific strength   The strength of a material divided by the specific gravity or weight of the material.

Surface veil   A thin sheet of reinforcement that is often applied to the top of the gel coat, thus preventing fiber blooming.

Tow   A bundle of fiber strands.

Unidirectional materials   Composite materials in which all the fibers are in one direction (parallel).

Vacuum bag   The airtight bag that is placed over the vacuum-bag materials to hold the vacuum and the bagging materials.

Wet-out   The coating of the reinforcement fibers by the resin matrix.

Whiskers   Very short-fiber reinforcements.

QUESTIONS

1. What is the purpose of the resin in parts with long fibers at high fiber concentrations?
2. Identify three advantages of composites over metals for structural applications.
3. Explain why very short-fiber composites tend to be made with thermoplastic matrices while longer-fiber composites use thermoset matrices predominately.
4. Discuss the use of sizing agents and coupling agents and distinguish between them.
5. Why are sizes and coupling agents generally not required for aramid fibers?
6. Why would a low-intensity mixer be used to make BMC rather than a high-intensity mixer, which would give better mixing and, presumably, better wet-out?
7. Indicate at least three problems of most composite processes that are solved in RTM.
8. Define a gel coat and give three reasons for its use.
9. A submarine driveshaft was made of epoxy and carbon fibers by filament winding. Indicate two reasons why composites have unique and excellent properties for this application. The part was cured in an oven at 250°F (120°C). On close examination, the shaft was found to spin irregularly and to cause vibrations. Identify four possible causes of this problem and four possible solutions.
10. You have been asked to manufacture a wiring junction box for connecting electrical wires in a fighter aircraft. The box sits under the pilot's seat and is part of the seat ejection system. Sufficient forces are exerted on the box that it must be made of a composite material using intermediate to long fibers. What fiber, matrix, and manufacturing method would you use for this part, and why?

REFERENCES

Modern Plastics Encyclopedia, New York: McGraw-Hill Book Co., Inc., 1994.

Strong, A. Brent, *Fundamentals of Composites Manufacturing,* Dearborn, MI: Society of Manufacturing Engineers (SME), 1988.

Strong, A. Brent, *High Performance and Engineering Thermoplastic Composites,* Lancaster, PA: Technomic Publishing Co., Inc., 1993.

Strong, A. Brent (ed.), *Composites in Manufacturing: Case Studies,* Dearborn, MI: Society of Manufacturing Engineers (SME), 1991.

CHAPTER TWENTY

RADIATION PROCESSES

CHAPTER OVERVIEW

This chapter examines the following concepts:

- Concepts and usefulness
- Equipment and process
- Properties, materials, and applications
- Plasma polymerization and reactions

INTRODUCTION

This chapter discusses the interactions of polymers with high-energy radiation. Although polymers certainly react with sunlight and other natural radiation sources, which generally cause unwanted reactions such as polymer degradation, the type of radiation of interest in the processing of polymers is from synthetic radiation sources that can be controlled and directed at making useful changes in material properties. The most important of these synthetic radiation sources for polymer processing is *electron beam radiation.*

Concepts

In electron beam radiation, the polymer material is bombarded with high-energy electrons that have been created in a large machine built specifically for that purpose. Such a machine is pictured in Photo 20.1. The photo shows a typical layout of an electron beam operation. Note that both of the major types of products (boxes and continuous extruded lengths) are shown being irradiated.

The high-energy electrons can cause several changes in the polymer. The nature of these changes depends on the nature of the polymer, the energy of the electrons, and other processing variables such as the time of exposure and the presence of modifying

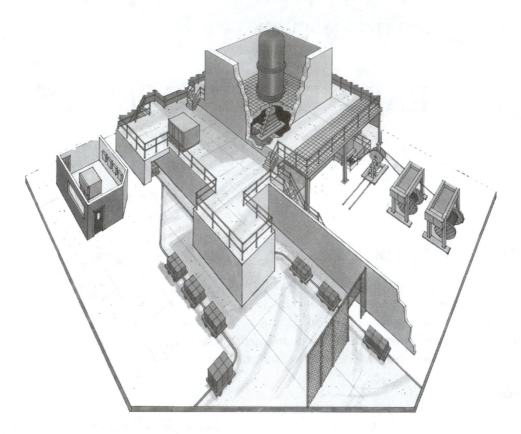

Photo 20.1   Electron irradiation machine (Courtesy: E-Beam Services, Inc.)

agents within the polymer material. In general, three main reactions can occur: crosslinking, scission, and molecular rearrangement. (*Scission,* or *chain scission,* is the breaking of the main backbone of the polymer and is a form of polymer degradation.)

Each of these reactions is initiated by the interaction of the polymer with the high-energy electrons. This interaction usually results in the removal (knocking off) of hydrogen atoms from the polymer chain, as shown in Figure 20.1. The removal of the hydrogens can result in the formation of unpaired electrons *(free radicals)* or charged atoms *(ions),* depending on whether the hydrogen atoms leave with or without an electron, which, in turn, depends on the energy of the electrons and the nature of the polymer. Whatever the form of the leaving hydrogen atoms, the sites on the polymer formerly occupied by the hydrogens become active sites. These active sites can bond to other active sites on another molecule to create a crosslink, or they can cause a local rearrangement of the bonding patterns in the polymer to cause either scission or molecular rearrangement.

These reactions can occur in both thermosets and thermoplastics. When a thermoset is irradiated by high-energy electrons, the thermoset can be crosslinked (cured), thus providing a method of crosslinking thermosets in addition to the normal chemical crosslink-

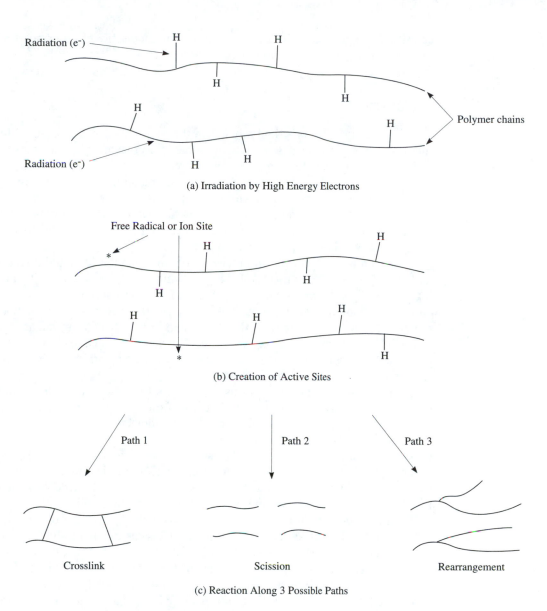

(a) Irradiation by High Energy Electrons

(b) Creation of Active Sites

(c) Reaction Along 3 Possible Paths

Figure 20.1 Initiation of reactions in a polymer from electron beam irradiation and the subsequent reactions of crosslinking, scission, and molecular rearrangement.

ing methods (peroxide-initiated free radical reactions, ring opening, molecular rearrangements, or standard reactions on multiple sites, as the case may be for each type of thermoset material). The thermosets can also experience scission or molecular rearrangement. If the reactions occur in thermoplastics, then thermoplastic materials can be crosslinked by this method, thus turning the thermoplastic into thermoset. Thermoplastics can also experience scission or rearrangement.

Usefulness

The ability to use some method other than the thermally activated method to crosslink a thermoset has advantages. For instance, some molecules may be very sensitive to degradation at even modest temperatures; others may require very high energies to crosslink. In either case, the use of electron beams for curing simplifies the processing. Since the temperature increase associated with electron beam reactions is less than that associated with thermal reactions, molds that may be weakened or degraded by thermal cures (such as a plastic mold) can be used with electron beam cures, provided that the material in the mold is not degraded by the electron beam.

The high level of control of the electron beam radiation level and direction is important in the choice of this method for crosslinking of polymers versus chemical crosslinking. The electron beam method allows the amount of crosslinking to be chosen by simply choosing the amount of radiation that is created. The electron beam method also allows much faster throughput because the crosslinking by electron beam radiation crosslinking is much faster than chemical crosslinking. A further advantage of electron beam radiation crosslinking is avoidance of the environmentally damaging and hazardous peroxides often required in conventional crosslinking.

The ability to use electron beams to crosslink thermoplastic polymers provides many unique processing opportunities. The thermoplastic material can be processed in the normal method, such as by extrusion or injection molding, and then subjected to the electron beam radiation to change the properties to those of a thermoset. In many cases the basic properties characteristic of a thermoplastic, such as toughness and elongation, are largely retained, while new properties characteristic of a thermoset, such as higher thermal resistance, are added. This irradiation process is used to make thermoplastic wire and cable insulation more resistant to environmental stress cracking, to raise the thermal stability of thermoplastic parts used in aerospace applications, to increase the toughness and resistance to cuts of golf ball covers, to make shrink-wrap and shrink-tubing, and to increase the hardness of rubber materials. All these attributes, and others discussed later in this chapter, reflect the presence of crosslinks made by the action of the electron radiation.

The second of the reaction paths for polymers irradiated by electrons is scission. Normally the breaking of bonds along the main polymer backbone has a negative effect on polymer properties and therefore is avoided. A few applications, such as the control of molecular weight for lubricants, viscosity modifiers, and intermediates for making synthetic fibers, are exceptions. These applications depend on the ability to carefully control the amount of radiation and, therefore, to control the amount of degradation that occurs within the polymer.

The third reaction path, molecular rearrangement, is used to add flexibility and useful life to polymers, especially hard, amorphous (glassy) polymers, that have been in service for an extended period of time and have begun to harden. This process, called *deaging,* reverses the aging process in these polymers. The electron beam radiation rearranges the molecules and, in the process, forces the polymer molecules apart, thus increasing the polymer's flexibility. Rearranging molecules by electron beam radiation is also used to decrease the number of defects in some crystal structures (such as semi-conductors, precious gems, and decorative glassware).

EQUIPMENT AND PROCESS

Machine

The electron beam machine can be compared to a giant TV picture tube. A power source converts normal ac power into rf power, which is then converted to very high-energy dc power (2.5 to 4.5 million volts) that is used to heat a filament (usually a tungsten wire). The hot filament emits electrons that enter a chamber where they are accelerated in speed by a magnetic field. The acceleration chamber is evacuated so that the electrons will not be deflected by gas molecules. The high-speed (high-energy) electrons then enter a chamber where the electrons are fanned out by a scanning device (somewhat like the way the electron gun in a TV picture tube sprays electrons across the face of the picture tube). The electrons flow through a delivery horn shaped to match the fan shape of the electrons. The horn has a window on the end that is covered by a thin metal sheet (usually titanium). This sheet is necessary to maintain the vacuum throughout the system. The electrons pass through the window and into the plastic part as the part passes under the window. Because of the radiation hazard present when the machine is on, the entire machine (see Photo 20.1 and Figure 20.2) is enclosed in some massive barrier, such as a concrete box.

The radiation energy of the electron beam machine is measured either in *grays (Gy)* or in *rads (rad)* where $1 \text{ G} = 1$ J/kg and $1 \text{ rad} = 10^{-2}$J/kg. Rads were formerly the most frequently used units, but grays are now favored as the official SI (international) unit of measure.

The dose received by a part is usually measured by *dosimetry*. This is done by placing a piece of radiation-sensitive tape on the polymer part as it goes through the machine. After exposure, the tape is removed and placed in a device *(dosimeter)* that compares some physical change (usually color) in the tape with a standard having a known dosage. The dosage is then determined as a function of the physical change.

Process Parameters

The principal control for the electron beam irradiation process is the amount of power supplied to the filament. As the power increases, the number of electrons increases and the amount of radiation impacting on the plastic part is increased. The dosage is therefore increased by increasing the voltage, assuming all other variables are constant. Another variable used to control dosage is the speed of travel of the part. If the part is left under the delivery horn for a longer period of time, the dosage increases. These two parameters (power and time) are mutually adjusted to give the desired effect in the polymer.

Time is a critical variable even after the part has passed through the irradiation process. The active sites can persist for some time and, as the polymer chains move and

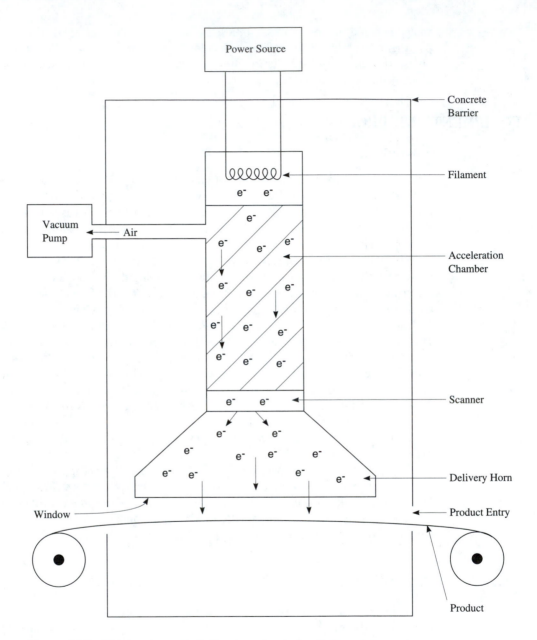

Figure 20.2 Electron beam radiation (accelerator) unit.

internally rearrange in response to stresses, can continue to react. The most common *postirradiation reactions* are cracking, embrittlement, and discoloration. These are all associated with scission, which would be favored by limited mobility of the polymer chains. Postirradiation reactions can be minimized by annealing because the heated polymer can move more freely and therefore react by crosslinking rather than by scission or rearrangement.

Only a small portion of the total amount of energy in the electron stream is absorbed by the polymer to produce the activated sites that create crosslinks, scission, or rearrangement. Some of the excess energy is absorbed by the polymer to give increased movement and vibrations within the polymer. These movements are registered as increased temperature. The typical temperature rise for a polymer after electron beam irradiation is 43°F (24°C) for each 100 kGy. At the dosages typical for polyethylene, the expected temperature rise is from 75 to 248°F (24 to 120°C). In some cases, the polymer is cooled by fans or water to prevent distortion from this increased temperature.

The depth of radiation penetration depends on the power and the nature of the polymer. For most polymers, a penetration of 0.16 inches (4 mm) is normal for most of the currently available electron acceleration machines.

Detection of Results

Dosage rates for crosslinking of polymers vary widely depending on the type of polymer. For polyethylene, one of the most commonly irradiated polymers, typical dosage is 10 kGy to 50 kGy, with significant degradation occurring above 1,000 kGy.

Even though the dosage can be detected quite accurately, the results of that dosage are somewhat difficult to determine. Since electron beam radiation causes three different reactions (crosslinking, scission, rearrangement), all of which can occur simultaneously or not at all, evaluating the extent to which each reaction occurred can be rather complicated. Tests exist that do measure each of the effects, but some distortion of the results from the other reactions can occur.

The amount of crosslinking is usually determined by a *gel fraction test* (ASTM D 2765). Because of the high molecular weight of the crosslinked material, the crosslinked portion of the polymer does not dissolve in some solvents, whereas the noncrosslinked portions of the polymer do dissolve. This allows the material to be separated into high and low molecular weight portions. Therefore, the fraction of crosslinked material can be determined by weighing the sample, dissolving out the noncrosslinked portion, and then filtering, drying, and weighing the undissolved portion.

Scission also causes changes in the molecular weight of the polymer, but, since all fractions are usually soluble, the gel fraction test cannot be used. Other tests that measure properties affected by changes in molecular weight include the melt index test and various viscosity tests. When scission alone occurs, these tests are quite reliable. However, when both scission and crosslinking occur simultaneously, the results can be misleading. For instance, the melt index test can be higher because of the existence of the low molecular materials created by scission, but it can also be lower because of the crosslinked materials. The same problem occurs with most of the viscosity tests.

Sedimentation and *gel chromatography* are testing methods that may be able to separate the low molecular weight and high molecular weight fractions. After separation, the fractions are compared with fractions from the polymer before irradiation to determine the relative amounts of newly formed small or large molecular weight fractions. Both sedimentation and gel chromatography depend on the natural tendency of materials of different molecular weights to separate themselves when subjected to stresses. The stress used for sedimentation is gravity; for gel chromatography the stress is absorptivity on a paper medium.

A process called *nuclear magnetic resonance (NMR)* can also be used to determine the amount of polymer that may have reacted to crosslinking, scission, or rearrangement. The NMR technique uses the inherent differences in electron density in various parts of the polymer to give different energy-absorption frequencies when the polymer is subjected to a varying magnetic field. Since the crosslinked molecules have different local electron densities than the noncrosslinked areas, the relative number of crosslinked polymers can be determined. Differences also exist between rearranged molecules and both crosslinked and noncrosslinked polymers.

The thermal transitions of a polymer depend on the molecular weight of the polymer and can be used for detecting the amount of crosslinking and scission. Some of the tests that measure these thermal transitions and can be used for monitoring radiation products include differential scanning calorimetry (DSC), thermal gravimetric analysis (TGA), and differential thermal analysis (DTA). In each of these tests, the nonirradiated and radiated samples are compared.

Scission may cause a change in color, as is often the case with degradation. This color change can be detected by comparing the color of the nonirradiated material with that of the irradiated material using a detection device such as a *reflectometer*. These devices compare the light reflected off the surfaces of the two materials and register the differences. The amount of reflected light is affected by the color of the sample.

Other Radiation Processes

Electron beam radiation processing is the most important process for irradiating polymers, but it is not the only process. Polymers are also subjected to γ-radiation from radioactive cobalt (and, less commonly, other radioactive elements), to high-energy microwaves, and to directed UV light for various processing purposes.

Gamma irradiation comes from radioactive cobalt atoms that are made by placing nonradioactive cobalt in nuclear reactors, where the cobalt receives neutron bombardment. The resulting radioactive cobalt then emits very high-energy photons (called γ-particles) that can be used to irradiate materials. The most important application for γ-particle radiation is the sterilization of materials used in medical applications. Electron beam radiation also is used to sterilize medical equipment and it competes strongly with γ-particle radiation in this application.

Although both electron beam irradiation and γ-radiation can be used for medical sterilization, the principal use of γ-radiation is for medical applications, and it has a dominant position in that application marketplace. However, the use of electron beam irradiation for this market is growing, and machines dedicated to this application have been commercialized. One such machine is shown in Photo 20.2.

Polymers are subjected to radiation during sterilization because polymers often form part or all of the medical device itself or the packaging of the device. In sterilization

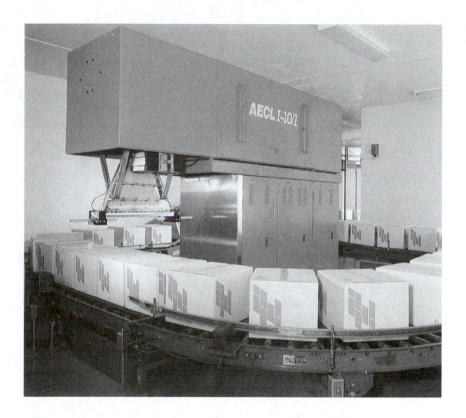

processing, the objective is to kill any live organisms present with as little change as possible in the polymers. Changes in the polymer can be minimized by controlling the radiation dose.

Cobalt or γ-particle radiation can also be used for crosslinking, scission, and rearrangement, although these applications are not as common. The disadvantages of γ-particle irradiation compared to electron beam irradiation diminish the use of γ-particles in these applications. The advantages and disadvantages of γ-particle irradiation are summarized in Table 20.1.

Table 20.1 Characteristics of γ-Particle (Cobalt) Radiation and Electron Beam Radiation

Property	γ-Particle Radiation	Electron Beam Radiation
Dose rate	Low	High
Penetration	High	Low
Temperature rise	Low	High
Directionality	Diffuse	Focused
Residual radiation	Abundant	Little
Controllability	Difficult	Easy

The most difficult problem to be overcome with γ-particle radiation is the residual radiation that exists because the source is radioactive with a long half-life. With electron beam radiation, the radiation drops to essentially zero as soon as the power to the electron beam acceleration machine is stopped. The other major benefit to electron beam irradiation is that it can be controlled, both in direction and in intensity, whereas the γ-particle radiation cannot be controlled. However, the advantage of γ-particle radiation is the much higher penetration that can be obtained because of the higher energy of the radiation particles. In very thick plastic samples, the use of γ-particle radiation to achieve high penetration may offset the operational problems.

High-energy microwaves are another radiation source used in polymer processing. The principal use of this radiation method is for curing (crosslinking). In this case, however, the microwaves are not used to create active sites on the polymer but rather to supply energy for the traditional chemical crosslinking mechanisms to occur, much like normal heating. The use of microwaves in this application is similar to their use in cooking, where microwaves can substitute for thermal heating. All of the normal components for traditional thermal curing (peroxides, accelerators, and so on) are present for microwave curing except, of course, the heat. A microwave generator is typically placed around the sample instead of placing the sample in an oven.

Microwave curing is much faster than thermal curing. In some processes, especially those where curing is done in-line in a continuous process (such as pultrusion), microwave curing can speed up the processing rate.

Directed UV light is used as an alternate to peroxide for initiating some crosslinking reactions and as a method for predictably degrading polymers. When used for crosslinking, the light from a UV lamp of the correct frequency range is directed onto the crosslinkable polymer. This UV light interacts with the π-electrons in the carbon-carbon double bonds to knock out one of the electrons in the bond and create a free radical. This free radical can then undergo all of the reactions that occur in normal crosslinking and polymerization processes. The advantage to using UV light rather than thermal cures is in thermal-sensitive materials and in avoidance of peroxides. Dentists use this technique in curing polymer teeth restorations.

PROPERTIES, MATERIALS, AND APPLICATIONS

Properties

Thermoset materials cured by high-energy electron beams γ-radiation, microwaves, or directed UV light have generally the same properties as thermosets cured by traditional thermal cures with peroxides or other thermal cures where peroxides are not required. The high-energy radiation methods may give some additional control capability or have some other benefit, but these benefits do not extend to product properties.

The properties of crosslinked thermoplastic materials should be compared to both the noncrosslinked analogue materials and to other crosslinked, more traditional, thermoset materials. The properties of the crosslinked thermoplastics are usually intermediate between the noncrosslinked thermoplastic and the typical thermoset. For instance, crosslinked polyethylene has the appearance and feel of noncrosslinked polyethylene. Flexural properties and most other readily discernable mechanical properties are only slightly affected by the crosslinking. Some improvements include increased impact toughness, improved environmental stress crack resistance, and increased abrasion resistance, but these properties are not readily apparent to the casual observer. The crosslinked polyethylene is not a hard and brittle material as would be common for a thermoset material like crosslinked polyester or epoxy. The changes in the crosslinked thermoplastic are subtle, but, as with the crosslinking of rotationally molded polyethylene, these changes can result in better overall performance in some applications.

Materials

Some monomers and polymers are more sensitive to radiation than are others (see Figure 20.3). The presence of benzene (aromatic) groups has a tendency to resist activation by radiation. Therefore, highly aromatic polymers such as PET, polycarbonate, and polystyrene are not generally susceptible to radiation processing. The resistant nature of these aromatic groups is so strong that their presence as additives, copolymers, or blends can desensitize other polymers to radiation processing.

To be effectively crosslinked by radiation, then, the polymer should contain one of the sensitive groups or be formed originally from monomers containing one of the sensitive groups. Therefore, polyesters that crosslink by peroxide activation of carbon-carbon double bonds and polymers with acrylic end groups (created from monomers having a carbon-carbon double bond) can be effectively crosslinked by radiation. Epoxies, on the other hand, are difficult to crosslink by radiation. However, when the epoxy end group is changed to an acrylic, crosslinking is excellent.

The internal movement of the molecules within a plastic has a strong influence on the ability of the material to be crosslinked, especially when converting a thermoplastic to a thermoset or in increasing the amount of crosslinking in a polymer. For instance, rubbery materials are affected much more strongly by radiation than are glassy (hard) amorphous materials, and amorphous materials are affected more strongly than are crystalline materials. The molecules in the rubbery and amorphous materials move freely so active sites approach each other and effect a bonding. In the glassy and crystalline materials, this movement is restricted, thus prohibiting the formation of crosslinks. The active sites in these motion-restricted molecules have several alternate paths. They can rearrange atomic

Figure 20.3 Molecular groups sensitive to radiation.

$C{=}C$	Carbon-Carbon Double Bond
$C{-}NH_2$	Amines
$C{-}SO_2$	Sulfonyl
$C{-}COOH$	Acids
$C{-}Halogen$	Halo-carbons (Where halogen is F, Cl, Br or I)

groups, but this is difficult because of the space restrictions. The active sites can also move electrons around to cause scission. The most probable occurrence is, however, that the hydrogen knocked off the backbone will reunite with the backbone at the same location. This results in a release of energy (the same amount absorbed when the radiation impacted the hydrogen) and the effect of the radiation is lost.

When highly crystalline and glassy materials are processed by radiation methods, an increase in sensitivity is gained by increasing the temperature of the material above the glass transition temperature (T_g). This gives greater internal movement to the material.

Some materials are more easily crosslinked by radiation, while others, also sensitive to radiation, favor scission or rearrangement. The choice of reactive path depends on the ease with which the active sites can react. If the sites cannot react quickly, the instability of the reactive site forces some other method of reaction, such as scission or rearrangement. This phenomena can be seen in the difference in reaction path between polyethylene and polypropylene.

Polyethylene crosslinks as the predominant reactive path after irradiation. The reactive sites formed on polyethylene react readily because of the flexibility of the polyethylene molecules and the relatively unhindered access that the sites have for each other. In polypropylene, however, the molecules are stiffer and the presence of the methyl pendant group restricts access between the active sites on adjacent chains. Therefore, polyethylene generally crosslinks and polypropylene generally has scission. The scission tendency of polypropylene is further enhanced by the greater stability of the active site in polypropylene because the free radical effect is spread over more carbon molecules. (The active site in polypropylene is spread over four carbons because the active-site carbon is attached to three other carbons. This type of carbon is called a *tertiary carbon*. The active-site carbon in polyethylene is attached to two other carbons and is called a *secondary carbon*.) Tertiary carbons are more stable than secondary carbons and therefore favor scission rather than crosslinking. This reactive difference is illustrated in Figure 20.4.

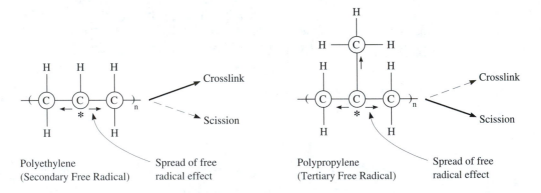

Figure 20.4 Differences in reaction path between polyethylene and polypropylene (secondary and tertiary carbons) showing stabilization from spreading the free radical charge.

Applications

The applications for radiation processing already discussed in this chapter are faster and more convenient crosslinking of thermosets, crosslinking of thermoplastics that would be difficult by conventional methods, control of molecular weight for viscosity modifiers, control of particle size, improved quality of crystalline structures due to atomic or molecular rearrangement, and medical sterilization. Several other applications are possible.

One important application is the modification of a polymer surface. The surface can be coated with a crosslinkable material and then crosslinked by radiative curing. This method avoids the problem that might be encountered when the polymer being coated softens or degrades with the heat normally required for crosslinking. This method of surface treatment can increase the toughness, scratch resistance, and stiffness of a material. Compact discs and computer floppy discs are improved by this technique.

A similar principle is involved in coating a low-priced material that forms the bulk of a product with a higher-priced material that can be crosslinked. This procedure cuts the cost of the overall product and still gives the required performance because the surface of the product may be the only part that requires the enhanced properties (such as plastic material that will be subjected to weathering).

A similar application is when some property is needed for the entire part (such as flexibility) but another property, such as abrasion resistance, is critical for the surface. This can be accomplished by making the bulk of the material from a polymer possessing the main property (flexibility) and then coating the outside with some other product that can be irradiated to obtain the desired surface property. Of course, the limited penetration of electron beam radiation allows this process to be accomplished without the need to coat the original material. Simply crosslinking the surface with radiation is sufficient to obtain the desired results.

The ability of radiation to cause molecular rearrangements has proven to be valuable in making polymer blends more stable. When these materials are irradiated, the molecules in the two materials migrate across their mutual polymer boundaries to form interpenetrating polymeric networks. These interpenetrating networks result in improved bonding between the two polymers (see Figure 20.5).

When radiation is used to join materials, the process is called *polymer grafting*. The result can be improved properties, stability of a blend of two or more polymers, or the joining of two polymers that would not otherwise join. The process for polymer grafting is the same as that shown in Figure 20.5.

If desired, two separate polymers can be joined by pressing them together and irradiating the junction. Bonding may or may not have already been done prior to irradiation. If the parts are already bonded, the irradiation is used to enhance the bonding. If they are not already bonded, the irradiation is used to create bonding. Usually, however, the bond strength obtained from bonding only with irradiation is not very strong.

PLASMA POLYMERIZATION AND REACTIONS

A highly ionized gas is called a *plasma*. Plasmas exist as the principal material in stars (such as the sun) and in fusion nuclear reactors. These plasmas are called *hot plasmas*

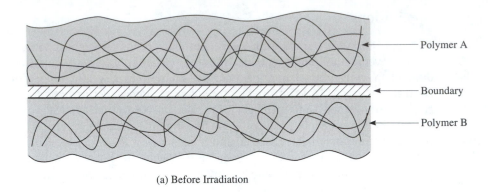

(a) Before Irradiation

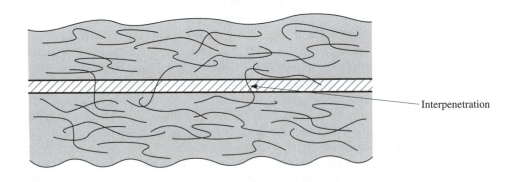

(b) After Irradiation

Figure 20.5   The increase in bonding in a polymer blend by interpenetrating polymer networks induced by radiation.

because they exist only at extremely high temperatures. Plasmas can also exist at or near room temperature. These *cold plasmas* are created by the ionizing effect of an rf (radio frequency) field on a gas. The rf field ionizes the gas molecules and imparts sufficient energy to the gas that it can be used as a source of energy for modifying monomer and polymer materials. The ionized gas particles are very reactive and create many unusual and useful reactions that cannot typically be done using standard chemical methods.

Equipment

The equipment needed to do cold plasma reactions is quite simple in concept, although the electronic stabilizing circuits and other refinements can be somewhat more complicated. The basic unit is shown in Figure 20.6.

The plasma machine consists of a chamber that can be evacuated so that only a few air molecules remain. The sample to be treated is placed inside the chamber, usually between

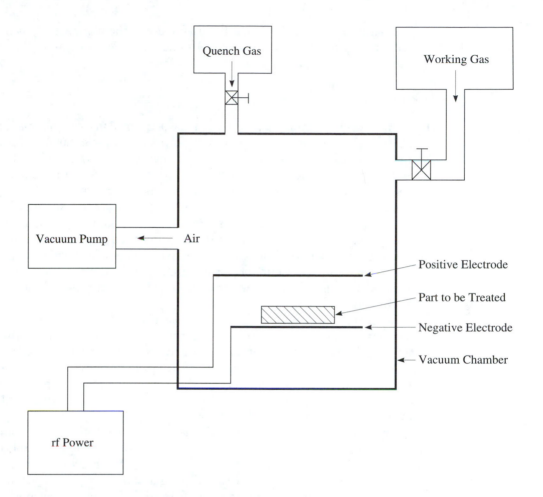

Figure 20.6 Plasma reaction vessel.

plates connected to the rf power source. A vacuum is created inside the chamber. (The creation of the vacuum gives more control over the system because only the desired gas will be ionized in the chamber.) After evacuating the chamber, the working gas is introduced. The pressure inside the chamber is still kept very low so that interactions between the gas molecules are minimized, thus maximizing the interactions with the part to be treated. The rf power is turned on and ionization of the gas molecules occurs. The ionized gas often develops a color (pink, blue, red, and so on) that is characteristic of the gas. The ionized gas interacts with the part in several ways. After these reactions have occurred, the rf power is turned off and the chamber is flooded with another gas (perhaps air) to quench the reactions. This relieves the vacuum, and the chamber can then be opened and the part removed.

Equipment with air locks are used to allow material to enter and leave the vacuum chamber on a continuous basis. These machines are used to treat fibers, films, and powders, as well as discrete parts conveyed through the chamber.

The nature of the reactions and the usefulness of the plasma equipment depend on several factors. These include the material to be treated, the nature of the working gas, and the nature of the quench gas.

Usefulness

One of the most important reactions that can be performed in a cold plasma is polymerization. One method of using plasma to form a polymer is to use a monomer as the working gas that is introduced into the chamber and then ionized by the rf field. The ionized monomer particles are very reactive and react in ways that reduce their energy by forming more stable species. One of the most effective reactions for forming a stable species is for the ionized monomer molecules to combine into a polymer on the surface of the part. The surface is then coated with the polymer formed by the recombining monomer particles.

Because of the high energy of the rf field, almost all gaseous molecules introduced into the rf field are ionized. Gases that would not otherwise be reactive can, therefore, be used as monomers. This feature of plasma reactions thus expands greatly the number of gases that can be used as monomers and also expands the number of polymer types that can be formed.

Polymers formed by plasma reactions are usually not the same as polymers formed by conventional methods, even when the same monomers are used. Plasma-formed polymers are almost always highly crosslinked, simply because the ionized gases that form the polymers are so highly reactive that each particle forms many bonds, thus resulting in a crosslinked network. Plasma polymers are therefore useful in forming tough and durable surfaces on materials. Some products that use this capability are compact discs, contact lenses, and scratch-resistant plastic windows.

A similar application is the coating of a metallic wafer with a polymer to form a capacitor or some other electrical device. Plasma polymerization is carefully controlled so that the polymer layer is very thin and virtually pinhole free. The ability to control the rate and thickness of the polymerization process is a major advantage of plasma polymerization.

Another application for plasma reactions is the modification of a polymer surface without the deposition of any material on the surface. This is accomplished by using a working gas that will not polymerize, such as the single gases oxygen, argon, nitrogen, ammonia, or some other simple gas. When a plastic part is subjected to one of these gases when it is ionized by the rf field, the gas reacts with the surface of the plastic part. This reaction etches the surface by lifting off the outer layers of plastic material. (This property is used to clean plastic surfaces.)

Ionized gases also combine with polymers to create a polymer surface with many small groups on the surface. For instance, a polymer in oxygen forms OH groups on the surface, where the O comes from the ionized gas and the H comes from the displacement of H atoms off the polymer. In a similar fashion, ammonia (NH_3) forms NH_2 groups on the surface plus release hydrogen molecules (H_2).

Some of the groups formed in the surface have a tendency to raise the energy of the surface of the plastic and make it more reactive in subsequent reactions. This property is used as a method of preparing the surface of a plastic for bonding with metals in a metallizing process or with adhesives in a bonding process. The amount of surface energy increase (called *surface activation*) can be determined by placing a drop of water on the surface of the material after it has been removed from the plasma chamber. If the water has a low angle of incidence on the surface, the water is said to wet the surface and the energy is high. If the angle is small, the surface is not wetted and the surface energy is low. Some gases (such as ammonia) lower the activity of the surface so that water will not wet the surface. These materials are said to give *surface pacification*.

Plasmas can, therefore, be used to modify the surfaces of plastic materials. This modification can be used to activate the surface in preparation for further surface reactions, to pacify the surface so that it will be less reactive, and to coat or polymerize the surface.

CASE STUDY 20.1

Making Shrink-Tubing Using Electron Beam Crosslinking

Shrink-tubing is used extensively in making electrical connections, but can be used in other applications where a connection or joint needs to be made between two thin parts or when a tight covering needs to be placed on a part. Shrink-tubing is used in joining two wires, for example, by placing a section of shrink-tubing onto one of the wires so that it is away from the end. The bare ends of the two wires are then twisted together to form a joint. The piece of shrink-tubing is then moved along the wire until the twisted joint is in the middle of the shrink-tubing section. The tubing is then heated, which causes it to shrink in diameter. If the diameter size of the shrink-tubing is appropriate, the tubing forms a tight seal against the two wires and completely encloses the twisted joint.

The ability of a material to shrink when heated is imparted by radiation processing. This is usually done with an electron beam accelerator.

The initial step in making the shrink-tubing is to extrude tubing in the diameter desired for the **final** diameter (the diameter it is to have **after** it has shrunk). After extrusion the tubing is passed through an electron beam irradiation device. This causes crosslinking in the polymer. Then, while still hot from the irradiation, or after heating again to give the material softness, the material is expanded in diameter and then quickly cooled to maintain the tubing in the larger diameter. The larger diameter is the size when the tubing is packaged and shipped. This is the size it has when placed over the parts to be joined **before** it shrinks.

The key to the shrinking of the tubing when heated is the crosslinks formed during the irradiation step. The crosslinks tie the polymer chains together in whatever configuration they may have at the time of crosslinking (irradiation). When the material is later expanded, this expansion stretches the crosslinks. The rapid cooling of the material in its expanded state freezes the crosslinks in the stretched positions. Then, when the material is reheated, the crosslinks are free to return to their normal, relaxed positions and they cause the material to contract. The lowest-energy position is the original size (the size of the material when the crosslinks were formed). This stretching of crosslinks is depicted in Figure 20.7.

Figure 20.7 Shrink-tubing showing crosslinking when the tubing has a small diameter and stretching of the crosslinks when the tubing is expanded.

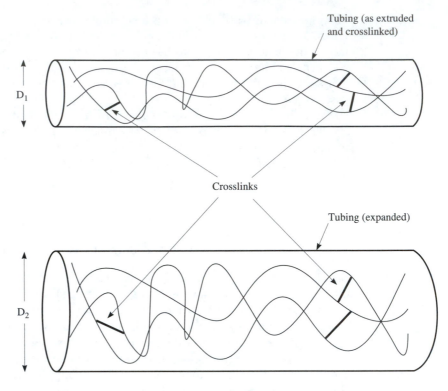

The most commonly used material for shrink-tubing is low-density polyethylene, although other polymers can be used also. Low-density polyethylene readily crosslinks with electron beam irradiation and therefore can be irradiated and crosslinked quickly. The expansion can be done simply by running the tubing through a vacuum system with sizing holes to define the larger diameter. Cooling can then be done by spraying with cold water. For electrical applications, polyethylene has the added benefit that it is an excellent electrical insulator.

SUMMARY

High-energy radiation is a method for processing polymers that causes several different changes in the nature of the polymeric structure. Under some conditions and with certain polymers, the irradiation of the material causes the polymer to crosslink. For thermosets, this crosslinking is an alternative method to the more common method of crosslinking with heat and/or peroxide initiators. After the thermosets are crosslinked, the properties are similar for radiation-crosslinked materials and materials crosslinked by the more traditional methods.

For thermoplastics, this crosslinking results in properties that are significantly different from other thermoplastics and other thermosets. The crosslinked thermoplastic has many properties that are still like those of the noncrosslinked analogue. In some properties, however, significant differences exist. For instance, the environmental stress-crack resistance (ESCR) is much better in the crosslinked material. Other significant property changes include abrasion resistance, toughness, heat stability, and nonmeltability.

In addition to crosslinking, the radiation causes the molecules to split apart, a phenomenon called scission. Scission dominates over crosslinking at very high radiation levels and when steric or some other interference prohibits the molecules from getting close enough to crosslink. Scission is a form of degradation, but can be used successfully to control particle sizes and polymer molecular weight.

Another possible reaction is molecular rearrangement. This reaction is favored when crosslinks are inhibited. The ability to get molecular rearrangement with radiation is used to remove defects in crystal structures.

The most important method used to radiate polymers is electron beam irradiation. In this method electrons are emitted from a hot filament. These electrons are then accelerated through a magnetic field and then fanned out so that they impact evenly over the surface of the part being irradiated.

Other high-energy radiation methods include directed UV and microwave. These are used mainly to cure thermosets. Radiation from nuclear sources (γ-radiation) is used mainly for sterilization but can be used for other purposes (crosslinking, scission, and rearrangement), although with some greater difficulty than electron irradiation. Electron radiation is used for sterilizing medical devices.

Cold plasma is another high-energy method to treat plastics. Plasma is used to polymerize new polymers on the surface of polymers and other materials. Plasma is also used to clean and etch surfaces and to change surface energy to improve bonding characteristics.

One of the most important uses for irradiated plastics is shrink-tubing. In shrink-tubing, the tubing material is crosslinked at a small diameter and then expanded and frozen in a larger diameter. This expansion stretches the crosslinks. When the material is reheated, the crosslinks relax and cause the material to shrink to the size it had when the crosslinks were formed.

GLOSSARY

Cold plasmas A plasma created as a result of exposure of a gas to an rf (radio frequency) field.

Deaging The use of molecular rearrangement from irradiation to reverse the normal aging (degradation) of some plastics.

Dosimetry The method of measuring the dosage received by a part subjected to radiation.

Electron beam radiation A process of inducing changes in polymers (usually crosslinking) by subjecting the polymer to a high-energy stream of electrons.

Free radicals Unpaired electrons.

Gamma radiation The use of radioisotopes that emit gamma particles (nuclear reactions) to irradiate materials.

Gel chromatography   A method to separate the fractions of a dissolved polymer according to the molecular weight of the fraction.

Gel fraction test   A test to determine the amount of crosslinking in a polymer.

Grays   The unit of radiation energy of an electron beam machine.

Hot plasma   The gas that forms as a result of nuclear fusion reactions.

Ions   Charged atoms.

Nuclear magnetic resonance (NMR)   A spectroscopic technique that allows identification of molecular species and can, therefore, assist in determining the amount of scission, crosslinking, or rearrangement that has occurred because of radiation.

Plasma   A highly ionized gas.

Postirradiation reactions   Molecular reactions that occur as a result of irradiation but that actually take place after the part has been removed from the radiation chamber.

Rads   A unit of radiation energy of an electron beam machine.

Reflectometer   A device used to detect a color change in a material.

Scission   The process of breaking the polymer chain through the use of radiation.

Sedimentation test   A method to separate low and high molecular weight fractions, thus permitting the detection of scission in a polymer.

Surface activation   Increasing the surface energy of a material by some method such as treatment with a plasma.

Surface pacification   Lowering of the surface energy of a material by some method, such as treatment with a plasma.

Tertiary carbon   A carbon atom bound to two other carbon atoms.

QUESTIONS

1. Why does crosslinking increase the resistance of golf ball covers to cuts?
2. Why does crosslinking increase the environmental stress-crack resistance of polymers?
3. Would HDPE, LDPE, or LLDPE be the most strongly affected (crosslinked) by electron beam radiation? Why?
4. Why do plastics heat up when they are irradiated?
5. What would be the result of a leakage of gas into the acceleration chamber of the electron beam irradiation machine?
6. Why do you think cobalt radiation can penetrate farther into plastics than can electron beam radiation?
7. What is the likely cause of the inability of shrink-tubing to shrink sufficiently to bind two wires tightly when it is shrunk onto a connection between the wires?
8. The surface of polyethylene after being treated with cold plasma using oxygen as the working gas is similar to the type of surface created when a spark is used to etch the surface of the same polyethylene material. Explain this similarity.
9. Studies have found that the rate of crosslinking of a thermoset material can be increased when both electron beam irradiation and thermal heating are used together. Explain why this is so.
10. Wire and cable material is often crosslinked by electron beam irradiation. Why is this material crosslinked?

REFERENCES

Boenig, Herman V., *Plasma Science and Technology*, Ithaca, NY: Cornell University Press, 1982.

———*Fundamentals of Plasma Chemistry and Technology*, Lancaster, PA: Technomic Publishing Co., Inc., 1988.

Frados, Joel, *Plastics Engineering Handbook* (5th ed.), Florence, KY: International Thomson Publishing, 1994.

Harris, Anthony R., "Electron Beam Irradiation Crosslinking of Polypropylene Panels," (thesis) Brigham Young University, 1994.

O'Donnell, James H., "Radiation Chemistry of Polymers," in *Effects of Radiation of High-Technology Polymers*, American Chemical Society, 1988.

Richardson, Terry L., *Industrial Plastics* (2nd ed.), Albany, NY: Delmar Publishers, Inc., 1989.

Tenney, Darrel B., and Wayne S. Slemp, "Radiation Durability of Polymeric Matrix Composites," in *Effects of Radiation on High-Technology Polymers*, American Chemical Society, 1988.

"Your Partner in Electron Beam Processing," E-Beam Services, Inc., SO-287-3M, 1987.

CHAPTER TWENTY ONE

FINISHING AND ASSEMBLY

CHAPTER OVERVIEW

This chapter examines the following concepts:

- Runner system trimming and flash removal
- Machining (cutting, drilling, milling, surface finishing)
- Nontraditional machining (lasers, water jet, hot wire)
- Shaping (postmold forming)
- Mechanical joining and assembly
- Adhesive bonding (materials, processes, designs)
- Nonadhesive bonding (fusion, ultrasonic, rf, spin, vibration, induction)
- Coating and decorating (painting, printing, metallizing, laminating, texturizing)

INTRODUCTION

Even though most plastics can easily be molded to the general shape desired, many plastic parts require additional finishing before they are in their final use form. For some parts, the additional finishing is as simple as removing the runners and sprue system. Additional finishing for other parts includes joining the parts with others into assemblies, either with simple snap fits or some other mechanical joining methods or by adhesive or nonadhesive bonding. One of the advantages of plastics over metals and ceramics is the multitude of ways in which assembly of plastic parts can be accomplished.

Plastics may also need to be coated or decorated for final use. This coating and decorating can be done by laminating plastics together with other materials (such as cloth, metal, or other plastics) or by painting, printing, metallizing, or just texturizing the surface. Several methods are available for accomplishing each of these tasks.

Finishing tasks often influence which material is used for a particular application. For instance, the use of plastics for automotive body panels was delayed for many years because the plastic panels needed extensive smoothing and finishing before they could be painted with a finish equivalent to the finish on the metal parts making up the remainder of the automotive body. Not until the development of low-profile SMC (which cures with a very smooth and defect-free surface) was plastic used extensively for automotive body panels. Even then some plastic formulations were excluded from use because the paint would not adhere well.

The choice of a material for some particular application should consider the full range of associated costs, including raw material price and availability, processing to final shape, assembly, and finishing. The assembly and finishing steps can involve costs associated with additional shaping or assembly steps, surface preparation, heat requirements, clamping fixtures, costs of additional finishing materials, and the time required to accomplish all these tasks. Plastics offer many advantages over metals and ceramics in these choices, but have some disadvantages as well. Each case should be decided specifically in light of these and the other economic and property considerations.

The choice of finishing can be quite complicated, depending on the material chosen. Some materials may need special finishing to prevent the effects of weathering, corrosion, oxidation, light (UV) degradation, moisture, and chemical attack from solvents, seawater, or acids. Almost every type of metal, plastic, and ceramic reacts differently to these environmental effects and needs to be considered specifically before being chosen for a particular application.

This chapter presents the major finishing and assembly methods for plastics, focusing on the plastic part as the material being acted on in the finishing, decorating, and assembly methods. Some consideration also is given to plastics as the agents of the various methods.

RUNNER SYSTEM TRIMMING AND FLASH REMOVAL

Runner System Removal

The most common method used to remove the runner system (which may include the sprue) is by trimming with hand tools such as wire cutters or small handsaws. This trimming can conveniently be done by an operator who is at the machine during the molding process. For many parts, the cycle time is sufficiently long that the operator can clip the runner system from the parts and also service the machine (load pellets, load inserts, and generally keep the machine within the desired operating limits). Having a single operator perform these tasks for two machines is common and for several machines is not unusual. The keys to having a single operator run several machines are, of course, the cycle time and the complexity of the other tasks that must be done.

Trimming off the runner system using manual tools can cause stresses in the plastic part, especially in brittle plastics such as polystyrene and rigid PVC. In some parts stresses can lead to premature failure. Annealing the parts after they are trimmed, however, often relieves the stresses.

Manual trimming is made easier by proper design of the gate. When the gate is small, the runner system breaks off more easily than when a large gate is used. The size of the gate depends on the type of material and flow characteristics needed in the cavity. These factors are discussed in more detail in the sections in the chapters on injection molding and transfer molding—processes that use gates and runner systems. Under ideal conditions, the gate is designed so that the runner system breaks off when the part is ejected from the mold. Stresses under these conditions are usually small. Some molds use hot-runner systems and need no trimming because the runners remain liquid. Only the part solidifies and so only the part is ejected from the mold.

Proper gate design also simplifies the automatic removal of runner systems. Injection molding machines that are run without operators often use automatic material feeders and robotic part removal. Separating the part from the runners typically involves using either a hot-runner system or a gate so small that the parts naturally separate at the time of ejection. Separating the gate from the part occasionally requires some mechanical shaking, but this can be automated.

After removal of the runner (by any of the aforementioned methods), a small piece of the gate or a rough spot may yet remain on the part. For thermoplastic materials that are generally not too brittle, this small piece can be removed by scraping along the edge of the part with either a file or a sharp tool. For thermosets and brittle plastics, removal methods include filing, buffing, grinding, and smoothing with sandpaper or emery cloth. These methods are not likely to cause stresses in the part.

Removing the runner system and the gate piece or roughness is often easier if done while the part is still warm. The warmth adds toughness to the part and thus diminishes any problems that might arise from internal stresses. This is especially important for brittle plastics.

Flash Removal

Flash is most easily removed when the part is very cold. The cold stiffens many materials and thus makes the flash brittle. The thinness of the flash means that its removal is not likely to add stresses to the part. Cooling the part to facilitate flash removal is called *cryogenic flash removal.*

After the part is cooled to the desired temperature (usually about the temperature of liquid nitrogen), it is placed into a rotating drum and tumbled. (The drum looks like a large, commercial-type clothes dryer.) Tumbling the parts causes the thin flash sections to break off. Some operations include other items in the tumbling step, such as frozen rubber discs or wood, which give slightly more mass to the impacts and therefore help break off the flash.

Flash can also be removed by traditional cutting or machining methods. This is often done by operators at the machine, but it is also commonly done in a separate operation, where the part is placed into a fixture and then rotated against a cutting blade or wheel. Sanding or polishing a part to obtain a very smooth surface is also done in similar fixture setups.

MACHINING

Machining is used to finish many plastic parts. The finishing may include minor shaping corrections to molded parts, trimming of excess materials surrounding the finished part (such as would be common in thermoforming and blow molding operations), making holes or other features that are needed for assembly or use, and surface preparation for other finishing steps. Machining is needed in some parts that require very close tolerances. These parts are molded oversize and then machined to exact dimensions that are tighter in tolerance than could be obtained by molding.

A number of different operations fall into the general category of machining, including cutting, sawing, drilling, tapping, milling, turning, routing, piercing, filing, sanding, buffing, and grinding. These machining operations were originally developed for metals, and some modifications in the standard processing methods should be made to allow for the differences between metals and plastics.

Thermal differences between metals and plastics are especially important in machining considerations. One thermal consideration is the fact that heated plastics expand much more than heated metals (up to 10 times more). This expansion can cause deviations from expected tolerances if the part is allowed to heat up during machining and if corrections are not made for subsequent contractions when the part cools. (For instance, drilled holes in plastics will tend to be undersized by roughly 0.002 inch (0.05 mm) because of the thermal contraction that occurs after drilling.)

Another thermal consideration is the relatively poor thermal conductivity of plastics, which makes heat dissipation difficult and (because of plastic's low melting point) may cause the plastic to melt or to soften and distort. Therefore, steps should be taken during machining to reduce heat generation. When intensive machining of metals causes them to heat up excessively, a cutting fluid is often used as a coolant or lubricant. Cutting fluids are far less commonly used with plastics. Many common cutting fluids are solvents that attack plastics. Some cutting fluids also accelerate stress cracking in the plastic, especially because of the inherent stresses that occur during machining operations. Furthermore, subsequent finishing steps may require clean surfaces and removing the cutting fluid may be difficult. Nevertheless, some machining operations can be carried out successfully with a cutting fluid. The fluid should not cause solvent reactions with the plastic. Some examples of cutting fluids for plastics include liquid air, high-pressure air, water, and nonreactive solvents.

The more common method to reduce heating of plastics during machining is simply to machine less intensively by extending the time to machine the part and by reducing the amount of material that is to be cut at any one moment (reducing the feed rate). Feed rates used for soft metals such as brass or for wood often are appropriate for plastics. For instance, a feed rate of 0.15 to 0.20 inch/rev (4 to 5 mm/rev) would be typical for drilling acrylics. For HDPE, a much softer and lower-melting plastic, drilling feed rates of 0.004 to 0.015 inch/rev (0.1 to 0.4 mm/rev) are typical. The slower feed in HDPE allows more time for heat dissipation.

In addition to machining more slowly, interrupting the machining process can reduce the amount of heating. For instance, drilling can be done by alternately drilling and then retracting to allow time for the part to cool. The retraction also removes the waste plastic created by the drilling process. (This interrupted process is called *peck drilling*.) When sawing, the sawtooth pattern on the blade can include gaps (missing teeth) that allow for cooling during the actual cutting process because the cutting is interrupted. These types of blades are called *skip-tooth cutters*.

Although it may seem strange, the heat buildup of the plastic can be reduced by increasing the cutting tool's speed of rotation, provided the feed rate is kept low. Because the tool carries away some of the heat from the cutting area, increasing the speed of the tool increases the amount of heat carried away. For instance, circular saws are commonly turned at 3000 rpm when cutting acrylic and HDPE. If scorching occurs, the turning rate can be increased to 4000 rpm.

Heating can also be reduced by shaping the cutting tools so that they slice into the plastic rather than impact against the surface. This is done by controlling the angle of the cutting tool relative to the part. The angle of the face of a cutting tool, measured relative to the radial lines of the tool, is called the *rake angle* and is illustrated in Figure 21.1 for saws, drills, and turning tools. This angle can also be expressed as the entrance angle of the tool into the workpiece.

The best sawing, cutting and drilling on plastics are done when the rake angle is slightly positive (0° to 5°). This positive rake angle not only gives the desired shearing action of the cutting surface against the part, it also allows for easier removal of the *chip* (debris that is cut away from the part). If the chip remains in the vicinity of the cut it may obstruct the cutter or heat up and melt. Both problems interfere with the cutting process. The positive rake angle is especially useful for soft materials (like plastics and brass or copper) because there is less of a tendency for the cutting tool to grab into the workpiece and damage or bind the workpiece in the vicinity of the cut.

Tools used to cut plastics should be kept sharp so that the cutting (slicing) action is maximized. Sharp cutting tools can also keep burrs and other surface defects to a minimum.

Another major difference between plastics and metals is that plastics generally are not as strong as metals. This lower strength can have a significant effect on the machining process. For instance, plastic materials should be supported when drilling or milling so that the material does not splinter or crack when the tool breaks out at the finish of the cut.

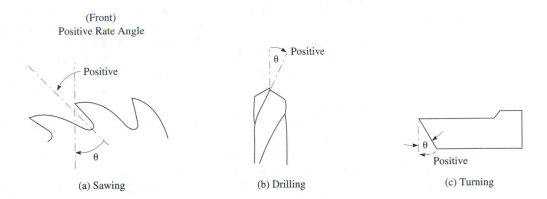

Figure 21.1   Rake angle characteristics for sawing, drilling, and turning.

Plastics can also be more brittle than metals. This brittleness causes special problems in machining because of the material's tendency to fracture during the machining process. Sharp cutting tools and low feed rates are especially important with brittle plastics. In some cases of machining brittle plastics, annealing may be required after cutting. Some brittle plastics, such as acrylics, are sold with a paper coating on the surface to give some backing during machining, to protect the surface, and to lubricate during the cutting process. Therefore, these materials should be machined with the paper in place. Thin brittle plastics can sometimes be straight-cut by scribing and then breaking, much as window glass is cut.

An especially difficult combination of materials to machine is composites in which a brittle matrix is combined with a tough or very hard reinforcement (such as epoxy combined with aramid fibers or carbon fibers). These composite materials require special tool shapes and special tool materials (such as diamonds) to effectively machine. Experience has shown that cutting composites with traditional tool steels is not acceptable because of wear. Carbide tools or other hardened surfaces should be used.

Filing, grinding, sanding, and other smoothing operations have some special considerations when working on plastics. The soft nature of plastics and their tendency to melt at low temperatures lead to rapid accumulations of materials in the finishing tool. These tools should be cleaned often during use so that the buildup does not affect the operation.

NONTRADITIONAL MACHINING

Especially difficult materials can often be cut more easily by using nonmechanical machining methods, such as water jet. In water jet cutting a very high-pressure water stream (about 50,000 psi or 350 MPa) is directed through a nozzle (approximately 0.005-inch or 0.13-mm diameter) and onto the part to be cut. Most materials, even composites, can be cut effectively using this technique, with stresses no higher than would occur during traditional machining operations. The amount of material removed in the cutting operation (called the *kerf*) is about the same in water jet cutting and traditional cutting.

Lasers are another nonmechanical machining method that is rapidly gaining popularity. Lasers cut by burning the material with a concentrated light beam of very high energy. One advantage of lasers is that they often vaporize the material when it is cut so that little if any residue is left after the machining operation. Lasers are used to make cuts (like a saw or a shear) and holes (such as a drill). The size of the hole can be varied by changing the laser's focus, which also determines the depth of the cut. By careful control over the focal length of the laser, holes can be cut through one side of a tube and not the other side, even though the tube is rather small in diameter. Another advantage of lasers is that they don't become dull.

One disadvantage of lasers is that their heat is sufficient to decompose plastic material and so create fumes. These fumes are hazardous, so ventilation is suggested. However, the advantage to vaporization of the material is that the cut is clean. Another problem with lasers is heating of the material in the vicinity of the cut. This edge heating usually is not severe, but it can result in changes in the surface characteristics of the material along the cut.

Table 21.1 Comparisons of Different Methods for Cutting Plastics

Cutting Method	Advantages	Disadvantages
Mechanical	• Well understood • Available tools and machines • Low capital costs	• Can cause internal stresses • May require hardened tools • High tool wear
Water jet	• Low tool wear • Cuts many materials • Moderate stresses	• High operating costs • High cost of the machine
Laser	• Clean cut (vaporizes plastic) • No tool wear • Cuts many materials • Can be focused to control cut	• Cannot cut thick materials well • Some damage zone around kerf • High machine cost • Possible harmful vapors
Hot wire	• Cuts many materials • Very low-cost machine	• Some decomposition zone • Possible harmful vapors • Materials may flow back together

Hot-wire cutting is a useful cutting method for many plastics, especially foams. Hot-wire cutting takes advantage of the low melting point of plastics to slice through simply by melting the material with a wire heated by electrical resistance. Some decomposition usually accompanies hot-wire cutting, so ventilation is suggested.

Table 21.1 shows a comparison of traditional and nontraditional cutting methods.

SHAPING (POSTMOLD FORMING)

One of the advantages of plastics over other materials is the ease of forming the plastic after it has been molded or given some intermediate shape. This postmold forming is important in several situations. For instance, acrylics often are formed after they have been cut from a sheet. This forming is much like thermoforming except that the part is already cut to the proper size and needs only some simple shaping. Heating can be done in an oven, but a strip heater is more commonly used when only a few bends are needed. A strip heater and a forming jig are illustrated in Figure 21.2.

Some parts must be made by cutting, drilling, machining, or otherwise finishing before they are formed into their final shape. These materials often are heated to a softening point and then manually formed, often with a simple forming jig.

Another situation where postmold forming is valuable is to introduce severe undercuts or other features that are difficult to mold. These can be added manually after the part is removed from the mold, usually with a forming jig.

Some applications require forming of a part or component after the other components in the assembly have been joined. It may be impossible to heat the entire assembly, but the part to be formed must be formed to the dimension required after the assembly has been made. An application of this type is a plastic retaining ring around a device.

Plastic parts can be reshaped to correct errors in molding. Although not very useful for small parts, reshaping can salvage some very large or complicated molded parts.

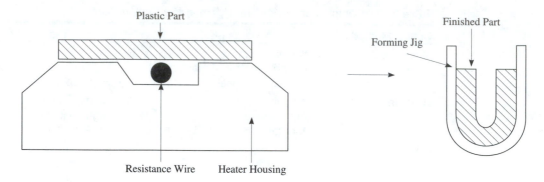

Figure 21.2   Postmold forming of a sheet material using a strip heater and a forming jig.

MECHANICAL JOINING AND ASSEMBLY

The principal methods of joining or bonding two materials together can be classified as mechanical joining, adhesive bonding, and nonadhesive bonding. Each of these methods is discussed in the following sections and then compared in a table at the end of the sections. Each of the methods has significant applications with plastics in which plastics are the materials that are joined or bonded and, in some cases, the materials that are the joining or bonding agents.

Mechanical joining and assembly methods for plastics include many methods also widely used for nonplastic parts, such as rivets and threads (developed for metals and modified for plastics). This section discusses traditional mechanical joining methods as they apply to plastics as well as mechanical joining methods that are unique to plastics. Most of these nontraditional joining methods rely on the flexibility of many plastics or some other property of plastics that allows them to be joined into assemblies without the need to bond the components together.

Traditional Joining and Assembly Methods

Joining two plastic parts with rivets is accomplished using the same basic procedure as used in joining metal parts. Holes are drilled in the parts to be joined, the holes are aligned, and a rivet is inserted into the hole. The rivet is then capped (flattened or headed) to secure it in place and lock the plastic parts together. If a traditional metal rivet is used as the joining device, head flattening is done mechanically, either on a cold or a heated rivet. If plastics are to be joined, heated rivets are generally not used because they melt the plastics.

Plastic rivets can also be used. In this case the rivet can be heated, provided that the temperature is lower than the softening temperature of the plastics to be joined or that the contact time at the high temperature is short enough that softening of the parts does not occur. Using a plastic rivet to join plastic parts means that the strength of the rivet will closely match the strength of the plastic. This reduces one of the major problems associated with the use of metal rivets on plastic parts: pulling out of the rivet. The high strength of metal compared to plastic means that any stress placed on the rivet zone may cause the metal rivet to rip through the surrounding plastic part. With plastic rivets, on the other hand, the rivet elongates slightly and may be the breaking point rather than the surrounding plastic material. In either case, rivets used to join plastic parts usually have very wide flanges so that stresses can be spread over a much wider area, helping the joint withstand the stresses without failure of the plastic material.

Plastic parts can also be joined using traditional metal screws. The threads are cut or molded in the plastic parts and then are joined using regular metal screws. Again, plastic screws can also be used. The relatively low strength of plastics can also be a factor in the strength of joints using screws, either metal or plastic. The threads in the plastic materials would be weaker than the screw and pullout by stripping the plastic threads could occur under stresses applied to the joint. This problem is reduced by using large, strong thread types. These heavy-duty threads often are widely spaced with heavy ridges, some even square. Although the strength of such heavy-duty threads is increased, the accuracy of the thread system and the tightness and retention of the screw is reduced. Hence, the chance of the thread loosening is much higher with plastic threads than with metal threads.

When long-term tightness is required, inserts with metal threads are used. These inserts are molded or, perhaps, bonded into the plastic material. When appropriately attached, these inserts give the proper anchor for metal screws so that retention and tightness are equivalent to full metal systems. Plastic screws can also be used with metal inserts, but the strength of the assembly would be lower because the plastic screw is generally weaker than the similar metal screw.

Self-threading screws can also be used with plastic parts. With these, the ease of joining (cutting the threads with the screw) is easier than with metals and about the same as with wood. The retention in plastics is about the same as in wood.

Plastic fasteners, either rivets, screws or other fasteners, have some advantages over metal analogues. The plastic fasteners are lighter, corrosion-resistant, insulative, flexible, durable, and usually less expensive. These also have the same or similar coefficients of thermal expansion as the plastic parts to be joined, thus eliminating a source of thermal stresses and possible joint movements. The disadvantages of plastic fasteners versus metal fasteners are the lower strength, shear resistance, torque strength, and heat resistance of the plastics.

Nontraditional (Unique to Plastics) Joining and Assembly Methods

The most important and most widely used joining and assembly method unique to plastics is the *snap joint,* or *interference joint.* This joint, several types of which are illustrated in Figure 21.3, uses the flexible nature of plastics to create a joint. Because such joints often can be opened and closed repeatedly, they are valuable as closure devices. The snap fit basically involves an undercut on one part that engages a molded lip on the other part.

Other types of fitting can be used because of the compressibility of plastics. Fittings for semirigid plastic tubes or hose can be of several types, as shown in Figure 21.4. The

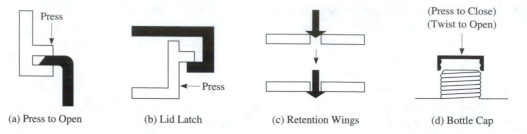

(a) Press to Open (b) Lid Latch (c) Retention Wings (d) Bottle Cap

Figure 21.3 Snap fits showing (a) a joint that opens repeatedly when pressed, (b) a lid latch, (c) a retention type with wings, and (d) a bottle cap that can be pressed to close but twisted to open.

Figure 21.4 Three different types of tubing fittings: (a) the outside-diameter (OD) fitting, (b) the inside-diameter (ID or barb) fitting, and (c) the ring-lock fitting.

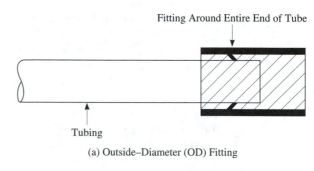

(a) Outside–Diameter (OD) Fitting

(b) Inside–Diameter (barb or ID) Fitting

(c) Ring-Lock Fitting

first type, called an outside-diameter (OD) fitting, fits over the end of the tubing. Two barbs are molded into the inside of the fitting. These barbs are sloped so that the tubing can be pressed into the fitting past the barbs. If the tubing is pulled so that it moves out of the fitting, the barbs dig into the surface of the tubing and retain it. The second type of fitting is inserted into the tubing and is called an inside-diameter (ID) or barb fitting. This fitting has barbs on its exterior that match or are slightly larger than the diameter of the tubing. When the ID fitting is forced into the tubing, the tubing expands around the barbs and the barb fitting is held in place. The third type of tubing fitting is called the ring-lock fitting. This fitting has a ring that is retained on a tapered section of the fitting. (The ring is slightly smaller than the largest dimension of the taper and is put in place by forcing the ring over the taper.) The taper is inserted into the tubing such that the tubing goes between the tapered section and the ring. Then, the tubing is pulled in one direction and the fitting is pulled in the other direction. This pulling action moves the ring against the taper and squeezes the pliable tubing between the ring and the taper, thus securing the fitting.

The *living hinge,* or *integral hinge,* is another important method for joining plastics that depends on the flexibility of the plastic materials. This hinge is a connecting strip of plastic between two larger plastic sections. The connecting strip and larger sections are made of the same material without joining; that is, they are molded or formed as a single unit. Immediately after molding, the part is flexed on the hinge to give the proper orientation to the molecules in the hinge region. The molecules should be oriented along the direction of the hinge for maximum hinge flexibility and long life. Figure 21.5 shows a box and lid connected by an integral, or living, hinge.

Although many flexible plastic materials can be used to form living hinges, the most commonly used plastic is polypropylene. Polypropylene has suburb flex life and good flexibility. Other candidate materials, such as polyethylene, tend to crack after repeated flexing. Still others, such as nylon or acetal, will form living hinges that are very stiff.

Figure 21.5 A small box and top using an integral, or living, hinge.

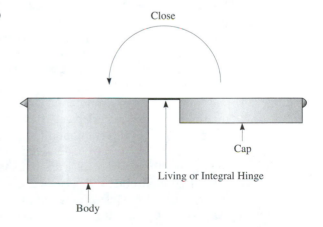

ADHESIVE BONDING

When two materials are joined together using adhesive bonding, those two materials are called the *adherends*. An alternate, name for an adherend is a *substrate*. The material that forms the bond between them is called the *adhesive*. This section begins with a discussion of adherends that are plastics, for it is the joining of plastic materials for finishing and assembly that is of primary concern. Later, this section examines those adhesive materials that are polymeric and, therefore, are an important application of some polymers.

Adherends

Not all plastics can be easily or, perhaps, even successfully bonded with adhesives. The differences between those plastics that are readily bonded and those that are not are based principally on surface characteristics.

A key feature of adherend surfaces that can be bonded easily is that the adhesive covers or coats the surface. This coating of the surface is called *wetting*. The controlling factors in whether a surface is wetted by a liquid or whether the liquid beads and doesn't wet the surface are the surface energies of the two materials. If the surface energy of the liquid is lower than the surface energy of the solid, then the solid surface will be wetted by that liquid. This rule arises from the free-energy equation introduced earlier in this text. Whenever the net free energy is negative, the phenomenon releasing that energy will occur naturally. Hence, when a solid surface is wetted by a liquid, the free energy is negative or energy of the system is lowered. This occurs when a high-energy surface is covered so that a lower energy surface results.

If an adherend surface is not wetted by the desired adhesive, then poor adhesive bonding will occur. This can sometimes be remedied by modifying the adherend surface to increase its energy and therefore allow wetting. The following are some ways to treat or prepare the surface of the adherend and promote adhesive bonding

Surface Cleaning.
Even plastic adherends that have high surface energies and are therefore easy to bond onto may have surface contaminants that reduce surface energy and interfere with bonding. These surface contaminants can come from the molding operation when mold releases are used to facilitate removal of the plastic part from the mold; from general handling, which imparts grease and oils onto the surface; from plant high-pressure air used to blow debris off the parts when the air has been contaminated with oil; and from general environmental contamination arising from a variety of sources, including the packaging materials and the general dusty conditions within the plant. Whatever the cause of the contamination, it should be removed.

The most common method of removing surface contamination is by wiping or dipping the material in a solvent. The solvent should be chosen to solvate the surface contamination without affecting the plastic adherend and should itself be easily removable. The solvent may also be required to remove other materials contaminating the surface. Water, for example, interferes with the cure mechanisms of some adhesives and must be removed for good bonding.

Chemical Etching.   When the clean surface of the adherend is still not sufficiently energetic to allow wetting by the adhesive, a chemical etch may be useful. In this situation, a chemical that will react with the adherend is applied to the surface of the adherend. This application changes the nature of the surface and, if the proper type of reaction occurs, the energy of the surface is raised and wetting by the adhesive becomes possible. The chemical etching material is usually removed before the adhesive is applied. If the etched surface is likely to change back or be altered in some way so that the activity is lost, a coating (called a *primer*) may be applied to preserve the surface characteristics.

Flame Treatment.   Many plastic materials, especially the polyolefins, can be treated with a flame to increase their surface energy. Brief treatment with an open flame often causes some oxidation of the surface of the adherend and thereby raises the surface energy. Flame treatment is frequently done on polyethylene and polypropylene to promote bonding and other adhesion-dependent phenomena such as printing.

Corona Treatment.   Oxidation of plastic surfaces can also be achieved by passing an electric spark, called a *corona,* over the surface of the adherend, which then becomes wettable. Corona treatment is generally more aggressive than flame treatment. After corona treatment, the surface may be both oxidized and physically eroded. The polyolefins are the most frequently corona-treated plastics.

Plasma Treatment.   Plasma (discussed in detail in the chapter on radiation processing) can be used to clean a surface, etch a surface, or react with a surface. It can also be used to cause a reaction (polymerization) to occur in the gaseous phase that would be deposited onto the surface. Plasma obviously has wide capabilities for surface modification of plastics, but the relatively high cost of the equipment and need for operating in a vacuum have limited its widespread use.

Coupling Agents.   Some surfaces can best be treated by coating the surface with some material that promotes bonding to the desired adhesive. Such materials are called *coupling agents* because they couple the surface and the adhesive. Typically, these materials are chemicals that have two different chemical natures, usually one nature on one end of the molecule and the other nature on the other end. When one end is chemically like the adherend and the other end is like the adhesive, then the molecule serves to link the adhesive and adherend together. Coupling agents are very common with ceramic adherends (such as fiberglass) but less so with plastic adherends. Nevertheless, coupling agents remain an interesting and potentially useful device for promoting adhesion.

Mechanical Abrasion.   When modifying the energy of the adherend surface is impossible, or when the modifications need enhancement, mechanical abrasion can further increase the ability of an adhesive to bond to an adherend. The surface imperfections (small cracks and pits) created by mechanical abrasion serve as locations where the adhesive can enter and then mechanically bond with the adherend. No thermodynamic change in the surface is made by this mechanical abrasion. Therefore, wetting is not increased. The bonding relies strictly on mechanical forces.

Some general comments about the chemical nature of surfaces and their energy may be useful in understanding the actions of the various surface treatments just described. In general, surfaces that have pendant oxygen atoms (usually as OH groups) are very energetic and, therefore, easily wettable and bondable. (The presence of OH and O groups on the surface of an adhesive is also favorable for adhesion promotion in many cases.) When oxidation occurs, such as with flame or corona treatments, O and OH groups are the chief products created on the surface. Other pendant groups that promote adhesion are groups containing oxygen, such as organic acids, esters, and ketones, NH, CN, Cl, and sulfur-containing groups.

Poor surface adhesion is often associated with having only hydrocarbon atoms on the surface, such as polyethylene and polypropylene, or with surfaces that have tightly held electrons, such as fluorocarbons or ethers. Common greases and oils are chemically similar to the polyolefins and, therefore, adversely affect the bondability of a surface when they are present. Hence, solvent removal is mostly to clean these olefinic (oil-like) materials from the surface.

The actual bonding at the surface can be of three types. Covalent bonds can be created between the adherend and the adhesive. The strongest bonding results when this occurs. The second type of bonding results from an attraction between the adherend and the adhesive such as hydrogen bonding, van der Waals forces, induced dipole attractions, and so on. These attractions can be quite strong, but may also be quite weak, depending on the nature of the materials. The third type of bonding is mechanical bonding where the adhesive fills into crevices on the surface of the adherend and thereby becomes mechanically entrained on the surface. Mechanical bonding can occur with both of the other types of bonding or it may be the only type of bond that is present.

Ideally the bond between the adherend and the adhesive will be so strong that if fracture occurs it will be either within the adhesive itself or within the adherend itself rather than at the junction between them. When failure occurs within a single material rather than at the boundary of the material, the failure is called *cohesive*. When the failure occurs at the interface or boundary between two different materials, such as between the adherend and the adhesive, the failure is called *adhesive*. The strongest bonds are those in which only cohesive failure occurs.

Adhesives

When applied to the adherend surface, adhesives are usually liquids but may also be pastes or solid films. In all cases, however, the adhesive goes through a liquid phase and wets the surface of the adherend or is somehow pressed against the surface so that interaction between the adhesive and the adherend occurs. After covering the surface of the adherend, most adhesives go through some chemical or physical change such that they become hard. (Exceptions to hardening adhesives would be those that are meant to stay flexible, such as rubber cement. In these cases the only change, if any, is a thickening of the material, normally from a solvent loss.) This hardening step can result from the loss of a solvent, cooling of the adhesive from a melt to a solid, chemical reaction within the adhesive

Table 21.2   Types of Adhesives and Common Polymers within each Type

| | Types of Adhesives | | | | |
	Structural	Hot Melt	Pressure Sensitive	Water-Base	Radiation Cured
Typical Polymers in the Adhesive Type	Epoxies Polyurethanes Acrylics Cyanoacrylates Anaerobics Silicones Phenolics	EVA PVA PE Nylons PP (amorphous) SBR	Rubber cement "Scotch®"-type	Starches Natural rubbers Casein PVAl PVA Amino resins	Acrylics Epoxies

such as crosslinking or polymerization, or pressure against the adhesive that causes some interaction to occur with the adherend.

A wide range of adhesives are available to meet the many applications that require adhesive bonding. Adhesives are used to bond plastics, metals, ceramics, composites, wood products, and various combinations of these. The adhesive is chosen for its ability to bond to the adherend and for its other properties such as structural strength, cost, ease of application, and environmental effects. For simplicity, adhesives can be divided into five common types (structural, hot melt, pressure sensitive, water base, and radiation cured), each comprised of many different specific materials, but all characterized generally as polymeric. Each of the five general adhesive groups is considered separately. These are outlined in Table 21.2.

Structural Adhesives.   These are adhesives that can be stressed mechanically and can withstand a variety of difficult environmental conditions such as high temperature, adverse solvents, and creep potential. Structural requirements dictate that most of these adhesives are thermosets that can be either one or two-component systems. The most important polymeric groups within this class of adhesives are epoxies, polyurethanes, modified acrylics, cyanoacrylates, anaerobics, silicones, phenolics, and various high-temperature polymers.

Epoxies (described in the chapter on thermosetting resins) are the most common of the structural adhesives. Epoxies are generally two-part adhesives, where one part is the epoxy and the other is the hardener (a chemical that will open the epoxy rings on two chains and bond between them). These adhesives can be cured either at room temperature or at elevated temperatures generally up to 350°F (175°C). They are generally brittle and relatively inexpensive.

Polyurethanes are tougher than epoxies, generally maintaining good flexibility even at low temperatures, although high-temperature stability is poor. Polyurethanes are made by combining two reactants, a polyol and an isocyanate, although in some formulations the polyol reactant can be water. Reaction begins immediately after mixing of the components and often occurs without the need for heating. Pot life is, therefore, limited.

Modified acrylics cure by a free-radical polymerization method. They differ from standard acrylics in that rubber or other tougheners have been added. The typical method for applying these materials is to apply one component to one adherend and the second component to the other adherend. The materials react when the adherends are pressed together. These adhesives can be used on most materials, even when the surface is slightly oily or improperly cleaned. Cure times are quite long.

Cyanoacrylates (superglues) are single-component systems that cure when in the presence of even small amounts of moisture, as is present in most ambient atmospheres. These adhesives bond well to most materials, including skin. This can present a safety hazard but is also an advantage wherein cyanoacrylates are used to suture skin together in surgical procedures. Cyanoacrylates are relatively expensive and strong but quite brittle.

Anaerobics are single-component systems that cure by free-radical polymerization. The polymerization reaction is inhibited by oxygen, so no reaction will take place until oxygen is removed from the system.

Silicones are available as either one- or two-component systems. The one-component systems cure on contact with atmospheric moisture. Silicones can be formulated to cure at room temperature (RTV) or at high temperature (HTV). They have good adhesion over a wide temperature range, -76 to $480°F$ ($-60°C$ to $250°C$), with capability to as high as $700°F$ ($370°C$) for some formulations. Silicones are tough and flexible with good solvent, moisture, and weathering resistance.

Phenolics and **urea formaldehydes** are low-cost thermosetting systems available in one- or two-component systems. They are generally strong but brittle. Phenolics are dark colored, which limits applications. These materials dominate the plywood and chipboard markets.

Hot-melt Adhesives.

These are thermoplastic materials that are heated to a melt, applied to the adherends, and then allowed to harden by cooling. The most important thermoplastics used for hot-melt adhesives include ethylene vinylacetate (EVA), polyvinyl acetates (PVA), polyethylene (PE), nylons, amorphous polypropylene, and various block copolymers, such as between styrene and butadiene. Melt viscosity is an important factor in the application of these materials. Therefore, a key to their successful use is a proper heating/dispenser unit. For industrial applications, the hot-air welding tool (shown in Figure 21.6) is very common. In this device a rod of the solid hot-melt resin is fed into a shaping head where it is melted by hot air that is passing through the machine. The molten material is applied to the adherend as the shaping head passes over the surface.

Automatic hot-melt dispensers are used in commercial applications where high speed and repeatable application are required. These automatic dispensers have a chamber where the plastic resin is placed and kept hot. The dispensers have a pumping system that moves the material from the reservoir in the heated chamber to the application nozzle. Then, when the signal is given (usually from some timing device), the pump moves the hot melt onto the surface in the quantity desired.

Pressure-sensitive Adhesives.

These materials, which are often supported on a tape or some other backing material, are tacky at room temperature and adhere when brought

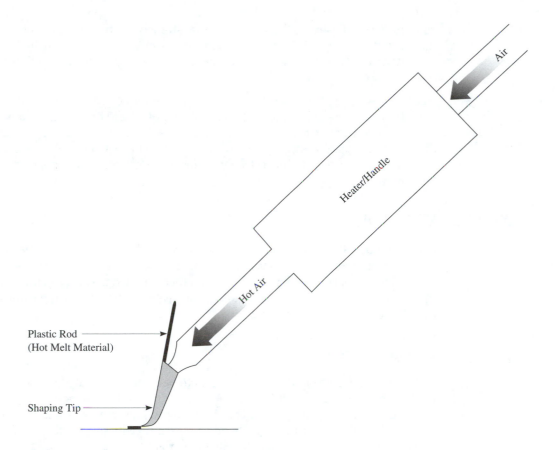

Figure 21.6 Application of a hot-melt adhesive.

into contact with the adherend through some pressure. These materials must be highly viscoelastic so that they flow when pressed together but they also must resist excess flowing, have some elastic nature, store bond-rupture energy to provide peel and tack, and dissipate energy during adhesion. The highly elastomeric nature of many of these materials has earned some of them the name *rubber cement*. The viscoelastic nature of the adhesives is clearly demonstrated when trying to remove a label that is bonded with a pressure-sensitive adhesive. If the label is pulled quickly, the adhesive resists the movement and the label breaks. If the label is pulled slowly, the adhesive gradually separates and the entire label can be removed. Hence, the time-dependent nature of the adhesive is demonstrated.

Water-base Adhesives. Many of these materials are natural products that are combined or occur naturally in combination with water (latex materials). Some typical examples of these are starch-based adhesives, latex rubber, casein (from milk products), animal glues, sodium carboxymethylcellulose, and sodium silicate. Nonnatural materials that are also water-base adhesives include polyvinyl alcohol, polyvinyl acetate, amino resins, and several rubbers. The water-base adhesives have lower strength than most of the other types of adhesives but are nontoxic, are nonpressure sensitive, and can occasionally be dissolved in solvents other than water. When used, the solvent must be removed to effect the bonding.

Radiation-Cured Adhesives.   These are generally free-radical cure materials that are placed between the adherends and then radiated to cure. The radiation source is usually either a UV light or an electron beam. One of the substrates must be transparent to the radiation. These materials can have strengths equivalent to structural adhesives. The most common applications are in electronic equipment, magnetic tapes, floppy disks, and dental and medical devices.

NONADHESIVE BONDING

Several methods of bonding plastic parts involve neither adhesives nor mechanical bonding methods. These nonadhesive bonding methods generally take advantage of the ability of plastic materials to soften and fuse with other softened plastics when pressed together in the softened state. The methods differ principally in the way energy is introduced into the plastic to cause the softening.

Fusion Bonding

In *fusion bonding* the portions of the parts to be joined are heated to softening and then pressed together. Care must be taken to ensure that only those portions of the plastic to be joined are heated and pressed together; otherwise, some deformation may occur. Fusion bonding differs from adhesive bonding in that no adhesive is used; it differs from hotmelt welding in that no separate bonding material is melted.

For best results, fusion bonding is done with materials that have a wide melting point and good melt strength. If a narrow-melting-point material is used, the material may become so low in viscosity that the fluid material runs out of the joint area.

The successful use of HDPE for gas distribution pipes is largely due to the ability of the pipe to be fusion bonded easily, with airtight bonds in the field environment. The polyethylene used for this application often is salmon colored and is sold in long lengths wound onto large reels. The installation often is done by placing the end of the new polyethylene pipe into the end of the larger old pipe (usually made from cast iron or steel). The thicker walls of the polyethylene pipe allow a smaller diameter of pipe to carry higher pressure and, therefore, an equivalent volume of gas. The polyethylene pipe is fed into the old pipe until that section of PE pipe runs out. At that point, two sections must be welded together. This is done using a portable heating and pressing unit. The ends to be bonded are each placed inside the heater and clamped in place. Then, when the pipe ends have softened from the heat, they are brought together and held until the fusion bond is formed. An example of fusion bonding of large pipe being done in the field is shown in Photo 21.1.

Some fusion bonds use a plate to heat the plastic parts. This type of fusion bonding, called *plate bonding,* is used at grocery store meat counters. The meat is placed onto a small plastic foam tray and then covered with a clear plastic film. This film usually is a

Photo 21.1   Field welding of PE pipe (Courtesy: Hoechst Corp.)

multilayered film having several films with different characteristics coextruded or laminated together. The outer layers are materials that have quite low melting points and good adhesive properties. After the meat is wrapped, the plastic is simply folded over itself and lightly heated with a plate heater attached to a heating iron. The outer layers quickly soften and fuse together, producing a watertight and airtight seal.

Ultrasonic Welding

The most important nonadhesive and nonmechanical method of joining plastics after fusion uses ultrasonics and is called *ultrasonic welding*. An ultrasonic signal generator generates the energy that is directed into the plastic. Sound energy normally is the vibrations of air molecules, but it can be the vibrations of any molecules at the frequencies generally associated with sound waves.

To fuse using ultrasonic vibrations, the ultrasonic signal generator (held in the operator's hand or more commonly mounted in a holding jig) is pressed against the two materials that are overlapped and touching in the region to be welded. The ultrasonic vibrations induce sympathetic vibrations in the plastic materials and cause some softening and, perhaps, some melting. When the plastics are pressed together (only moderate pressure is required), they fuse.

The efficiency of the ultrasonic welding method is substantially improved if the vibration energy is concentrated into a small region rather than across the entire surface to be welded. By concentrating the energy into a small region, the small region melts and spreads over the entire bonding area, thus creating the bond desired and doing it with less energy. The small region for energy concentration is created by making a small peak or protrusion, called an *energy director,* on the surface of the plastic in the area to be bonded (see Figure 21.7a).

Figure 21.7b illustrates the principle of *staking.* This method of joining uses ultrasonic energy to soften the end or a post on a plastic part that protrudes through a hole in a part to which it is to be joined. After the parts are put together, the ultrasonic signal generator is brought close to the end to be softened and the energy is applied for sufficient time to soften the end. While the end is soft, a forming tool is pressed against it and the plastic is squeezed outward so that it overlaps onto the second part. The forming tool can be the same as the end of the ultrasonic generator. This process is similar to riveting using heated rivets that are capped or headed by a pressure/forming tool.

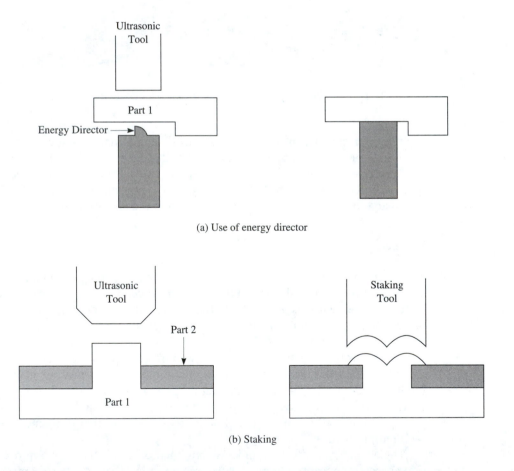

Figure 21.7   Ultrasonic welding showing (a) the use of an energy director to obtain efficient ultrasonic welding and (b) the principle of staking.

RF Welding

Radio frequency (rf) welding is similar to ultrasonic welding except in the frequency and power of the energy signal. Some plastics are highly susceptible to induced vibrations in the radio frequency range (a higher-energy frequency than ultrasonic signals). Frequencies in the rf range induce a back and forth movement in the plastic molecules as the alternating current switches polarity. Some molecules attempt to align themselves with these alternating signals and therefore heat up when rf energy is induced on them. Plastics having this property are said to have a *dissipation factor* or a high *loss rate*. Common plastics with high dissipation factors include ABS, PVC, and cellulosics (but not PE, PP, and PS).

Spin, or Friction, Welding

The necessary heat required to soften and fuse some plastics can be achieved by simple friction when the two parts to be joined are moved rapidly while in mutual contact. The setup most common for this technique is when a circular part is to be welded to a flat plate or a circular rod is to be welded inside a hole (see Figure 21.8).

For best efficiency, the surfaces to be friction welded should be relatively smooth and free of contamination. Very large surfaces are difficult to weld by this method because of the large amount of energy that is required to melt or soften the entire surface simultaneously. Further, *spin, or friction, welding* works best when the melting points of the parts to be joined are low.

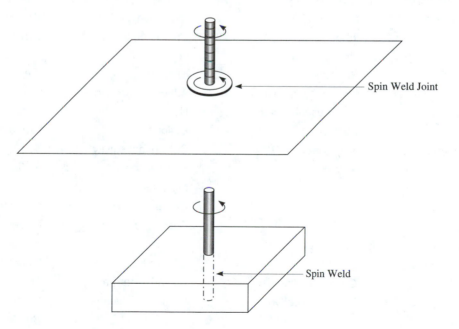

Figure 21.8 Spin, or friction, welding.

Induction Welding

Induction welding, or *electromagnetic heating,* heats by utilizing the property of some metals to vibrate in a magnetic field. If some metallic powders or filings having this property are placed along the region to be bonded and the region is subjected to a magnetic field, the metal particles begin to vibrate, heating the surrounding plastic molecules and causing them to soften or melt slightly so that fusion occurs. Another technique for induction welding is simply to place the plastic materials to be joined around a metal part (such as a rod) and then pass the rod and material through a magnetic field. Sufficient heat can be generated to fuse the plastic materials.

Summary of Joining and Bonding

The general characteristics of three general methods of joining and bonding—mechanical joining, adhesive bonding, and nonadhesive bonding—are compared in Table 21.3.

Table 21.3.   Comparison of Mechanical Joining, Adhesive Bonding and Nonadhesive Bonding

Property	Mechanical Joining	Adhesive Bonding	Nonadhesive Bonding
Cost	High	Low	Low
Time required	Several steps, each rapid	Few steps, long cure	Few steps, rapid
Surface preparation	Minimal	Extensive, critical	Moderate
Thin sections	To be avoided	Can be done	Can be done
Joint weight	Heavy	Light	Very light
External surface	Protrusions	Smooth	Can be smooth
Temperature limits	Limited by the plastic	Adhesive may limit	Limited by the plastic
Inherent damage	Holes usually made	None	None
Ability to inspect	Easy	Difficult	Very difficult
Material interactions	May corrode	Solvent sensitivity	None
Moisture penetration	Significant	Self-sealing	Self-sealing
Stress concentrations	Significant	Minimal	Minimal
Long-term loads	Fatigue	Creep	Minimal
Peel resistance	Resistant	Susceptible	Moderate
Tensile resistance	Susceptible	Resistant	Resistant
Vibration damping	Minimal	Inherent	Minimal
Health and safety	Cutting, drilling danger	Solvent, thermal	Thermal
Materials	Most plastics OK	Not for polyolefins	Restricted by method
Equipment cost	Low	Very low	Can be high

JOINT DESIGN

Joints must be able to withstand four types of stresses: (1) normal or tensile stresses that are perpendicular (normal) to the plane of the bond; (2) shear stresses that are parallel to the plane of the bond; (3) peel stresses that are a combination of shear and normal and that act when one of the adherends is flexible; and (4) cleavage that is a combination of shear and normal and that occurs when the forces are not symmetrical on the bond. These stress types are shown in Figure 21.9 for adhesive bonds, although other bonding systems experience the same stresses. In strict terms, only the normal and shear stresses are needed to characterize the stresses, but since all four types are reflected in tests for bond strengths they are all commonly used to characterize the bond strengths.

Properly designed bonds spread the stresses over as much of the bond area as possible and minimize the amount of force that might arise from lever effects on the bond. Figure 21.10 shows some lap joints that reflect various degrees of strength from joint design. Note that the best of the joints shown has both symmetry and a large surface area for the bond. Others that are not as good lack one or both of these features.

Bonds used for attachment of two materials using angle joints are shown in Figure 21.11. Again, the principles of low leverage and large surface areas apply.

In designing joints, the primary consideration is the required joint strength, which dictates the type of adhesive material (at least the class of adhesives) that can be used.

Stress Types

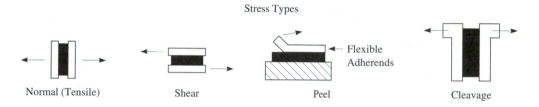

Normal (Tensile)   Shear   Peel   Flexible Adherends   Cleavage

Figure 21.9 Types of stresses in bonded structures.

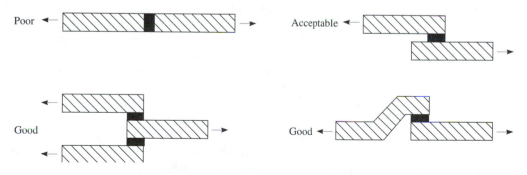

Poor   Acceptable   Good   Good

Figure 21.10 Various joint designs illustrating good and poor practices.

Figure 21.11   Attachment bonds showing good and poor practices.

Other factors are the environment and the method to be used to make the bond. Then, the type of bond can be chosen and the forces calculated to determine if the bond meets the minimum requirements.

COATING AND DECORATING

Sometimes a designer wants to add decorative effects to a plastic product that cannot be achieved through regular molding methods. For this reason there are many secondary operations in use just for decorating and coating. These methods can be used principally for adding aesthetic value to the product but can also enhance performance, usually by protecting the material from unwanted environmental effects.

Coatings are thin layers of materials that are applied to the surfaces of other materials (called the substrates) so that the thin layer material adheres to the substrate. If no adhesion occurs, as when the surface of the substrate material is merely etched or embossed, the process is called *decorating*. This section considers the coating and decorating methods used when the substrate material is a plastic.

Painting

Because of the ease of coloring plastics by simply adding pigment or dyes to the plastic material itself, a logical question to ask is why would someone want to paint the surface of a plastic part. Although the answers may vary widely, the principal reason given is that painting imparts a more aesthetically pleasing and richer color than can be obtained with internal colorants. Another reason is that the plastic might be part of an assembly, perhaps with metal, and the manufacturer wants all the parts to be the same color (such as with automobile exterior panels).

Painting is the most commonly used method to coat plastics. This section treats the case where the paint is a liquid that is applied to the surface of the plastic. A later section discusses the case where the paint is a powder.

Most paints consist of a polymer dissolved in a solvent and solidify by evaporation of the solvent. The solvent can be either organic or water, and both types are in wide use. Other paints are liquid polymers that crosslink when mixed together and may or may not include a solvent.

The plastic is painted along with the metal so that all the exterior parts have the same color intensity and shade. This requirement means that plastics used in the exterior panels of automobiles must be as smooth as the metal parts surrounding them and that their surfaces must accept the paint in much the same way that metals do. Some plastic materials do not meet these requirements and therefore are not used for automotive exterior panels or other applications that require this high level of paint compatibility. For instance, paint does not adhere readily to polyethylene, and so it is not used for these high-performance automotive applications. In general, plastics with a high surface energy do adhere to paints and can be used.

The painting of polyethylene and other plastics resistant to coatings is often a problem. The paintability of these low-energy surfaces can be improved by activating the surface with chemicals, plasma, corona, or flame etching. Adding a primer coat helps protect the activated surface and promotes additional adhesion with the coating.

Liquid Coating.   Spraying is the most versatile and most common method of painting. Spraying can easily be automated and generally gives an excellent surface quality—full coverage, uniform thickness, few runs, and few imperfections. Of course, the technology needed to achieve these results has been developed for many years.

Spray guns usually use air pressure mixed with the liquid stream or hydraulic pressure on the paint itself to create the spray. The distance from the part and the speed of application are important factors to be worked out for each situation. Masking of areas that are not to be painted or are to be painted at a different time (perhaps with a different color) is easily done, although usually with considerable manual labor. Overspray is inherent in the spraying method, and so the process is usually conducted in booths so that the overspray can be confined and the solvent, if any, can be captured and not released to the atmosphere.

Other traditional painting methods include rolling, brushing, or wiping the paint onto the surface of the plastic. Although used less frequently than spraying, these techniques have some advantages, especially for small lots and for special effects that can be created. For instance, a woodgrain effect can be created by using a roller with an engraved surface or by wiping or brushing the surface after a suitable paint or ink has been applied to an already painted surface.

Parts that are very irregular can be painted by dip-coating. This is done by immersing the parts into the paint. Spinning or rotating of the parts after dip-coating is often done to assist in eliminating bubbles and promote uniform drainage of excess paint from the part.

Another common method of painting is *silk-screening*. In this method a series of screens is prepared (assuming that several colors are to be printed on the part), with one screen for each color. Each screen masks a different portion of the overall picture to be painted. The part is passed under each screen in turn as paint is forced (usually with a roller) through the unmasked portion of the screen and onto the surface of the plastic part. Some drying time is needed between each paint application (each screen), but that time can be shortened by using photocurable inks and passing the part under a UV lamp after each screen.

Manufacturers may have problems getting paint to adhere to the plastic. This is usually because the plastic part was not properly prepared prior to coating. The techniques already discussed for ensuring good adhesion also apply to painting. The presence of mold release is an especially troublesome problem in painting. The mold release should be carefully removed before the part is painted.

Powder Coating and Electrostatic Coating

Whereas painting applies materials to the surface of a substrate as liquids, *powder coating* applies a powdered solid coating (usually a plastic) to a substrate. The part to be coated is immersed in a fluidized bed of plastic particles, which adhere to the substrate, or it is sprayed with the powder much as would be done in liquid painting. After the part is thoroughly coated, it is baked at a high-enough temperature to fuse the coating particles but not so hot that the part is damaged. Metals, ceramics, and plastic parts all can be coated using powder coating. With all types of materials, concern should be taken to ensure that the surface of the substrate bonds well to the coating. When the part to be coated is a plastic, special care should be exercised that the fusion step does not cause degradation or softening that will deform the plastic part.

Another method of applying a plastic coating to a substrate using powders is called **electrostatic coating.** In this method the part to be coated is placed inside a chamber and connected to an electrode so that surface charge is developed on the part. The coating material is then sprayed into the chamber, which has been evacuated and into which an electric field has been imposed by placement of an electrode having the opposite charge to the part. The sprayed plastic particles pick up an electrostatic charge from the electric field inside the chamber and coat the plastic parts. By this method the particles are directed toward the part to be coated. If the plastic part does not readily accept a charge, the surface of the plastic can be dipped in a metal solution to give a very light surface coating that attracts the charged particles. Electrostatic coatings are especially useful for applying a very thin coating to a plastic. Overspray is minimal (less than 5%) in most well-run electrostatic coating operations.

Metallizing

In vacuum metallizing a pure metal or metal alloy is evaporated at high temperature in a closed, vacuumized chamber and then allowed to condense on a plastic part in the chamber, which is usually at room temperature. The plastic parts can be coated in batches, often several parts at one time, or in a continuous process that is especially appealing for coating plastic films. Some common parts that are metallized by the batch process include compact discs (CDs), toys (such as plastic cap pistols and race cars), car dash and grill parts, water-fountain buttons, reflective mirrors (such as for headlights), cosmetic cases, furniture, electrical capacitors, holiday decorations, sunglasses, and traffic signs.

Metallized plastic films are a major plastic product that has found many applications, both because of the low permeability of the metal (such as balloons and potato-chip bags) and because of the reflective, electrical, or thermal properties of the metal. Some other specific applications include greenhouse shades, magnetic tape, copying film, photograph film, optical filters, multilayer flexible circuit boards, agricultural films, and thermal protective gear (such as for firefighters and space suits).

Several types of plastic film are commonly metallized, including PET, PE, PP, PVC, PS, nylon, polyimide, and PC, but the most common, by far, is PET. The most common metal used for coating the films is aluminum, although selenium, cadmium, silver, copper, gold, chromium, nickel chromium, palladium, and titanium are also deposited. Even some nonmetals such as silicon monoxide and magnesium fluoride are coated. The deposited coating is typically about 1 to 2 microinches (0.005 to 0.01 μm) thick; because the metal layer is so thin, the metal mimics the surface of the plastic. Typically the coated film has 1,000 times more plastic than metal.

The thickness and uniformity of the metal coating are determined by the rate of deposition of the metal, which is affected by the temperature of the evaporation electrode (how much metal is vaporized), the rate of movement of the film through the chamber, the distance between the metal electrode and the film, the angle between the film and the electrode, and the degree of vacuum in the chamber. The pressure also affects adherence of the film, with lower pressures being better.

A common measure of coating thickness is optical density, which measures the amount of light that passes through the coated film. Therefore, optical scanning can detect differences in coating thickness (uniformity) as well as total amount deposited.

The metallizing process can also be done in solution where the plastic is charged by one electrode and the other electrode carries an opposite charge. The metal ions can be in the solution. With the imposition of current through the electrochemical cell, the metal ions can be made to plate out on the plastic part, a process called **electroplating.**

Printing

Printing, putting ink on a surface, is similar in concept to painting except that the printing does not cover the entire surface. As with painting, printing requires that the surfaces of some plastics be activated for improved adhesion. Polyethylene and polypropylene, for example, both need surface treatment.

Printing can be done either with a roller or pad to mark portions of a plastic surface or with a spray (ink jet), as is often used to mark manufacturing codes or other short identifications. In these applications, the solvent content is kept quite low so that drying occurs rapidly. Applicators for the ink can be timed to the flow of product down a manufacturing line so that the product indexes with the printer.

When a roller or pad is used, two methods are common. The first is simply applying the ink to a raised portion of the surface of the plastic part. The second method is accomplished by transferring the ink from another surface onto the plastic. Generally this transfer method is done by using a master roll on which the desired pattern has been engraved. Ink is then applied to this master roll so that the ink sits on only the raised portions of the roll. Then, a pad is brought against the master roll and the ink is transferred to the pad. The pad is then pressed against the surface of the plastic and the ink again transfers. The nature of the pad (usually a silicone) and the types of ink are important in achieving clean transfers.

Thermostatic Printing

This process involves the use of a special paper that is printed, using special inks, with the image that is to be applied to the plastic surface. The paper is placed on the surface of the plastic part and heat is applied. This heating causes the inks to flow onto the plastic and simultaneously causes the plastic surface to soften, thereby increasing the reception of the ink. Upon cooling, the ink is strongly bonded to the plastic. The printing is relatively unaffected by moderate abrasion, UV light, oils, solvents, detergents, heat, humidity, and stains.

This process has found widespread application in precise printing applications requiring high durability, such as the numbers and letters on computer keyboards, appliance dials and indicators, and sports equipment, such as skis and snowboards. One drawback to the process is that not all plastics can be printed.

Laminating and Related Processes

Films of various types can be coated onto plastic parts by several processes, the most important of which is *laminating*. The coating film, usually a plastic, is brought into a roller assembly into which the material to be coated is also brought. Then with pressure and, usually, heat, the two materials are pressed together such that a bond forms between them and the coating is mated to the part. The heating can be done either at the roller assembly or before.

Laminating is also useful for bonding several layers of plastic material together at one time. When several layers are laminated, some of the layers are almost always heated. The several layers may be useful for making barrier films and packaging materials. These may also be made by coextruding, but that process is not always possible. Laminating allows customized multilayer materials to be made without the high cost of coextrusion.

Laminating has an advantage in that it is done without the use of adhesives. The materials can therefore move rapidly through the laminating process, usually much more quickly than if an adhesive is applied.

Laminating can be used to transfer a film coating from one surface to another surface, a process called *transfer coating*. A film coating is placed onto a substrate on which a release material has been previously applied. The coating adherence to this substrate is slight so care must be taken not to disturb the material by touching it or by mechanically working it. This coated substrate is then carefully rolled and pressed against the substrate to be coated using a lamination technique. When the two substrates are pressed together with the coating between them, the coating leaves the release material and bonds to the new substrate. Some pressure from the laminating rolls usually accomplishes this task, although heating of the new substrate may also be required.

Lamination can also be used to coat a liquid plastic onto a film or sheet substrate. The liquid plastic can be a polymer solution or a thermoset that is not fully cured. In this method a doctor blade is used to spread the plastic liquid material over the surface of the substrate. Then, after the coating is spread, the substrate and the coating are run through laminating rollers to ensure that the liquid coating is uniformly spread across the substrate. If the lamination process includes heating, the layer can be cured at the same time.

If the liquid material is a thermoplastic that is heated and applied to a substrate, the process is called *extrusion coating*. An extruder is placed so that its nozzle applies the resin to the substrate (which is usually a film or sheet). The substrate is then run through lamination rollers to spread the coating. In this case, the laminating rolls may be chilled so that the coating solidifies rapidly.

In all these processes where liquids are coated onto substrates, the lamination process is slowed somewhat when compared to the application of a film onto the surface. The slower speed is required to ensure uniform distribution of the liquid on the substrate.

Many products use lamination. These products include vinyl flooring and countertops (Formica) where the plastic coats a paper substrate.

Decorating

Decorating differs from coating in that only a portion of the substrate surface is covered in decorating. Masking or differentiating the surface of the substrate often is required to limit the region to which the decorating is applied.

Perhaps the simplest decoration method is the application of decals. These printed plastic sheets can be applied to many plastic surfaces, often with only minor heating to bond them. The heat should not, of course, cause distortion of either the decal or the plastic substrate. The printing allows great flexibility in design and use of color. Moreover, with some care, many contoured and even complex surfaces can be decorated with decals. However, some plastics do not bond well to the decals, so surface treatment or heating must be done with great care to achieve a permanent bond.

One very useful decorating technique is *hot stamping*. In this decorating method a film is placed inside a mold (usually a blow mold or a thermoforming mold, although it would work with other processes under restricted conditions). The part is then pressed against the film and the film attaches to the plastic part. If desired, the film can be pressed against only raised portions of the surface, thus giving an appearance that is highly desired by some marketers. Some common applications also include simulated wood grain and holographic images.

When metal foil is coated onto a substrate (which can be a plastic), the process is called *hot-foil stamping*. This process is similar to laminating except that the metal foil is generally not coated over the entire surface of the plastic but, rather, onto raised areas so that a surface design is made. This technique is useful for decorating small bottles.

The primary advantage of hot stamping is the look of the finished decoration. A foil-stamp image can have a clean, metallic look similar to gold leaf. It is the only method where permanent gold and silver metallic graphics can be conveniently produced. Another advantage is that it is a dry process. There are no solvents or wet inks and no smearing, and parts can be handled immediately after decorating. Also, there is minimal training needed for an operator to do the actual transfer. The setup costs are not as high as many of the other printing or coating processes and the cost of the film/foil is fairly moderate. Both thermoplastics and thermosets can be hot stamped, and the foil can be applied to contoured surfaces. Due to the thermal bonding involved in the process, the foil has good abrasion resistance as well.

The primary disadvantage is the limitation in print quality, especially with multicolor jobs. Decal heat transfers look far better and have more application options, including multiple colors, scratch and fade resistance, and the ability to transfer on many substrate surfaces and textures. Hot-stamping equipment is moderately expensive to purchase and there is some downtime while changing foils. Aside from contoured surfaces, complex 3D surfaces are not a hot-stamping specialty.

Texturizing the surface of a plastic part involves forming of a pattern on the surface of the part without molding. The pattern can be either a repeat pattern or a singular pattern. When the pattern is a repeat, it is usually applied with a heated, embossed roller. As the roller moves across the surface of the plastic, it melts the plastic slightly and creates the pattern. Singular patterns are usually stamped onto the plastic surface, although almost any method that melts the plastic surface can be used to create a pattern. After being textured by either method, the plastic surface can be printed (touching only the high parts of the pattern) or otherwise decorated to give emphasis to the texturing.

CASE STUDY 21.1

Comparison of Adhesive-Bonded and Metal Attachments

The aerospace industry has traditionally secured parts together using metal attachments such as rivets, bolts, and screws with their accompanying securing part, such as nuts. These attachments have a long history of success and therefore have also been used to secure plastic or composite parts that are increasingly used in place of the metal parts to save weight and to give other operational and manufacturing advantages. However, the thin nature of many of these plastic and composite aerospace parts makes the use of metal fasteners difficult. The metal fasteners have a tendency to tear out of the hole because of the plastic part's comparatively lower strength. To solve this problem, aerospace manufacturers have experimented with a new type of fastener. This new fastener uses adhesives to bond the fastener to the surface of the plastic and composite part, without the need for a hole to be drilled (see Figure 21.12).

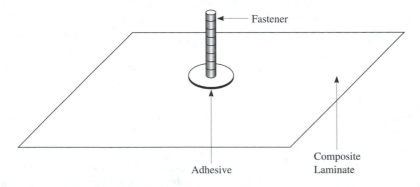

Figure 21.12 Plastic fastening fixture showing the adhesive bonding surface and the insert.

The need for a strong fastener that does not require holes being made in the substrate has been made critical by the development of composite materials for aerospace. These aerospace materials are designed for the lowest possible weight while still maintaining the strength and stiffness required in tension. This optimization of these specific properties has often resulted in parts that are very thin laminates of thermoset resin (usually epoxy) and carbon fibers. These laminates have good strength and stiffness in the plane of the laminate (commonly called the x-y plane) but very little strength and stiffness perpendicular to the plane (the z direction). This low stiffness and strength perpendicular to the plane of the laminate leads to weakness around fasteners, especially a tendency to tear around the head of the fastener. This tendency is especially strong when a hole has been made in the laminate because the hole serves as a starting point for the tearing.

The existence of holes can also result in thicker and heavier structures because the strength of the laminate in the x-y plane is compromised by the existence of a hole. The hole serves as a starting point for tensile failure, just as it does for shear failure. To compensate, the laminate must be made thicker and stronger.

Another potentially serious problem with holes in the structure is corrosion. Penetrating fasteners can have an electrolytic couple with the laminate material, thus producing corrosion at the fastener site. Although carbon-fiber epoxy is not as severely corroded as metals, some potential does exist for corrosion with, for example, aluminum. This corrosion problem can be solved by coating the fastener with a plastic or by using a noncarbon-fiber reinforcement in the vicinity of the hole. Both these solutions are costly and the coating is only temporary since it can wear off.

The extra time required to make (drill) the holes may also increase the cost of the assembly. Drilling in composite materials is difficult because of the widely different nature of the two materials (matrix and reinforcement). Therefore, eliminating the need to drill holes produces a significant savings in labor and material potentially lost because of poorly drilled holes or backface damage.

Many types of adhesive fasteners are available to remedy the problem with penetrating fasteners. These can have threaded inserts, rivetless nutplates, and tie-wrap mounts, just to name a few. The one feature that seems to be common in all these adhesive fasteners is the baseplate. The baseplate provides enough surface area to allow for effective adhesive bonding. Baseplate size, which determines bond strength, can be increased to satisfy the load requirement. Standard sizes range from 1.6 cm to 6.0 cm in diameter and provide as much as 20 cm^2 of surface area.

The fasteners can be made from a variety of materials. Aluminum, titanium, stainless steel, composites, and plastics are common. The baseplates and the attachment device can be made of different materials to meet specific mechanical and physical property requirements.

Whenever bonding is planned, adhesive selection becomes a primary concern. The chemical industry provides a multitude of adhesives for any given application. Two-part, medium-viscosity, room-temperature-cure adhesives often are the most successful for aerospace applications because of their user-friendly properties. Two properties rank as the most desirable. The first is a 1:1 mix ratio of two different color components that combine to give a third distinct color, signifying a complete mix. The other is a quick cure for rapid-handling strength. The adhesive must possess a pot life sufficient to allow time for

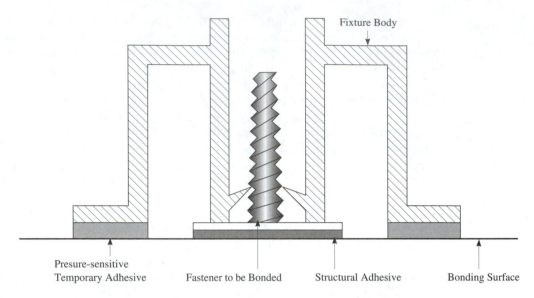

Fixture Body

Presure-sensitive
Temporary Adhesive   Fastener to be Bonded   Structural Adhesive   Bonding Surface

Figure 21.13   Temporary installation fixture for securing adhesive fasteners.

completing the application without rushing. There is typically a trade-off between quick cure and long pot life.

Epoxies, urethanes, polysulfides, and acrylics are adhesives that have been successfully proven in structural applications in the aerospace industry. Ultimately the choice of adhesive depends on the performance characteristics necessary to make an effective attachment. Characteristics such as installation environment, end-use environment, cure time, pot life, application equipment, desired strength level, optimal failure mode, and cost should be considered when identifying an appropriate adhesive.

When applying adhesive-bonded fasteners, surface preparation becomes the dominant concern in making a successful joint. The amount of adhesive required for a reliable bond depends on the material being bonded and the adhesive used. From a manufacturing standpoint, it is desirable to perform the least amount of surface preparation possible while still creating a reliable bond.

There are standard actions that always will enhance bonding regardless of the adhesive used. A good solvent wipe of the substrate and the fastener base is considered to be the minimum surface preparation required. Abrasion of the surfaces with a scuff pad, sandpaper, or grit blasting, prior to solvent wiping, provides a surface condition that is generally accepted as excellent for successful adhesive bonding.

To make installation of adhesive-bonded fasteners quick, easy, and less expensive than traditional fastening, a disposable installation fixture is employed. The fixture, which is held in place by a temporary pressure-sensitive adhesive, accurately locates the fastener and provides a constant down force on it. The action squeezes out the adhesive to a thickness consistent from fastener to fastener, while holding the fastener in place until the adhesive reaches handling strength. A typical installation fixture is pictured in Figure 21.13.

The strengths of the adhesively bonded fasteners have proven to be acceptable for many applications, especially light loads or where shear is likely to be a major factor acting on the fastener. The surface area of the fastener base can be increased to handle the high shear loads.

SUMMARY

Finishing and assembly are important in many applications because the plastic part that has been molded may still need some work done on it to make it suitable for use or for inclusion in some assembly. The most common finishing step is the removal of the runner system and flash. These excess materials often are part of the molding process and their routine removal is common. In most cases the excess material can be mechanically removed quite easily.

Machining (cutting, drilling, turning, and so on) is generally to be avoided if possible because of the additional cost of performing these operations. When machining cannot be avoided, the major consideration is reduction of heat so that the plastic material does not melt or become distorted. Plastics melt at lower temperatures than metals and have relatively poor thermal conductivity, making heat problems potentially severe. Consequently, machining should be done with small feed increments. Methods to conduct away the heat, such as interrupted teeth on saw blades, peck drilling, and coolants, should be considered.

Nonmechanical machining has proven successful, but some machines that do this are expensive. The most common nonmechanical machining methods are water jet cutting, laser cutting, and hot wire. The quality of the nonmechanical machining can be as good as or superior to traditional machining methods.

Because plastics often are flexible, nonbonded assembly is widely used. Snap fits and other latching methods take advantage of the ability of plastics to be deformed slightly and to return to their original shape and position. This flexibility allows the use of plastics as living or integral hinges, at least in some plastics where stress cracking is not a problem.

Many plastics can be joined readily by adhesives. These materials have high surface energies that are readily wetted by the adhesive. To obtain the best adhesive bond possible, the surfaces of the materials to be joined should be prepared. The preparation usually includes a solvent wipe (cleaning) and may also include some abrasion. Plastics that are not wetted by the adhesive may require additional surface preparation such as chemical, flame, corona, or plasma etching.

Plastics can also be joined directly, without the use of mechanical fasteners or adhesives. Fusion bonding is used to join plastics as they are heated and melted slightly on their surfaces. In addition to the normal heating that is used in fusion bonding, the heat can be created by spinning or by vibrating two parts against each other. Ultrasonic and rf welding are much the same except that ultrasonic or rf impulses are used to melt the surfaces prior to joining. Induction welding uses the heat generated by a metal filler in the plastic moving through a magnetic field to heat the plastic material.

Plastics are coated for both decoration and protection. The coatings can be applied as a solvent spray (paint), as plastic particles in a fluidized bed, or as films or melts that are pressed onto the substrate to be coated. Although not very conductive, plastics can also be coated by electroplating of metals onto the surface.

GLOSSARY

Adhesive failure   Failure that occurs between two different materials at their bonding interface.

Chip   The debris that is cut away during a machining operation.

Cohesive failure   Failure that occurs within a single material.

Corona   Treatment of a material with an electric spark in order to increase the bondability of the surface.

Coupling agents   Chemicals that attach to a surface and then to the coating to be put onto the surface, thus linking the coating and the surface.

Cryogenic flash removal   A process in which the parts are cooled, often with liquid air, and then tumbled or otherwise subjected to some mechanical impact to remove flash.

Dissipation factor   A property of a material in which the molecules of the material attempt to align with an electric field, especially an alternating field, and thus create internal heating in the material; also called loss rate.

Electroplating   The coating of a plastic with a metal, wherein the metal is in a solution and the plastic is charged opposite to the metal so that the metal will adhere.

Electrostatic coating   A powder-coating technique in which the powder is applied to the surface by charging the surface with an electric charge that is opposite of that applied to the powder.

Energy director   A small protrusion of plastic material on a surface that is to be ultrasonically bonded that provides a place for the energy to concentrate, which results in a melting of the energy director and greater energy efficiency in the bonding operation.

Extrusion coating   Application of a coating material by extruding the coating onto a roller and then laminating the melt onto the substrate.

Friction welding   Another name for spin welding.

Fusion bonding   A joining method in which the adhesive and the adherend molecules are thermally softened and then pressed together, thus causing the molecules to intermingle across the boundary.

Hot-foil stamping   A process of applying a metal decoration to a plastic part by pressing the metal against the hot plastic.

Induction welding   A joining technique in which the energy to melt the surface of the material to be joined is supplied from a magnetic field that interacts with the metal particles imbedded in the surface or located near the surface; also called electromagnetic heating.

Kerf   The width of the material removed due to the cutting tool (such as a saw blade).

Laminating   The application of a plastic film to a substrate.

Peck drilling   Alternating advancement and retraction of a drill during a drilling operation to allow for cooling of the part and removal of the chips.

Plate bonding   Fusion bonding in which a plate is used to heat the materials.

Powder coating   Coating a material by covering the surface with a powder and then fusing the powder into a continuous film.

Primer   A coating that is applied under paint to promote adhesion.

Radio frequency (rf) welding A joining process in which the energy from an rf generator is used to melt a plastic, which then fuses with another material to form a bond.

Rake angle The angle between the cutting surface of a tool and the radial line of the tool.

Rubber cement A common name for several types of pressure-sensitive adhesives.

Skip-tooth cutters Saw blades and other continuous cutting tools that have skips in the teeth to allow for cooling of the part being cut.

Snap joint A joining system that employs the natural ability of a plastic material to deform slightly and to snap back (recover elastically) to its original position; also called interference joints.

Spin welding A joining technique in which the surfaces of the materials to be joined are softened or melted by the friction created between the surfaces as one material is spun against the other.

Staking A joining method in which ultrasonic energy is applied to a surface, which is thereby softened and then mechanically deformed to overlap another surface and create a mechanical fastening, somewhat like riveting.

Transfer coating A method of coating in which a film material is placed on a roller (which is coated with a release material) and then transferred to another surface.

Ultrasonic welding Adhesion caused by application of ultrasonic energy, which causes the molecules of the adherend and/or the adhesive to vibrate and fuse together.

Vacuum metallizing A technique in which a metal is coated onto the surface of a plastic by passing the plastic through a vacuum chamber in which atomized metal particles have been created.

QUESTIONS

1. Identify five factors that can be varied to optimize the machining of plastic parts.
2. Identify four advantages of laser cutting plastics over traditional cutting (machining) methods.
3. Discuss the effect of plastic brittleness in snap-fit joints.
4. Discuss the value of a primer coat when painting a plastic part.
5. Name and contrast four treatment methods for increasing the adhesive capability of a plastic surface.
6. Explain staking, and explain the effect of its use with large-diameter posts in large-diameter holes.
7. Identify three methods for coating a plastic with a metal.
8. How can a two-part adhesive be monitored to ensure proper mixing of the two components?
9. What advantage does a two-part adhesive system have over a one-part system?
10. What is the curing agent (that is, the material that initiates the reaction) for cyano-acrylates?

REFERENCES

Adhesives, Vol 3, Engineering Materials Handbook, Metals Park, OH: ASM, International.

Baird, Ronald J., and David T. Baird, *Industrial Plastics,* South Holland, IL: The Goodheart-Willcox Co., Inc., 1986.

Chanda, Manas, and Salil K. Roy, *Plastics Technology Handbook,* New York: Marcel Dekker, Inc., 1987.

Frados, Joel (ed.), *Plastics Engineering Handbook* (5th ed.), Florence, KY: International Thomson Publishing, 1994.

Milby, Robert V., *Plastics Technology,* New York: McGraw-Hill Book Co., 1973.

"Plexiglas® Fabrication Manual," AtoHaas North America Inc., PLA-21b, Dec 1992.

"Plexiglas® Forming Manual," AtoHaas North America Inc., PLA-53b, Jan 1993.

Richardson, Terry L., *Industrial Plastics* (2nd ed.), Albany, NY: Delmar Publishers, Inc., 1989.

Strong, A. Brent, "Composites in Manufacturing: Case Studies," Dearborn, MI: Society of Manufacturing Engineers (SME), 1991.

CHAPTER TWENTY-TWO

ENVIRONMENTAL ASPECTS OF PLASTICS

CHAPTER OVERVIEW

This chapter examines the following concepts:

- Source reduction
- Recycling (collection, handling/sorting, reclamation/cleaning, end uses)
- Regeneration
- Degradation
- Landfill
- Incineration
- Total product life cycle
- Future

INTRODUCTION

The presence of humanity has had a strong impact on the world environment throughout history, and this impact seems to be increasing with our modern society. The impact is most commonly seen in the many wastes created by society. Gaseous wastes pollute the air and contribute to increased carbon dioxide in the atmosphere, which is associated with the "greenhouse effect" and is purported to be increasing the temperature of the world. Air pollution also is a major source of acid rain, which is destroying forests. In addition, animals suffer from lung disorders and other diseases caused by air pollution. Liquid wastes pollute streams, rivers, and groundwater, causing fish to die and other animals to build up toxins in their bodies. Solid wastes clutter the roadside, beaches, and oceans, and accumulate in an ever-increasing number of landfills. These problems have often been blamed on technology. However, just as technology has led to increased industrialization and the extent of human impact on the environment, technology can be the tool to reduce the negative impact.

In particular, this chapter examines environmental aspects of plastics. Although air and liquid pollution have important consequences for the environment, the primary focus of this chapter is on solid waste pollution.

Plastics have become common materials in everyday life and, along with other materials such as paper, are often used in disposable (one-time use) applications that are a major contributor to solid waste. While the use of plastics in disposables is still much less than paper-based products, the wide use and growth of plastics in these applications elevates concern about plastics as a serious pollution problem. When not disposed of properly, the plastic materials are widely seen and often criticized, in part because of their long life and obviousness. The disposal problem is not simply technical, but has important social, economic, and political aspects. All of these aspects should be brought together to work on finding the most intelligent method of using plastics as well as other materials.

Plastic materials can be used intelligently for many applications, but are not appropriate for other applications. While the initial choice of material for each application is typically made on a technical basis, economic, social, and political considerations are frequently important inputs in the decision. Social and political considerations, however, should be based on well-reasoned arguments, hopefully with sufficient data that the conclusions are valid technically as well. Too often these social and political considerations are based on emotion, short-term views, and a lack of overall understanding. This text does not purport to be the authoritative source for solving waste or pollution problems, but an attempt is made to present a broad perspective and to point out both the benefits and problems associated with plastics and their relative effect on the environment.

This chapter begins by examining several different options that have been identified and tried for solving the problems associated with solid wastes handling and disposal: (1) source reduction, (2) recycling, (3) regeneration, (4) degradation, (5) landfill, and (6) incineration. This section includes a discussion of the problems and benefits associated with the use of plastics in each option. The chapter then turns to an evaluation of the proper material from a total life-cycle basis. This basis includes the environmental and economic costs associated with the product from its inception to its final disposition, rather than just the costs associated with the original manufacture of the product as has been common in the past. Finally, the chapter considers what the future might bring for environmental protection and the use of plastics.

SOURCE REDUCTION

Source reduction refers to a reduction in the amount of material that is used in any application and, therefore, a reduction in the amount of material potentially discarded when that use is completed. The simplest methods to employ source reduction are to use fewer products that cause waste, to choose sizes and types of products whereby the waste is minimized, and for manufacturers to reduce the material requirements of the product. For instance, the amount of packaging material in a 16-ounce (1-L) soda can is 40% less than the material in two 8-ounce (0.5-L) cans. Larger sizes are almost always more efficient in using materials. Additionally, the large sizes are often more economical for the consumer.

Another source reduction method is to decrease the thickness (amount) of the material in the application. Trash bags are a good example of how this principle works. When polyethylene trash bags were first introduced, the bags were generally about 0.003 inches or 3 mils (0.08 mm) thick. These bags were made of traditional low-density polyethylene (LDPE). With improvements in the process, the thickness was reduced to about 2 mils (0.05 mm). Then, with the introduction of linear low-density polyethylene (LLDPE), this inherently stronger and tougher material allowed the thickness to be reduced to 1 mil (0.025 mm). The use of coextrusion and high-density polyethylene (HDPE) in these applications has further reduced the thickness so that some trash bags of only 0.7 mils (0.017 mm) are now available. These data mean that the original bags were more than 400% thicker than the equivalent bag made using current technology.

The reduction in thickness (weight) of trash bags is only one of many examples of the reduction of weight in plastic products through increased material and manufacturing capability. For instance, 1-gallon (4-L) milk jugs have dropped from typical weights of 0.25 pound (95 g) to 0.15 pound (60 g) over the past few years.

These reductions in product thickness have principally been driven by economic factors. (Competition between part manufacturers and various resin producers.) This competition has been beneficial for the environment, however, because the bulk and weight of the trash bags and milk jugs now being used are about half of what they were just a few years ago, without any significant decrease in performance.

Suppliers of other materials such as metals and paper have also reduced the thickness and weight of their products to help alleviate the solid waste problem. However, at this time, plastics technology and the development of new plastics seems to be a more fruitful field for overall source reduction than other materials that have been in use considerably longer.

Substituting plastics for other materials has also been a method of achieving source reduction. An excellent example of this reduction is the widespread substitution of plastic grocery sacks for paper sacks. As shown in Figure 22.1, the volume (height) of 1,000 paper sacks is over 13 times greater than the same number of plastic grocery sacks. The weight comparison is 140 pounds (63.5 kg) for the 1,000 paper sacks and only 15.6 pounds (7.0 kg) for the plastic sacks. Secondary implications such as freight, fuel, storage, and so on make the argument in favor of plastic bags even more compelling.

Plastics obviously have a volume and weight advantage over many other packaging materials. In fact, Germany's Society for Research into the Packaging Market estimates that without plastics, the volume of currently used packaging materials would increase by 250% and the cost of packaging would more than double. The substitution of plastics for other materials raises the obvious question of whether that substitution brings problems for the environment that might be more serious than the weight or volume reduction. Three potential problems seem obvious. The first is difficulties that might arise from the recycling of plastics. The second is the resistance of most plastics to most forms of natural degradation. The third is potentially harmful off-gases when plastics are incinerated. These problems are discussed in later sections since they are aspects of the overall challenge of handling solid waste.

Figure 22.1   Comparison of the volume (height) and weight of 1,000 grocery sacks.

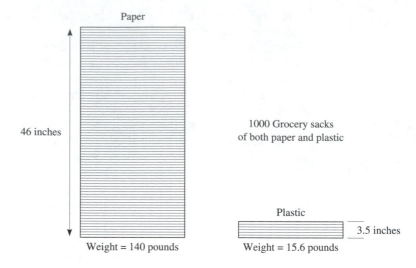

Paper

46 inches

1000 Grocery sacks
of both paper and plastic

Plastic

3.5 inches

Weight = 140 pounds   Weight = 15.6 pounds

RECYCLING OF PLASTICS

Recycling refers to the reprocessing and refabrication of a material that has been used and discarded by a consumer and that otherwise would be destined for disposal as solid waste. This type of recycle is called *postconsumer recycle (PCR)*, as opposed to recycle that is created as a normal part of the scrap from a manufacturing process (generally called *regrind*.) Regrind is normally consumed (reused) within a manufacturing process and so is not part of the solid waste problem.

The reprocessing and refabrication of PCR materials into useful products requires several steps: collection, handling/sorting, reclamation/cleaning, and end-use fabrication. Each of these steps is costly and so recycling is not as inherently inviting, either economically or from an environmental impact viewpoint, as is source reduction. Still, even with tremendous steps being taken in source reduction, materials will continue to be discarded and recycling will continue to be important as a means of reducing the amount of material in landfills and other waste streams.

The amounts of all materials being recycled continue to increase. In the United States 24% of all waste was recycled in 1990, up 10% in just five years. Some materials, like aluminum soda cans, have been recycled for many years and a high percentage (over 50%) of the total are recycled. Paper, glass, and plastics, too, are beginning to be recycled more extensively. Reasonable estimates are that between 8% and 17% of plastic products are recycled. Plastics recycling principally involves two products—PET soda bottles and HDPE milk jugs. These two products represent 26% of all waste packaging and (since packaging is 30% of all waste) they are approximately 8% of the total waste stream. Because PET and HDPE bottles are as convenient to recycle as aluminum cans, their recycle percentages of total production, 57% of HDPE bottles and 31% of PET bottles, already approach those of aluminum cans.

Collection

The favorable recycling percentages for aluminum cans, HDPE jugs, and PET soda bottles suggest that consumer convenience is an important part of recycling success. These plastics products are easy for the consumer to identify and place in separate recycling bins provided by the waste collector or to save in their own bin and take to a recycling center. In general, voluntary recycling by the consumer has proven to be the most important single factor in improving recycling of all materials.

An incentive for consumer collection, sorting, and conveyance to a recycling center is the fee paid by aluminum companies when they buy back the cans. This fee is economically justified because of the high cost (especially the high energy requirement) of producing new aluminum (from ore) versus the cost of reprocessing aluminum scrap. For many plastics, the cost of the virgin plastic material is about the same as the costs involved in recycling and so little or no fee can be economically justified for buying scrap plastics. One exception is PET soda bottles. The costs of collecting and bulk bundling PET is estimated to be 7 cents/pound and the cost of reprocessing PET into a clean and usable resin form is about 30 cents/pound, for a total recycling cost of 37 cents/pound. New resin is approximately 55 cents/pound, giving a recycling advantage of 18 cents/pound.

Municipalities have realized that recycling economics are not as favorable for plastics as for aluminum and so many municipalities have imposed a fee on the consumer (typically about 7¢ in the United States) that can be reclaimed when the plastic bottle is returned. This fee is intended to give the consumer an incentive to collect and convey PET bottles to recycling centers. The fee can be used to offset some of the costs of recycling. In some locations, machines accept the recycled plastic products (usually bottles) and pay the consumer the fee.

Generally, however, consumers do not sort their solid wastes but rather mix all materials together. This means that municipalities must handle and sort the products if recycling is to be done. Some municipalities have mandated that consumers sort their solid wastes by type of material, such as metals, plastics, newspapers, glass, and mixed. These are either picked up on separate collection runs or picked up in separate bins and kept separated in the collection trucks. The costs of separate collections are considerably higher than those of traditional mixed collections.

Handling/Sorting

All recycled materials must be handled. Sometimes the handling involves simply conveying the materials from the pickup point (usually the consumer's house or job location, but occasionally the recycle center) to the reclamation facility. If economic recycling is to be done, some sorting of the materials is also necessary. The maximum economic benefit is obtained when each recycled material is sorted by specific product type—aluminum cans, PET soda bottles, HDPE milk jugs, and so on. In some cases, slightly less restrictive separations are also economically viable. Some broader groupings are all aluminum alloys, all PET bottles and other PET waste, and all HDPE waste. Even broader material groupings (such as all metals, all plastics, and so on) usually require further separations in order to be viable. One exception is the case when all plastic parts are grouped. Under certain conditions these collections of several types of plastics can be recycled. The mixture of several plastic types is called *mixed recycle* or *comingled recycle*.

The handling and sorting steps in the recycling process can be relatively simple or relatively difficult and costly, depending on the degree of consumer participation and the level of sorting required. If consumers carefully sort the recyclables by specific product type (aluminum cans, PET bottles, HDPE milk jugs, newspapers, and so on) and convey them to the recycling center, then the handling consists simply of bundling the products into some convenient shipping form (such as crushing the bottles and pressing them together into a bale) and then conveying the materials to the reclamation location. However, if the materials must be separated from a general waste stream as would be created if consumers did no sorting, considerable labor is required to pick out the recyclables by hand. Even if consumers have separated the materials into general material groups (such as metals, plastics, and paper products), some additional sorting at the collection point is required. For the highest economic benefit, the aluminum must be separated from the iron and steel, the HDPE and PET and other recyclable plastics must be separated from the plastics that are not to be recycled, and so on.

In the case of metals, some separation by metal family can be done by machine because iron and steel are magnetic. Aluminum cannot readily be separated from other metals such as zinc and brass, but generally at this stage, aluminum cans simply are separated by sight and the rest of the metal is sent to a metal reclaimer.

Some sorting of plastics can also be done by machine. These machines rely on inherent characteristics of the various plastic resins that distinguish one resin from the others. For instance, the light absorption of a resin can be used because each resin type has a different absorption pattern. A machine has been developed that shines light through the plastics and detects the pattern of light absorbed (or transmitted). This detection then triggers a separation mechanism to route the plastic part in a particular direction along the conveying system according to the detected pattern. This process has been demonstrated in test runs but is not as yet widely used.

The most widely used method of sorting plastics is by sight. Some plastic products, such as PET soda bottles and HDPE jugs, are easy to identify. Others, however, are much more difficult to distinguish, especially when they are not easily identified by shape. To assist both consumers and sorters at the collection facility, a system was developed by the Society for the Plastic Industry (SPI) for identifying the resin type or family. A number and a recycling symbol have been assigned to each of the major resin types and that number is molded onto each plastic product (usually on the bottom). The major resin types and their numbers are given in Table 22.1. For instance, PET soda bottles and other PET waste is abbreviated PETE and given the number 1. HDPE is number 2, and polyvinyl chloride is number 3 (V). (PETE is used instead of PET and V is used instead of PVC because of possible trademark infringements if the more common abbreviations were used.)

Even with the several special categories for the most common plastics, many plastic types are assigned to the "other" category (7). These "other" products present special processing problems and must be handled as comingled recycle. Plastics that are unknown or that have several different types of plastic in them are classed as "other." The large

Table 22.1 Numbering System for Plastics Recycling

Plastic Type	Recycling Symbol	Recycling Number
Polyethylene terephthalate	PETE	1
High-density polyethylene	HDPE	2
Polyvinyl chloride	V	3
Low-density polyethylene	LDPE	4
Polypropylene	PP	5
Polystyrene	PS	6
Other	OTHER	7

number of plastics that fall into this "other" category is a problem in the economics of recycling.

The many coextruded plastics and multilayered plastics are all "other." Sometimes these multilayered materials are needed for proper product performance. For instance, small lunchbox drink boxes are made of six layers, each with a particular purpose. The outer layer is polyethylene, which serves as a water-resistant coating to keep the package dry and structurally sound. The second layer is paper, which provides stiffness, strength, and shape. The third layer is polyethylene, which bonds the second and fourth layers together. The fourth layer is aluminum foil, which forms a barrier against light and oxygen and prevents spoilage or taste change. The fifth layer is another polyethylene adhesive. The sixth layer is a polyethylene film, which provides an inert food contact surface that prevents spoilage, contamination, or leakage.

We can see, therefore, that some plastics are easy to sort by shape, number, or physical characteristic and can be reformed into the original application, just as is done with aluminum cans. The relatively pure (one type of plastic resin) materials can also be used to make products that are not the same as the original but are nevertheless quite valuable. For instance, PET soda bottles can be recycled into fibers, strapping, and reinforcement for concrete; HDPE milk bottles can be recycled into flowerpots, pipe, toys, trash cans, soft-drink-bottle carriers, pails, and drums; and vinyl can be recycled into drainpipe, vinyl floor tile, outdoor furniture, and truck bed liners.

Reclamation/Cleaning

After the recycle materials have been sorted, they must be chopped or shredded so that they can be further processed. The plastic materials are usually chopped into small flakes about the size of a fingernail. These flakes are then treated with solvents and washed to remove any residual contaminants (such as the original contents and paper labels).

The mixed or comingled recycle materials are usually sent to the fabricators as cleaned flakes. The carefully sorted recycle materials are often extruded into pellets so that the refabricator will be able to use the traditional processing equipment for making the finished parts.

A final separation step is sometimes done with well-sorted flakes. After the flakes have been cleaned, the flakes are introduced into a water bath. The flakes that have a density lower than water (chiefly polyolefins) float, whereas the heavier plastics sink. This method is used to separate PET bottles from the HDPE bottom caps used by some molders to give the soda bottles a stronger and more stable bottom. The HDPE has a specific gravity of about 0.96 and so it floats after agitation. The PET has a specific gravity of about 1.2 and so it sinks.

End Uses—Sorted PCR

The ideal use for recycled material is the same application that was the original use or some other high-value application. For many plastics that have been carefully sorted, the reuse in the original application is perfectly acceptable. Examples include recycled LDPE for new bags and films, recycled PS in insulation and delicate instrument packaging, PP in auto parts and industrial fibers, and vinyl in detergent bottles and pipe. These examples are just a few of the hundreds of products that are and will be made from recycled materials that are reformed into their original uses. Because thermoplastic materials can generally be reheated and reprocessed many times without significant degradation, reusing these materials in almost any application is possible.

For most plastic resins, however, some minor changes in resin properties are to be expected. A typical change is a reduction in molecular weight with an accompanying drop in resin melt viscosity and an increase in the melt index. However, these changes are minor and probably reflect random molecular scission rather than a widespread phenomenon throughout the melt. This conclusion arises from the apparent retention of many physical and mechanical properties such as yield strength and elongation.

Some fabricators of products made from postconsumer recycle have noted that the small changes in molecular weight and other properties cause processing difficulties. Finely tuned injection molding operations find that the optimum cycles are altered and sometimes the filling of the mold is much more difficult. Rotational molders also have found that postconsumer recycle can alter their cycles and cause disruption in the uniformity of their products. Blow molders generally have found fewer problems in their processing but have found some limitations on the use of PCR. These molders are able to use PCR by blending it with virgin material. Blend ratios of 20% PCR are typical, although some have found that up to 50% PCR can be used without significant changes in processing or product performance.

However, recyclers and molders of PCR plastics are reluctant to reuse the recycled materials in medical and food-contacting applications because of the danger of contamination and disease. Therefore, PET from soda bottles typically is not reused in bottles but is made into nonfood-contacting applications such as carpets, textile fibers, and fiberfill for sleeping bags and winter coats. HDPE from milk jugs is reused in nonfood applications such as motor oil bottles, recycling bins, trash carts, and laundry soap packages.

End Uses—Comingled PCR

This category of materials includes all of the plastic materials that are normally included in the "other" category in the recycling separation system. These "other" plastics include several types that should be considered separately: thermoplastic materials that cannot be separated into a single resin type or family, thermoset materials, and composite materials.

Thermoplastic materials are the most abundant of the three "other" types. Some success has been found in recycling these materials, but they cannot be recycled back into their original products because they usually come from many different products. Moreover, they cannot be processed as a single material because a single batch of comingled materials likely contains materials with varying melting points and processing characteristics. Therefore, different processing methods are needed for comingled materials. New products are also required.

The value of products made with comingled recycle will undoubtedly be less than the value of the original products made with virgin materials. The focus on economics should therefore shift to a comparison with products made from nonplastic sources. The objective should be to find applications and value where products made of generic plastics materials have inherent advantages over other materials. Several of these types of products have been identified.

The most common product made from comingled PCR is plastic wood. Plastic wood has its highest value in applications where natural wood rapidly decays or rots. These applications have high maintenance and would, therefore, justify a higher price if the maintenance were eliminated as it would be if plastic were used. Another advantage of the plastic wood is the absence of the highly toxic antifungal and antirot additives that must be pressurized into natural wood to allow them to be used in the high rot- and degradation-sensitive applications. Wood docks and pilings are submerged in seawater and have high maintenance problems, so they are prime target markets for plastic wood. Other applications include park benches and tables, signposts, playground equipment, and landscape timbers when weathering is a problem, especially when access is difficult. Some manufacturers have had success in making plastic wood for home fences, simply because of the consumer's desire for low maintenance. Plastic wood splinters less than natural wood and is therefore valuable as a decking material and has shown promise in playground equipment and reusable pallets.

The plastic wood is generally produced by pressing the comingled flakes together at a temperature sufficient to melt the majority of the flakes. These flakes then bind the non-melted flakes together. Color additives mask the multicolored appearance of most comingled PCR. The process can be designed like ram extrusion which is inexpensive and well suited to making single parts that do not have to be melted together (as would be the case if they were extruded). Alternately, an adiabatic (low thermal input and high mechanical heating) extruder with little filtration and a very wide die orifice can be used to melt most of the plastic and form it into wood-shaped pieces.

Plastic wood is not without its problems, however. The dissimilar hygroscopic characteristics of some components (for instance, nylon versus PE) cause differential water absorption and dimensional changes. This problem can be reduced by controlling the source streams and ensuring that highly hygroscopic materials are not mixed with hydrophobic materials. Plastic lumber falls short as a structural material compared to natural wood. This can be designed around by using thicker and heavier plastic wood beams or by placing some natural wood in critical load-bearing locations.

The most important drawback to plastic wood is its higher cost (about 20%) compared to untreated lumber. Therefore, its major uses continue to be in applications where treated lumber would be used, especially those in which an additional cost for maintenance would be high.

Some researchers have demonstrated that a woodlike product can be created by combining comingled PCR with waste wood fibers. The materials are mixed in a twin-screw extruder (wide clearance) and then formed by a die.

Another technique for avoiding the processing problems with comingled PCR is to shear the flakes with very high-intensity shearing and simultaneously remove the heat. This can be done in special twin-screw extruders. Processors have found that the resulting granule materials can be processed much like a virgin resin.

Mixed recycle material has also been used for applications where the principal value is the lighter weight of the plastic versus a traditional product. One example is parking stops, which typically are made of concrete. These products are there simply to take up space and so almost any material that can withstand the environment will work. These applications are the lowest valued and should be pursued only to use excess capacity in the recycling operations.

Thermoset materials can also be part of the "other" category in the plastics recycling system. Most people consider thermosets to be nonrecyclable because they cannot be remelted and therefore cannot be reformed into their original product. However, as already discussed, many materials are recycled into products other than their original form and, moreover, some are recycled without ever being melted. The possibility exists, therefore, for recycling of thermoset plastics.

This type of recycling has been demonstrated using several different types of thermoset materials. In these applications the thermosets are ground into powders and then used as fillers in other products. For instance, ground rubber tires are used successfully in polyolefin blends and crosslinked PE is used with virgin HDPE to make tote bins, wheels for carts, trays, and pallets. The mechanical properties of the products with thermoset fillers are only slightly less than or superior to the properties of virgin materials. These results indicate that the thermoset particles bond with the virgin to create a stronger material then would have been created if nonplastic fillers had been used. Alternatively, in some cases the filler adds toughness or some other beneficial property to the virgin material.

Composite materials can also be recycled even though composites are often made from a thermoset resin and have the added complication of a reinforcing material. However, parts made from sheet molding compound (SMC) and bulk molding compound (BMC), which are made of thermoset polyester with glass reinforcement and inorganic fillers, have been recycled by grinding the parts and then using the resultant granules as fillers in other SMC/BMC mixes or in polyolefin molding compounds. The presence of the reinforcement does not seem to present any special problems except that the grinding of the cured material is somewhat more difficult than would be the case with nonreinforced thermosets.

The properties of the filled parts are generally the same as the conventional molding compound, provided the amount of recycle is 20% or less. Problems that have not as yet been fully resolved include processing materials with paint or adhesive contaminants, control of moisture in the granulated material, segregation of various SMC/BMC formulations to provide material consistency, and development of viable SMC/BMC collection and recycled product marketing scenarios.

Other composite products, such as advanced composites and nonpolyester glass-reinforced plastics, are recycled by grinding the composites and then using the granules in molding compounds. These molding compounds are processed by compression molding and by injection molding.

REGENERATION

Regeneration is the process of breaking down the polymer molecules in the plastic material into more basic chemicals. These basic chemicals can then be used to create new polymers or other usable materials. Regeneration can be called *chemical recycling* because it deals with plastics at a chemical level rather than at a polymeric level.

The easiest polymers to regenerate are the condensation polymers such as PET and nylon. With the addition of heat and pressure in the presence of a reactive agent, the polymers depolymerize (commonly called *unzipping*) and regenerate the monomers. Addition polymers such as PS, PE, and PP are more difficult to depolymerize. Regeneration of these polymers requires much higher temperatures than condensation polymers and generally is done in a facility much like an oil refinery, where *cracking* is used to break apart large crude oil molecules into simpler forms. Thermosets are the most difficult of all to regenerate and typically are done by *pyrolysis,* which is the decomposition of a material using heat processes in the absence of oxygen.

The advantage of regeneration over conventional recycling is that regeneration may be more effective for mixed recycle. Each of the fractions (condensation polymers, addition polymers, and thermosets) may be selectively regenerated. This allows a much higher value for the waste than would be obtained from conversion to a product that does not have a high intrinsic value.

One problem with regeneration, when compared to traditional recycling, is the greater potential for air and water pollution. The regenerated monomers are liquids or gases that can be pollutants. The processes themselves also create by-products that are potentially polluting. A further problem is that considerable thermal and chemical energy was already expended to get the monomers into the polymer form. Regeneration requires additional thermal and chemical energy to break them apart and then requires more energy to put them back together.

Even with these difficulties, several large resin manufacturers are exploring ways to regenerate. They view the enormous backlog of landfill plastics as potential feedstock that could be valuable if crude oil sources become scarce because of war, catastrophe, or simply dwindling supplies.

DEGRADATION

An alternate tactic to deal with solid plastic wastes is to make plastics degradable. The term *degradable* means that the plastic can break down into smaller molecules by natural means, usually by some biological agent or by sunlight. The hope of those who favor degradability as the answer to plastic solid waste disposal is that the plastics can be left wherever is convenient and they will eventually disappear into some innocuous material that becomes part of the dirt. In support of this point of view is the fact that all materials, including plastics, eventually degrade. (Even rocks degrade.)

In reality, however, degradable materials have many problems. Some materials degrade very slowly. Others degrade into substances that are hazardous and therefore are polluting. Furthermore, some applications require that the material not degrade, so considerable effort has been made to prolong the life of these materials. For instance, packaging materials are intended to hold and protect their contents for the entire shelf life of the product. Because shelf life could be a long time, package life must also be long.

Still, some applications could be well served by a degradable product. For instance, sutures in medical applications could be absorbed by the body. Another application is the plastic rings used to join a six-pack of beverage cans. The product does not actually touch the food item and so would not be catastrophic should it degrade before use was completed. Moreover, when disposed of improperly, the rings are a nuisance and can harm some animals.

Several degradable polymers have been developed and are now being used for applications in which degradability makes sense. These degradable polymers include polyvinyl alcohol, polyvinyl ethanoate, many aliphatic polyesters and polyurethanes, and several others that are based on natural products.

LANDFILLS

The use of sanitary landfill is the most commonly used means of solid waste disposal—as much as 90% of all waste in the United States, 85% in the United Kingdom, and 60 to 70% in European countries. Essentially, the rubbish is simply buried in the ground. Landfills are popular because they are less expensive than any other method of waste disposal. No sorting is required, no processing, and no worry about molding the residue into useful products. Some costs, however, are inherent in modern landfills. Careful control of the landfill process is required in order to protect the site and its surroundings from problems of odor, fire, vermin, and pollution from seepage. The seepage problem is controlled by lining the landfill, usually with an impermeable plastic sheet that is designed to last indefinitely and to retain the water and other liquids within the landfill, thus preventing the seepage from contaminating groundwater.

Plastics in landfills are fairly inert though not completely so. The majority of common polymers are not readily degradable and do not contribute to water-soluble residues, but their components such as plasticizers may do so. This overall inertness is an advantage in landfills where considerable effort goes into not allowing the landfill to degrade.

Perhaps the most serious problem with degradable plastics and other degradable materials is that degradation generally requires specific environmental conditions to occur, and these conditions may or may not be present. The most common example of this is in landfills where supposedly degradable materials, such as newspapers, are perfectly readable even after 50 years. Even vegetables and other food products have been found to be essentially intact after 10 to 15 years in a landfill. The problem is that the air and organisms necessary for degradation are not present. Hence, degradation is not only unwanted in landfills, it is uncertain.

The biggest problem with landfills is that the space available for them is dwindling. Landfills are reaching capacity, and alternatives for the disposal of waste are needed. These alternatives are explored in the other sections of this chapter, at least as they can be applied to plastics.

A reasonable question to ask is, How much of the landfill volume (or weight) is plastics and is that percentage increasing rapidly? The answer is illustrated in Figure 22.2.

It is obvious from these data that plastics are a fairly large component of the waste (19.9% by volume and 8.0% by weight). The largest component of waste is paper products (34.1% by volume and 40.0% by weight). Recent excavations into municipal landfills show that over the past 20 years, the percentage of plastics in landfills has not grown significantly relative to the other materials. Probable reasons for this surprising result (in light of our increasing use of plastics) are that source reduction is working for plastics and that the use of other materials is growing about as fast as plastics. (Consider the increase in the amount of paper used with the advent of computers and copy machines.) Nevertheless, plastics are under considerable scrutiny and ways are actively being sought to reduce the amounts of plastics placed in landfills.

Figure 22.2 Materials in municipal solid waste landfills by volume and weight (in a recent year).

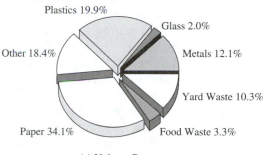

(a) Volume Percentages

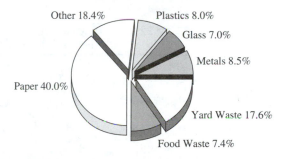

(b) Weight Percentages

INCINERATION

Incineration, or controlled burning, is another option for disposing of a large percentage of municipal solid waste. In this reuse scheme, plastic wastes can be used along with paper and other flammable waste materials. The paper, plastics, and other flammables are separated from the rest of the waste stream. In practice, high-value plastics that are easily sorted, such as PET bottles, are manually separated. The remaining plastics, paper, and other flammable materials are chopped into flakes and pressed into pellets for ease of handling and economic transport from the recycling center to the burn location.

The burning should, of course, have some value in addition to the elimination of solid wastes. The most common purpose for burning is the generation of electricity. The efficiency of the burning of municipal waste is related to the energy content of the wastes versus the energy content of more traditional fuels such as natural gas and oil. The energy content of various materials is given in Table 22.2.

The issues of concern with incineration include the creation of hazardous air emissions, the problems of ash disposal, and the release of carbon dioxide gases that could hasten global warming. These concerns relate to all burned materials; plastics have potential problems for the environment as do the other components that are burned.

In regard to emissions, the focus has been on the following pollutants: particulates; heavy metals such as lead, cadmium, zinc, chromium, and mercury; acid gases that lead to acid rain (trace concentrations of sulfur, chlorine, and fluorine yield these acids); NOx (also an acid-forming gas); and trace organics, especially chlorinated dioxins and furans which may be carcinogenic. Experts on incineration technology assert that most of the hazardous components and emissions can be efficiently limited to safe levels by means of available technologies. In European countries that are very environmentally conscious, such as Denmark and Sweden, a much larger percentage of municipal waste is incinerated than in the United States. This good performance depends on proper combustion conditions, excellent off-gas and other emission-control equipment, responsible management, and workers who carefully maintain and improve the system.

Still, the risks associated with incineration are real. All things considered, source reduction and recycling are preferred over incineration and regeneration (also potentially polluting) for managing solid waste. Incineration, however, may be preferred over landfill.

Table 22.2   Energy Content of Various Solid Waste Materials and Conventional Fuels Burned to Generate Electricity

Material	Energy Values (BTU/pound)
PET	10,900
HDPE	18,700
Rubber	12,800
Newspaper	8,000
Wood	7,300
Yard waste	2,900
Fuel oil	20,900
Coal	9,600

TOTAL PRODUCT LIFE CYCLE

For many consumers, environmentalists, government entities, and companies, a two-part question concerning the environment needs to be answered. That question is: What is the total impact of a particular product or product type on the environment over the entire life of the product, and how does that impact compare with other products that might be competitive in use? This question is not easy to answer, for several reasons.

The environmental impact of any product is difficult to assess because the origin of the product's life is difficult to define. In the case of a metal part, for instance, would the origin of the part be when the part is formed, or when the metal is mined, or when the mine is first developed? And, what about the origin of the manufacturing plant in which the metal is formed? Obviously some origin limits need to be established so that rational assessments of the total environmental impact of a particular part can be made.

Assessing the total life of a product and the product's environmental impact involves more than simply looking at the impact made by the disposal of the product. The entire life must account for an extended period of time, perhaps through several uses, and must consider the impact of the manufacturing process originally used to form the product as well as the disposal impact. The cost might also consider the most likely disposal method, such as recycling, and attempt to determine the costs when the end of the product's life is as a recycled entity.

Another problem in comparing two products and their environmental impacts is the wide differences in processes that are likely with products made of different materials. Again, some reasonable assumptions need to be made or a comparison would be meaningless.

Some assessments have been made based on assumptions that have attempted to set reasonable boundaries for origins and comparison bases, but these assessments always are open to challenge by those who may want to make a particular point that differs from the conclusions drawn. Nevertheless, the concept of using total product life as a basis for determining the real environmental impact of a product is sound, and, on that basis, a few comparisons are made in this section. These comparisons show how viewing the entire life of a product can alter the perception of the environmental impact that a product may have.

Total Product Costs for Plastic and Paper Hot-Drink Cups

One of the most visible pollutants and, therefore, a product that is under pressure for environmental impact assessment is the hot-drink cup. These cups typically are made from polystyrene foam or from paper. A total product life comparison of these two products is therefore pertinent and demonstrates the method that might be followed in determining the true environmental impact of various products. The results of the comparison are given in Table 22.3.

An analysis of the data in Table 22.3 leads to the following conclusions:

- The energy requirements to produce the paper cups are much greater than (approximately double) the amounts needed for the plastic cups.

Table 22.3   Total Product Life Costs and Environmental Impact for Plastic and Paper Hot-Drink Cups

Item		Paper Cup	Plastic Foam Cup
Per Cup:			
Raw materials	Wood and bark	25–17 g	0 g
	Petroleum products	1.5–2.9 g	3.4 g
	Other chemicals	1.1–1.7 g	0.07–0.12 g
Finished weight		10.1 g	1.5 g
Per metric ton:			
Utilities	Steam	9,000–12,000 kg	5,500–7,000 kg
	Power	980 kWh	260–300 kWh
	Cooling water	50 m^3	130–140 m^3
Water effluent	Volume	50–190 m^3	1–4 m^3
	Suspended solids	4–16 kg	0.4–0.6 kg
	Products of degradation	2–20 kg	0.2 kg
	Organochlorines	2–4 kg	0 kg
	Metal salts	40–80 kg	10–20 kg
Air emissions	Chlorine	0.2 kg	0 kg
	Chlorine dioxide	0.2 kg	0 kg
	Reduced sulfides	1–2 kg	0 kg
	Particulates	2–3 kg	0.3–0.5 kg
	Chlorofluorocarbons	0 kg	0 kg
	Pentane	0 kg	35–50 kg
	Sulphur dioxide	10 kg	3–4 kg
Recycle potential:	To primary user	Washing can destroy	Easy to reuse
	Postconsumer	Difficulties with coating	Resin reuse easy
Ultimate disposal:	Proper incineration, heat recovery	Clean, 20 MJ/kg	Clean, 40 MJ/kg
	Mass to landfill	10.1 g	1.5 g
	Biodegradable under proper conditions	Yes. Degradable leachate to water	No. Inert.

- The amount of water needed to produce the paper cups is much greater (approximately ten times) than the amount needed for the plastic cups.
- Air emissions are much greater for the paper cups except for pentane.
- Recycle potential, a method of source reduction, is much better for the plastic cup than for the nonreusable paper cup. Postconsumer recycling potential is also much better for the plastic cup.
- Ultimate disposal to either incineration or a landfill is about the same for both paper and plastic. The paper is biodegradable, but, as mentioned previously, the proper conditions must be met and those are not common in landfills.

The overall conclusion is, therefore, that the plastic cups are less damaging to the environment than are the paper cups.

Energy Requirements

The energy requirements needed to make various products, plastic or otherwise, are very important because of the potential for pollution from any of the sources used to create the energy. The generally rising cost of energy puts those products that require high energy to make at an initial disadvantage. However, as the product is recycled, the energy requirements overall drop, sometimes favoring one product over another. Such a trend is seen in Figure 22.3, where the energy requirement decreases as the amount of recycling increases, just as would be expected because recycling takes less energy then does making the product originally. The paper sack originally takes more energy than either the 2.1-mil PE sack or the 1.5-mil PE sack, but as the amount of recycling increases, the energy of the paper sack drops faster than the PE sacks so that at 100% recycling, the paper is about the same as the 2.1-mil PE sack. Hence, over the total life of the products, the paper sack becomes more favorable as recycling increases although still not as good as the low-weight plastic.

The decrease in energy required also affects the cost of the products from a total-cost standpoint. Hence, in the analysis given in Table 22.3, the energy requirement should really be reduced by the amount of recycling that is done. In the case of paper cups, however, the cups are almost never recycled and so the numbers shown in the table do not need the correction for recycling and the proper comparison is made near the 0% recycling rate.

In terms of energy consumption, plastics have some inherent advantages over most other materials, chiefly because the temperatures required in processing plastics are so much lower than temperatures for molding other materials and because plastics are lighter in weight than many competitive materials. Some examples may illustrate these favorable energy factors.

Figure 22.3 Energy requirements for grocery sacks as a function of recycling.

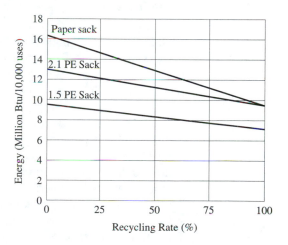

Refrigerators and Freezers. Plastics are used extensively in household refrigerators and freezers. It is estimated that the 326 million pounds of plastics used in 1978 would have required 2.6 times that weight if produced in the alternate materials (metal and glass). The production of the plastic components consumed a total of 15.8 trillion BTUs, compared to 23 billion BTUs that would have been required to produce the components in glass and metal. The use of plastic thus saved 7.2 trillion BTUs.

Plastic Pipe. In construction, more tonnage of plastics goes into piping products than for any other single use. In 1977, the weight of plastic pipe produced was 2.0 billion pounds, while the weight of the equivalent metal pipe would have been 17.5 billion pounds. The energy consumption for the plastic was 84 trillion BTUs and for the metal would have been 408 trillion BTUs, for an energy savings of 324 trillion BTUs.

Beverage Bottles. When the PET soda bottle (2-L size) was originally introduced in the mid-1970s, the entire market for family-sized bottles was glass. The energy consumption was 34.2 trillion BTUs. Assuming the same market size, if PET bottles had been used, the energy consumption would have been 18.2 trillion BTUs, a savings of 16 trillion BTUs, or the energy equivalent of 2.8 million barrels of crude oil.

Automobiles. Plastics are increasingly being used as components in new automobiles because of all-around performance and because of the lower weight of plastics versus traditional materials. For every 400 pounds of weight savings, the average car increases its gas mileage by one mile per gallon. It is estimated the plastics saved 275 pounds in each car in 1977, giving a total energy savings of 80.5 trillion BTUs, or the equivalent of 14 million barrels of crude oil.

Total Product Life Accountability

Some governments (especially Germany) have mandated or are working on legislation that will require that the original manufacturer of a product be ultimately accountable for that product throughout its total life cycle, including disposal. Under these rules, consumers are obligated to return worn-out products to the point of purchase, where the products are then forwarded to the original manufacturer. Although some serious problems have yet to be resolved with the implementation of such a total product life accountability, the move in this direction has made manufacturers aware of this potential responsibility and has spurred many to develop systems for recycling of their products.

One of the most common solutions is to develop methods for recycling of the raw materials. For instance, molders of polystyrene foam packing materials are developing programs to densify the foam for shipping and then sell the material to extruders and compounders who will convert the material into polystyrene products or pellets.

In a more complicated case, automobile manufacturers are developing methods to reclaim all automobile parts. Markets for used metals, which are remelted and then reformed, have existed for many years. Recently methods for recycling plastic body panels have been developed (as already discussed in this chapter). Other parts of the car that can be recycled include foam seats, which can be chopped and reused in carpet padding, and fabrics, which can be remelted and then reformed into either fibers or molded parts. Glass windows can be remelted and reformed, and even lubricating fluids can be cleaned and reused. Automobile manufacturers are also changing the design and specifications for several parts so that they can be recycled more easily.

FUTURE

Some people have suggested that all single-use disposable items that cannot be recycled easily should be eliminated because of their negative impact on the environment. These people especially target packaging materials because they so often are improperly disposed of and tend to clutter roadsides and parks and eventually find their way into the landfills. This extreme approach is not likely to be adopted since packaging materials are needed, sometimes for special applications with special requirements that diminish the likelihood of recycling. However, the pressure to improve the recycling capability of disposable items is felt by most manufacturers of these items and they generally have responded by improving the recycle capability of their products.

Items that are not defined as disposable, such as refrigerators, automobiles, furniture, and almost everything manufactured, will eventually wear out and become disposable. As already discussed, the manufacturers of many of these items are also changing designs to make the components of these products recyclable whenever possible.

Of the options for reducing solid waste that were introduced at the beginning of this chapter (source reduction, recycling, regeneration, degradation, landfill, and incineration), source reduction and recycling seem to be gaining the most popularity among the public. These methods are sure to be the least polluting and, from that standpoint, are the most acceptable. However, the other methods of waste disposal have some advantages and should, under some circumstances, be considered as viable alternatives.

Governments will continue to impose recycling and to set standards for products. It behooves all manufacturers and others to properly educate the public and, simultaneously, to take the appropriate steps to develop products that have the lowest environmental impact possible throughout the entire life cycle of the product.

The raw materials from which the various products are made will increasingly come under scrutiny for environmental impact. Most of the raw materials are nonrenewable, including metals from mines and crude oil and coal for plastics. Great care must be taken to ensure that these raw material sources are developed under an environmentally sound policy. Even supposedly renewable resources, such as trees, are now seen as having an environmental impact. Old growth forests are not renewable, and efforts to get timber from these forests are meeting strong resistance.

The best solution for the environmental problem of product use and disposal seems to be a rational analysis of each product type and its own particular use and disposal characteristics. Manufacturers should be open to suggestions from groups who lobby for environmental policy. Similarly, these groups should be willing to put aside emotional issues in favor of reasoned positions based on scientific principles and to modify their positions as new scientific information and solutions become available.

CASE STUDY 22.1

Recycling Solid Wastes

One of the major problems in recycling of municipal solid wastes is that the endeavor is not always commercially successful. Many companies have built extensive recycling facilities only to find that they could not operate the facilities and make money. To continue operation they therefore required extensive grants from federal, state, or municipal governments, but these grants could not be continued indefinitely and so the facilities closed.

One company, Reuter, Incorporated, built and is operating two recycling facilities that employ two different recycling concepts, and is making a success out of both. One facility is located in Eden Prairie, Minnesota, near the corporate headquarters in Hopkins, Minnesota. This facility was designed to recover recyclable materials and manufacture a densified Refuse Derived Fuel (dRDF) that would be sold to the municipal waste-to-energy power generation facilities as a fuel. In this process municipal and commercial waste haulers bring their loads to the facility where the loads are weighed and the haulers pay a usage (tipping) fee. The haulers then dump their trash onto a large, enclosed tipping floor where the large, nonprocessable items such as mattresses, furniture, and so on are removed. Metals and wood are removed for recycling. The amount of nonprocessable materials is less than 2% of the total. The bulk of the material is pushed onto a conveyor system where small material (less than two inches) falls through holes and is sent to a composting site. The conveyor then enters a picking station where high-value recyclable items (PET soda bottles, milk jugs, and aluminum cans, for example) and difficult-to-process materials such as textiles are removed by hand. From the picking room the waste material is sent to a hammer mill where it is reduced to less than eight inches in size. The stream then passes under strong magnets where additional magnetic materials are removed. An air separator removes large items, which are then sent to landfill or recycling. The small materials are sent to a flaking unit and a densifier where they are compressed into fuel pellets that are then sold to the power plant. A breakdown of the fractions recycled is given in Table 22.4.

Table 22.4   Eden Prairie Recycling Plant Waste Reduction (Abatement)

	Tons Processed (per month)	% of Total	% of Residuals	% of Material Abated
Amount processed	7,891	100.0		
Recyclables	361	4.6	0	4.6
Fuel produced	3,779	47.9	0	47.9
Fine residue composted	647	8.2	17	6.81
Fine residue landfilled	869	11.0	100	0.0
Heavy residue combusted	1,889	23.9	8	22.0
Tip floor residues combusted	188	2.4	15	2.4
Tip floor residues landfilled	125	1.6	100	0.0
Percentage of original amount abated				83.35

Table 22.5 Pembroke Pines Recycling Plant Waste Reduction (Abatement)

	Tons Processed (per month)	% of Total	% of Residuals	% of Material Abated
Amount processed	11,460	100.0		
Recyclables	844	7.4	0	7.4
Tip floor rejects	202	1.8	15	1.53
Coarse treatment rejects incinerated	3,192	27.8	8	25.58
To compost	7,222	63.0	2	61.0
Percentage of original amount abated				95.3

The second Reuter facility is located in Pembroke Pines, Florida. This facility was built to use composting as the principal recycling technique and has been equally successful. The process begins much like the Eden Prairie process with a weighing and tipping of trash onto a tipping floor, where large, nonprocessable materials are removed. The material is pushed onto a conveyor where magnetic materials are removed and small materials less than 2 inches in size are screened out. A second separation removes materials that are less than 8 inches in size. This medium separation and the remaining large materials are conveyed separately through handpicking stations where high-value recyclables and additional nonprocessables are removed. Beyond the picking stations the medium and large fractions are again combined and fed by conveyor to a large shredder. This shredder reduces the material to a size of 8 inches or smaller. This material is then conveyed to a mixing drum, where it is recombined with the small fraction and water is added. The noncompostable and larger items are directed to a residue trailer, from which they are sent to an incinerator. The smaller materials are then conveyed to a compost building, where they are spread into 600-foot (180-m) long rows that are 13-feet (4-m) wide. The floor is built with aeration ducts that draw air through the piles and carry the air to large earth filters, used to reduce the odors created by the composting process. Once each week, for 6 weeks, the pile is turned. At the end of the sixth week, the composted material is loaded onto another conveyor and taken into the fine-treatment building, where glass, stones, and other heavy objects are removed. The compost is then sold as a fertilizer product. The operations of the Pembroke plant waste-reduction system are summarized in Table 22.5.

SUMMARY

Some surveys of public perceptions regarding plastics and their impact on the environment have indicated that the public believes that plastics are more detrimental to the environment than perhaps any other type of material. These perceptions are based largely on the obvious presence of discarded plastics along roadsides and on reports that the amount of discarded plastics is growing and is therefore an ever-increasing component of municipal solid waste. These perceptions have some basis in truth, but are not entirely correct.

Plastics are a significant component of solid waste, but they are neither the largest component nor the fastest-growing component. The relatively slow growth of plastics in solid wastes being sent to landfills is primarily due to the strong efforts within the plastics manufacturing community to reduce the amount of plastic material in the solid waste stream. These efforts include five ways of handling solid waste: source reduction, recycling, regeneration, degradation, and landfills.

Source reduction involves finding methods to cut down on the amount of plastic used. This can mean making thinner or smaller products, thus using less plastic material, or it can mean asking the consumer to reuse some plastic parts, thus reducing the amount of single-use disposable items.

Recycling is an important technique for reducing the amount of waste that must be disposed of in a landfill. Some plastics such as PET soda bottles and HDPE milk jugs, are readily recycled. Other plastics can also be recycled, but they are less obvious as unique shapes and therefore are less easily sorted. Automated sorting systems help solve some of the sorting problems and therefore improve the capability to sort and then recycle plastics economically. Almost every type of plastic can be recycled, even those that cannot be melted. These can be recycled as comingled (mixed) recycle or as fillers in other plastics. In these materials especially, the costs of recycling often are high when compared to the costs of virgin plastic materials. This differential in cost has inhibited recycling of many plastics and will continue to be a problem in the foreseeable future.

Another method of reducing the amount of plastics in landfills is to regenerate the original monomers or other chemicals that were used to make the plastics in the original chemical processes. Regeneration has the advantage of being able to obtain a higher value for the mixed cycle material than could be obtained from recycling. Regeneration might also be a method of attacking the reservoir of plastics already in landfills. However, regeneration requires that energy be placed into the plastic material so that it can be broken into its chemical components, and that energy requirement might be quite high.

Many people believe that degradation is the ideal way to reduce solid waste. However, in actual practice, most materials, even those normally thought to be degradable (such as paper and foods) do not easily degrade in landfills because the conditions for degradation are not met. Furthermore, the ideal of degradation is that the material will degrade and disappear without any adverse affect on the environment, which rarely happens. The degradation products of many materials, even paper and foods, often are detrimental to the environment and they contaminate groundwater. When placed in landfills, degradation, if it occurs, is a serious problem. Landfills are best if stable over long periods of time.

Landfill is the most widely used method of disposing of solid waste. Plastics represent a significant portion of landfill materials, but they are not the largest segment and they are growing at a slower rate than some other materials (such as paper).

Incineration is yet another method of disposing of solid waste. Many people worry about the potential for air pollution from the burning of waste, and those worries are well founded. However, if operated properly, incineration facilities produce very little pollution. The advantages of incineration include the high-energy content of the plastic material and the ability to burn both paper and plastics together without sorting either.

A strong movement, spurred somewhat by legislative action, is underway to consider total product life when determining the environmental impact of the product. This total life concept can also include the manufacturer's responsibility for the product and its proper disposal. In the light of the total life of a product and the total environmental impact, plastic materials have a surprisingly low impact. For instance, the decision of some fast-food companies to eliminate foamed plastics in favor of paper was probably bad for the environment from a total life viewpoint. The pressures to make these short-sighted and, apparently, incorrect decisions can be alleviated by proper education of the public and diligence on the part of the plastics industry to identify and practice good procedures for reducing the amount of plastics in the waste stream.

GLOSSARY

Comingled recycle   A mixture of several types of plastic recycled products.
Cracking   The process of breaking apart a polymer into basic constituents with very high heat and pressure.
Degradable   A material that breaks down by natural means (usually biological or environmental).
Incineration   Controlled burning.
Mixed recycle   Recycling material that is a mixture of several types of plastics; also referred to as comingled recycle.
Postconsumer recycle (PCR)   Material discarded by the consumer that might be recycled.
Pyrolysis   The decomposition of a material using heat processes in the absence of oxygen.
Recycling   The reprocessing and refabrication of a material that has been used.
Regeneration   The process of breaking down the plastic material into more basic materials, which can then be used to make new (virgin) materials; also called chemical recycling.
Regrind   The reprocessing and refabrication of a material that has been molded and then rejected for some manufacturing reason but has not been used by the consumer.
Source reduction   The reduction of the amount of solid waste generated by reducing the amount of material in any application.
Unzipping   The process by which some plastics depolymerize when heated.

QUESTIONS

1. Identify three reasons why plastics have been slower to be recycled than aluminum cans.
2. List five methods of disposing of municipal solid wastes.
3. What are the three major components (material types) of a landfill?
4. What are two major disadvantages to degradable plastics?
5. What is regeneration, and what is its potential value?
6. Why are most plastics actually beneficial in a landfill?
7. What is the chief economic disadvantage of recycling plastics?
8. To what does total product life-cycle costing refer?

9. What is the biggest problem associated with incineration as a waste disposal method?
10. Compare the overall environmental impact of plastic and paper grocery sacks.

REFERENCES

"Answers to Your Questions About Plastics in the Environment," Washington, DC: American Plastics Council.

Blankenship, Joseph E., "Achieving Maximum Waste Reduction by Integration of Compositing, Fuel Production and Incineration," Hopkins, MN.: Reuter, Inc.

"The Blueprint for Plastics Recycling," Washington, DC: The Council for Solid Waste Solutions.

"Crosslinked PE Can Be Recycled," *Plastics Technology,* June 1992, 43–45.

"Estimates of the Volume of Municipal Solid Waste and Selected Components," Franklin Associates, Ltd., October 19, 1989.

Henshaw, John W., Alan D. Owens, Dan Q. Houston, Irvin T. Smith, and Todd Cook, "Recycling of a Cyclic Thermoplastic Composite Material by Injection and Compression Molding," *Journal of Thermoplastic Composite Materials,* Vol. 7, January 1994, 14–29.

"Implications of Post-Consumer Plastic Waste," *Plastics Engineering,* September and October 1990.

Inoh, Takashi, Toshio Yokoi, Ken-ichi Sekiyama, Norihisa Kawamura, and Yasuhiro Mishima, "SMC Recycling Technology," *Journal of Thermoplastic Composite Materials,* Vol. 7, January 1994, 42–55.

Jutte, Ralph B., and W. David Graham, "Recycling SMC Scrap as a Reinforcement," *Plastics Engineering,* May 1991, 13–16.

Leaversuch, Robert D., "Chemical Recycling Brings Real Versatility to Solid-Waste Management," *Modern Plastics,* July 1991, 40–43.

"Recycling Faces Reality as Bottom Line Looms," *Modern Plastics,* July 1994, pp. 48D–49.

Lindsay, Karen F., "'Truly Degradable' Resins are Now Truly Commercial," *Modern Plastics,* February, 1992.

Maldas, D., and B. V. Kokta, "Effect of Recycling on the Mechanical Properties of Wood Fiber-Polystyrene Composites. Part 1: Chemithermomechanical Pulp as a Reinforcing Filler," *Polymer Composites,* Vol. 11, No. 2, April 1990, 77–83.

Nir, Moira Marx, "Implications of Post-consumer Plastic Waste," *Plastics Engineering,* October 1990.

Oliphant, K., and W. E. Baker, "The Use of Cryogenically Ground Rubber Tires as a Filler in Polyolefin Blends," *Polymer Engineering and Science,* Mid-February 1993, 166–174.

Peterson, Joakim, and Peter Nilsson, "Recycling of SMC and BMC in Standard Process Equipment," *Journal of Thermoplastic Composite Materials,* Vol. 7, January 1994, 56–63.

"Plastic Facts," Society of Plastics Engineers, Inc.

"Plastics in Perspective: Answers to Your Questions About Plastics and the Environment," American Plastics Council, Society of the Plastics Industry, Inc., Washington, DC.

Rathje, William L., "Once and Future Landfills," *National Geographic,* May 1991, 117–133.

"Recycle: Be Part of the Solution, Not Part of the Problem," Springfield, IL: Illinois Department of Energy and Natural Resources.

"Recycling SMC Scrap as a Reinforcement," *Plastics Engineering,* May 1991, 13–16.

"Resource and Environmental Profile Analysis of Foam Polystyrene and Bleached Paperboard Containers," Washington, DC: The Council for Solid Waste Solutions.

"Resource and Environmental Profile Analysis of Polyethylene and Unbleached Paper Grocery Sacks," Washington, DC: The Council for Solid Waste Solutions.

"Resource and Environmental Profile Analysis of Polyethylene Milk Bottles and Polyethylene-Coated Paperboard Milk Cartons," Washington, DC: The Council for Solid Waste Solutions.

"The Solid Waste Management Dilemma," Washington, DC: The Council for Solid Waste Solutions, June 1990.

"SPI's Voluntary National Container Material Code System," Washington, DC: The Society of the Plastics Industry, Inc., AU-164.

Strong, A. Brent, and Norman C. Lee, "Performance Evaluation of Refuse Carts Using Post Consumer Regrind Material," *Antec 93,* New Orleans, May 9–13, 1993.

Testin, Robert F., and Peter J. Vergano, "Plastic Packaging: Opportunities and Challenges," Washington, DC: The Society of the Plastics Industry, Inc.

"The Urgent Need to Recycle," A Special Advertising Section in *Time Magazine,* July 17, 1989, Sponsored by The Council for Solid Waste Solutions.

Walling, Ronald L., "Incorporating Post-consumer HDPE in Large Blow-Molded Applications," *Plastics Engineering,* January 1993, pp. 28–29.

"Waste Solutions," *Modern Plastics,* April 1990.

Wendorf, Mark A. "Blends of Polypropylene with Recycled HDPE," *PRD RETEC,* Chicago Section and the State of Illinois, Department of Energy and Natural Resources, Schaumburg, IL, June 14–16, 1993.

Yam, Kit L., Binoy K. Gogoi, Christopher C. Lai, and Susan E. Selke, "Composites From Compounding Wood Fibers with Recycled High Density Polyethylene," *Polymer Engineering and Science,* Vol. 30, No. 11, Mid-June 1990, 693–699.

CHAPTER TWENTY-THREE

OPERATIONS

CHAPTER OVERVIEW

This chapter examines the following concepts:

- Safety and cleanliness
- Plastic resin handling and conveying
- Plant layout
- Quality assurance

INTRODUCTION

Many important activities in plastic-part manufacturing are associated with operations. These activities occur principally on the shop floor and in areas that surround and support the basic manufacturing process. These activities include safety and cleanliness and handling of resins and other materials (such as solvents), and may involve special conditions for special plastic products (such as medical and food grades). Many of these activities are overlooked in typical discussions of plastic-manufacturing processes, but their importance to the good quality and functioning of the manufacturing plant suggest that some comments about them should be given. These are not afterthoughts but rather introductions to concepts that can mean the difference between good products and bad products.

SAFETY AND CLEANLINESS

Safety and plant cleanliness (including neatness and proper organization) are important operations activities, as well as major social considerations. Therefore, they are addressed in this text as they apply to plastics manufacturing.

Plastics manufacturing occurs in two major steps. The first step is the manufacture of the basic plastic resins from the raw materials; the second step is the conversion of the resin into a molded shape. Each of these steps is discussed separately.

Production of the Basic Plastic Material

The raw materials, which are commonly derived from crude oil, usually are liquids or gases. These raw materials are then converted into the polymeric resin, which is usually in the form of a powder or granules (somewhat like the consistency of laundry soap), a liquid, or pellets. The conversion of ethylene gas to polyethylene and the combination of an amine and an acid to form nylon are examples of the formation of basic resins. This resin formation is usually done by large chemical companies using very large, high-throughput equipment. Some of the large companies that produce the basic resins include DuPont, Union Carbide, Exxon, Huntsman, Dow, Allied Signal, BASF, Hoechst, Ciba Geigy, Hercules, Imperial Chemical Industries, Toray, and Solvay. Most of these companies operate in many parts of the world.

Safety is an important consideration in these plants and is watched closely because of the potential for catastrophe. The plants are so large and the temperatures, pressures, and fire danger occasionally so high that an accident could kill many people. Moreover, the raw materials used to make the basic resins can pollute and, in some cases, cause disease. The sheer volume of materials handled by these plants raises the awareness of the people running the plants and others in the nearby communities. Fortunately, most of these large plastics manufacturing companies are socially responsible and have installed extensive safety programs and environmental controls that have resulted in an excellent overall industry performance. The chemical industry has a much better safety record than does industry in general. Still, these companies and the public need to remain aware of the dangers and ensure that correct safety practices are followed.

One of the major problems with the manufacture of the basic resins is the potential toxicity of the chemicals involved in these processes. As already indicated, large chemical and petroleum companies are the primary producers of these basic materials. In some cases the polymers are liquids. These liquid chemicals must be handled carefully, with full understanding of the potential dangers. To ensure this, *material safety data sheets (MSDS)* should be consulted before any use of the materials. MSDS sheets are sent with the chemicals and must be stored in convenient locations so that any person handling the chemicals can inspect them. Training in the use and reading of MSDS sheets is mandated.

Care should be taken to avoid all of the potential dangers of liquid handling—vapor inhalation, skin irritation, splashes, and prolonged exposure in general. In every case the proper protective equipment should be worn and the environment (room) should be properly vented. Eye washes and emergency showers should be available and convenient. Disposal of waste liquids must be done by proper methods.

Production of the Plastic Part—Thermosets

Thermoset resins often are liquids and sometimes have highly volatile solvents added, so all the precautions concerning the handling of liquid materials should be observed by those who mold thermoset parts. Proper equipment and ventilation should be used, and MSDS sheets should be consulted so that dangers are well understood before proceeding to mold the plastic part.

The processing of thermoset materials often involves the evolution of heat. When this occurs, several precautions should be taken. First, the materials should be lightly covered so that any splashing or boiling that occurs will not be a problem. Second, because the heat evolved can cause burns, protective clothing, safety glasses with side shields, and gloves should be worn.

Peroxides are chemicals that are added to initiate some polymerization or curing reactions. These materials are extremely hazardous because of the ease with which they react. Peroxides should be stored in special containers, usually in a special refrigerator, and should be handled only according to the manufacturer's MSDS and other accepted procedures. In some cases an accelerator might also be used in curing reactions. Never mix accelerators directly with peroxides. Each should be mixed with the resin or solvent prior to any mixing of these two reactive materials.

Solvents of any kind should be handled safely and only after consultation of the MSDS for potential hazards. In addition to the toxicity danger of some solvents, the potential for flammability should also be considered. Emissions of solvent vapors should be appropriately controlled.

Production of the Plastic Part—Thermoplastics

The molding of thermoplastics into a plastic part is usually done in facilities that are much smaller than the large resin-manufacturing plants. Because these manufacturing facilities are smaller than those involved in producing the basic resins, safety aspects are sometimes overlooked. This is a serious mistake. There is the potential for serious injury and even loss of life in any manufacturing facility, no matter what its size. In most cases, injuries can be prevented by following simple safety practices. (Environmental considerations, while important in general, are less critical for these plants because the materials used are almost all solids and are less polluting than liquids and gases.)

Perhaps the greatest hazard in plants where there are thermoplastic-molding operations is the potential for spills of the thermoplastic material on the floor. These materials, often in the form of pellets, can be very slippery and have been known to cause serious falls, because they slide under a person's shoes. This spillage occurs frequently during transport of the material and, in a good plant, are always cleaned up promptly.

Solvents may sometimes be used, however, so the precautions already discussed for solvent use and disposal should be carefully followed.

General Safety Rules in Plastic-Molding Operations

The specific safety considerations for each of the principal plastics part-manufacturing processes are discussed in detail in each of the chapters dealing with the particular processes. However some overall safety concepts are worth considering as they concern all thermoplastic part-manufacturing processes.

1. Much plastic-manufacturing equipment is mechanical and moves under pressure or creates pressure. When manufactured commercially, the equipment usually has safety guards and other safety interlocks. Many accidents in the plastics industry result from operating the equipment with these safety devices either removed or bypassed. The

reason usually given is to increase production time or for ease of operation. These are not good reasons for ignoring or superseding safety equipment or rules.

2. The plastic is usually formed into its new shape by heating the plastic and then introducing it into a mold while it is a liquid or soft semisolid. Therefore, relatively high temperatures are usually present. Caution should be exercised around all equipment so that burns or hot leaks of the plastic do not occur.

3. Many plastics are brittle and can be easily broken, especially when cut, trimmed, or sanded. Eye protection with side shields should always be worn to avoid potential damage from your own work or from that of your co-worker.

4. Use the proper tools. The use of "expedient" tools can often result in injury and in damage to the machines.

5. Develop a set of workable safety rules for your work facility, make sure everyone agrees with them and understands them, and then follow the rules carefully.

6. Understand and comply with governmental safety and work rules.

7. Wear gloves for protection from hot plastic.

8. When working near heated surfaces, such as molds for themosets, wear arm thermal protectors or, at least, long sleeves.

9. The emergency shop procedures should be well known.

10. A basic training course be completed before operating any machine.

Cleanliness and Neatness

In all plastic-molding operations cleanliness and neatness are good practices because they promote safety and quality. In some molding operations, cleanliness is critically important. For instance, molding of plastics for medical, electronics, and food applications often requires clean-room operations in which the air is specially filtered to remove dust, and operators are required to wear special protective clothing to prevent contamination of the parts. Much effort goes into keeping these facilities clean, and even small spills or carelessness can result in lost product and, perhaps, a shutdown of the facility until the problem is corrected. In some cases, government inspectors insist on cleanliness standards.

Routine safety and cleanliness inspections are wise. These inspections can avoid potential safety problems and improve general productivity.

PLASTIC RESIN HANDLING, CONVEYING, AND DRYING

Resin pellets, powders, and flake are usually shipped to the molding facility either as loose bulk material (usually in rail cars but occasionally in special hopper trucks) or in packages (usually in 1000-pound (500-kg) capacity boxes called gaylords, or in 50-pound or (20-kg) sacks). Both bulk and packaged shipments present problems in material handling.

Bulk materials must be unloaded from the bulk carrier. This unloading is almost always done by pneumatic transfer. Figure 23.1 shows a system used to accomplish this transfer.

In the system shown, the valve on the bottom of the rail hopper car is connected to a pneumatic hose line. Air is brought into the system and that air sweeps the resin out of

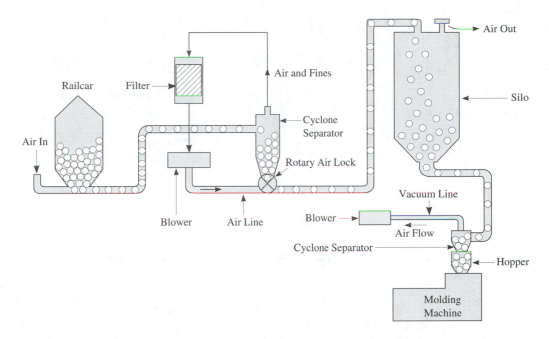

Figure 23.1 Pneumatic system for unloading bulk materials from a rail car into a silo.

the bottom of the rail car and into a cyclone separator. The cyclone separator works by bringing the air and resin into the top of the separator at a tangential direction. The airflow circles inside the separator and loses energy through friction. At some point, the resin falls out of the airstream and drops to the bottom of the separator. The air and any fine particles that may be present exit the top of the separator. This stream of air and fines goes through a filter, where the fines are removed. The airstream then flows into the intake side of a blower. The outlet side of the blower then conveys the air through a rotary air lock on the bottom of the cyclone separator, where the pellets are again picked up. The airstream and pellets are then directed into the top of a silo, where the pellets fall into the silo and the air exits at the top.

The resin can be picked up by a vacuum system from the bottom of the silo and conveyed to another cyclone separator that sits on top of the hopper of a molding machine. A separate blower supplies the vacuum for this system.

The pneumatic conveying system for packaged resins is not too different from that shown for bulk resins. In the system for packages, shown in Figure 23.2, the resin is picked up out of the gaylord by a suction wand and moved to a cyclone separator, where the resin drops out and the fines and the air exist at the top. The cyclone separator sits on top of a hopper on the molding machine. The airflow and the fines move to a filter and then to the inlet side of a blower.

When bags are used rather than gaylords, the bags can be emptied into a receiving bin that acts like a gaylord. The bags can also be emptied directly into the hopper over the molding machine. However, this practice is not preferred because of the potential for con-

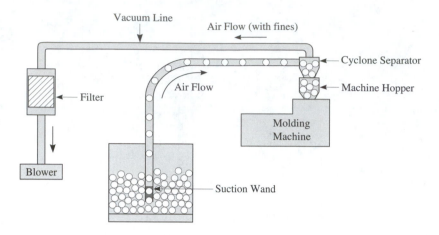

Figure 23.2   Pneumatic conveying system for packaged resins.

tamination. Loose paper from the bags, dirt, and foreign objects can be accidentally introduced. If the potential for sabotage exists, these open hoppers are likely places for problems to occur.

Some resins need to be dried before molding. This drying is usually done by blowing warm air through the resin for a few hours. The resin becomes dry because the warm air has a greater moisture-carrying capacity and therefore causes the moisture on the pellets to evaporate. The opposite effect is seen when air gets cooler. The cool air cannot hold as much water and so droplets of water (dew) form and coat solid surfaces in contact with the cool air.

Several arrangements are possible to accomplish this drying and still convey the resin. The simplest is to blow the warm air through the hopper on top of the molding machine. This requires no special equipment other than a special drying hopper. If the residence time in the hopper is not sufficient to properly dry the resin, then a separate dryer is placed in the conveying system. This dryer can be fed just like a small silo. The resin can be picked up from the dryer by a vacuum system and conveyed to the hopper as would occur from a silo.

PLANT LAYOUT

For large plants, the layout of the plant for efficient movement of material is critically important to overall efficiency. Without this layout, movements of the materials from receiving to production and then to shipping or inventory could interfere with each other with crossing patterns. If the plant is to have some automated material movement, such as conveyor belts, the layout is even more critical. Even in small plants, much difficulty can be avoided by simply thinking through the patterns of movement within the plant.

In large plants where the motions are complicated and, occasionally, even in small plants, a computer simulation method *(factory simulation)* is used to graphically depict the movement of the material around the plant. These programs also allow each manufacturing step to be simulated, thus looking for material buildups at points of inefficiencies (pinch points in the process). Such simulations are especially helpful when considerable assembly or multiple manufacturing steps must be carried out.

Ideally, of course, the plant should be simulated before it is even built. More common, however, is to establish the plant in an existing building that can be only slightly modified to accommodate the plastic-molding operation. When this occurs, simulation techniques can still be used to test several equipment and department arrangement plans within the restrictions imposed by the building.

QUALITY ASSURANCE

This text is not a quality text. However, quality is an important part of all manufacturing and often plays a critical role in plastics molding. In establishing a quality assurance system for the plant, several key factors should be considered:

- Studies over many years have shown that quality in a part is best produced by controlling the process rather than trying to separate out the good parts from the bad by frequent (usually 100%) inspection. Not only is 100% inspection highly labor intensive, it leads to high amounts of waste and just doesn't work as well as controlling the process. The system of controlling the process to control product quality is called *statistical process control (SPC)*.

- Process control is best done by the operators of the production line itself. They need to accept the responsibility for keeping the line within the parameters that will result in good product. The job of the quality assurance department is then to assist them in this task by defining the machine setting ranges that are most likely to result in good-quality parts, providing the inspection tools and methods for correctly measuring the important dimensions of the part, and assembling the statistics to assist the operators in taking long-term actions for improvement.

- Improvement needs to be continuous. The process of quality improvement employs many tools that can look at processes and products over many days, months, and years to help in detecting trends that cause the quality to decline or to be less than optimal.

- Everyone in the organization needs to be quality conscious and dedicated to its improvement. With this type of attitude, the entire organization—as well as those who interact with it (suppliers and customers)—will actively work to improve quality and, therefore, to improve profitability.

- As products become more sophisticated, the methods for determining quality and for controlling the process also need to become more sophisticated. Real-time controllers for many plastic-manufacturing processes are now available, although they should be used with caution if the process is not in a steady-state condition of good-quality operation.

CASE STUDY 23.1

Establishing QC for a PET Bottle Plant

The manufacture of PET soft-drink bottles has become a high-speed, high-technology operation in which a typical blow molding machine can produce up to 20,000 bottles per hour. At such high production rates, the need for careful quality control at all phases of the bottle-making process is critical or many reject bottles could be made before problems are detected. The highly competitive nature of the business requires that bottles of the highest quality be manufactured.

The resin supplied to bottle fabricators is a special grade of PET that has been formulated specifically for the bottles. Because the resin manufacturers have strong quality standards, the bottle manufacturers make few quality checks on a resin manufacturer's product after it has been qualified and used for a reasonable period of time. Some spot checks are made, but these are as much to characterize the resin as to check its quality.

Most bottle-manufacturing facilities are high-volume production facilities that receive the resin by rail cars that hold about 180,000 pounds (80,000 kg) each. The resin is unloaded and conveyed pneumatically into silos. The resin is then dried and conveyed automatically to the feed hoppers of the injection molding machines. The color concentrate, when used, is metered automatically by weight into the hoppers. The preforms are injection molded in large presses (about 200 tons or 1800 kN) that have molds with typically 64 (8 × 8) cavities. All the molds are automatically unloaded, and the gates are all submarine type so that no operator clipping is required. The runners are reground and reused at a small concentration level with the virgin resin.

The preforms are sampled using a MIL Standard sampling technique and checked for quality. The most important quality characteristic of the preforms is the flow pattern of the resin during the molding. This pattern is detected by observing the preform under polarized light. Patterns of acceptable quality preforms are used as visual comparisons so that the trained quality control operator can identify preforms that are not being made correctly. Adjustments to the injection molding machines can be made, although everything must be done quickly because the preforms move by pneumatic conveyance directly from the injection molding machine to the blow molding operation. The emphasis in injection molding is, therefore, process stability. The machine parameters are tightly controlled within long-established limits. Statistical process control (SPC) is a required method of operation. Control charts that focus on the quality of the preform, as measured by the polarized light test, are the major source of process monitoring. The thread dimensions, and other dimensional characteristics, are also measured. In general, the injection molding operation runs without problems for most bottle manufacturers. Small deviations in part size, except for the threads, are not critical to the successful making of a bottle.

The blow molding operation is done by large machines that are divided into two interconnected sections. In the first section the preforms are individually captured and held by the indexing part of the machine and then moved in a continuous belt process through an oven for heating. The residence time and the oven temperature are set to standard limits established by experimental design procedures. One interesting operational requirement, in addition to the obvious need to properly heat the preform, is the need to keep from heating the threaded portion of the preform. The threads were molded precisely in injection molding, and they should not be reheated or their dimensions could change.

After the individually indexed preforms have been heated, they are moved on a continuous line into the blow molding part of the machine. By passing through several sections, the bottles are stretched and then fully blown. The total time for blowing is only a few seconds.

The blowing operation is more likely to have quality problems than is the injection molding operation. The machine is much more mechanical, so deviations can occur as the machine parts wear or deviate from their positions. Therefore, several quality control checks are made to verify that the molding was done correctly.

One quality check is the bottle-pressurization test. Bottles are inflated to 400 psi (2.8 MPa) to ensure that they will not break under the filling pressure. Filling pressure is typically 150 psi (1.0 MPa), so the higher test standard gives some margin of safety in this critical property. Bottles are sectioned so that the wall thicknesses can be verified. The bottles are cut so that tensile properties can be run to verify minimum tensile modulus requirements. Bottles are also compressed to verify that they are stiff enough to withstand the pressures of filling without collapsing. Another quality check is done by filling a bottle and dropping it from a set height, usually 6 feet (2 m). This drop is considered to be the worst-case in-use problem for the bottles.

As you can see, these tests are a combination of standard tests and product-specific tests. This mixture of standard and special tests is typical throughout the plastics industry. Standard tests are used as much as possible so that the carefully designed testing procedures can be followed and standard testing equipment can be used. The standard tests also allow designers to anticipate test results. Special tests typically reflect in-use conditions and are, therefore, product specific. These test procedures are often set by the manufacturer to ensure that the product will meet performance standards. The manufacturer usually uses the results from these special tests as a sales aid, thereby demonstrating not only that the product performs well, but that the manufacturer is responsible in assuring good performance. Some trade associations adopt tests to ensure that all manufacturers are producing acceptable product. For instance, the plastic pipe manufacturers must subject their product to rigorous testing so that they can carry a certification of the association, a certification that is required by most purchasers.

The blow molded bottles are also checked for surface imperfections. Unless they are very large, these imperfections are rarely a reason for rejecting a bottle. Small nicks and imperfections are common because the preforms bang against each other when they are conveyed from injection molding to blow molding. Small nicks are difficult to see in the preforms. When expansion occurs during blow molding, however, the nicks also expand and become much more apparent. Fortunately, these nicks have little effect on performance, and most are covered by the label.

The labels are put on the blown bottles in a separate machine. The bottles are transferred from the blowing machine to the labeling operation by a continuous conveyor, with the bottles still individually indexed. Each bottle is labeled and then sent to an automated palletizer and stacker. An automatic visual detecting device monitors that the label has been properly applied. After palletizing, the bottles are wrapped with shrink-film for shipping to the bottler, where they are filled.

SUMMARY

The operational activities conducted on the shop floor to support the manufacturing effort can be very important in determining the overall quality and performance of the products and of the plant. All employees should be aware of these activities and work to improve them.

Safety and cleanliness are of utmost importance in maintaining a good workforce and in improving the efficiency of the facility. Safety also has a strong psychological effect on the workforce.

Resins are conveyed within the manufacturing facility from storage to the molding machines. The conveying system, when automated or semiautomated, is usually pneumatic. These automated and semiautomated systems move the pellets efficiently and cleanly. Provision for resin drying can also be included within the conveying system.

Some manufacturers make little effort to have good support systems. These manufacturers are usually less able to compete. Long-term costs are actually higher when these support systems are ignored.

Proper plant layout and design are important for efficient operations and can be improved by simulating and animating the process. Computer software packages, which easily accomplish this simulation, are readily available. If the plant is so small that such a system is not needed, at least some careful thought should be given to minimizing cross traffic and insuring that the material flows through the plant in an organized fashion.

Quality is important in all manufacturing operations. The most effective method for ensuring that the product is of good quality is to control the process making the part. Efforts to separate good parts from bad through 100% inspection are highly labor intensive, result in high amounts of waste, and are simply not effective. All persons in the plant should be quality conscious, with the operator of the line having primary responsibility for assuring quality.

GLOSSARY

Factory simulation   A computer software system that allows the operations in a plant and the design of the plant (equipment layout, etc.) to be viewed in animated simulation.

Material safety data sheet (MSDS)   A sheet that lists the ingredients and dangers of a particular chemical or plastic material.

Quality assurance   The process of controlling all aspects of the process and product to assure that quality is built into the product.

Statistical process control (SPC)   The method of controlling the quality of the part by controlling the operation of the process.

QUESTIONS

1. What is an MSDS?
2. List three critical elements in maintaining a safe workplace.

REFERENCES

ASM International, *Engineering Materials Handbook, Vol. 2—Engineering Plastics,* Metals Park, OH, 1988.

Braun, Dietrich, *Simple Methods for Identification of Plastics* (2nd ed.), Munich: Hanser Publishers, 1986.

Shah, Vishu, *Handbook of Plastics Testing Technology,* New York: John Wiley and Sons, Inc., 1984.

Appendix 1   Cost Estimating Form for Injection Molding

Part name:		Customer:				Date:	
Part number:		Part print/revision:				Estimator:	
Request for quote reference:						Approved:	
Resin and Additive Costs	Resin and Grade	Source	Price	Additives	Additives Costs	Additive Content per Part	Total Material Cost
Part Cost	Part Weight (grams)	Waste Factor	Adjusted Weight	Material Costs		Factory Cost Contribution per 1,000 Parts (Part Cost)	
Tooling Costs	Tool Description	External Cost	Internal Cost	Total Cost	Expected Quantity	Factory Cost Contribution per 1,000 Parts (Tooling)	
Machine Costs	Cavities	Cycle Time (s)		Parts/h	Auto/ Operator	Machine Rate	Factory Cost Contribution per 1,000 Parts (Machine)
Secondary Operation Costs	Operation	Cycle Time	Rate	Operation	Cycle Time	Rate	Factory Cost Contribution per 1,000 Parts (Secondary)
Purchased Items Costs	Item	Source	Cost/ 1,000 Parts	Item	Source	Cost/ 1,000 Parts	Factory Cost Contribution per 1,000 Parts (Purchased)
Packaging and Shipping Costs	Material	Source	Cost/ 1,000 Parts	Material	Source	Cost/ 1,000 Parts	Factory Cost Contribution per 1,000 Parts (Packaging)
Total Factory Cost per 1,000 Parts							
General and Administrative Costs per 1,000 Parts							
Marketing Costs per 1,000 Parts							
Total Cost per 1,000 Parts							

Form created by A. Brent Strong for *Plastics: Materials and Processing*

Appendix 2 Plastics Design/Selection Matrix

Procedures:
1. After confirming the need for the new product, develop a set of functional specifications which define **what the part should do in actual use.**
2. Select the key properties (usually 3 or 4) which must be met in order that the part perform its functions properly. List these on the Design/Selection Matrix at the tops of the appropriate columns.
3. Give numerical weighting factors to the properties (usually numbers from 1 to 10 with the most important being 10). The weighting factors should reflect the importance of each particular property or characteristic in achieving part performance. Put the weighting factors into the appropriate columns in the matrix.
4. Investigate the performance of each resin for each of the properties and list the value in the appropriate Value/Rank column. Then, rank the resins according to their performance value for that property with the best resin receiving the highest rank. Proceed through all of the properties so that each resin is ranked according to each property.
5. Multiply the rank by the weighting factor for each property and each resin to obtain the score. Total the scores to see which resin has the highest performance score overall.

Note: For additional information on the use of this chart, see the example in the chapter on design of plastic materials.

Resins	Properties									Total Score
	Property 1:		Property 2:		Property 3:		Property 4:			
	Value/ Rank	WF = Score	Value/ Rank	WF = Score	Value/ Rank	WF = Score	Value/ Rank	WF = Score		Total Score

* WF is the weighting factor

Answers to Selected Questions

CHAPTER 1

1. What are the three necessary and sufficient criteria that must be satisfied by all plastics materials?

   The material must be (a) composed of very large (polymeric) molecules, (b) must be synthetic, and (c) must be formable.

3. Identify and describe the forms of plastic resins.

   Resins are usually liquids (like honey), granules (like laundry soap powder), flakes (like uncooked oatmeal), or pellets (like very short spaghetti).

5. What is the first modern plastic that was synthesized with a specific set of properties in mind, who sponsored the work, and when was it done?

   *One correct answer would be Celluloid, invented by James Wesley Hyatt in 1868. However, the first truly synthetic plastic that was made to a **specification** was nylon. The work was sponsored by the DuPont Company (Wallace Carothers was the chief scientist) and the work was done in the 1930s.*

7. Define the term "polymer" and relate it to the term "plastics."

   Polymer is a very long molecule made by joining many small molecules together. All plastics are polymers, but some polymers, for instance, natural polymers such as fur and leather, are not plastics.

CHAPTER 2

1. Describe the differences in the carbon and oxygen atoms.

   Carbon has six protons in the nucleus and six electrons. (The number of neutrons determines the isotope of carbon but is unimportant for the differences in behavior between carbon and oxygen.) Oxygen has eight protons and eight electrons.

3. Identify the type of bond and the product formula expected between potassium (K) and bromine (Br) and explain the basic nature of this bond. Show the resulting electron configurations of K and Br after the bond is formed.

The materials will form an ionic bond since one is a metal (loses electrons) and the other is a nonmetal (gains electrons). The product formed with be KBr. After the bond is formed, K will have transferred its outer electron to Br. Potassium will then have a stable octet and a net charge of +1. Bromine will have a stable octet and a net charge of −1.

5. Describe the type of bonding between carbon and chlorine.

Carbon and chlorine are both nonmetals and will therefore share electrons to form covalent bonds. Carbon will always form four bonds by sharing its outer four electrons. Since each chlorine needs only one electron to complete its octet, four chlorines are required to give carbon its four bonds. The resulting material is CCl_4.

7. Why is it harder to make very long polymer chains using the condensation polymerization method than using the addition polymerization method?

In the condensation reaction every monomer molecule can react with equal probability upon encountering any monomer of the opposite type. This situation favors the formation of many small chains which grow simultaneously. As the chains get larger, their ability to continue reacting is slightly less than small chains because the large chains have less mobility and, therefore, less chance to encounter a monomer with which it can react. In addition polymerization, encounters between monomers do not form bonds unless one of the monomers has been initiated, that is, reacted with an initiator so that the monomer contains a free radical. Therefore, in addition polymerization, since only a few of the monomers are initiated, the growth of the chains will be largely at the sites that are on the already formed polymers, thus making those polymer chains even longer.

9. Define monomer and polymer. Write typical polymeric repeating unit structures for both addition and condensation polymerization and explain the various symbols contained therein.

A monomer is a simple molecule that can be chemically reacted to form a long chain. A polymer is that long chain of monomer units. In a typical repeating unit structure for a polymer made by addition polymerization, the monomer is represented without the carbon-carbon double bond and with two bonds extending beyond the basic monomer unit. These bonds indicate that the basic unit bonds to additional units on either side of the unit shown. The entire basic unit is enclosed in parentheses (which normally go through the lateral bonds) to indicate that the basic unit exists as a group that is repeated many times down the chain. A subscripted n follows the second parenthesis and represents the number of times the basic unit is repeated down the chain. For polymers made by the condensation polymerization method, the basic unit is the combination of the two monomers with the monomers joined at the active sites but with the small, condensate molecule removed. All the other symbols are the same as for the addition polymerization case.

11. Describe crosslinking and the resultant properties that it will create.

Crosslinking is the process of connecting polymer chains with covalent bonds. When this occurs, the total length of the polymer chains increases tremendously which raises the melting point of the polymer, usually beyond the decomposition temperature. Therefore, polymers that have been crosslinked will decompose before they melt.

13. What is the molecular difference between thermoset and thermoplastic materials?

The thermoset materials are crosslinked and thermoplastics are not. This crosslinking raises the total length (molecular weight) of the thermoset and thereby prevents the material from being remeltable.

CHAPTER 3

1. Contrast the interatomic or intermolecular forces present in solids, liquids and gases. Explain the consequences of these forces and how these forces are normally overcome.

   In solids the interatomic and intermolecular forces are very strong. The particles (atoms or molecules) are relatively fixed in relation to all the surrounding particles. These materials resist movements of the particles. Hence, solids are rigid. In liquids the interparticle forces are less strong than in solids, but there is still some association between the particles. The particles can move freely to take the shape of the container but will not fill the entire volume of the container. In gases the interparticle forces are nearly non-existent. Gases can take both the shape and fill the volume of the container. The interparticle forces are normally overcome by adding thermal energy to the material. This thermal energy causes the particles to vibrate, rotate, and translate, thus eventually breaking the interparticle forces.

3. Name three methods of reducing the amount of thermal degradation that might occur in a heat-sensitive plastic. How does each reduce thermal degradation?

   (1) Reduce the amount of regrind (as a percentage) that is used during processing. This assures that the heat history of the total material will never become large, thus reducing the likelihood that degradation from accumulated thermal processing will occur. (2) Add thermal stabilizers—materials that preferentially accept the heat and, therefore, reduce the amount of heat encountered by the polymer. (3) Add processing aids—materials that facilitate processing, generally by allowing the processing to occur at lower temperatures. Hence, the total amount of thermal energy added to the system is reduced. (4) Use some mechanical energy to melt the plastic material so that all of the energy is not from thermal sources.

5. What are thermal stresses and how are they caused in plastic materials? How can they be relieved?

   Thermal stresses are energy concentrations that are retained in a material resulting from the inability of the material to change shape as would normally occur upon cooling or heating. These thermal stresses can be relieved by slowly heating the material to a moderate temperature (annealing), thus allowing the molecules to move and the stresses to relax.

7. Why must average molecular weights be used for polymers rather than exact molecular weights based upon the molecular formula as is done for small molecules?

   Small molecules have exact formulas which always apply. In polymers, the chains can be of varying length and so a general formula must be used. This general formula represents the number of monomer units by an unknown (usually an n). Hence, if the exact number of units in each of the chains is not known, some statistical measure of the polymer chain length must be used. The average molecular weight is a convenient statistical measure to use.

9. Contrast and discuss the difference between amorphous and crystalline regions in a polymer.

   In crystalline regions the polymer chains are tightly packed together which results in a high density. The polymers in these crystalline regions have relatively strong intermolecular bonds which increase the stability of the polymer and increase most mechanical and physical properties. Amorphous regions are zones within the polymer structure where no crystallinity exists. Even highly crystalline polymers usually have some amorphous regions. The physical and mechanical properties for largely amorphous polymers are lower than analogous crystalline polymers.

11. Discuss how you would expect the glass transition to be affected if a large, pendent atomic group were added to the monomer unit.

    Most large pendent groups will require more thermal energy to achieve long-range chain movements. Therefore, the glass transition temperature would be expected to increase.

13. Discuss the implications in MWD when the number average molecular weight and the weight average molecular weight are widely different.

    If all the polymers were the same length, the number and weight average molecular weights would be the same. Therefore, when the number and weight averages are vastly different, it implies that the molecular weights are also vastly different and that the MWD is wide.

15. Discuss how copolymerization would affect crystallinity.

    The nature of copolymerization would cause more randomness to occur along the molecular chain than would the corresponding homopolymers. Therefore, crystallinity would be lower.

17. Why does processing of a crystalline material like nylon require more critical thermal control as opposed to an amorphous material like ABS?

    Nylon, like all crystalline materials, has a sharp melting point. Processing below this temperature is virtually impossible. Processing much above this temperature can make the melt excessively runny (low viscosity). Amorphous materials (like ABS) have a broad melting range and processing 50 to 75°F below T_{mp} is possible. Most grades of nylon, for example, do not process below 508°F, the point where it loses crystallinity and is capable of flow. ABS can be processed from 420 to 550 °F.

CHAPTER 4

1. Explain what is meant by a viscoelastic material and relate its response to applied stresses by comparing the material with an elastic solid and a viscous fluid.

   Viscoelastic materials have some properties that are similar to elastic solids and some properties that are similar to viscous fluids. When a stress is applied to a viscoelastic material that is a solid, the solid will deform much like an elastic solid but will not immediately recover its original shape when the stress is removed. The viscoelastic material has used up some of the applied stress through internal heating and, hence, not all of the energy is available to assist in the shape recovery.

3. Using a molecular view, explain why compression strength is generally less than tensile strength in polymers.

   When a compressive load is applied to a polymer, the molecules are forced closer together. If the polymer is highly crystalline, this force will simply press on the crystals with little damage, much as would happen in a crystalline solid. However, in an amorphous material, such as most plastics, there is considerable space between the polymer chains, which will allow deformation of the polymers at lower stress levels. Furthermore, the spaces allow the polymers to collapse into the holes, thus reducing the ultimate compressive strength. Hence, the compressive forces are lower for polymers than for highly crystalline materials.

5. Discuss how creep is likely to be affected by aromatic pendant groups.

   Large pendant groups such as the aromatics will decrease creep because they cause interference between the molecular chains. This interference raises the energy required to move the chains relative to each other and, as a result, creep is decreased.

7. What is ASTM?

The American Society for Testing and Materials (ASTM) is an organization that publishes uniform testing procedures for common tests.

9. A stress-strain experiment was done on a plastic and the following data were noted on the plot of the experiment: ultimate tensile strength = 9000 psi, yield strength = 5000 psi, proportional limit = 4000 psi, strain-to-failure = .025, strain at yield = .020, strain at proportional limit = .015. The original length of the sample was 4 inches. What was the modulus?

    The modulus is calculated by examining only the slope of the stress-strain curve before yield, that is, before any non-linear behavior is seen. Hence, the slope at the beginning of the experiment is stress at yield (5000 psi) divided by strain at proportional limit (.015).

11. What is the likely affect on modulus of adding a filler? Why?

    Modulus will likely increase by adding most common fillers. The reasons are that the modulus is itself a stiff material. Secondly, the filler will likely fill into the spaces between the molecules and thus inhibit the molecules from moving over each other. This resistance to intermolecular motion is seen as an increase in modulus.

CHAPTER 5

1. Describe how ultraviolet light degrades plastics and why uv-light does not generally degrade metals.

    Ultraviolet (UV) light interacts with the electrons in covalent bonds, often raising the energy of the electrons so they are no longer in a favorable bonding energy level. When this happens, the bond is broken. This rarely occurs in metals because the energy of the sea of electrons can be excited with little effect on any individual bond between the atoms. Hence, the bonds remain intact even when some of the electrons are excited to higher energy levels.

3. Explain the relationship between plastic crystallinity and sensitivity to solvent attack.

    Highly crystalline polymers have strong secondary bonds that resist the formation of solvent-polymer bonds or, perhaps, require the formation of more or stronger solvent-polymer bonds to achieve solvent penetration into the polymer network. Hence, crystallinity reduces solvent sensitivity.

5. Explain in thermodynamic terms and molecular structure terms why small molecules are more readily solvated than large molecules.

    In thermodynamic terms, solvation will occur when the free energy for the solvation is negative, that is, the energy is more favorable for solvation than for staying in the solid state. Either the enthalpy term (ΔH), the entropy term ($T\Delta S$), or both, must be negative and dominate whichever of the terms might be positive. The enthalpy term will only be negative when the solvation bonds are stronger than the polymer bonds. This rarely happens. Hence, the entropy term must be more negative than the enthalpy term is positive. This can happen at high temperatures because T becomes very large. It can also happen when the entropy, ΔS, is large. The entropy will be large when great randomness is associated with the system. Because small polymer molecules can become more random when solvated than can large molecules, small molecules are more favorable to a dominating entropy term and will therefore solvate more easily. (The increased randomness with small molecules arises from the inherent structure that is always associated with large molecules, just because they are large and structurally more defined.) The molecular structure view of solvation is simply that small polymer molecules require fewer solvent secondary bonds to equal the secondary bonds between the polymer molecules. Hence, solvation is easier than for large polymer molecules.

7. Describe the key features of a polymer that would make it electrically conductive.

Electrical conductivity results from the ability of electrons to move freely within the material's structure. This can occur in polymers when the electrons are delocalized. One method to achieve delocalization is when the bonds in the polymer are conjugated double bonds. Another method to achieve conductivity is to add a conductive filler to the plastic material.

9. What is the limiting oxygen index (LOI)?

The LOI is a test for flammability of a plastic sample. In this test the sample is supported in a candle-like arrangement and ignited. The composition of the gases surrounding the burning sample is controlled and gradually the amount of oxygen is reduced or increased until the sample will just barely burn. That level of oxygen is called the limiting oxygen index.

CHAPTER 6

1. Explain the cause of the differences in structure in LDPE, HDPE, and LLDPE.

The structures are dependent on the amount of branching in the molecules. When the branches are long and frequent, the molecules cannot pack tightly together thus preventing crystallization. This is LDPE. When the branches are short and infrequent, the molecules can pack together and crystals can form. This is HDPE. When the structure is similar to HDPE but with slightly longer and more frequent branches, the polymers cannot pack together and the density is low like LDPE, but other properties are like HDPE.

3. Explain why isotactic polypropylene can crystallize while the other polypropylene stereoisomers cannot crystallize.

In isotactic polypropylene all of the methyl groups are on the same side of the tertiary carbon. This reduces the intermolecular interference between these groups and allows the molecules to pack tightly together, thus forming crystals.

5. Why should the concentration of regrind PVC be kept low in any rigid PVC extrusion operation?

Regrind material is rejected product that has been chopped into small particles so that it can be reprocessed to make new parts. This is an advantage of thermoplastics. With PVC, however, the tendency to degrade is so pronounced that each time the PVC is heated, the degrading gets worse. To further complicate the matter, the degradation of PVC is autocatalytic, meaning that the decomposition happens faster as more is formed (like a chain reaction). Therefore, if the concentration of regrind PVC is too high, the accumulation of degrading material will be too high in one place and the material will degrade rapidly.

7. Why is crystal polystyrene clear?

The clarity comes from the ability of light to pass through the material unimpeded by any internal barriers. Crystal polystyrene has no crystalline structure to interfere with the passage of light (it is amorphous) and generally has few additives. It is therefore transparent.

9. Would you expect ABS to be clear or opaque? Why?

Clarity comes only if light transmission is unimpeded in passing through the plastic. ABS is usually multiphasic and so every phase boundary is a potential light-scattering location. Hence, ABS is generally opaque.

CHAPTER 7

1. Which type of backbone substituents stiffen the polymer chains?

Functional groups (a) and (d) restrict the motion of the polymer chain and therefore result in chain stiffening while (b) and (c) give only minor chain restriction.

3. Compare the water absorptivity of nylon (6/6) to nylon (6/12) and explain the differences that are noted.

    Nylon (6/6) will have more polyamide groups per length of chain which will increase the polar nature of the polymer. This will cause the water absorption to be higher than in the (6/12) product.

5. Discuss why acetals are generally not made into fibers.

    The very high crystallinity of acetals makes them brittle and, therefore, not suitable for fibers.

7. If nylon is sensitive to moisture absorption, why can it still function well as a tow rope for water skiers?

    The absorption of cold water is very slow in nylon. Furthermore, tow ropes are usually not totally submerged for long periods of time and therefore do not have high absorptions. Furthermore, nylon ropes are seldom pulled to their limit in this application so the small decrease in tensile strength is not critical.

9. Why is a polyimide film ideal in a very high-temperature electrical application, even one in which the film may be slowly degraded?

    Polyimide film has a very high temperature capability. Then as it might degrade, rather than melt, it remains a solid that keeps the electrical conductors separated. The char that is formed may eventually conduct, but char formation is usually a slow process.

CHAPTER 8

1. What is meant by a stage B resin?

    These are thermoset resins which have been polymerized but not crosslinked. The resins are B-staged to improve the handling and control the cure.

3. What are the structural features in the melamine monomer that lead to the very high hardness of cured melamine resins?

    The melamine monomer has three amine groups attached to the ring, with each amine group containing a nitrogen and two hydrogens. Each of these nitrogens is capable of being a crosslinking site. Therefore, the hardness in melamine plastics comes from the very high crosslink density that is possible because of the multiple crosslink sites.

5. Discuss two methods that might be employed to make an epoxy resin tougher.

One method would be to add a rubber toughener to the epoxy resin. Another method would be to change the nature of the epoxy so that it is more aliphatic and less aromatic. (This would, of course, mean changing the type of resin being used.)

7. What is the origin of the variety in the properties of polyurethanes?

The urethane bond is easily formed with little consequence of the groups that are between the active groups on the polymers. Therefore, a wide variety of structural shapes are possible for these groups. These groups determine the basic properties of the polyurethane and, therefore, give rise to the variability.

9. If a polyol has four active sites per molecule and an isocyanate also has four active sites per molecule, why isn't a 1:1 mixture by weight a likely ideal concentration mixture? What should be the correct ratio?

Although the number of active sites is in the ratio of 1:1, the differences in molecular weight of the two components must be taken into account. For instance, assume that the polyol is a long-chain polymer with a molecular weight of 10,000 and that the diisocyanate is a small molecule with a molecular weight of only 100. To achieve complete reaction of all the reactive groups, there would need to be the same number of groups, not the same weight. Therefore, the weight of polyol should be 100 times the weight of diisocyanate to give the same number of polymers and the same number of reactive sites, a weight ratio of 100:1.

CHAPTER 9

1. What are the key polymeric structural features common to most elastomeric materials?

Elastomers are generally linear, aliphatic materials with structures that do not pack together easily and are therefore noncrystalline. These materials have highly random structures in their relaxed state but will become more ordered if stretched.

3. Why is the cis form of polyisoprene softer than the trans form?

The cis form will not pack together as easily as the trans form. Hence, there are fewer intermolecular attractions in the cis form and the molecules can slide over each other more easily. This ease of sliding and absence of intermolecular attractions means that the cis form is softer than the trans.

5. Explain why silicones have higher gas permeability than do carbon-based molecules with equivalent pendent groups.

The bonds along the silicone backbone (Si—O—Si) are longer than the bonds in the carbon polymer (C—C—C). This increase in bond length means that the silicone polymer has more space between atoms. This increase in space means that a diffusing atom or molecule can pass through the mass of molecules more easily in silicones than in the equivalent carbon polymer. Of course, if the pendent groups of the polymers are vastly different, the effects from these pendent groups on space between molecules may overwhelm the bond-distance effect. Likewise, crystallinity could overwhelm the bond-distance effect.

7. What is the effect of crosslinks on an elastomer and what is the structural explanation for their effect? What would happen to the amount of stretch, hardness, strength, and creep in an elastomer if the crosslink density were increased?

Crosslinks prevent the elastomer from creeping at high temperature. Also, crosslinks give some limitations to the range of elasticity for the polymer. These effects arise from the limiting effects of the crosslinks. The crosslinks bind the molecules together and do not permit unlimited stretching. Also, during creep, the crosslinks act as limiting forces to the continuation of creeping motion. If the crosslink density were increased, the amount of stretching would be lower, the strength higher, the hardness higher and the amount of creep would be lower.

9. Explain what may happen to the properties of a rubber material if the carbon black filler is poorly mixed into the batch.

The carbon black serves two main functions in a rubber compound. The first of these is protection against degradation from UV radiation as it depends upon the carbon black to absorb the UV radiation. This function is improved with dispersion of the carbon black, simply because many small and distributed particles can absorb more effectively than fewer, larger particles. However, this effect is not great. The other function of carbon black is to strengthen the elastomer. For carbon black to properly perform this function, the carbon black particles need to be widely distributed and intimately mixed with the elastomeric polymers because this strengthening comes from the interactions between the carbon black particles and the molecules. These interactions occur at the molecular level, rather than at the macro level. Therefore, if the carbon black is poorly distributed, the interactions will not, in general, be at the molecular level and will only be with a few of the molecules. This will have a significant effect on the strengthening of the polymer that would is for from the addition of the carbon black.

11. (a) Explain why stretching an elastomer causes an unstable state and, therefore, when the stretching force is removed, the elastomer recovers. (b) Explain why stretching of some polyethylene molecules results in "blushing" and this situation is not eliminated when the stretching force is removed.

(a) The entropy of the system decreases when the polymer is stretched because the stretching causes the molecules to become aligned and, therefore, more ordered. Increases in order cause decreases in entropy which causes a net increase in free energy and, therefore, the state is unstable. When the stretching force is removed, the entropy can increase and thus the energy will decrease, causing a return to the original state. (b) In stretching some polymers, the ordering of the molecules causes crystalline bonds to form. These bonds overcome the forces of entropy decrease and result in a net decrease in free energy, thus creating a more stable state. The crystalline structures will, therefore, persist even when the stretching force is removed. The blushing is caused by light refracting off the crystalline structures.

CHAPTER 10

1. Define simultaneous engineering, and point out a problem that might arise if it is not done.

Simultaneous engineering is the concurrent development of the part design and the process design. A problem that could arise if they are not done at the same time is that a part is designed that would be very difficult to make, where with just a minor recognition of the capabilities of a process, the design could be changed and the problem avoided.

3. Indicate why design rules must be done after a preliminary choice is made on both process and material.

The design rules are both process and material specific. For instance, the shrinkage that is to be allowed would vary depending on the material and might be different if the process chosen does not melt the resin, say with rotomolding instead of a melting process such as injection molding.

5. Discuss the advantages of rapid prototyping methods versus conventional prototyping.

Rapid prototyping allows a part to be made directly from a CAD rendering. This ensures that the prototype is the same as the drawing. Also, the time to create the prototype is much shorter. In some rapid prototyping methods, the material of the prototype can even be the same as the material of the part, thus allowing mechanical prove-out as well as dimensional verification.

CHAPTER 11

1. What are the differences between adiabatic heating and thermal heating as applied to extruders and why are both important?

   Adiabatic heating is caused by the mechanical action of the screw on the resin inside the extruder. Thermal heating is caused by the heaters that surround the extruder barrel. Both are important because they combine to give the internal energy needed for melting. Some especially thermal-sensitive polymers cannot tolerate much thermal heating (such as PVC) and therefore need to have more mechanical energy.

3. Discuss the temperature profile shapes normally used for nylon and polyethylene when using a general-purpose screw. Which is higher in actual value? What are the slopes of each?

   The nylon temperature profile is higher because nylon melts at a much higher temperature than does polyethylene. However, when nylon reaches that temperature, it melts over a narrower range. With a general purpose screw, normally the feed section would be shorter than with the ideal nylon-type screw. Therefore, the nylon temperature in the feed zone would be higher to compensate for the reduction in the adiabatic heating and the time in the feed zone. The compression zone in the general purpose screw is longer than in the nylon-type screw and therefore the heat can be reduced in the compression zone. The net result is a declining heat profile for nylon when a general purpose screw is used. The opposite is true for polyethylene. When using a general purpose screw, the heat profile for polyethylene increases.

5. Discuss three operational steps that can be taken to reduce the thickness of an extruded part.

   The drawdown can be increased by speeding up the puller with respect to the speed of the extruder. (Note that this will also increase molecular orientation.) The gap between the outlet of the die and the cooling tank can be increased which will allow some additional drawdown before the part enters the cooling tank. The sizing rings inside the cooling tank can be reduced or the vacuum inside the tank can be reduced to give a smaller diameter.

7. Discuss the relationship between the melt temperature and the location of the frost line in blown film manufacturing.

   As the temperature of the melt increases, the frost line rises farther up the bubble, that is, it gets farther away from the die. This is because the frost line is the point of crystallization of the resin and, with a higher melt temperature, the time required for the melt temperature to fall to the crystallization temperature will be longer.

9. During the operation of an extruder in making thin-walled tubing, several defects have been encountered that appear to be small globules of resin. Investigation of these globules reveals that they will soften but not melt when heated. Explain what these globules might be and suggest two possible remedies for eliminating them.

   The globules are likely crosslinked resin that has, for some reason, been created within the extruder. One likely cause is that the extruder may be overheating the resin, thus causing some of the material to crosslink. A solution for this would be to lower the temperature. Another solution would be to add thermal stabilizer additive to the resin which could also prevent this crosslinking from occurring. Another cause of the formation of crosslinked material could be nonstreamlined flow in the die. This can be investigated by stopping the extruder and carefully examining the die to see if material seems to be hanging up in some location. If that is the case, the die should be smoothed.

11. Why do the cross-sectional shapes of the die orifice and the extruded part sometimes differ?

The polymer material flows unevenly through thick and thin sections. Therefore, if a sharp corner is cut in the orifice, the corner will restrict the flow and the part will have a rounded corner. To obtain a part with a sharp corner, the orifice should have some restriction in the flow of the middle of the part that will the push flow to the corner, relatively.

CHAPTER 12

1. Why are injection molding machines not as effective for mixing additives or other resins as are traditional extrusion machines?

The length-to-diameter ratio (L/D), which is a measure of the ability of the machines to mix thoroughly, is much lower for injection molding than for extrusion. Injection molding machines are simply not long enough. Furthermore, the mixing intensity of injection molding machines is lower than in extruders.

3. What feature in a mold will allow a hollow, cylindrical part to be made?

Hollow parts are made by inserting a core into the mold so that the material will flow around the core, displacing the resin from the volume to be hollow. When the mold is openend and the part is removed, the core is withdrawn and the solid, hollow part is created. Normally the withdrawal of the core occurs simultaneously with the opening of the mold.

5. What is packing the mold and why is it important in obtaining good injection molded parts?

Packing is the process of continuing to inject resin into the cavity even after the cavity is full. As the plastic shrinks from thermal processes (thermal shrinkage), the part will reduce in size. This reduction can be compensated for by continuing the injection of resin as the cooling and shrinking occurs. Packing can ensure proper part size and can help eliminate sink marks which can occur in thick sections that shrink more than thinner sections within the part.

7. How does high crystallinity in a resin affect the way the resin is injection molded, including any postmolding considerations which might be made?

Crystallization normally gives materials a sharper melting point. That is favorable in injection molding because it makes the material easier to heat to a uniform viscosity. Furthermore, the crystallinity usually means that the material will cool and solidify more sharply, thus decreasing the time that the material must be cooled. After the part is removed, residual crystallization is common. This continuing crystallization can distort the part. Therefore, cooling fixtures are often needed to ensure that the part will crystallize and complete cooling without part distortion.

9. Why is it important to have the sections of the molded part as uniform in thickness as possible?

Thick sections cool more slowly than thin sections because of the time required for the heat to move through the plastic material. The cooling occurs from the outside toward the middle because the heat is transferred most easily from the outside. Therefore, the inside is liquid after the outside is solid. Because of the natural shrinkage that occurs upon cooling, the liquid can still shrink whereas the outside, which is solid, cannot. This may cause a sink mark over the center section as it cools and shrinks. This problem can be solved by making all the sections of the same thickness.

11. Assume that you are assigned to determine the minimum clamping force for a part to be molded out of polystyrene. The part cross-sectional area is 10×14 inches. What is the clamping force required if, as a general rule, 2.5 tons of force are needed for each square inch of cross-sectional area?

The clamping force can be calculated from the equation:

Clamping Force = (Injection Pressure) (Total Cavity Projected Area). In this case the total cavity projected area is 140 square inches. If we use the assumption of 2.5 tons of pressure (force) for each square inch of area we get a total clamping force by multiplying $140 \times 2.5 = 350$ tons clamping force.

13. Why is low specific heat capacity desired in a mold cavity material for some applications and a high specific heat capacity desired in others?

Specific heat determines the amount of heat necessary to raise the temperature of the material. For an injection mold that is cooled, a high specific heat is desired so that the heat from the part that is being molded will not cause the mold to heat. This will minimize the work of the coolant. In a mold that is heated, such as a thermoset mold, a low specific heat will allow the mold to be heated more easily.

15. Identify four advantages and two disadvantages of using aluminum rather than tool steel to make cavities for injection molds.

<u>Aluminum advantages</u>
1. Lower initial cost due to lower machining time.
2. Shorter delivery time.
3. More even heat distribution.
4. Shorter cycle from higher thermal conductivity.
5. Lighter weight.
6. Easier to cast, should that be the preferred fabrication method.
<u>Aluminum disadvantages</u>
1. Must be anodized to get hardness and anodizing may wear off.
2. Long-term wear is high.

17. What is the purpose of cold-well extensions?

These extensions of the runners are placed at points where the runners change directions and are a means of capturing the cold, leading front of the resin. By capturing the cold end of the resin flow, some assurance is given that the resin will flow smoothly through the gate without plugging.

19. Identify two advantages of using inserts in injection molds.

1. Inserts allow a single base to be used for multiple parts.
2. Inserts can be changed and modified easily should they be damaged.

CHAPTER 13

1. What is the principal problem in forming a bottle using injection molding?

If the diameter of the mouth of the bottle is smaller than the diameter of the body, some method must be provided to remove the core pin. The core pin will have the diameter of the body and, therefore, must collapse or be reduced in diameter by some other method. This is a difficult mechanical problem that is eliminated when blow molding is the method for forming the bottle.

3. List three advantages of the moveable mold system versus the rotary system for continuous extrusion blow molding.

The moveable mold system has fewer molds than the rotary system and therefore less investment. The moveable mold system is less complicated mechanically and is therefore less costly to buy and to maintain. The moveable system need not be synchronized as precisely as the rotary system.

5. Why can blow molding molds be made out of aluminum whereas injection molding molds are usually made out of tool steel?

The pressures associated with blow molding are far less than the pressures in injection molding. Therefore, deformation is less likely. Another reason for the ability to use aluminum in blow molding is that the material does not flow along the mold surface but rather is pressed against the surface. The flow of the plastic along the surface, as occurs in injection molding, is abrasive and wears out the mold. One further reason that aluminum can be used is the lower temperature of the material when it touches the mold. The high temperatures of injection molding resins when they enter the molds cause thermal shrinkage and contribute to abrasive wear.

7. What method is used in injection blow molding to achieve the type of part wall thickness control that can be obtained in programmable parisons?

Programmable parisons only have meaning in extrusion blow molding because they are made with a movable mandrel in the extrusion die. However, a similar effect can be achieved in injection blow molding if the core pin or the parison mold cavity is shaped. This shaping will create some thick section of the parison which can be placed in the locations that will travel the farthest in the mold.

9. What factors are likely to determine the maximum size part that can be blow molded?

The parison of the part must be able to be formed without excessive deformation under the weight of the parison itself. This is determined by the melt strength of the resin, the speed of formation of the parison, and the temperature. Also important are the wall thickness and the blow-up ratio.

CHAPTER 14

1. Discuss the pressure capabilities of vacuum forming relative to pressure forming.

With vacuum forming the maximum pressure available is the difference between the pressure in the evacuated chamber and atmospheric pressure. Hence, the pressure is limited to a max of one atmosphere. With pressure forming any pressure can be applied, although normally very high pressures are not needed. The pressures used in pressure forming, however, would always be higher than those used in vacuum forming.

3. What is backdrilling of a vent hole and why is it done?

Backdrilling is cutting a larger hole behind the part of the vent hole that touches the plastic sheet. This larger hole gives greater air flow and allows faster application of the vacuum.

5. Discuss the differences between blister packaging and skin packaging.

Blister packaging forms a pocket of stiff plastic material in a standard thermoforming method. The pocket is filled with the product and then sealed. Skin packs are made with a flexible plastic that is formed over the top of the part and the display card by drawing a vacuum through holes in the card. The skin pack is, therefore, tight against the product and the card whereas the blister pack is not tight against the product, but allows some movement of the product inside the pocket (blister). Normally the blister pack is preferred aesthetically but the skin pack is less expensive.

7. A deep cup (20 cm deep × 10 cm diameter) is to be formed. Compare thermoforming, blow molding, and injection molding as processes for making the cup. What technical and economic considerations should be considered in determining which method is best?

   To make such a long, narrow part by injection molding the mold must be quite large and the flow to fill the mold would be difficult. Blow molding, on the other hand, would be ideally suited technically because the parison would be long already and forming to the shape indicated would be fairly simple. Thermoforming, too, would be technically simple except for the possible problem of uneven sides where such a deep draw relative to the diameter is required (Depth ratio = 2:1).

   The economic considerations between blow molding and thermoforming are largely based upon the higher capital costs associated with the blow molding process. Therefore, justification for the higher costs must come from higher volume. Specific costs of equipment should be weighed against the number of parts to be made to give some measure of the break-even point.

9. What is the purpose of a vacuum surge tank?

   The surge tank, when evacuated, provides a location for the air to flow when the tank is opened to a closed chamber. The atmospheric pressure pushes the material out of the chamber and into the surge tank. Without the surge tank, the chamber would have to be evacuated by a pump which can be a rather slow process.

CHAPTER 15

1. What constraints are necessary for biaxial rotation to evenly coat the mold in a rotational molding process?

   Biaxial rotation must have both axes rotating simultaneously. The frequency of rotation should be different for each axis, and the rotational frequencies should not be even multiples of each other. The rotational speeds should be slow.

3. Explain how polyethylene is crosslinked in a rotomolding operation.

   Normally thermoplastics such as polyethylene cannot be crosslinked. However, if a peroxide or some other free-radical-forming material is added to the polyethylene and the temperature is raised, some hydrogens in the polyethylene can be removed and free radicals formed in their place. These free radicals can then react with crosslinking agents containing carbon-carbon double bonds which have also been added to the mixture. These crosslinking agents bond to the free radical locations on the polyethylene. Because the crosslink agents have at least two of these reactive carbon-carbon double bond sites, they can react with more than one free radical location and therefore tie together multiple polyethylene chains. Increased heat is needed to give the molecules the mobility to move together to form the crosslink bonds.

5. Why is there a limit on the thickness of parts that can be reasonably made by rotomolding?

   The thickness of the material determines, to some extent, the length of the heating cycle. If too much time is required to properly heat and fuse all of the material, some resin may begin to degrade. Hence, the limit on thickness is imposed by the maximum length of time before degradation occurs.

7. What is the major limitation on the distance between walls in a double-walled carrying case made by rotational molding?

The resin powder must be able to flow freely between the walls or nonuniform walls will result.

9. Describe why the weight of the resin charge is important in part performance.

    The weight of the resin determines the wall thickness. The wall thickness has a major effect on the length of the heating cycle. Because plastic is not a good thermal conductor, even small increases in thickness required disproportionate times for heating. If the heating is too long, degradation can occur. If the thickness increases without the knowledge of the machine operator, the last material to adhere to the surface may not get hot enough to fully fuse. Therefore, small changes in weight can affect the heat cycle and part physical properties.

CHAPTER 16

1. Briefly describe five different casting techniques.

    (1) Mold casting is when the casting resin is poured into a mold and forms the entire part. (2) Embedded is when an article is to be totally encased by the resin. The embedding technique uses a mold, similar to that used in mold casting, and suspends the article in the mold so that the resin fills the mold and completely surrounds the article. (3) Encapsulation is when an article is to be coated with resin but the final shape of the part is to be roughly the same as the coated article. Examples of this are painting and dip coating. (4) Cell casting is when the casting resin is poured between parallel glass plates separated by a gasket to retain the resin and set the thickness of the part. Sheets are typically made by this method or by the closely associated continuous casting method that uses two continuous stainless steel belts to define the surfaces of the sheets. (5) Solvent casting is done by applying a solution of a resin in a volatile solvent onto a belt, causing the solvent to evaporate, and then pulling the solidified resin off the belt.

3. What are plastisols and organisols?

    Plastisols are generally over 90% solids where the solid material is a powdered resin (such as PVC) and the nonsolid material is a plasticizer. Organisols are also mixtures of powdered resin and plasticizer but with the addition of up to 40% volatile solvent.

5. Discuss the advantages of casting over other plastic molding operations in equipment size and cost.

    Cell casting does not use high pressures and can, therefore, use a much wider variety of materials for molds and other processing equipment. Hence, the capital costs associated with casting are usually much less than would be required with other plastic processing methods. The absence of a need to withstand high pressures also means that casting can be used for making very large parts.

7. Why does casting have more design flexibility than other plastic manufacturing processes?

    Because of the ability to use flexible molds in casting, deep undercuts, high surface details and other design features that would be difficult to mold with other processes can be done easily by casting.

9. Identify three methods for controlling an exotherm in casting.

    Three methods for controlling an exotherm are making all sections of a part thin, casting thick parts in multiple pours, and minimizing the heat created by lengthening the curing time and eliminating the need to heat the material.

11. Define and describe slush molding.

    Slush molding is a process that uses a powdered resin suspended in a nonvolatile solvent as the starting material. The mold and processing equipment can be similar, at least in function, to standard rotomolding equipment. The mold is filled with the liquid material and is then heated and rotated so that the walls are uniformly coated with material that sticks as it becomes hot. When the material stuck to the walls of the mold has reached the desired thickness, the mold rotation is stopped and the excess liquid is poured out of the mold. The mold is then returned to the oven and rotated until fused. The cycle is completed by cooling and then unloading the part.

CHAPTER 17

1. Describe the four basic methods of creating a foam.

   Mechanical—vigorously mix the resin with a gas (usually air) to create a foam; chemical—create a gas by some chemical process within the mixture of resin and foaming agent so that the gas foams the resin; physical—create a gas by some change in the physical conditions of the resin and foaming agent mixture; hollow spheres—mix glass microballoons into the resin.

3. What is meant by "blowing the cells open," and how can this event be controlled?

   "Blowing the cells open" is the conversion of closed cells to open cells by the rupturing of the cell walls. This can be controlled by controlling the amount and rate of expansion of the gas inside the cells. This control can be achieved by limiting the expansion of the mold in high-pressure foam molding. It can also be controlled by reducing the temperature at which the foaming occurs. Other methods that limit the expansion of a gas, such as increasing the pressure under that expansion occurs, can also prevent blowing the cells open.

5. What is an integral skin, and how is it controlled?

   An integral skin is a tough outer layer of the foam that is formed when the cells collapse due to contact with the mold, with air, or with any substance other than the foam. The amount of cellular collapse to create the integral foam can be controlled by adjusting the pressure and temperatures surrounding the foam, the temperature of the foreign material touched by the foam, and the presence of any other foreign substance, such as water or dust that might cause the cells to collapse.

7. Explain how foam-in-place and RIM differ.

   In both foam-in-place and RIM the material is squirted into a closed mold until the mold is filled. The foaming usually occurs at the same time as the filling. The major difference is that in RIM the foamed part is removed after it has solidified and the mold is reused. In foam-in-place, the mold is the sides of the part that is to contain the foam and so the foam is not removed.

9. Predict how the acceleration curves versus stress curves would appear for different thicknesses of the same material. Draw curves for 2 cm, 4 cm, and 6 cm thicknesses.

   The curves are shown in the following figure:

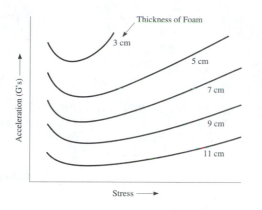

Art Guide: UAA17-01

— CHAPTER 18 ———————————————————————

1. Calculate the size of the press required to mold a part using a multicavity transfer mold in which the part size is 2.5 inches by 2 inches. The mold contains 4 cavities. There are 4 runners, each of which has a diameter of 0.375 inches and a length of 1.5 inches. The sprue has a projected area of 0.25 square inches. Assume that a transfer pressure 8,000 psi is needed.

   The parts and runner system are depicted in Figure 18.10.

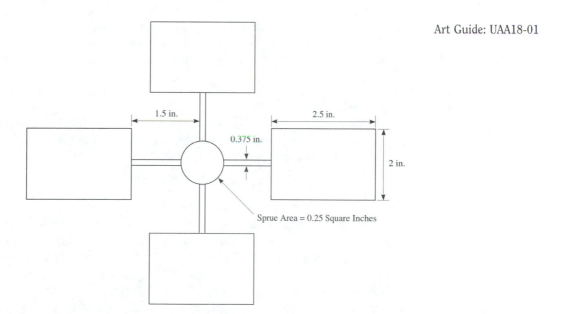

Art Guide: UAA18-01

Figure 18.10   Parts in a multiequity mold showing important dimensions for the calculation of press size.

$$Part\ dimensions = 2.5 \times 2\ in. = 5.0\ in.^2;$$
$$number\ of\ cavities = 4;$$
$$total\ cavity\ area = 4 \times 5.0 = 20.0\ in.^2;$$
$$runner\ area = 0.375 \times 1.5 = 0.562\ in.^2;$$
$$total\ runner\ area = 4 \times 0.562 = 2.25\ sq.\ in.;$$
$$sprue\ area = 0.25\ in.^2;$$
$$total\ molded\ area = 20.00 + 2.25 + 0.25 = 22.50\ in.^2;$$
$$press\ capacity\ required = total\ molded\ area \times transfer\ pressure\ required$$
$$= 22.50\ in.^2 \times 8000\ psi$$
$$= 180,000\ lb\ or\ 90\ tons$$

3. Explain how both undercuring and overcuring can cause part cracking on removal from the compression molding process.

   Overcuring can result in brittle parts that crack when the ejection system pushes the part out of the mold. Overcured parts may also stick to the mold. Undercuring can produce cracked parts because the strength of the part is lower than desired and may not be sufficient to withstand the pressures of ejection.

5. Describe poker chipping, and explain how it can be prevented.

   Poker chipping occurs in the ram extrusion process when the charges of compacted resin are not well bonded together. The billet made by the ram extrusion process can therefore be in separate charge portions. These are compacted together like discs or poker chips. This problem can be prevented by improving the consolidation between charges, usually done most easily by increasing the fusion temperature.

7. Why can't the cull, sprue, and runners from a transfer molding operation be reprocessed as regrind like runners in an injection molding operation?

   In a transfer molding operation the cull, runners, and sprue are made from thermoset materials that cannot be remelted, which is required for reprocessing as regrind material. In injection molding the runners and sprue are made from thermoplastics that can be remelted and reprocessed as regrind.

9. What is the major advantage of a flash type mold closure system in compression molding? What are the disadvantages?

   The flash type closure system is much simpler than the positive and semipositive systems. Therefore, mold costs are lower. However, the part dimensions are less well defined and, of course, flash must always be removed from the parts.

CHAPTER 19

1. What is the purpose of the resin in parts with long fibers at high fiber concentrations?

   The plastic resin is the matrix in these composites and its purpose is to bond the fibers together, give shape to the part, and protect the fibers from adverse environmental conditions.

3. Explain why very short fiber composites tend to be made with thermoplastic matrices while longer fiber composites use thermoset matrices predominately.

   The very short fiber composites are made with thermoplastic matrices because of the low cost of manufacture if traditional thermoplastic manufacturing methods are used. Furthermore, the applications for these materials are generally well suited to the thermoplastic composites, with additional strength or stiffness as provided by the fibers. On the other hand, composites with long fibers are made from thermoset because of the difficulty of wetting-out the fibers. The additional stiffness and higher temperature capabilities of thermosets also contribute to their use.

5. Why are sizes and coupling agents generally not required for aramid fibers?

Aramid fibers are inherently tough and therefore resistant to damage from later processing steps. Hence, a size is rarely needed. Aramid fibers are organic and therefore have good compatibility with the organic resins that are most often used for matrices in composites. Therefore, coupling agents are not needed.

7. Indicate at least three problems of most composite processes that are solved in RTM.

Most composite processes are open to the atmosphere, thus leading to escape of solvents and other volatiles and, therefore, pollution. Secondly, the RTM process can be easily automated whereas most composite processes cannot. Third, RTM permits both short and long fibers to be used.

9. A submarine driveshaft was made of epoxy and carbon fibers by filament winding. Indicate two reasons why composites have unique and excellent properties for this application. The part was cured in an oven at 250°F (120°C). On close examination, the shaft was found to spin irregularly and to cause vibrations. Identify four possible causes of this problem and four possible solutions.

Composites are excellent as driveshafts for submarines because they reduce vibrations more so than other materials, they are lightweight and so the forces to change the speed or direction of the shafts are lower than metal, and they resist corrosion. The cause of the vibration could be one or more of the following: (1) If cured horizontally, the resin may have flowed to the lower side of the shaft. This can be corrected by rotating the shaft during cure. (2) The mandrel on which the shaft was wound may not have been round or straight. Solve by truing the shaft. (3) The turning of the shaft could be at the natural frequency of the shaft. This can be solved by changing the shaft so that it vibrates at some other frequency. (4) The shaft may not be strong or stiff enough and the turning could be causing deflections. This can be solved by adding more windings to the shaft. (5) The winding pattern may not be balanced; that is, the windings may be predominantly in one direction so that when the part cures and thermal expansion and contraction occurs, the part is warped. This can be solved by making sure that the windings are symmetrical about a mirror plane in the middle of the winding pattern.

CHAPTER 20

1. Why does crosslinking increase the resistance of golf ball covers to cuts?

Crosslinking toughens the plastic material. This toughness arises from the greater ability of the crosslinked polymer to distribute energy throughout its structure because the crosslinks are natural connections between the polymer chains, thus transmitting the energy directly.

3. Would HDPE, LDPE, or LLDPE be the most strongly affected (crosslinked) by electron beam radiation? Why?

LDPE would be the most strongly affected by the radiation. LDPE is much more flexible than the other two materials. This flexibility indicates high internal motion of the molecules and, therefore, the ability to move to accommodate crosslink formation.

5. What would be the result of a leakage of gas into the acceleration chamber of the electron beam irradiation machine?

Gas molecules within the acceleration chamber would interfere with the electrons being moved through the chamber. This interference may cause scattering of the electrons or, perhaps, reaction with the electrons. In either case, the intensity of the electron beam would be reduced.

7. What is the likely cause of the inability of shrink-tubing to shrink sufficiently to bind two wires tightly when it is shrunk onto a connection between the wires?

    If the diameter of the shrink-tubing when it was crosslinked is too large, the tubing diameter may be too large to tightly capture the wires. The diameter can never be made to go smaller than the diameter at the time of crosslinking.

9. Studies have found that the rate of cross-linking of a thermoset material can be increased when both electron beam irradiation and thermal heating are used together. Explain why this is so.

    The thermal heating gives the molecules more movement and this movement can make the crosslinking reactions more effective and more rapid because there are more collisions between the molecules and, therefore, more reactive occurrences.

CHAPTER 21

1. Identify five factors that can be varied to optimize the machining of plastic parts.

    The machining of plastics can be optimized by varying tool shape (design), feed rate, cutting speed, process conditions (cutting fluid, and so on), and depth of cut.

3. Discuss the effect of plastic brittleness in snap-fit joints.

    Snap-fit joints depend on the ability of the plastic material to deform slightly without breakage or deformation. If the plastic part is so stiff that small amounts of movement will cause it to break, snap fits cannot be used. This breakage might also occur in the mold if a snap fit depends on an undercut in the part that must be stripped out of the mold.

5. Name and contrast four treatment methods for increasing the adhesive capability of a plastic surface.

    The surface can be treated with chemicals, plasma, flame, or corona. Chemical etching is not effective on all plastics and so its application is limited. Plasma is almost always effective, but plasma machines are expensive and some experimentation is needed to develop the correct gas and machine parameters to get optimal results. Both corona and flame are inexpensive and usually effective. Corona is somewhat faster than flame, but corona often etches the surface.

7. Identify three methods for coating a plastic with a metal.

    Electroplating, foil transfer, and vacuum deposition.

9. What advantage does a two-part adhesive system have over a one-part system?

    The two-part system is stable until the parts are mixed. Therefore shelf life is like to be much longer than for a one-part system.

CHAPTER 22

1. Identify three reasons why plastics have been slower to be recycled than aluminum cans.

    The greater variety of plastic materials has led to difficulty in sorting; the greater value of recycled aluminum relative to the cost of making new aluminum and the low differential between plastic virgin and recycled materials; the ease of separating aluminum cans (obvious shape) and the corresponding difficulty in separating some types of plastics by shape and type.

3. What are the three major components (material types) of a landfill?

    Paper, yard wastes, and plastics.

5. What is regeneration, and what is its potential value?

    Regeneration is the conversion of plastic material back into the chemicals used to make them originally. The advantage of this method is that the chemicals have a higher value than some of the mixed recycle of plastics from which the regenerated chemicals were derived.

7. What is the chief economic disadvantage of recycling plastics?

    The costs associated with recycling raise the cost of recycled materials too close to the value of the original virgin materials.

9. What is the biggest problem associated with incineration as a waste disposal method?

    Incineration has a great potential for air pollution from the waste gases of the burning process.

CHAPTER 23

1. What is an MSDS?

    A MSDS is a Material Safety Data Sheet that is sent by the manufacturer of the raw material. The MSDS indicates the chemical components in a liquid material and any potential hazards. It also indicates parameters such as boiling point, flash point, chemical reactivity, and toxicity.

3. List two specific dangers associated with molding of thermoplastics.

    Slipping on spilled pellets and falling; and burns from the hot areas used in melting the thermoplastic.

Index

Abrasion, 146, 200, 232, 715
ABS (*see* Acrylonitrile butadiene styrene)
Accelerators, 281
Accumulator, 488
Acetal, 239–241
Acrylic, 157, 179, 246, 691, 718
Acrylic syrup, 247, 575, 580
Acrylonitrile butadiene styrene (ABS), 148, 223, 311
Active site, 265, 273, 279, 292
Addition polymerization, 51–56, 57, 60–62, 193, 207, 211, 217, 248, 291
Additives, 147
Adherends, 714
Adhesive, 714, 716
Adhesive bonding, 714–720
Adiabatic heating, 359
Advanced composites, 644
Advanced neural networks (ANN), 458
Advanced thermoplastics, 653
Aging, 92
Aliphatic, 46–47, 194, 282, 304
Aliphatic thermoset elastomers, 307–312
Alkyd, 284
Alligator hide, 379
Alloys, 219
Allylic, 284
Alpha (α)—cellulose, 277
Alpha (α)—olefins, 200
Alternating copolymer, 64
Aluminum molds, 496
Amino plastics, 276–279
Amorphous, 76–79, 85, 164, 178, 213, 242, 246, 304, 313, 691
Anaerobic adhesives, 718
Anisotropic, 648

Annealing, 96, 635
Antioxidants, 159
Aramid, 110, 238, 653–656
Arc resistance, 177, 277
Area ratio, 530
Aromatic, 46–47, 175, 182, 238, 252, 289
Arrhenius equation, 96, 158, 170
ASTM, 334
Atactic, 210
Atom, 26
Autocatalytic decomposition, 213
Autoclave, 280, 664, 666

B-staging, 271, 272, 274, 287, 290, 291, 581, 629
Backdrilling, 527
Balanced mold, 434
Balatta, 309
Ballistics particle manufacturing (BPM), 344
Bambooing, 379
Banbury mixer, 318
Bank (material), 365, 399
Barcol hardness, 146
Barrel, 356, 377
Barrier cloth, 664
Barrier property, 169–173, 405
Barrier screw, 367
Bending force (*see* Flexural force)
Bent strip test, 168
Beryllium-copper, 440
Biaxial rotation, 546
Biaxially oriented, 396, 398, 505
Bifunctional, 57
Billets, 634
Billow forming, 515
Bimodel molecular weight distribution, 106
Binder, 683

Bleeder cloth, 664
Bleed-out, 216
Blends, 219
Blister pack, 218, 522
Block copolymer, 64
Blow molding, 483–507
Blowing agents, 591
Blow-up ratio, 398, 501
Blushing, 141, 169, 210
BMI, 291
Bonding, 31–39, 195
Branching, 54, 196
Break point, 625
Breakdown voltage, 176
Breaker plate, 360, 377
Breathing the mold, 275, 630
Bridge molecule, 268, 280
Bridging, 359, 375
Brittleness, 144
Brookfield viscosity, 129
Bulk density, 77, 553
Bulk factor, 77
Bulk material, 634
Bulk molding compound (BMC), 284–286, 629, 659, 748
Bulk properties, 4
Bumping the mold (*see* Breathing the mold)
Bun, 598
Buna, 311
Butadiene rubber (BR), 221, 310
Butyl rubber, 221, 311

Calendaring, 166, 216, 319, 399
Capacitive coupling, 256
Capacity, 369
Caprolactam, 238, 580
Carbon atom bonding, 40–48
Carbon black, 156, 187–188, 311
Carbon fibers, 222, 246, 653
Carbon-carbon double bond (*see* Double bond)
Carcinogen, 211
Cast molds, 442
Casting foams, 597
Casting, 247, 567
Catalyst, 8, 51–52, 55–56, 265
Cauterizing, 255
Cavities, 438
Cavity mold, 511
Cell casting, 575
Cellular plastics (*see* Foams)
Cellulose (cellulosics), 7, 161, 254, 402
Ceramics, 16, 28, 94, 643
Chain-growth polymerization (*see* Addition polymerization)

Char, 88, 92, 177, 182, 252, 256, 275
Charpy test, 144
Check valve, 425
Chemical etching, 442, 715
Chemical foaming, 591
Chemical properties, 3, 159
Chemical recycling, 749
Chloroprene rubber (CR), 312
Chopped fibers, 652, 657
Cis, 308
Clam shell mold, 667
Clamping, 443
Class A surface, 287, 637
Cling, 206, 216
Closed cell foams, 589
Clothes hanger die, 394
Coating, 398, 726
Coblow molding, 499
Coefficient of thermal expansion (CTE), 81, 94, 232
Coextrustion, 403
Cohesive, 716
Coining, 635
Coinjection molding, 459
Cold forming, 634
Cold well extensions, 434
Collapsing guide, 397
Collective properties, 4
Colligative properties, 72
Color, 147, 180
Comingled recycle, 744, 746
Commodity thermoplastics, 193, 232
Comonomer, 200, 205
Compatibility agents, 221
Composite materials, 9, 147, 282, 643
Compounding, 13–14, 220, 317, 354, 366
Compression molding, 276–277, 285, 319, 615, 659
Compression or melting section, 359
Compression press, 616
Compression properties tests, 139, 604
Compression ratio, 360, 423, 634
Compression set, 140, 607
Compressive force, 123–125, 139, 306
Computer-aided design (CAD), 330, 342, 441
Computer-aided mold-flow analysis, 432, 466
Concentrates, 382
Concrete stakes (case study), 344
Condensation polymerization, 6–8, 56–63, 274, 278, 291
Conduction, 610
Conductivity, 172–175
Cone and plate viscometer, 129
Conjugated bonds, 175

Constrained geometry polymers, 218
Construction, 17
Continuous casting, 575
Continuous extrusion blow molding, 486
Continuous fibers, 652
Convection, 610
Converter, 354
Copolyester elastomers, 374
Copolymer, 63–65, 201, 205, 219, 222, 240, 250, 310, 313
Core pin, 432, 447
Cored screws, 359
Corona, 715
Corotating, 366
Corrosion, 214
Corrugated sheet/pipe, 403
Costs of plastics, 11, 469
Counterrotating screws, 366
Coupling agents, 655, 715
Covalent bonding, 34–36, 717
Cram feeding, 380, 386
Crazing, 168, 211, 277
Creep, 81–83, 141–143, 232, 245, 250, 272, 306, 313
Crosshead or offset die, 391, 393
Crosslink density, 284, 289, 307
Crosslinking, 63, 168, 173, 188, 203, 265–272, 287, 306, 553, 682, 690, 697
Cryogenic flash removal, 630, 705
Crystal polystyrene, 217
Crystalline, 32, 34, 39, 76–79, 85, 144, 163, 180, 240, 398
Cull, 625
Curing, 63, 263, 289, 653
Curing agent, 265, 275
Cutting fluid, 706
Cyanoacrylates, 718

D-orbital, 29
DAP, 284
Dashpot, 130
Daylight opening, 616
Deaging, 684
Debulking, 663
Decals, 731
Decomposition point or decomposition temperature, 87–88, 250, 263
Decorating, 726
Deflection temperature under load, 83
Degassing, 680
Degradation, 91, 156, 240, 381, 447, 749
Degree of crystallinity, 77
Degree of polymerization, 100
Delocalized electrons, 175, 181

Density, 77, 111–113, 173, 196, 202, 249, 594, 604
Density gradient column, 77
Depolymerization, 240
Design rules, 327–336
Die, 361, 372, 377, 389–396, 551–56
Die swell, 372, 389, 391
Dielectric constant, 177
Dielectric strength, 176
Differential calorimetry, 78, 84, 88, 688
Diffusion/diffusivity constant, 170
Diisocyanate, 293
Dilatant, 127
Dilatometer, 95
Diluent, 271, 280
Dimensions, 706
Dip casting, 572
Dipole, 36–37
Dipping (see also Dip casting), 319
Directed UV light, 690
Dissipation factor, 177, 178, 723
Domains, 295
Dosimetry, 685
Double cond, 42–48, 193, 268, 279, 289, 291, 311, 313
Dough molding compound (see Bulk molding compound)
Downstroke machine, 616
Draft angle, 438, 454, 527, 551
Drag flow, 369
Drape forming, 518
Draw down ratio, 168, 373, 391, 530
Drawing, 396, 401
Drive motor, 355, 377
Drop box method, 556
Dry spinning, 403
Dual cure system (epoxy), 289
Durometer, 146
Dyes, 180

Edge gate, 436
Ejector system, 442, 676
Elastic elongation, 304
Elastic limit, 304
Elastic solids, 122–125
Elastomers, 208, 221, 294, 303–307
Electrical discharge machining (EDM), 441, 625
Electromagnetic welding, 724
Electron beam radiation, 187, 204, 681
Electron configuration, 29
Electronegativity, 30, 36–37, 68, 206, 248
Electroplating, 729
Electostatic attraction, 32
Electrostatic coating, 728
Electrosurgical blades, 255

Elongation, 124, 137, 188
Embedding, 571
Encapsulation, 572
End-capping, 60, 240, 291
Endotherm, 270
Energy director, 722
Energy requirements, 755
Engineering composites, 644
Engineering stress-strain, 137
Engineering thermoplastics, 231–235
Entanglement, 103
Entertainment, 17
Enthalpy, 164–165
Entropy, 164–165, 306
Environmental aspects, 155
Environmental stress-crack resistance (ESCR),
    167, 203, 374, 410
Epoxy, 287–291, 570, 581, 691, 717
Ergonomic, 331
Ethylene acrylic acid (EAA), 206
Ethylene propylene (EPM), 208, 313
Ethylene propylene diene rubber (EPDM), 208,
    313
Ethylene vinylacetate (EVA), 206, 718
Exotherm, 270, 570, 583
Expanded beads, 601
Expanded foam molding, 600
Expanded plastics, 592
Expanded polystyrene foam, 218, 600
Expansion processes, 168, 438, 592
Extrudate, 353, 371, 378
Extrudate bank, 365, 399
Extruder burn, 356
Extrusion, 105, 128, 147, 168, 187, 351–355
Extrusion blow molding, 485–490, 493
Extrusion coating, 731
Extrusion foaming, 596

F-orbital, 29
Falling dart impact test, 144
Family mold, 439
Fan gate, 436
Fatigue, 232
Feed section, 359
Feed throat, 356
Female mold (see Cavity mold)
Fiber blooming, 663
Fiber placement, 665
Fiberglass, 169, 296, 367, 653
Fiberglass reinforced plastics (FRP), 282, 321, 652
Fibers, 400
Fick's law of diffusion, 171
Filament winding, 665
Fill rate, 457

Fillers, 147, 183, 213, 311
Film, 242, 394, 395–397, 576
Finite element analysis (FEA), 332, 466
Fisheyes, 382
Flame retardance, 192, 283
Flame treatment, 715
Flammability, 182–185, 295
Flash, 621, 630
Flash-type mold closure, 621
Flexible foams, 294, 589
Flexural force, 123–125, 140
Flexural property tests, 140
Flight depth, 357
Flight of screw, 357
Flow-control injection molding, 464
Fluidized bed coating, 573
Fluoroelastomers, 314
Fluoropolymers, 248, 256
Fluorinated ethylene propylene (FEP), 250
Foam, 219, 294, 589
Foam spraying, 598
Foaming agent, 294, 591
Foaming processes, 592–602
Foam-in-place, 596
Force feeding, 380, 386
Ford cup viscometer, 130
Form, fill and seal, 536
Formaldehyde, 240
Forming jig, 710
Fourier transform infra-red (FTIR), 185
Fractional melt materials, 387
Free energy, 164–165
Free forming, 517
Free fusion, 635
Free radical, 42, 51–55, 196, 279, 682
Free radical polymerization (see Addition
    polymerization)
Freeze line, (see Frost line)
Friction welding, 723
Frost line, 398
Functional group, 44–48, 57
Functional specification, 328
Fundamental vacuum forming, 511
Furniture, 17
Fused deposition modeling (FDM), 344
Fusion, 195, 216, 250, 545, 553, 573, 635, 720

Gamma irradiation, 688
Gap distance, 373
Gas injection molding, 465
Gate, 436, 452
Gated casting, 571
Gaylord, 470, 769
Gear pump, 367

Gel coat, 282, 663
Gel fraction test, 687
Gel permeation chromatography (GPC), 102, 687
Gel spinning, 403
Gels, 92
General-purpose screw, 360, 374
Glass spheres, 592
Glass transition temperature (T_g), 84–91, 109, 134, 166, 213, 215, 306, 692
Glassy region, 90, 165
Gloss, 246
Go/no-go gauges, 392
Graft copolymer, 65, 222, 693
Graphical data presentation, 458
Graphite fibers, (*see* Carbon fibers)
Green compact, 634
Gum space, 394
Gutta percha, 309
Gypsum, 320

Halogens, 182, 691
Hand lay-up, 663
Hard rubber, 308
Hardener or hardening agent, 265, 273, 288
Hardness tests, 146
HDPE, 199
Head zone, 360
Heat capacity, 84
Heat distortion temperature (HDT), 82–84, 222
Heat history (*see* Thermal history)
Heat of reaction, 270
Hevea rubber (*see* Natural rubber)
Hexa, 275
High-energy microwaves, 690
High-impact polystyrene (HIPS), 221
High-performance thermoplastics, 233, 243, 252, 663, 669
High-pressure foam molding, 593
High-selectivity catalyst, 210
High-temperature vulcanization (HTV), 317, 718
Histogram, 98
History of plastics, 5–9
Hobbing, 442, 623
Hollow fill fibers, 390
Hollow spheres, 592
Homogeneous, 219
Homopolymer, 63, 201, 205, 240
Hookean, 125, 135
Hot stamping, 731
Hot-runner molds, 436
Hot-wire cutting, 599, 709
Hot-melt plastics (adhesives) 568, 582, 718
HTV (*see* High-temperature vulcanization)
Hybrid composites, 653

Hybrid orbital, 40
Hydrated, 183
Hydraulic clamping, 444
Hydrogen bond, 37, 76, 81, 103, 168, 236, 716
Hygroscopic, 357, 366, 747
Hysteresis, 136

Identification, 185
Immersion test, 163
Immiscible, 220
Impact modifiers (*see* Toughness modifiers)
Impact polystyrene (*see* High-impact polystyrene)
Impact strength (*see* Impact toughness)
Impact toughness, 78, 103, 143–146, 204, 222
Incineration, 752, 758
Index of refraction, 181
Induced dipole bond, 37
Induction welding, 724
Industry (plastics), 12–14, 17
Infrared spectroscopy, 77–78, 185
Inhibitors, 281, 386
Initiator, 51–56, 265, 279, 555
Injection blow molding, 490–494
Injection compression molding, 465
Injection molding, 105, 276, 277, 419–470, 490, 658
Insulator, 173, 176
Integral hinge, 211, 713
Integral skin, 594
Interference joint, 711
Intermittent extrusion blow molding, 486
Intermolecular atttractions, 103
Internal mold release, 275, 278
Internal sizing mandrel, 392
Intrinsic viscosity [η], 129
Invert forming, 515
Ion, 32, 68, 682
Ionic bond, 32–33, 68, 207
Ionic character, 50
Ionomer, 207
Irrigation tubing, 406
Iso resin, 282
Isocyanate, 293
Isoprene rubber (IR), 308–309
Isotactic, 209
Isotope, 218
Isotropic, 452, 648
Izod test, 144

Jacketed rotational molding, 550
Joint design, 725

K factor, 94

Kelvin model, 132
Kerf, 527, 708
Knitting, 453, 464
Knockout pins, 442, 616

L/D (*see* Length/diameter)
Lamellar injection molding, 464
Laminate design theory, 649
Laminated object manufacturing (LOM), 342
Laminates, 649
Laminating, 730
Laminating resin, 281
Land, 363, 372, 391
Landfill, 750
Laser, 708
Latex, 308, 319
LDPE, 168, 198
Leakage flow, 369
Leathery region, 90, 165
Length/diameter, 357, 382, 423
Leveling, 255
Light transmission, 179, 246
Limiting oxygen index (LOI), 183
Liquid crystals, 110
Liquid resins, 568, 624
Living hinge, 211, 713
LLDPE, 168, 200, 555
Long-range movements, 84, 134
Loss factor, 723
Low-density polyethylene (*see* LDPE)
Low-pressure foam molding, 593
Low-pressure injection molding, 464
Low-pressure molds, 578
Low-profile system, 287
Low-shrink system (*see* Low-profile system)
Lubricants, 225, 387
Lubricity, 249
Luminous transmittance, 180

Machining, 441, 705
Maintenance, 377, 459
Macro viewpoint, 115
Macromolecules (definition), 3
Maddox section, 367
Male mold (*see* Plug mold)
Mandrel, 390, 665
Mass cast molds, 579
Mass spectroscopy, 185
Master batch, 187
Mat, 657
Matched die molding, (*see* Compression molding)
Matched die forming, 521
Material bank, 366
Material safety data sheet (MSDS), 766

Matrix, 643, 651
Maximum-use temperature (*see* Thermal stability temperature)
Maxwell model, 131
Mechanical foaming, 591
Mechanical forming, 522
Mechanical joining, 710
Medical, 17
Melamines, 278
Melt fracture, 378
Melt index, 106–108, 111–113, 387, 534, 628
Melt spinning, 401
Melt strength, 105, 203, 387, 528
Melting point or temperature (T_m), 85–90, 104–105, 165, 202
Memory, 204
Mesh, 553
Metal oxide, 207, 287, 312
Metallic alloys, 34, 219
Metallic bond, 33–34
Metallizing, 728
Metallocenes, 218
Metals, 16, 30–34, 94, 173
Metering section, 360
Micro viewpoint, 75–118
Microballoons, 592, 604
Microorganisms, 158
Mill, 222
Miscible, 220
Mixed recycle, 744
Modified acrylics, 118
Modulus (E), 125, 304
Mold, 290, 427, 494, 510, 551, 567, 592, 616
Mold base, 429–432
Mold casting, 569
Mold cavities, 429
Mold construction, 441
Mold costs, 471, 474
Mold inserts, 439, 458
Mold materials, 440, 494–496, 525, 567, 578, 623
Mold release, 275, 458, 593
Molding compound or molding resin, 271, 281, 284
Molecular orbitals, 40–44
Molecular viewpoint, 25
Molecular weight, 97–108, 129, 144, 202
Molecular weight distribution (MWD), 98–106, 202, 387
Molecule, 1–3, 35
Monomers, 1, 48, 205, 211, 217, 232, 246, 250, 292
Mooney viscometer, 130
Moveable platen, 427
Multicavity mold, 429, 438

Multiple injection points, 453, 464
Multiple-mold system, 486
Multiple-pour method, 574

Names, 234
Natural rubber, 308–309
Neat resin, 281
Necking, 137
Neoprene, 312
Newtonian fluids, 127
Nitrile butadiene rubber, 312
Nonflammable, 249, 275, 312
Nonmetals, 32, 34
Non-Newtonian, 127
Nonstick, 248, 255
Notch sensitive, 239
Novolacs, 275
Nuclear magnetic resonance (NMR), 185, 688
Nucleus, 28
Nylon, 8, 20–21, 58–60, 160, 235, 276–277, 359, 429, 536, 580

Octet rule, 29, 34
Offset die (see Crosshead die)
Ohmic resistance, 174
Oil resistant elastomers, 312
Olefins, 194
Oligomer, 3, 271
One-stage resins, 275
Opaque, 147, 179
Open cell foams, 589
Optical properties, 178–182, 245, 246, 576
Orange peel effect, 379, 497, 559
Organic chemistry, 39–48
Organic reinforcement fibers, 656
Organisol, 568, 573
Orientation effects, 242, 452
Orthotropic, 648
Oxidation, 93, 158, 313
Oxygen index test, 183
Ozone, 64

P-orbital, 29
Packaging, 16
Painting, 247, 573, 726
Pan casting, 597
Paper cup, 753
Parison, 484
Parison programmming, 495
Parison transfer system, 486
Parting agent, 583
Parting line, 429, 438, 496
Pattern, 595
Payoff, 666

Peck drilling, 706
Pendant group, 49–50, 76, 108, 194, 206, 217, 308
Perfluoroalkoxy (PFA), 251
Periodic table, 26–28, 97
Permanent deformation, 136
Permanent set, 140, 141
Permeability, 169–173, 206
Permitivity, 177
Peroxide (see Initiator)
PETE, 744
Phenolic, 7, 182, 243, 274–276, 581, 718
Phenyl, 316
Photopolymer, 342, 690
Physical foaming, 591
Pi (π) orbital, 43–48, 53, 268
Pigments, 180
Pillar packs, 470
Pillow forming, (see Reverse draw forming)
Pitch, 358, 655
Plasma, 693, 715
Plastic (definition), 1–6
Plastic memory, 373
Plastic region, 136
Plasticization, 161
Plasticizer, 165–167, 212, 214–216
Plastics industry, 13
Plastisol, 216, 568, 573
Plate bonding, 720
Plug assist forming, 513
Plug flow, 395
Plug mold, 518
Pneumatic feeding, 769
Pocket knife case study, 468
Poker chipping, 636
Polar bond, 36, 160, 235
Polyacetylene, 176
Polyacrylonitrile (PAN), 222, 246, 655
Polyamide (PA), 235–238
Polyaryletherketones, 252
Polyarylesters, 243
Polycarbonate (PC), 179, 182, 244
Polyester, 241, 279, 580, 584
Polyetheretherketone, 252
Polyethylene (PE), 187–188, 194–205, 359
Polyethyleneterephthalate (PET), 173, 243, 503, 729, 772
Polyhexafluoropropylene (PHFP), 250
Polyimide (PI), 182, 254, 291
Polymer (definition), 1–4, 48
Polymerization, 2–4, 12–15
Polymethylmethacrylate (PMMA), 157, 180, 182, 246
Polyol, 292
Polyolefins, 194

Polyoxymethylene (POM), 239
Polyphenylene oxide (PPO), 252
Polyphenylenes, 252
Polypropylene (PP), 208–211, 313
Polysiloxanes (*see* Silicones)
Polystyrene (PS), 180, 182, 217–219
Polysulfone (PSU), 253
Polytetrafluoroethylene (PTFE), 65–67, 182, 248
Polyurethane (PUR), 292–295, 314, 581, 633, 717
Polyvinyl alcohol (PVAL), 161
Polyvinyl acetate(PVA), 718
Polyvinyl chloride (PVC or vinyl), 161, 166, 182, 211, 224
Polyvinyl chloride/polyvinyl acetate copolymer, 216
Polyvinylidene chloride, 216
Porous-well tubing, 403
Positive pumping, 367
Positive type mold closure, 621
Postconsumer recycle, 499, 742
Postextrusion forming, 403
Postmold shrinkage, 452
Potting, 285, 571
Pourability, 77, 553
Powder coating, 728
Preexpansion, 600
Preform, 319, 491, 629
Premix, (*see* Bulk molding compound)
Preplasticizing machine, 427
Prepolymers (*see* Oligomers)
Prepreg, 290, 657
Press capacity, 617
Press thermoforming, 670
Pressure flow, 369
Pressure forming, 511
Pressure relief valves, 367
Pressure sensitive adhesives, 718
Prices, 11
Primer, 715
Printing, 729
Processing aids, 93
Processing temperature, 88
Profiles, 388
Projected area, 443, 620
Proportional limit, 135
Prototyping, 330, 341
Pseudoplastic, 127
Puller, 364
Pultrusion, 668
Pump, 320
Purging, 240, 371, 386
Push-pull injection molding, 464

Quench, 53, 60, 272
Quality assurance, 771

R-value, 610
Radiant panel test, 184
Radiation, 610, 681, 720
Radio frequency (rf) welding, 723
Rake angle, 707
Ram extrusion, 250, 636
Ram injection, 423
Random copolymer, 64
Rapid prototyping, 341
Recreation, 17
Reaction injection molding (RIM), 294, 594, 633
Reactive diluents, 271, 281
Reactive group (*see* Active site)
Rebond, 603
Reciprocating screw, 423–427, 488
Recycling, 742, 758
Reflectometer, 182, 688
Refracture index, 181
Regeneration, 402, 749
Regrind, 91, 214, 405, 742
Regular copolymer, 64
Reinforcements, 146, 275, 653
Relative viscosity, 129
Release cloth, 664
Relief valves, 367
Repeating unit, 53, 97
Resilience, 306
Resin, 13, 22
Resin distributor, 13
Resin manufacturers, 13
Resin transfer molding (RTM), 661
Resistivity, 174
Resoles, 275, 581
Reverse draw forming, 515
Rheometry, 126, 130
Rigid foams, 294, 589
Ring gate, 437
Ring opening, 62, 287
Ripening, 287
Rising mold system, 486
Rivet, 710
Rockwell hardness, 146
Roll wrapping, 665
Room temperature vulcanization (RTV), 317, 581, 718
Root diameter of screw, 357
Rotating spindle viscometer, 130
Rotational molding, 543–566
Rotomolding (*see* Rotational molding)
RTV (*see* Room temperature vulcanization)

Rubber, 148, 306
Rubber cement, 716, 719
Runners, 429, 432, 624, 704
Rupture, 136

S-orbital, 29
Safety, 378, 446, 560, 609, 632, 765
Sag point, 511
Sandwich composite, 665
Saturated, 42
Scission, 314, 682, 692
Scorching, 318
Screen pack, 360, 377, 384
Sea of electrons, 33
Secondary bond, 37–39, 162–164, 168, 236
Secondary carbons, 209, 692
Sedimentation, 687
Selected laser sintering (SLS), 343
Self-extinguishing, 182, 212
Semipositive mold type closure, 623
Shaping, 709
Sharkskin, 379
Shear force, 123–125, 140
Shear rate, 126
Shear-thinning, 127, 669
Sheet molding compound (SMC), 284–287, 659,
    704, 748
Shore hardness, 146
Short-range movements, 84, 134
Short shots, 457
Shot size, 425
Shrink tubing, 697
Shrinkage, 203, 449, 451, 570
Sieve analysis (particle size), 77
Sigma (σ) orbital, 43–48
Silanes, 365
Silica, 316
Silicones, 315–317, 570, 578, 581, 584,
    718
Silk screening, 727
Silly putty, 122, 135
Simultaneous engineering, 328
Sink marks, 287, 449
Sintering, 66, 200, 250, 575, 634
Sizing, 654
Sizing plates, 363
Skewed MWD, 100–102
Skin pack, 522
Skip tooth cutters, 706
Skiving, 598
Sliding/compression, 497
Slug, 528
Slurry, 320
Slush casting, 573

Smoke density test, 184
Snap fit, 528
Snap joint, 711
Snap-back forming, 519
Soda pop bottle (case study), 503
Softening temperature (see Glass transition
    temperature)
Solid casting, 569
Solid-phase forming, 634
Solids, liquids, gases, 79
Solute, 159–165
Solvent, 159–165, 214
Solvent casting, 576
Solvent crazing, 169, 277
Solvent resistance, 159
Solvent welding, 167, 214, 235, 246
Source reduction, 740
Spandex fibers, 295
Specific gravity, 77, 162
Specific stiffness, 650
Specific strength, 650
Specular gloss, 182
Spider die, 391
Spin welding, 723
Spinneret, 401
Spinning, 400
Spray-up, 662
Spring equation, 125, 131
Springback, 452
Sprue, 428, 624
Sprue bushing, 428
Sprue puller, 431, 625
Stacked mold, 429
Stage resins, 271
Stain resistance, 163
Staking, 722
Staple, 401
Startup, 371
Starve feeding, 380
Static mixers, 367
Static powder molding, 574
Stationary platen, 427
Step-growth polymerization (see Condensation
    polymerization)
Stereoisomers, 208
Stereolithography, 342
Stereoregular catalyst, 210
Steric effects, 108–110, 194
Stiffness (see Modulus)
Straight vacuum forming, 511
Strain (ε), 124, 141
Strain to failure, 136
Stress (σ), 124
Stress strain curve, 124, 141

Stress-induced crystallization (*see also* Blushing), 140, 505
Stretch blow molding, 492
String up, 371
Stripping, 454, 585, 593
Structural adhesives, 717
Structural reaction injection molding (SRIM), 596
Styrene acrylonitrile (SAN), 222
Styrene butadiene rubber (SBR), 221, 310
Submarine gate, 436
Substrate, 714, 726
Surface activation, 697
Surface casting, 571
Surface cleaning, 714
Surface pacification, 697
Surface resistivity, 174
Surface veil, 663
Surging, 380
Syndiotactic, 210
Syntactic foams, 592, 604
Synthetic fibers, (*see* Fibers)
Synthetic rubber (*see* Isoprene rubber)
Syrups, 568, 580
Systematic step-wise identification, 185

Tab gate, 436
Tensile force, 123–124, 138
Tensile properties tests, 138
Tensile strength, 103, 111–113, 137
Tent frame, 397
Tenter hooks, 396
Terpolymer, 64, 208, 224, 313
Tertiary carbons, 209, 692
Textiles, 17
Texturizing, 732
Thermal aging (*see* Aging)
Thermal conductivity (K), 93–94, 610
Thermal decomposition temperature, 212
Thermal expansion, 94
Thermal heating, 359
Thermal history, 92, 166, 213
Thermal resistivity (R), 610
Thermal stabilizers, 93
Thermal stresses, 95
Thermal transitions, 81–88
Thermoforming, 509
Thermogravimetric analysis (TGA), 88
Thermomechanical analysis (TMA), 85, 88
Thermoplastic, 62–63, 85–86, 629
Thermoplastic olefin elastomers (TPO), 208, 313
Thermoplastic polyesters, 241
Thermoset, 62–63, 85–88, 263–264, 615, 629, 748
Thermostatic printing, 730
Thixotropic, 128

Thrust bearing, 356, 377
Time-temperature superposition, 134
Toggle, 445
Tolling, 354
Torpedo or spreader, 426
Torsion force, 123–125, 140
Total product life cycle, 753
Toughness, 104, 143–146, 203
Toughness modifiers or enhances, 148, 310
Toughness tests, 111–113, 143
Tow, 401, 653
Trans, 309
Transfer chamber, 624
Transfer coating, 730
Transfer molding, 276, 277, 624
Transfer coating, 730
Transition section (*see* Compression section)
Translucent, 179
Transparent, 179
Transportation, 17
Trash carts (case study), 561
Triallylcyanurate (TAC), 204, 555
Trouble shooting, 383, 459, 501, 535, 559, 583, 608, 631, 671
True stress-strain, 137
Tubular die, 396, 404
Tup, 144
Twin-screw extruders, 214, 357, 365
Two-stage clamping, 446, 630
Two-stage resins, 275
Two-stage screws, 366

UL 94 flammability tests, 185
UL temperature index, 83–84
Ultimate strength, 136
Ultrahigh-molecular-weight polyethylene (UHMWPE), 105, 200, 656
Ultralow density polyethylene, 202
Ultrasonic welding, 216, 721
Ultraviolet (UV) light, 156–158, 179, 187–188, 217, 690
Undercut, 454, 528, 557
Unidirectional fibers, 657
Universal thickener, 255
Unsaturated, 42, 279
Unzipping, 240, 749
Upstroke machine, 616
Urea formaldehyde (UF), 278
Utility poles (case study), 295

Vacuum bag, 664
Vacuum forming, 511
Vacuum metallizing, 728
Valence, 30

Van der Waals bonds, 37, 160, 716
Velocity gradient, 125–126
Vent, 354, 359, 438
Vibration damping, 311, 607
Vicat softening temperature, 83
Vinyl, 166, 212, 581
Vinyl chloride, 211
Vinyl copolymers, 216
Vinyl dispersion, 216
Vinyl esters, 283, 652
Vinylidene fluoride, 314
Virgin plastic, 96, 214
Viscoelastic, 131–134
Viscosity, 122, 125–130, 268, 318, 450, 629
Viscous fluids, 122, 580
Voigt model, 132
Volume resistivity, 174
Vulcanization, 6, 308

Warpage, 447
Water absorption, 162–163, 236

Water base adhesives, 719
Water jet cutting, 708
Weathering, 155–159
Weight average molecular weight, 100
Weighting factor, 340
Weld line, 453, 570
Wet spinning, 402
Wet-out, 651
Wetting, 714
Whiskers, 651
Wood (plastic), 747

X ray, 77–78

Yarn, 401
Yield point, 135, 304
Young's modulus (*see* Modulus)

Z-average molecular weight, 101
Zahn cup viscometer, 130
Ziegler-Natta catalyst, 9, 56, 199, 210, 309

THANK YOU FOR YOUR PURCHASE! THIS BOOK ENTITLES YOU TO 15 FREE LOGINS TO PROSPECTOR WEB™, THE PLASTIC MATERIALS DATABASE BY IDES, Inc.

Prospector Web™ is an interactive database used to find and compare plastic materials. You can specify your application requirements to search a catalog of global plastics data. Search the IDES database utilizing over 200 engineering and processing properties. Test data is available in both ASTM and ISO format and can be viewed in either English or Metric units. IDES Inc. also offers quoting software, Prospector Desktop, and a free data sheet look up service at http://www.freemds.com.

Dr. Brent Strong of Brigham Young University says, "**Prospector Web™** is an integral tool in our classes. I expect quality from my students and quality starts with the superior material selection tools provided by IDES, Inc. When it comes to plastic materials information there is simply no other choice."

To access your **15 FREE LOGINS** point your Internet browser to http://www.idesinc.com, click on **Prospector Web** and enter the following access codes:

> **Username:** *7512915*
>
> **Password: student**

A little about IDES, Inc...

IDES Inc., *"The World's Source for Plastic Materials Information"* provides information and software tools to improve and expedite information needs in the plastics industry. We produce state of the art products that unite plastic information databases with computer technology. IDES' excellence in developing leading edge technology, coupled with many years of knowledge in working with materials information make us the world leader in this specialized field.

209 Grand Ave
Laramie, WY 82070 USA
Phone: 800-788-4668 • Fax: 307-745-9339
http://www.idesinc.com
sales@idesinc.com